Mary Jones, Rich
Jennifer Gregory and Dennis Taylor

Cambridge International AS and A Level

Biology

Coursebook

Fourth Edition

CAMBRIDGE
UNIVERSITY PRESS

University Printing House, Cambridge CB2 8BS, United Kingdom
One Liberty Plaza, 20th Floor, New York, NY 10006, USA
477 Williamstown Road, Port Melbourne, VIC 3207, Australia
4843/24, 2nd Floor, Ansari Road, Daryaganj, Delhi – 110002, India
79 Anson Road, #06–04/06, Singapore 079906

Cambridge University Press is part of the University of Cambridge.

It furthers the University's mission by disseminating knowledge in the pursuit of education, learning and research at the highest international levels of excellence.

Information on this title: www.cambridge.org

First published 2003
Second edition 2007
Third edition 2013
Fourth edition 2014
20 19 18 17 16 15 14 13 12 11 10 9 8

Printed in the United Kingdom by Latimer Trend

A catalogue record for this publication is available from the British Library

ISBN978-1-107-63682-8 Paperback with CD-ROM for Windows® and Mac®
ISBN 978-1-316-63770-8 Paperback with CD-ROM for Windows ®
and Mac ® Paperback + Cambridge Elevate enhanced edition, 2 years
ISBN 978-1-107-70045-1 Cambridge Elevate enhanced edition, 2 years

Contents

AS Level

A Level

How to use this book

Each chapter begins with a short list of the facts and concepts that are explained in it.

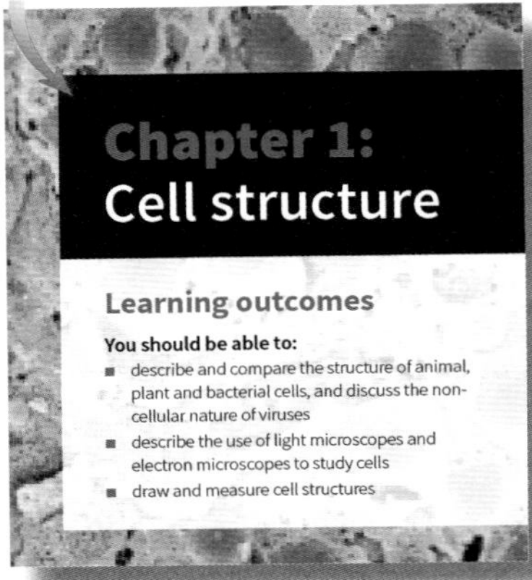

Chapter 1:
Cell structure

Learning outcomes

You should be able to:

- describe and compare the structure of animal, plant and bacterial cells, and discuss the non-cellular nature of viruses
- describe the use of light microscopes and electron microscopes to study cells
- draw and measure cell structures

There is a short context at the beginning of each chapter, containing an example of how the material covered in the chapter relates to the 'real world'.

Where biology meets psychology

We have five senses: touch, sight, hearing, taste and smell. It's a controversial view, but some people believe in extrasensory perception (ESP), telepathy and having premonitions as a 'sixth sense'. Recent research suggests that we detect subtle changes, which we cannot put into words, so imagine it is an extra sense. Some people also have synaesthesia – a condition where stimulation of, say, hearing also produces a visual response (Figure 15.1).

But we do have a genuine sixth sense, one which we take for granted. In his essay 'The Disembodied Lady', the neurologist Oliver Sacks relates the story of a woman who woke up one day to find she had lost any sense of having a body. All the sensory neurones from the receptors in her muscles and joints had stopped sending impulses. She had no feedback from her muscles and could not coordinate her movements. The only way she could live without this sixth sense was to train herself to rely entirely on her eyesight for coordinating her muscles. A man with the same condition describes the efforts needed to do this as equivalent to running a marathon every day. Curiously, the night before Oliver Sacks's patient found she had total loss of body awareness, she dreamt about it.

Figure 15.1 Crossed wires? By studying electrical activity in the brain, researchers have found that some people do indeed hear colour and see sound.

Questions throughout the text give you a chance to check that you have understood the topic you have just read about. You can find the answers to these questions on the CD-ROM.

This book does not contain detailed instructions for doing particular experiments, but you will find background information about the practical work you need to do in these boxes. There are also two chapters, P1 and P2, which provide detailed information about the practical skills you need to develop during your course.

BOX 4.4: Investigating osmosis in plant cells

1 Observing osmosis in plant cells

Epidermal strips are useful material for observing plasmolysis. Coloured sap makes observation easier. Suitable sources are the inner surfaces of the fleshy storage leaves of red onion bulbs, rhubarb petioles and red cabbage.

The strips of epidermis may be placed in a range of molarities of sucrose solution (up to $1.0\,\text{mol dm}^{-3}$) or sodium chloride solutions of up to 3%. Small pieces of the strips can then be placed on glass slides, mounted in the relevant solution, and observed with a microscope. Plasmolysis may take several minutes, if it occurs.

2 Determining the water potential of a plant tissue

The principle in this experiment is to find a solution of known water potential which will cause neither a gain nor a loss in water of the plant tissue being examined. Samples of the tissue – for example, potato – are allowed to come into equilibrium with a range of solutions (for example, sucrose solutions) of different water potentials, and changes in either mass or volume are recorded. Plotting a graph of the results allows the solution that causes no change in mass or volume to be determined. This solution will have the same water potential as the plant tissue.

QUESTION

4.8 Two neighbouring plant cells are shown in Figure 4.16.

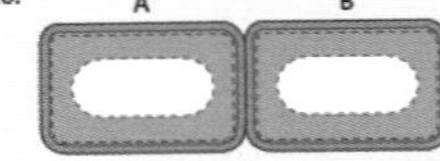

Figure 4.16

a In which direction would there be net movement of water molecules?

b Explain what is meant by **net movement**.

c Explain your answer to a.

d Explain what would happen if both cells were placed in

i pure water

ii a $1\,\text{mol dm}^{-3}$ sucrose solution with a water potential of –3510 kPa.

The text and illustrations describe and explain all of the facts and concepts that you need to know. The chapters, and often the content within them as well, are arranged in the same sequence as in your syllabus.

Active transport

If the concentration of particular ions, such as potassium and chloride, inside cells is measured, it is often found that they are 10–20 times more concentrated inside than outside. In other words, a concentration gradient exists, with a lower concentration outside and a higher concentration inside the cell. The ions inside the cell originally came from the external solution, therefore diffusion cannot be responsible for this gradient because, as we have seen, ions diffuse from high concentration to low concentration. The ions must therefore accumulate **against** a concentration gradient.

The process responsible is called active transport. It is achieved by carrier proteins, each of which is specific for a particular type of molecule or ion. However, unlike facilitated diffusion, active transport requires energy, because movement occurs up a concentration gradient. The energy is supplied by the molecule ATP (adenosine triphosphate) which is produced during respiration inside the cell. The energy is used to make the carrier protein change its shape, transferring the molecules or ions across the membrane in the process (Figure 4.17).

An example of a carrier protein used for active transport is the sodium–potassium ($Na^+ – K^+$) pump (Figure 4.18 and page 272). Such pumps are found in the

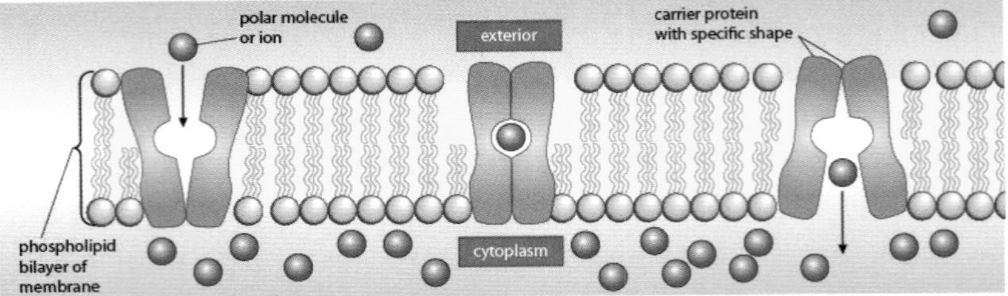

Figure 4.17 Changes in the shape of a carrier protein during active transport. Here, molecules or ions are being pumped into the cell against a concentration gradient. (Compare Figure 4.9.)

Important equations and other facts are shown in highlight boxes.

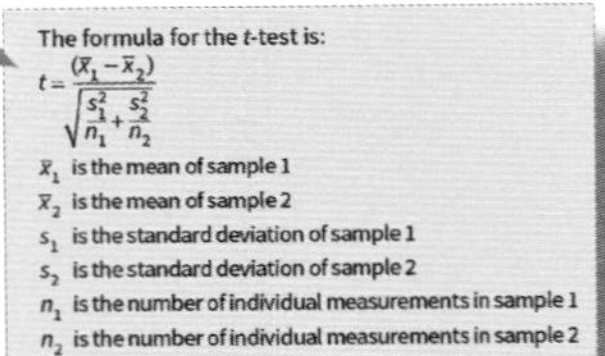

The formula for the *t*-test is:

$$t = \frac{(\bar{x}_1 - \bar{x}_2)}{\sqrt{\frac{s_1^2}{n_1} + \frac{s_2^2}{n_2}}}$$

$\bar{x}_1$ is the mean of sample 1

$\bar{x}_2$ is the mean of sample 2

s_1 is the standard deviation of sample 1

s_2 is the standard deviation of sample 2

n_1 is the number of individual measurements in sample 1

n_2 is the number of individual measurements in sample 2

Wherever you need to know how to use a formula to carry out a calculation, there are worked example boxes to show you how to do this.

WORKED EXAMPLE 2

Calculating the magnification of a photograph or image

To calculate M, the magnification of a photograph or an object, we can use the following method.

Figure 1.9 shows two photographs of a section through the same plant cells. The magnifications of the two photographs are the same. Suppose we want to know the magnification of the plant cell labelled P in Figure 1.9b. If we know its actual (real) length we can calculate its magnification using the formula

$$M = \frac{I}{A}$$

The real length of the cell is 80 µm.

Step 1 Measure the length in mm of the cell in the photograph using a ruler. You should find that it is about 60 mm.

Step 2 Convert mm to µm. (It is easier if we first convert all measurements to the same units – in this case micrometres, µm.)

$$1\,\text{mm} = 1000\,\mu\text{m}$$

so $60\,\text{mm} = 60 \times 1000\,\mu\text{m} = 60\,000\,\mu\text{m}$

Step 3 Use the equation to calculate the magnification.

$$\text{magnification, } M = \frac{\text{image size, } I}{\text{actual size, } A} = \frac{60\,000\,\mu\text{m}}{80\,\mu\text{m}} = \times 750$$

The multiplication sign in front of the number 750 means 'times'. We say that the magnification is 'times 750'.

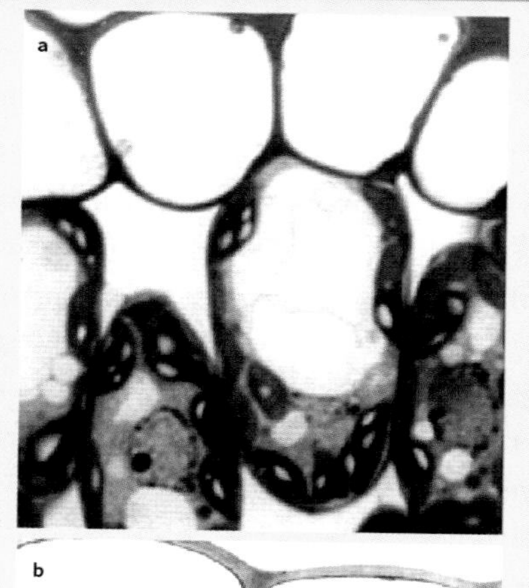

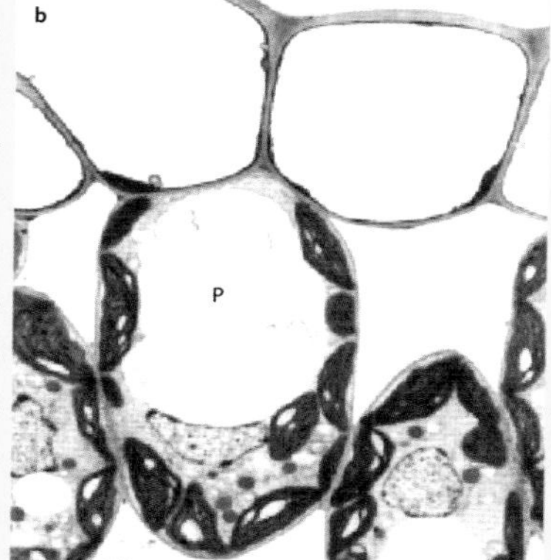

Figure 1.9 Photographs of the same types of plant cells seen **a** with a light microscope, **b** with an electron microscope, both shown at a magnification of about ×750.

Definitions that are required by the syllabus are shown in highlight boxes.

A **macromolecule** is a large biological molecule such as a protein, polysaccharide or nucleic acid.

A **monomer** is a relatively simple molecule which is used as a basic building block for the synthesis of a polymer; many monomers are joined together to make the polymer, usually by condensation reactions; common examples of molecules used as monomers are monosaccharides, amino acids and nucleotides.

A **polymer** is a giant molecule made from many similar repeating subunits joined together in a chain; the subunits are much smaller and simpler molecules known as monomers; examples of biological polymers are polysaccharides, proteins and nucleic acids.

Key words are highlighted in the text when they are first introduced.

An example of a carrier protein used for active transport is the sodium–potassium ($Na^+ – K^+$) pump (Figure 4.18 and page 272). Such pumps are found in the

You will also find definitions of these words in the Glossary.

sodium–potassium pump a membrane protein (or proteins) that moves sodium ions out of a cell and potassium ions into it, using ATP

There is a summary of key points at the end of each chapter. You might find this helpful when you are revising.

Summary

- The basic unit of life, the cell, can be seen clearly only with the aid of microscopes. The light microscope uses light as a source of radiation, whereas the electron microscope uses electrons. The electron microscope has greater resolution (allows more detail to be seen) than the light microscope, because electrons have a shorter wavelength than light.
- With a light microscope, cells may be measured using an eyepiece graticule and a stage micrometer. Using the formula $A = \frac{I}{M}$ the actual size of an object (A) or its magnification (M) can be found if its observed (image) size (I) is measured and A or M, as appropriate, is known.

Questions at the end of each chapter begin with a few multiple choice questions, then move on to questions that will help you to organise and practise what you have learnt in that chapter. Finally, there are several more demanding exam-style questions, some of which may require use of knowledge from previous chapters. Answers to these questions can be found on the CD-ROM.

11 a The fruit fly, *Drosophila melanogaster*, feeds on sugars found in damaged fruits. A fly with normal features is called a wild type. It has a grey striped body and its wings are longer than its abdomen. There are mutant variations such as an ebony-coloured body or vestigial wings. These three types of fly are shown in the figure.

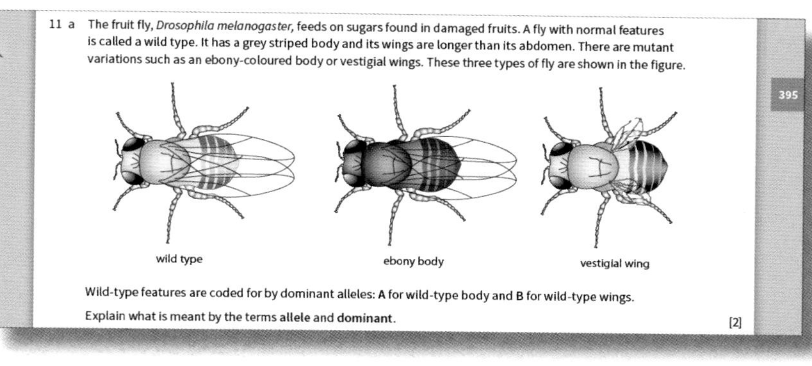

Wild-type features are coded for by dominant alleles: **A** for wild-type body and **B** for wild-type wings.

Explain what is meant by the terms **allele** and **dominant**. [2]

395

Introduction

This fourth edition of *Cambridge International AS and A Level Biology* provides everything that you need to do well in your Cambridge International Examinations AS and A level Biology (9700) courses. It provides full coverage of the syllabus for examinations from 2016 onwards.

The chapters are arranged in the same sequence as the material in your syllabus. Chapters 1 to P1 cover the AS material, and Chapters 12 to P2 cover the extra material you need for the full A level examinations. The various features that you will find in these chapters are explained on the next two pages.

In your examinations, you will be asked many questions that test deep understanding of the facts and concepts that you will learn during your course. It's therefore not enough just to learn words and diagrams that you can repeat in the examination; you need to ensure that you really understand each concept fully. Trying to answer the questions that you will find within each chapter, and at the end, should help you to do this. There are answers to all of these questions on the CD-ROM that comes with this book.

Although you will study your biology as a series of different topics, it's very important to appreciate that all of these topics link up with each other. Some of the questions in your examination will test your ability to make links between different areas of the syllabus. For example, in the AS examination you might be asked a question that involves bringing together knowledge about protein synthesis, infectious disease and transport in mammals. In particular, you will find that certain key concepts come up again and again. These include:

- cells as units of life
- biochemical processes
- DNA, the molecule of heredity
- natural selection
- organisms in their environment
- observation and experiment

As you work through your course, make sure that you keep on thinking about the work that you did earlier, and how it relates to the current topic that you are studying. On the CD-ROM, you will also find some suggestions for other sources of particularly interesting or useful information about the material covered in each chapter. Do try to track down and read some of these.

Practical skills are an important part of your biology course. You will develop these skills as you do experiments and other practical work related to the topic you are studying. Chapters P1 (for AS) and P2 (for A level) explain what these skills are, and what you need to be able to do to succeed in the examination papers that test these skills.

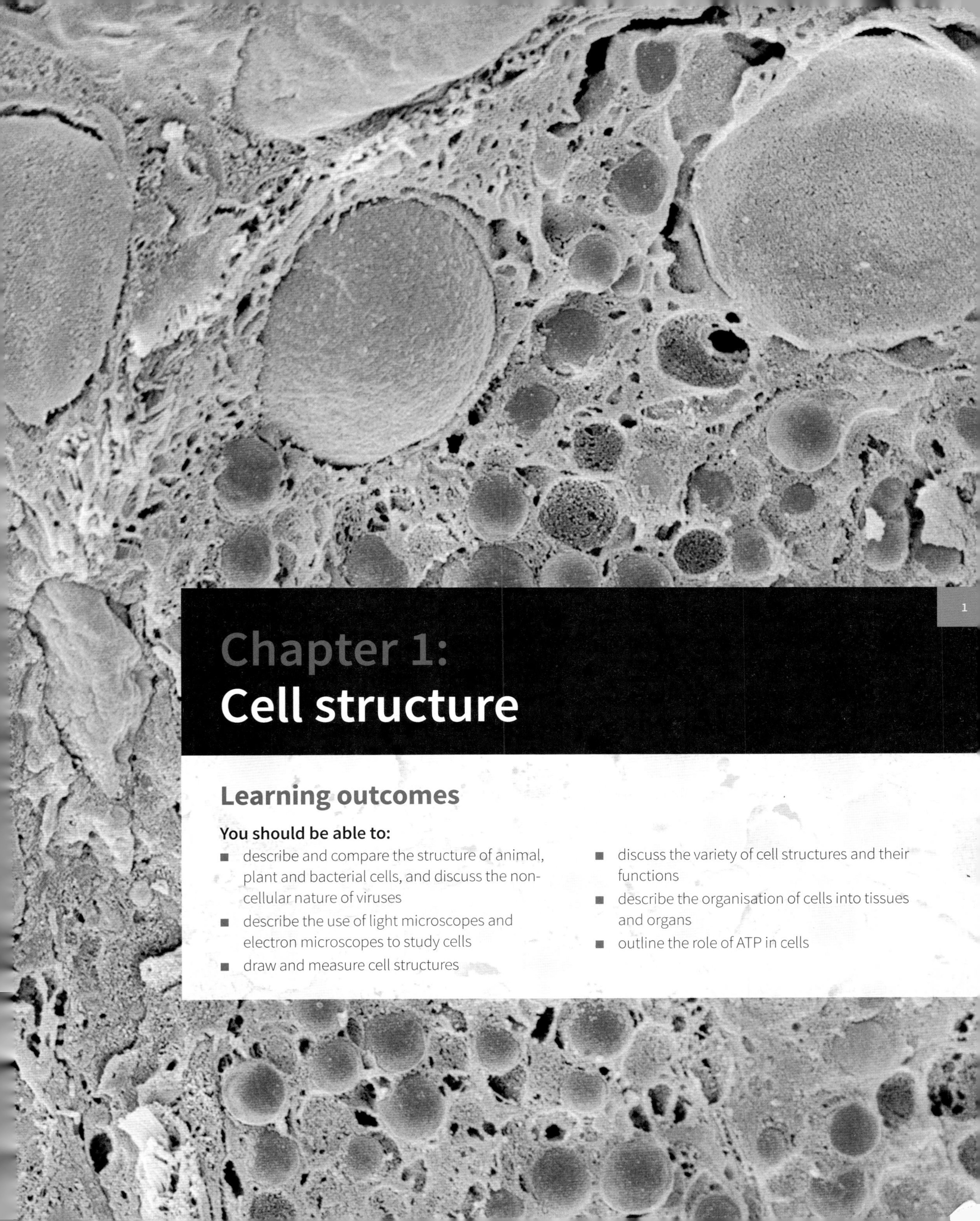

Chapter 1: Cell structure

Learning outcomes

You should be able to:

- describe and compare the structure of animal, plant and bacterial cells, and discuss the non-cellular nature of viruses
- describe the use of light microscopes and electron microscopes to study cells
- draw and measure cell structures
- discuss the variety of cell structures and their functions
- describe the organisation of cells into tissues and organs
- outline the role of ATP in cells

Thinking outside the box

Progress in science often depends on people thinking 'outside the box' – original thinkers who are often ignored or even ridiculed when they first put forward their radical new ideas. One such individual, who battled constantly throughout her career to get her ideas accepted, was the American biologist Lynn Margulis (born 1938, died 2011: Figure 1.1). Her greatest achievement was to use evidence from microbiology to help firmly establish an idea that had been around since the mid-19th century – that new organisms can be created from combinations of existing organisms which are not necessarily closely related. The organisms form a symbiotic partnership, typically by one engulfing the other – a process known as endosymbiosis. Dramatic evolutionary changes result.

The classic examples, now confirmed by later work, were the suggestions that mitochondria and chloroplasts were originally free-living bacteria (prokaryotes) which invaded the ancestors of modern eukaryotic cells (cells with nuclei). Margulis saw such symbiotic unions as a major driving cause of evolutionary change. She continued to challenge the Darwinian view that evolution occurs mainly as a result of competition between species.

Figure 1.1 Lynn Margulis: 'My work more than didn't fit in. It crossed the boundaries that people had spent their lives building up. It hits some 30 sub-fields of biology, even geology.'

In the early days of microscopy an English scientist, Robert Hooke, decided to examine thin slices of plant material. He chose cork as one of his examples. Looking down the microscope, he was struck by the regular appearance of the structure, and in 1665 he wrote a book containing the diagram shown in Figure 1.2.

If you examine the diagram you will see the 'pore-like' regular structures that Hooke called 'cells'. Each cell appeared to be an empty box surrounded by a wall. Hooke had discovered and described, without realising it, the fundamental unit of all living things.

Although we now know that the cells of cork are dead, further observations of cells in living materials were made by Hooke and other scientists. However, it was not until almost 200 years later that a general cell theory emerged from the work of two German scientists. In 1838 Schleiden, a botanist, suggested that all plants are made of cells, and a year later Schwann, a zoologist, suggested the same for animals. The **cell theory** states that the basic unit of structure and function of all living organisms is the cell. Now, over 170 years later, this idea is one of the most familiar and important theories in biology. To it has been added Virchow's theory of 1855 that all cells arise from pre-existing cells by cell division.

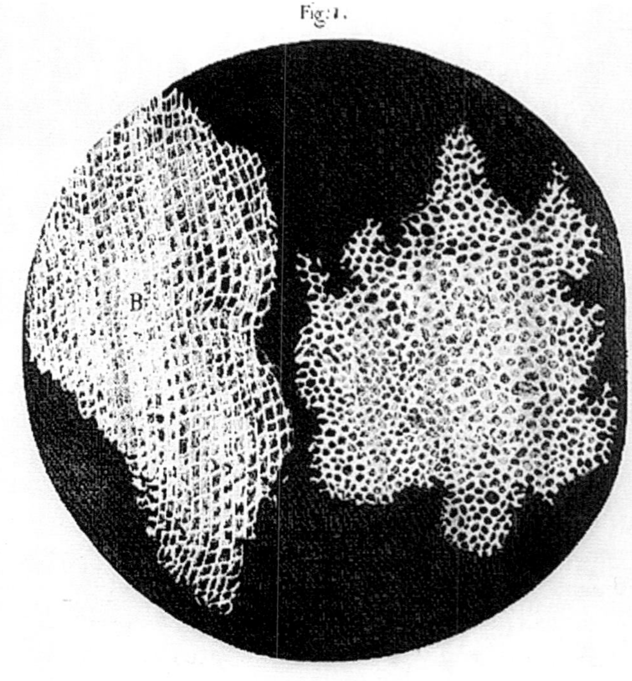

Figure 1.2 Drawing of cork cells published by Robert Hooke in 1665.

Why cells?

A cell can be thought of as a bag in which the chemistry of life is allowed to occur, partially separated from the environment outside the cell. The thin membrane which surrounds all cells is essential in controlling exchange between the cell and its environment. It is a very effective barrier, but also allows a controlled traffic of materials across it in both directions. The membrane is therefore described as **partially permeable**. If it were **freely permeable**, life could not exist, because the chemicals of the cell would simply mix with the surrounding chemicals by diffusion.

Cell biology and microscopy

The study of cells has given rise to an important branch of biology known as **cell biology**. Cells can now be studied by many different methods, but scientists began simply by looking at them, using various types of microscope.

There are two fundamentally different types of microscope now in use: the light microscope and the electron microscope. Both use a form of radiation in order to create an image of the specimen being examined. The light microscope uses light as a source of radiation, while the electron microscope uses electrons, for reasons which are discussed later.

Light microscopy

The 'golden age' of light microscopy could be said to be the 19th century. Microscopes had been available since the beginning of the 17th century but, when dramatic improvements were made in the quality of glass lenses in the early 19th century, interest among scientists became widespread. The fascination of the microscopic world that opened up in biology inspired rapid progress both in microscope design and, equally importantly, in preparing material for examination with microscopes. This branch of biology is known as **cytology**. Figure 1.3 shows how the light microscope works.

By 1900, all the structures shown in Figures 1.4 and 1.5 had been discovered. Figure 1.4 shows the structure of a generalised animal cell and Figure 1.5 the structure of a generalised plant cell as seen with a light microscope. (A generalised cell shows **all** the structures that are typically found in a cell.) Figure 1.6 shows some **actual** human cells and Figure 1.7 shows an actual plant cell taken from a leaf.

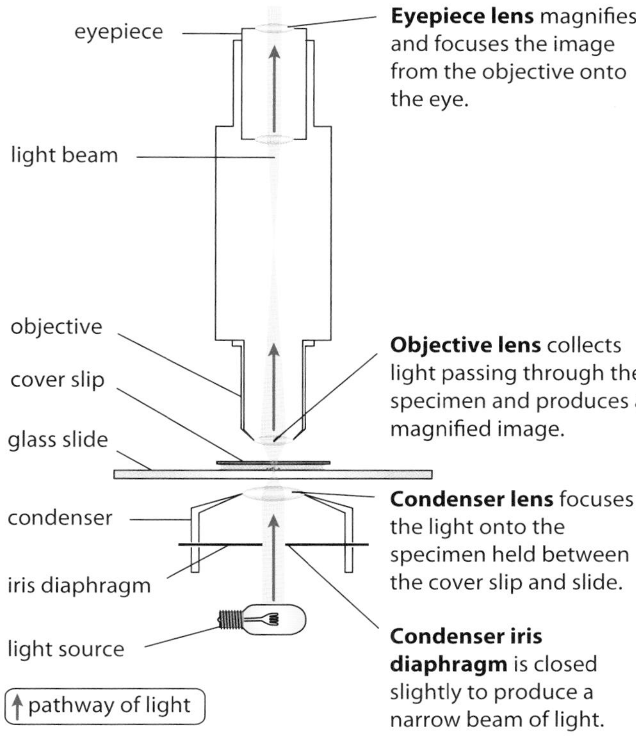

Figure 1.3 How the light microscope works.

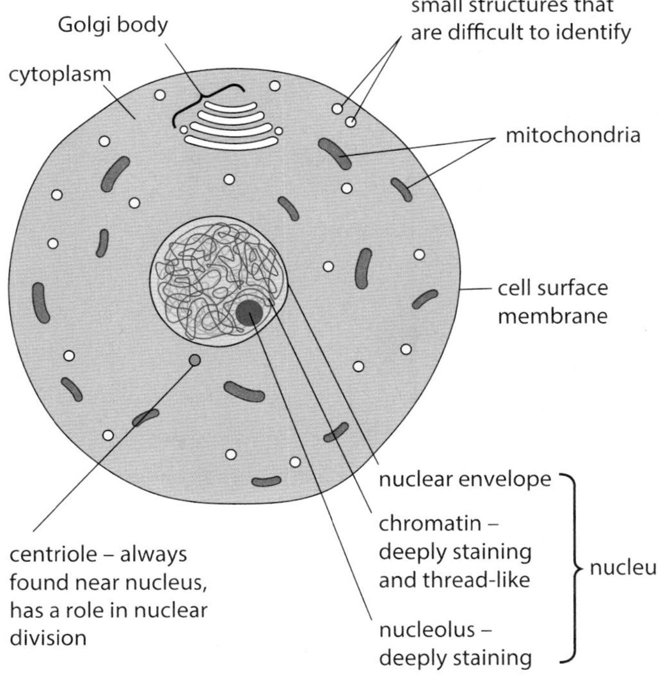

Figure 1.4 Structure of a generalised animal cell (diameter about 20 μm) as seen with a very high quality light microscope.

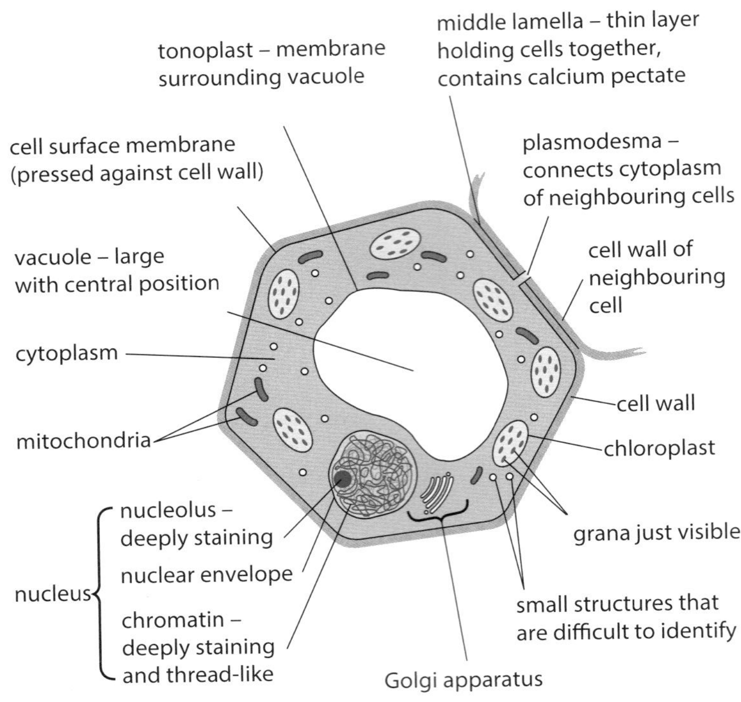

Figure 1.5 Structure of a generalised plant cell (diameter about 40 µm) as seen with a very high quality light microscope.

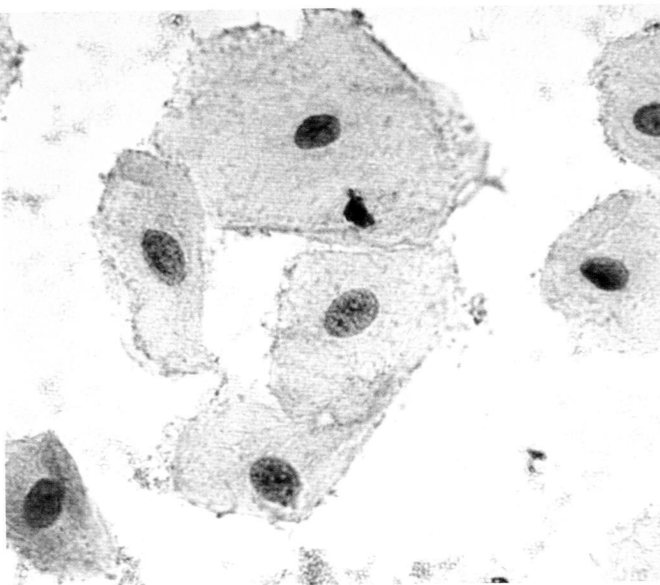

Figure 1.6 Cells from the lining of the human cheek (×400), each showing a centrally placed nucleus, which is a typical animal cell characteristic. The cells are part of a tissue known as squamous (flattened) epithelium.

QUESTION

1.1 Using Figures 1.4 and 1.5, name the structures that animal and plant cells have in common, those found in only plant cells, and those found only in animal cells.

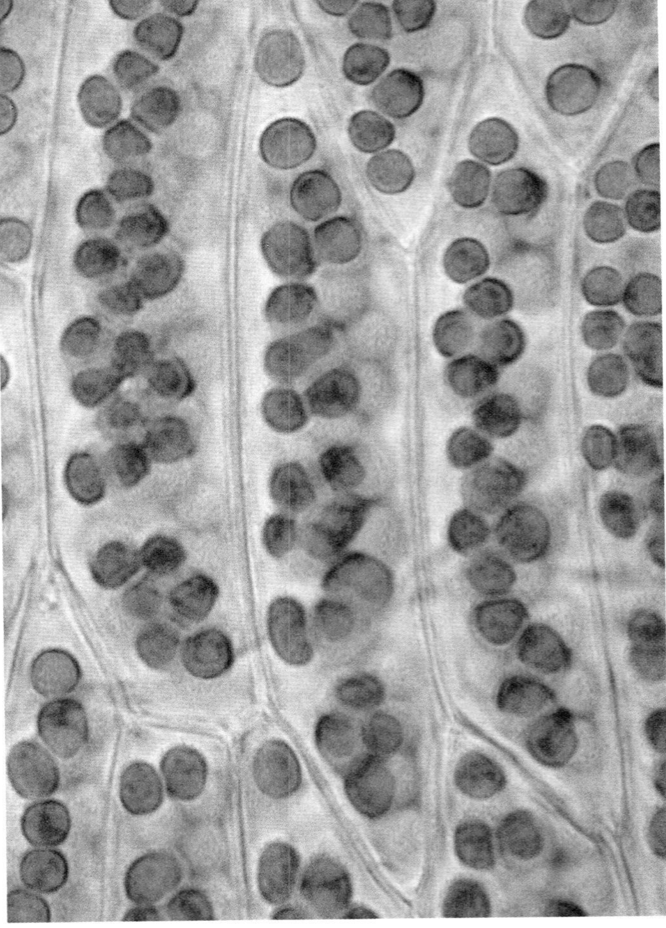

Figure 1.7 Photomicrograph of a cells in a moss leaf (×400).

Animal and plant cells have features in common

In animals and plants each cell is surrounded by a very thin cell surface membrane. This is also sometimes referred to as the plasma membrane.

Many of the cell contents are colourless and transparent so they need to be stained to be seen. Each cell has a nucleus, which is a relatively large structure that stains intensely and is therefore very conspicuous. The deeply staining material in the nucleus is called chromatin and is a mass of loosely coiled threads. This material collects together to form visible separate chromosomes during nuclear division (page 98). It contains DNA (deoxyribonucleic acid), a molecule which contains the instructions that control the activities of the cell (see Chapter **6**). Within the nucleus an even more deeply staining area is visible, the nucleolus, which is made of loops of DNA from several chromosomes. The number of nucleoli is variable, one to five being common in mammals.

The material between the nucleus and the cell surface membrane is known as cytoplasm. Cytoplasm is an aqueous (watery) material, varying from a fluid to a jelly-like consistency. Many small structures can be seen within it. These have been likened to small organs and hence are known as organelles. An organelle can be defined as a functionally and structurally distinct part of a cell. Organelles themselves are often surrounded by membranes so that their activities can be separated from the surrounding cytoplasm. This is described as **compartmentalisation**. Having separate compartments is essential for a structure as complex as an animal or plant cell to work efficiently. Since each type of organelle has its own function, the cell is said to show **division of labour**, a sharing of the work between different specialised organelles.

The most numerous organelles seen with the light microscope are usually mitochondria (singular: mitochondrion). Mitochondria are only just visible, but films of living cells, taken with the aid of a light microscope, have shown that they can move about, change shape and divide. They are specialised to carry out aerobic respiration.

The use of special stains containing silver enabled the Golgi apparatus to be detected for the first time in 1898 by Camillo Golgi. The Golgi apparatus is part of a complex internal sorting and distribution system within the cell (page 15). It is also sometimes called the **Golgi body** or **Golgi complex**.

Differences between animal and plant cells

The only structure commonly found in animal cells which is absent from plant cells is the centriole. Plant cells also differ from animal cells in possessing cell walls, large permanent vacuoles and chloroplasts.

Centrioles

Under the light microscope the centriole appears as a small structure close to the nucleus (Figure 1.4, page 3). Centrioles are discussed on page 18.

Cell walls and plasmodesmata

With a light microscope, individual plant cells are more easily seen than animal cells, because they are usually larger and, unlike animal cells, surrounded by a cell wall outside the cell surface membrane. This is relatively rigid because it contains fibres of cellulose, a polysaccharide which strengthens the wall. The cell wall gives the cell a definite shape. It prevents the cell from bursting when water enters by osmosis, allowing large pressures to develop inside the cell (page 84). Cell walls may also be reinforced with extra cellulose or with a hard material called lignin for extra strength (page 141). Cell walls are freely permeable, allowing free movement of molecules and ions through to the cell surface membrane.

Plant cells are linked to neighbouring cells by means of fine strands of cytoplasm called plasmodesmata (singular: plasmodesma), which pass through pore-like structures in their walls. Movement through the pores is thought to be controlled by the structure of the pores.

Vacuoles

Although animal cells may possess small vacuoles such as phagocytic vacuoles (page 87), which are temporary structures, mature plant cells often possess a large, permanent, central vacuole. The plant vacuole is surrounded by a membrane, the tonoplast, which controls exchange between the vacuole and the cytoplasm. The fluid in the vacuole is a solution of pigments, enzymes, sugars and other organic compounds (including some waste products), mineral salts, oxygen and carbon dioxide.

Vacuoles help to regulate the osmotic properties of cells (the flow of water inwards and outwards) as well as having a wide range of other functions. For example, the pigments which colour the petals of certain flowers and parts of some vegetables, such as the red pigment of beetroots, may be located in vacuoles.

Chloroplasts

Chloroplasts are found in the green parts of the plant, mainly in the leaves. They are relatively large organelles and so are easily seen with a light microscope. It is even possible to see tiny 'grains' or grana (singular: granum) inside the chloroplasts using a light microscope. These are the parts of the chloroplast that contain chlorophyll, the green pigment which absorbs light during the process of photosynthesis, the main function of chloroplasts. Chloroplasts are discussed further on page 19.

Points to note

- You can think of a plant cell as being very similar to an animal cell, but with extra structures.
- Plant cells are often larger than animal cells, although cell size varies enormously.
- Do not confuse the cell **wall** with the cell surface **membrane**. Cell walls are relatively thick and physically strong, whereas cell surface membranes are very thin. Cell walls are freely permeable, whereas cell surface membranes are partially permeable. **All** cells have a cell surface membrane.
- Vacuoles are not confined to plant cells; animal cells may have small vacuoles, such as phagocytic vacuoles, although these are not usually permanent structures.

We return to the differences between animal and plant cells as seen using the **electron microscope** on page 13.

Units of measurement

In order to measure objects in the microscopic world, we need to use very small units of measurement, which are unfamiliar to most people. According to international agreement, the International System of Units (SI units) should be used. In this system, the basic unit of length is the **metre** (symbol, **m**). Additional units can be created in multiples of a thousand times larger or smaller, using standard prefixes. For example, the prefix **kilo** means **1000** times. Thus 1 kilometre = 1000 metres. The units of length relevant to cell studies are shown in Table 1.1.

It is difficult to imagine how small these units are, but, when looking down a microscope and seeing cells clearly, we should not forget how amazingly small the cells actually are. The smallest structure visible with the human eye is about 50–100 μm in diameter. Your body contains about 60 million million cells, varying in size from about 5 μm to 40 μm. Try to imagine structures like mitochondria, which have an average diameter of 1 μm. The smallest cell organelles we deal with in this book, ribosomes, are only about 25 nm in diameter! You could line up about 20 000 ribosomes across the full stop at the end of this sentence.

Electron microscopy

As we said on page 3, by 1900 almost all the structures shown in Figures 1.4 and 1.5 (pages 3 and 4) had been discovered. There followed a time of frustration for microscopists, because they realised that no matter how much the design of light microscopes improved, there was a limit to how much could ever be seen using light.

In order to understand why this is, it is necessary to know something about the nature of light itself and to understand the difference between **magnification** and **resolution**.

Magnification

Magnification is the number of times larger an image is, than the real size of the object.

$$\text{magnification} = \frac{\text{observed size of the image}}{\text{actual size}}$$

or

$$M = \frac{I}{A}$$

Here I = observed size of the image (that is, what you can measure with a ruler) and A = actual size (that is, the real size – for example, the size of a cell before it is magnified).

If you know two of these values, you can work out the third one. For example, if the observed size of the image and the magnification are known, you can work out the actual size: $A = \frac{I}{M}$. If you write the formula in a triangle

Fraction of a metre	Unit	Symbol
one thousandth = 0.001 = 1/1000 = 10^{-3}	millimetre	mm
one millionth = 0.000 001 = 1/1000 000 = 10^{-6}	micrometre	μm
one thousand millionth = 0.000 000 001 = 1/1000 000 000 = 10^{-9}	nanometre	nm

Table 1.1 Units of measurement relevant to cell studies: μ is the Greek letter mu; 1 micrometre is a thousandth of a millimetre; 1 nanometre is a thousandth of a micrometre.

as shown on the right and cover up the value you want to find, it should be obvious how to do the right calculation.

Some worked examples are now provided.

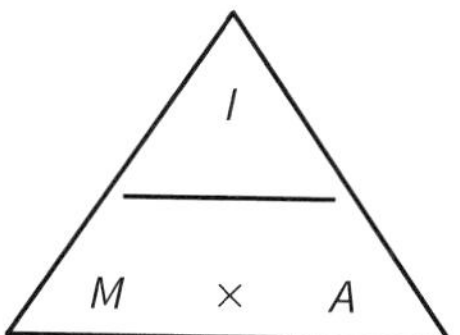

WORKED EXAMPLE 1

Measuring cells

Cells and organelles can be measured with a microscope by means of an **eyepiece graticule**. This is a transparent scale. It usually has 100 divisions (see Figure 1.8a). The eyepiece graticule is placed in the microscope eyepiece so that it can be seen at the same time as the object to be measured, as shown in Figure 1.8b. Figure 1.8b shows the scale over a human cheek epithelial cell. The cell lies between 40 and 60 on the scale. We therefore say it measures 20 eyepiece units in diameter (the difference between 60 and 40). We will not know the actual size of the eyepiece units until the eyepiece graticule scale is calibrated.

To calibrate the eyepiece graticule scale, a miniature transparent ruler called a **stage micrometer** scale is placed on the microscope stage and is brought into focus. This scale may be etched onto a glass slide or printed on a transparent film. It commonly has subdivisions of 0.1 and 0.01 mm. The images of the two scales can then be superimposed as shown in Figure 1.8c.

In the eyepiece graticule shown in the figure, 100 units measure 0.25 mm. Hence, the value of each eyepiece unit is:

$$\frac{0.25}{100} = 0.0025\,\text{mm}$$

Or, converting mm to μm:

$$\frac{0.25 \times 1000}{100} = 2.5\,\mu\text{m}$$

The diameter of the cell shown superimposed on the scale in Figure 1.8b measures 20 eyepiece units and so its actual diameter is:

$$20 \times 2.5\,\mu\text{m} = 50\,\mu\text{m}$$

This diameter is greater than that of many human cells because the cell is a flattened epithelial cell.

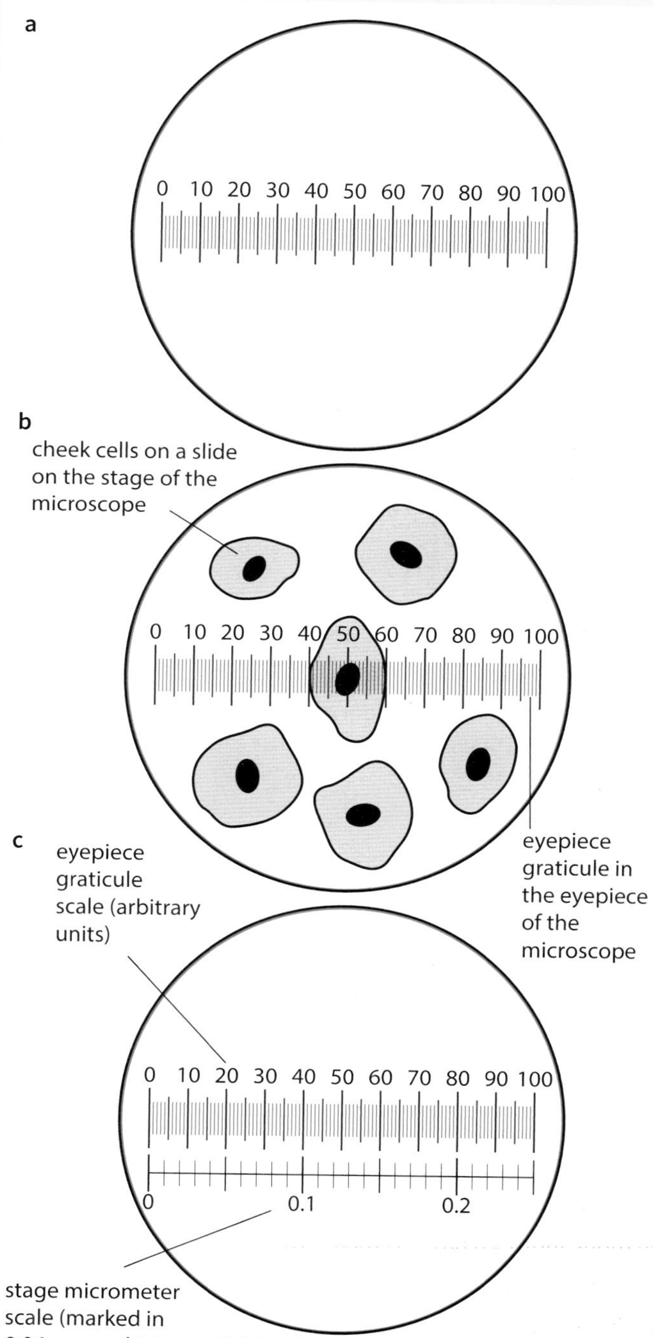

Figure 1.8 Microscopical measurement. Three fields of view seen using a high-power (×40) objective lens. **a** An eyepiece graticule scale. **b** Superimposed images of human cheek epithelial cells and the eyepiece graticule scale. **c** Superimposed images of the eyepiece graticule scale and the stage micrometer scale.

WORKED EXAMPLE 2

Calculating the magnification of a photograph or image

To calculate M, the magnification of a photograph or an object, we can use the following method.

Figure 1.9 shows two photographs of a section through the same plant cells. The magnifications of the two photographs are the same. Suppose we want to know the magnification of the plant cell labelled P in Figure 1.9b. If we know its actual (real) length we can calculate its magnification using the formula

$$M = \frac{I}{A}$$

The real length of the cell is 80 µm.

Step 1 Measure the length in mm of the cell in the photograph using a ruler. You should find that it is about 60 mm.

Step 2 Convert mm to µm. (It is easier if we first convert all measurements to the same units – in this case micrometres, µm.)

$$1\,\text{mm} = 1000\,\mu\text{m}$$

so

$$60\,\text{mm} = 60 \times 1000\,\mu\text{m} = 60\,000\,\mu\text{m}$$

Step 3 Use the equation to calculate the magnification.

$$\text{magnification, } M = \frac{\text{image size, } I}{\text{actual size, } A} = \frac{60\,000\,\mu\text{m}}{80\,\mu\text{m}} = \times 750$$

The multiplication sign in front of the number 750 means 'times'. We say that the magnification is 'times 750'.

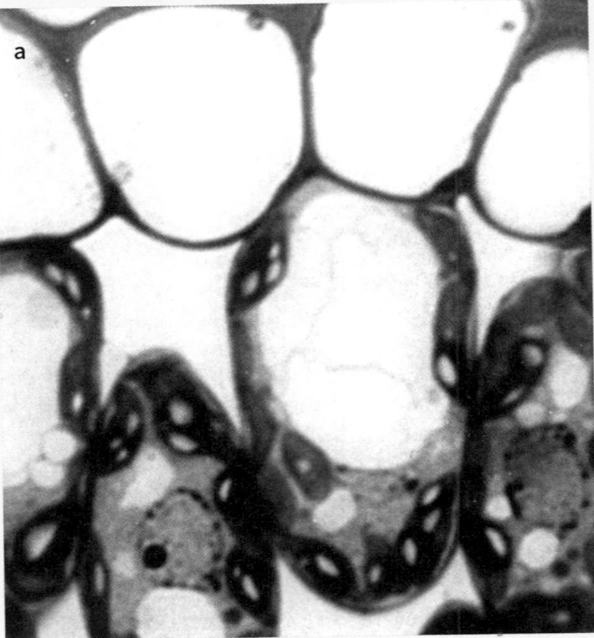

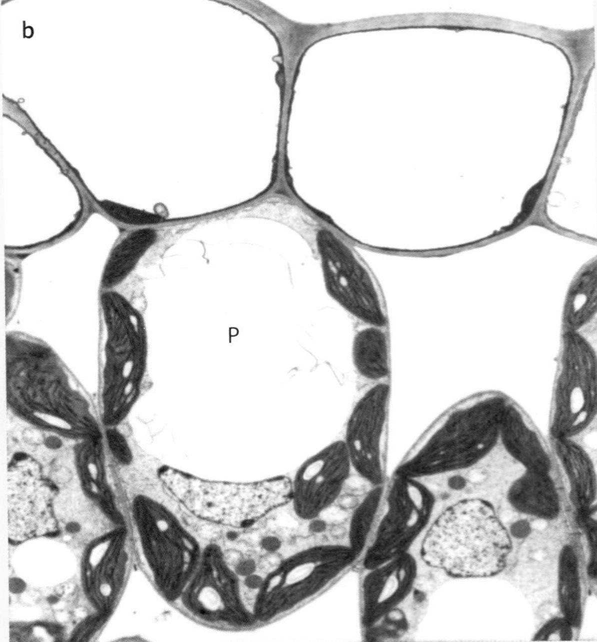

Figure 1.9 Photographs of the same types of plant cells seen **a** with a light microscope, **b** with an electron microscope, both shown at a magnification of about ×750.

QUESTION

1.2 a Calculate the magnification of the drawing of the animal cell in Figure 1.4 on page 3.

b Calculate the actual (real) length of the chloroplast labelled **X** in Figure 1.29 on page 21.

WORKED EXAMPLE 3

Calculating magnification from a scale bar

Figure 1.10 shows a lymphocyte.

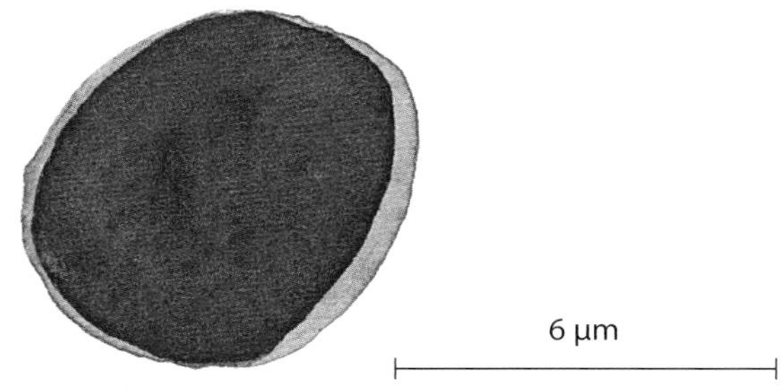

Figure 1.10 A lymphocyte.

We can calculate the magnification of the lymphocyte by simply using the scale bar. All you need to do is measure the length of the scale bar and then substitute this and the length it represents into the equation.

Step 1 Measure the scale bar. Here, it is 36 mm.

Step 2 Convert mm to µm:

$36\,\text{mm} = 36 \times 1000\,\mu\text{m} = 36\,000\,\mu\text{m}$

Step 3 Use the equation to calculate the magnification:

$$\text{magnification, } M = \frac{\text{image size, } I}{\text{actual size, } A} = \frac{36\,000\,\mu\text{m}}{6\,\mu\text{m}} = \times 6000$$

WORKED EXAMPLE 4

Calculating the real size of an object from its magnification

To calculate A, the real or actual size of an object, we can use the following method.

Figure 1.27 on page 19 shows parts of three plant cells magnified ×5600. One of the chloroplasts is labelled 'chloroplast' in the figure. Suppose we want to know the actual length of this chloroplast.

Step 1 Measure the observed length of the image of the chloroplast (I), in mm, using a ruler. The maximum length is 40 mm.

Step 2 Convert mm to µm:

$40\,\text{mm} = 40 \times 1000\,\mu\text{m} = 40\,000\,\mu\text{m}$

Step 3 Use the equation to calculate the actual length:

$$\text{actual size, } A = \frac{\text{image size, } I}{\text{magnification, } M} = \frac{40\,000\,\mu\text{m}}{5600} = 7.1\,\mu\text{m (to one decimal place)}$$

BOX 1.1: Making temporary slides

Background information

Biological material may be examined live or in a preserved state. Prepared slides contain material that has been killed and preserved in a life-like condition. This material is often cut into thin sections to enable light to pass through the structures for viewing with a light microscope. The sections are typically stained and 'mounted' on a glass slide, forming a permanent preparation.

Temporary preparations of fresh material have the advantage that they can be made rapidly and are useful for quick preliminary investigations. Sectioning and staining may still be carried out if required. Sometimes macerated (chopped up) material can be used, as when examining the structure of wood (xylem). A number of temporary stains are commonly used. For example, iodine in potassium iodide solution is useful for plant specimens. It stains starch blue-black and will also colour nuclei and cell walls a pale yellow. A dilute solution of methylene blue can be used to stain animal cells such as cheek cells.

Viewing specimens yourself with a microscope will help you to understand and remember structures more fully. This can be reinforced by making a pencil drawing on good quality plain paper, using the guidance given later in Chapter 7 (Box 7.1, page 129). Remember always to draw what you see, and not what you think you should see.

Procedure

The material is placed on a clean glass slide and one or two drops of stain added. A cover slip is carefully lowered over the specimen to protect the microscope lens and to help prevent the specimen from drying out. A drop of glycerine mixed with the stain can also help prevent drying out.

Suitable animal material: human cheek cells

Suitable plant material: onion epidermal cells, lettuce epidermal cells, *Chlorella* cells, moss leaves

Resolution

Look again at Figure 1.9 (page 8). Figure 1.9a is a **light micrograph** (a photograph taken with a light microscope, also known as a **photomicrograph**). Figure 1.9b is an **electron micrograph** of the same specimen taken at the same magnification (an electron micrograph is a picture taken with an electron microscope). You can see that Figure 1.9b, the electron micrograph, is much clearer. This is because it has greater resolution. Resolution can be defined as the ability to distinguish between two separate points. If the two points cannot be **resolved**, they will be seen as one point. In practice, resolution is the amount of detail that can be seen – the greater the resolution, the greater the detail.

The maximum resolution of a light microscope is 200 nm. This means that if two points or objects are closer together than 200 nm they cannot be distinguished as separate.

It is possible to take a photograph such as Figure 1.9a and to magnify (enlarge) it, but we see no more detail; in other words, we do not improve resolution, even though we often enlarge photographs because they are easier to see when larger. With a microscope, magnification up to the limit of resolution can reveal further detail, but any further magnification increases blurring as well as the size of the image.

Resolution is the ability to distinguish between two objects very close together; the higher the resolution of an image, the greater the detail that can be seen.

Magnification is the number of times greater that an image is than the actual object;
magnification = image size ÷ actual (real) size of the object.

The electromagnetic spectrum

How is resolution linked with the nature of light? One of the properties of light is that it travels in waves. The length of the waves of visible light varies, ranging from about 400 nm (violet light) to about 700 nm (red light). The human eye can distinguish between these different wavelengths, and in the brain the differences are converted to colour differences. (Colour is an invention of the brain!)

The whole range of different wavelengths is called the **electromagnetic spectrum**. Visible light is only one part of this spectrum. Figure 1.11 shows some of the parts of the electromagnetic spectrum. The longer the waves, the lower their frequency (all the waves travel at the same speed, so imagine them passing a post: shorter waves pass at higher frequency). In theory, there is no limit to how short or how long the waves can be. Wavelength changes with energy: the greater the energy, the shorter the wavelength.

Now look at Figure 1.12, which shows a mitochondrion, some very small cell organelles called ribosomes (page 15) and light of 400 nm wavelength, the shortest visible wavelength. The mitochondrion is large enough to interfere with the light waves. However, the ribosomes are far too small to have any effect on the light waves. The general rule is that the limit of resolution is about one half the wavelength of the radiation used to view the specimen. In other words, if an object is any smaller than half the wavelength of the radiation used to view it, it cannot be seen separately from nearby objects. This means that the best resolution that can be obtained using a microscope that uses visible light (a light microscope) is 200 nm, since the shortest wavelength of visible light is 400 nm (violet light). In practice, this corresponds to a maximum useful magnification of about 1500 times. Ribosomes are approximately 25 nm in diameter and can therefore never be seen using light.

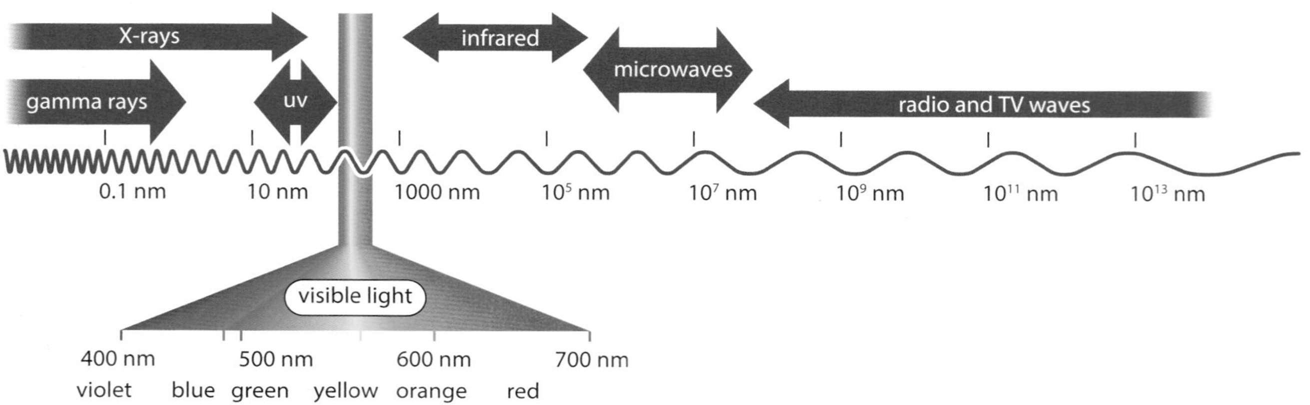

Figure 1.11 Diagram of the electromagnetic spectrum (the waves are not drawn to scale). The numbers indicate the wavelengths of the different types of electromagnetic radiation. Visible light is a form of electromagnetic radiation. The arrow labelled uv is ultraviolet light.

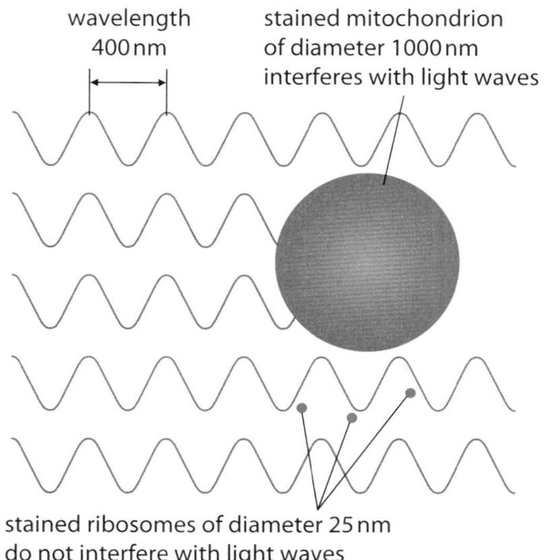

Figure 1.12 A mitochondrion and some ribosomes in the path of light waves of 400 nm length.

If an object is transparent, it will allow light waves to pass through it and therefore will still not be visible. This is why many biological structures have to be stained before they can be seen.

QUESTION

1.3 Explain why ribosomes are not visible using a light microscope.

The electron microscope

Biologists, faced with the problem that they would never see anything smaller than 200 nm using a light microscope, realised that the only solution would be to use radiation of a shorter wavelength than light. If you study Figure 1.11, you will see that ultraviolet light, or better still X-rays, look like possible candidates. Both ultraviolet and X-ray microscopes have been built, the latter with little success partly because of the difficulty of focusing X-rays. A much better solution is to use **electrons**. Electrons are negatively charged particles which orbit the nucleus of an atom. When a metal becomes very hot, some of its electrons gain so much energy that they escape from their orbits, like a rocket escaping from Earth's gravity. Free electrons behave like electromagnetic radiation. They have a very short wavelength: the greater the energy, the shorter the wavelength. Electrons are a very suitable form of radiation for microscopy for two major reasons. Firstly, their wavelength is extremely short (at least as short as that of X-rays). Second, because they are negatively charged, they can be focused easily using electromagnets (a magnet can be made to alter the path of the beam, the equivalent of a glass lens bending light).

Using an electron microscope, a resolution of 0.5 nm can be obtained, 400 times better than a light microscope.

Transmission and scanning electron microscopes

Two types of electron microscope are now in common use. The **transmission electron microscope**, or **TEM**, was the type originally developed. Here the beam of electrons is passed **through** the specimen before being viewed. Only those electrons that are **transmitted** (pass through the specimen) are seen. This allows us to see thin sections of specimens, and thus to see inside cells. In the **scanning electron microscope** (SEM), on the other hand, the electron beam is used to scan the **surfaces** of structures, and only the **reflected** beam is observed.

An example of a scanning electron micrograph is shown in Figure 1.13. The advantage of this microscope is that surface structures can be seen. Also, great depth of field is obtained so that much of the specimen is in focus at the same time and a three-dimensional appearance is achieved. Such a picture would be impossible to obtain with a light microscope, even using the same magnification and resolution, because you would have to keep focusing up and down with the objective lens to see different parts of the specimen. The disadvantage of the SEM is that it cannot achieve the same resolution as a TEM. Using an SEM, resolution is between 3 nm and 20 nm.

Figure 1.13 False-colour scanning electron micrograph of the head of a cat flea (×100).

Viewing specimens with the electron microscope

Figure 1.14 shows how an electron microscope works and Figure 1.15 shows one in use.

It is not possible to see an electron beam, so to make the image visible the electron beam has to be projected onto a fluorescent screen. The areas hit by electrons shine brightly, giving overall a black and white picture. The stains used to improve the contrast of biological specimens for electron microscopy contain heavy metal atoms, which stop the passage of electrons. The resulting picture is like an X-ray photograph, with the more densely stained parts of the specimen appearing blacker. 'False-colour' images can be created by colouring the standard black and white image using a computer.

To add to the difficulties of electron microscopy, the electron beam, and therefore the specimen and the fluorescent screen, must be in a vacuum. If electrons collided with air molecules, they would scatter, making it impossible to achieve a sharp picture. Also, water boils at room temperature in a vacuum, so all specimens must be dehydrated before being placed in the microscope. This means that only dead material can be examined. Great efforts are therefore made to try to preserve material in a life-like state when preparing it for electron microscopy.

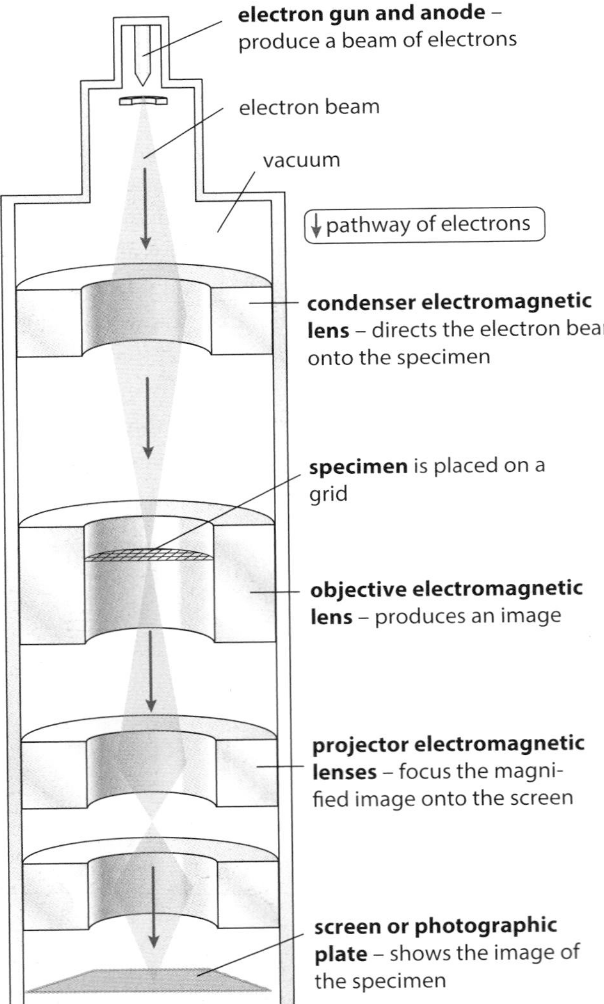

Figure 1.14 How an electron microscope (EM) works.

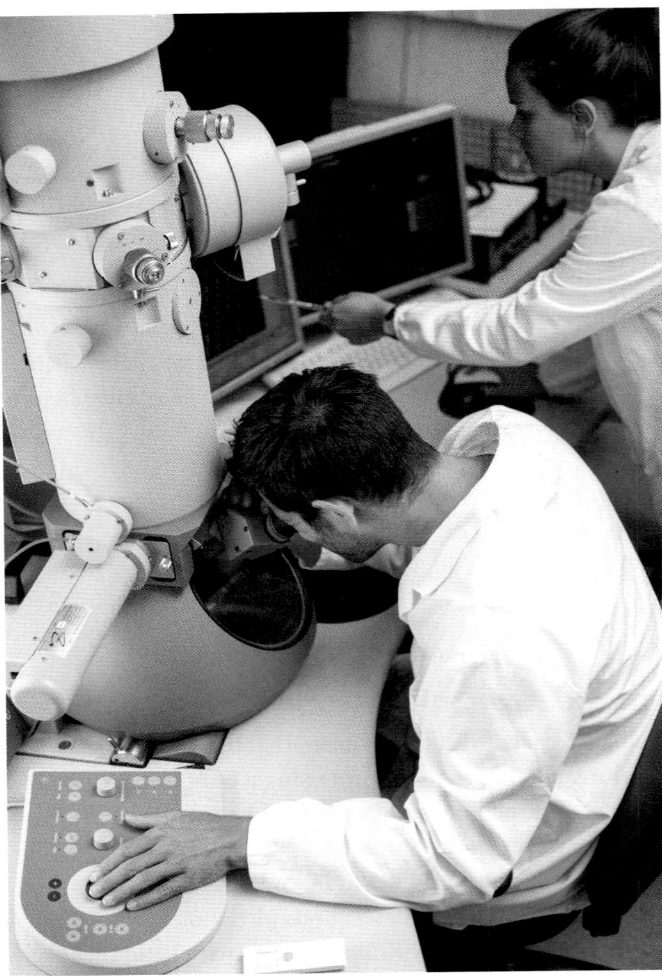

Figure 1.15 A transmission electron microscope (TEM) in use.

Ultrastructure of an animal cell

The fine (detailed) structure of a cell as revealed by the electron microscope is called its **ultrastructure**.

Figure 1.16 shows the appearance of typical animal cells as seen with an electron microscope, and Figure 1.17 is a diagram based on many other such micrographs.

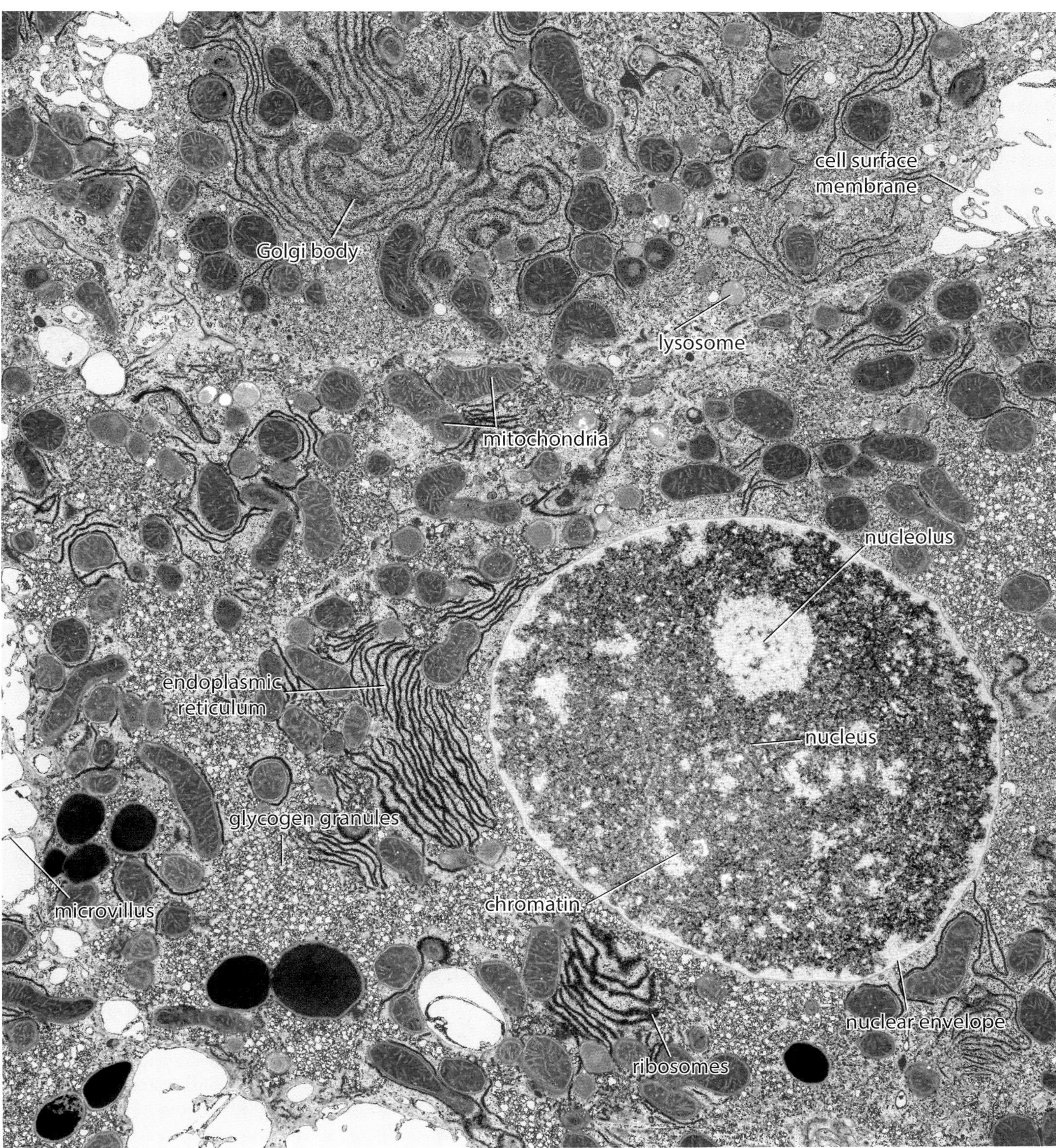

Figure 1.16 Representative animal cells as seen with a TEM. The cells are liver cells from a rat (×9600). The nucleus is clearly visible in one of the cells.

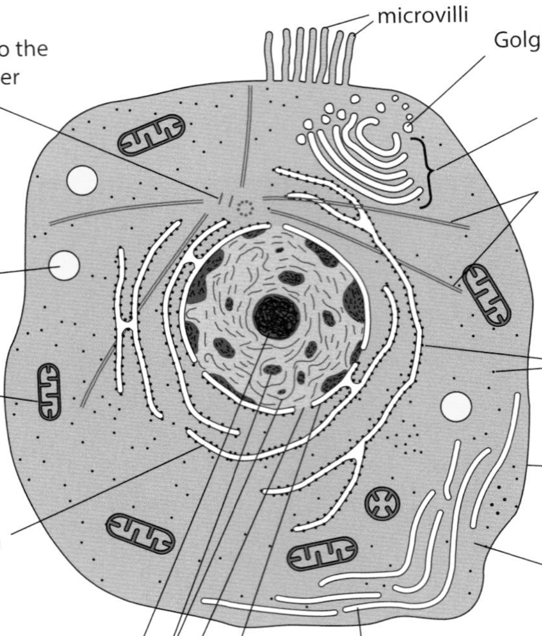

microvilli
Golgi vesicle
centrosome with two centrioles close to the nucleus and at right angles to each other
Golgi body
microtubules radiating from centrosome
lysosome
ribosomes
mitochondrion
cell surface membrane
rough endoplasmic reticulum
cytoplasm
nucleus { nucleolus
chromatin
nuclear pore
nuclear envelope (two membranes)
smooth endoplasmic reticulum

Figure 1.17 Ultrastructure of a typical animal cell as seen with an electron microscope. In reality, the ER is more extensive than shown, and free ribosomes may be more extensive. Glycogen granules are sometimes present in the cytoplasm.

QUESTION

1.4 Compare Figure 1.17 with Figure 1.4 on page 3. Name the structures in an animal cell which can be seen with the electron microscope but not with the light microscope.

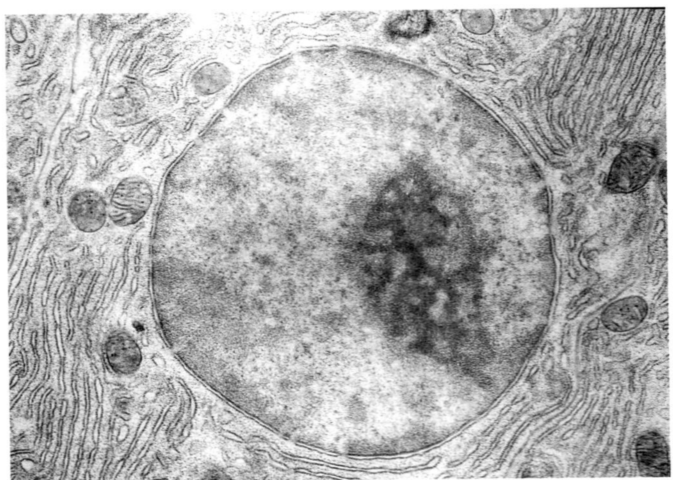

Figure 1.18 Transmission electron micrograph of the nucleus of a cell from the pancreas of a bat (×7500). The circular nucleus is surrounded by a double-layered nuclear envelope containing nuclear pores. The nucleolus is more darkly stained. Rough ER (page 15) is visible in the surrounding cytoplasm.

Structures and functions of organelles

Compartmentalisation and division of labour within the cell are even more obvious with an electron microscope than with a light microscope. We will now consider the structures and functions of some of the cell components in more detail.

Nucleus

The **nucleus** (Figure 1.18) is the largest cell organelle. It is surrounded by two membranes known as the nuclear envelope. The outer membrane of the nuclear envelope is continuous with the endoplasmic reticulum (Figure 1.17).

The nuclear envelope has many small pores called nuclear pores. These allow and control exchange between the nucleus and the cytoplasm. Examples of substances leaving the nucleus through the pores are mRNA and ribosomes for protein synthesis. Examples of substances entering through the nuclear pores are proteins to help make ribosomes, nucleotides, ATP (adenosine triphosphate) and some hormones such as thyroid hormone T3.

Within the nucleus, the chromosomes are in a loosely coiled state known as chromatin (except during nuclear division, Chapter 5). Chromosomes contain DNA, which is organised into functional units called genes. Genes control the activities of the cell and inheritance; thus the nucleus controls the cell's activities. When a cell is about to divide, the nucleus divides first so that each new cell will have its own nucleus (Chapters 5 and 16). Also within the nucleus, the **nucleolus** makes ribosomes, using the information in its own DNA.

Endoplasmic reticulum and ribosomes

When cells were first seen with the electron microscope, biologists were amazed to see so much detailed structure. The existence of much of this had not been suspected. This was particularly true of an extensive system of membranes running through the cytoplasm, which became known as the endoplasmic reticulum (ER) (Figures 1.18, 1.19 and 1.22). The membranes form an extended system of flattened compartments, called sacs, spreading throughout the cell. Processes can take place inside these sacs, separated from the cytoplasm. The sacs can be interconnected to form a complete system (reticulum) – the connections have been compared to the way in which the different levels of a parking lot are connected by ramps. The ER is continuous with the outer membrane of the nuclear envelope (Figure 1.17).

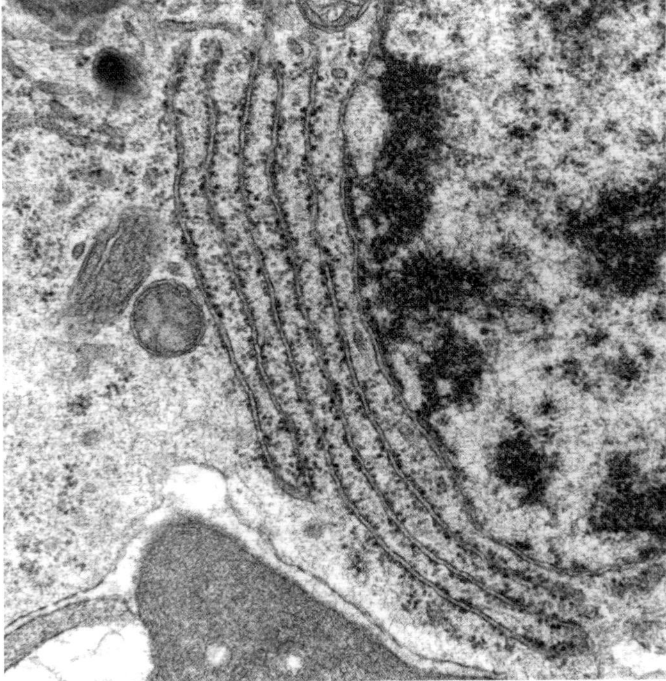

Figure 1.19 Transmission electron micrograph of rough ER covered with ribosomes (black dots) (×17 000). Some free ribosomes can also be seen in the cytoplasm on the left.

There are two types of ER: rough ER and smooth ER. **Rough ER** is so called because it is covered with many tiny organelles called ribosomes. These are just visible as black dots in Figures 1.18 and 1.19. At very high magnifications they can be seen to consist of two subunits: a large and a small subunit. Ribosomes are the sites of protein synthesis (page 119). They can be found free in the cytoplasm as well as on the rough ER. They are very small, only about 25 nm in diameter. They are made of RNA (ribonucleic acid) and protein. Proteins made by the ribosomes on the rough ER enter the sacs and move through them. The proteins are often modified in some way on their journey. Small sacs called vesicles can break off from the ER and these can join together to form the Golgi body. They form part of the secretory pathway because the proteins can be exported from the cell via the Golgi vesicles (Figure 1.2).

Smooth ER, so called because it lacks ribosomes, has a completely different function. It makes lipids and steroids, such as cholesterol and the reproductive hormones oestrogen and testosterone.

Golgi body (Golgi apparatus or Golgi complex)

The **Golgi body** is a stack of flattened sacs (Figure 1.20). More than one Golgi body may be present in a cell. The stack is constantly being formed at one end from vesicles which bud off from the ER, and broken down again at the other end to form **Golgi vesicles**. The stack of sacs together with the associated vesicles is referred to as the Golgi apparatus or Golgi complex.

The Golgi body collects, processes and sorts molecules (particularly proteins from the rough ER), ready for transport in Golgi vesicles either to other parts of the cell or out of the cell (secretion). Two examples of protein processing in the Golgi body are the addition of sugars to proteins to make molecules known as glycoproteins, and the removal of the first amino acid, methionine, from newly formed proteins to make a functioning protein. In plants, enzymes in the Golgi body convert sugars into cell wall components. Golgi vesicles are also used to make lysosomes.

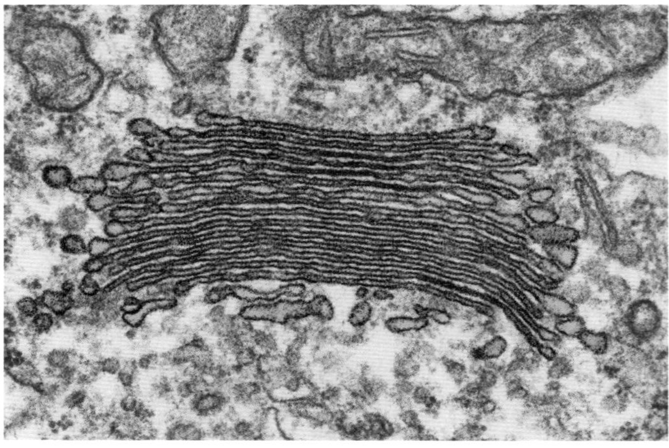

Figure 1.20 Transmission electron micrograph of a Golgi body. A central stack of saucer-shaped sacs can be seen budding off small Golgi vesicles (green). These may form secretory vesicles whose contents can be released at the cell surface by exocytosis (page 87).

Lysosomes

Lysosomes (Figure 1.21) are spherical sacs, surrounded by a single membrane and having no internal structure. They are commonly 0.1–0.5 µm in diameter. They contain digestive (hydrolytic) enzymes which must be kept separate from the rest of the cell to prevent damage from being done. Lysosomes are responsible for the breakdown (digestion) of unwanted structures such as old organelles or even whole cells, as in mammary glands after lactation (breast feeding). In white blood cells, lysosomes are used to digest bacteria (see endocytosis, page 87). Enzymes are sometimes released outside the cell – for example, in the replacement of cartilage with bone during development. The heads of sperm contain a special lysosome, the acrosome, for digesting a path to the ovum (egg).

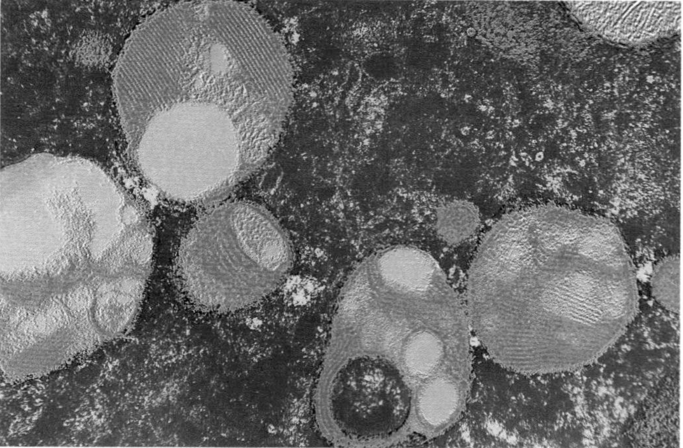

Figure 1.21 Lysosomes (orange) in a mouse kidney cell (×55 000). They contain cell structures in the process of digestion, and vesicles (green). Cytoplasm is coloured blue here.

Mitochondria

Structure

The structure of the mitochondrion as seen with the electron microscope is visible in Figures 1.16, 1.22, 12.13 and 12.14. Mitochondria (singular: mitochondrion) are usually about 1 µm in diameter and can be various shapes, often sausage-shaped as in Figure 1.22. They are surrounded by two membranes (an envelope). The inner of these is folded to form finger-like cristae which project into the interior solution, or **matrix**. The space between the two membranes is called the **intermembrane** space. The outer membrane contains a transport protein called **porin**, which forms wide aqueous channels allowing easy access of small, water-soluble molecules from the surrounding cytoplasm into the intermembrane space. The inner membrane is a far more selective barrier and controls precisely what ions and molecules can enter the matrix.

The number of mitochondria in a cell is very variable. As they are responsible for aerobic respiration, it is not surprising that cells with a high demand for energy, such as liver and muscle cells, contain large numbers of mitochondria. A liver cell may contain as many as 2000 mitochondria. If you exercise regularly, your muscles will make more mitochondria.

Function of mitochondria and the role of ATP

As we have seen, the main function of mitochondria is to carry out aerobic respiration, although they do have other functions, such as the synthesis of lipids. During

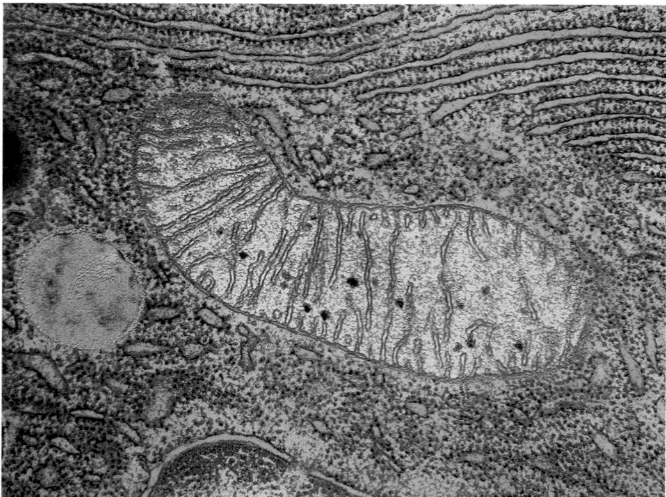

Figure 1.22 Mitochondrion (orange) with its double membrane (envelope); the inner membrane is folded to form cristae (×20 000). Mitochondria are the sites of aerobic cell respiration. Note also the rough ER.

respiration, a series of reactions takes place in which energy is released from energy-rich molecules such as sugars and fats. Most of this energy is transferred to molecules of ATP. ATP (adenosine triphosphate) is the energy-carrying molecule found in all living cells. It is known as the universal energy carrier.

The reactions of respiration take place in solution in the matrix and in the inner membrane (cristae). The matrix contains enzymes in solution, including those of the Krebs cycle (Chapter 12) and these supply the hydrogen and electrons to the reactions that take place in the cristae. The flow of electrons along the precisely placed electron carriers in the membranes of the cristae is what provides the power to generate ATP molecules, as explained in Chapter 12. The folding of the cristae increases the efficiency of respiration because it increases the surface area available for these reactions to take place.

Once made, ATP leaves the mitochondrion and, as it is a small, soluble molecule, it can spread rapidly to all parts of the cell where energy is needed. Its energy is released by breaking the molecule down to ADP (adenosine diphosphate). This is a hydrolysis reaction. The ADP can then be recycled into a mitochondrion for conversion back to ATP during aerobic respiration.

The endosymbiont theory

In the 1960s, it was discovered that mitochondria and chloroplasts contain ribosomes which are slightly smaller than those in the cytoplasm and are the same size as those found in bacteria. The size of ribosomes is measured in 'S units', which are a measure of how fast they sediment in a centrifuge. Cytoplasmic ribosomes are 80S, while those of bacteria, mitochondria and ribosomes are 70S. It was also discovered in the 1960s that mitochondria and chloroplasts contain small, circular DNA molecules, also like those found in bacteria. It was later proved that mitochondria and chloroplasts are, in effect, ancient bacteria which now live inside the larger cells typical of animals and plants (see prokaryotic and eukaryotic cells, page 21). This is known as the **endosymbiont theory**. 'Endo' means 'inside' and a 'symbiont' is an organism which lives in a mutually beneficial relationship with another organism. The DNA and ribosomes of mitochondria and chloroplasts are still active and responsible for the coding and synthesis of certain vital proteins, but mitochondria and chloroplasts can no longer live independently.

Mitochondrial ribosomes are just visible as tiny dark orange dots in the mitochondrial matrix in Figure 1.22.

Cell surface membrane

The cell surface membrane is extremely thin (about 7 nm). However, at very high magnifications, at least ×100 000, it can be seen to have three layers, described as a **trilaminar appearance**. This consists of two dark lines (heavily stained) either side of a narrow, pale interior (Figure 1.23). The membrane is partially permeable and controls exchange between the cell and its environment. Membrane structure is discussed further in Chapter 4.

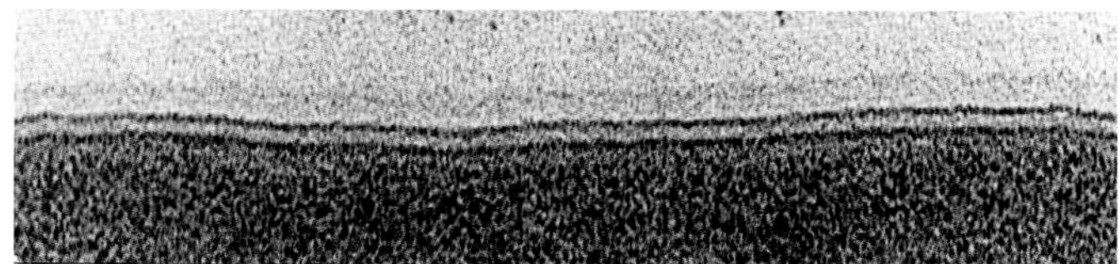

Figure 1.23 Cell surface membrane (×250 000). At this magnification the membrane appears as two dark lines at the edge of the cell.

Microvilli

Microvilli (singular: microvillus) are finger-like extensions of the cell surface membrane, typical of certain epithelial cells (cells covering surfaces of structures). They greatly increase the surface area of the cell surface membrane (Figure 1.17 on page 14). This is useful, for example, for absorption in the gut and for reabsorption in the proximal convoluted tubules of the kidney (page 308).

Microtubules and microtubule organising centres (MTOCs)

Microtubules are long, rigid, hollow tubes found in the cytoplasm. They are very small, about 25 nm in diameter. Together with actin filaments and intermediate filaments (not discussed in this book), they make up the cytoskeleton, an essential structural component of cells which helps to determine cell shape.

Microtubules are made of a protein called tubulin. Tubulin has two forms, α-tubulin (alpha-tubulin) and β-tubulin (beta-tubulin). α- and β-tubulin molecules combine to form dimers (double molecules). These dimers are then joined end to end to form long 'protofilaments'. This is an example of polymerisation. Thirteen protofilaments then line up alongside each other in a ring to form a cylinder with a hollow centre. This cylinder is the microtubule. Figure 1.24 (overleaf) shows the helical pattern formed by neighbouring α- and β-tubulin molecules.

Apart from their mechanical function of support, microtubules have a number of other functions. Secretory vesicles and other organelles and cell components can be moved along the outside surfaces of the microtubules, forming an intracellular transport system. Membrane-bound organelles are held in place by the cytoskeleton. During nuclear division (Chapter 5), the spindle used for the separation of chromatids or chromosomes is made of microtubules, and microtubules form part of the structure of centrioles.

The assembly of microtubules from tubulin molecules is controlled by special locations in cells called microtubule organising centres (MTOCs). These are discussed further in the following section on centrioles. Because of their simple construction, microtubules can be formed and broken down very easily at the MTOCs, according to need.

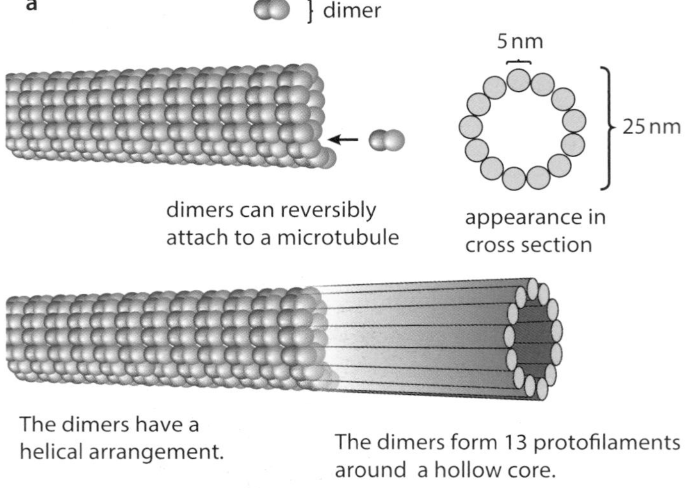

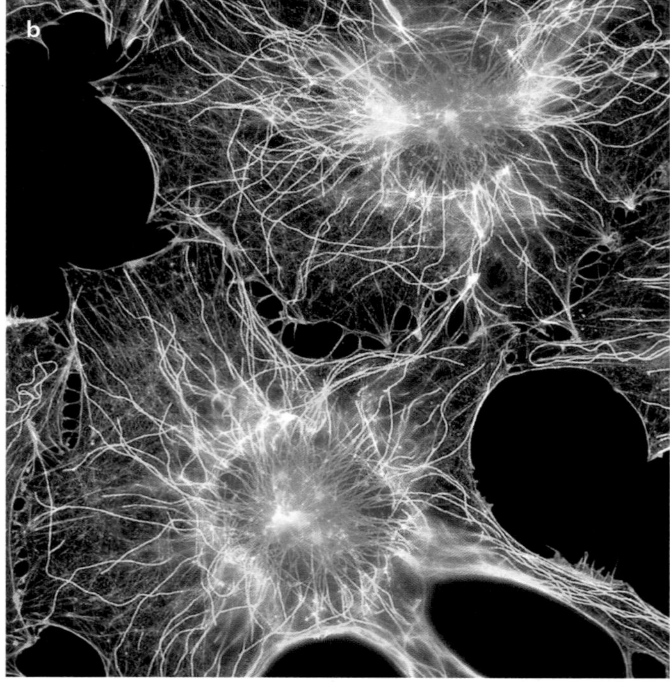

Figure 1.24 **a** The structure of a microtubule and **b** the arrangement of microtubules in two cells. The microtubules are coloured yellow.

Centrioles and centrosomes

The extra resolution of the electron microscope reveals that just outside the nucleus of animal cells there are really two centrioles and not one as it appears under the light microscope (compare Figures 1.4 and 1.17). They lie close together and at right angles to each other in a region known as the centrosome. Centrioles and the centrosome are absent from most plant cells.

A centriole is a hollow cylinder about 500 nm long, formed from a ring of short microtubules. Each centriole contains nine triplets of microtubules (Figures 1.25 and 1.26).

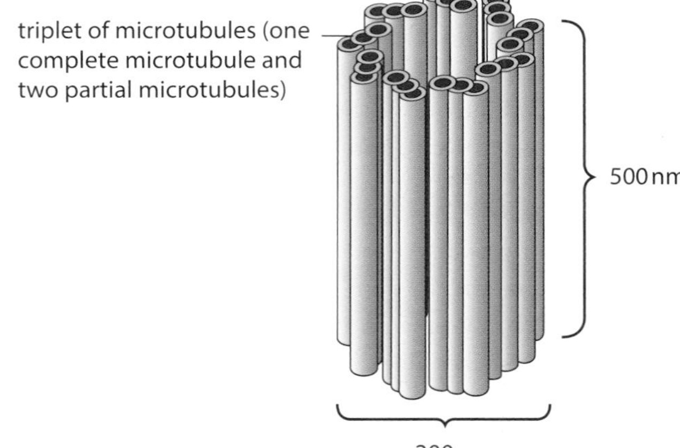

Figure 1.25 The structure of a centriole. It consists of nine groups of microtubules arranged in triplets.

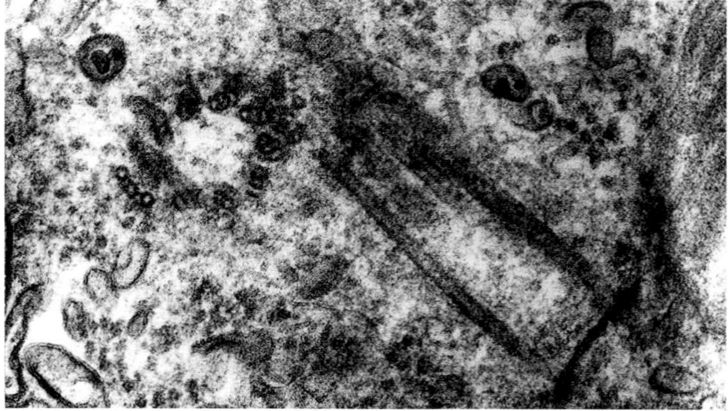

Figure 1.26 Centrioles in transverse and longitudinal section (TS and LS) (×86 000). The one on the left is seen in TS and clearly shows the nine triplets of microtubules which make up the structure.

The function of the centrioles remains a mystery. Until recently, it was believed that they acted as MTOCs for the assembly of the microtubules that make up the spindle during nuclear division (Chapter 5). It is now known that this is done by the centrosome, but does not involve the centrioles.

Centrioles found at the bases of cilia (page 189) and flagella, where they are known as basal bodies, do act as MTOCs. The microtubules that extend from the basal bodies into the cilia and flagella are essential for the beating movements of these organelles.

Ultrastructure of a plant cell

All the structures so far described in animal cells are also found in plant cells, with the exception of centrioles and microvilli. The plant cell structures that are not found in animal cells are the cell wall, the large central vacuole, and chloroplasts. These are all shown clearly in Figures 1.27 and 1.28. The structures and functions of cell walls and vacuoles have been described on page 5.

Chloroplasts

The structure of the chloroplast as seen with the electron microscope is visible in Figures 1.27–1.29 and at a higher resolution in Figure 13.6. Chloroplasts tend to have an elongated shape and a diameter of about 3 to 10 μm (compare 1 μm diameter for mitochondria). Like mitochondria, they are surrounded by two membranes, forming the chloroplast envelope. Also like mitochondria, chloroplasts replicate themselves independently of cell division by dividing into two.

The main function of chloroplasts is to carry out photosynthesis. Chloroplasts are an excellent example of how structure is related to function, so a brief understanding of their function will help you to understand their structure.

During the first stage of photosynthesis (the light dependent stage) light energy is absorbed by photosynthetic pigments, particularly the green pigment chlorophyll. Some of this energy is used to manufacture ATP from ADP. An essential stage in the process is the

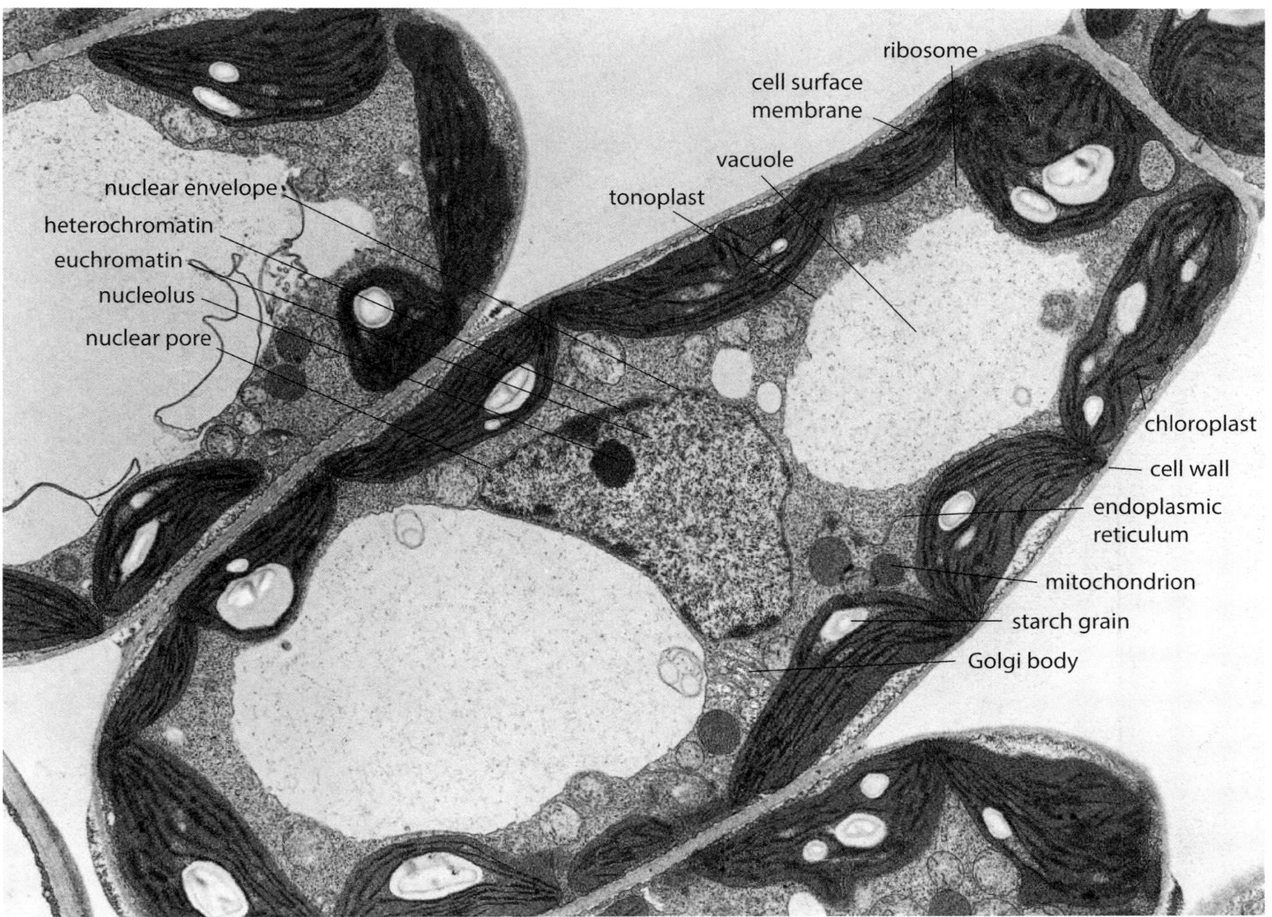

Figure 1.27 A representative plant cell as seen with a TEM. The cell is a palisade cell from a soya bean leaf (×5600).

splitting of water into hydrogen and oxygen. The hydrogen is used as the fuel which is oxidised to provide the energy to make the ATP. This process, as in mitochondria, requires electron transport in membranes. This explains why chloroplasts contain a complex system of membranes.

The membrane system is highly organised. It consists of fluid-filled sacs called thylakoids which spread out like sheets in three dimensions. In places, the thylakoids form flat, disc-like structures that stack up like piles of coins many layers deep, forming structures called grana (from their appearance in the light microscope; 'grana' means grains). These membranes contain the photosynthetic pigments and electron carriers needed for the light dependent stage of photosynthesis. Both the membranes and whole chloroplasts can change their orientation within the cell in order to receive the maximum amount of light.

The second stage of photosynthesis (the light independent stage) uses the energy and reducing power generated during the first stage to convert carbon dioxide into sugars. This requires a cycle of enzyme-controlled reactions called the Calvin cycle and takes place in solution in the stroma (the equivalent of the matrix in mitochondria). The sugars made may be stored in the form of starch grains in the stroma (Figures 1.27 and 13.6). The lipid droplets also seen in the stroma as black spheres in electron micrographs (Figure 1.29) are reserves of lipid for making membranes or from the breakdown of membranes in the chloroplast.

Like mitochondria, chloroplasts have their own protein synthesising machinery, including 70S ribosomes and a circular strand of DNA. In electron micrographs, the ribosomes can just be seen as small black dots in the stroma (Figure 13.6, page 291). Fibres of DNA can also sometimes be seen in small, clear areas in the stroma.

As with mitochondria, it has been shown that chloroplasts originated as endosymbiotic bacteria, in this case photosynthetic blue-green bacteria. The endosymbiont theory is discussed in more detail on page 17.

QUESTION

1.5 Compare Figure 1.28 with Figure 1.5 on page 4. Name the structures in a plant cell which can be seen with the electron microscope but not with the light microscope.

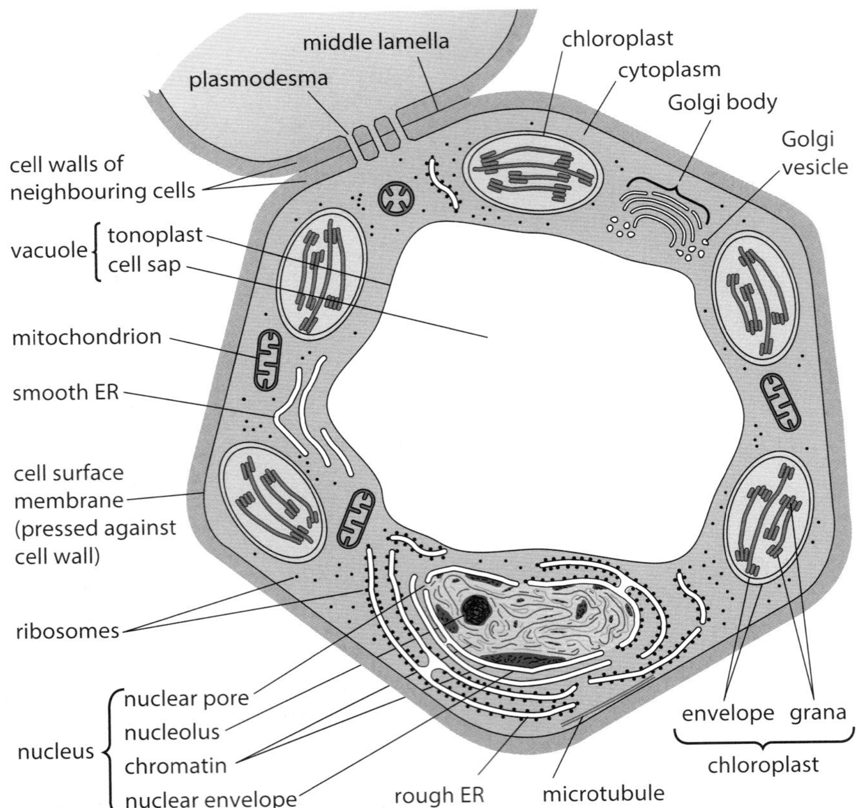

Figure 1.28 Ultrastructure of a typical plant cell as seen with the electron microscope. In reality, the ER is more extensive than shown. Free ribosomes may also be more extensive.

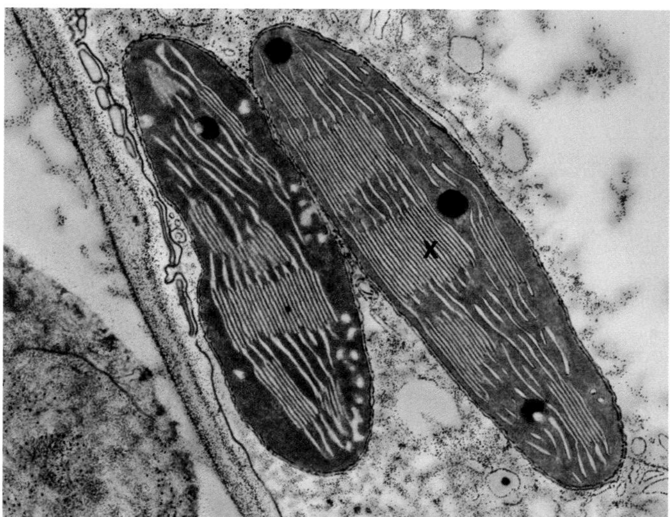

Figure 1.29 Chloroplasts (×16 000). Thylakoids (yellow) run through the stroma (dark green) and are stacked in places to form grana. Black circles among the thylakoids are lipid droplets. See also Figure 13.6, page 291. Chloroplast X is referred to in Question 1.2.

Two fundamentally different types of cell

At one time it was common practice to try to classify **all** living organisms as either animals or plants. With advances in our knowledge of living things, it has become obvious that the living world is not that simple. Fungi and bacteria, for example, are very different from animals and plants, and from each other. Eventually it was discovered that there are two fundamentally different types of cell. The most obvious difference between these types is that one possesses a nucleus and the other does not.

Organisms that lack nuclei are called prokaryotes ('pro' means before; 'karyon' means nucleus). They are, on average, about 1000 to 10 000 times smaller in **volume** than cells with nuclei, and are much simpler in structure – for example, their DNA lies free in the cytoplasm.

Organisms whose cells possess nuclei are called **eukaryotes** ('eu' means true). Their DNA lies inside a nucleus. Eukaryotes include **animals**, **plants**, **fungi** and a group containing most of the unicellular eukaryotes known as **protoctists**. Most biologists believe that eukaryotes evolved from prokaryotes, 1500 million years after prokaryotes first appeared on Earth. We mainly study animals and plants in this book, but **all** eukaryotic cells have certain features in common.

A generalised prokaryotic cell is shown in Figure 1.30. A comparison of prokaryotic and eukaryotic cells is given in Table 1.2.

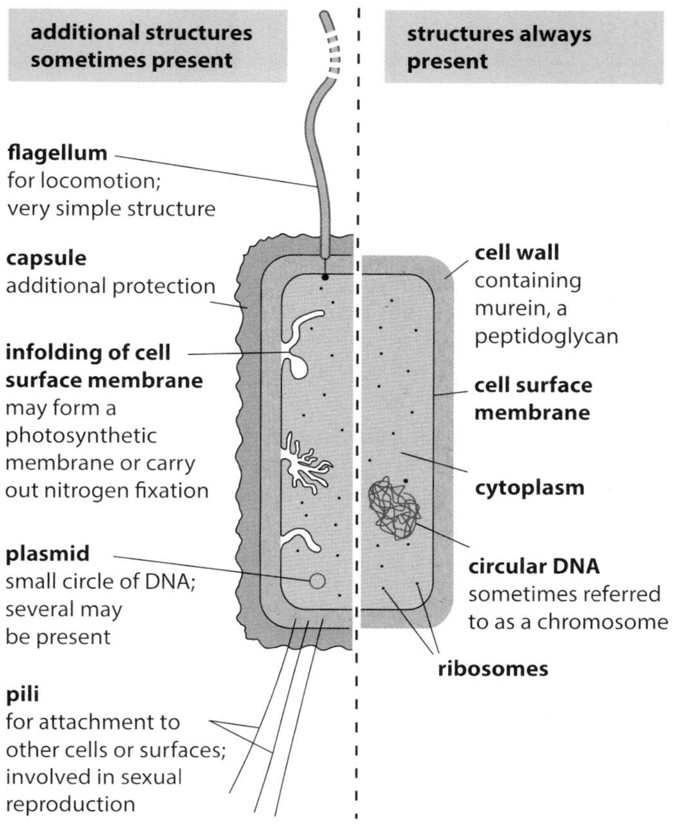

Figure 1.30 Diagram of a generalised bacterium showing the typical features of a prokaryotic cell.

QUESTION

1.6 List the structural features that prokaryotic and eukaryotic cells have in common. Briefly explain why each of the structures you have listed is essential.

Viruses

In 1852, a Russian scientist discovered that certain diseases could be transmitted by agents that, unlike bacteria, could pass through the finest filters. This was the first evidence for the existence of viruses, tiny 'organisms' which are much smaller than bacteria and are on the boundary between what we think of as living and non-living. Unlike prokaryotes and eukaryotes, viruses do not have a cell structure. In other words, they are not surrounded by a partially permeable membrane containing cytoplasm with ribosomes. They are much simpler in structure. Most consist only of:

- a self-replicating molecule of DNA or RNA which acts as its genetic code
- a protective coat of protein molecules.

Prokaryotes	Eukaryotes
average diameter of cell is 0.5–5 µm	cells commonly up to 40 µm diameter and commonly 1000–10 000 times the volume of prokaryotic cells
DNA is circular and lies free in the cytoplasm	DNA is not circular and is contained in a nucleus – the nucleus is surrounded by an envelope of two membranes
DNA is naked	DNA is associated with protein, forming structures called chromosomes
slightly smaller (70S) ribosomes (about 20 nm diameter) than those of eukaryotes	slightly larger (80S) ribosomes (about 25 nm diameter) than those of prokaryotes
no ER present	ER present, to which ribosomes may be attached
very few cell organelles – no separate membrane-bound compartments unless formed by infolding of the cell surface membrane	many types of cell organelle present (extensive compartmentalisation and division of labour): ■ some organelles are bounded by a single membrane, e.g. lysosomes, Golgi body, vacuoles ■ some are bounded by two membranes (an envelope), e.g. nucleus, mitochondrion, chloroplast ■ some have no membrane, e.g. ribosomes, centrioles, microtubules
cell wall present – wall contains murein, a peptidoglycan (a polysaccharide combined with amino acids)	cell wall sometimes present, e.g. in plants and fungi – contains cellulose or lignin in plants, and chitin (a nitrogen-containing polysaccharide similar to cellulose) in fungi

Table 1.2 A comparison of prokaryotic and eukaryotic cells.

Figure 1.31 shows the structure of a simple virus. It has a very symmetrical shape. Its protein coat (or **capsid**) is made up of separate protein molecules, each of which is called a **capsomere**.

Viruses range in size from about 20–300 nm (about 50 times smaller on average than bacteria).

All viruses are parasitic because they can only reproduce by infecting and taking over living cells. The virus DNA or RNA takes over the protein synthesising machinery of the host cell, which then helps to make new virus particles.

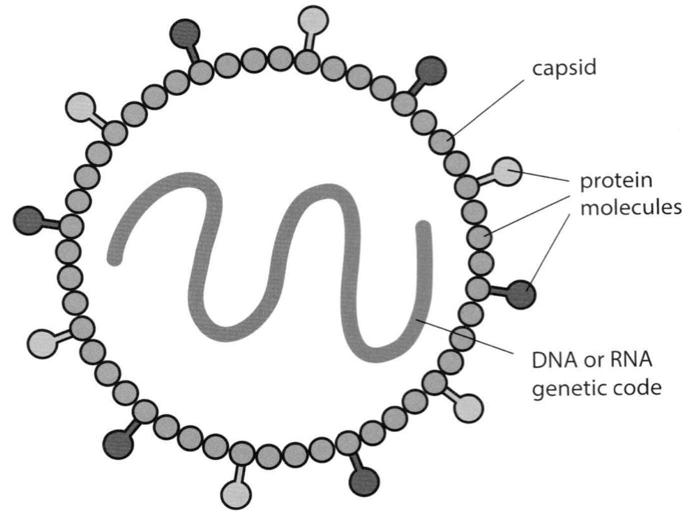

Figure 1.31 The structure of a simple virus.

Summary

- The basic unit of life, the cell, can be seen clearly only with the aid of microscopes. The light microscope uses light as a source of radiation, whereas the electron microscope uses electrons. The electron microscope has greater resolution (allows more detail to be seen) than the light microscope, because electrons have a shorter wavelength than light.
- With a light microscope, cells may be measured using an eyepiece graticule and a stage micrometer. Using the formula $A = \frac{I}{M}$ the actual size of an object (A) or its magnification (M) can be found if its observed (image) size (I) is measured and A or M, as appropriate, is known.
- All cells are surrounded by a partially permeable cell surface membrane that controls exchange between the cell and its environment. All cells contain genetic material in the form of DNA, and ribosomes for protein synthesis.

- The simplest cells are prokaryotic cells, which are thought to have evolved before, and given rise to, the much more complex and much larger eukaryotic cells. Prokaryotic cells lack a true nucleus and have smaller (70S) ribosomes than eukaryotic cells. They also lack membrane-bound organelles. Their DNA is circular and lies naked in the cytoplasm.
- All eukaryotic cells possess a nucleus containing one or more nucleoli and DNA. The DNA is linear and bound to proteins to form chromatin.
- The cytoplasm of eukaryotic cells contains many membrane-bound organelles providing separate compartments for specialised activities (division of labour). Organelles of eukaryotic cells include endoplasmic reticulum (ER), 80S ribosomes, mitochondria, Golgi apparatus and lysosomes. Animal cells also contain a centrosome and centrioles. Plant cells may contain chloroplasts, often have a large, permanent, central vacuole and have a cell wall containing cellulose.

End-of-chapter questions

1 Which **one** of the following cell structures can be seen with a light microscope?

A mitochondrion
B ribosome
C rough ER
D smooth ER [1]

2 The use of electrons as a source of radiation in the electron microscope allows high resolution to be achieved because electrons:

A are negatively charged.
B can be focused using electromagnets.
C have a very short wavelength.
D travel at the speed of light. [1]

3 Which one of the following structures is found in animal cells, but not in plant cells?

A cell surface membrane
B centriole
C chloroplast
D Golgi body [1]

4 Copy and complete the following table, which compares light microscopes with electron microscopes. Some boxes have been filled in for you.

Feature	Light microscope	Electron microscope
source of radiation		
wavelength of radiation used		about 0.005 nm
maximum resolution		0.5 nm in practice
lenses	glass	
specimen		non-living or dead
stains	coloured dyes	
image	coloured	

[8]

5 List **ten** structures you could find in an electron micrograph of an animal cell which would be absent from the cell of a bacterium. [10]

6 ***Advice on answering question 6***: If you are asked to distinguish between two things, it is likely that it is because they have certain things in common and that they may even be confused with each other. In your answer it is helpful where relevant to point out similarities as well as differences. Remember that for organelles there may be differences in both structure and function.

Distinguish between the following pairs of terms:

- a magnification and resolution [3]
- b light microscope and electron microscope [2]
- c nucleus and nucleolus [4]
- d chromatin and chromosome [3]
- e membrane and envelope [3]
- f smooth ER and rough ER [4]
- g prokaryote and eukaryote [4]

[Total: 23]

7 List:

- a **three** organelles each lacking a boundary membrane
- b **three** organelles each bounded by a single membrane
- c **three** organelles each bounded by two membranes (an envelope) [9]

8 Identify each cell structure or organelle from its description below.

- a manufactures lysosomes
- b manufactures ribosomes
- c site of protein synthesis
- d can bud off vesicles which form the Golgi body
- e can transport newly synthesised protein round the cell
- f manufactures ATP in animal and plant cells
- g controls the activity of the cell, because it contains the DNA
- h carries out photosynthesis
- i can act as a starting point for the growth of spindle microtubules during cell division
- j contains chromatin
- k partially permeable barrier only about 7 nm thick
- l organelle about 25 nm in diameter [12]

9 The electron micrograph on page 25 shows part of a secretory cell from the pancreas. The secretory vesicles are Golgi vesicles and appear as dark round structures. The magnification is ×8000.

- a Copy and complete the table. Use a ruler to help you find the actual sizes of the structures. Give your answers in micrometres.

Structure	Observed diameter (measured with ruler)	Actual size
maximum diameter of a Golgi vesicle		
maximum diameter of nucleus		
maximum length of the labelled mitochondrion		

[9]

b Make a fully labelled drawing of **representative** parts of the cell. You do not have to draw everything, but enough to show the structures of the main organelles. Use a full page of plain paper and a sharp pencil. Use Figures 1.16 and 1.17 in this book and the simplified diagram in **d** below to help you identify the structures. [14]

c The mitochondria in pancreatic cells are mostly sausage-shaped in three dimensions. Explain why some of the mitochondria in the electron micrograph below appear roughly circular. [1]

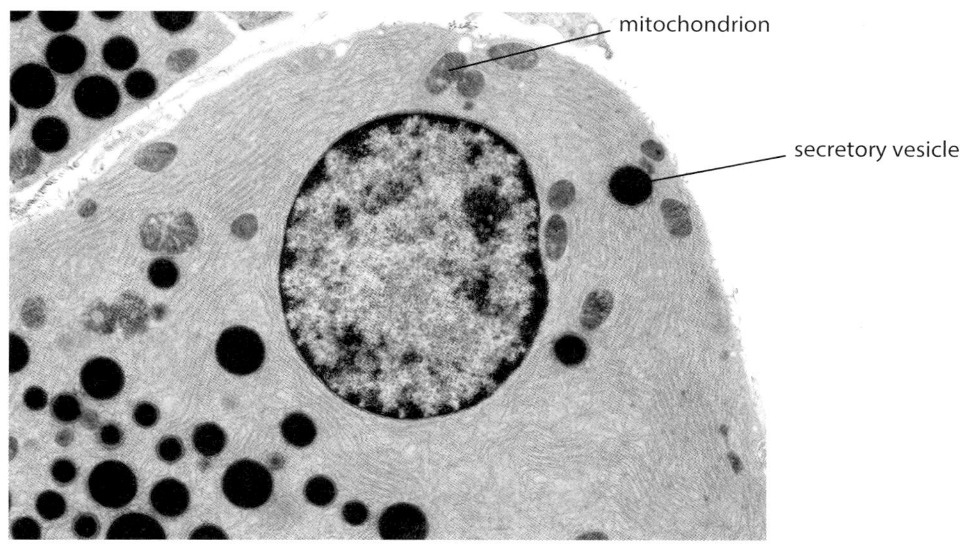

d The figure below shows a diagram based on an electron micrograph of a secretory cell from the pancreas. This type of cell is specialised for secreting (exporting) proteins. Some of the proteins are digestive enzymes of the pancreatic juice. The cell is very active, requiring a lot of energy. The arrows show the route taken by the protein molecules.

A magnified

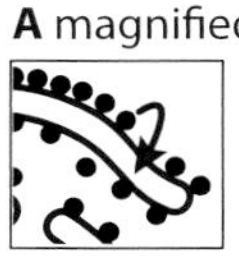

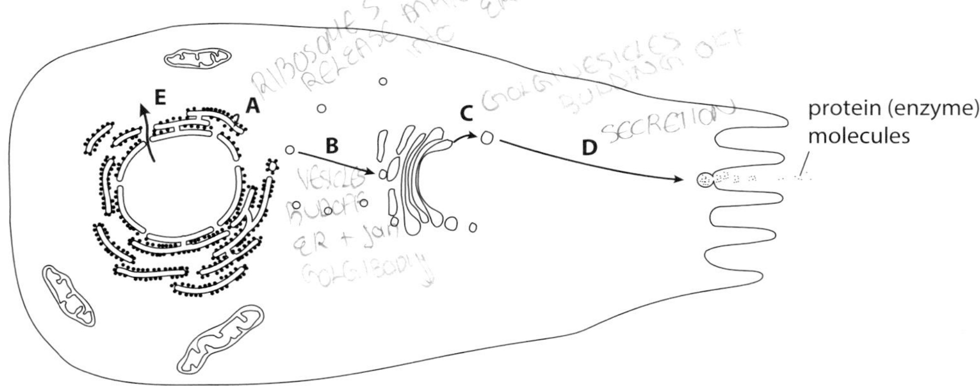

i Describe briefly what is happening at each of the stages **A**, **B**, **C** and **D**. [8]

ii Name **one** molecule or structure which leaves the nucleus by route **E**. [1]

iii Through which structure must the molecule or structure you named in **ii** pass to get through the nuclear envelope? [1]

iv Name the molecule which leaves the mitochondrion in order to provide energy for this cell. [1]

[Total: 35]

10 One technique used to investigate the activity of cell organelles is called differential centrifugation. In this technique, a tissue is homogenised (ground in a blender), placed in tubes and spun in a centrifuge. This makes organelles sediment (settle) to the bottom of the tubes. The larger the organelles, the faster they sediment. By repeating the process at faster and faster speeds, the organelles can be separated from each other according to size. Some liver tissue was treated in this way to separate ribosomes, nuclei and mitochondria. The centrifuge was spun at 1000 *g*, 10 000 *g* or 100 000 *g* ('*g*' is gravitational force).

a In which of the three sediments – 1000 g, 10 000 g or 100 000 g – would you expect to find the following?

i ribosomes

ii nuclei

iii mitochondria [1]

b Liver tissue contains many lysosomes. Suggest why this makes it difficult to study mitochondria using the differential centrifugation technique. [4]

[Total: 5]

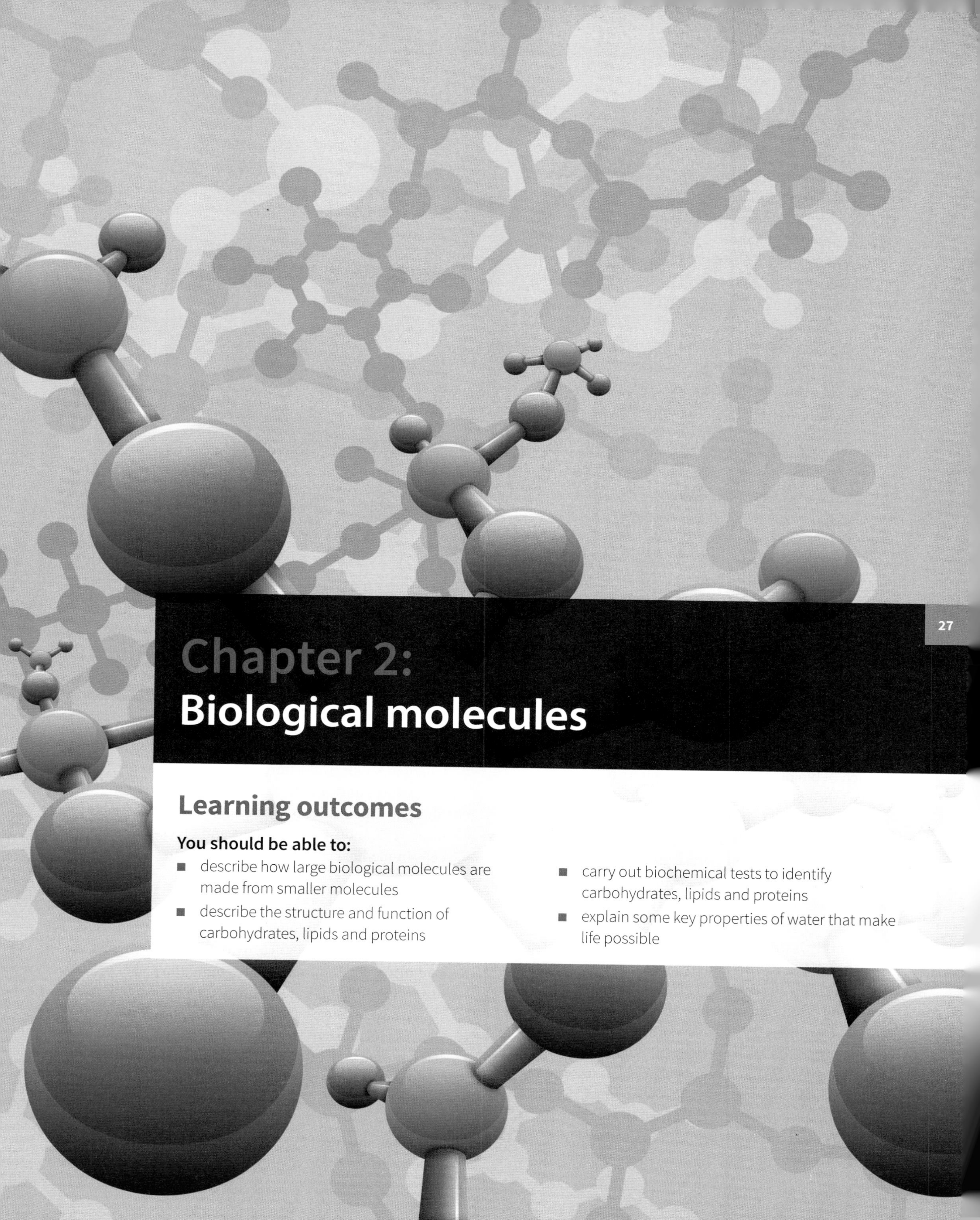

Chapter 2: Biological molecules

Learning outcomes

You should be able to:

- describe how large biological molecules are made from smaller molecules
- describe the structure and function of carbohydrates, lipids and proteins
- carry out biochemical tests to identify carbohydrates, lipids and proteins
- explain some key properties of water that make life possible

'And the winner is ...'

Nobel prizes were first awarded in 1901. The prizes were founded by Alfred Nobel, the inventor of dynamite. The winning scientists are referred to as Nobel laureates.

The study of biological molecules has been so important in the last 100 years that it has inevitably led to the award of many Nobel prizes. Many of the winners have been associated with the University of Cambridge.

For example, William and Lawrence Bragg (father and son) won the Physics prize in 1915 for work on X-ray crystallography, which was to lead to the discovery of the structure of key biological molecules. Frederick Sanger won prizes in 1958 and 1980 for work on sequencing the subunits of proteins and nucleic acids. James Watson and Francis Crick, along with Maurice Wilkins from King's College London, won the 1962 prize for Physiology and Medicine for their discovery of the structure of DNA in 1953, arguably one of the most important scientific discoveries of all time. John Kendrew and Max Perutz received the Chemistry prize in the same year for their work on the three-dimensional structure of the proteins myoglobin (Figure 2.1) and haemoglobin, essential for an understanding of how proteins function.

Not surprisingly, Cambridge has become a centre of excellence for technologies associated with biology, particularly in the pharmaceutical and computing industries. Scientists from many disciplines and from all over the world have the opportunity to work together in a close-knit and highly productive community.

Figure 2.1 Kendrew's original model of the myoglobin molecule, made in 1957.

The study of biological molecules forms an important branch of biology known as **molecular biology**. The importance of the subject is clear from the relatively large number of Nobel prizes that have been awarded in this field. It has attracted some of the best scientists, even from other disciplines like physics and mathematics.

Molecular biology is closely linked with **biochemistry**, which looks at the chemical reactions of biological molecules. The sum total of all the biochemical reactions in the body is known as **metabolism**. Metabolism is complex, but it has an underlying simplicity. For example, there are only 20 common amino acids used to make naturally occurring proteins, whereas theoretically there could be millions. Why so few? One possibility is that all the manufacture and reactions of biological molecules must be controlled and regulated and, the more there are, the more complex the control becomes. (Control and regulation by enzymes is examined in Chapter 3.)

Another striking principle of molecular biology is how closely the structures of molecules are related to their functions. This will become clear in this chapter and in Chapter 3. Our understanding of how structure is related to function may lead to the creation of a vast range of 'designer' molecules to carry out such varied functions as large-scale industrial reactions and precise targeting of cells in medical treatment.

The building blocks of life

The four most common elements in living organisms are, in order of abundance, hydrogen, carbon, oxygen and nitrogen. They account for more than 99% of the atoms found in all living things. Carbon is particularly important because carbon atoms can join together to form long chains or ring structures. They can be thought of as the basic skeletons of organic molecules to which groups of other atoms are attached. Organic molecules always contain carbon and hydrogen.

It is believed that, before **life** evolved, there was a period of **chemical** evolution in which thousands of carbon-based molecules evolved from the more simple molecules that existed on the young planet Earth. Such an effect can be artificially created reasonably easily today given similar raw ingredients, such as methane (CH_4), carbon dioxide (CO_2), hydrogen (H_2), water (H_2O), nitrogen (N_2), ammonia (NH_3) and hydrogen sulfide (H_2S), and an energy source – for example, an electrical discharge. These simple but key biological molecules, which are relatively limited in variety, then act as the building blocks for larger molecules. The main ones are shown in Figure 2.2.

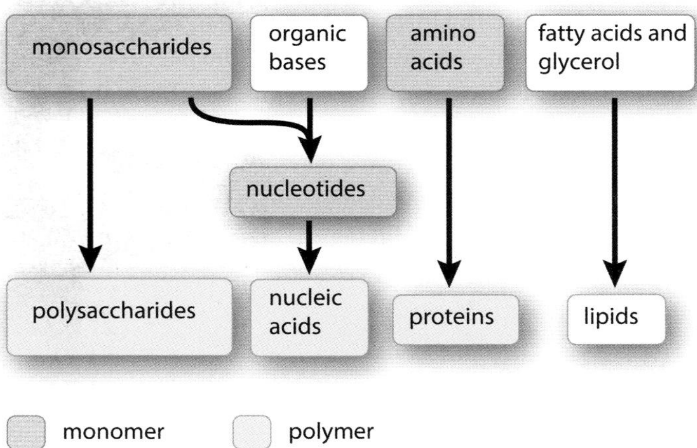

Figure 2.2 The building blocks of life.

Monomers, polymers and macromolecules

The term **macromolecule** means giant molecule. There are three types of macromolecule in living organisms, namely polysaccharides, proteins (polypeptides) and nucleic acids (polynucleotides). The prefix 'poly' means many, and these molecules are **polymers**, meaning that they are made up of many repeating subunits that are similar or identical to each other. These subunits are referred to as **monomers**. They are joined together like beads on a string. Making such molecules is relatively simple because the same reaction is repeated many times.

The monomers from which polysaccharides, proteins and nucleic acids are made are monosaccharides, amino acids and nucleotides respectively, as shown in Figure 2.2. Figure 2.2 also shows two types of molecule which, although not polymers, are made up of simpler biochemicals. These are lipids and **nucleotides**.

Natural examples of polymers are cellulose and rubber. There are many examples of industrially produced polymers, such as polyester, polythene, PVC (polyvinyl chloride) and nylon. All these are made up of carbon-based monomers and contain thousands of carbon atoms joined end to end.

We shall now take a closer look at some of the small biological molecules and the larger molecules made from them. Organic bases, nucleotides and nucleic acids are dealt with in Chapter 6.

Carbohydrates

All carbohydrates contain the elements carbon, hydrogen and oxygen. The 'hydrate' part of the name comes from the fact that hydrogen and oxygen atoms are present in the ratio of 2 : 1, as they are in water ('hydrate' refers to water). The **general formula** for a carbohydrate can therefore be written as $C_x(H_2O)_y$.

Carbohydrates are divided into three main groups, namely monosaccharides, disaccharides and polysaccharides. The word 'saccharide' refers to a sugar or sweet substance.

Monosaccharides

Monosaccharides are **sugars**. Sugars dissolve easily in water to form sweet-tasting solutions. Monosaccharides have the general formula $(CH_2O)_n$ and consist of a **single** sugar molecule ('mono' means one). The main types of monosaccharides, if they are classified according to the number of carbon atoms in each molecule, are **trioses** (3C), **pentoses** (5C) and **hexoses** (6C). The names of all sugars end with **-ose**. Common hexoses are glucose, fructose and galactose. Two common pentoses are ribose and deoxyribose.

A **macromolecule** is a large biological molecule such as a protein, polysaccharide or nucleic acid.

A **monomer** is a relatively simple molecule which is used as a basic building block for the synthesis of a polymer; many monomers are joined together to make the polymer, usually by condensation reactions; common examples of molecules used as monomers are monosaccharides, amino acids and nucleotides.

A **polymer** is a giant molecule made from many similar repeating subunits joined together in a chain; the subunits are much smaller and simpler molecules known as monomers; examples of biological polymers are polysaccharides, proteins and nucleic acids.

Molecular and structural formulae

The formula for a hexose can be written as $C_6H_{12}O_6$. This is known as the **molecular formula**. It is also useful to show the arrangements of the atoms, which can be done using a diagram known as the **structural formula**. Figure 2.3 shows the structural formula of glucose, a hexose, which is the most common monosaccharide.

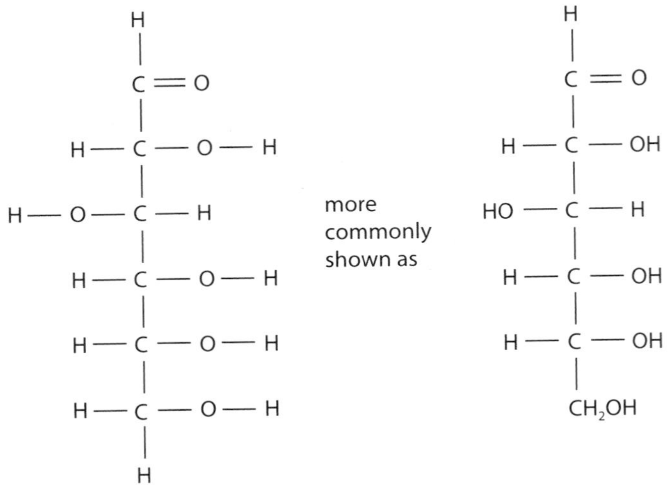

Figure 2.3 Structural formula of glucose. –OH is known as a hydroxyl group. There are five in glucose.

Ring structures

One important aspect of the structure of pentoses and hexoses is that the chain of carbon atoms is long enough to close up on itself and form a more stable ring structure. This can be illustrated using glucose as an example. When glucose forms a ring, carbon atom number **1** joins to the oxygen on carbon atom number **5** (Figure 2.4). The ring therefore contains oxygen, and carbon atom number **6** is not part of the ring.

You will see from Figure 2.4 that the hydroxyl group, –OH, on carbon atom **1** may be **above** or **below** the plane of the ring. The form of glucose where it is below the ring is known as **α-glucose (alpha-glucose)** and the form where it is above the ring is **β-glucose (beta-glucose)**. The same molecule can switch between the two forms. Two forms of the same chemical are known as **isomers**, and the extra variety provided by the existence of α- and β-isomers has important biological consequences, as we shall see in the structures of starch, glycogen and cellulose.

QUESTION

2.1 The formula for a hexose is $C_6H_{12}O_6$ or $(CH_2O)_6$. What would be the formula of:

a a triose?

b a pentose?

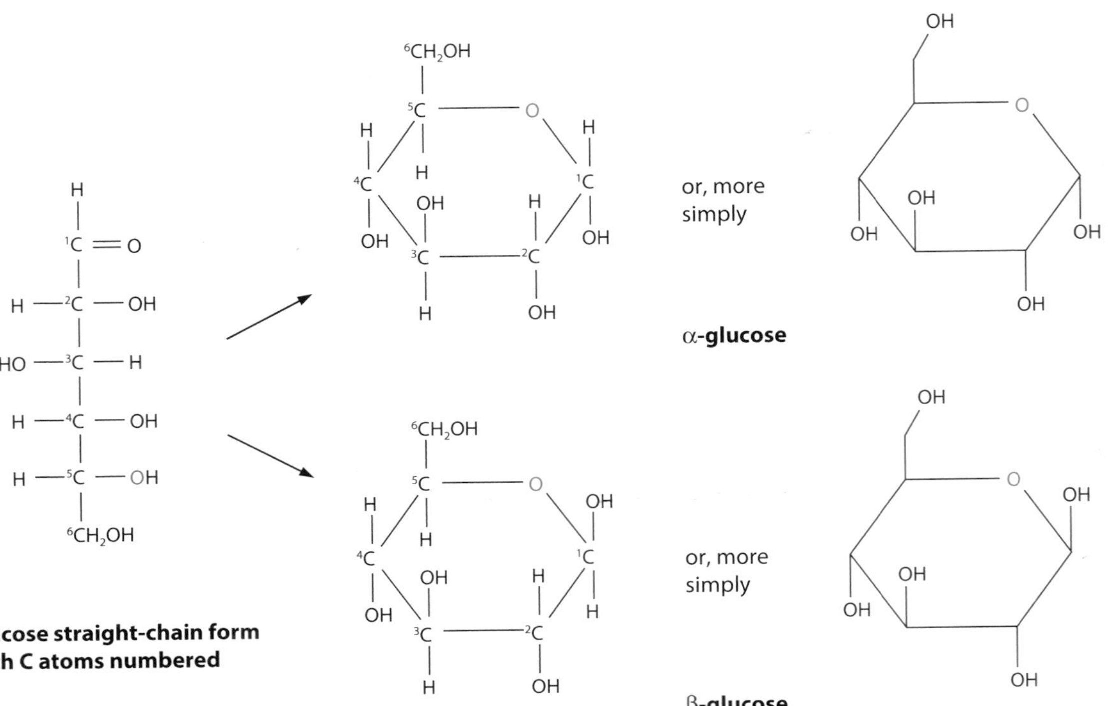

Figure 2.4 Structural formulae for the straight-chain and ring forms of glucose. Chemists often leave out the C and H atoms from the structural formula for simplicity.

Roles of monosaccharides in living organisms

Monosaccharides have two major functions. First, they are commonly used as a source of energy in respiration. This is due to the large number of carbon–hydrogen bonds. These bonds can be broken to release a lot of energy, which is transferred to help make ATP (adenosine triphosphate) from ADP (adenosine diphosphate) and phosphate. The most important monosaccharide in energy metabolism is glucose.

Secondly, monosaccharides are important as building blocks for larger molecules. For example, glucose is used to make the polysaccharides starch, glycogen and cellulose. Ribose (a pentose) is one of the molecules used to make RNA (ribonucleic acid) and ATP. Deoxyribose (also a pentose) is one of the molecules used to make DNA (Chapter 6).

Disaccharides and the glycosidic bond

Disaccharides, like monosaccharides, are sugars. They are formed by two monosaccharides joining together. The three most common disaccharides are maltose (glucose + glucose), sucrose (glucose + fructose) and lactose (glucose + galactose). Sucrose is the transport sugar in plants and the sugar commonly bought in shops. Lactose is the sugar found in milk and is therefore an important constituent of the diet of young mammals.

The joining of two monosaccharides takes place by a process known as **condensation**. Two examples are shown in Figure 2.5. In Figure 2.5a two molecules of α-glucose combine to make the disaccharide maltose. In Figure 2.5b α-glucose and β-fructose combine to make the disaccharide sucrose. Notice that fructose has a different ring structure to glucose.

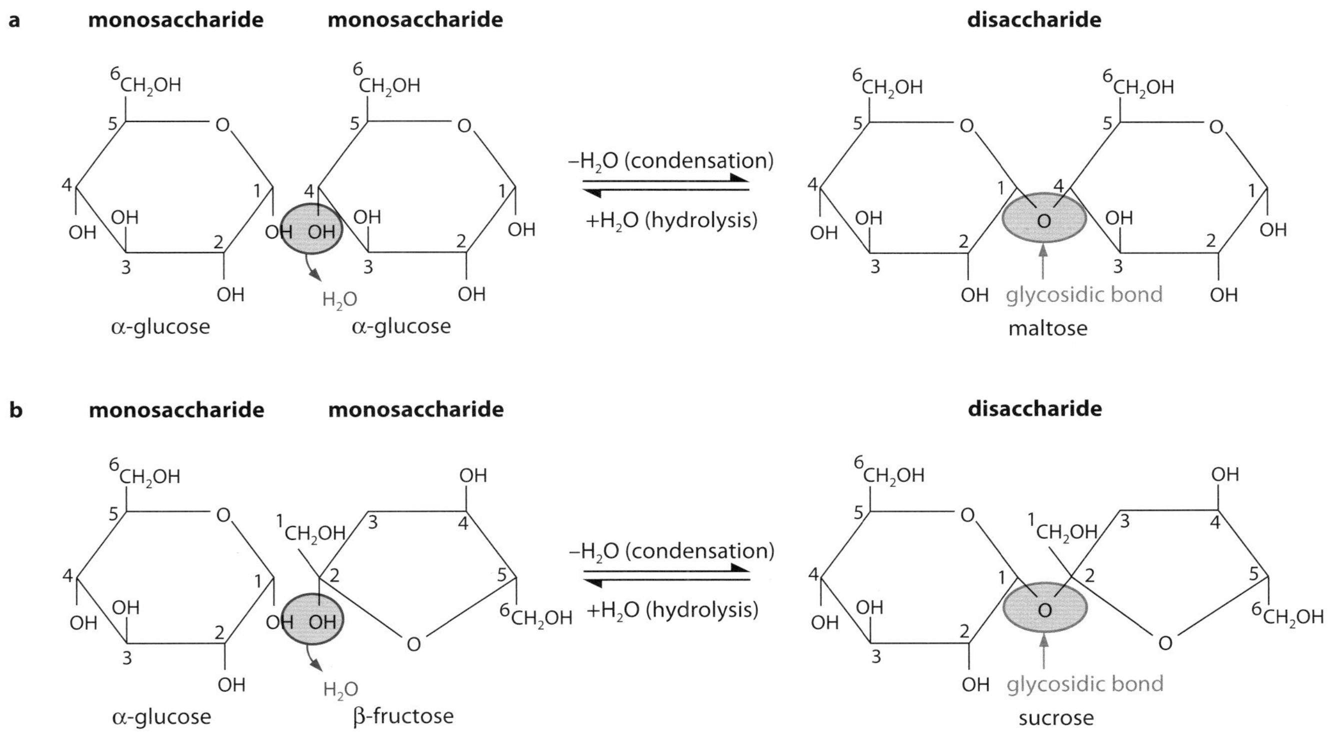

Figure 2.5 Formation of a disaccharide from two monosaccharides by condensation. **a** Maltose is formed from two α-glucose molecules. This can be repeated many times to form a polysaccharide. Note that in this example the glycosidic bond is formed between carbon atoms **1** and **4** of neighbouring glucose molecules. **b** Sucrose is made from an α-glucose and a β-fructose molecule.

A **monosaccharide** is a molecule consisting of a single sugar unit with the general formula $(CH_2O)_n$.

A **disaccharide** is a sugar molecule consisting of two monosaccharides joined together by a glycosidic bond.

A **polysaccharide** is a polymer whose subunits are monosaccharides joined together by glycosidic bonds.

For each condensation reaction, two hydroxyl (–OH) groups line up alongside each other. One combines with a hydrogen atom from the other to form a water molecule. This allows an oxygen 'bridge' to form between the two molecules, holding them together and forming a **disaccharide** ('di' means two). The bridge is called a glycosidic bond.

In theory any two –OH groups can line up and, since monosaccharides have many –OH groups, there are a large number of possible disaccharides. The shape of the enzyme controlling the reaction determines which –OH groups come alongside each other. Only a few of the possible disaccharides are common in nature.

The reverse of condensation is the **addition** of water, which is known as hydrolysis (Figure 2.5). This takes place during the digestion of disaccharides and polysaccharides, when they are broken down to monosaccharides.

BOX 2.1: Testing for the presence of sugars

1 Reducing sugars – background information

The reducing sugars include all monosaccharides, such as glucose, and some disaccharides, such as maltose. The only common non-reducing sugar is sucrose.

Reducing sugars are so called because they can carry out a type of chemical reaction known as **reduction**. In the process they are **oxidised**. This is made use of in the Benedict's test using **Benedict's reagent**. Benedict's reagent is copper(II) sulfate in an alkaline solution and has a distinctive blue colour. Reducing sugars reduce soluble blue copper sulfate, containing copper(II) ions, to insoluble brick-red copper oxide, containing copper(I). The copper oxide is seen as a brick-red precipitate.

$$\underset{\text{}}{\text{reducing sugar}} + \underset{\text{blue}}{Cu^{2+}} \rightarrow \text{oxidised sugar} + \underset{\text{red-brown}}{Cu^{+}}$$

Procedure

Add Benedict's reagent to the solution you are testing and heat it in a water bath. If a reducing sugar is present, the solution will gradually turn through green, yellow and orange to red-brown as the insoluble copper(I) oxide forms a precipitate. As long as you use excess Benedict's reagent (more than enough to react with all of the sugar present), the intensity of the red colour is related to the concentration of the reducing sugar. You can then estimate the concentration using colour standards made by comparing the colour against the colours obtained in tests done with reducing sugar solutions of known concentration. You could also measure the time taken for the colour to change.

Alternatively, you can use a colorimeter to measure subtle differences in colour precisely.

2 Non-reducing sugars – background information

Some disaccharides, such as sucrose, are **not** reducing sugars, so you would get a negative result from Benedict's test. In such a case, a brick-red precipitate in the test described below will tell you that a non-reducing sugar is present. If both a reducing sugar and a non-reducing sugar are present, the precipitate obtained in the test below will be heavier than the one obtained in Benedict's test.

In the non-reducing sugars test, the disaccharide is first broken down into its two monosaccharide constituents. The chemical reaction is hydrolysis and can be brought about by hydrochloric acid. The constituent monosaccharides will be reducing sugars and their presence can be tested for using Benedict's test after the acid has been neutralised.

Procedure

Heat the sugar solution with hydrochloric acid. This will release free monosaccharides. Benedict's reagent needs alkaline conditions to work, so you need to neutralise the test solution now by adding an alkali such as sodium hydroxide. Add Benedict's reagent and heat as before and look for the colour change. If the solution goes red now but didn't in the first stage of the test, there is non-reducing sugar present. If there is **still** no colour change, then there is no sugar of any kind present.

QUESTION

2.2 a Why do you need to use **excess** Benedict's reagent if you want to get an idea of the concentration of a sugar solution?

b Outline how you could use the Benedict's test to estimate the concentration of a solution of a reducing sugar.

Polysaccharides

Polysaccharides are polymers whose subunits (monomers) are monosaccharides. They are made by joining many monosaccharide molecules by condensation. Each successive monosaccharide is added by means of a glycosidic bond, as in disaccharides. The final molecule may be several thousand monosaccharide units long, forming a macromolecule. The most important polysaccharides are starch, glycogen and cellulose, all of which are polymers of glucose. Polysaccharides are **not** sugars.

Since glucose is the main source of energy for cells, it is important for living organisms to store it in an appropriate form. If glucose itself accumulated in cells, it would dissolve and make the contents of the cell too concentrated, which would seriously affect its osmotic properties (page 82). Glucose is also a reactive molecule and would interfere with normal cell chemistry. These problems are avoided by converting glucose, by condensation reactions, to a storage polysaccharide, which is a convenient, compact, inert (unreactive) and insoluble molecule. The storage polysaccharide formed is starch in plants and glycogen in animals. Glucose can be made available again quickly by an enzyme-controlled reaction.

Starch and glycogen

Starch is a mixture of two substances – amylose and amylopectin. Amylose is made by condensations between α-glucose molecules, as shown in Figure 2.5a. In this way, a long, unbranching chain of several thousand 1,4 linked glucose molecules is built up. ('1,4 linked' means they are linked between carbon atoms **1** and **4** of successive glucose units.) The chains are curved (Figure 2.6) and coil up into helical structures like springs, making the final molecule more compact. Amylopectin is also made of many 1,4 linked α-glucose molecules, but the chains are shorter than in amylose, and branch out to the sides. The branches are formed by 1,6 linkages, as shown in Figure 2.7.

Mixtures of amylose and amylopectin molecules build up into relatively large starch grains, which are commonly found in chloroplasts and in storage organs such as potato tubers and the seeds of cereals and legumes (Figure 2.8). Starch grains are easily seen with a light microscope, especially if stained; rubbing a freshly cut potato tuber on a glass slide and staining with iodine–potassium iodide solution (Box 2.2) is a quick method of preparing a specimen for viewing.

Starch is never found in animal cells. Instead, a substance with molecules very like those of amylopectin is

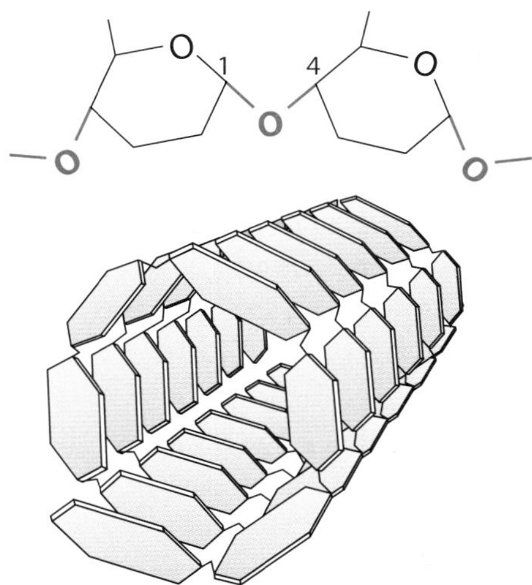

Figure 2.6 Arrangement of α-glucose units in amylose. The 1,4 linkages cause the chain to turn and coil. The glycosidic bonds are shown in red and the hydroxyl groups are omitted.

QUESTION

2.3 What type of chemical reaction would be involved in the formation of glucose from starch or glycogen?

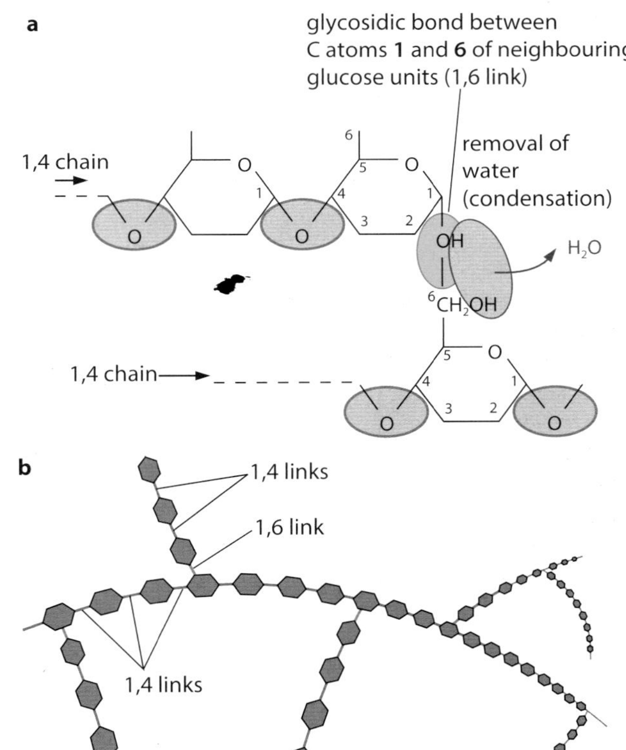

Figure 2.7 Branching structure of amylopectin and glycogen. **a** Formation of a 1,6 link, a branchpoint, **b** Overall structure of an amylopectin or glycogen molecule. Amylopectin and glycogen only differ in the amount of branching of their glucose chains.

used as the storage carbohydrate. This is called glycogen. Glycogen, like amylopectin, is made of chains of 1,4 linked α-glucose with 1,6 linkages forming branches (Figure 2.7b). Glycogen molecules tend to be even more branched than amylopectin molecules. Glycogen molecules clump together to form granules, which are visible in liver cells and muscle cells, where they form an energy reserve.

QUESTION

2.4 List **five** ways in which the molecular structures of glycogen and amylopectin are similar.

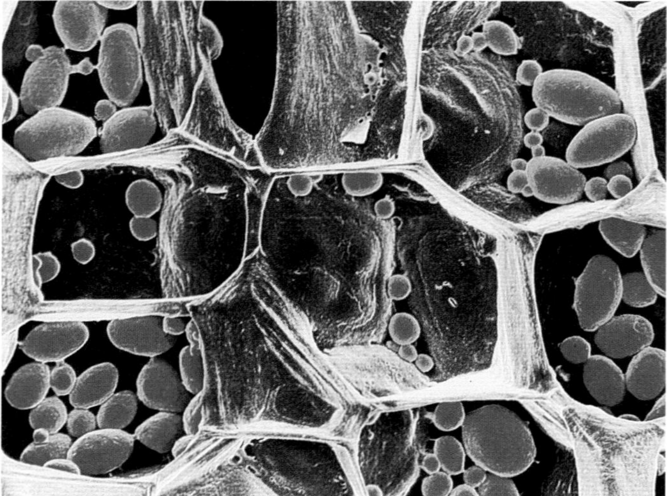

Figure 2.8 False-colour scanning electron micrograph of a slice through a raw potato showing cells containing starch grains or starch-containing organelles (coloured red) (×260).

BOX 2.2: Testing for the presence of starch

Background information

Starch molecules tend to curl up into long spirals. The hole that runs down the middle of this spiral is just the right size for iodine molecules to fit into. To test for starch, you use something called 'iodine solution'. (In fact, iodine won't dissolve in water, so the 'iodine solution' is actually iodine in potassium iodide solution.) The starch–iodine complex that forms has a strong blue-black colour.

Procedure

Iodine solution is orange-brown. Add a drop of iodine solution to the solid or liquid substance to be tested. A blue-black colour is quickly produced if starch is present.

Cellulose

Cellulose is the most abundant organic molecule on the planet, due to its presence in plant cell walls and its slow rate of breakdown in nature. It has a structural role, being a mechanically strong molecule, unlike starch and glycogen. However, the only difference between cellulose and starch and glycogen is that cellulose is a polymer of β-glucose, not α-glucose.

Remember that in the β-isomer, the –OH group on carbon atom 1 projects **above** the ring (Figure 2.4 on page 30). In order to form a glycosidic bond with carbon atom 4, where the –OH group is **below** the ring, one glucose molecule must be upside down (rotated 180°) relative to the other. Thus successive glucose units are linked at 180° to each other, as shown in Figure 2.9.

This arrangement of β-glucose molecules results in a strong molecule because the hydrogen atoms of –OH groups are weakly attracted to oxygen atoms in the same cellulose molecule (the oxygen of the glucose ring) and also to oxygen atoms of –OH groups in neighbouring molecules. These hydrogen bonds (page 35) are individually weak, but so many can form, due to the large number of –OH groups, that collectively they provide enormous strength. Between 60 and 70 cellulose molecules become tightly cross-linked to form bundles called **microfibrils**. Microfibrils are in turn held together in bundles called **fibres** by hydrogen bonding.

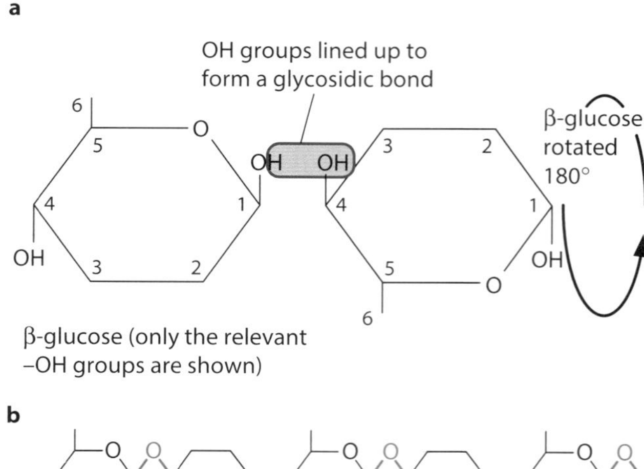

Figure 2.9 **a** Two β-glucose molecules lined up to form a 1,4 link. Note that one glucose molecule must be rotated 180° relative to the other, **b** Arrangement of β-glucose units in cellulose: glycosidic bonds are shown in red and hydroxyl groups are omitted.

A cell wall typically has several layers of fibres, running in different directions to increase strength (Figure 2.10). Cellulose makes up about 20–40% of the average cell wall; other molecules help to cross-link the cellulose fibres, and some form a glue-like matrix around the fibres, which further increases strength.

Cellulose fibres have a very high tensile strength, almost equal to that of steel. This means that if pulled at both ends they are very difficult to stretch or break, and makes it possible for a cell to withstand the large pressures that develop within it as a result of osmosis (page 82). Without the wall, the cell would burst when in a dilute solution. These pressures help provide support for the plant by making tissues rigid, and are responsible for cell expansion during growth. The arrangement of fibres around the cell helps to determine the shape of the cell as it grows.

Despite their strength, cellulose fibres are freely permeable, allowing water and solutes to reach or leave the cell surface membrane.

QUESTION

2.5 Make a table to show **three** ways in which the molecular structures of amylose and cellulose differ.

Dipoles and hydrogen bonds

When atoms in molecules are held together by covalent bonds, they share electrons with each other. Each shared pair of electrons forms one **covalent bond**. For example, in a water molecule, two hydrogen atoms each share a pair of electrons with an oxygen atom, forming a molecule with the formula H_2O.

oxygen atom
covalent bond
hydrogen atom

However, the electrons are not shared absolutely equally. In water, the oxygen atom gets slightly more than its fair share, and so has a small negative charge, written δ– (delta minus). The hydrogen atoms get slightly less than their fair share, and so have a small positive charge, written δ+ (delta plus).

This unequal distribution of charge is called a **dipole**.

In water, the negatively charged oxygen of one molecule is attracted to a positively charged hydrogen of another, and this attraction is called a **hydrogen bond** (see diagram below). It is much weaker than a covalent bond, but still has a very significant effect. You will find out how hydrogen bonds affect the properties of water on pages 46–47.

hydrogen bond

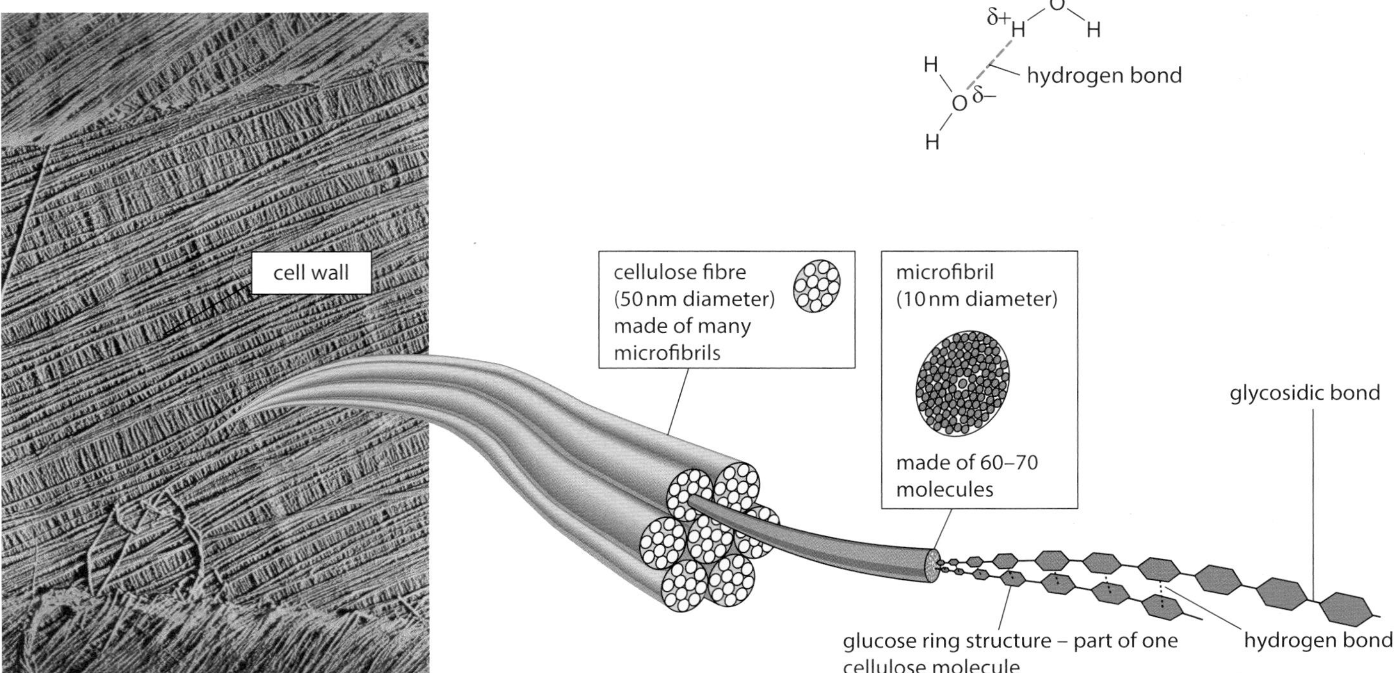

Figure 2.10 Structure of cellulose.

Dipoles occur in many different molecules, particularly wherever there is an –OH, –CO or –NH group. Hydrogen bonds can form **between** these groups, because the negatively charged part of one group is attracted to the positively charged part of another. These bonds are very important in the structure and properties of carbohydrates and proteins.

$$>C=O^{\delta-}----{}^{\delta+}H-N<$$

Molecules which have groups with dipoles, such as sugars, are said to be **polar**. They are attracted to water molecules, because the water molecules also have dipoles. Such molecules are said to be **hydrophilic** (water-loving), and they tend to be soluble in water. Molecules which do not have dipoles are said to be **non-polar**. They are not attracted to water, and they are **hydrophobic** (water-hating). Such properties make possible the formation of cell membranes (Chapter 4).

Lipids

It is difficult to define precisely what we mean by a 'lipid' because lipids are a very varied group of chemicals. They are all organic molecules which are insoluble in water. The most familiar lipids are fats and oils. Fats are solid at room temperature and oils are liquid at room temperature – chemically they are very similar. We could say that true lipids are esters formed by fatty acids combining with an alcohol.

Fatty acids

Fatty acids are a series of acids, some of which are found in fats (lipids). They contain the acidic group –COOH, known as a carboxyl group. The larger molecules in the series have long hydrocarbon tails attached to the acid 'head' of the molecule (Figure 2.11). As the name suggests, the hydrocarbon tails consist of a chain of carbon atoms combined with hydrogen. The chain is often 15 or 17 carbon atoms long.

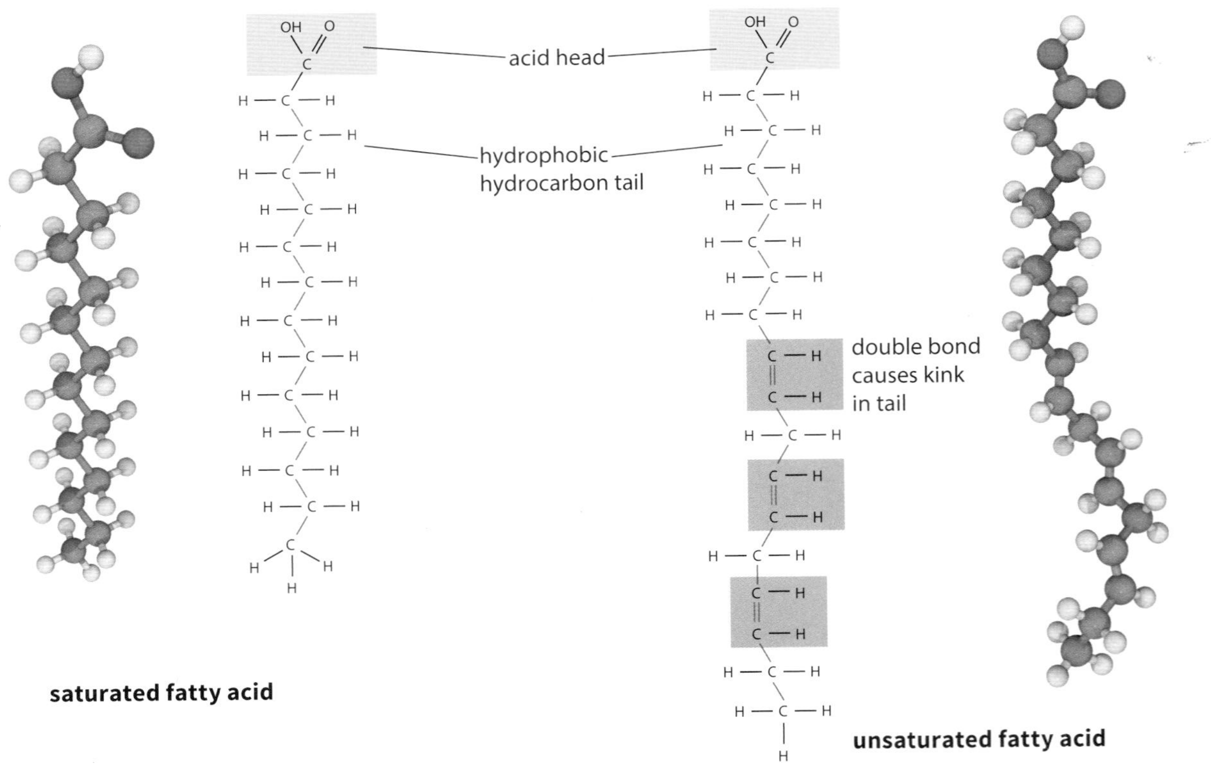

Figure 2.11 Structure of a saturated and an unsaturated fatty acid. Photographs of models are shown to the sides of the structures. In the models, hydrogen is white, carbon is grey and oxygen is red.

The tails of some fatty acids have double bonds between neighbouring carbon atoms, like this: –C=C–. Such fatty acids are described as **unsaturated** because they do not contain the maximum possible amount of hydrogen. They form **unsaturated lipids**. Double bonds make fatty acids and lipids melt more easily – for example, most oils are unsaturated. If there is more than one double bond, the fatty acid or lipid is described as **polyunsaturated**; if there is only one it is **monounsaturated**.

Animal lipids are often saturated (no double bonds) and occur as fats, whereas plant lipids are often unsaturated and occur as oils, such as olive oil and sunflower oil.

Alcohols and esters

Alcohols are a series of organic molecules which contain a hydroxyl group, –OH, attached to a carbon atom. Glycerol is an alcohol with three hydroxyl groups (Figure 2.12).

The reaction between an acid and an alcohol produces a chemical known as an ester. The chemical link between the acid and the alcohol is known as an **ester bond** or an ester linkage.

$$-\overset{|}{\underset{|}{C}}-COOH + HO-\overset{|}{\underset{|}{C}}- \longrightarrow -\overset{|}{\underset{|}{C}}-COO\overset{|}{\underset{|}{C}}- + H_2O$$

acid alcohol ester

The –COOH group on the acid reacts with the –OH group on the alcohol to form the ester bond, –COO–. This is a condensation reaction because water is formed as a product. The resulting ester can be converted back to acid and alcohol by the reverse reaction of adding water, a reaction known as hydrolysis.

Triglycerides

The most common lipids are triglycerides (Figure 2.13). These are fats and oils. A glyceride is an ester formed by a fatty acid combining with the alcohol glycerol. As we have seen, glycerol has three hydroxyl groups. Each one is able to undergo a condensation reaction with a fatty acid. When a triglyceride is made, as shown in Figure 2.12, the final molecule contains three fatty acids tails and three ester bonds ('tri' means three). The tails can vary in length, depending on the fatty acids used.

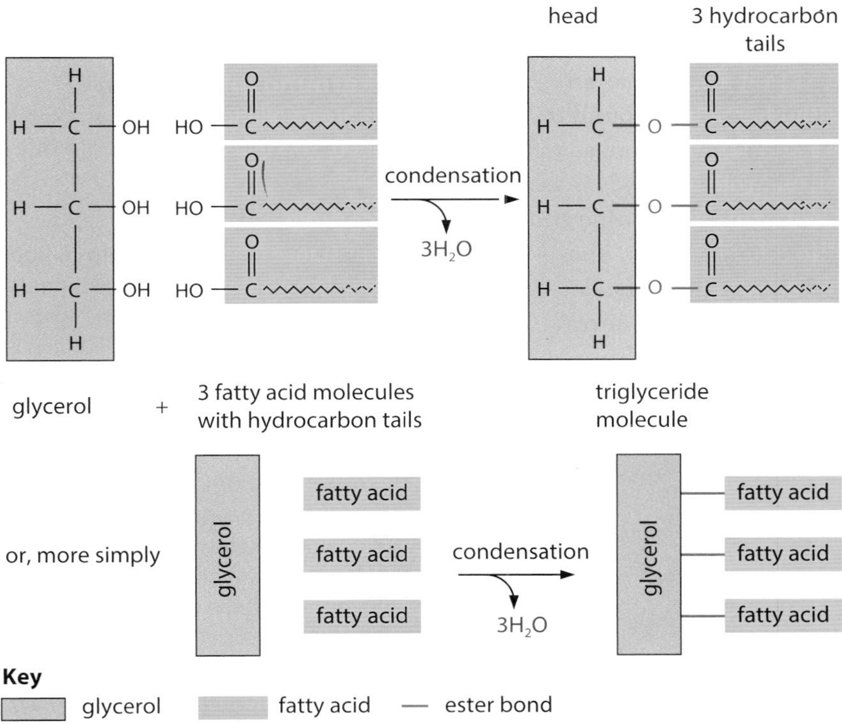

Figure 2.12 Formation of a triglyceride from glycerol and three fatty acid molecules.

Triglycerides are insoluble in water but are soluble in certain organic solvents, including ether, chloroform and ethanol. This is because of the non-polar nature of the hydrocarbon tails: they have no uneven distribution of electrical charge. Consequently, they will not mix freely with water molecules and are described as hydrophobic (water-hating). Figure 2.13 shows a simplified diagram of a triglyceride.

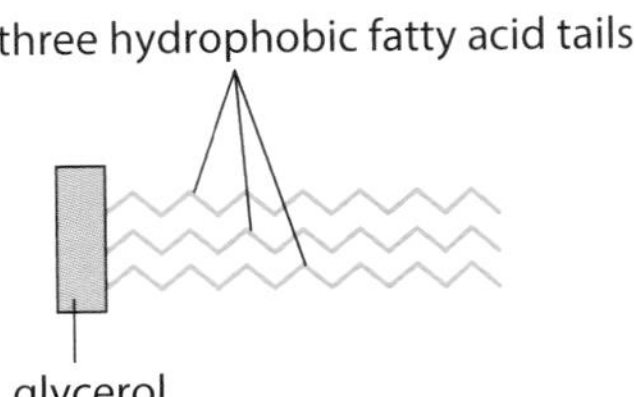

Figure 2.13 Diagrammatic representation of a triglyceride molecule.

Roles of triglycerides

Lipids make excellent **energy reserves** because they are even richer in carbon–hydrogen bonds than carbohydrates. A given mass of lipid will therefore yield more energy on oxidation than the same mass of carbohydrate (it has a higher calorific value), an important advantage for a storage product.

Fat is stored in a number of places in the human body, particularly just below the dermis of the skin and around the kidneys. Below the skin it also acts as an **insulator** against loss of heat. Blubber, a lipid found in sea mammals like whales, has a similar function, as well as providing buoyancy. An unusual role for lipids is as a **metabolic source of water.** When oxidised in respiration they are converted to carbon dioxide and water. The water may be of importance in very dry habitats. For example, the desert kangaroo rat (Figure 2.14) never drinks water and survives on metabolic water from its fat intake.

Figure 2.14 The desert kangaroo rat uses metabolism of food to provide most of the water it needs.

Phospholipids

Phospholipids are a special type of lipid. Each molecule has the unusual property of having one end which is soluble in water. This is because one of the three fatty acid molecules is replaced by a phosphate group, which is polar (page 35) and can therefore dissolve in water. The phosphate group is **hydrophilic** (water-loving) and makes the head of a phospholipid molecule hydrophilic, although the two remaining tails are still hydrophobic (Figure 2.15). This allows the molecules to form a membrane around a cell, where the hydrophilc heads lie in the watery solutions on the outside of the membrane, and the hydrophobic tails form a layer that is impermeable to hydrophilic substances. The biological significance of this will become apparent when we study membrane structure (Chapter 4).

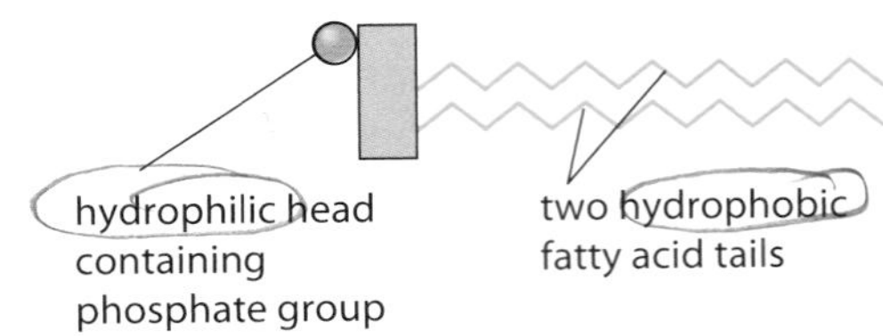

Figure 2.15 Diagrammatic representation of a phospholipid molecule. Compare with Figure 2.13.

BOX 2.3: Testing for the presence of lipids

Background information

Lipids are insoluble in water, but soluble in ethanol (alcohol). This fact is made use of in the **emulsion test** for lipids.

Procedure

The substance that is thought to contain lipids is shaken vigorously with some absolute ethanol (ethanol with little or no water in it). This allows any lipids in the substance to dissolve in the ethanol. The ethanol is then poured into a tube containing water. If lipid is present, a cloudy white suspension is formed.

Further information

If there is no lipid present, the ethanol just mixes into the water. Light can pass straight through this mixture, so it looks completely transparent. But if there is lipid dissolved in the ethanol, it cannot remain dissolved when mixed with the water. The lipid molecules form tiny droplets throughout the liquid. This kind of mixture is called an **emulsion**. The droplets reflect and scatter light, making the liquid look white and cloudy.

Proteins

Proteins are an extremely important class of macromolecule in living organisms. More than 50% of the dry mass of most cells is protein. Proteins have many important functions. For example:

- all enzymes are proteins
- proteins are essential components of cell membranes – their functions in membranes, such as receptor proteins and signalling proteins, are discussed in Chapter 4
- some hormones are proteins – for example, insulin and glucagon
- the oxygen-carrying pigments haemoglobin and myoglobin are proteins
- antibodies, which attack and destroy invading microorganisms, are proteins
- collagen, another protein, adds strength to many animal tissues, such as bone and the walls of arteries
- hair, nails and the surface layers of skin contain the protein keratin
- actin and myosin are the proteins responsible for muscle contraction
- proteins may be storage products – for example, casein in milk and ovalbumin in egg white.

Despite their tremendous range of functions, all proteins are made from the same basic monomers. These are **amino acids**.

Amino acids

Figure 2.16 shows the general structure of all amino acids and of glycine, the simplest amino acid. They all have a central carbon atom which is bonded to an **amine** group, $-NH_2$, and a **carboxylic acid** group, $-COOH$. It is these two groups which give amino acids their name. The third component that is always bonded to the carbon atom is a hydrogen atom.

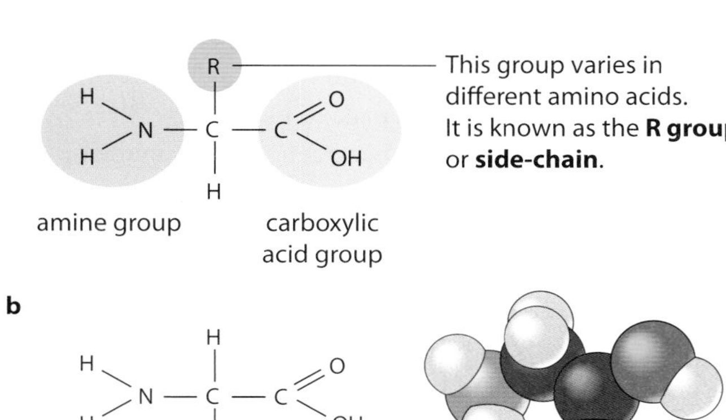

Figure 2.16 **a** The general structure of an amino acid. **b** Structure of the simplest amino acid, glycine, in which the R group is H, hydrogen. R groups for the 20 naturally occurring amino acids are shown in Appendix 1.

The only way in which amino acids differ from each other is in the remaining, fourth, group of atoms bonded to the central carbon. This is called the **R group**. There are 20 different amino acids which occur in the proteins of living organisms, all with a different R group. You can see their molecular formulae in Appendix 1. (You do not need to remember all the different R groups.) Appendix 1 also shows the three-letter abbreviations commonly used by scientists for the names of the amino acids. Many other amino acids have been synthesised in laboratories.

The peptide bond

Figure 2.17 shows how two amino acids can join together. One loses a hydroxyl (–OH) group from its carboxylic acid group, while the other loses a hydrogen atom from its amine group. This leaves a carbon atom of the first amino acid free to bond with the nitrogen atom of the second. The link is called a **peptide bond**. The oxygen and two hydrogen atoms removed from the amino acids

amino acid | amino acid | $-H_2O$ (condensation) | $+H_2O$ (hydrolysis) | dipeptide | peptide bond

Figure 2.17 Amino acids link together by the loss of a molecule of water to form a peptide bond.

form a water molecule. We have seen this type of reaction, a condensation reaction, in the formation of glycosidic bonds (Figure 2.5 on page 31) and in the synthesis of triglycerides (Figure 2.12 on page 37).

The new molecule which has been formed, made up of two linked amino acids, is called a **dipeptide**. Any number of extra amino acids could be added to the chain in a series of condensation reactions. A molecule made up of many amino acids linked together by peptide bonds is called a **polypeptide**. A polypeptide is another example of a polymer and a macromolecule, like a polysaccharide. A complete **protein** molecule may contain just one polypeptide chain, or it may have two or more chains which interact with each other.

In living cells, **ribosomes** are the sites where amino acids are joined together to form polypeptides. The reaction is controlled by enzymes. You can read more about this on pages 119–121.

Polypeptides can be broken down to amino acids by breaking the peptide bonds. This is a hydrolysis reaction, involving the addition of water (Figure 2.17), and happens naturally in the stomach and small intestine during digestion. Here, protein molecules in food are hydrolysed into amino acids before being absorbed into the blood.

Primary structure

A polypeptide or protein molecule may contain several hundred amino acids linked into a long chain. The particular amino acids contained in the chain, and the sequence in which they are joined, is called the **primary structure** of the protein. Figure 2.18 shows the primary structure of the protein ribonuclease, an enzyme.

There are an enormous number of different **possible** primary structures. Even a change in one amino acid in a chain made up of thousands may completely alter the properties of the polypeptide or protein.

Secondary structure

The amino acids in a polypeptide chain have an effect on each other even if they are not directly next to each other. A polypeptide chain, or part of it, often coils into a corkscrew shape called an **α-helix** (Figure 2.19a). This **secondary structure** is due to hydrogen bonding between the oxygen of the –CO– group of one amino acid and the hydrogen of the –NH– group of the amino acid four places ahead of it. Each amino acid has an –NH– and a –CO– group, and Figure 2.19a shows that all these groups are involved in hydrogen bonding in the α-helix, holding the structure firmly in shape. Hydrogen bonding is a result of the polar characteristics of the –CO– and –NH– groups (page 36).

Sometimes hydrogen bonding can result in a much looser, straighter shape than the α-helix, which is called a

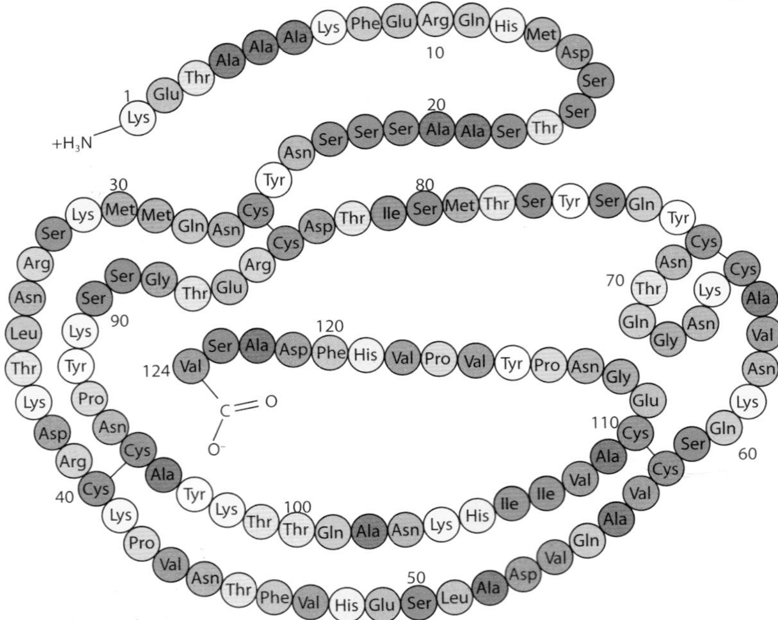

Figure 2.18 The primary structure of ribonuclease. Ribonuclease is an enzyme found in pancreatic juice, which hydrolyses (digests) RNA (Chapter 6). Notice that at one end of the amino acid chain there is an $-NH_3^+$ group, while at the other end there is a $-COO^-$ group. These are known as the amino and carboxyl ends, or the N and C terminals, respectively. Note the use of three-letter abbreviations for the amino acids. These are explained in Appendix 1.

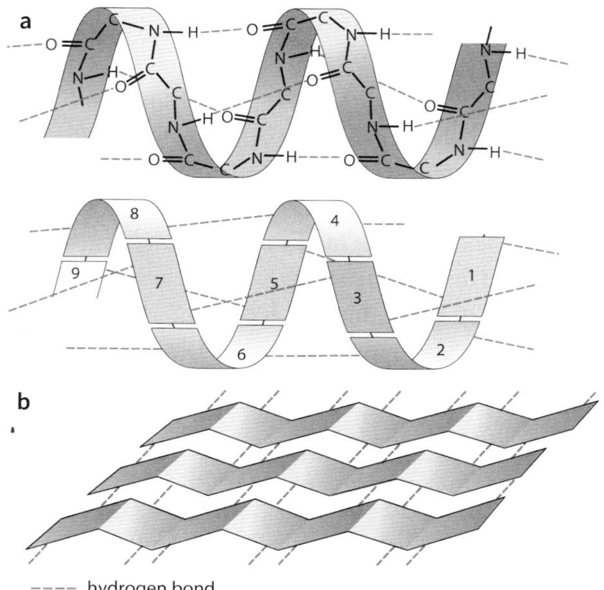

Figure 2.19 Protein secondary structure. **a** Structure of the α-helix. The R groups are not shown. **b** Another common arrangement is the β-pleated sheet. Both of these structures are held in shape by hydrogen bonds between the amino acids.

β-pleated sheet (Figure 2.19b). Hydrogen bonds, although strong enough to hold the α-helix and β-pleated sheet structures in shape, are easily broken by high temperatures and pH changes. As you will see, this has important consequences for living organisms.

Some proteins or parts of proteins show no regular arrangement at all. It all depends on which R groups are present and therefore what attractions occur between amino acids in the chain.

In diagrams of protein structure, α-helices can be represented as coils or cylinders, β-sheets as arrows, and random coils as ribbons (Figures 2.20 and 2.21).

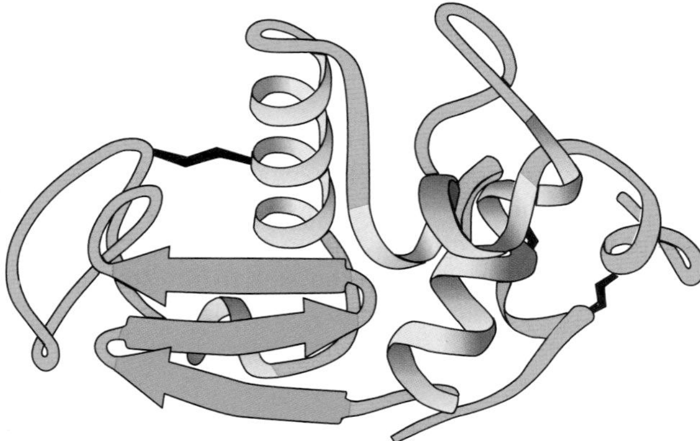

Figure 2.20 Secondary and tertiary structure of lysozyme. α-helices are shown as blue coils, β-sheets as green arrows, and random coils as red ribbons. The black zig-zags are disulfide bonds.

Tertiary structure

In many proteins, the secondary structure itself is coiled or folded. Figure 2.20 shows the complex way in which a molecule of the protein lysozyme folds. Here the α-helices are represented as coils, while in Figure 2.21, which shows the secondary and tertiary structure of myoglobin, the α-helices are shown as cylinders.

At first sight, the myoglobin and lysozyme molecules look like disorganised tangles, but this is not so. The shape of the molecules is very precise, and the molecules are held in these exact shapes by bonds between amino acids in different parts of the chain. The way in which a protein coils up to form a precise three-dimensional shape is known as its **tertiary structure**.

Primary structure is the sequence of amino acids in a polypeptide or protein.

Secondary structure is the structure of a protein molecule resulting from the regular coiling or folding of the chain of amino acids, e.g. an α-helix or β-pleated sheet.

Tertiary structure is the compact structure of a protein molecule resulting from the three-dimensional coiling of the already-folded chain of amino acids.

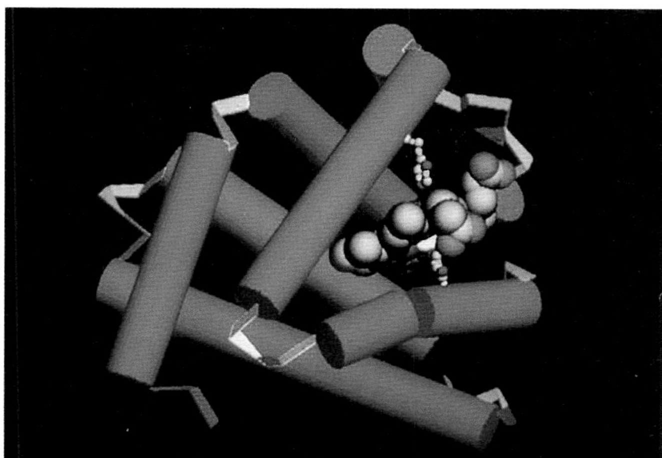

Figure 2.21 A computer graphic showing the secondary and tertiary structures of a myoglobin molecule. Myoglobin is the substance which makes meat look red. It is found in muscle, where it acts as an oxygen-storing molecule. The blue sections are α-helices and are linked by sections of polypeptide chain which are more stretched out – these are shown in red. At the top right is an iron-containing haem group (page 43).

Figure 2.22 shows the four types of bond which help to keep folded proteins in their precise shapes. **Hydrogen bonds** can form between a wide variety of R groups. **Disulfide bonds** form between two cysteine molecules, which contain sulfur atoms. (Can you spot the four disulfide bonds in ribonuclease in Figure 2.18?) **Ionic bonds** form between R groups containing amine and carboxyl groups. (Which amino acids have these?) **Hydrophobic interactions** occur between R groups which are non-polar, or hydrophobic.

a Hydrogen bonds form between strongly polar groups – for example, –NH–, –CO– and –OH groups.

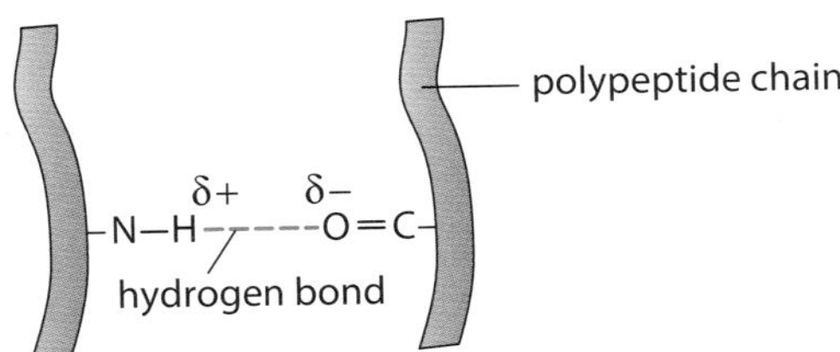

b Disulfide bonds form between cysteine molecules. They are strong covalent bonds. They can be broken by reducing agents.

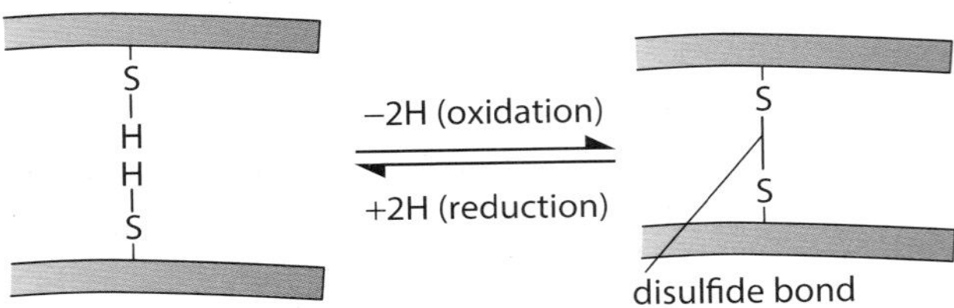

c Ionic bonds form between ionised amine (NH_3^+) groups and ionised carboxylic acid (COO^-) groups. They can be broken by pH changes.

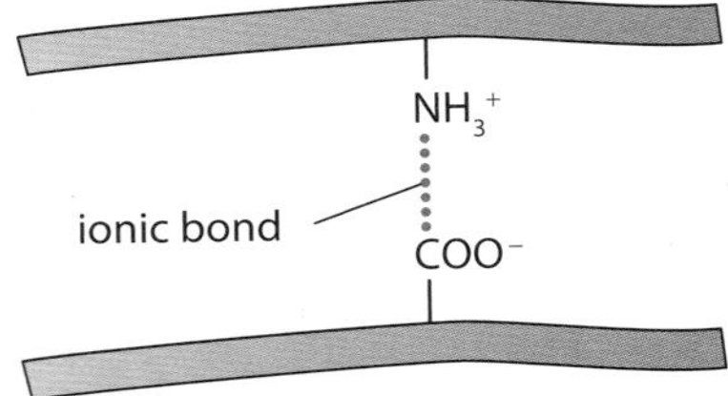

d Weak hydrophobic interactions occur between non-polar R groups. Although the interactions are weak, the groups tend to stay together because they are repelled by the watery environment around them.

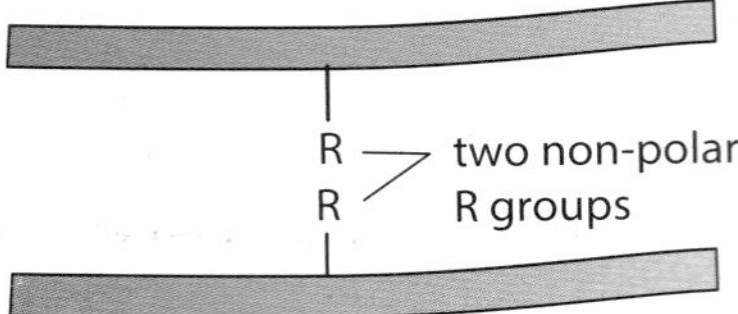

Figure 2.22 The four types of bond which are important in protein tertiary structure: **a** hydrogen bonds, which are also important in secondary structure, **b** disulfide bonds, **c** ionic bonds and **d** hydrophobic interactions.

Quaternary structure

Many protein molecules are made up of two or more polypeptide chains. Haemoglobin is an example of this, having four polypeptide chains in each molecule (Figure 2.23). The association of different polypeptide chains is called the **quaternary structure** of the protein. The chains are held together by the same four types of bond as in the tertiary structure. More details of haemoglobin are given in the next section.

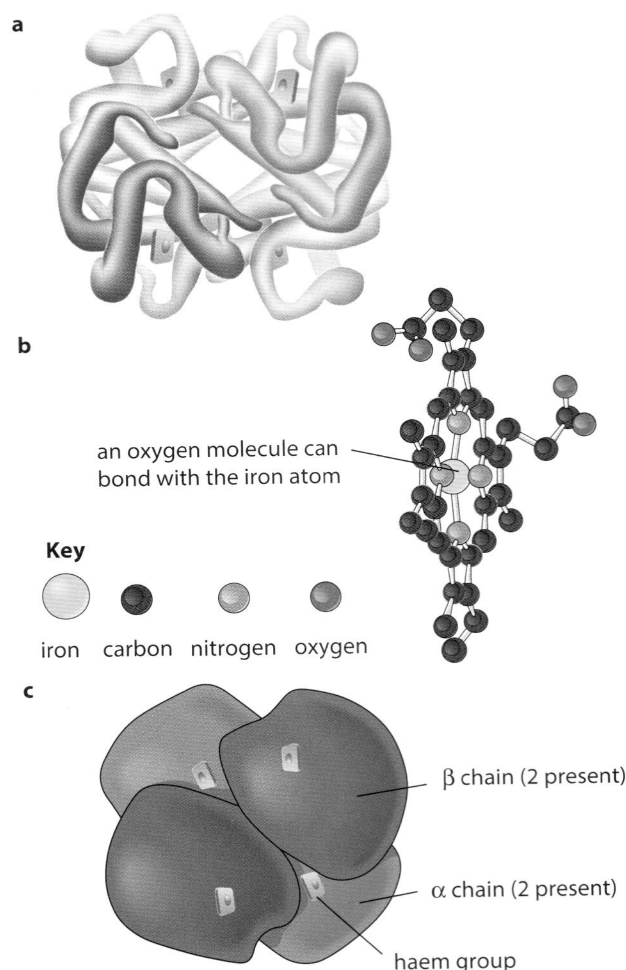

Figure 2.23 Haemoglobin. **a** Each haemoglobin molecule contains four polypeptide chains. The two α chains are shown in purple and blue, and the two β chains in brown and orange. Each polypeptide chain contains a haem group, shown in yellow and red. **b** The haem group contains an iron atom, which can bond reversibly with an oxygen molecule. **c** The complete haemoglobin molecule is nearly spherical.

Quaternary structure is the three-dimensional arrangement of two or more polypeptides, or of a polypeptide and a non-protein component such as haem, in a protein molecule.

Globular and fibrous proteins

A protein whose molecules curl up into a 'ball' shape, such as myoglobin or haemoglobin, is known as a globular protein. In a living organism, proteins may be found in cells and in other aqueous environments such as blood, tissue fluid and in phloem of plants. Globular proteins usually curl up so that their non-polar, hydrophobic R groups point into the centre of the molecule, away from their watery surroundings. Water molecules are excluded from the centre of the folded protein molecule. The polar, hydrophilic R groups remain on the outside of the molecule. Globular proteins, therefore, are usually soluble, because water molecules cluster around their outward-pointing hydrophilic R groups (Figure 2.24).

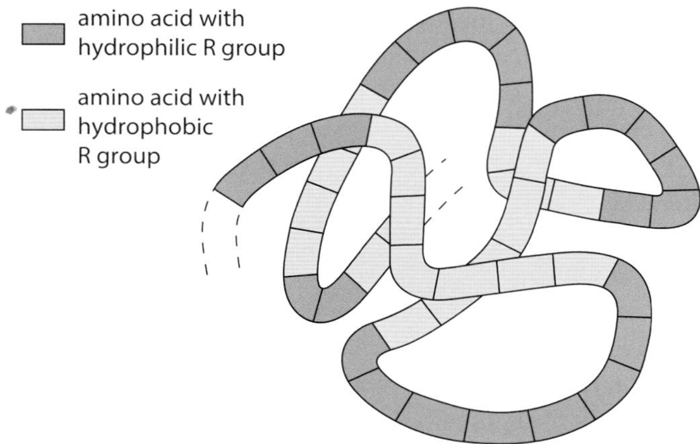

Figure 2.24 A section through part of a globular protein molecule. The polypeptide chain coils up with hydrophilic R groups outside and hydrophobic ones inside, which makes the molecule soluble.

Many globular proteins have roles in metabolic reactions. Their precise shape is the key to their functioning. Enzymes, for example, are globular proteins.

Many other protein molecules do not curl up into a ball, but form long strands. These are known as fibrous proteins. Fibrous proteins are not usually soluble in water and most have structural roles. For example, **keratin** forms hair, nails and the outer layers of skin, making these structures waterproof. Another example of a fibrous protein is **collagen** (pages 44–45).

Haemoglobin – a globular protein

Haemoglobin is the oxygen-carrying pigment found in red blood cells, and is a globular protein. We have seen that it is made up of four polypeptide chains, so it has a quaternary structure. Each chain is itself a protein known as globin. Globin is related to myoglobin and so has a very similar tertiary structure (Figures 2.21 and 2.23). There are many types of globin – two types are used to make haemoglobin, and these are known as alpha-globin (α-globin) and beta-globin (β-globin). Two of the haemoglobin chains, called α chains, are made from α-globin, and the other two chains, called β chains, are made from β-globin.

The haemoglobin molecule is nearly spherical (Figure 2.23). The four polypeptide chains pack closely together, their hydrophobic R groups pointing in towards the centre of the molecule, and their hydrophilic ones pointing outwards.

The interactions between the hydrophobic R groups inside the molecule are important in holding it in its correct three-dimensional shape. The outward-pointing hydrophilic R groups on the surface of the molecule are important in maintaining its solubility. In the genetic condition known as sickle cell anaemia, one amino acid which occurs in the **surface** of the β chain is replaced with a different amino acid. The correct amino acid is glutamic acid, which is polar. The substitute is valine, which is non-polar. Having a non-polar R group on the outside of the molecule makes the haemoglobin much less soluble, and causes the unpleasant and dangerous symptoms associated with sickle cell anaemia in anyone whose haemoglobin is all of this 'faulty' type (Figure 2.25).

Each polypeptide chain of haemoglobin contains a **haem group**, shown in Figure 2.23b. A group like this, which is an important, permanent, part of a protein molecule but is not made of amino acids, is called a prosthetic group.

Each haem group contains an iron atom. One oxygen molecule, O_2, can bind with each iron atom. So a complete haemoglobin molecule, with four haem groups, can carry four oxygen molecules (eight oxygen atoms) at a time.

It is the haem group which is responsible for the colour of haemoglobin. This colour changes depending on whether or not the iron atoms are combined with oxygen. If they are, the molecule is known as **oxyhaemoglobin**, and is bright red. If not, the colour is purplish.

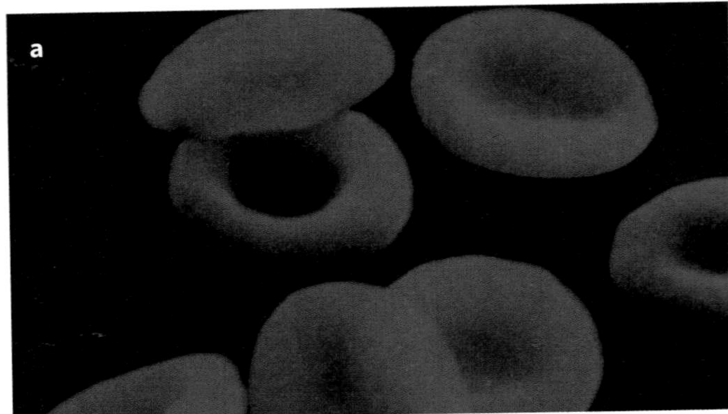

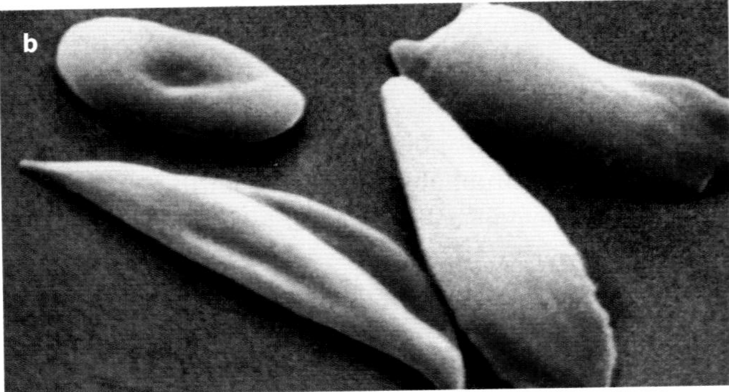

Figure 2.25 **a** Scanning electron micrograph of human red blood cells (× 3300). Each cell contains about 250 million haemoglobin molecules. **b** Scanning electron micrograph of red blood cells from a person with sickle cell anaemia. You can see a normal cell and three or four sickled cells (×3300).

BOX 2.4: Testing for the presence of proteins

Background information

All proteins have peptide bonds, containing nitrogen atoms. These form a purple complex with copper(II) ions and this forms the basis of the biuret test.

The reagent used for this test is called **biuret reagent**. You can use it as two separate solutions: a dilute solution of potassium hydroxide or sodium hydroxide, and a dilute solution of copper(II) sulfate. Alternatively, you can use a ready-made biuret reagent that contains both the copper(II) sulfate solution and the hydroxide ready mixed. To stop the copper ions reacting with the hydroxide ions and forming a precipitate, this ready-mixed reagent also contains sodium potassium tartrate or sodium citrate.

Procedure

The biuret reagent is added to the solution to be tested. No heating is required. A purple colour indicates that protein is present. The colour develops slowly over several minutes.

Collagen – a fibrous protein

Collagen is the most common protein found in animals, making up 25% of the total protein in mammals. It is an insoluble fibrous protein (Figure 2.26) found in skin (leather is preserved collagen), tendons, cartilage, bones, teeth and the walls of blood vessels. It is an important **structural protein**, not only in humans but in almost all animals, and is found in structures ranging from the body walls of sea anemones to the egg cases of dogfish.

As shown in Figure 2.26b, a collagen molecule consists of three polypeptide chains, each in the shape of a helix. (This is not an α-helix – it is not as tightly wound.) These three helical polypeptides are wound around each other, forming a three-stranded 'rope' or 'triple helix'. The three strands are held together by hydrogen bonds and some covalent bonds. Almost every third amino acid in each polypeptide is glycine, the smallest amino acid. Glycine is found on the insides of the strands and its small size allows the three strands to lie close together and so form a tight coil. Any other amino acid would be too large.

Each complete, three-stranded molecule of collagen interacts with other collagen molecules running parallel to it. Covalent bonds form between the R groups of amino acids lying next to each other. These cross-links hold many collagen molecules side by side, forming **fibrils**. The ends of the parallel molecules are staggered; if they were not, there would be a weak spot running right across the collagen fibril. Finally, many fibrils lie alongside each other, forming strong bundles called **fibres**.

The advantage of collagen is that it is flexible but it has tremendous tensile strength, meaning it can withstand large pulling forces without stretching or breaking. The human Achilles tendon, which is almost pure collagen fibres, can withstand a pulling force of 300 N per mm^2 of cross-sectional area, about one-quarter the tensile strength of mild steel.

Collagen fibres line up according to the forces they must withstand. In tendons they line up in parallel bundles along the length of the tendon, the direction of tension. In skin, they may form layers, with the fibres running in different directions in the different layers, like cellulose in cell walls. In this way, they resist tensile (pulling) forces from many directions.

a

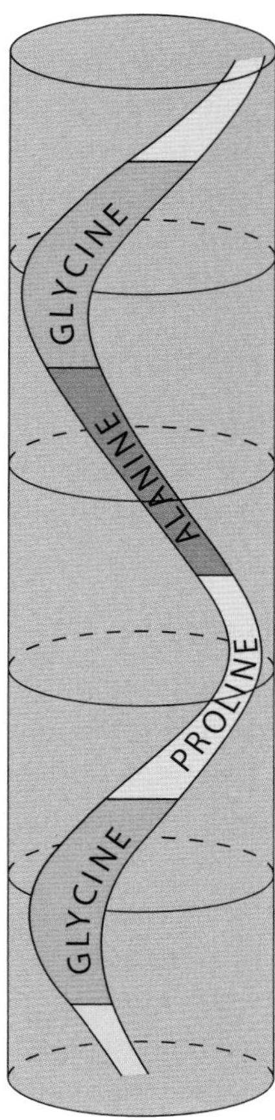

The polypeptides which make up a collagen molecule are in the shape of a stretched-out helix. Every third amino acid is glycine.

b

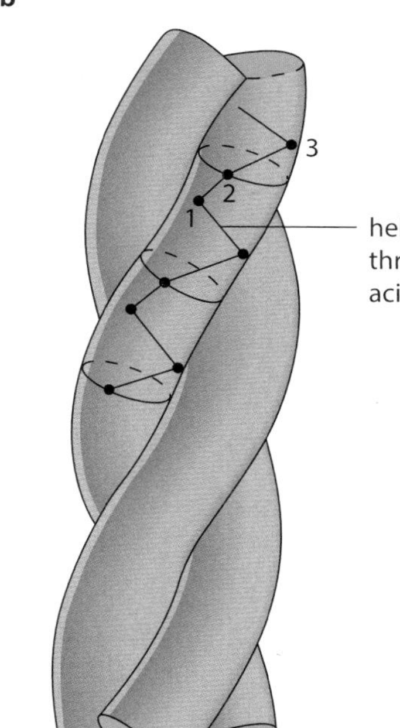

Three helices wind together to form a collagen molecule. These strands are held together by hydrogen bonds and some covalent bonds.

c

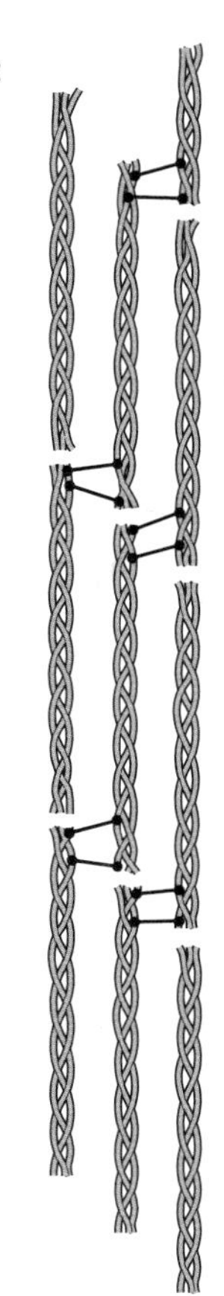

Many of these triple helices lie side by side, linked to each other by covalent cross-links between the side chains of amino acids near the ends of the polypeptides. Notice that these cross-links are out of step with each other; this gives collagen greater strength.

d

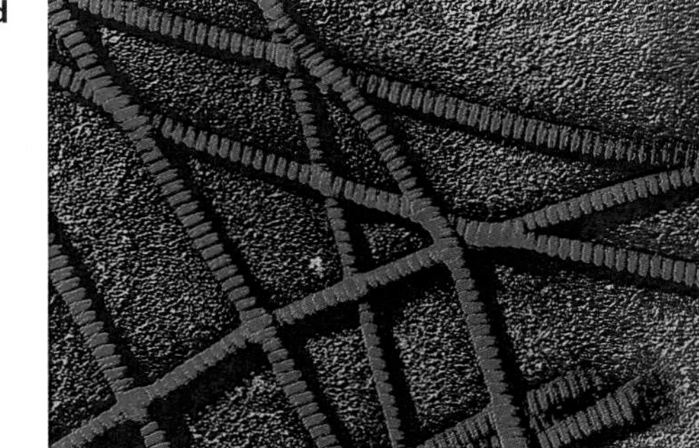

A scanning electron micrograph of collagen fibrils (×17 000). Each fibril is made up of many triple helices lying parallel with one another. The banded appearance is caused by the regular way in which these helices are arranged, with the staggered gaps between the molecules (shown in **c**) appearing darker.

e

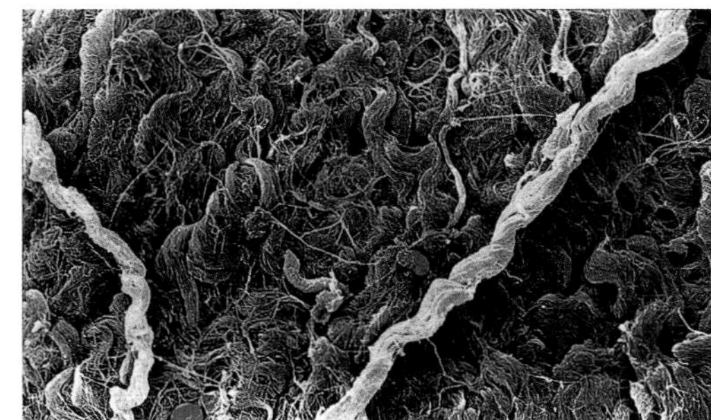

A scanning electron micrograph of human collagen fibres (×2000). Each fibre is made up of many fibrils lying side by side. These fibres are large enough to be seen with an ordinary light microscope.

Figure 2.26 Collagen. The diagrams and photographs begin with the very small and work up to the not-so-small. Thus three polypeptide chains like the one shown in **a** make up a collagen molecule, shown in **b**; many collagen molecules make up a fibril, shown in **c** and **d**; many fibrils make up a fibre, shown in **e**.

Water

Water is arguably the most important biochemical of all. Without water, life would not exist on this planet. It is important for two reasons. First, it is a major component of cells, typically forming between 70% and 95% of the mass of the cell. You are about 60% water. Second, it provides an environment for those organisms that live in water. Three-quarters of the planet is covered in water.

Although it is a simple molecule, water has some surprising properties. For example, such a small molecule would exist as a gas at normal Earth temperatures were it not for its special property of hydrogen bonding to other water molecules (page 35). Also, because water is a liquid, it provides a medium for molecules and ions to mix in, and hence a medium in which life can evolve.

The hydrogen bonding of water molecules makes the molecules more difficult to separate and affects the physical properties of water. For example, the energy needed to break the hydrogen bonds makes it more difficult to convert water from a liquid to a gas than to convert similar compounds which lack hydrogen bonds, such as hydrogen sulfide (H_2S), which is a gas at normal air temperatures.

Water as a solvent

Water is an excellent solvent for ions and polar molecules (molecules with an uneven charge distribution, such as sugars and glycerol) because the water molecules are attracted to the ions and polar molecules and therefore collect around and **separate** them (Figure 2.27). This is what happens when a chemical dissolves in water. Once a chemical is in solution, it is free to move about and react with other chemicals. Most processes in living organisms take place in solution in this way.

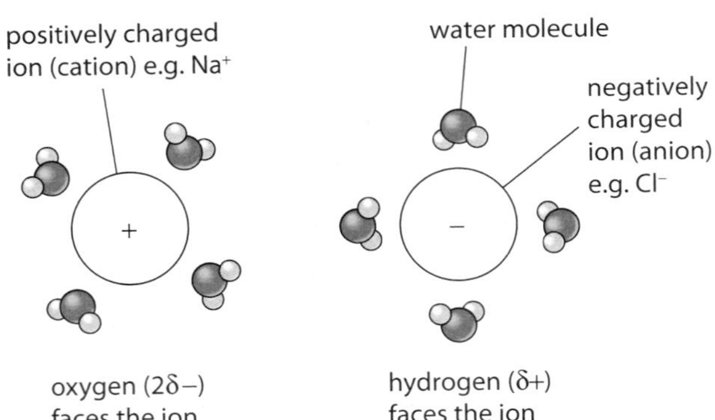

Figure 2.27 Distribution of water molecules around ions in a solution.

By contrast, non-polar molecules such as lipids are insoluble in water and, if surrounded by water, tend to be **pushed together** by the water, since the water molecules are **attracted to each other**. This is important, for example, in hydrophobic interactions in protein structure and in membrane structure (Chapter 4), and it increases the stability of these structures.

Water as a transport medium

Water is the transport medium in the blood, in the lymphatic, excretory and digestive systems of animals, and in the vascular tissues of plants. Here again its solvent properties are essential.

High specific heat capacity

The heat capacity of a substance is the amount of heat required to raise its temperature by a given amount. The specific heat capacity of water (or simply the specific heat) is the amount of heat energy required to raise the temperature of 1 kg of water by 1 °C.

Water has a relatively high heat capacity. In order for the temperature of a liquid to be raised, the molecules must gain energy and consequently move about more rapidly. The hydrogen bonds that tend to make water molecules stick to each other make it more difficult for the molecules to move about freely; the bonds must be broken to allow free movement. This explains why more energy is needed to raise the temperature of water than would be the case if there were no hydrogen bonds. Hydrogen bonding, in effect, allows water to store more energy than would otherwise be possible for a given temperature rise.

The high heat capacity of water has important biological implications because it makes water more resistant to changes in temperature. This means that the temperature within cells and within the bodies of organisms (which have a high proportion of water) tends to be more constant than that of the air around them. Biochemical reactions therefore operate at relatively constant rates and are less likely to be adversely affected by extremes of temperature. It also means that large bodies of water such as lakes and oceans are slow to change temperature as environmental temperature changes. As a result they provide more stable habitats for aquatic organisms.

High latent heat of vapourisation

The latent heat of vapourisation is a measure of the heat energy needed to vaporise a liquid (cause it to evaporate), changing it from a liquid to a gas. In the case of water, it involves the change from liquid water to water vapour.

Water has a relatively high latent heat of vapourisation. This is a consequence of its high heat capacity. The fact that water molecules tend to stick to each other by hydrogen bonds means that relatively large amounts of energy are needed for vapourisation to occur, because hydrogen bonds have to be broken before molecules can escape as a gas. The energy transferred to water molecules during vapourisation results in a corresponding loss of energy from their surroundings, which therefore cool down. This is biologically important because it means that living organisms can use evaporation as a cooling mechanism, as in sweating or panting in mammals. A large amount of heat energy can be lost for relatively little loss of water, reducing the risk of dehydration. It can also be important in cooling leaves during transpiration.

The reverse is true when water changes from liquid to solid ice. This time the water molecules must lose a relatively large amount of energy, making it less likely that the water will freeze. This is an advantage for aquatic organisms and makes it less likely that their bodies will freeze.

Density and freezing properties

Water is an unusual chemical because the solid form, ice, is less dense than its liquid form. Below 4 °C, the density of water starts to decrease. Ice therefore floats on liquid water and insulates the water under it. This reduces the tendency for large bodies of water to freeze completely, and increases the chances of life surviving in cold conditions.

Changes in the density of water with temperature cause currents, which help to maintain the circulation of nutrients in the oceans.

QUESTION

2.6 State the property of water that allows each of the following to take place and, in each case, explain its importance:

- **a** the cooling of skin during sweating
- **b** the transport of glucose and ions in a mammal
- **c** much smaller temperature fluctuations in lakes and oceans than in terrestrial (land-based) habitats.

High surface tension and cohesion

Water molecules have very high cohesion – in other words they tend to stick to each other. This explains why water can move in long, unbroken columns through the vascular tissue in plants (Chapter 7), and is an important property in cells. High cohesion also results in high surface tension at the surface of water. This allows certain small organisms, such as pond skaters, to exploit the surface of water as a habitat, allowing them to settle on or skate over its surface (Figure 2.28).

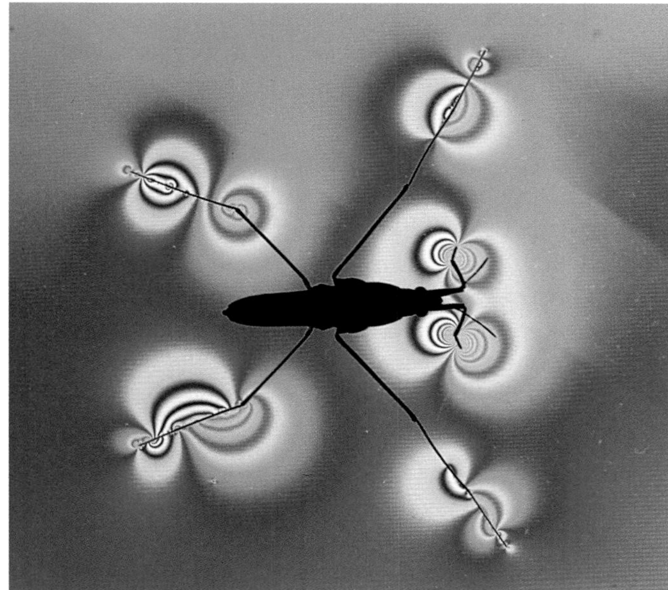

Figure 2.28 A pond skater standing on the surface of pond water. This was photographed through an interferometer, which shows interference patterns made by the pond skater as it walks on the water's surface. The surface tension of the water means the pond skater never breaks through the surface.

Water as a reagent

Water takes part as a reagent in some chemical reactions inside cells. For example, it is used as a reagent in photosynthesis. During photosynthesis, energy from sunlight is used to separate hydrogen from the oxygen in water molecules. The hydrogen is then effectively used as a fuel to provide the energy needs of the plant – for example, by making glucose, an energy-rich molecule. The waste oxygen from photosynthesis is the source of the oxygen in the atmosphere which is needed by aerobic organisms for respiration. Water is also essential for all hydrolysis reactions. Hydrolysis is the mechanism by which large molecules are broken down to smaller molecules, as in digestion.

Summary

- The larger biological molecules are made from smaller molecules. Polysaccharides are made from monosaccharides, proteins from amino acids, nucleic acids from nucleotides, lipids from fatty acids and glycerol. Polysaccharides, proteins and nucleic acids are formed from repeating identical or similar subunits called monomers, and are therefore polymers. These build up into giant molecules called macromolecules.
- The smaller units are joined together by condensation reactions. Condensation involves removal of water. The reverse process, adding water, is called hydrolysis and is used to break the large molecules back down into smaller molecules.
- The linkages that join monosaccharides are called glycosidic bonds. Carbohydrates have the general formula $C_x(H_2O)_y$ and comprise monosaccharides, disaccharides and polysaccharides. Monosaccharides (e.g. glucose) and disaccharides (e.g. sucrose) are very water-soluble and together are known as sugars. They are important energy sources in cells and also important building blocks for larger molecules like polysaccharides.
- Monosaccharides may have straight-chain or ring structures and may exist in different isomeric forms such as α-glucose and β-glucose. Benedict's reagent can be used to test for reducing and non-reducing sugars. The test is semi-quantitative.
- Polysaccharides include starch, glycogen and cellulose. Starch is an energy storage compound in plants. 'Iodine solution' can be used to test for starch. Starch is made up of two types of molecule, amylose and amylopectin, both made from α-glucose. Amylose is an unbranching molecule, whereas amylopectin has a branching structure.
- Glycogen is an energy storage compound in animals, which is also made from α-glucose. Its structure is similar to that of amylopectin, but with more branching. Cellulose is a polymer of β-glucose molecules. The molecules are grouped together by hydrogen bonding to form mechanically strong fibres with high tensile strength that are found in plant cell walls.
- Lipids are a diverse group of chemicals, the most common of which are triglycerides (fats and oils). Triglycerides are made by condensation between three fatty acid molecules and glycerol. They are hydrophobic and do not mix with water, acting as energy storage compounds in animals, as well as having other functions such as insulation and buoyancy in marine mammals. Phospholipids have a hydrophilic phosphate head and two hydrophobic fatty acid tails. This is important in the formation of membranes. The emulsion test can be used to test for lipids.
- Proteins are long chains of amino acids which fold into precise shapes. Biuret reagent can be used to test for proteins. The linkages that join amino acids are called peptide bonds. The sequence of amino acids in a protein, known as its primary structure, determines the way that it folds and hence determines its three-dimensional shape and function.
- Many proteins contain areas where the amino acid chain is twisted into an α-helix; this is an example of secondary structure. The structure forms as a result of hydrogen bonding between the amino acids. Another secondary structure formed by hydrogen bonding is the β-pleated sheet. Further folding of proteins produces the tertiary structure. Often, a protein is made from more than one polypeptide chain. The association between the different chains is the quaternary structure of the protein. Tertiary and quaternary structures are very precise and are held in place by hydrogen bonds, disulfide bonds (which are covalent), ionic bonds and hydrophobic interactions.
- Proteins may be globular or fibrous. A molecule of a globular protein – for example haemoglobin – is roughly spherical. Most globular proteins are soluble and metabolically active. Haemoglobin contains a non-protein (prosthetic) group, the haem group, which contains iron. This combines with oxygen. A molecule of a fibrous protein – for example, collagen – is less folded and forms long strands. Fibrous proteins are insoluble. They often have a structural role. Collagen has high tensile strength and is the most common animal protein, being found in a wide range of tissues.
- Water is important within plants and animals, where it forms a large part of the mass of each cell. It is also an environment in which organisms can live. Extensive hydrogen bonding gives water unusual properties.

- Water is liquid at most temperatures on the Earth's surface. It has a high specific heat capacity, which makes liquid water relatively resistant to changes in temperature. Water acts as a solvent for ions and polar molecules, and causes non-polar molecules to group together. Water has a relatively high latent heat of vapourisation, meaning that evaporation has a strong cooling effect. It has high cohesion and surface tension which affects the way it moves through narrow tubes such as xylem and allows it to form a surface on which some organisms can live. Water acts as a reagent inside cells, as in hydrolysis reactions, and in photosynthesis as a source of hydrogen.

End-of-chapter questions

1 Which term describes both collagen and haemoglobin?

A enzymes

B fibrous proteins

C globular proteins

D macromolecules [1]

2 What type of chemical reaction is involved in the formation of disulfide bonds?

A condensation

B hydrolysis

C oxidation

D reduction [1]

3 Which diagram best represents the arrangement of water molecules around sodium (Na^+) and chloride (Cl^-) ions in solution?

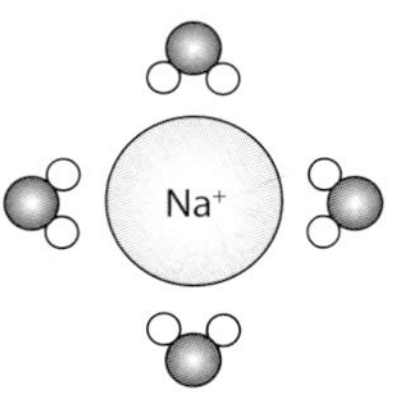

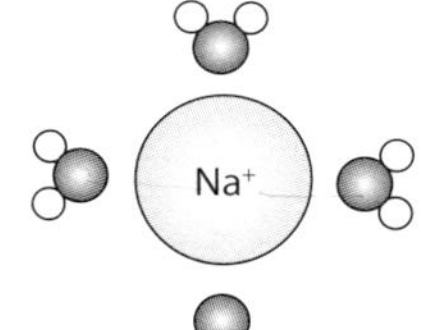

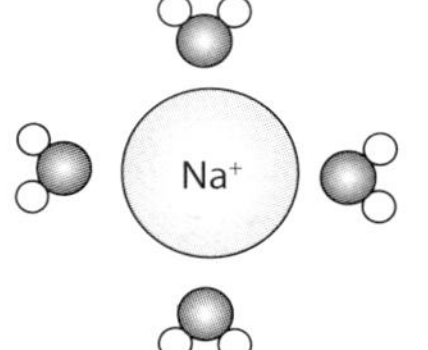

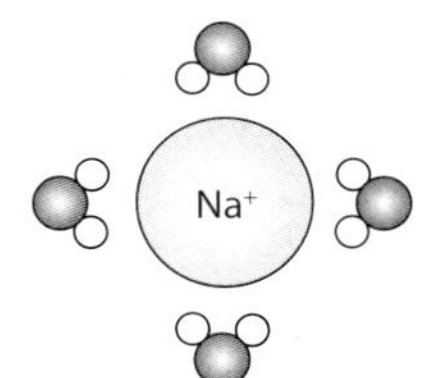

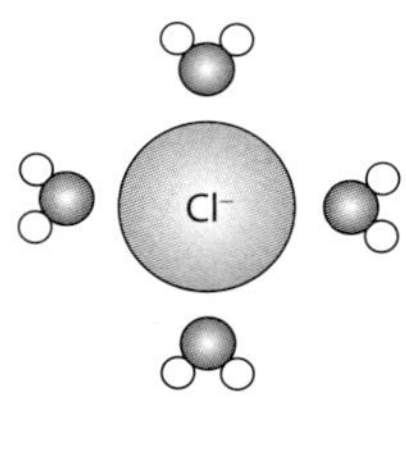

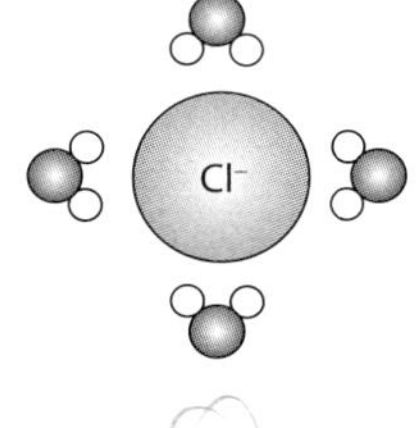

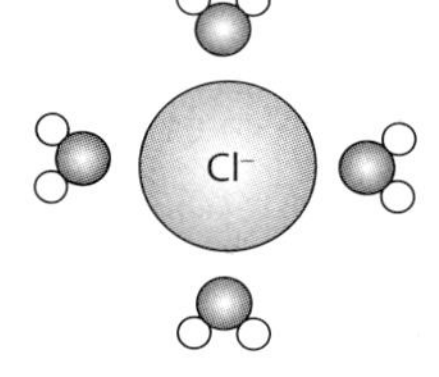

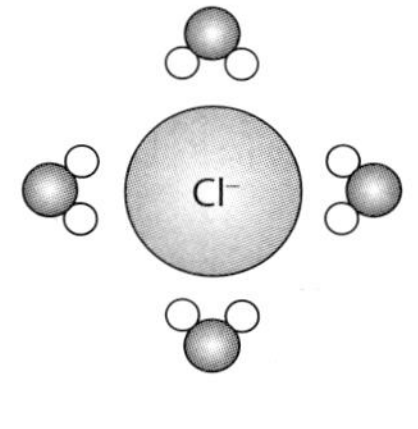

A B C D

[1]

4 Copy and complete the following table. Place a tick or a cross in each box as appropriate.

	Globular protein, e.g. haemoglobin	Fibrous protein, e.g. collagen	Monosaccharide	Disaccharide	Glycogen	Starch	Cellulose	Lipid
monomer			✓					
polymer								
macromolecule								
polysaccharide					✓	✓	✓	
contains subunits that form branched chains					✓	✓		
contains amino acids	✓	✓						
made from organic acids and glycerol								✓
contains glycosidic bonds				✓	✓	✓	✓	
contains peptide bonds	✓	✓						
one of its main functions is to act as an energy store								✓
usually insoluble in water								✓
usually has a structural function		✓					✓	
can form helical or partly helical structures		✓						
contains the elements carbon, hydrogen and oxygen only			✓	✓	✓	✓	✓	✓

1 mark for each correct column [8]

5 Copy and complete the table below, which summarises some of the functional categories into which proteins can be placed.

Category	Example
structural	1. Collagen 2. Keratin
enzyme	Amylase
hormones	insulin
transport	haemoglobin and myoglobin
defensive	white blood cells
muscle contraction	actin and myosin
storage	ovalbumin & casein

[8]

6 State **three** characteristics of monosaccharides. [3]

→ general formula $(CH_2O)_n$
→ source of E in respiration
→ building blocks for larger molecule

7 The diagram below shows a disaccharide called lactose. The carbon atoms are numbered. You are not expected to have seen this structure before. Lactose is a reducing sugar found in milk. It is made from a reaction between the two monosaccharides glucose and galactose.

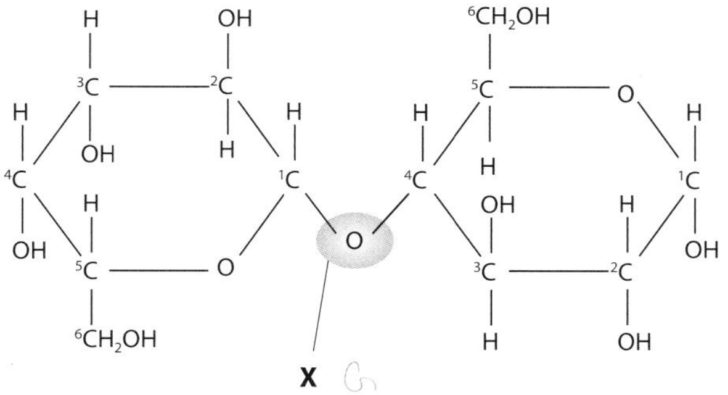

- **a** Suggest **two** functions that lactose could have. [2]
- **b** What is the name given to the reaction referred to above that results in the formation of lactose? [1]
- **c** Identify the bond labelled **X** in the diagram. [1]
- **d** Draw diagrams to show the structures of separate molecules of glucose and galactose. [2]
- **e** Using the information in the diagram, is the alpha or beta form of glucose used to make lactose? Explain your answer. [2]
- **f** Like lactose, sucrose is a disaccharide. If you were given a solution of lactose and a solution of sucrose, state briefly how you could distinguish between them. [2]

[Total: 10]

8 **a** The diagram below shows the structures of three amino acids.

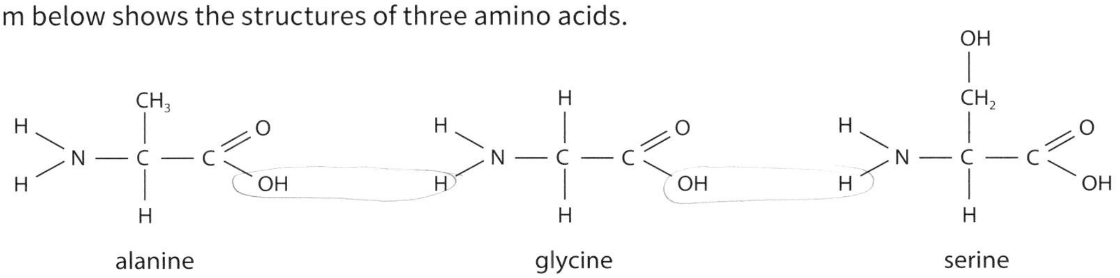

alanine　　glycine　　serine

- **i** Draw a diagram to show the structure of the tripeptide with the following sequence: alanine–glycine–serine. [3]
- **ii** What is the name given to the sequence of amino acids in a protein? [1]
- **iii** What substance, apart from the tripeptide, would be formed when the three amino acids combine? [1]
- **iv** Draw a ring around an atom or group of atoms making up an R group that could hydrogen bond with a neighbouring R group. [1]
- **v** Draw a ring around **and label** the peptide bond(s) you have drawn in the diagram. [1]
- **vi** Draw a ring around a group of atoms which could hydrogen bond with a –CO – group in an alpha helix (α-helix). **Label** this group **A**. [1]

b State **three** features that α-helices and beta sheets (β-sheets) have in common. [3]

c A protein can be described as a polymer. State the meaning of the term **polymer**. [2]

d X and Y represent two different amino acids.

- **i** Write down the sequences of all the possible tripeptides that could be made with just these two amino acids. [1]
- **ii** From your answer to **d i**, what is the formula for calculating the number of different tripeptides that can be formed from two different amino acids? [1]

[Total: 15]

9 Copy the diagrams below.

A

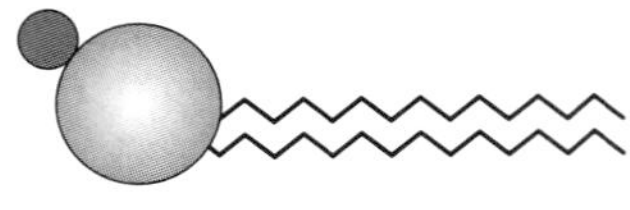

B

a Identify with labels which one represents a lipid and which a phospholipid. [1]

b i For molecule **A**, indicate on the diagram where hydrolysis would take place if the molecule was digested. [2]

ii Name the products of digestion. [2]

c Each molecule has a head with tails attached. For molecule **B**, label the head to identify its chemical nature. [1]

d i Which of the two molecules is water-soluble? [1]

ii Explain your answer to **d i**. [1]

e State **one** function of each molecule. [2]

[Total: 10]

10 a Copy the following table to summarise some differences between collagen and haemoglobin.

Collagen	Haemoglobin

[5]

Use the following to guide you.

Row 1: State whether globular or fibrous.

Row 2: State whether entirely helical or partly helical.

Row 3: State the type of helix.

Row 4: State whether a prosthetic group is present or absent.

Row 5: State whether soluble in water or insoluble in water.

b State **one** way in which the structure of haemoglobin is related to its function. [2]

c Haemoglobin possesses a quaternary structure. What does this mean? [1]

d Name the **five** elements found in haemoglobin. [2]

[Total: 10]

Chapter 3: Enzymes

Learning outcomes

You should be able to:

- explain how enzymes work
- describe and explain the factors that affect enzyme activity
- use V_{max} and K_m to compare the affinity of different enzymes for their substrates
- explain how reversible inhibitors affect the rate of enzyme activity
- carry out experiments with enzymes under controlled conditions
- explain the advantages of using immobilised enzymes

The best means of defence is attack

If you are a beetle and about to be eaten by a predator such as a spider or a frog, how do you escape? Bombardier beetles have evolved a spectacularly successful strategy (Figure 3.1). It makes use of the fantastic speeds of enzyme-controlled reactions. When threatened by a predator, the beetle uses the extended tip of its abdomen to squirt a boiling hot chemical spray at its attacker.

The release of the spray is accompanied by a loud popping sound. The beetle can swivel the tip of the abdomen to spray accurately in almost any direction. With the potential predator reeling from this surprise attack, the beetle makes good its escape.

How are enzymes involved? Inside the beetle's abdomen is a chemical mixing chamber into which hydrogen peroxide and hydroquinone are released.

The chamber contains two enzymes, catalase and peroxidase, which make the reactions they catalyse proceed several million times faster than normal. Hydrogen peroxide is broken down into oxygen and water and the oxygen is used to oxidise the hydroquinone into quinone. The reactions are violent and release a great deal of heat, vaporising about 20% of the resulting liquid. Within a fraction of a second a boiling, foul-smelling gas and liquid mix is explosively discharged through an outlet valve.

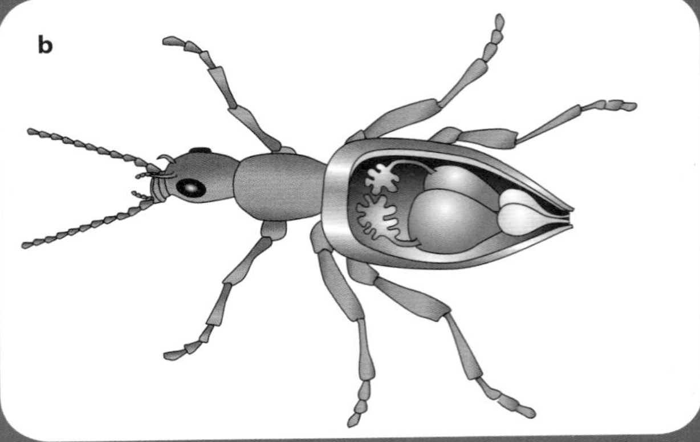

Figure 3.1 **a** A bombardier beetle sprays a boiling chemical spray at an annoying pair of forceps. **b** Abdominal organs generating the spray.

Mode of action of enzymes

Enzymes are protein molecules which can be defined as **biological catalysts.** A catalyst is a molecule which speeds up a chemical reaction but remains unchanged at the end of the reaction. Virtually every metabolic reaction which takes place within a living organism is catalysed by an enzyme and enzymes are therefore essential for life to exist. Many enzyme names end in **-ase** – for example amylase and ATPase.

Intracellular and extracellular enzymes

Not all enzymes operate within cells. Those that do are described as **intracellular**. Enzymes that are secreted by cells and catalyse reactions outside cells are described as **extracellular.** Digestive enzymes in the gut are an example. Some organisms secrete enzymes outside their bodies. Fungi, for example, often do this in order to digest the substrate on which they are growing.

Lock and key and induced fit hypotheses

Enzymes are globular proteins. Like all globular proteins, enzyme molecules are coiled into a precise three-dimensional shape, with hydrophilic R groups (side-chains) on the outside of the molecule ensuring that they are soluble. Enzyme molecules also have a special feature in that they possess an active site (Figure 3.2). The active site of an enzyme is a region, usually a cleft or depression, to which another molecule or molecules can bind. This molecule is the **substrate** of the enzyme. The shape of the active site allows the substrate to fit perfectly. The idea that the enzyme has a particular shape into which the substrate fits exactly is known as the lock and key hypothesis. The substrate is the **key** whose shape fits the **lock** of the enzyme. The substrate is held in place by temporary bonds which form between the substrate and some of the R groups of the enzyme's amino acids. This combined structure is termed the **enzyme–substrate complex**.

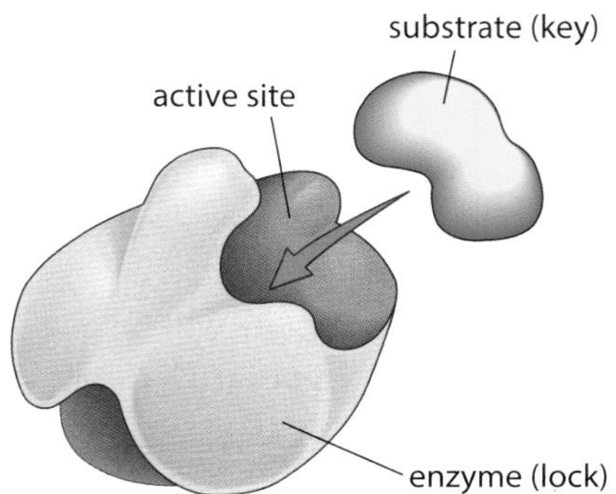

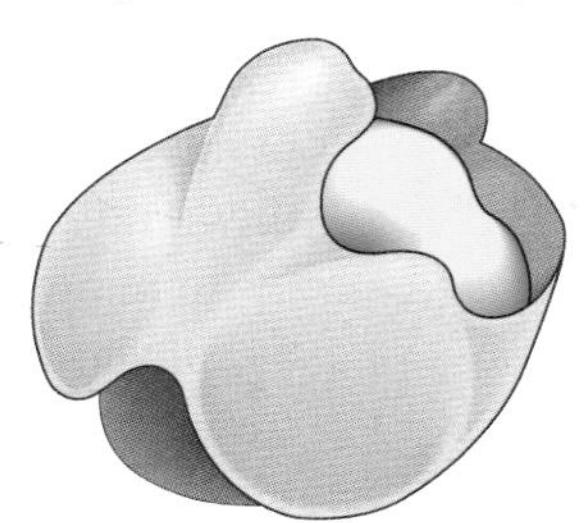

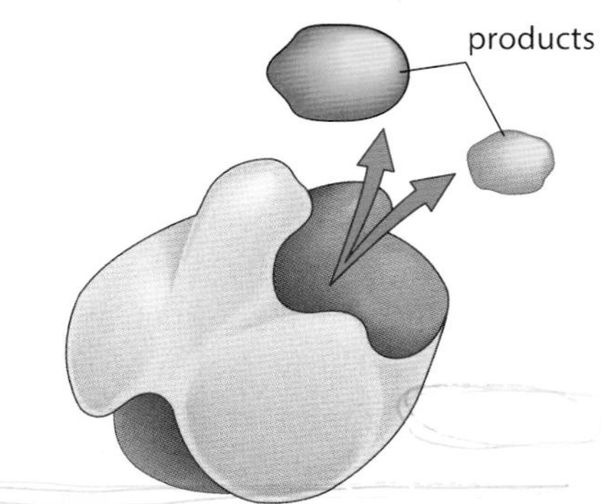

a An enzyme has a cleft in its surface, called the active site. The substrate molecule has a complementary shape.

b Random movement of enzyme and substrate brings the substrate into the active site. An enzyme–substrate complex is temporarily formed. The R groups of the amino acids in the active site interact with the substrate.

c The interaction of the substrate with the active site breaks the substrate apart. An enzyme–product complex is briefly formed, before the two product molecules leave the active site, leaving the enzyme molecule unchanged and ready to bind with another substrate molecule.

Figure 3.2 How an enzyme catalyses the breakdown of a substrate molecule into two product molecules.

Each type of enzyme will usually act on only one type of substrate molecule. This is because the shape of the active site will only allow one shape of molecule to fit. The enzyme is said to be **specific** for this substrate.

In 1959 the lock and key hypothesis was modified in the light of evidence that enzyme molecules are more flexible than is suggested by a rigid lock and key. The modern hypothesis for enzyme action is known as the **induced fit hypothesis**. It is basically the same as the lock and key hypothesis, but adds the idea that the enzyme, and sometimes the substrate, can change shape slightly as the substrate molecule enters the enzyme, in order to ensure a perfect fit. This makes the catalysis even more efficient.

An enzyme may catalyse a reaction in which the substrate molecule is split into two or more molecules, as shown in Figure 3.2. Alternatively, it may catalyse the joining together of two molecules, as when making a dipeptide. A simplified diagram is shown in Figure 3.3. This diagram also shows the enzyme–product complex which is briefly formed before release of the product. Interaction between the R groups of the enzyme and the atoms of the substrate can break, or encourage formation of, bonds in the substrate molecule, forming one, two or more **products**.

When the reaction is complete, the product or products leave the active site. The enzyme is unchanged by this process, so it is now available to receive another substrate molecule. The rate at which substrate molecules can bind to the enzyme's active site, be formed into products and leave can be very rapid. The enzyme catalase, for example,

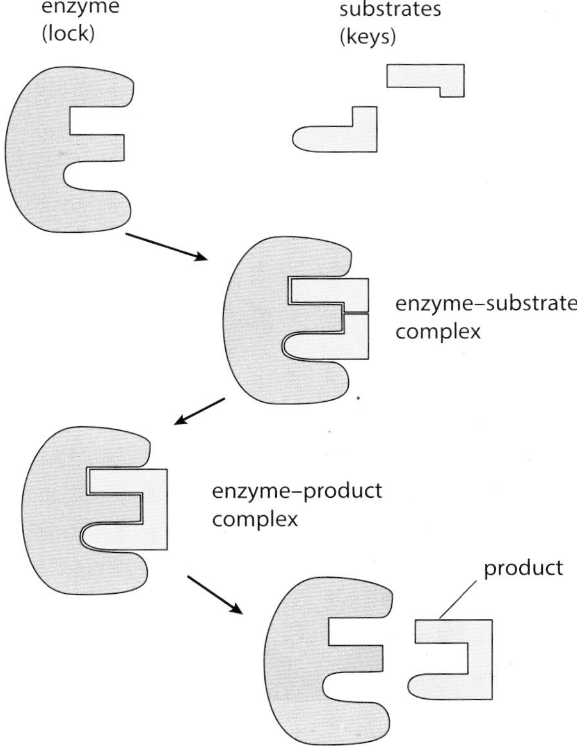

Figure 3.3 A simplified diagram of enzyme function. Note that in this example the enzyme is catalysing the joining together of two molecules.

can bind with hydrogen peroxide molecules, split them into water and oxygen, and release these products at a rate of 10 million molecules per second.

The interaction between the substrate and the active site, including the slight change in shape of the enzyme (induced fit) which results from the binding of the substrate, is clearly shown by the enzyme lysozyme. Lysozyme is a natural defence against bacteria that is found in tears, saliva and other secretions. It breaks the polysaccharide chains that form the cell walls of bacteria. The tertiary structure of the enzyme has already been shown in Figure 2.20 (page 41). Figure 3.4 shows how part of the polysaccharide substrate is broken down in the active site.

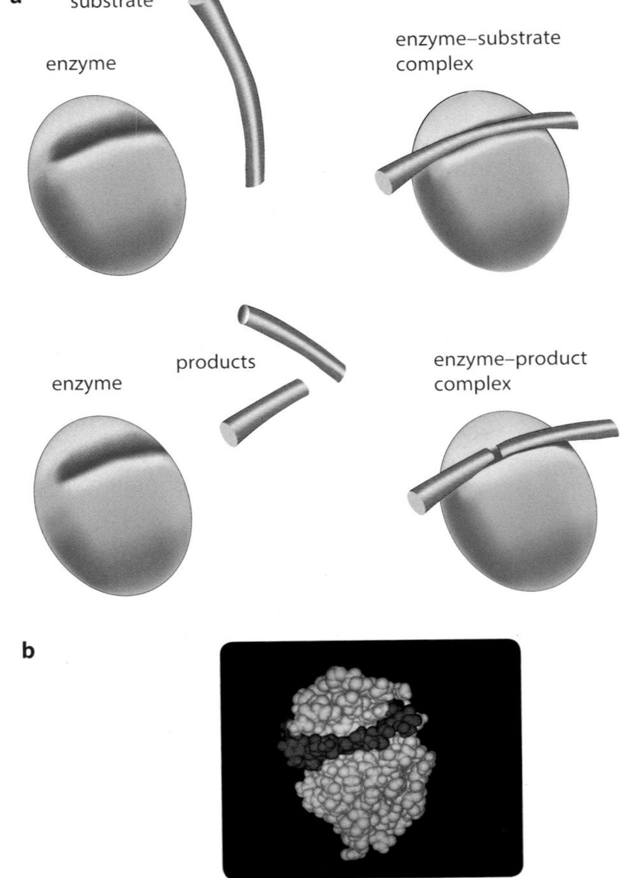

Figure 3.4 Lysozyme breaking a polysaccharide chain. This is a hydrolysis reaction. **a** Diagram showing the formation of enzyme–substrate and enzyme–product complexes before release of the products. **b** Space-filling model showing the substrate in the active site of the enzyme. The substrate is a polysaccharide chain which slides neatly into the groove (active site) and is split by the enzyme. Many such chains give the bacterial cell wall rigidity. When the chains are broken, the wall loses its rigidity and the bacterial cell explodes as a result of osmosis.

Enzymes reduce activation energy

As catalysts, enzymes increase the rate at which chemical reactions occur. Most of the reactions which occur in living cells would occur so slowly without enzymes that they would virtually not happen at all.

In many chemical reactions, the substrate will not be converted to a product unless it is temporarily given some extra energy. This energy is called activation energy (Figure 3.5a).

One way of increasing the rate of many chemical reactions is to increase the energy of the reactants by heating them. You have probably done this by heating substances which you want to react together. In the Benedict's test for reducing sugar, for example, you need to heat the Benedict's reagent and sugar solution together before they will react (page 32).

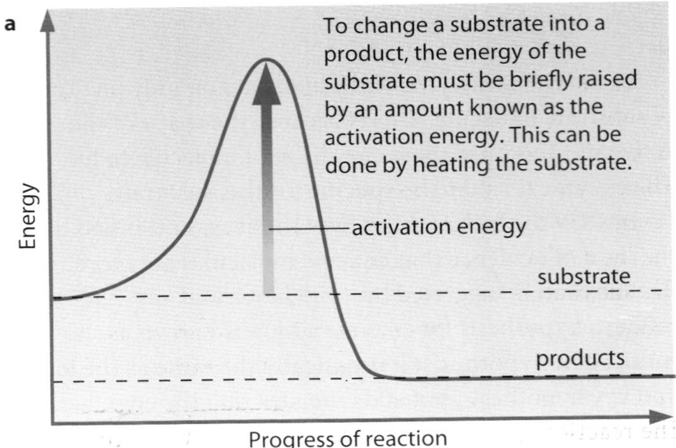

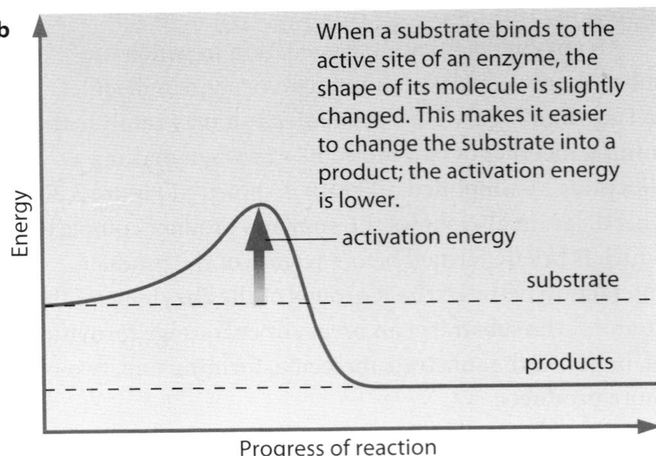

Figure 3.5 Activation energy **a** without enzyme; **b** with enzyme.

Mammals such as humans also use this method of speeding up their metabolic reactions. Our body temperature is maintained at 37 °C, which is usually much warmer than the temperature of the air around us. But even raising the temperature of cells to 37 °C is not enough to give most substrates the activation energy which they need to change into products. Enzymes avoid this problem because they **decrease** the activation energy of the reaction which they catalyse (Figure 3.5b). They do this by holding the substrate or substrates in such a way that their molecules can react more easily. Reactions catalysed by enzymes will take place rapidly at a much lower temperature than they otherwise would.

The course of a reaction

You may be able to carry out an investigation into the rate at which substrate is converted into product during an enzyme-controlled reaction. Figure 3.6 shows the results of such an investigation using the enzyme catalase. This enzyme is found in the tissues of most living things and catalyses the breakdown of hydrogen peroxide into water and oxygen. (Hydrogen peroxide is a toxic product of several different metabolic reactions, and so it must be got rid of quickly.) It is an easy reaction to follow, as the oxygen that is released can be collected and measured.

The reaction begins very swiftly. As soon as the enzyme and substrate are mixed, bubbles of oxygen are released quickly. A large volume of oxygen is collected in the first minute of the reaction. As the reaction continues, however, the rate at which oxygen is released gradually slows down. The reaction gets slower and slower, until it eventually stops completely.

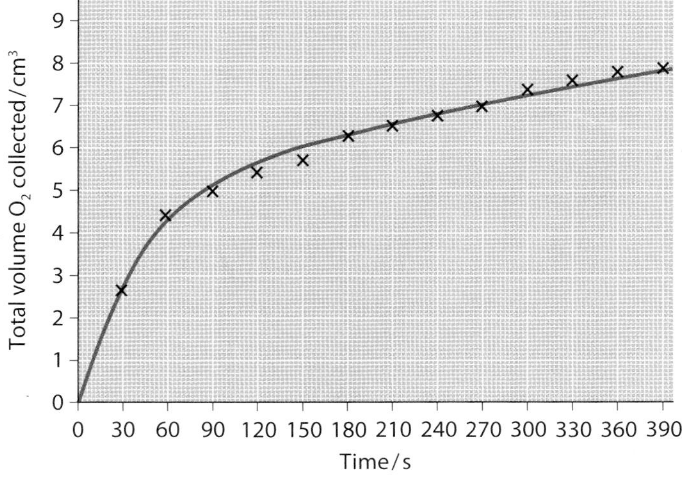

Figure 3.6 The course of an enzyme-catalysed reaction. Catalase was added to hydrogen peroxide at time 0. The gas released was collected in a gas syringe, the volume being read at 30 s intervals.

The explanation for the course of the reaction is quite straightforward. When the enzyme and substrate are first mixed, there are a large number of substrate molecules. At any moment, virtually every enzyme molecule has a substrate molecule in its active site. The rate at which the reaction occurs depends only on how many enzyme molecules there are and the speed at which the enzyme can convert the substrate into product, release it, and then bind with another substrate molecule. However, as more and more substrate is converted into product, there are fewer and fewer substrate molecules to bind with enzymes. Enzyme molecules may be 'waiting' for substrate molecules to hit their active sites. As fewer substrate molecules are left, the reaction gets slower and slower, until it eventually stops.

The curve of a graph such as the one in Figure 3.6 is therefore steepest at the beginning of the reaction: the rate of an enzyme-controlled reaction is always fastest at the beginning. This rate is called the **initial rate of reaction**. You can measure the initial rate of the reaction by calculating the slope of a tangent to the curve, as close to time 0 as possible (see Figure P1.15, page 260, for advice on how to do this). An easier way of doing this is simply to read off the graph the amount of oxygen given off in the first 30 seconds. In this case, the rate of oxygen production in the first 30 seconds is 2.7 cm^3 of oxygen per 30 seconds, or 5.4 cm^3 per minute.

Factors that affect enzyme action

The effect of enzyme concentration

Figure 3.7a shows the results of an investigation in which different concentrations of catalase solution (from celery extract) were added to the same volumes of hydrogen peroxide solution. Concentration was varied by varying the initial volume of extract and then making up to a standard volume. You can see that the shape of all five curves is similar. In each case, the reaction begins very quickly (steep curve) and then gradually slows down (curve levels off). Because the quantity of hydrogen peroxide is the same in all five reactions, the total amount of oxygen eventually produced will be the same; so, if the investigation goes on long enough, all the curves will meet.

To compare the rates of these five reactions, in order to look at the effect of enzyme concentration on reaction rate, it is fairest to look at the rate **right at the beginning of the reaction**. This is because, once the reaction is under way, the amount of substrate in each reaction begins to vary, because substrate is converted to product at different rates in each of the five reactions. It is only at the very beginning of the reaction that we can be sure that differences in reaction rate are caused only by differences in enzyme concentration.

To work out the initial rate for each enzyme concentration, we can calculate the slope of the curve 30 seconds after the beginning of the reaction, as explained earlier. Ideally, we should do this for an even earlier stage of the reaction, but in practice this is impossible. We can then plot a second graph, Figure 3.7b, showing the initial rate of reaction against enzyme concentration.

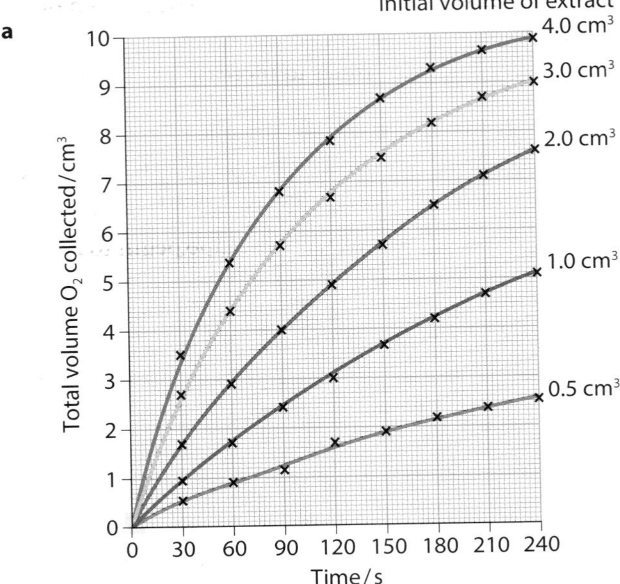

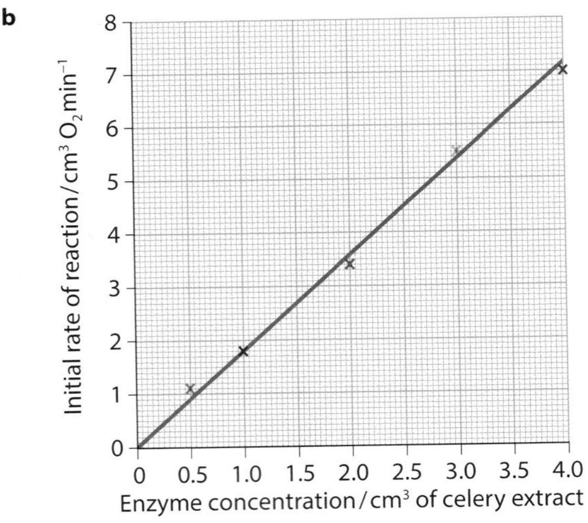

Figure 3.7 The effect of enzyme concentration on the rate of an enzyme-catalysed reaction. **a** Different volumes of celery extract, which contains catalase, were added to the same volume of hydrogen peroxide. Water was added to make the total volume of the mixture the same in each case. **b** The rate of reaction in the first 30 s was calculated for each enzyme concentration.

This graph shows that the initial rate of reaction increases linearly. In these conditions, reaction rate is directly proportional to the enzyme concentration. This is just what common sense says should happen. The more enzyme present, the more active sites will be available for the substrate to slot into. As long as there is plenty of substrate available, the initial rate of a reaction increases linearly with enzyme concentration.

BOX 3.1: Measuring reaction rate

It is easy to measure the rate of the catalase–hydrogen peroxide reaction, because one of the products is a gas, which is released and can be collected. Unfortunately, it is not always so easy to measure the rate of a reaction. If, for example, you wanted to investigate the rate at which amylase breaks down starch, it would be very difficult to observe the course of the reaction because the substrate (starch) and the product (maltose) remain as colourless substances in the reaction mixture.

The easiest way to measure the rate of this reaction is to measure the rate at which starch disappears from the reaction mixture. This can be done by taking samples from the mixture at known times, and adding each sample to some iodine in potassium iodide solution. Starch forms a blue-black colour with this solution. Using a colorimeter, you can measure the intensity of the blue-black colour obtained, and use this as a measure of the amount of starch still remaining. If you do this over a period of time, you can plot a curve of 'amount of starch remaining' against 'time'. You can then calculate the initial reaction rate in the same way as for the catalase–hydrogen peroxide reaction.

It is even easier to observe the course of this reaction if you mix starch, iodine in potassium iodide solution and amylase in a tube, and take regular readings of the colour of the mixture in this one tube in a colorimeter. However, this is not ideal, because the iodine interferes with the rate of the reaction and slows it down.

QUESTION

3.1 a In the breakdown of starch by amylase, if you were to plot the amount of starch remaining against time, sketch the curve you would expect to obtain.

b How could you use this curve to calculate the initial reaction rate?

QUESTION

3.2 Why is it better to calculate the initial rate of reaction from a curve such as the one in Figure 3.6, rather than simply measuring how much oxygen is given off in 30 seconds?

The effect of substrate concentration

Figure 3.8 shows the results of an investigation in which the amount of catalase was kept constant and the amount of hydrogen peroxide was varied. Once again, curves of oxygen released against time were plotted for each reaction, and the initial rate of reaction calculated for the first 30 seconds. These initial rates of reaction were then plotted against substrate concentration.

As substrate concentration increases, the initial rate of reaction also increases. Again, this is only what we would expect: the more substrate molecules there are around, the more often an enzyme's active site can bind with one. However, if we go on increasing substrate concentration, keeping the enzyme concentration constant, there comes a point where every enzyme active site is working continuously. If more substrate is added, the enzyme simply cannot work faster; substrate molecules are effectively 'queuing up' for an active site to become vacant. The enzyme is working at its maximum possible rate, known as $\mathbf{V_{max}}$. V stands for velocity.

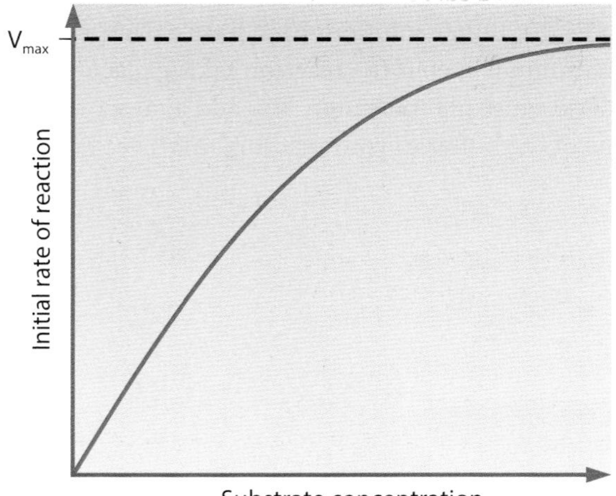

Figure 3.8 The effect of substrate concentration on the rate of an enzyme-catalysed reaction.

QUESTION

3.3 Sketch the shape that the graph in Figure 3.7b would have if excess hydrogen peroxide were not available.

Temperature and enzyme activity

Figure 3.9 shows how the rate of a typical enzyme-catalysed reaction varies with temperature. At low temperatures, the reaction takes place only very slowly. This is because molecules are moving relatively slowly. Substrate molecules will not often collide with the active site, and so binding between substrate and enzyme is a rare event. As temperature rises, the enzyme and substrate molecules move faster. Collisions happen more frequently, so that substrate molecules enter the active site more often. Moreover, when they do collide, they do so with more energy. This makes it easier for bonds to be formed or broken so that the reaction can occur.

As temperature continues to increase, the speed of movement of the substrate and enzyme molecules also continues to increase. However, above a certain temperature, the structure of the enzyme molecule vibrates so energetically that some of the bonds holding the enzyme molecule in its precise shape begin to break. This is especially true of hydrogen bonds. The enzyme molecule begins to lose its shape and activity, and is said to be **denatured**. This is often irreversible. At first, the substrate molecule fits less well into the active site of the enzyme, so the rate of the reaction begins to slow down. Eventually the substrate no longer fits at all, or can no longer be held in the correct position for the reaction to occur.

The temperature at which an enzyme catalyses a reaction at the maximum rate is called the **optimum temperature**. Most human enzymes have an optimum temperature of around 40 °C. By keeping our body temperatures at about 37 °C, we ensure that enzyme-catalysed reactions occur at close to their maximum rate.

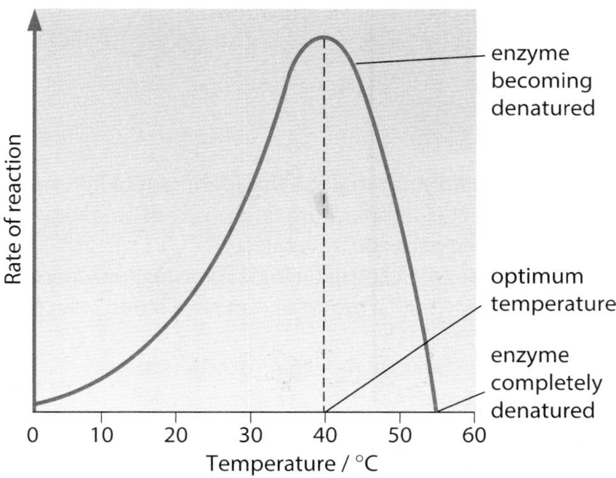

Figure 3.9 The effect of temperature on the rate of an enzyme-controlled reaction.

It would be dangerous to maintain a body temperature of 40 °C, as even a slight rise above this would begin to denature enzymes.

Enzymes from other organisms may have different optimum temperatures. Some enzymes, such as those found in bacteria which live in hot springs (Figure 3.10), have much higher optimum temperatures. Some plant enzymes have lower optimum temperatures, depending on their habitat.

pH and enzyme activity

Figure 3.11 shows how the activity of an enzyme is affected by pH. Most enzymes work fastest at a pH of somewhere around 7 – that is, in fairly neutral conditions. Some, however, such as the protease pepsin, which is found in the acidic conditions of the stomach, have a different optimum pH.

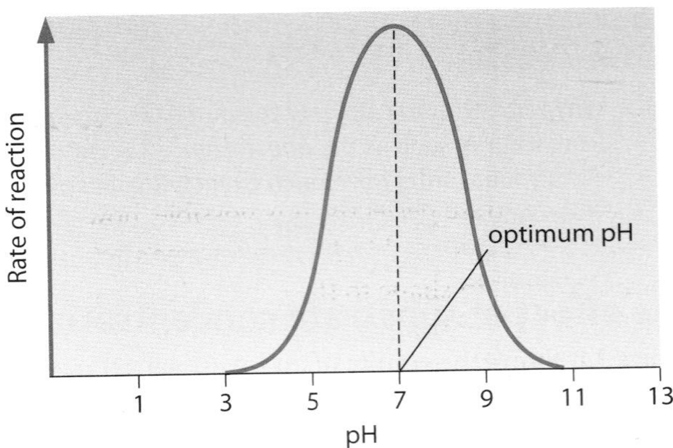

Figure 3.11 The effect of pH on the rate of an enzyme-controlled reaction.

Figure 3.10 Not all enzymes have an optimum temperature of 40 °C. Bacteria and algae living in hot springs such as this one in Yellowstone National Park, USA, are able to tolerate very high temperatures. Enzymes from such organisms are proving useful in various industrial applications.

pH is a measure of the concentration of hydrogen ions in a solution. The lower the pH, the higher the hydrogen ion concentration. Hydrogen ions can interact with the R groups of amino acids – for example, by affecting ionisation (the negative or positive charges) of the groups. This affects the ionic bonding between the groups (page 42), which in turn affects the three-dimensional arrangement of the enzyme molecule. The shape of the active site may change and therefore reduce the chances of the substrate molecule fitting into it. A pH which is very different from the optimum pH can cause denaturation of an enzyme.

When investigating pH, you can use buffer solutions (Chapter P1). Buffer solutions each have a particular pH and maintain it even if the reaction taking place would otherwise cause pH to change. You add a measured volume of the buffer to your reacting mixture.

QUESTIONS

3.4 How could you carry out an experiment to determine the effect of temperature on the rate of breakdown of hydrogen peroxide by catalase?

3.5 Proteases are used in biological washing powders.

a How would a protease remove a blood stain on clothes?

b Most biological washing powders are recommended for use at low washing temperatures. Why is this?

c Washing powder manufacturers have produced proteases which can work at temperatures higher than 40 °C. Why is this useful?

QUESTION

3.6 Trypsin is a protease secreted in pancreatic juice, which acts in the duodenum. If you add trypsin to a suspension of milk powder in water, the enzyme digests the protein in the milk, so that the suspension becomes clear.

How could you carry out an investigation into the effect of pH on the rate of activity of trypsin? (A suspension of 4 g of milk powder in 100 cm^3 of water will become clear in a few minutes if an equal volume of a 0.5% trypsin solution is added to it.)

Enzyme inhibitors

Competitive, reversible inhibition

As we have seen, the active site of an enzyme fits one particular substrate perfectly. It is possible, however, for some other molecule to bind to an enzyme's active site if it is very similar in shape to the enzyme's substrate. This would then **inhibit** the enzyme's function.

If an **inhibitor** molecule binds only briefly to the site, there is competition between it and the substrate for the site. If there is much more of the substrate present than the inhibitor, substrate molecules can easily bind to the active site in the usual way, and so the enzyme's function is unaffected. However, if the concentration of the inhibitor rises, or that of the substrate falls, it becomes less and less likely that the substrate will collide with an empty site. The enzyme's function is then inhibited. This is therefore known as **competitive inhibition** (Figure 3.12a). It is said to be **reversible** (not permanent) because it can be reversed by increasing the concentration of the substrate.

An example of competitive inhibition occurs in the treatment of a person who has drunk ethylene glycol. Ethylene glycol is used as antifreeze, and is sometimes drunk accidentally. Ethylene glycol is rapidly converted in the body to oxalic acid, which can cause irreversible kidney damage. However, the active site of the enzyme which converts ethylene glycol to oxalic acid will also accept ethanol. If the poisoned person is given a large dose of ethanol, the ethanol acts as a competitive inhibitor, slowing down the action of the enzyme on ethylene glycol for long enough to allow the ethylene glycol to be excreted.

Non-competitive, reversible inhibition

A different kind of reversible inhibition takes place if a molecule can bind to another part of the enzyme rather than the active site. While the inhibitor is bound to the enzyme it can seriously disrupt the normal arrangement of hydrogen bonds and hydrophobic interactions holding the enzyme molecule in its three-dimensional shape (Chapter 2). The resulting distortion ripples across the molecule to the active site, making the enzyme unsuitable for the substrate. While the inhibitor is attached to the enzyme, the enzyme's function is blocked no matter how much substrate is present, so this is an example of **non-competitive inhibition** (Figure 3.12b).

Inhibition of enzyme function can be lethal, but in many situations inhibition is essential. For example, metabolic reactions must be very finely controlled and balanced, so no single enzyme can be allowed to 'run wild', constantly churning out more and more product.

One way of controlling metabolic reactions is to use the **end-product** of a chain of reactions as a non-competitive, reversible inhibitor (Figure 3.13). As the enzyme converts substrate to product, it is slowed down because the end-

a Competitive inhibition

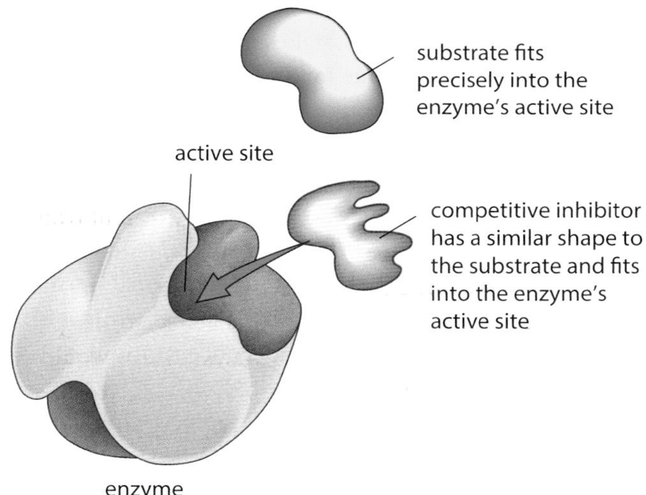

b Non-competitive inhibition

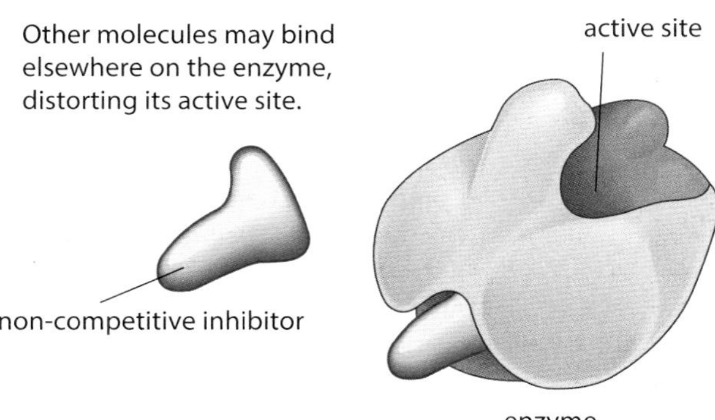

Figure 3.12 Enzyme inhibition. **a** Competitive inhibition. **b** Non-competitive inhibition.

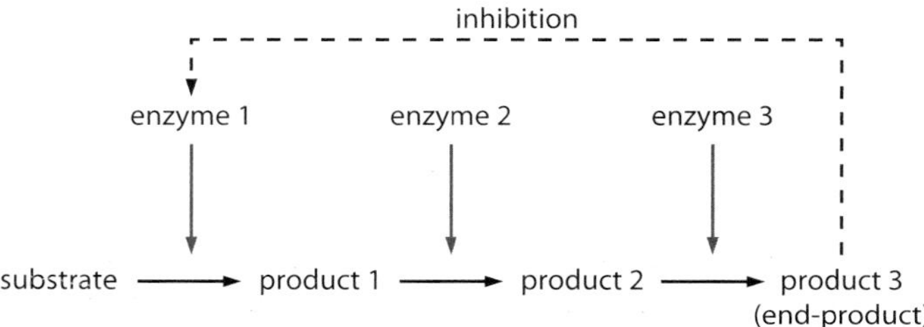

Figure 3.13 End-product inhibition. As levels of product 3 rise, there is increasing inhibition of enzyme 1. So, less product 1 is made and hence less product 2 and 3. Falling levels of product 3 allow increased function of enzyme 1 so products 1, 2 and 3 rise again and the cycle continues. This end-product inhibition finely controls levels of product 3 between narrow upper and lower limits, and is an example of a **feedback mechanism**.

product binds to another part of the enzyme and prevents more substrate binding. However, the end-product can lose its attachment to the enzyme and go on to be used elsewhere, allowing the enzyme to reform into its active state. As product levels fall, the enzyme is able to top them up again. This is termed **end-product inhibition**.

Comparing enzyme affinities

There is enormous variation in the speed at which different enzymes work. A typical enzyme molecule can convert around one thousand substrate molecules into product per second. This is known as the **turnover rate**. The enzyme carbonic anhydrase (Chapter 12) is one of the fastest enzymes known. It can remove 600 000 molecules of carbon dioxide from respiring tissue per second, roughly 10^7 times as fast as the reaction would occur in the absence of the enzyme. It has presumably evolved such efficiency because a build-up of carbon dioxide in tissues would quickly become lethal. Speeds such as these are only possible because molecules within cells move about very quickly by diffusion over short distances, with tens or hundreds of thousands of collisions per second occurring between enzyme and substrate molecules.

Simple measurements of the rate of activity of enzymes can be carried out. See for example, Box 3.1, Figures 3.6 and 3.7 and Questions 3.4 and 3.6. More precise measurements of the rate at which enzymes work are difficult and complex to make, but are important for our understanding of how enzymes work together to control cell metabolism. One of the key steps towards understanding how well an enzyme performs is to measure the theoretical maximum rate (velocity), V_{max}, of the reaction it catalyses. At V_{max} all the enzyme molecules are bound to substrate molecules – the enzyme is saturated with substrate. The principle of how V_{max} is measured is described on page 58. To summarise, the reaction rate is measured at different substrate concentrations while keeping the enzyme concentration constant. As substrate concentration is increased, reaction rate rises until the reaction reaches its maximum rate, V_{max}.

QUESTION

3.7 For each substrate concentration tested, the rate should be measured as soon as possible. Explain why.

The initial rate for each substrate concentration is plotted against substrate concentration, producing a curve like that shown in Figure 3.8. This type of curve is described as asymptotic and such curves have certain mathematical properties. In particular, the curve never completely flattens out in practice. In theory, it does so at infinite substrate concentration, but this is obviously impossible to measure. This makes it impossible to accurately read off the value for V_{max} from the graph. There is, however, a way round this problem. Instead of plotting substrate concentration, [S], on the *x*-axis and velocity (rate) on the *y*-axis, we can plot 1/[S] (the inverse of substrate concentration) and 1/velocity (the inverse of velocity) respectively. Such a plot is called a double-reciprocal plot. (Remember, the word 'reciprocal' means 'inverse'.) One advantage of doing this is that while it is impossible to plot infinite substrate concentration, 1/infinity is zero, which can be plotted, so V_{max} can be found accurately. Also, the resulting graph is a straight line. It is easier to understand this if you use some specimen

QUESTION

3.8 a Copy the table and complete it by calculating the remaining values for 1/[S] and 1/v to one decimal place.

[S] / arbitrary units	1/[S] / arbitrary units	v / arbitrary units	1/v / arbitrary units
0.02	50.0	0.025	40.0
0.04		0.041	
0.06		0.052	
0.08		0.061	
0.10		0.067	
0.20		0.085	

Table 3.1 Some results from an experiment to determine the effect of substrate concentration on the velocity of an enzyme reaction. [S] = substrate concentration, v = velocity (rate of reaction)

b Draw two graphs, one with [S] on the *x*-axis and v on the *y*-axis, and a second with 1/[S] on the *x*-axis and 1/v on the *y*-axis. (The second graph is a double-reciprocal plot.) Compare the two graphs. Note that the double-reciprocal plot is a straight line.

c After reading the rest of this section, calculate the values for V_{max} and K_m using the data from your double-reciprocal plot.

results to plot the two types of graph. The table in Question 3.8 gives you some results and it is worth spending some time answering the question before proceeding.

Figure 3.14a shows a double-reciprocal plot. Note that it is a straight line. Using this graph, we can find V_{max} in the following way. First, we find $1/V_{max}$. This is the point where the line crosses (intersects) the *y*-axis because this is where 1/[S] is zero (and therefore [S] is infinite). Once we know $1/V_{max}$ we can calculate V_{max}.

Another useful value can be obtained from the double-reciprocal plot, namely the **Michaelis–Menten constant**, K_m. The Michaelis–Menten constant is the substrate concentration at which an enzyme works at half its maximum rate ($\frac{1}{2}V_{max}$). At this point half the active sites of the enzyme are occupied by substrate. The higher the affinity of the enzyme for the substrate, the lower the substrate concentration needed for this to happen. Thus the Michaelis–Menten constant is a measure of the affinity of the enzyme for its substrate. The higher the affinity, the lower the Michaelis–Menten constant and the quicker the reaction will proceed to its maximum rate, although the maximum rate itself is not affected by the Michaelis–Menten constant. V_{max} and K_m therefore provide two different ways of comparing the efficiency of different enzymes. V_{max} gives information about the maximum rate of reaction that is possible (though not necessarily the rate under cell conditions) while K_m measures the affinity of the enzyme for the substrate. The higher the affinity, the more likely the product will be formed when a substrate molecule enters the active site, rather than the substrate simply leaving the active site again before a reaction takes place. These two aspects of efficiency are rather like using the maximum speed and acceleration to measure the efficiency of a car.

How can we find K_m from a double-reciprocal plot? The answer is that the point where the line of the graph intersects the *x*-axis is $-1/K_m$ (note that it is in the negative region of the *x*-axis). Figure 3.14a shows this point. From the value for $-1/K_m$, we can calculate K_m. Figure 3.14b shows the normal plot (as in Figure 3.7 and your first graph in Question 3.8). The relationship between $\frac{1}{2}V_{max}$ and K_m is shown in Figure 3.14b.

The value of K_m for a particular enzyme can vary, depending on a number of factors. These include the identity of the substrate, temperature, pH, presence of particular ions, overall ion concentration, and the presence of poisons, pollutants or inhibitors.

Turnover numbers, which are related to V_{max} and K_m values, for four enzymes are shown in Table 3.2. This shows the great variation in efficiency that is possible between enzymes, and the fact that V_{max} and K_m are independent of each other.

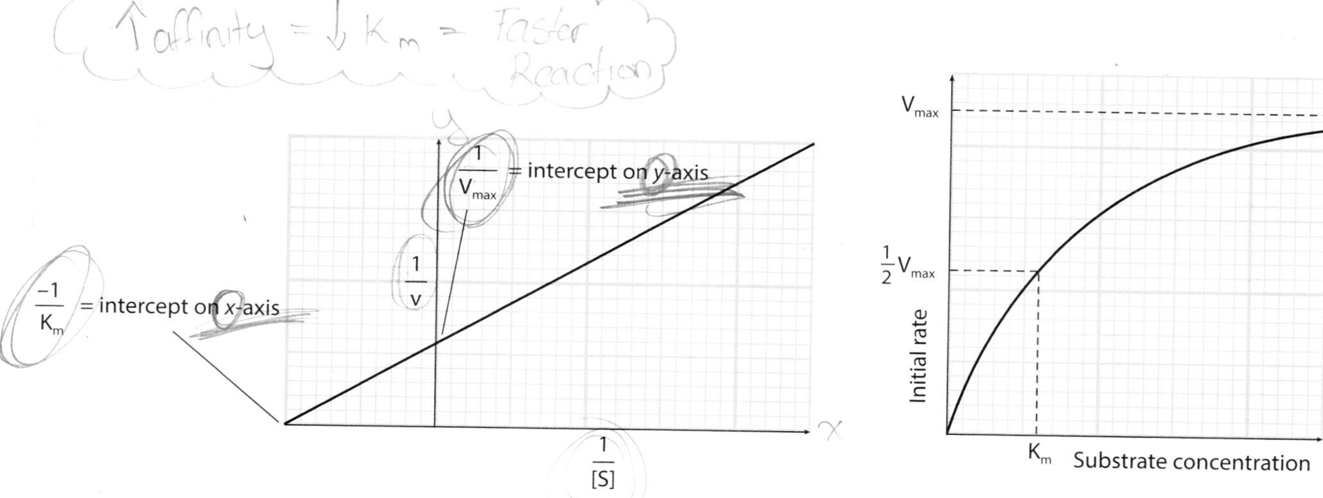

Figure 3.14 **a** A double-reciprocal plot of substrate concentration against initial rate: **b** A graph showing the effect of substrate concentration on initial rate, with V_{max}, ½ V_{max} and K_m values shown.

Enzyme	Substrate	Maximum turnover number / per second	K_m / $\mu mol\,dm^{-3}$
carbonic anhydrase	carbon dioxide	600 000	8000
penicillinase	penicillin	2000	50
chymotrypsin	protein	100	5000
lysozyme	acetylglucosamine	0.5	6

Table 3.2 Turnover numbers and K_m values for four enzymes. Note that the unit for K_m is a concentration. (The turnover number per second is the number of molecules of substrate that one molecule of an enzyme converts to product per second. This is related to V_{max}.)

The significance of V_{max} and K_m values

Knowing the values of V_{max} and K_m has a number of applications.

- It enables scientists to make computerised models of biochemical pathways or even the behaviour of whole cells because it helps to predict how each reaction in a proposed pathway will proceed and therefore how the enzymes will interact. The consequences of changing conditions such as temperature, pH or the presence of inhibitors can be built into the models.
- An enzyme's preference for different substrates can be compared quantitatively.
- By understanding what affects enzyme efficiency, scientists may in future be able to design better catalysts, linking this to genetic engineering.
- For a commercially important enzyme, the performance of the same enzyme from different organisms can be compared.
- The calculations involved can be applied to other fields of biochemistry, such as antibody–antigen binding.
- Knowing K_m means the proportion of active sites occupied by substrate molecules can be calculated for any substrate concentration.

QUESTION

3.9 Which of the four enzymes in Table 3.2 has the highest affinity for its substrate? Briefly explain your answer.

Immobilising enzymes

Enzymes have an enormous range of commercial applications – for example, in medicine, food technology and industrial processing. Enzymes are expensive. No company wants to have to keep buying them over and over again if it can recycle them in some way. One of the best ways of keeping costs down is to use **immobilised enzymes**.

The enzyme lactase can be immobilised using alginate beads (Box 3.2). Milk is then allowed to run through the column of lactase-containing beads. The lactase hydrolyses the lactose in the milk to glucose and galactose. The milk is therefore lactose-free, and can be used to make lactose-free dairy products for people who cannot digest lactose.

You can see that enzyme immobilisation has several obvious advantages compared with just mixing up the enzyme with its substrate. If you just mixed lactase with milk, you would have a very difficult task to get the lactase back again. Not only would you lose the lactase, but also you would have milk contaminated with the enzyme. Using immobilised enzymes means that you can keep and re-use the enzymes, and that the product is enzyme-free.

Another advantage of this process is that the immobilised enzymes are more tolerant of temperature changes and pH changes than enzymes in solution. This may be partly because their molecules are held firmly in shape by the alginate in which they are embedded, and so do not denature as easily. It may also be because the parts of the molecules that are embedded in the beads are not fully exposed to the temperature or pH changes.

BOX 3.2: Immobilised enzymes

Figure 3.15 shows one way in which enzymes can be immobilised. The enzyme is mixed with a solution of sodium alginate. Little droplets of this mixture are then added to a solution of calcium chloride. The sodium alginate and calcium chloride instantly react to form jelly, which turns each droplet into a little bead. The jelly bead contains the enzyme. The enzyme is held in the bead, or **immobilised**.

These beads can be packed gently into a column. A liquid containing the enzyme's substrate can be allowed to trickle steadily over them (Figure 3.16).

As the substrate runs over the surface of the beads, the enzymes in the beads catalyse a reaction that converts the substrate into product. The product continues to trickle down the column, emerging from the bottom, where it can be collected and purified.

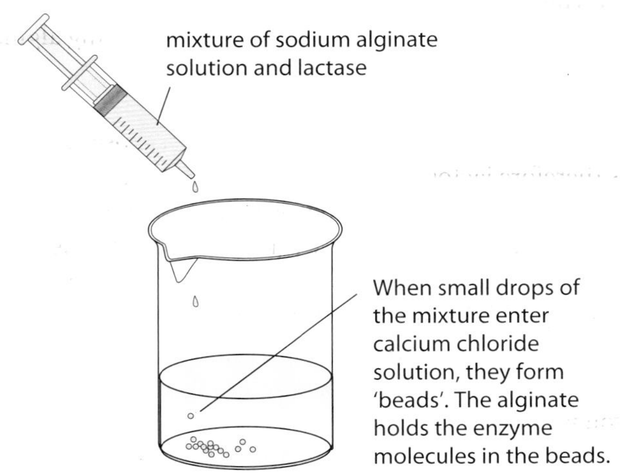

Figure 3.15 Immobilising enzyme in alginate.

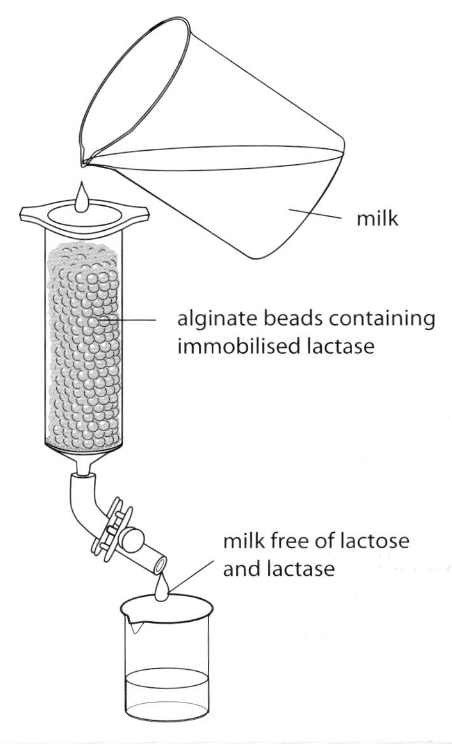

Figure 3.16 Using immobilised enzyme to modify milk.

QUESTIONS

3.10 **a** Outline an investigation you could carry out to compare the temperature at which the enzyme lactase is completely denatured within 10 minutes

i when free in solution,

ii when immobilised in alginate beads.

b Outline an experiment you could carry out to investigate how long it takes the enzyme lactase to denature at 90 °C

i when free in solution,

ii when immobilised in alginate beads.

c Outline how you would determine the optimum pH of the enzyme lactase

i when free in solution,

ii when immobilised in alginate beads.

3.11 Summarise the advantages of using immobilised enzymes rather than enzyme solutions.

Summary

- Enzymes are globular proteins which catalyse metabolic reactions. Each enzyme has an active site with a flexible structure which can change shape slightly to fit precisely the substrate molecule. This is called the induced fit hypothesis. When the substrate enters the active site, an enzyme–substrate complex is temporarily formed in which the R groups of the amino acids in the enzyme hold the substrate in place.
- Enzymes may be involved in reactions which break down molecules or join molecules together. They work by lowering the activation energy of the reactions they catalyse.
- The course of an enzyme reaction can be followed by measuring the rate at which a product is formed or the rate at which a substrate disappears. A progress curve, with time on the *x*-axis, can be plotted. The curve is steepest at the beginning of the reaction, when substrate concentration is at its highest. This rate is called the initial rate of reaction.
- Various factors affect the rate of activity of enzymes. Four important factors are enzyme concentration, substrate concentration, temperature and pH. The greater the concentration of the enzyme, the faster the rate of reaction, provided there are enough substrate molecules present. The greater the concentration of the substrate, the faster the rate of reaction, provided enough enzyme molecules are present. During enzyme reactions, rates slow down as substrate molecules are used up.
- Each enzyme has an optimum temperature at which it works fastest. As temperature increases above the optimum temperature, the enzyme gradually denatures (loses its precise tertiary structure). When an enzyme is completely denatured, it ceases to function, but denaturation is sometimes reversible.
- Each enzyme has an optimum pH. Some enzymes operate within a narrow pH range; some have a broad pH range.
- Enzymes are also affected by the presence of inhibitors, which slow down their rate of reaction or stop it completely. Competitive inhibitors are molecules which are similar in shape to the normal substrate molecules. They compete with the substrate for the active site of the enzyme. Competitive inhibition is reversible because the inhibitor can enter and leave the active site.
- Non-competitive inhibitors either bind permanently to the active site or bind at a site elsewhere on the enzyme, causing a change in shape of the active site. Binding of non-competitive inhibitors may or may not be reversible.
- The efficiency of an enzyme can be measured by finding the value known as the Michaelis–Menten constant, K_m. To do this the maximum rate of reaction, V_{max}, must first be determined. Determination of V_{max} involves finding the initial rates of reactions at different substrate concentrations while ensuring that enzyme concentration remains constant.
- Enzymes can be immobilised – for example by trapping them in jelly (alginate) beads. This is commercially useful because the enzyme can be re-used and the product is separate from (uncontaminated by) the enzyme. Immobilisation often makes enzymes more stable.

End-of-chapter questions

1 The diagram below shows an enzyme and two inhibitors of the enzyme, **X** and **Y**. Which of the following describes the two inhibitors?

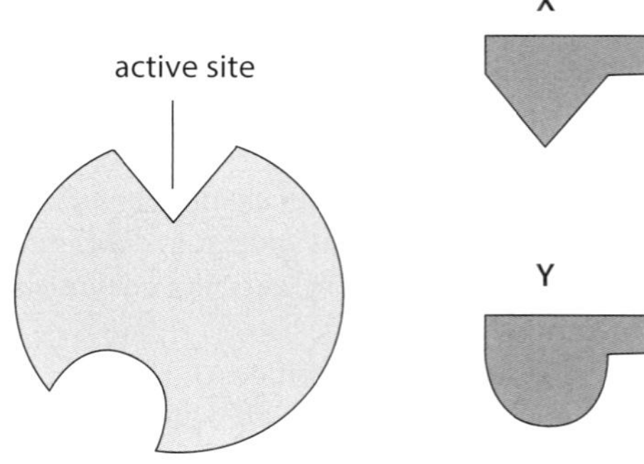

A **X** and **Y** are competitive inhibitors.
B **X** and **Y** are non-competitive inhibitors.
C **X** is a competitive inhibitor and **Y** is a non-competitive inhibitor.
D **X** is a non-competitive inhibitor and **Y** is a competitive inhibitor.

[1]

2 In a reaction controlled by an enzyme, which of the following graphs shows the effect of substrate concentration on the rate of the reaction?

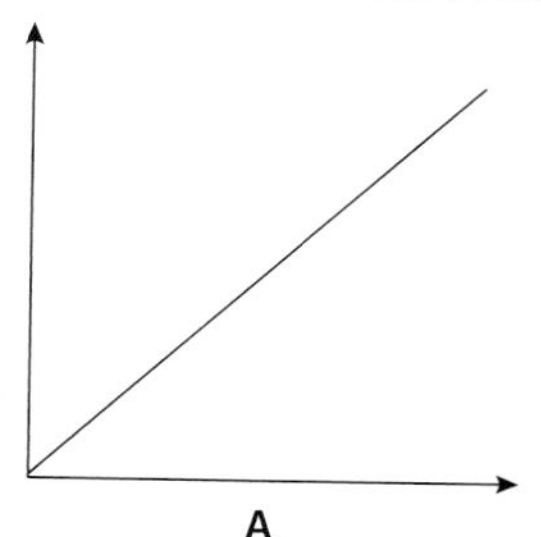

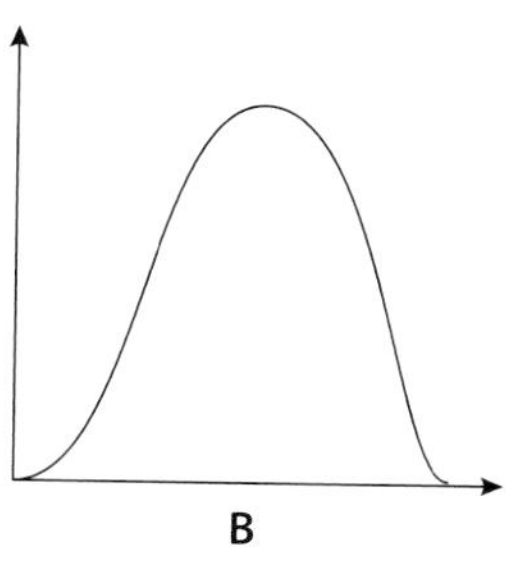

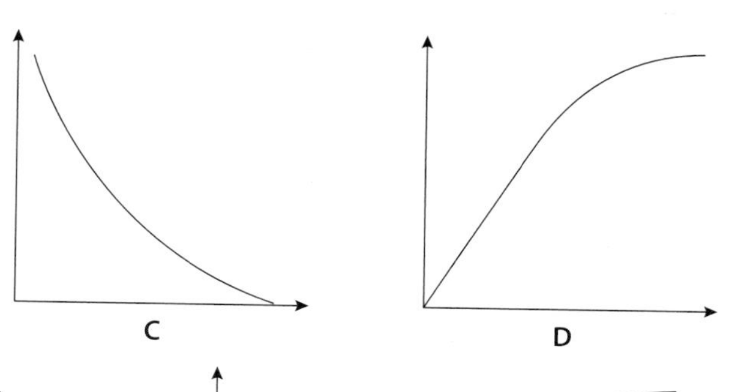

[1]

3 The graph shows the progress of the digestion of starch by the enzyme salivary amylase. Why does the reaction slow down?

A End-product inhibition by maltose.

B The salivary amylase is becoming denatured.

C The salivary amylase is gradually becoming saturated with starch.

D There are fewer and fewer substrate molecules left to bind with the salivary amylase.

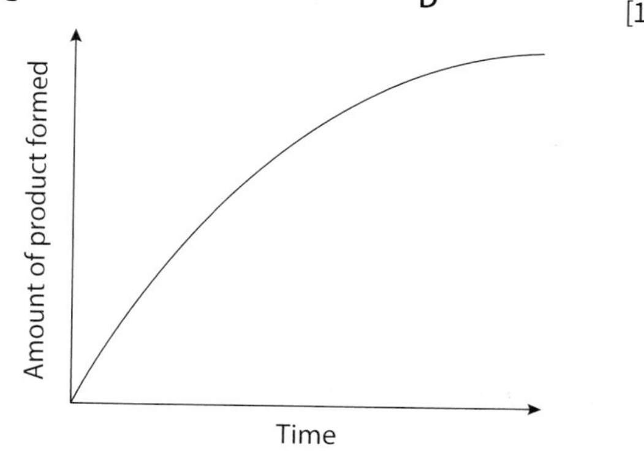

[1]

4 If methylene blue dye is added to a suspension of yeast cells, living cells do not take up the stain, and they remain colourless. However, dead cells are stained blue. This fact was used to carry out an investigation into the rate at which yeast cells were killed at two different temperatures (at high temperatures the yeast enzymes will be denatured). The results are shown in the diagram below.

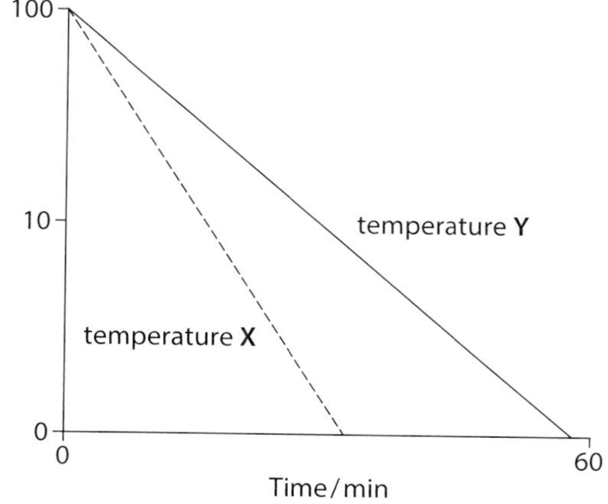

Which of the following is correct?

	The higher temperature is	The vertical axis (y-axis) should be labelled
A	X	% coloured cells
B	Y	% colourless cells
C	X	% colourless cells
D	Y	% coloured cells

[1]

5 Copy the graph in question **3** and draw a line from which the initial rate of reaction could be calculated. [1]

6 The graph shows the effect of changes in pH on the activity of the enzyme lysozyme.

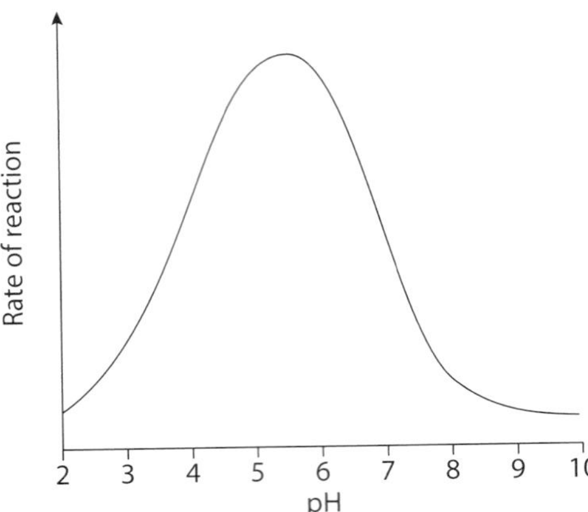

a Describe the effect of pH on this enzyme. [2]

b Explain why pH affects the activity of the enzyme. [4]

[Total: 6]

7 The graph below shows the effect of temperature on the rate of reaction of an enzyme.

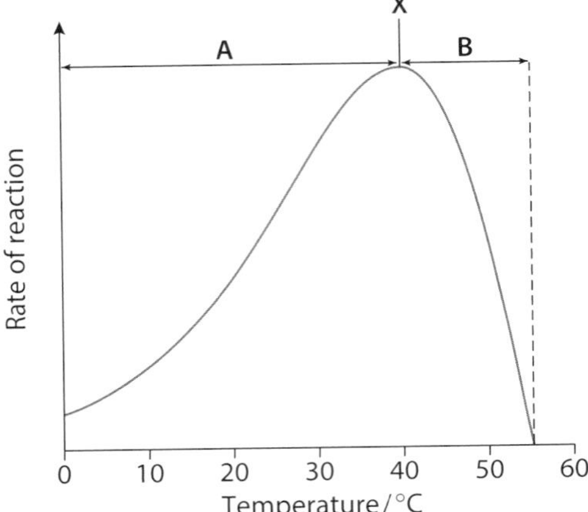

a What is indicated by **X**? [1]

b What temperature would **X** be for a mammalian enzyme? [1]

c Explain what is happening in region **A**. [3]

d Explain what is happening in region **B**. [3]

e Enzymes are effective because they lower the activation energy of the reactions they catalyse. Explain what is meant by 'activation energy'. [2]

[Total: 10]

8 The reaction below occurs during aerobic respiration. The reaction is catalysed by the enzyme succinate dehydrogenase.

a Name the substrate in this reaction. [1]

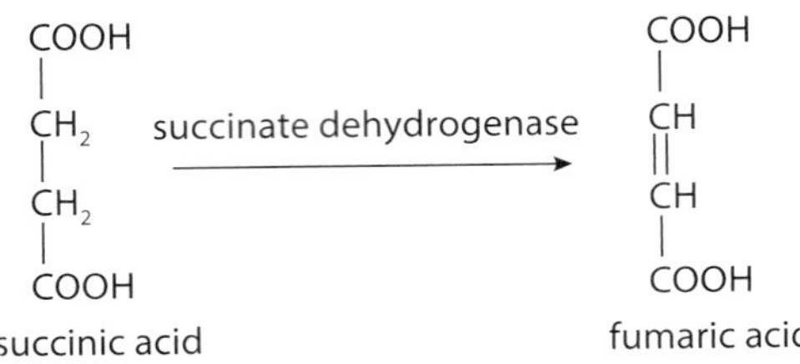

b The molecule malonic acid, which is shown here, inhibits this reaction. It does not bind permanently to the enzyme. Describe how malonic acid inhibits the enzyme succinate dehydrogenase. [3]

COOH
|
CH_2
|
COOH

malonic acid

c Heavy metals such as lead and mercury bind permanently to –SH groups of amino acids present in enzymes. These –SH groups could be in the active site or elsewhere in the enzyme.

i Name the amino acid which contains –SH groups. [1]

ii Explain the function of –SH groups in proteins and why binding of heavy metals to these groups would inhibit the activity of an enzyme. [4]

iii What type of inhibition would be caused by the heavy metals? [1]

[Total: 10]

9 You are provided with three solutions: **A**, **B** and **C**. One solution contains the enzyme amylase, one contains starch and one contains glucose. Starch is the substrate of the enzyme. The product is the sugar maltose. You are provided with only one reagent, Benedict's solution, and the usual laboratory apparatus.

a Outline the procedure you would follow to identify the three solutions. [6]

b What type of reaction is catalysed by the enzyme? [1]

[Total: 7]

10 The activity of the enzyme amylase can be measured at a particular temperature by placing a sample into a Petri dish containing starch-agar ('a starch-agar plate'). Starch-agar is a jelly containing starch. One or more 'wells' (small holes) are cut in the agar jelly with a cork borer, and a sample of the enzyme is placed in each well. The enzyme molecules then diffuse through the agar and gradually digest any starch in their path. At the end of the experiment, iodine in potassium iodide solution is poured over the plate. Most of the plate will turn blue-black as iodine reacts with starch, but a clear 'halo' (circle) will be seen around the well where starch has been digested. Measuring the size of the halo can give an indication of the activity of the enzyme.

A student decided to investigate the rate at which a mammalian amylase is denatured at 60 °C. She heated different samples of the enzyme in a water bath at 60 °C for 0, 1, 5, 10 and 30 minutes. She then allowed the samples to cool down to room temperature and placed samples of equal volume in the wells of five starch-agar plates, one plate for each heating period. She then incubated the plates in an oven at 40 °C for 24 hours.

The results of the student's experiment are shown on the next page. A diagram of one dish is shown, and the real size of one halo from each dish is also shown.

a Why did the student cut four wells in each dish rather than just one? [1]

b One dish contained samples from amylase which was not heated (time zero). This is a control dish. Explain the purpose of this control. [1]

c Explain why the starch-agar plates were incubated at 40 °C and not room temperature. [1]

d Describe what was happening in the dishes during the 24 hours of incubation. [4]

e Why was it important to add the same volume of amylase solution to each well? [1]

f Measure the diameter in mm of the representative halo from each dish. Record the results in a suitable table. [4]

g Only one halo from each dish is shown in the diagrams. In practice there was some variation in the diameters of the four halos in each dish. How would you allow for this when processing your data? [1]

h Plot a graph to show the effect of length of time at 60 °C on the activity of the enzyme. [5]

i Describe and explain your results. [4]

continued ...

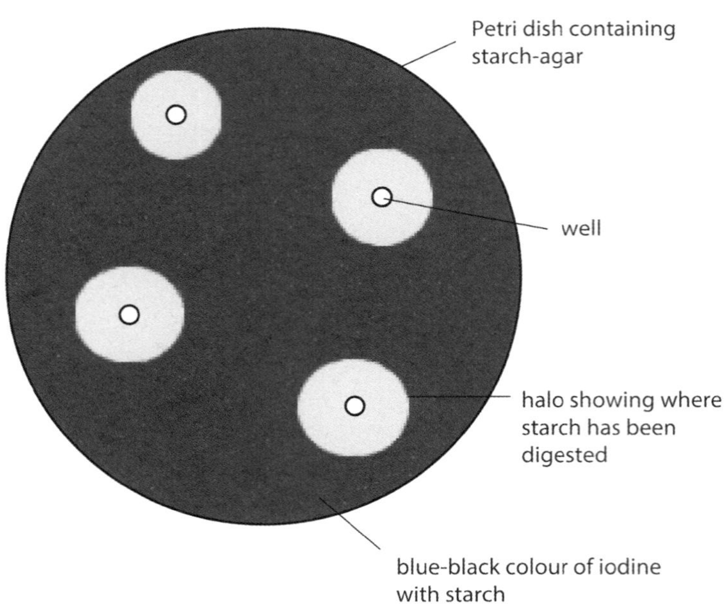

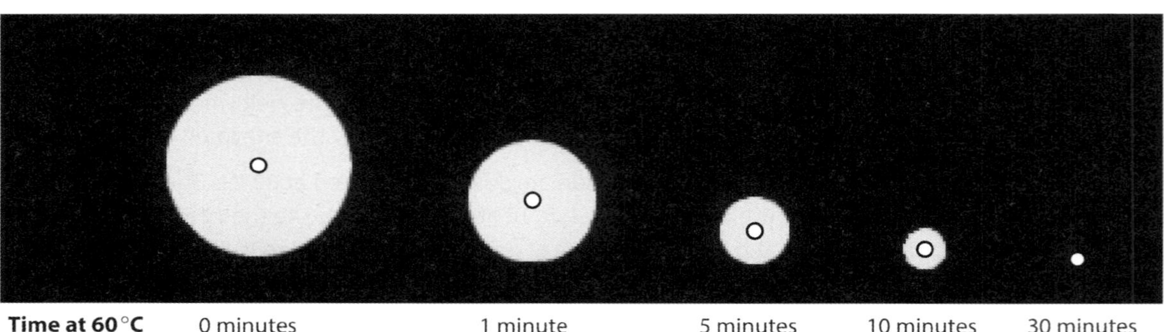

j Another student discovered that amylases from fungi and bacteria are more resistant to high temperatures than mammalian amylases. Using starch-agar plates as a method for measuring the activity of an amylase at 40 °C, outline an experiment that the student could perform to discover which amylase is most resistant to heat. Note that temperatures up to 120 °C can be obtained by using an autoclave (pressure cooker). [5]

k Enzymes are used in many industrial processes where resistance to high temperatures is an advantage. State **three** other variables apart from temperature which should be controlled in an industrial process involving enzymes. [3]

[Total: 30]

11 Two inhibitors of the same enzyme, inhibitor **A** and inhibitor **B**, were investigated to discover if they were competitive or non-competitive. In order to do this, the rate of reaction of the enzyme was measured at different concentrations of substrate without inhibitor, with inhibitor **A** and with inhibitor **B**. Graphs of the data were plotted as shown on the next page. The graphs showed that one inhibitor was competitive and the other non-competitive.

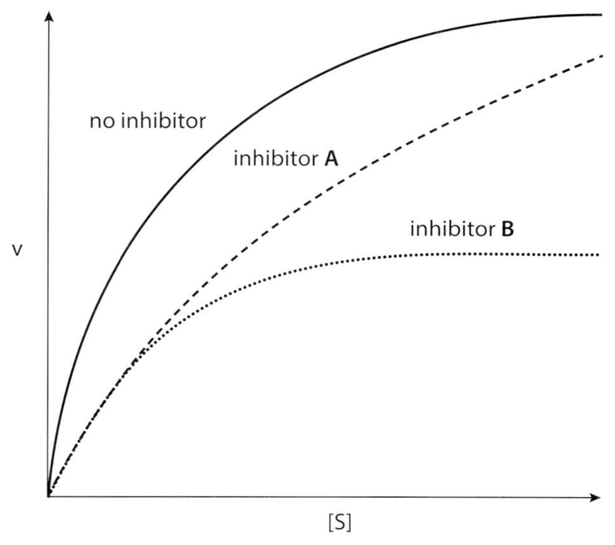

v = velocity (rate)

[S] = substrate concentration

Copy the graphs.

a Label the graph for 'no inhibitor' to show the position of V_{max}, ½V_{max} and K_m. [3]

b State the effect that inhibitor **A** had on V_{max} and K_m of the enzyme. [2]

c State the effect that inhibitor **B** had on V_{max} and K_m of the enzyme. [2]

d Which inhibitor is competitive and which is non-competitive? Explain your answer. [4]

Double-reciprocal plots of the data obtained produced the following graphs.

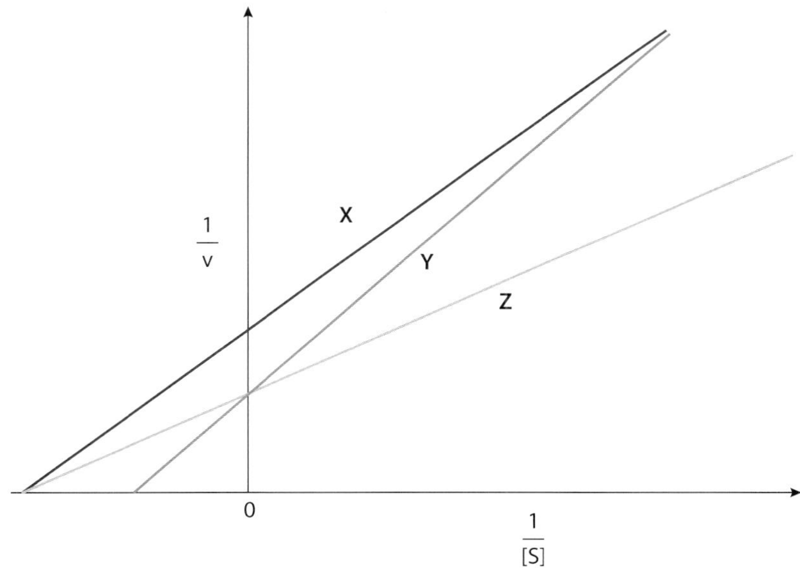

e Identify which line, **X**, **Y** or **Z**, corresponds to each of the following experiments:

i inhibitor absent

ii competitive inhibitor present

iii non-competitive inhibitor present

Briefly explain your answers to **ii** and **iii**. [5]

[Total: 16]

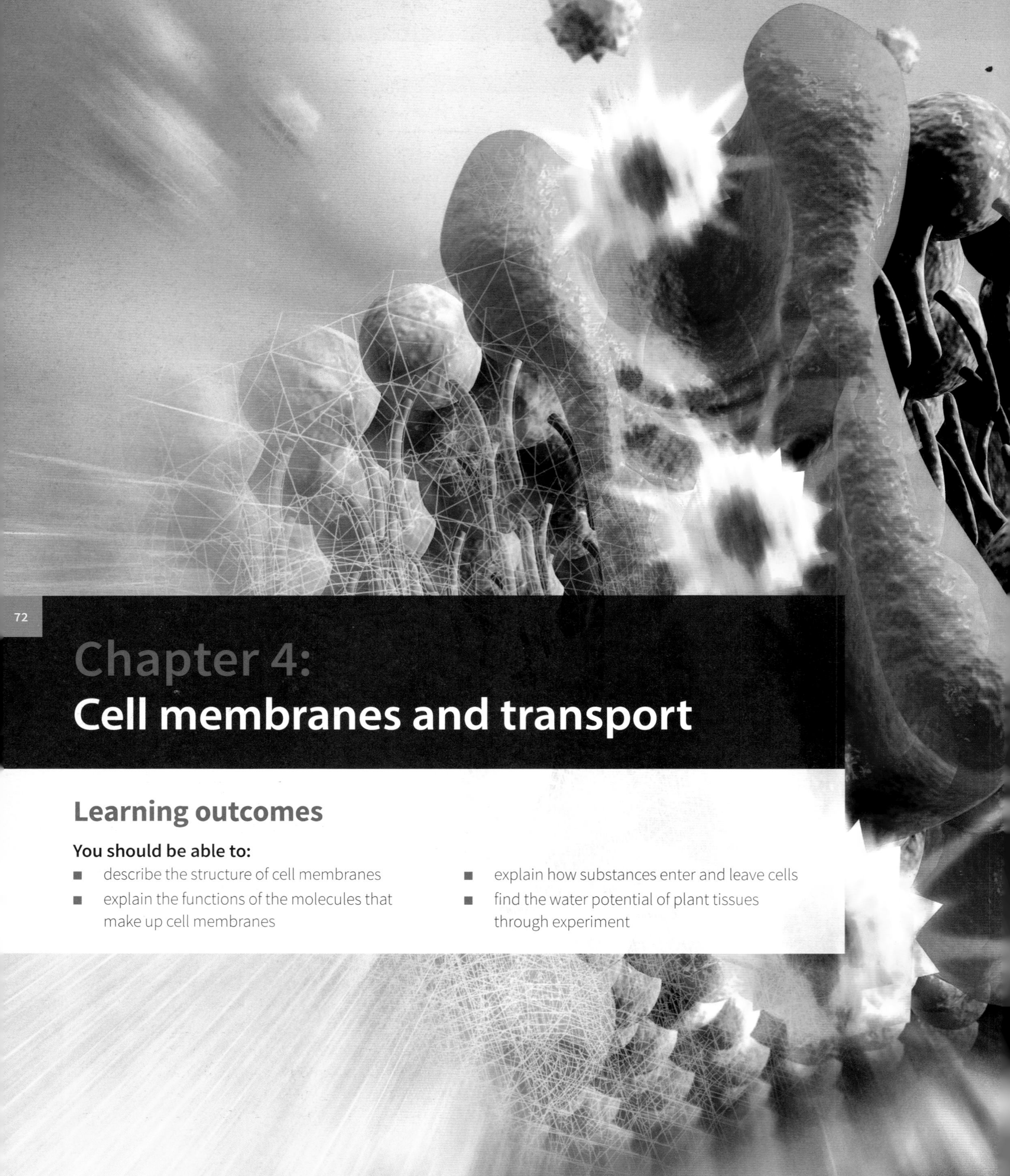

Chapter 4: Cell membranes and transport

Learning outcomes

You should be able to:

- describe the structure of cell membranes
- explain the functions of the molecules that make up cell membranes
- explain how substances enter and leave cells
- find the water potential of plant tissues through experiment

Drug runners

Liposomes are artificially prepared membrane-bound compartments (vesicles). They can be prepared by breaking up biological membranes into pieces, some of which re-seal themselves into balls resembling empty cells, though much smaller on average. Like intact cells, they are surrounded by a phospholipid bilayer and the interior is usually aqueous. They were first described in 1961. Since then they have been used as artificial models of cells and, more importantly, for medical applications. In particular, they have been used to deliver drugs.

To do this, the liposome is made while in a solution of the drug, so the drug is inside the liposome. The liposome is then introduced into the body and when it reaches a target cell, such as a cancer cell or other diseased cell, it fuses with that cell's surface membrane, delivering the drug inside the cell. Precise targeting can be achieved by inserting the correct recognition molecule – for example an antigen or antibody – into the liposome membrane. Other targeting methods also exist.

A recent (2013) discovery illustrating the usefulness of liposomes is that they provide a safe way of delivering the powerful anti-cancer drug staurosporine. Although this drug has been available since 1977, it kills any cells, including healthy ones that it comes into contact with because it interferes with several cell signalling pathways. Disguising agents have been added to the outer surfaces of liposomes carrying the drug which hide the drug from the immune system and allow it to target cancer cells only.

Liposomes have many other uses. For example, they are used in the cosmetics industry to deliver skin care products, such as aloe vera, collagen, elastin and vitamins A and E, when rubbed on skin. Liposome delivery of food supplements by mouth has also been tried with some success – absorption rates can be much higher than with traditional tablets.

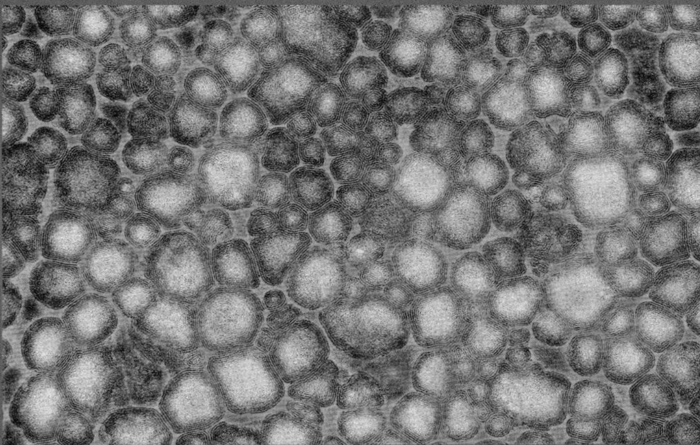

Figure 4.1 Liposomes.

In Chapter 1, you saw that **all** living cells are surrounded by a very thin membrane, the cell surface membrane. This controls the exchange of materials such as nutrients and waste products between the cell and its environment. Inside cells, regulation of transport across the membranes of organelles is also vital. Membranes also have other important functions. For example, they enable cells to receive hormone messages. It is important to study the structure of membranes if we are to understand how these functions are achieved.

Phospholipids

An understanding of the structure of membranes depends on an understanding of the structure of phospholipids (page 38). From phospholipids, little bags can be formed inside which chemicals can be isolated from the external environment. These bags are the membrane-bound compartments that we know as cells and organelles.

Figure 4.2a shows what happens if phospholipid molecules are spread over the surface of water. They form a single layer with their heads in the water, because these are polar (hydrophilic), and their tails projecting out of the water, because these are non-polar (hydrophobic). The term 'polar' refers to the uneven distribution of charge which occurs in some molecules. The significance of this is explained on pages 35–36.

If the phospholipids are shaken up with water, they can form stable ball-like structures in the water called **micelles** (Figure 4.2b). Here all the hydrophilic heads face outwards into the water, shielding the hydrophobic tails, which point in towards each other. Alternatively, two-layered structures, called **bilayers**, can form in sheets (Figure 4.2c). It is now known that this phospholipid bilayer is the basic structure of membranes (Figure 4.2d).

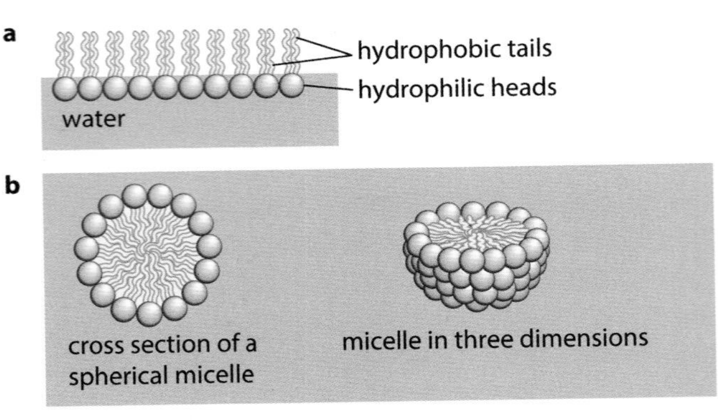

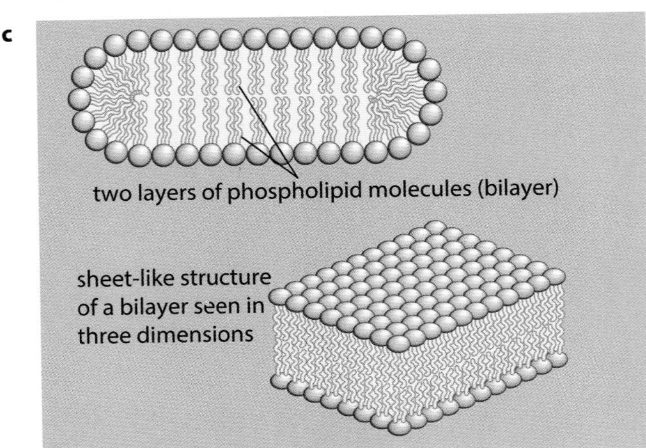

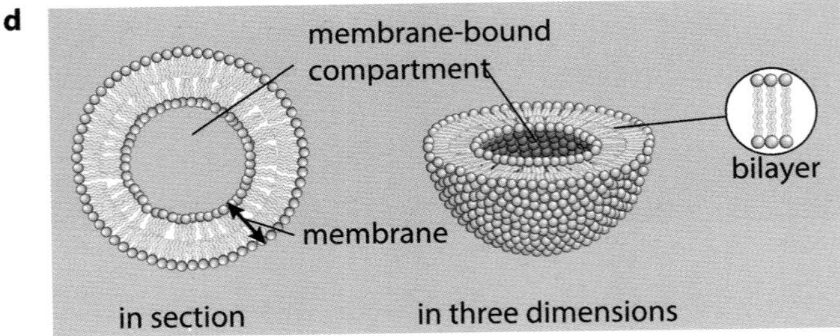

Figure 4.2 Phospholipids in water: **a** spread as a single layer of molecules (a monolayer) on the surface of water **b** forming micelles surrounded by water **c** forming bilayers **d** bilayers forming membrane-bound compartments.

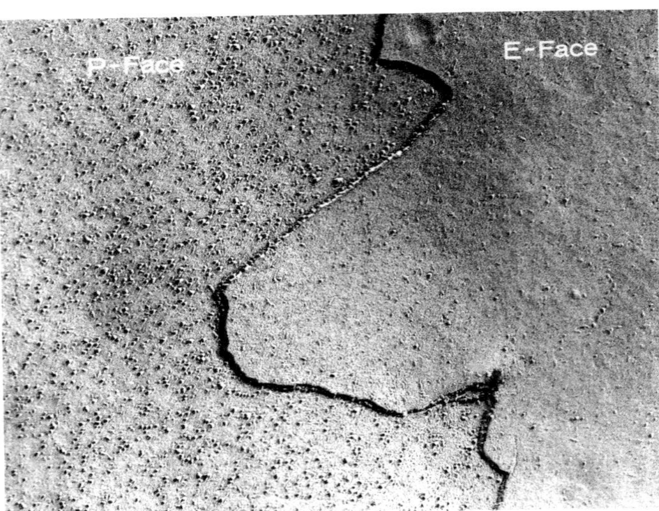

Figure 4.3 Scanning electron micrograph of a cell surface membrane. The membrane has been prepared by freeze-fracturing, which has split open the bilayer. The P-face is the phospholipid layer nearest the inside of the cell and shows the many protein particles embedded in the membrane. The E-face is part of the outer phospholipid layer (×50 000).

Structure of membranes

The phospholipid bilayer is visible using the electron microscope at very high magnifications of at least ×100 000 (Figure 1.23 on page 17). The double black line visible using the electron microscope is thought to show the hydrophilic heads of the two phospholipid layers; the pale zone between is the hydrophobic interior of the membrane. The bilayer (membrane) is about 7 nm wide.

Membranes also contain proteins. These can be seen in certain electron micrographs, such as Figure 4.3.

In 1972, two scientists, Singer and Nicolson, used all the available evidence to put forward a hypothesis for membrane structure. They called their model the **fluid mosaic model**. It is described as 'fluid' because both the phospholipids and the proteins can move about by diffusion. The phospholipid bilayer has the sort of fluidity we associate with olive oil. The phospholipids move sideways, mainly in their own layers. Some of the protein molecules also move about within the phospholipid bilayer, like icebergs in the sea. Others remain fixed to structures inside or outside the cell. The word 'mosaic' describes the pattern produced by the scattered protein molecules when the surface of the membrane is viewed from above.

Figures 4.4 and 4.5 are diagrams of what we imagine a membrane might look like if we could see the individual molecules.

Features of the fluid mosaic model

The membrane is a double layer (**bilayer**) of phospholipid molecules. The individual phospholipid molecules move about by diffusion within their own monolayers.

The phospholipid tails point inwards, facing each other and forming a non-polar hydrophobic interior. The phospholipid heads face the aqueous (water-containing) medium that surrounds the membranes.

Some of the phospholipid tails are saturated and some are unsaturated. The more unsaturated they are, the more fluid the membrane. This is because the unsaturated fatty acid tails are bent (Figure 2.11, page 36) and therefore fit together more loosely. Fluidity is also affected by tail

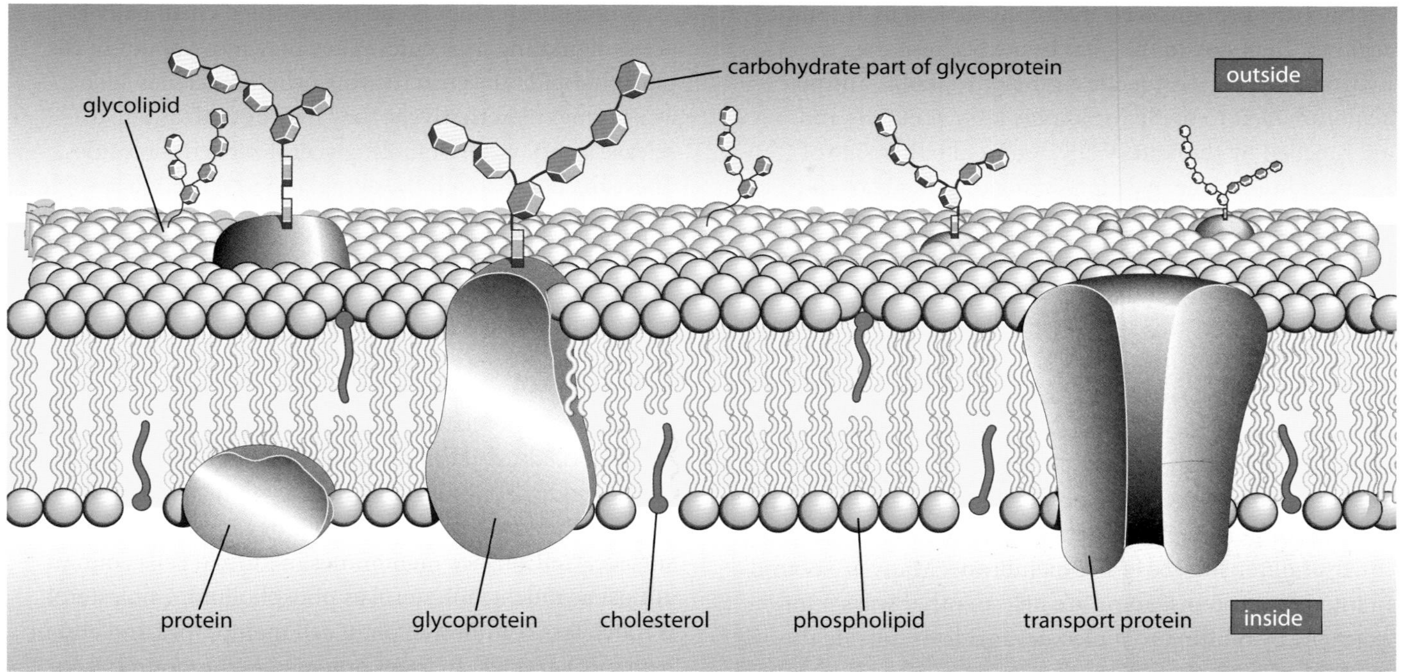

Figure 4.4 An artist's impression of the fluid mosaic model of membrane structure.

length: the longer the tail, the less fluid the membrane. As temperature decreases, membranes become less fluid, but some organisms which cannot regulate their own temperature, such as bacteria and yeasts, respond by increasing the proportion of unsaturated fatty acids in their membranes.

Two types of protein are recognised, according to their position in the membrane.

Proteins that are found embedded within the membrane, such as those in Figure 4.5, are called **intrinsic proteins** (or **integral proteins**). Intrinsic proteins may be found in the inner layer, the outer layer or, most commonly, spanning the whole membrane, in which case they are known as **transmembrane proteins**. In transmembrane proteins, the hydrophobic regions which cross the membrane are often made up of one or more α-helical chains.

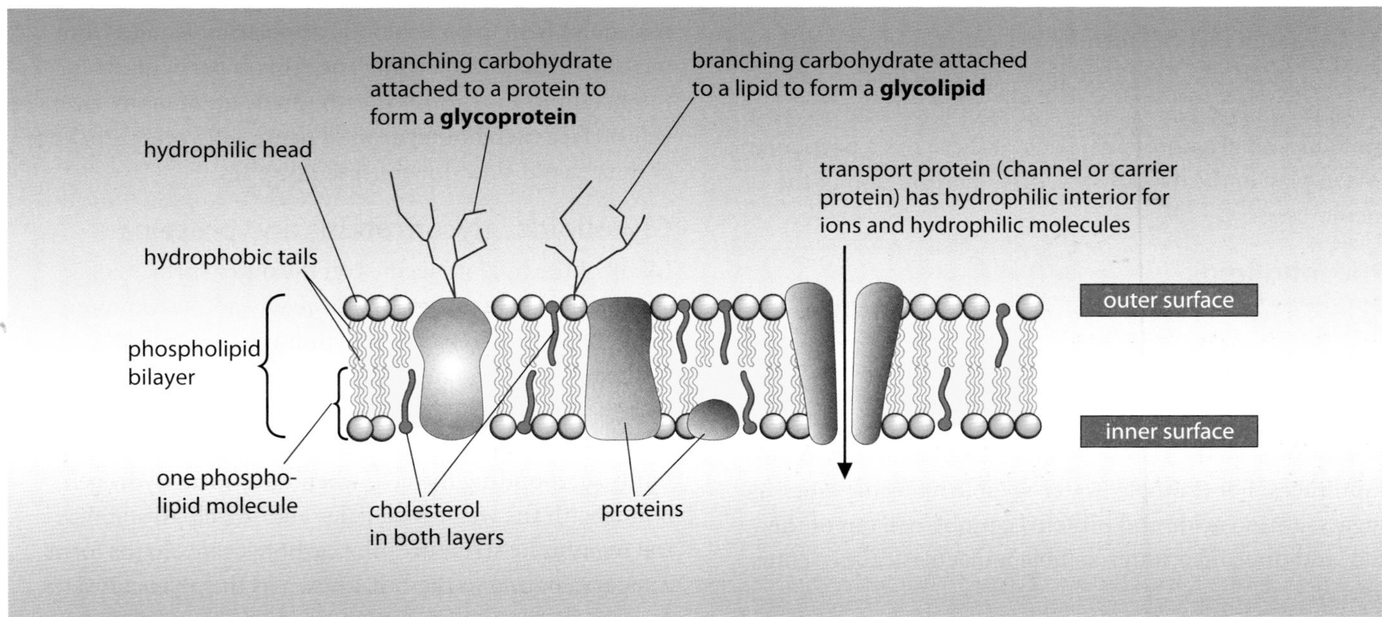

Figure 4.5 Diagram of the fluid mosaic model of membrane structure.

Intrinsic proteins have hydrophobic and hydrophilic regions. They stay in the membrane because the hydrophobic regions, made from hydrophobic amino acids, are next to the hydrophobic fatty acid tails and are repelled by the watery environment either side of the membrane. The hydrophilic regions, made from hydrophilic amino acids, are repelled by the hydrophobic interior of the membrane and therefore face into the aqueous environment inside or outside the cell, or line hydrophilic pores which pass through the membrane.

Most of the intrinsic protein molecules float like mobile icebergs in the phospholipid layers, although some are fixed like islands to structures inside or outside the cell and do not move about.

A second type of protein molecule is the **extrinsic protein** (or **peripheral protein**). These are found on the inner or outer surface of the membrane. Many are bound to intrinsic proteins. Some are held in other ways – for example, by binding to molecules inside or outside the cell, or to the phospholipids.

All the proteins referred to from now on in this chapter are intrinsic proteins.

Many proteins and lipids have short, branching carbohydrate chains attached to that side of the molecule which faces the outside of the membrane, thus forming glycoproteins and glycolipids, respectively.

The total thickness of the membrane is about 7 nm on average. Molecules of cholesterol are also found in the membrane.

Roles of the components of cell membranes

We have seen that cell membranes contain several different types of molecule. There are three types of lipid, namely phospholipids, cholesterol and glycolipids. There are also proteins and glycoproteins. Each of these has a particular role to play in the overall structure and function of the membrane.

Phospholipids

As explained on pages 73–76, phospholipids form the bilayer, which is the basic structure of the membrane. Because the tails of phospholipids are non-polar, it is difficult for polar molecules, or ions, to pass through membranes, so they act as a barrier to most water-soluble substances. For example, water-soluble molecules such as sugars, amino acids and proteins cannot leak out of the cell, and unwanted water-soluble molecules cannot enter the cell.

Some phospholipids can be modified chemically to act as signalling molecules. They may move about in the phospholipid bilayer, activating other molecules such as enzymes. Alternatively, they may be hydrolysed to release small, water-soluble, glycerol-related molecules. These diffuse through the cytoplasm and bind to specific receptors (page 78). One such system results in the release of calcium ions from storage in the ER, which in turn brings about exocytosis of digestive enzymes from pancreatic cells as described on page 87.

Cholesterol

Cholesterol is a relatively small molecule. Like phospholipids, cholesterol molecules have hydrophilic heads and hydrophobic tails, so they fit neatly between the phospholipid molecules with their heads at the membrane surface. Cell surface membranes in animal cells contain almost as much cholesterol as phospholipid. Cholesterol is much less common in plant cell membranes and absent from prokaryotes. In these organisms, compounds very similar to cholesterol serve the same function.

At low temperatures, cholesterol increases the fluidity of the membrane, preventing it from becoming too rigid. This is because it prevents close packing of the phospholipid tails. The increased fluidity means cells can survive colder temperatures. The interaction of the phospholipid tails with the cholesterol molecules also helps to stabilise cells at higher temperatures when the membrane could otherwise become **too** fluid. Cholesterol is also important for the mechanical stability of membranes, as without it membranes quickly break and cells burst open. The hydrophobic regions of cholesterol molecules help to prevent ions or polar molecules from passing through the membrane. This is particularly important in the myelin sheath (made up of many layers of cell surface membrane) around nerve cells, where leakage of ions would slow down nerve impulses.

Glycolipids, glycoproteins and proteins

Many of the lipid molecules on the outer surfaces of cell surface membranes, and probably all of the protein molecules, have short carbohydrate chains attached to them. These 'combination' molecules are known as glycolipids and glycoproteins, respectively. The carbohydrate chains project like antennae into the watery fluids surrounding the cell, where they form hydrogen bonds with the water molecules and so help to stabilise the membrane structure. The carbohydrate chains form a sugary coating to the cell, known as the **glycocalyx**. In animal cells, the glycocalyx is formed mainly from glycoproteins; in plant cells it mainly comprises glycolipids.

The carbohydrate chains help the glycoproteins and glycolipids to act as **receptor molecules**, which bind with particular substances at the cell surface. Different cells have different receptors, depending on their function. There are three major groups of receptor.

One group of receptors can be called 'signalling receptors', because they are part of a signalling system that coordinates the activities of cells. The receptors recognise messenger molecules like hormones and neurotransmitters. (Neurotransmitters are the chemicals that cross synapses, allowing nerve impulses to pass from one cell to another, and are discussed in Chapter 15.) When the messenger molecule binds to the receptor, a series of chemical reactions is triggered inside the cell. An example of a signalling receptor is the glucagon receptor in liver cells (Figure 14.24, page 317). Cells that do not have glucagon receptors are not affected by glucagon. Signalling is discussed in the next section.

A second group of receptors are involved in endocytosis (page 87). They bind to molecules that are parts of the structures to be engulfed by the cell surface membrane.

A third group of receptors is involved in binding cells to other cells (cell adhesion) in tissues and organs of animals.

Some glycolipids and glycoproteins act as cell markers or antigens, allowing cell–cell recognition. Each type of cell has its own type of antigen, rather like countries with different flags. For example, the ABO blood group antigens are glycolipids and glycoproteins which have small differences in their carbohydrate portions.

Many proteins act as **transport proteins**. These provide hydrophilic channels or passageways for ions and polar molecules to pass through the membrane. There are two types of transport protein: channel proteins and carrier proteins. Their roles are described on pages 82 and 86. Each transport protein is specific for a particular kind of ion or molecule. Therefore the types of substances that enter or leave the cell can be controlled.

Other membrane proteins may be **enzymes** – for example, the digestive enzymes found in the cell surface membranes of the cells lining the small intestine. These catalyse the hydrolysis of molecules such as disaccharides.

Some proteins on the inside of the cell surface membrane are attached to a system of protein filaments inside the cell, known as the cytoskeleton. These proteins help to maintain and decide the shape of the cell. They may also be involved in changes of shape when cells move.

Proteins also play important roles in the membranes of organelles. For example, in the membranes of mitochondria and chloroplasts they are involved in the processes of respiration and photosynthesis. (You will find out much more about this if you continue your biology course to A Level.)

Cell signalling

Cell signalling is an important, rapidly expanding area of research in modern biology, with wide applications. It is important because it helps to explain how living organisms control and coordinate their bodies. In this chapter, we concentrate on a few basic principles of signalling, highlighting the importance of membranes. As with other areas of biology, such as biochemistry, many of the fundamental principles and mechanisms are shared between all living organisms – plants, animals, fungi, protoctists and bacteria.

What is signalling? Basically, signalling is getting a message from one place to another.

Why do living organisms need signalling? All cells and organisms must be able to respond appropriately to their environments. This is made possible by means of a complex range of signalling pathways which coordinate the activities of cells, even if they are large distances apart in the same body. The basic idea of a signalling pathway can be summarised in a simple diagram (Figure 4.6). You will meet examples of cell signalling throughout this book and this diagram is a useful starting point for analysing the various pathways.

As Figure 4.6 shows, a signalling pathway includes receiving a stimulus or signal, transmitting the message and making an appropriate response. Conversion of the original signal to a message that is then transmitted is called transduction. Transmitting the message involves crossing barriers such as cell surface membranes. Signalling molecules are usually very small for easy transport.

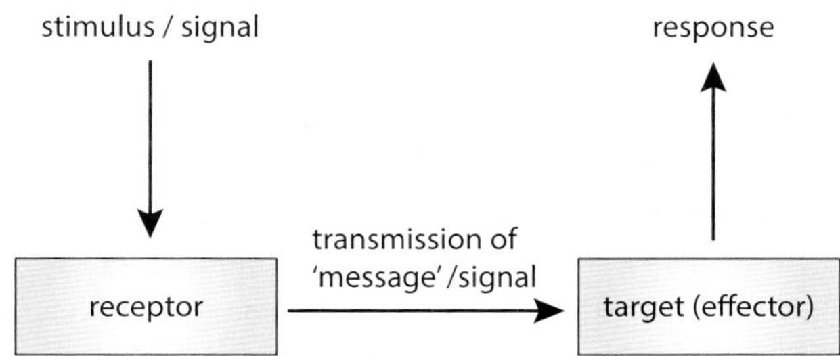

Figure 4.6 Basic components of a signalling pathway.

Distances travelled may be short, as with diffusion within one cell, or long, as with long-distance transport in blood (animals) or phloem (plants). There are usually many components and different mechanisms along the route. Signalling includes both electrical and chemical events and their interactions with each other – for example, the events associated with the nervous and hormonal systems in animals. These events involve a wide range of molecules produced by cells within the body (e.g. hormones and neurotransmitters) as well as outside stimuli (e.g. light, drugs, pheromones and odours).

The cell surface membrane is a critical component of most signalling pathways because it is a barrier to the movement of molecules, controlling what moves between the external and internal environments of the cell. In a typical signalling pathway, molecules must cross or interact with cell surface membranes.

Signalling molecules are very diverse. If they are hydrophobic, such as the steroid hormones (e.g. oestrogen), they can diffuse directly across the cell surface membrane and bind to receptors in the cytoplasm or nucleus.

QUESTION

4.1 Why does the cell surface membrane not provide a barrier to the entry of hydrophobic molecules into the cell?

More commonly, the signalling molecule is water-soluble. In this case, a typical signalling pathway starts with the signal arriving at a protein receptor in a cell surface membrane. The receptor is a specific shape which recognises the signal. Only cells with this receptor can recognise the signal. The signal brings about a change in the shape of the receptor, and since this spans the membrane, the message is in effect passed to the inside of the cell (signal transduction). Changing the shape of the receptor allows it to interact with the next component of the pathway, so the message gets transmitted.

This next component is often a 'G protein', which acts as a switch to bring about the release of a 'second messenger', a small molecule which diffuses through the cell relaying the message. (G proteins are so-called because the switch mechanism involves binding to GTP molecules. GTP is similar to ATP, but with guanine in place of adenine.)

Many second messenger molecules can be made in response to one receptor molecule being stimulated. This represents an amplification (magnification) of the original signal, a key feature of signalling. The second messenger typically activates an enzyme, which in turn activates further enzymes, increasing the amplification at each stage. Finally, an enzyme is produced which brings about the required change in cell metabolism.

The sequence of events triggered by the G protein is called a signalling cascade. Figure 4.7 is a diagram of a simplified cell signalling pathway involving a second messenger. Examples of such a pathway involving the hormones adrenaline and glucagon are discussed in Chapter 14.

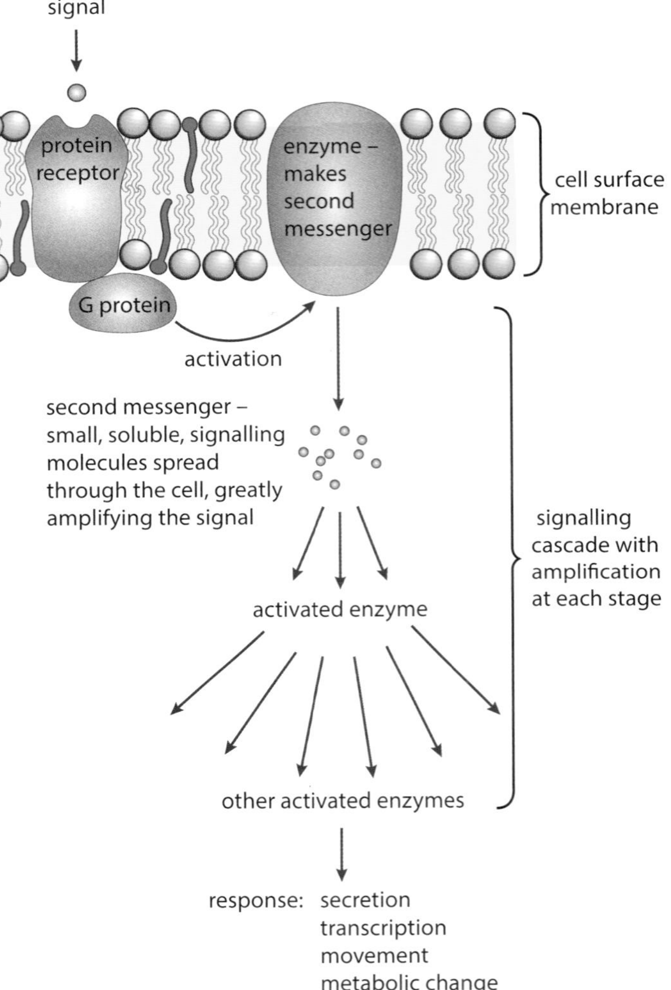

Figure 4.7 A simplified cell signalling pathway involving a second messenger.

Besides examples involving second messengers, there are three other basic ways in which a receptor can alter the activity of a cell:

- opening an ion channel, resulting in a change of membrane potential (e.g. nicotine-accepting acetylcholine receptors, Chapter 15)
- acting directly as a membrane-bound enzyme (e.g. insulin receptor)

- acting as an intracellular receptor when the initial signal passes straight through the cell surface membrane. For example, the oestrogen receptor is in the nucleus and directly controls gene expression when combined with oestrogen.

Figure 4.8 summarises some typical signalling systems. Note that apart from the secretion of chemical signals, direct cell-cell contact is another mechanism of signalling. This occurs, for example, during embryonic development and when lymphocytes detect foreign antigens on other cells.

Movement of substances into and out of cells

We have seen that a phospholipid bilayer around cells makes a very effective barrier, particularly against the movement of water-soluble molecules and ions. The aqueous contents of the cell are therefore prevented from escaping. However, some exchange between the cell and its environment is essential. There are five basic mechanisms by which this exchange is achieved: diffusion, facilitated diffusion, osmosis, active transport and bulk transport.

QUESTION

4.2 Suggest **three** reasons why exchange between the cell and its environment is essential.

Diffusion

If you open a bottle of perfume in a room, it is not long before molecules of scent spread to all parts of the room (and are detected when they fit into membrane receptors in your nose). This will happen, even in still air, by the process of diffusion. Diffusion can be defined as the net movement, as a result of random motion of its molecules or ions, of a substance from a region of its higher concentration to a region of its lower concentration. The molecules or ions move down a **concentration gradient**. The random movement is caused by the natural kinetic energy (energy of movement) of the molecules or ions. As a result of diffusion, molecules or ions tend to reach an equilibrium situation, where they are evenly spread within a given volume of space. The phenomenon of diffusion can be demonstrated easily using non-living materials such as glucose and Visking tubing (Box 4.1) or plant tissue (Box 4.2).

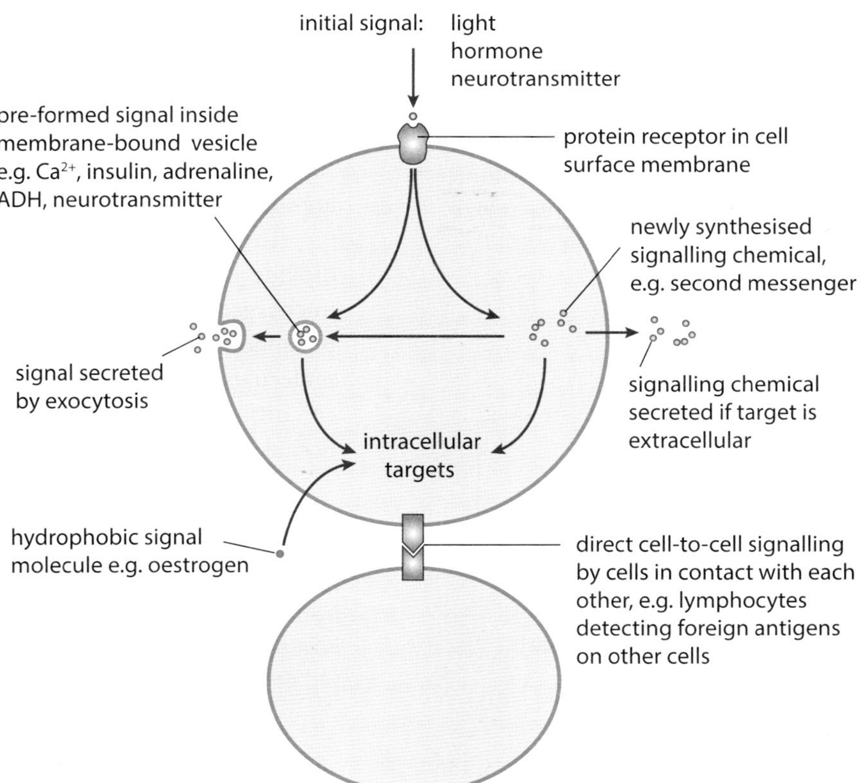

Figure 4.8 A few of the possible signalling pathways commonly found in cells. Note the role of membranes in these pathways.

Some molecules or ions are able to pass through living cell membranes by diffusion. Temporary staining of plant cells, e.g. adding iodine solution to epidermal cells, shows that this is possible. The rate at which a substance diffuses across a membrane depends on a number of factors, including the following.

- **The 'steepness'** of the concentration gradient – that is, the difference in the concentration of the substance on the two sides of the surface. If there are, for example, many more molecules on one side of a membrane than on the other, then at any one moment more molecules will be moving (entirely randomly) from this side than from the other. The greater the difference in concentration, the greater the difference in the number of molecules passing in the two directions, and hence the faster the rate of diffusion.
- **Temperature**. At high temperatures, molecules and ions have much more kinetic energy than at low temperatures. They move around faster, and thus diffusion takes place faster.
- **The surface area** across which diffusion is taking place. The greater the surface area, the more molecules or ions can cross it at any one moment, and therefore the faster diffusion can occur. The surface area of cell membranes can be increased by folding, as in microvilli in the intestine and kidneys or the cristae inside mitochondria. The larger the cell, the smaller its surface area in relation to its volume. This can easily be demonstrated by studying the diagram in Question 4.3. To make the calculations easier, cells are shown as cubes, but the principle remains the same – volume increases much more rapidly than surface area as size increases. (See also Box 4.3.)

QUESTION

4.3 The diagram shows three cubes.

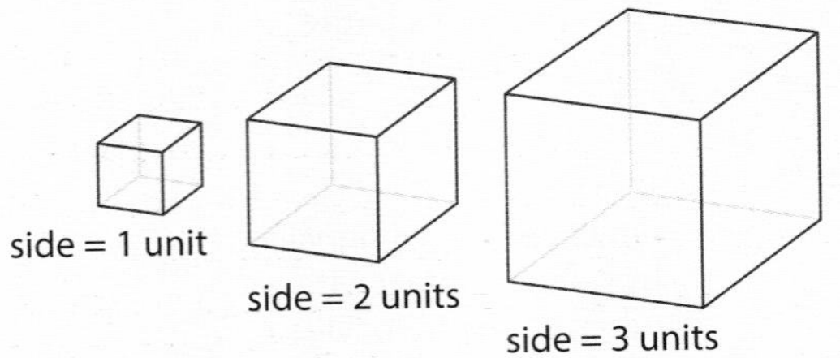

Calculate the surface area, volume and surface area : volume ratio of each of the cubes.

Cells rely on diffusion for internal transport of molecules. This results in a limit on the size of cells, because once inside a cell, the time it takes a molecule to reach a certain destination by diffusion increases rapidly with distance travelled. In fact, the rate falls in proportion to the square of the distance. Diffusion is therefore only effective over very short distances, such as the 7 nm across a membrane. An amino acid molecule, for example, can travel a few micrometres in several seconds, but would take several hours to diffuse a centimetre. An aerobic cell would quickly run out of oxygen and die if it were too large. Most cells are no larger than about 50 μm in diameter.

The surface area : volume ratio decreases as the size of any three-dimensional object increases.

QUESTION

4.4 The fact that surface area : volume ratio decreases with increasing size is also true for whole organisms. Explain the relevance of this for transport systems within organisms.

- **The nature of the molecules or ions**. Large molecules require more energy to get them moving than small ones do, so large molecules tend to diffuse more slowly than small molecules. Non-polar molecules, such as glycerol, alcohol and steroid hormones, diffuse much more easily through cell membranes than polar ones, because they are soluble in the non-polar phospholipid tails.

  The respiratory gases – oxygen and carbon dioxide – cross membranes by diffusion. They are uncharged and non-polar, and so can cross through the phospholipid bilayer directly between the phospholipid molecules. Water molecules, despite being very polar, can diffuse rapidly across the phospholipid bilayer because they are small enough.

Diffusion is the net movement of molecules or ions from a region of higher concentration to a region of lower concentration down a gradient, as a result of the random movements of particles.

BOX 4.1: Demonstrating diffusion using Visking tubing

Visking tubing (also known as dialysis tubing) is a partially permeable, non-living membrane made from cellulose. It possesses molecular-sized pores which are small enough to prevent the passage of large molecules, such as starch and sucrose, but will allow the passage of smaller molecules by diffusion, such as glucose.

This can be demonstrated by filling a length of Visking tubing (about 15 cm) with a mixture of starch and glucose solutions. If the tubing is suspended in a boiling tube of water for a period of time, the presence of starch and glucose outside the tubing can be tested for at intervals to monitor whether diffusion out of the tubing has occurred. The results should indicate that glucose, but not starch, diffuses out of the tubing.

This experiment can be made more quantitative. It would be interesting, for example, to try to estimate the concentration of glucose at each time interval by setting up separate tubes, one for each planned time interval, and using a semi-quantitative Benedict's test each time. A colorimeter would be useful for this. Alternatively, a set of colour standards could be prepared. A graph could be drawn showing how the rate of diffusion changes with the concentration gradient between the inside and outside of the tubing.

Further experiments could be designed if sucrose and an enzyme that breaks down sucrose (sucrase) are added to the Visking tubing. Experiments involving amylase, which breaks down starch, could also be designed.

BOX 4.3: Investigating the effect of size on diffusion

The effect of size on diffusion can be investigated by timing the diffusion of ions through blocks of agar of different sizes.

Solid agar is prepared in suitable containers such as ice cube trays. If the agar is made up with very dilute sodium hydroxide solution and Universal Indicator, it will be coloured purple. Cubes of the required dimensions (for example, sides of 2 cm × 2 cm, 1 cm × 1 cm, 0.5 cm × 0.5 cm) can be cut from the agar, placed in a container and covered with a diffusion solution such as dilute hydrochloric acid. (The acid should have a higher molarity than the sodium hydroxide so that its diffusion can be monitored by a change in colour of the indicator. Alternatively, the agar can be made up with Universal Indicator only, although its colour will be affected by the pH of the water used.)

Either the time taken for the acid to completely change the colour of the indicator in the agar blocks, or the distance travelled into the block by the acid in a given time (e.g. 5 minutes), can be measured. The times can be converted to rates.

Finally, the rate of diffusion (rate of colour change) can be plotted against the surface area : volume ratio.

Using the same techniques, you may be able to design further experiments. For example, you could investigate the effect on the rate of diffusion of the steepness of the concentration gradient.

BOX 4.2: Demonstrating diffusion using plant tissue

An experiment showing how the permeability of membranes is affected by environmental factors such as chemicals and temperature can be performed with beetroot.

Pieces of beetroot can be placed into water at different temperatures or into different alcohol concentrations. Any damage to the cell membranes results in the red pigment, which is normally contained within the large central vacuole, leaking out of the cells by diffusion. Changes in the colour of the surrounding solution can be monitored qualitatively or quantitatively. As in the experiment in Box 4.1, a colorimeter or a set of colour standards could be used. Alternatively, you could simply put the tubes in order and make up a colour scale (e.g. from 0 to 10), using water as 0 and the darkest solution as 10. There is an opportunity to design your own experiment.

What is being observed is diffusion of the red dye from a region of high concentration in the vacuoles to a region of low concentration in the solution outside the pieces of beetroot. Diffusion is normally prevented by the partially permeable nature of the cell membranes. After reading how molecules cross membranes, you may like to think about how the dye gets into the vacuoles in the first place.

Facilitated diffusion

Large polar molecules, such as glucose and amino acids, cannot diffuse through the phospholipid bilayer. Nor can ions such as sodium (Na^+) or chloride (Cl^-). These can only cross the membrane with the help of certain protein molecules. Diffusion that takes place in this way is called facilitated diffusion. 'Facilitated' means made easy or made possible, and this is what the proteins do.

There are two types of protein involved, namely **channel proteins** and **carrier proteins**. Each is highly specific, allowing only one type of molecule or ion to pass through it.

Channel proteins are water-filled pores. They allow charged substances, usually ions, to diffuse through the membrane. Most channel proteins are 'gated'. This means that part of the protein molecule on the inside surface of the membrane can move to close or open the pore, like a gate. This allows control of ion exchange. Two examples are the gated proteins found in nerve cell surface membranes. One type allows entry of sodium ions, which happens during the production of an action potential (page 335). Another allows exit of potassium ions (K^+) during the recovery phase, known as repolarisation. Some channels occur in a single protein; others are formed by several proteins combined.

Whereas channel proteins have a fixed shape, carrier proteins can flip between two shapes (Figure 4.9). As a result, the binding site is alternately open to one side of the membrane, then the other. If the molecules are diffusing across the membrane, then the direction of movement will normally depend on their relative concentration on each side of the membrane. They will move down a concentration gradient from a higher to a lower concentration. However, the rate at which this diffusion takes place is affected by how many channel or carrier protein molecules there are in the membrane, and, in the case of channel proteins, on whether they are open or not. For example, the disease cystic fibrosis is caused by a defect in a channel protein that should be present in the cell surface membranes of certain cells, including those lining the lungs. This protein normally allows chloride ions to move out of the cells. If the channel protein is not correctly positioned in the membrane, or if it does not open the chloride channel as and when it should, then the chloride ions cannot move out.

Facilitated diffusion is the diffusion of a substance through transport proteins in a cell membrane; the proteins provide hydrophilic areas that allow the molecules or ions to pass through the membrane which would otherwise be less permeable to them.

Osmosis

Osmosis is a special type of diffusion involving water molecules only. In the explanations that follow, remember that:

solute + solvent = solution

In a sugar **solution**, for example, the **solute** is sugar and the **solvent** is water.

In Figure 4.10 there are two solutions separated by a **partially permeable membrane**. This is a membrane that allows only certain molecules through, just like membranes in living cells. In the situation shown in Figure 4.10a, solution **B** has a higher concentration of

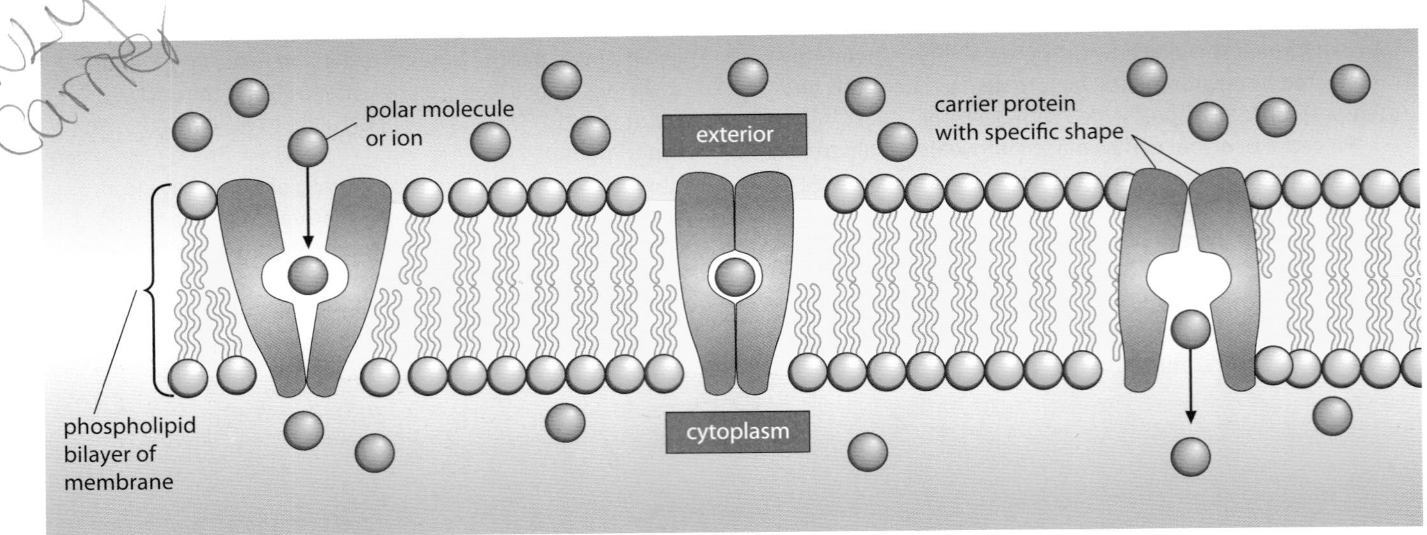

Figure 4.9 Changes in the shape of a carrier protein during facilitated diffusion. Here, there is a net diffusion of molecules or ions **into** the cell down a concentration gradient.

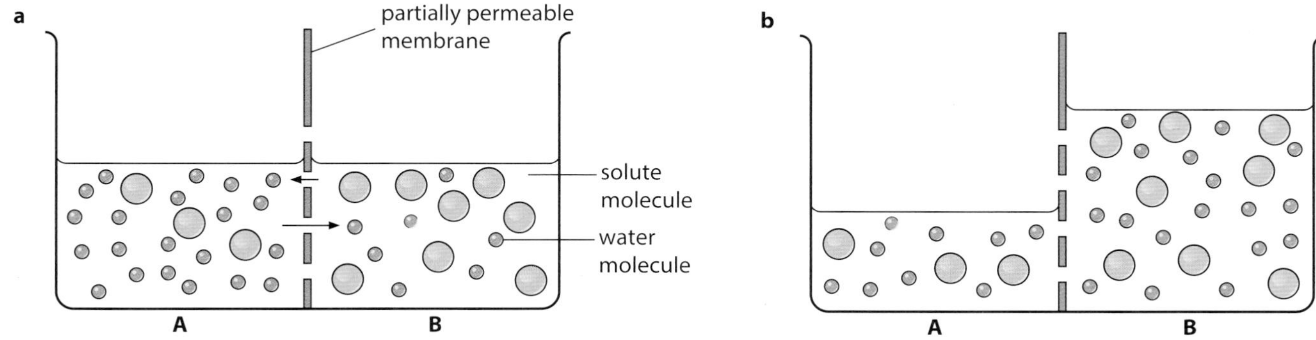

Figure 4.10 Two solutions separated by a partially permeable membrane. **a** Before osmosis. The solute molecules are too large to pass through the pores in the membrane, but the water molecules are small enough. **b** As the arrows show, more water molecules moved from A to B than from B to A, so the net movement has been from A to B, raising the level of solution in B and lowering it in A.

solute molecules than solution **A**. Solution **B** is described as more **concentrated** than solution **A**, and solution A as more **dilute** than solution **B**.

First, imagine what would happen if the membrane was not present. Both solute molecules and water molecules are free to move anywhere within the solutions. As they move randomly, both water molecules and solute molecules will tend to spread themselves evenly throughout the space available, by diffusion.

Now consider the situation where a partially permeable membrane is present, as shown in Figure 4.10. The solute molecules are too large to get through the membrane. Only water molecules can pass through. The solute molecules move about randomly, but as they hit the membrane they simply bounce back. The numbers of solute molecules each side of the membrane stay the same. The water molecules also move about randomly, but they are able to move both from **A** to **B** and from **B** to **A**. Over time, the water molecules will tend to spread themselves out more evenly between **A** and **B**.

This means that **A** will end up with fewer water molecules, so that the solution becomes more concentrated with solute. **B** will end up with more water molecules, so that it becomes more dilute. We will also find that the volume of liquid in **B** will increase, because it now contains the same number of solute molecules, but more water molecules.

This movement of water molecules from a dilute solution to a concentrated solution, through a partially permeable membrane, is called osmosis.

Water potential

The term water potential is very useful when considering osmosis. The Greek letter psi, ψ, can be used to mean water potential.

You can think of water potential as being the tendency of water to move out of a solution. This depends on two factors:

- how much water the solution contains in relation to solutes, and
- how much pressure is being applied to it.

Water always moves from a region of high water potential to a region of low water potential. We say water always moves down a water potential gradient. This will happen until the water potential is the same throughout the system, at which point we can say that equilibrium has been reached.

For example, a solution containing a lot of water (a dilute solution) has a higher water potential than a solution containing only a little water (a concentrated solution). In Figure 4.10a, solution **A** has a higher water potential than solution **B**, because solution **A** is more dilute than solution **B**. This is why the net movement of water is from **A** to **B**.

Now look again at Figure 4.10b. What would happen if we could press down very hard on side **B** (Figure 4.11)?

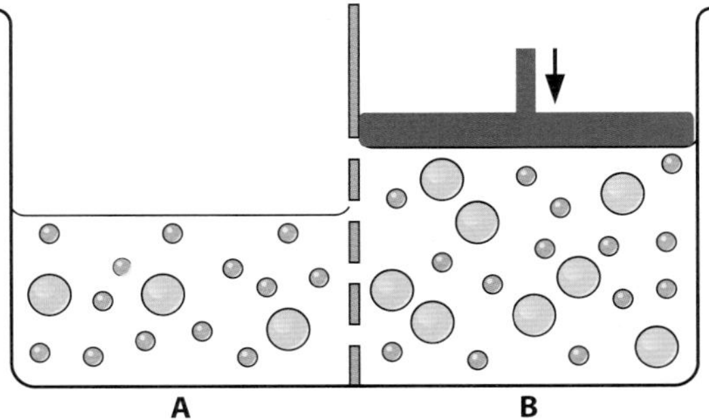

Figure 4.11 Applying pressure to a solution increases the tendency of water to move out of it – that is, it increases its water potential. Here, water molecules move from **B** to **A**.

It would be possible to 'squeeze' some of the water back into **A**. By increasing the pressure on the liquid in **B**, we are increasing the tendency for water to move out of it – that is, we are increasing its water potential, until it is higher than the water potential in **A**. Pressure on a liquid increases water potential.

By definition, the water potential of pure water at atmospheric pressure is 0. This means that a solution (water with a solute or solutes dissolved in it) must have a water potential less than 0 – that is, a negative value.

Osmosis is the net movement of water molecules from a region of higher water potential to a region of lower water potential, through a partially permeable membrane, as a result of their random motion (diffusion).

QUESTION

4.5 In Figure 4.10b, imagine that the solutions in **A** and **B** are in equilibrium – that is, there is no net movement of water molecules. What can you say about the water potentials of the two solutions?

Solute potential and pressure potential

We have seen that there are two factors that determine the water potential of a solution – the concentration of the solution, and the pressure applied to it.

The contribution of the concentration of the solution to water potential is called **solute potential**. We can think of solute potential as being the extent to which the solute molecules decrease the water potential of the solution. The more solute there is, the lower the tendency for water to move out of the solution.

Just like water potential, solute potential is 0 for pure water, and has a negative value for a solution. Adding more solute to a solution decreases its water potential. So the greater the concentration of the solute, the more negative the value of the solute potential. The psi symbol can be used to show the solute potential, but this time with the subscript s – ψ_s.

The contribution of pressure to the water potential of a solution is called **pressure potential**. We can see in Figure 4.11 that increasing the pressure on **B** increases the tendency of water to move out of it – that is, it increases its water potential. Pressure potential can be shown using the symbol ψ_p.

Osmosis in animal cells

Figure 4.12 shows the effect of osmosis on an animal cell. A convenient type of animal cell to study in practical work is the red blood cell. A slide of fresh blood viewed with a microscope will show large numbers of red blood cells. Different samples of blood can be mixed with solutions of different water potential. Figure 4.12a shows that if the water potential of the solution surrounding the cell is too high, the cell swells and bursts. If it is too low, the cell shrinks (Figure 4.12c). This shows one reason why it is important to maintain a constant water potential inside the bodies of animals.

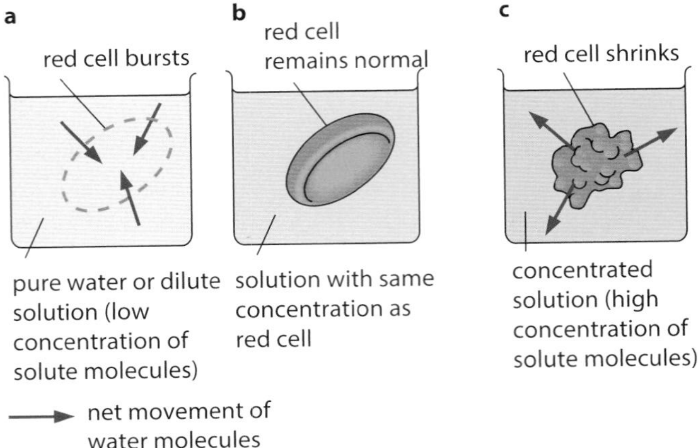

Figure 4.12 Movement of water into or out of red blood cells by osmosis in solutions of different concentration.

QUESTION

4.6 In Figure 4.12:

a which solution has the highest water potential?

b which solution has the lowest solute potential?

c in which solution is the water potential of the red cell the same as that of the solution?

Osmosis in plant cells

Unlike animal cells, plant cells are surrounded by cell walls, which are very strong and rigid (page 5). Imagine a plant cell being placed in pure water or a dilute solution (Figure 4.13a). The water or solution has a higher water potential than the plant cell, and water therefore enters the cell through its partially permeable cell surface membrane by osmosis. Just like in the animal cell, the volume of the cell increases, but in the plant cell the the cell wall pushes back against the expanding **protoplast** (the living part of the cell inside the cell wall), and pressure starts to build up rapidly. This is the pressure potential, and it increases the water potential of the cell until the water potential inside the cell equals the water potential outside the cell, and equilibrium is reached (Figure 4.13b). The cell wall is so inelastic that it takes very little water to enter the cell to achieve this. The cell wall prevents the cell from bursting, unlike the situation when an animal cell is placed in pure water or a dilute solution. When a plant cell is fully inflated with water it is described as fully **turgid**. For plant cells, then, water potential is a combination of solute potential and pressure potential. This can be expressed in the following equation:

$$\psi = \psi_s + \psi_p$$

Figure 4.13c shows the situation where a plant cell is placed in a solution of lower water potential. An example of the latter would be a concentrated sucrose solution. In such a solution, water will **leave** the cell by osmosis. As it does so, the protoplast gradually shrinks until it is exerting no pressure at all on the cell wall. At this point the pressure potential is zero, so the water potential of the cell is equal to its solute potential (see the equation above). Both the solute molecules and the water molecules of the external solution can pass through the freely permeable cell wall, and so the external solution remains in contact with the shrinking protoplast. As the protoplast continues to shrink, it begins to pull away from the cell wall (Figure 4.14). This process is called **plasmolysis**, and a cell in which it has happened is said to be **plasmolysed** (Figures 4.13c and 4.14). The point at which pressure potential has just reached zero and plasmolysis is **about** to occur is referred to as **incipient plasmolysis**. Eventually, as with the animal cell, an equilibrium is reached when the water potential of the cell has decreased until it equals that of the external solution.

The changes described can easily be observed with a light microscope using strips of epidermis peeled from rhubarb petioles or from the swollen storage leaves of onion bulbs and placed in a range of sucrose solutions of different concentration (Figure 4.15).

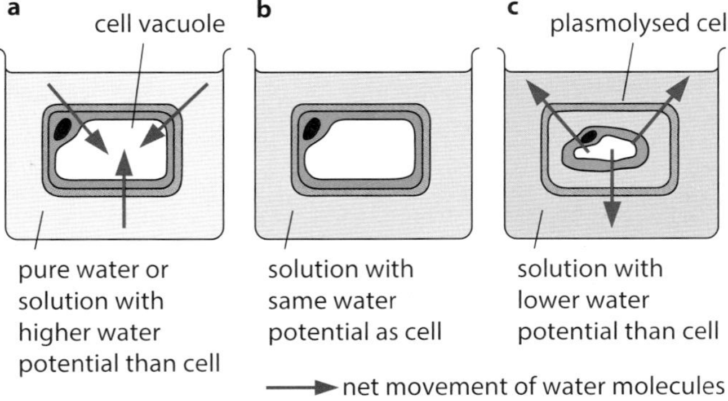

Figure 4.13 Osmotic changes in a plant cell in solutions of different water potential.

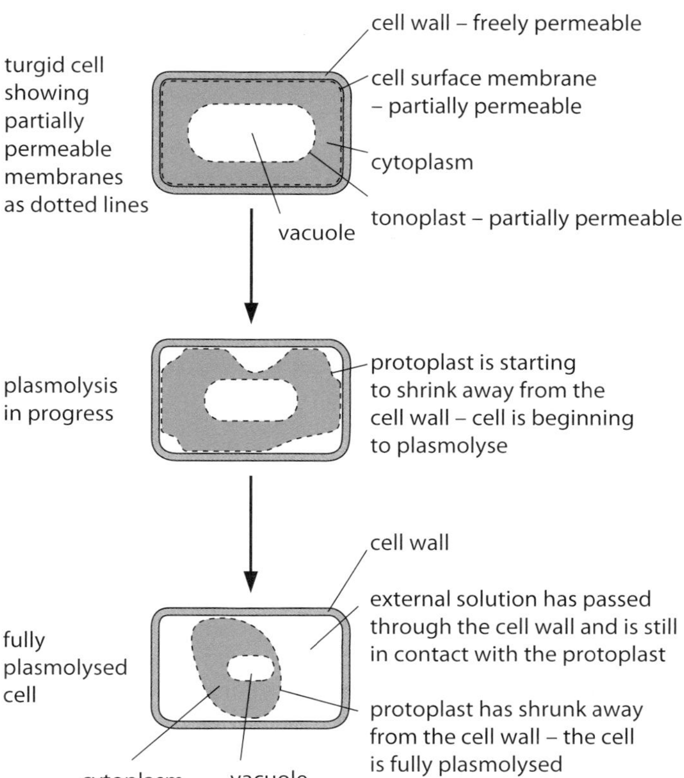

Figure 4.14 How plasmolysis occurs.

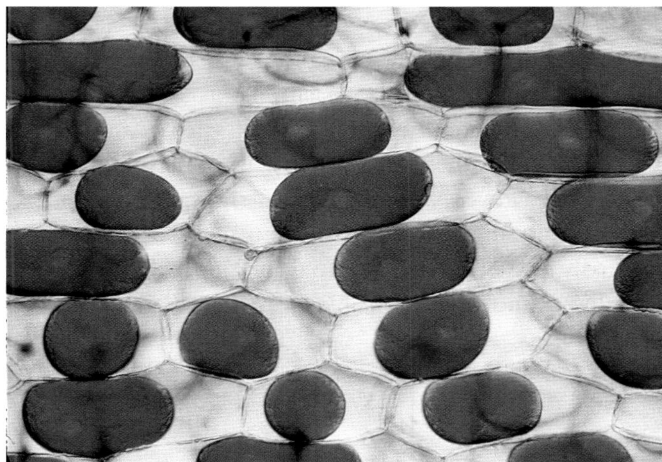

Figure 4.15 Light micrograph of red onion cells that have plasmolysed (×100).

QUESTION

4.7 Figures 4.14 and 4.15 shows a phenomenon called plasmolysis. Why can plasmolysis not take place in an animal cell?

BOX 4.4: Investigating osmosis in plant cells

1 Observing osmosis in plant cells

Epidermal strips are useful material for observing plasmolysis. Coloured sap makes observation easier. Suitable sources are the inner surfaces of the fleshy storage leaves of red onion bulbs, rhubarb petioles and red cabbage.

The strips of epidermis may be placed in a range of molarities of sucrose solution (up to 1.0 mol dm^{-3}) or sodium chloride solutions of up to 3%. Small pieces of the strips can then be placed on glass slides, mounted in the relevant solution, and observed with a microscope. Plasmolysis may take several minutes, if it occurs.

2 Determining the water potential of a plant tissue

The principle in this experiment is to find a solution of known water potential which will cause neither a gain nor a loss in water of the plant tissue being examined. Samples of the tissue – for example, potato – are allowed to come into equilibrium with a range of solutions (for example, sucrose solutions) of different water potentials, and changes in either mass or volume are recorded. Plotting a graph of the results allows the solution that causes no change in mass or volume to be determined. This solution will have the same water potential as the plant tissue.

Active transport

If the concentration of particular ions, such as potassium and chloride, inside cells is measured, it is often found that they are 10–20 times more concentrated inside than outside. In other words, a concentration gradient exists, with a lower concentration outside and a higher concentration inside the cell. The ions inside the cell originally came from the external solution, therefore diffusion cannot be responsible for this gradient because, as we have seen, ions diffuse from high concentration to low concentration. The ions must therefore accumulate **against** a concentration gradient.

The process responsible is called **active transport**. It is achieved by carrier proteins, each of which is specific for a particular type of molecule or ion. However, unlike facilitated diffusion, active transport requires energy, because movement occurs **up** a concentration gradient. The energy is supplied by the molecule ATP (adenosine triphosphate) which is produced during respiration inside the cell. The energy is used to make the carrier protein change its shape, transferring the molecules or ions across the membrane in the process (Figure 4.17).

An example of a carrier protein used for active transport is the **sodium–potassium (Na^+ – K^+) pump** (Figure 4.18). Such pumps are found in the cell surface

QUESTION

4.8 Two neighbouring plant cells are shown in Figure 4.16.

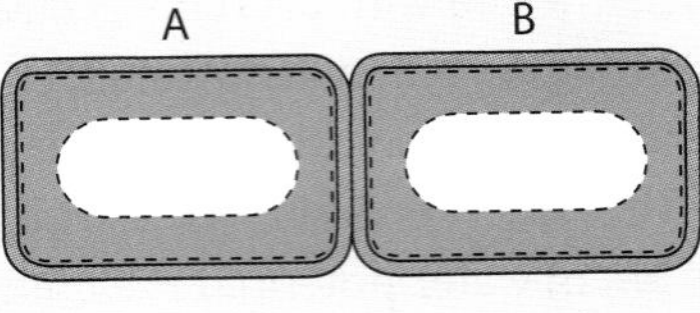

Figure 4.16

a In which direction would there be net movement of water molecules?

b Explain what is meant by **net movement**.

c Explain your answer to **a**.

d Explain what would happen if both cells were placed in

i pure water

ii a 1 mol dm^{-3} sucrose solution with a water potential of −3510 kPa.

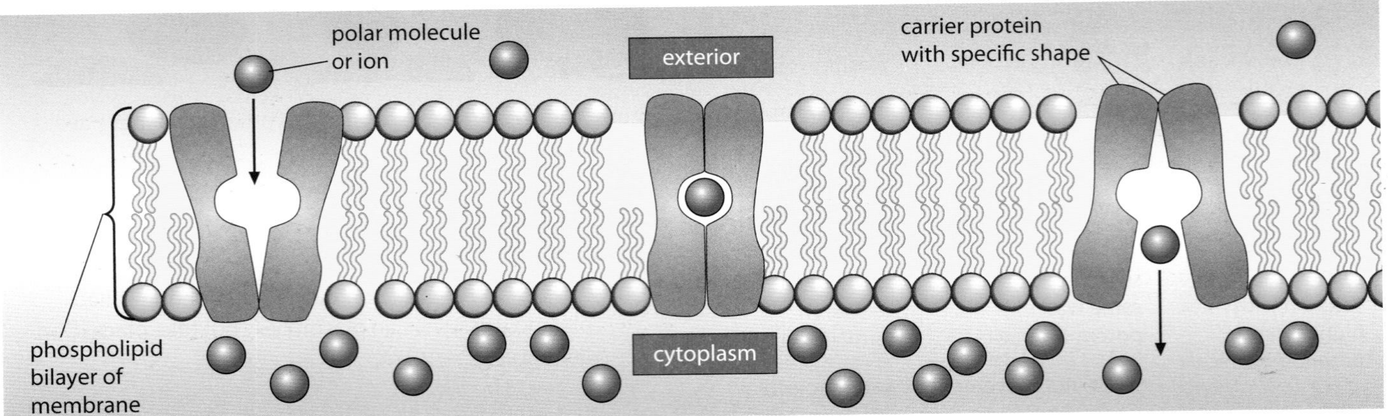

Figure 4.17 Changes in the shape of a carrier protein during active transport. Here, molecules or ions are being pumped into the cell **against** a concentration gradient. (Compare Figure 4.9.)

membranes of all animal cells. In most cells, they run all the time, and it is estimated that on average they use 30% of a cell's energy (70% in nerve cells).

The role of the $Na^+ - K^+$ pump is to pump three sodium ions out of the cell at the same time as allowing two potassium ions into the cell for each ATP molecule used. The ions are both positively charged, so the net result is that the inside of the cell becomes more negative than the outside – a potential difference (p.d.) is created across the membrane. The significance of this in nerve cells is discussed in Chapter 15 (pages 333–334).

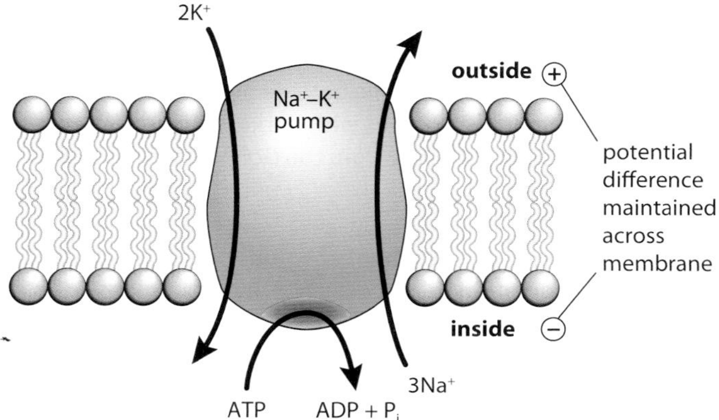

Figure 4.18 The $Na^+ - K^+$ pump.

In Figure 4.18, you can see that the pump has a receptor site for ATP on its inner surface. It acts as an ATPase enzyme in bringing about the hydrolysis of ATP to ADP (adenosine diphosphate) and phosphate to release energy.

Active transport can therefore be defined as the energy-consuming transport of molecules or ions across a membrane against a concentration gradient (from a lower to a higher concentration). The energy is provided by ATP from cell respiration. Active transport can occur either into or out of the cell.

Active transport is important in reabsorption in the kidneys, where certain useful molecules and ions have to be reabsorbed into the blood after filtration into the kidney tubules. It is also involved in the absorption of some products of digestion from the gut. In plants, active transport is used to load sugar from the photosynthesising cells of leaves into the phloem tissue for transport around the plant (Chapter 7), and to load inorganic ions from the soil into root hairs.

Bulk transport

So far we have been looking at ways in which **individual** molecules or ions cross membranes. Mechanisms also exist for the bulk transport of large quantities of materials into cells (**endocytosis**) or out of cells (**exocytosis**). Large molecules such as proteins or polysaccharides, parts of cells or even whole cells may be transported across the membrane. This requires energy, so it is a form of active transport.

Endocytosis involves the engulfing of the material by the cell surface membrane to form a small sac, or 'endocytic vacuole'. It takes two forms.

- Phagocytosis or 'cell eating' – this is the bulk uptake of **solid** material. Cells specialising in this are called phagocytes. The process is called **phagocytosis** and the vacuoles **phagocytic vacuoles.** An example is the engulfing of bacteria by certain white blood cells (Figure 4.19).
- Pinocytosis or 'cell drinking' – this is the bulk uptake of **liquid**. The vacuoles (vesicles) formed are often extremely small, in which case the process is called **micropinocytosis.**

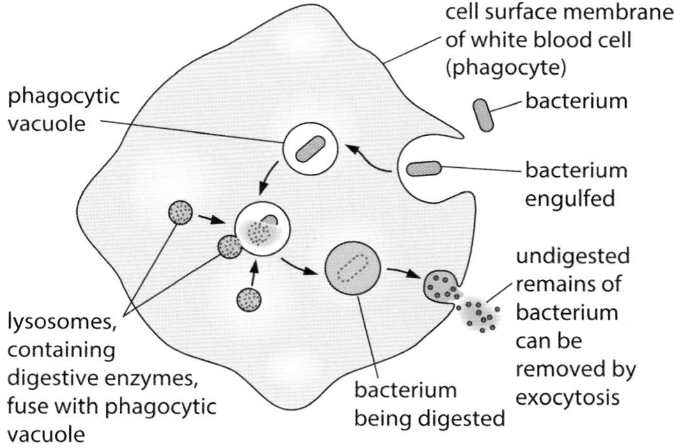

Figure 4.19 Stages in phagocytosis of a bacterium by a white blood cell.

Exocytosis is the reverse of endocytosis and is the process by which materials are removed from cells (Figure 4.20). It happens, for example, in the secretion of digestive enzymes from cells of the pancreas (Figure 4.21). Secretory vesicles from the Golgi body carry the enzymes to the cell surface and release their contents. Plant cells use exocytosis to get their cell wall building materials to the outside of the cell surface membrane.

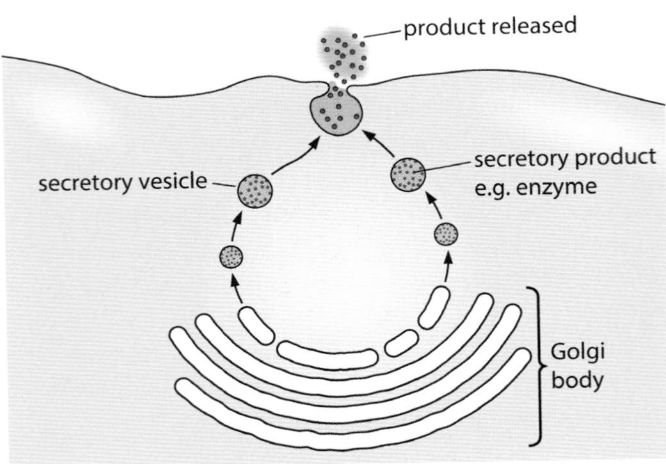

Figure 4.20 Exocytosis in a secretory cell. If the product being secreted is a protein, the Golgi body is often involved in chemically modifying the protein before it is secreted, as in the secretion of digestive enzymes by the pancreas.

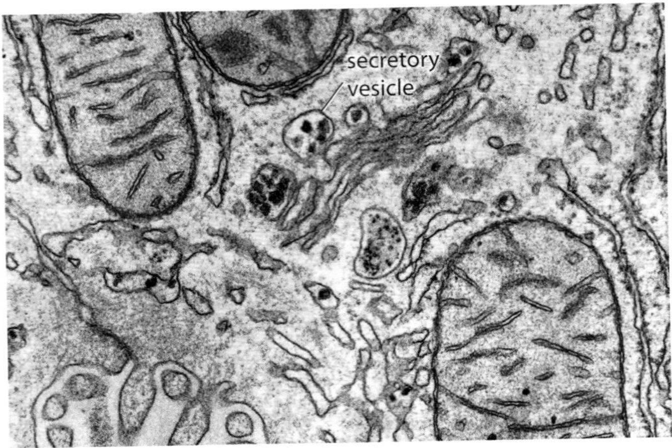

Figure 4.21 Transmission electron micrograph of pancreatic acinar cell secreting protein. The outside of the cell is coloured green. Golgi (secretory) vesicles with darkly stained contents can be seen making their way from the Golgi body to the cell surface membrane.

Active transport is the movement of molecules or ions through transport proteins across a cell membrane, against their concentration gradient, using energy from ATP.

Endocytosis is the bulk movement of liquids (pinocytosis) or solids (phagocytosis) into a cell, by the infolding of the cell surface membrane to form vesicles containing the substance; endocytosis is an active process requiring ATP.

Exocytosis is the bulk movement of liquids or solids out of a cell, by the fusion of vesicles containing the substance with the cell surface membrane; exocytosis is an active process requiring ATP.

Summary

- The basic structure of a membrane is a 7 nm thick phospholipid bilayer with protein molecules spanning the bilayer or within one or other layer. Phospholipids and some proteins move within the layers. Hence the structure is described as a fluid mosaic – the scattered protein molecules resemble pieces of a mosaic. Phospholipid bilayers are a barrier to most water-soluble substances because the interior of the membrane is hydrophobic. Cholesterol is needed for membrane fluidity and stability.
- Some proteins are transport proteins, transporting molecules or ions across the membrane. They may be either channel proteins or carrier proteins. Channel proteins have a fixed shape; carrier proteins change shape. Some proteins act as enzymes – for example, in the cell surface membranes of microvilli in the gut.
- Glycolipids and glycoproteins form receptors – for example, for hormones or neurotransmitters. Glycolipids and glycoproteins also form antigens, which are cell recognition markers. Membranes play an important role in cell signalling, the means by which cells communicate with each other.
- The cell surface membrane controls exchange between the cell and its environment. Some chemical reactions take place on membranes inside cell organelles, as in photosynthesis and respiration.
- Diffusion is the net movement of molecules or ions from a region of their higher concentration to one of lower concentration. Oxygen, carbon dioxide and water cross membranes by diffusion through the phospholipid bilayer. Diffusion of ions and larger polar molecules through membranes is allowed by transport proteins. This process is called facilitated diffusion.
- Water moves from regions of higher water potential to regions of lower water potential. When water moves from regions of higher water potential to regions of lower water potential through a partially permeable membrane, such as the cell surface membrane, this diffusion is called osmosis.
- Pure water has a water potential (Ψ) of zero. Adding solute reduces the water potential by an amount known as the solute potential (Ψ_s), which has a negative value. Adding pressure to a solution increases the water potential by an amount known as the pressure potential (Ψ_p), which has a positive value. The following equation is used: $\Psi = \Psi_s + \Psi_p$

- In dilute solutions, animal cells burst as water moves into the cytoplasm from the solution. In dilute solutions, a plant cell does not burst, because the cell wall provides resistance to prevent it expanding. The pressure that builds up as water diffuses into a plant cell by osmosis is the pressure potential. A plant cell in this state is turgid. In concentrated solutions, animal cells shrink, while in plant cells the protoplast shrinks away from the cell wall in a process known as plasmolysis.
- Some ions and molecules move across membranes by active transport, against the concentration gradient. This needs a carrier protein and ATP to provide energy. Exocytosis and endocytosis involve the formation of vacuoles to move larger quantities of materials respectively out of, or into, cells by bulk transport. There are two types of endocytosis, namely phagocytosis (cell eating) and pinocytosis (cell drinking).

End-of-chapter questions

1 What are the most abundant molecules in the cell surface membranes of plant cells?

A cholesterol
B glycolipids
C phospholipids
D proteins [1]

2 Where are the carbohydrate portions of glycolipids and glycoproteins located in cell surface membranes?

A the inside and outside surfaces of the membrane
B the inside surface of the membrane
C the interior of the membrane
D the outside surface of the membrane [1]

3 The cells of the myelin sheath are wrapped in layers around nerve cell axons. Freeze-fractured preparations of the myelin sheath cell surface membranes show very few particles. This indicates that myelin membranes contain relatively few of which type of molecule?

A cholesterol
B glycolipids
C polysaccharides
D proteins [1]

4 Prepare a table to summarise briefly the major functions of phospholipids, cholesterol, glycolipids, glycoproteins and proteins in cell surface membranes. [15]

5 a Describe fully what will occur if a plant cell is placed in a solution that has a higher water potential than the cell. Use the following terms in your answer.

cell wall, freely permeable, partially permeable, cell surface membrane, vacuole, tonoplast, cytoplasm, solute potential, pressure potential, water potential, turgid, osmosis, protoplast, equilibrium [14]

b Describe fully what will occur if a plant cell is placed in a solution that has a lower water potential than the cell. Use the following terms in your answer.

cell wall, freely permeable, partially permeable, cell surface membrane, vacuole, tonoplast, cytoplasm, solute potential, pressure potential, water potential, incipient plasmolysis, plasmolysed, osmosis, protoplast, equilibrium [15]

[Total: 29]

6 The diagram shows part of a membrane containing a channel protein. Part of the protein molecule is shaded.

A
B
C
7 nm

a Identify the parts labelled **A**, **B** and **C**. [3]

b For each of the following, state whether the component is hydrophilic or hydrophobic:

i A ii B iii darkly shaded part of protein

iv lightly shaded part of protein. [2]

c Explain how ions would move through the channel protein. [3]

d State **two** features that the channel proteins and carrier proteins of membranes have in common. [2]

e State **one** structural difference between channel and carrier proteins. [1]

f Calculate the magnification of the drawing. Show your working. [4]

[Total: 15]

7 Copy the table below and place a tick or cross in each box as appropriate.

Process	Uses energy in the form of ATP	Uses proteins	Specific	Controllable by cell
diffusion				
osmosis				
facilitated diffusion				
active transport				
endocytosis and exocytosis				

[20]

8 Copy and complete the table below to compare cell walls with cell membranes.

Feature	Cell wall	Cell membrane
is the thickness normally measured in nm or μm?		
location		
chemical composition		
permeability		
function		
fluid or rigid		

[6]

9 A cell with a water potential of –300 kPa was placed in pure water at time zero. The rate of entry of water into the cell was measured as the change in water potential with time. The graph shows the results of this investigation.

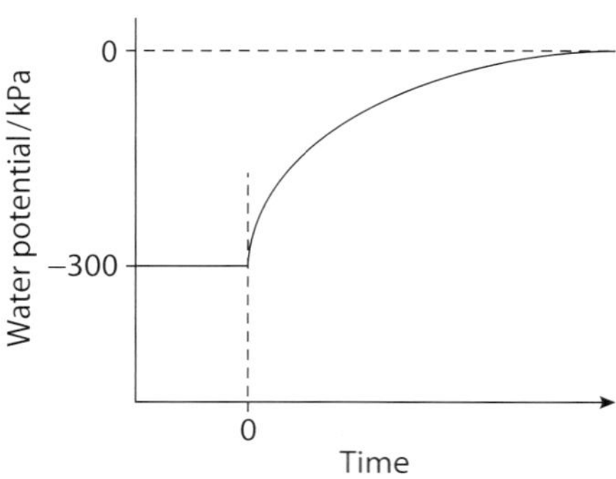

Describe and explain the results obtained. [8]

10 The rate of movement of molecules or ions across a cell surface membrane is affected by the relative concentrations of the molecules or ions on either side of the membrane. The graphs below show the effect of concentration difference (the steepness of the concentration gradient) on three transport processes, namely diffusion, facilitated diffusion and active transport.

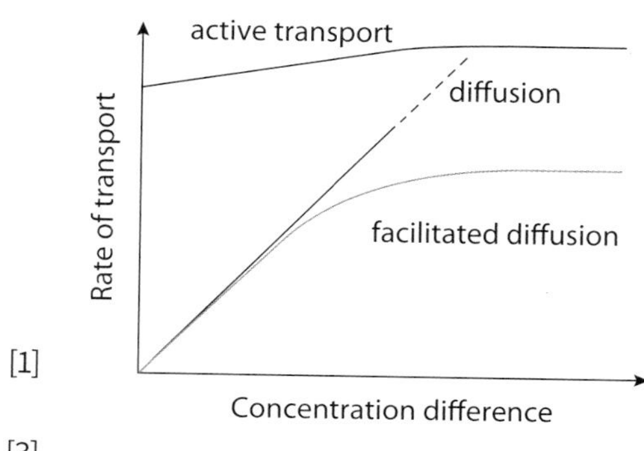

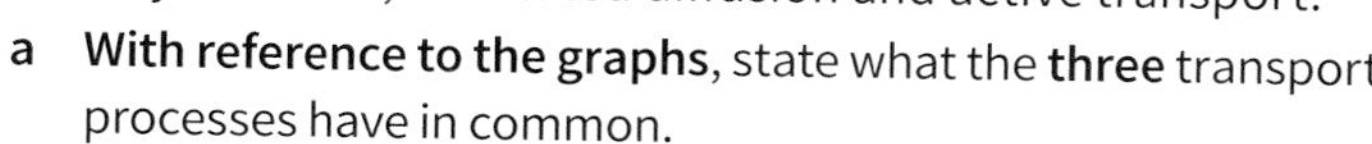

a **With reference to the graphs**, state what the **three** transport processes have in common. [1]

b Explain the rates of transport observed when the concentration difference is zero. [3]

c i Which **one** of the processes would stop if a respiratory inhibitor were added? [1]

ii Explain your answer. [2]

d Explain the difference between the graphs for diffusion and facilitated diffusion. [5]

[Total: 12]

11 When a cell gains or loses water, its volume changes. The graphs show changes in the water potential (ψ), pressure potential (ψ_p) and solute potential (ψ_s) of a plant cell as its volume changes as a result of gaining or losing water. (Note that 80% relative cell volume means the cell or protoplast has shrunk to 80% of the volume it was at 100% relative cell volume.)

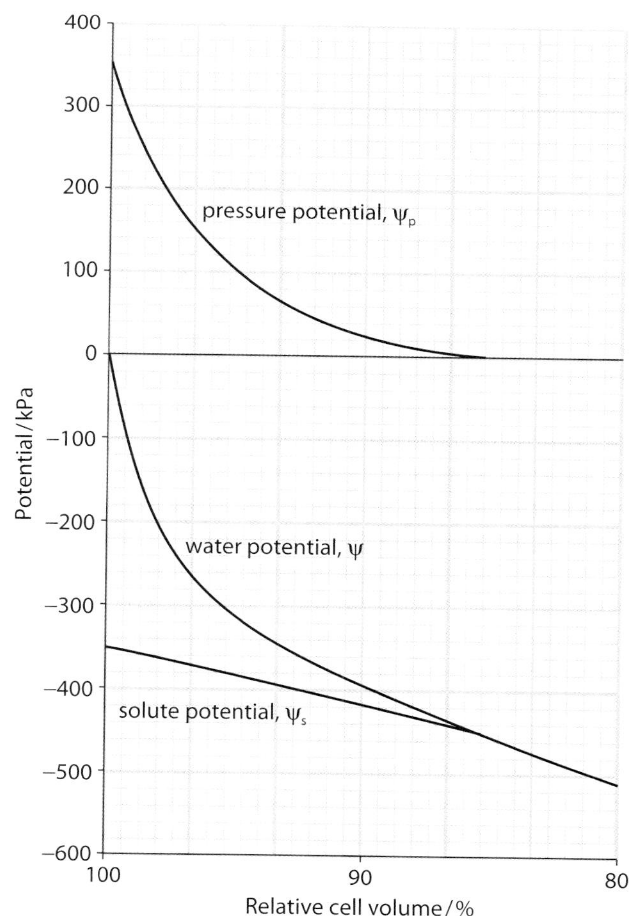

a What is a protoplast? [1]

b i What is the pressure potential at 90%, 95% and 100% relative cell volume? [1]

ii Calculate the change in pressure potential between 90% and 95% relative cell volume and between 95% and 100% relative cell volume. [2]

iii Explain why the pressure potential curve is not linear. [2]

iv State the water potential when the cell reaches maximum turgidity. [1]

The graph above shows that as the cell loses water, pressure potential falls and the relative cell volume decreases (the cell shrinks).

c i What is the minimum value of the pressure potential? [1]

ii In a shrinking cell, what is the relative cell volume when the minimum value of the pressure potential is reached? [1]

iii What is the term used to describe the state of the cell at this point? [1]

iv What happens to the values of water potential and solute potential at this point? [1]

v State the equation which links ψ_p, ψ_s and ψ. [1]

vi Describe what is happening to the cell between the point identified in **c ii** and **c iii** above and 80% relative cell volume. [5]

d As the cell changes volume, the change in solute potential is much less than the change in pressure potential. Suggest an explanation for this. [3]

[Total: 20]

12 The diagram shows the concentration in mmol dm^{-3} of two different ions inside a human red blood cell and in the plasma outside the cell.

ion	blood plasma	red blood cell
Na^+	144	15
K^+	5	150

a Explain why these concentrations could not have occurred as a result of diffusion. [1]

b Explain how these concentrations could have been achieved. [2]

c If respiration of red blood cells is inhibited, the concentrations of potassium ions and sodium ions inside the cells gradually change until they come into equilibrium with the plasma. Explain this observation. [4]

[Total: 7]

Chapter 5: The mitotic cell cycle

Learning outcomes

You should be able to:

- describe the structure of chromosomes
- describe the cell cycle – the cycle of events by which body cells grow to a certain size and then divide into two
- explain how a nucleus divides into two genetically identical nuclei by mitosis
- prepare and observe a root tip squash in order to see stages of mitosis with a light microscope
- explain the significance of mitosis
- explain the significance of telomeres
- explain the significance of stem cells
- outline how uncontrolled cell division can lead to cancer

Why grow old?

Is it useful to prolong human life? The forerunners of modern chemists, the alchemists, thought so (Figure 5.1). They had two main aims: first, the ability to transform 'base' metals, such as lead, into the 'noble metals' (gold and silver) and second, to discover the elixir of life, which would confer eternal youth.

By the early 20th century, scientists had relegated these aims to impossible dreams. Now, however, we are once again challenging the idea that the process of ageing is inevitable.

Why **do** organisms grow old and die? Interest in the process of ageing was rekindled with the discovery of telomeres in 1978. These are protective sequences of nucleotides found at the ends of chromosomes, which become shorter every time a cell divides. A gradual degeneration of the organism occurs, resulting in ageing.

Some cells are able to replenish their telomeres using the enzyme telomerase. It is thought that cancer cells can do this and so remain immortal. It may therefore be possible to prevent the ageing of normal cells by keeping the enzyme telomerase active.

If the ageing process could be slowed or prevented, this would raise some important moral and ethical issues. Should the treatment be universally available? If not, who should benefit? What if you could live for 600 years? Should you be entitled to so many years of healthy life before the drug was withdrawn? If so, would this create a black market for the drug?

Figure 5.1 A 19th century oil painting showing an alchemist at work.

All living organisms grow and reproduce. Since living organisms are made of cells, this means that cells must be able to grow and reproduce. Cells reproduce by dividing and passing on copies of their genes to 'daughter' cells. The process must be very precisely controlled so that no vital genetic information is lost. In Chapter 6, we discuss how DNA can copy itself accurately. In this chapter, we consider how whole cells can do the same.

In Chapter 1, we saw that one of the most conspicuous structures in eukaryotic cells is the nucleus. Its importance has been obvious ever since it was realised that the nucleus always divides before a cell divides. Each of the two daughter cells therefore contains its own nucleus. This is important because the nucleus controls the cell's activities. It does this through the genetic material, DNA, which is able to act as a set of instructions, or code, for life (Chapter 6).

So, nuclear division combined with cell division allows cells, and therefore whole organisms, to reproduce themselves. It also allows multicellular organisms to grow. The cells in your body, for example, are all genetically identical (apart from the gametes – reproductive cells); they were all derived from one cell, the zygote, which was the cell formed when two gametes from your parents fused.

Chromosomes

Just before a eukaryotic cell divides, a number of thread-like structures called chromosomes gradually become visible in the nucleus. They are easily seen, because they stain intensely with particular stains. They were originally termed chromosomes because 'chromo' means coloured and 'somes' means bodies.

The number of chromosomes is characteristic of the species. For example, in human cells there are 46 chromosomes, and in fruit fly cells there are only eight. Figure 5.2 is a photograph of a set of chromosomes in the nucleus of a human cell.

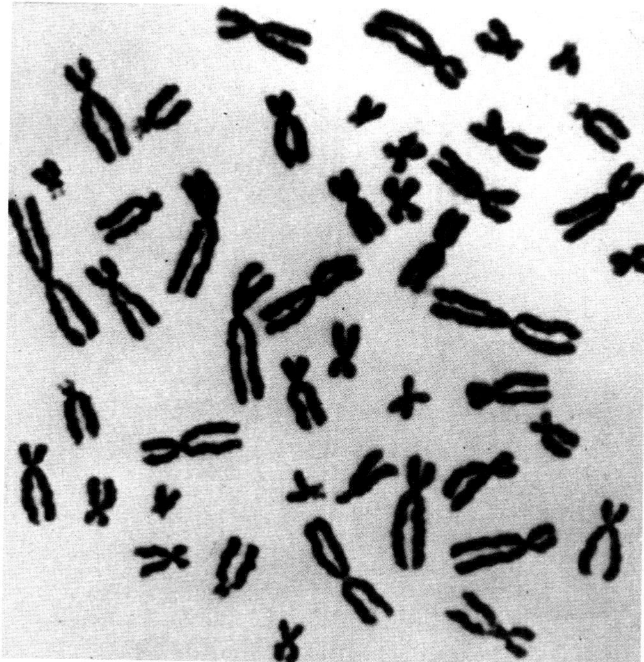

Figure 5.2 Photograph of a set of chromosomes in a human male, just before cell division. Each chromosome is composed of two chromatids held at the centromere. Note the different sizes of the chromosomes and the different positions of the centromeres.

The structure of chromosomes

Before studying nuclear division, you need to understand a little about the structure of chromosomes. Figure 5.3 is a simplified diagram of the structure of a chromosome just before cell division. You can see that the chromosome at this stage is a double structure. It is made of two identical structures called chromatids, joined together. This is because during the period between nuclear divisions, which is known as **interphase**, each DNA molecule in a nucleus makes an identical copy of itself (Chapter 6). Each chromatid contains one of these DNA copies, and the two chromatids are held together by a narrow region called the **centromere**, forming a chromosome. The centromere can be found anywhere along the length of the chromosome, but the position is characteristic for a particular chromosome.

Each chromatid contains one DNA molecule. DNA is the molecule of inheritance and is made up of a series of genes. Each gene is one unit of inheritance, coding for one polypeptide that is involved in a specific aspect of the functioning of the organism. The fact that the two DNA molecules in sister chromatids, and hence their genes, are identical is the key to precise nuclear division. When cells divide, one chromatid goes into one daughter cell and one goes into the other daughter cell, making the daughter cells genetically identical.

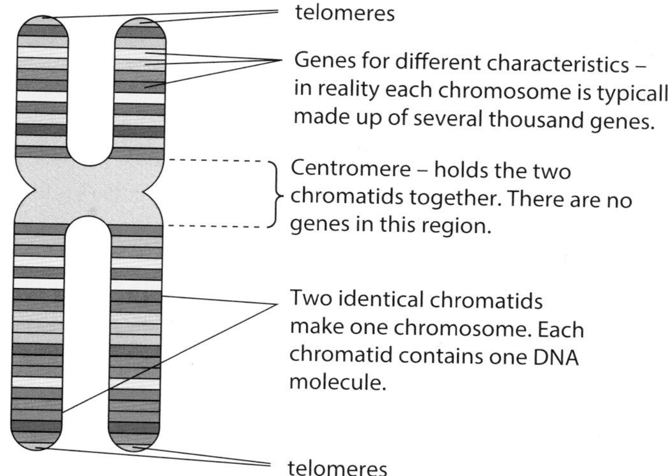

Figure 5.3 Simplified diagram of the structure of a chromosome.

So much information is stored in DNA that it needs to be a very long molecule. Although only 2 nm wide, the total length of DNA in the 46 chromosomes of an adult human cell is about 1.8 metres. This has to be packed into a nucleus which is only 6 μm in diameter. This is the equivalent of trying to get an 18 km length of string into a ball which is only 6 cm in diameter! In order to prevent the DNA getting tangled up into knots, a precise scaffolding made of protein molecules is used. The DNA is wound around the outside of these protein molecules. The combination of DNA and proteins is called chromatin. Chromosomes are made of chromatin. Chemically speaking, most of the proteins are basic (the opposite of acidic) and are of a type known as histones. Because they are basic, they can interact easily with DNA, which is acidic.

The precise details of chromatin structure are complex and you do not need to remember them, but they provide you with useful background knowledge. The solution to the packing problem is controlled coiling of the DNA. Coils can themselves be coiled to form 'supercoils'; these may then be looped, coiled or folded in precise ways which are still not fully understood. We do, however, understand the basic unit of structure. This is called a nucleosome (Figure 5.4). (Although you do not need to know about nucleosomes, this will help you to understand how DNA forms chromosomes.) The nucleosome is cylindrical in shape, about 11 nm wide by 6 nm long. It is made up of eight histone molecules. The DNA is wrapped around the outside of the cylinder, making 1⅔ turns (equivalent to 147 base pairs) before linking to the next nucleosome. The DNA between the nucleosomes (linker DNA, 53 base pairs in length) is also held in place by a histone molecule. Nucleosomes line up like a string of beads to form a fibre 10 nm wide. This string can be further coiled and supercoiled, involving some non-histone proteins.

The extent of coiling varies during the cell cycle, the period between one cell division and the next. The chromosomes seen just before nuclear division represent the most tightly coiled (condensed) form of DNA. Between nuclear divisions, some uncoiling occurs. In fact, chromatin exists in two forms – euchromatin and heterochromatin. Euchromatin is loosely coiled, whereas heterochromatin is tightly coiled, as in the chromosomes seen at nuclear division. During the period between divisions (interphase) the majority is in the form of euchromatin. This is where the active genes are located. The genes in the heterochromatin are mostly inactive. Chromatin is easily stained – the more tightly coiled it is, the more densely it stains. It is therefore easy to see chromatin in the light microscope (nuclei stain easily) and in the electron microscope the two forms of chromatin are clearly visible. This is obvious, for example, in Figure 1.27.

Chromatin is at its most condensed in chromosomes at the metaphase stage of mitosis (the stages of mitosis are dealt with later in this chapter). The fact that all the information of the cell is tightly packed makes it easier to separate the information into two new cells. This is one of the main functions of chromosomes. Chromosomes also possess two features essential for successful nuclear division, namely centromeres and telomeres. Centromeres are visible in Figures 5.2 and 5.3. Telomeres are visible if chromosomes are stained appropriately (Figure 5.5). Centromeres are discussed with mitosis on page 100 and the significance of telomeres is discussed on page 104.

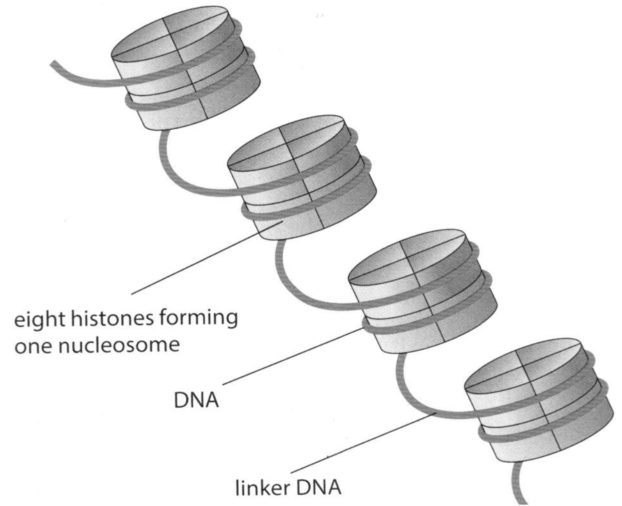

Figure 5.4 How nucleosomes are involved in DNA coiling. Note that you do not need to learn about nucleosomes.

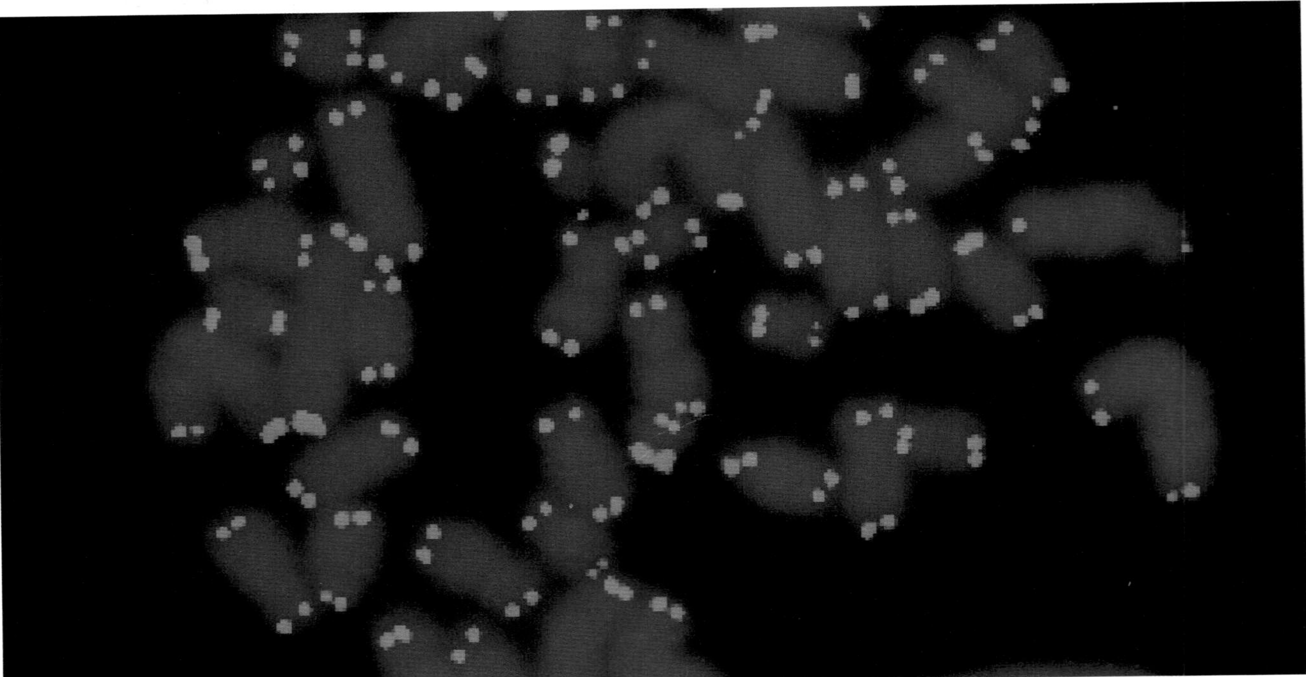

Figure 5.5 Fluorescent staining of human chromosome telomeres as seen with a light microscope. Chromosomes appear blue and telomeres appear pink. (×4000)

Mitosis

Mitosis is nuclear division that produces two genetically identical daughter nuclei, each containing the same number of chromosomes as the parent nucleus. Mitosis is part of a precisely controlled process called the cell cycle.

The cell cycle

The cell cycle is the regular sequence of events that takes place between one cell division and the next. It has three phases, namely **interphase**, **nuclear division** and **cell division**. These are shown in Figure 5.6.

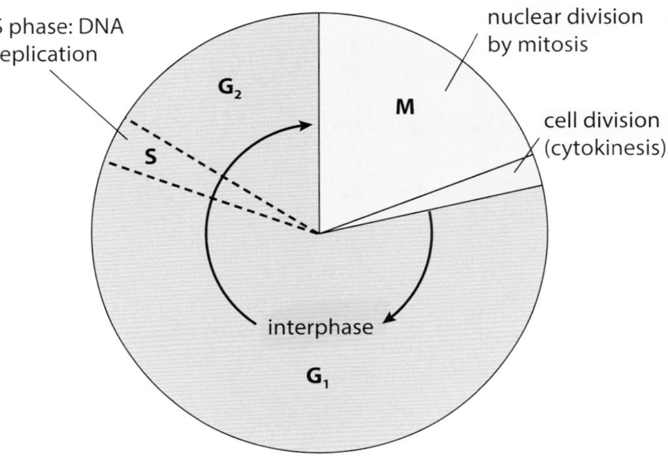

Figure 5.6 The mitotic cell cycle. DNA replication takes place during interphase, the period between cell division and the next nuclear division: **S** = synthesis (of DNA); **G** = gap; **M** = mitosis.

During interphase, the cell grows to its normal size after cell division and carries out its normal functions, synthesising many substances, especially proteins, in the process. At some point during interphase, a signal may be received that the cell should divide again. If this happens, the DNA in the nucleus replicates so that each chromosome consists of two identical chromatids. This phase of the cell cycle is called the **S phase** – S stands for synthesis (of DNA). This is a relatively short phase. The gap after cell division and before S phase is called the **G_1 phase** (G for gap). The gap after S phase and before cell division is called the **G_2 phase**. Interphase therefore consists of G_1, S and G_2. During G_1, cells make the RNA, enzymes and other proteins needed for growth. At the end of G_1, the cell becomes committed to dividing or not dividing.

During G_2 the cell continues to grow and new DNA is checked and any errors are usually repaired. Preparations are also made to begin the process of division. For example, there is a sharp increase in production of the protein tubulin which is needed to make microtubules for the mitotic spindle.

Nuclear division follows interphase. This may be referred to as the **M phase** (M for mitosis). Growth stops temporarily during mitosis. After the M phase, when the nucleus has divided into two, the whole cell divides to create two genetically identical cells.

The length of the cell cycle is very variable, depending on environmental conditions and cell type. On average, root tip cells of onions divide once every 20 hours; epithelial cells in the human intestine every 10 hours.

In animal cells, cell division involves constriction of the cytoplasm between the two new nuclei, a process called **cytokinesis**. In plant cells, it involves the formation of a new cell wall between the two new nuclei.

Mitosis

The process of mitosis is best described by annotated diagrams as shown in Figure 5.7. Although in reality the process is continuous, it is usual to divide it into four main stages for convenience, like four snapshots from a film. The four stages are called **prophase**, **metaphase**, **anaphase** and **telophase**.

Most nuclei contain many chromosomes, but the diagrams in Figure 5.7 show a cell containing only four chromosomes for convenience. Colours are used to show whether the chromosomes are from the female or male parent. An animal cell is used as an example.

The behaviour of chromosomes in plant cells is identical to that in animal cells. However, plant cells do not contain centrosomes and, after nuclear division, a new cell wall must form between the daughter nuclei. It is **chromosome** behaviour, though, that is of particular interest. Figure 5.7 summarises the process of mitosis diagrammatically. Figure 5.8 (animal) and Figure 5.9 (plant) show photographs of the process as seen with a light microscope.

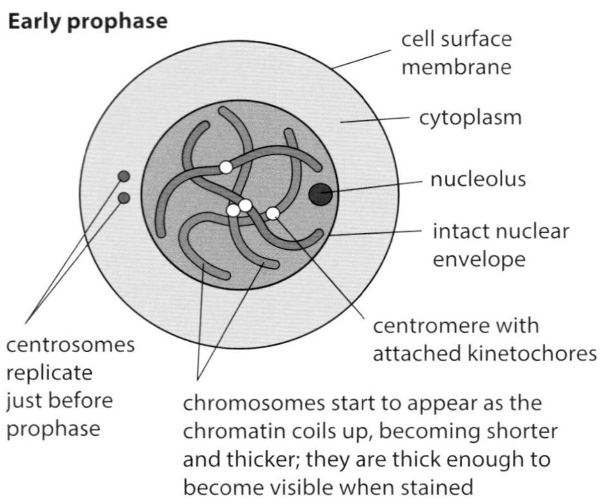

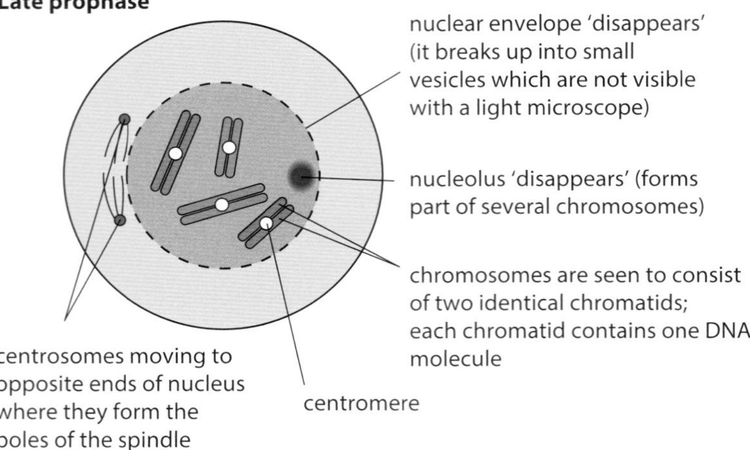

At the end of prophase a spindle is formed (see below).

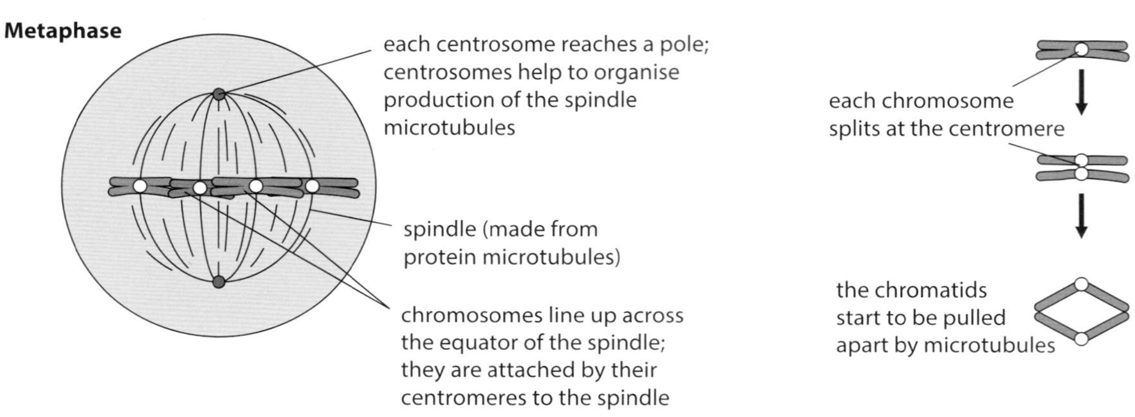

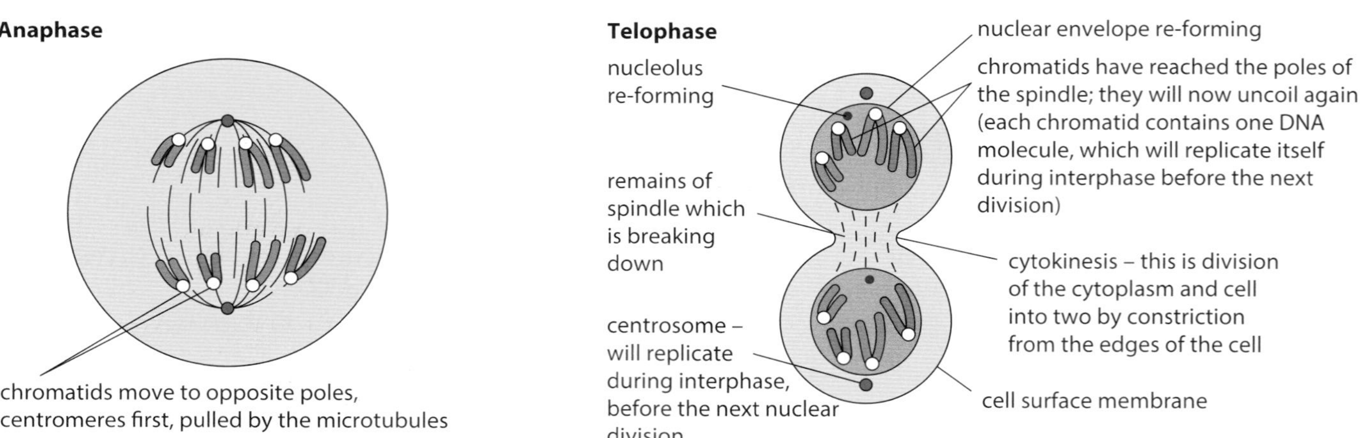

Figure 5.7 Mitosis and cytokinesis in an animal cell.

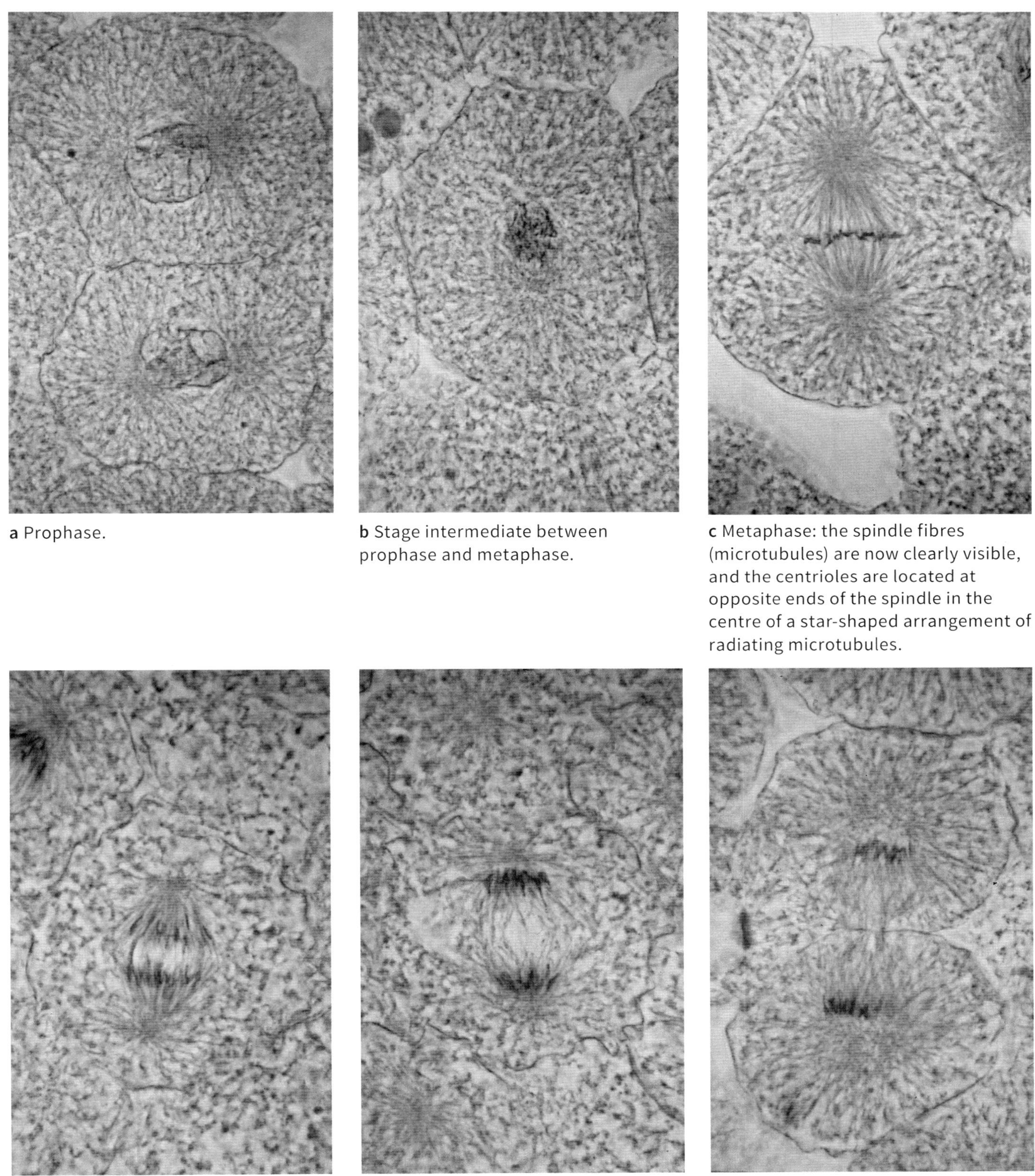

a Prophase.

b Stage intermediate between prophase and metaphase.

c Metaphase: the spindle fibres (microtubules) are now clearly visible, and the centrioles are located at opposite ends of the spindle in the centre of a star-shaped arrangement of radiating microtubules.

d Early anaphase.

e Anaphase.

f Telophase and cell division (cytokinesis).

Figure 5.8 Stages of mitosis and cell division in an animal cell (whitefish) (×900). Chromosomes are stained darkly.

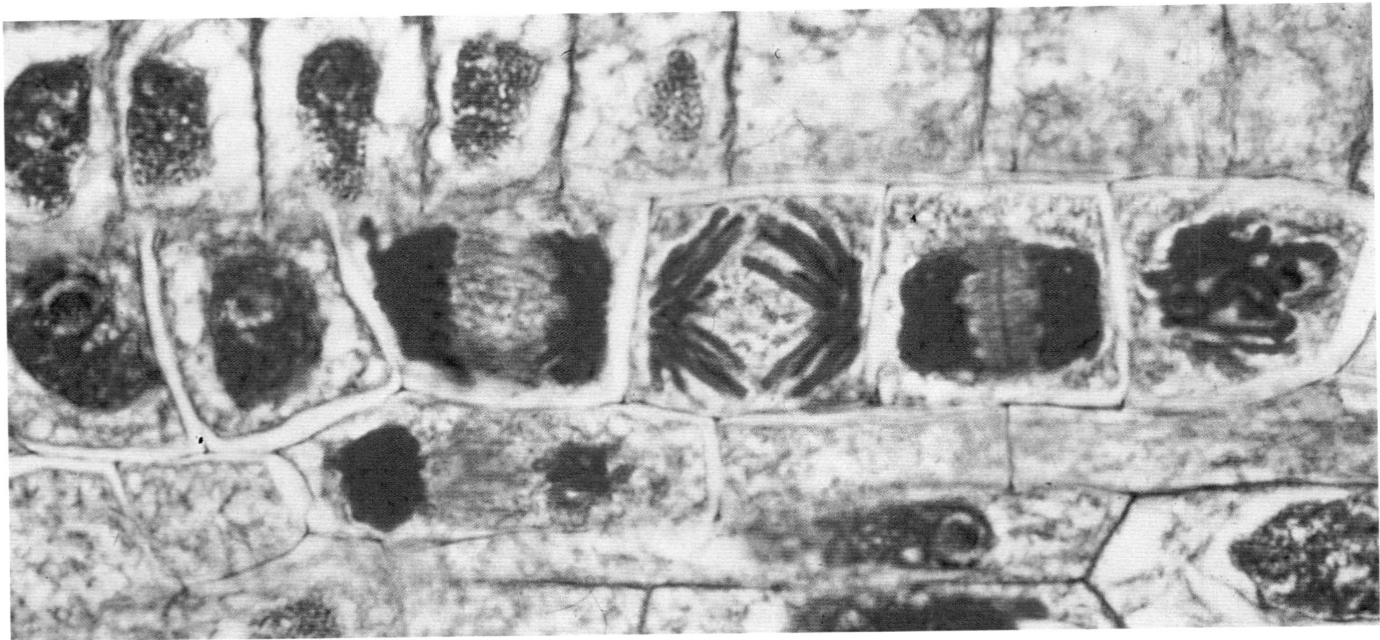

Figure 5.9 Longitudinal section of onion root tip showing stages of mitosis and cell division typical of plant cells (×400). Try to identify the stages based on information given in Figure 5.7.

Centromeres, centrosomes and centrioles

The centromere is needed for the separation of chromatids during mitosis. It is visible as a constriction (Figures 5.2 and 5.3) and is the site of attachment of spindle microtubules. Each metaphase chromosome has two **kinetochores** at its centromere, one on each chromatid (Figure 5.10). These are made of protein molecules which bind specifically to the DNA in the centromere and also bind to microtubules. Bundles of microtubules called spindle fibres extend from the kinetochores to the poles of the spindle during mitosis. Construction of kinetochores begins before nuclear division starts (during the S phase of the cell cycle) and they are lost again afterwards.

The microtubules attached to a given kinetochore pull the kinetochore, with the rest of its chromatid dragging behind, towards the pole. This is achieved by shortening of the microtubules, both from the pole end and from the kinetochore end.

The poles of the spindle are where the centrosomes are located, one at each pole. As noted in Chapter 1, the centrosome is an organelle found in animal cells that acts as the microtubule organising centre (MTOC) for construction of the spindle. Each centrosome consists of a pair of centrioles surrounded by a large number of proteins. It is these proteins that control production of the microtubules, not the centrioles. Plant mitosis occurs without centrosomes.

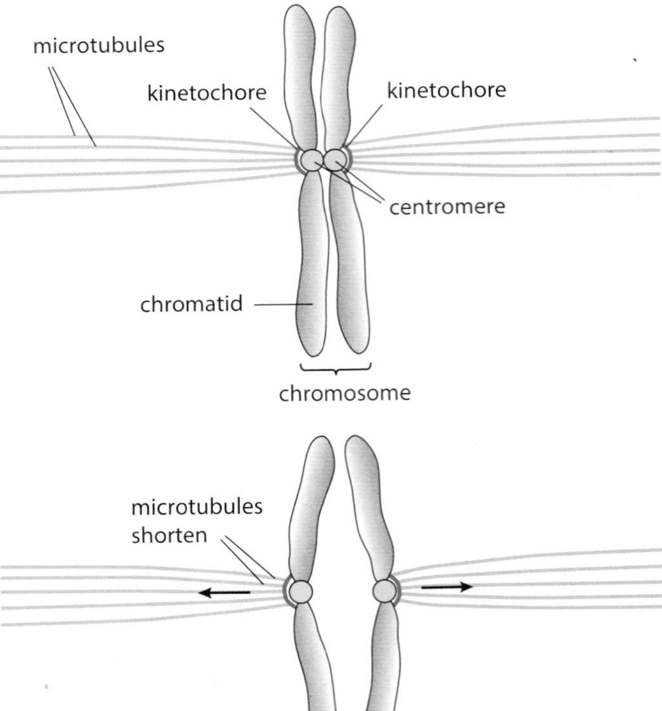

Figure 5.10 Role of the centromere, kinetochores and microtubules during mitosis.

QUESTION

5.1 How can the microtubules be shortened? (See Chapter 1.)

About 60–70% of the spindle microtubules that extend from the centrosome are not attached to kinetochores, but have free ends. Some of these are microtubules that have not found kinetochores to attach to, since this is a random process. Some may be involved in chromatid movement in ways other than that described.

Biological significance of mitosis

- **Growth**. The two daughter cells formed have the same number of chromosomes as the parent cell and are genetically identical (clones). This allows growth of multicellular organisms from unicellular zygotes. Growth may occur over the entire body, as in animals, or be confined to certain regions, as in the meristems (growing points) of plants.
- **Replacement of cells and repair of tissues**. This is possible using mitosis followed by cell division. Cells are constantly dying and being replaced by identical cells. In the human body, for example, cell replacement is particularly rapid in the skin and in the lining of the gut. Some animals are able to regenerate whole parts of the body; for example, starfish can regenerate new arms.
- **Asexual reproduction**. Mitosis is the basis of asexual reproduction, the production of new individuals of a species by a single parent organism. The offspring are genetically identical to the parents. Asexual reproduction can take many forms. For a unicellular organism such as *Amoeba*, cell division inevitably results in reproduction. For multicellular organisms, new individuals may be produced which bud off from the parent in various ways (Figure 5.11). Budding is particularly common in plants; it is most commonly a form of vegetative propagation in which a bud on part of the stem simply grows a new plant. The new plant eventually becomes detached from the parent and lives independently. The bud may be part of the stem of an overwintering structure such as a bulb or tuber. The ability to generate whole organisms from single cells or small groups of cells is important in biotechnology and genetic engineering, and is the basis of cloning.
- **Immune response**. The cloning of B- and T-lymphocytes during the immune response is dependent on mitosis (Chapter 11).

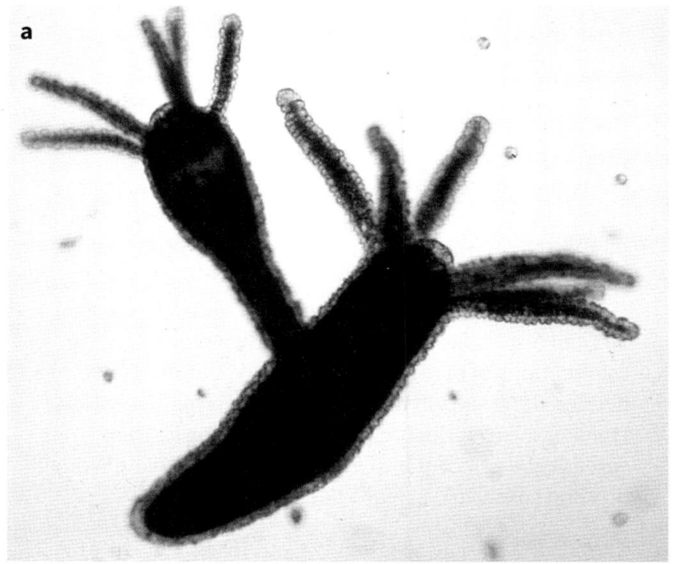

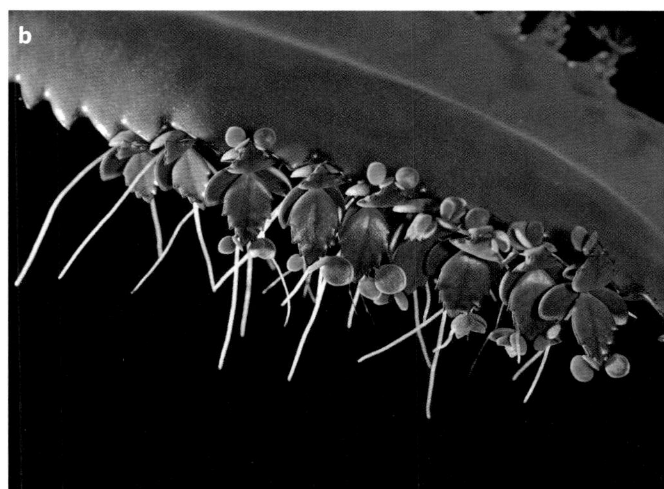

Figure 5.11 **a** Asexual reproduction by budding in *Hydra* (×60). *Hydra* lives in fresh water, catching its prey with the aid of its tentacles. The bud growing from its side is genetically identical to the parent and will eventually break free and live independently. **b** Asexual reproduction in *Kalanchoe pinnata*. The plant produces genetically identical new individuals along the edges of its leaves.

QUESTION

5.2 In the mitotic cell cycle of a human cell:

- **a** how many chromatids are present as the cell enters mitosis?
- **b** how many DNA molecules are present?
- **c** how many kinetochores are present?
- **d** how many chromatids are present in the nucleus of each daughter cell after mitosis and cell division?
- **e** how many chromatids are present in the nucleus of a cell after replication of DNA?

BOX 5.1: Investigating mitosis using a root tip squash

Background

Growth in plants is confined to regions known as meristems. A convenient example to study is the root tip meristem. This lies just behind the protective root cap. In this meristem there is a zone of cell division containing small cells in the process of mitosis.

You may be able to study prepared slides of root tips. You can also make your own temporary slides. Cutting thin sections of plant material is tricky, but this is not needed if the squash technique is used. This involves staining the root tip, then gently squashing it. This spreads the cells out into a thin sheet in which individual dividing cells can be clearly seen.

Procedure

The root tips of garlic, onion, broad bean and sunflower provide suitable material. Bulbs or seeds can be grown suspended by a pin over water for a period of a week or two. The tips of the roots (about 1 cm) are removed and placed in a suitable stain such as warm, acidified acetic orcein. This stains chromosomes a deep purple. The stained root tip can be squashed into a sheet of cells on a glass slide, using a blunt instrument such as the end of the handle of a mounted needle.

You should be able to see, and draw, cells similar to those shown in Figure 5.9 (but note that Figure 5.9 shows a longitudinal **section** of a root tip, not a squash). It would also be useful to use Figure 5.9 to make some annotated drawings of the different stages of mitosis.

QUESTIONS

5.3 Draw a simple diagram of a cell which contains only one pair of chromosomes:
- **a** at metaphase of mitosis
- **b** at anaphase of mitosis.

5.4 State **two** functions of centromeres during nuclear division.

5.5 Thin sections of adult mouse liver were prepared and the cells stained to show up the chromosomes. In a sample of 75 000 cells examined, nine were found to be in the process of mitosis. Calculate the length of the cell cycle in days in liver cells, assuming that mitosis lasts one hour.

The significance of telomeres

The ends of chromosomes are 'sealed' by structures called telomeres (they have been compared with the plastic tips on the ends of shoe laces – see Figure 5.12). These are made of DNA with short base sequences that are repeated many times (multiple repeat sequences). In telomeres, one strand of the DNA is rich in the base guanine (G) and the other strand is rich in the complementary base cytosine (C) (Chapter 6). Their main function is to ensure that when DNA is replicated, the ends of the molecule are included in the replication and not left out. The copying enzyme cannot run to the end of the DNA and complete the replication – it stops a little short of the end. (It is not possible to understand the reason for this without a detailed knowledge of replication.) If part of the DNA is not copied, that piece of information is lost. At each subsequent division, another small section of information would be lost. Eventually, the loss of vital genes would result in cell death.

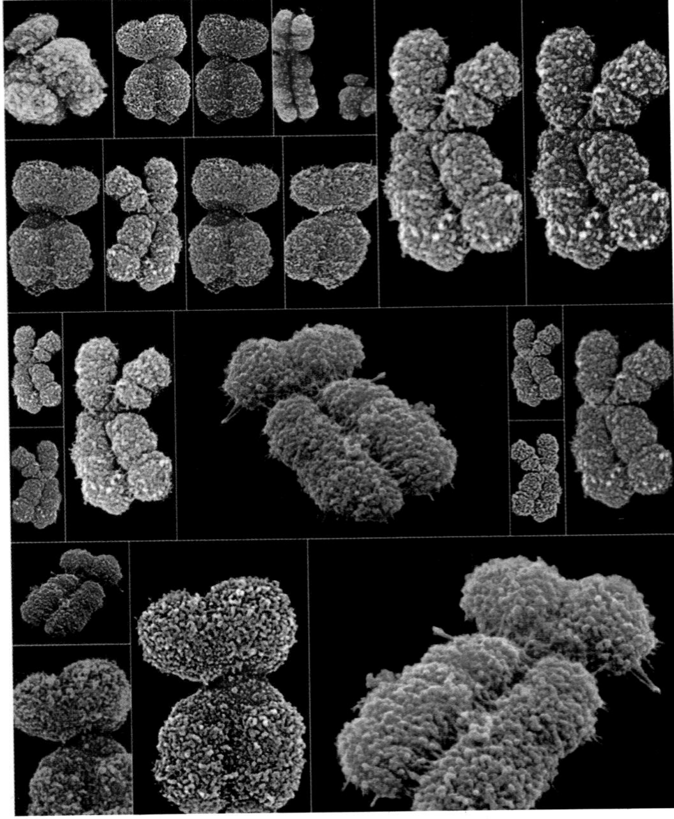

Figure 5.12 Coloured scanning electron micrographs of human chromosomes showing the location of telomeres at the ends of the chromosomes. Chromatids and centromeres are also clearly visible. Telomeres consist of short repeated sequences of DNA and proteins. As cells replicate and age the telomeres gradually get shorter. Stem cells are an exception.

The solution is to make the DNA a bit longer by adding some more bases. They have no useful information, but allow the copying enzyme to complete copying the meaningful DNA. As long as extra bases are added during each cell cycle, no vital information will be lost and the cell will be able to continue dividing successfully. The enzyme that performs this role is called telomerase. The extra DNA it adds is the telomere. The main function of telomeres is therefore to prevent the loss of genes during cell division and to allow continued replication of a cell.

It has been shown that some cells do not 'top up' their telomeres at each division. These tend to be fully differentiated (specialised) cells. With each division, their telomeres get a little shorter until the vital DNA is no longer protected and the cell dies. This could be one of the mechanisms of ageing, by which we grow old and die. This, of course, suggests that by somehow preventing the loss of telomeres we might be able to slow down or even prevent the process of ageing.

Stem cells

A stem cell is a cell that can divide an unlimited number of times (by mitosis). When it divides, each new cell has the potential to remain a stem cell or to develop (differentiate) into a specialised cell such as a blood cell or muscle cell.

The extent of the power of a stem cell to produce different cell types is variable and is referred to as its **potency**. Stem cells that can produce any type of cell are described as **totipotent**. The zygote formed by the fusion of a sperm with an egg at fertilisation is totipotent, as are the cells up to the 16-cell stage of development in humans. After that, some cells become specialised to form the placenta, while others lose this ability but can form all the cells that will lead to the development of the embryo and later the adult. These **embryonic stem cells** are described as **pluripotent**.

As tissues, organs and systems develop, cells become more and more specialised. There are more than 200 different cell types in an adult human body.

QUESTION

5.6 As a result of mitosis, all 200+ types of cell contain the same set of genes as the zygote. What does this suggest about the mechanism by which cells become different?

The more 'committed' cells become to particular roles, the more they lose the ability to divide until, in the adult, most cells do not divide. However, for growth and repair it is essential that small populations of stem cells remain which can produce new cells. Adult stem cells have already lost some of the potency associated with embryonic stem cells and are no longer pluripotent. They are only able to produce a few types of cell and may be described as **multipotent**. For example, the stem cells found in bone marrow are of this type. They can replicate any number of times, but can produce only blood cells, such as red blood cells, monocytes, neutrophils and lymphocytes. Mature blood cells have a relatively short life span, so the existence of these stem cells is essential. For example, around 250 thousand million (250 billion) red blood cells and 20 billion white blood cells are lost and must be replaced each day.

In the adult, stem cells are found throughout the body – for example in the bone marrow, skin, gut, heart and brain. Research into stem cells has opened up some exciting medical applications. **Stem cell therapy** is the introduction of new adult stem cells into damaged tissue to treat disease or injury. Bone marrow transplantation is the only form of this therapy that has progressed beyond the experimental stage into routine medical practice, but in the future it is hoped to be able to treat conditions like diabetes, muscle and nerve damage, and brain disorders such as Parkinson's and Huntington's diseases. Experiments with growing new tissues, or even organs, from isolated stem cells in the laboratory have also been conducted.

Cancer

Cancer is one of the most common diseases of developed countries, accounting for roughly one in four deaths, similar to the number of deaths from strokes and coronary heart disease combined. Lung cancer alone caused about 1 in 16 of all deaths in the UK in 2009 (one person every 15 minutes). Although prostate cancer is the most common form of cancer in men (24% of cases), lung cancer causes more deaths. Breast cancer is the most common form of cancer in women (31% of cases). There are, in fact, more than 200 different forms of cancer, and the medical profession does not think of it as a single disease.

Cancers show us the importance of controlling cell division precisely, because cancers are a result of uncontrolled mitosis. Cancerous cells divide repeatedly and form a tumour, which is an irregular mass of cells (Figure 5.13). The cells usually show abnormal changes in shape (Figure 5.14).

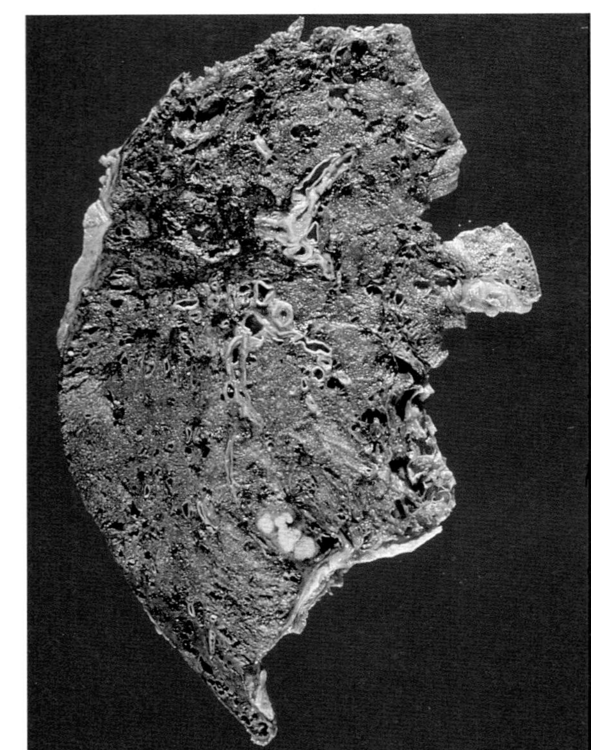

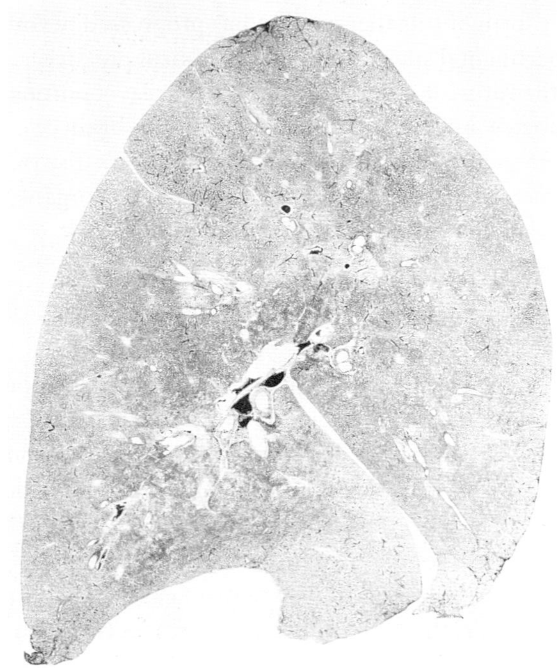

Figure 5.13 **a** Lung of a patient who died of lung cancer, showing rounded deposits of tumour (bottom, white area). Black tarry deposits throughout the lung show the patient was a heavy smoker. **b** Section of a healthy human lung. No black tar deposits are visible.

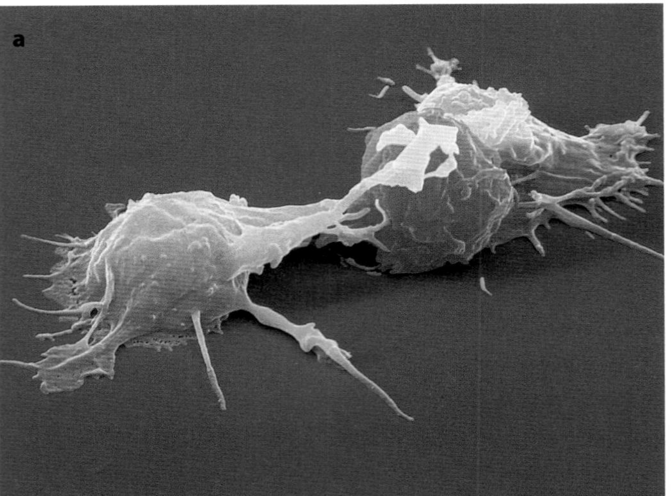

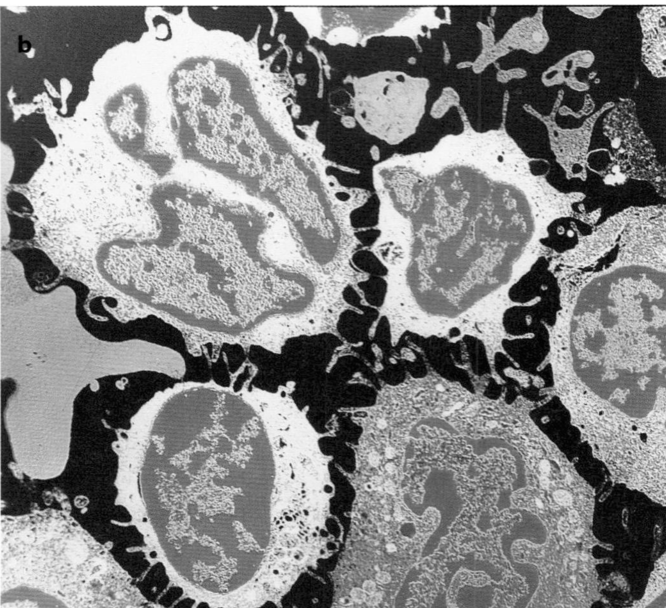

Figure 5.14 **a** False-colour scanning electron micrograph of a cancer cell (red) and white blood cells (orange and yellow). White blood cells gather at cancerous sites as an immune response. They are beginning to flow around the cancer cell, which they will kill using toxic chemicals (×4500). **b** False-colour transmission electron micrograph of abnormal white blood cells isolated from the blood of a person with hairy-cell leukaemia. The white blood cells are covered with characteristic hair-like cytoplasmic projections. Leukaemia is a disease in which the bone marrow and other blood-forming organs produce too many of certain types of white blood cells. These immature or abnormal cells suppress the normal production of white and red blood cells, and increase the patient's susceptibility to infection (×6400).

Cancers are thought to start when changes occur in the genes that control cell division. A change in any gene is called a **mutation**. The particular term for a mutated gene that causes cancer is an **oncogene**, after the Greek word 'onkos', meaning bulk or mass. Mutations are not unusual events, and most of the time they don't lead to cancer. **Most** mutated cells are either affected in some way that results in their early death or are destroyed by the body's immune system. Since most cells can be replaced, mutation usually has no harmful effect on the body. Cancerous cells, however, manage to escape both possible fates, so, although the mutation may originally occur in only one cell, it is passed on to all that cell's descendants. By the time it is detected, a typical tumour usually contains about a thousand million cancerous cells. Any agent that causes cancer is called a **carcinogen** and is described as **carcinogenic**.

Although you do not need to know about different types of tumours, you may be interested to know that not all tumours are cancerous. Some tumours do not spread from their site of origin, and are known as benign tumours; warts are a good example. It is only tumours that spread through the body, invade other tissues and destroy them that cause cancer, and these are known as **malignant tumours**. Malignant tumours interfere with the normal functioning of the area where they have started to grow. They may block the intestines, lungs or blood vessels. Cells can break off and spread through the blood and lymphatic system to other parts of the body to form **secondary growths**. The spread of cancers in this way is called **metastasis**. It is the most dangerous characteristic of cancer, since it can be very hard to find secondary cancers and remove them.

The steps involved in the development of cancer are shown in Figure 5.15.

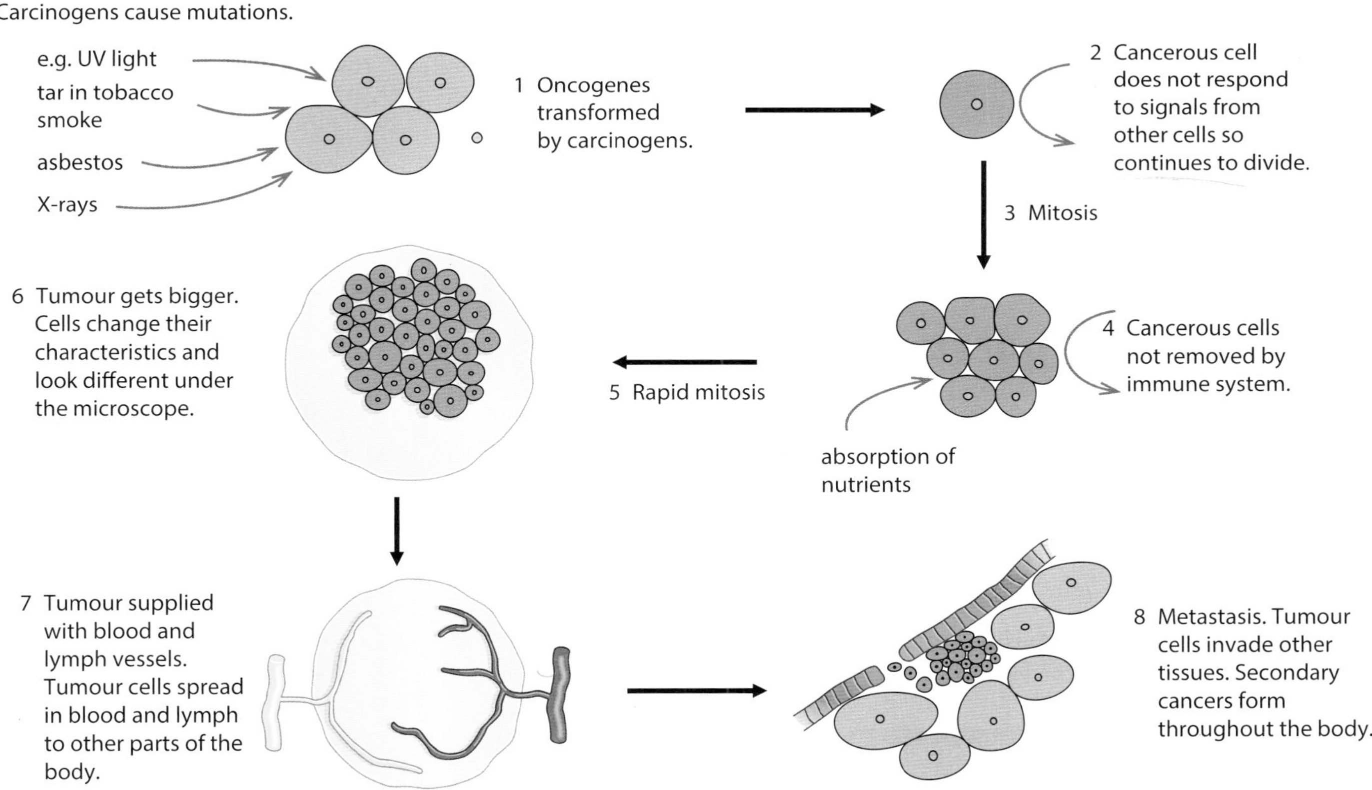

Figure 5.15 Stages in the development of cancer.

Summary

- Growth of a multicellular organism is a result of parent cells dividing to produce genetically identical daughter cells.
- During cell division, the nucleus divides first, followed by division of the whole cell. Division of a nucleus to produce two genetically identical nuclei is achieved by the process of mitosis. Mitosis is used in growth, repair, asexual reproduction and cloning of cells during an immune response.
- Although a continuous process, for convenience, mitosis can be divided into four phases: prophase, metaphase, anaphase, telophase. The period between successive nuclear and cell divisions is called interphase. During interphase chromosomes are visible only as loosely coiled material called chromatin. Chromatin consists of DNA held in position by basic proteins called histones in subunits known as nucleosomes.
- During prophase of mitosis, chromatin condenses (coils up more tightly) to form chromosomes which are easily visible in the light microscope when stained. Each chromosome consists of two identical chromatids, held together by a centromere. Each chromatid contains a single DNA molecule, formed when DNA replicates during interphase.
- The period from one cell division to the next is called the cell cycle. This has four phases: G_1 is the first growth phase after cell division; S phase is when the DNA replicates (S for synthesis); G_2 is a second growth phase; M phase (M for mitosis) is when nuclear division takes place (followed by cell division). G stands for 'gap' (between the S and M phases).
- The ends of chromosomes are capped with special regions of DNA known as telomeres. Telomeres are needed to prevent the loss of genes from the ends of chromosomes during replication of DNA.
- Many specialised cells lose the ability to divide, but certain cells known as stem cells retain this ability. Stem cells are essential for growth from zygote to adult and for cell replacement and tissue repair in the adult.
- The behaviour of chromosomes during mitosis can be observed in stained preparations of root tips, either in section or in squashes of whole root tips.
- Cancers are tumours resulting from repeated and uncontrolled mitosis. They are thought to start as the result of mutation.

End-of-chapter questions

1 During prophase of mitosis, chromosomes consist of two chromatids. At which stage of the cell cycle is the second chromatid made?

A cytokinesis
B G_1
C G_2
D S [1]

2 Growth of cells and their division are balanced during the cell cycle. Which column shows the consequences that would follow from the two errors shown in the table?

Error	Consequence			
	A	B	C	D
speeding up the growth rate without speeding up the cell cycle	larger and larger cells	larger and larger cells	smaller and smaller cells	smaller and smaller cells
speeding up the cell cycle without speeding up the growth rate	larger and larger cells	smaller and smaller cells	larger and larger cells	smaller and smaller cells

[1]

3 A cell with four chromosomes undergoes a cell cycle including mitosis. Which diagram correctly shows the changes in chromatid number during interphase?

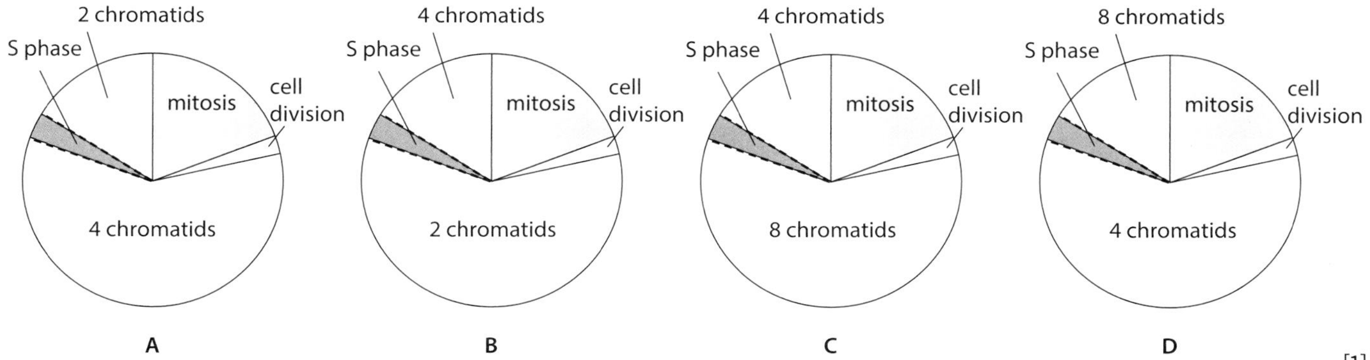

[1]

4 Distinguish between the following terms: **centrosome**, **centriole** and **centromere**. [6]

5 The diagram shows three cells (labelled **A**, **B** and **C**) from a root tip which have been stained to show chromosomes.

a Identify the stage of mitosis shown by each cell. [3]

b Describe what is happening at each stage. [3]

[Total: 6]

6 a Diagram **1** below shows a plant cell dividing by mitosis. Only two of the many chromosomes are shown for simplicity.

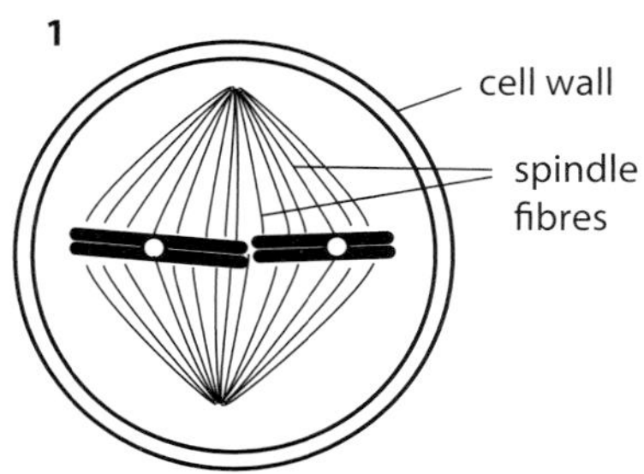

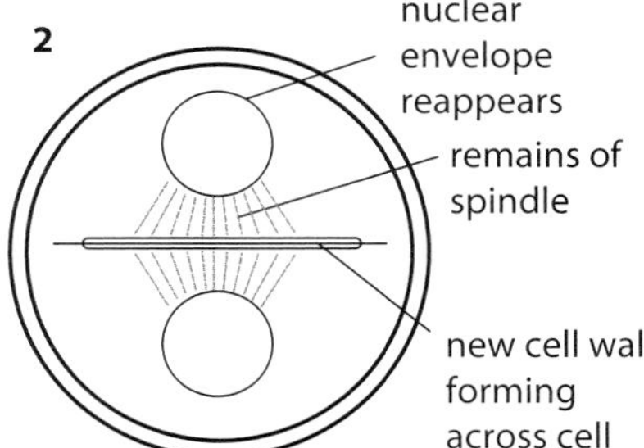

i What stage of mitosis is shown? [1]

ii Draw prophase for the same cell assuming the cell has only two chromosomes, as in diagram **1**. [1]

b Diagram **2** shows the same cell at telophase. The cell is beginning to divide and a new cell wall is forming, spreading out from the middle of the cell. Copy the diagram and add drawings of the chromosomes as they would appear at this stage. [1]

c Diagram **3** shows chromosomes in the nucleus of an animal cell.

Draw a diagram to show what the nucleus would look like in anaphase of mitosis. [3]

[Total: 6]

7 In Chapter 1 it was noted that microtubules are tiny tubes made of protein subunits which join together and the protein is called tubulin. Colchicine is a natural chemical which binds to tubulin molecules, thus preventing the formation of microtubules.

a Why should the binding of colchicine to tubulin molecules interfere with the formation of microtubules? [2]

b What structure or structures involved in mitosis are made of microtubules? [2]

c When cells treated with colchicine are observed, the dividing cells are all seen to be in the same stage of mitosis. Suggest with reasons the identity of this stage. [3]

[Total: 7]

8 If a 10 cm long piece of string was packed into a 5 cm long tube, the packing ratio would be 2 (2 cm of string per cm of tube). The packing ratio is a useful measure of the degree of compactness achieved.

The same idea can be applied to the problem of packing DNA into chromosomes.

a A chromosome 10 μm long was found to contain 8.7 cm of DNA. What is the packing ratio of DNA in this chromosome? Show your working. [3]

b There are 46 chromosomes in an adult human cell. Their average length is about 6 μm. The total length of DNA in the 46 chromosomes is 1.8 m. What is the overall packing ratio for DNA in human chromosomes? Show your working. [4]

c A nucleosome packages 50 nm of DNA into a diameter of 11 nm. What is the packing ratio of DNA in nucleosomes? [1]

d The linker DNA between nucleosomes is 53 base pairs long. Each base pair occupies a 0.34 nm length of the DNA double helix. Nucleosomes plus their linker DNA form a 10 nm wide fibre (the 'beads-on-a string' form of chromatin). What is the packing ratio of DNA in this form of chromatin? Show your working. [5]

e Nucleosomes and their linker DNA (the beads-on-a-string form of chromatin) are coiled into a helix to produce a 30 nm wide fibre called a solenoid. This has about six nucleosomes per turn of the helix. 40 000 nm of DNA are packed into a 1 μm length of this solenoid. What is the packing ratio of 30 nm fibres? [1]

f Each nucleosome is made of eight histone protein molecules. These molecules are highly conserved during evolution, meaning there are extremely few changes in their amino acid sequences over time (far fewer than is usual). What does this suggest about their functioning? [2]

g As stated in Question **8b**, the total length of DNA in the 46 chromosomes of an adult human cell is 1.8 m. The average adult human contains about 100 trillion cells (100 million million, or 10^{14}).

i What is the total length of DNA in a human? Express your answer in km. [1]

ii How many times would this stretch to the Sun and back?
(Average distance to Sun = 149 600 000 km = about 150 million km = 1.5×10^8 km.) [1]

[Total: 18]

9 Which of the following statements are true and which are false?

A Centrosomes are replicated before M phase of the cell cycle begins.

B Sister chromatids contain identical DNA.

C The microtubules attached to a given kinetochore extend to both poles of the spindle.

D Microtubule polymerisation and depolymerisation is a feature of the S phase of the cell cycle.

E Kinetochores are found in the centrosomes.

F Telomeres are the sites of attachment of microtubules during mitosis.

G Sister chromatids remain paired as they line up on the spindle at metaphase. [1 mark each]

[Total: 7]

Chapter 6: Nucleic acids and protein synthesis

Learning outcomes

You should be able to:

- describe the structures of nucleotides and nucleic acids
- describe the semi-conservative replication of DNA
- explain how the sequence of nucleotides in DNA codes for the sequence of amino acids in a polypeptide
- explain the roles of DNA and RNA in transcription and translation
- discuss the effect of gene mutation

Chemical factories of the future?

All of us – and all other living organisms – have cells that contain DNA. DNA is constructed from a chain of smaller molecules called nucleotides, and the sequence of the bases in these nucleotides acts as a genetic code, determining the proteins that are made in the cell and hence the organism's characteristics. The genetic code is universal. It is the same in all organisms.

But recently, the genetic code of a bacterium, ***Escherichia coli*** (Figure 6.1), has been deliberately modified. One of the three-letter 'words' of its genetic code, which originally told the bacterium's ribosomes to stop making a protein, has been changed to code for an amino acid that is not found in nature. The new code word now instructs the bacterium to insert the unnatural amino acid into a protein.

These modified bacteria can be made to take up different unnatural amino acids. This means that these bacteria can be used to produce new proteins with specific, unusual properties that could be of use to us. The possibilities are almost endless. For example, a completely new structural protein could be made that is able to bind to a metal, which could be used to build new structures. A new enzyme could be produced that is only active in the presence of another molecule, which could be used as a therapeutic drug to treat human diseases.

The fact that the novel amino acids are not found in nature means that these modified bacteria can only make the new proteins in laboratory conditions, where they are supplied with these amino acids. There is no chance of them surviving if they were to escape into the environment. Nevertheless, our ability to make such fundamental changes to an organism is thought-provoking. The chemistry of proteins will look very different in the future.

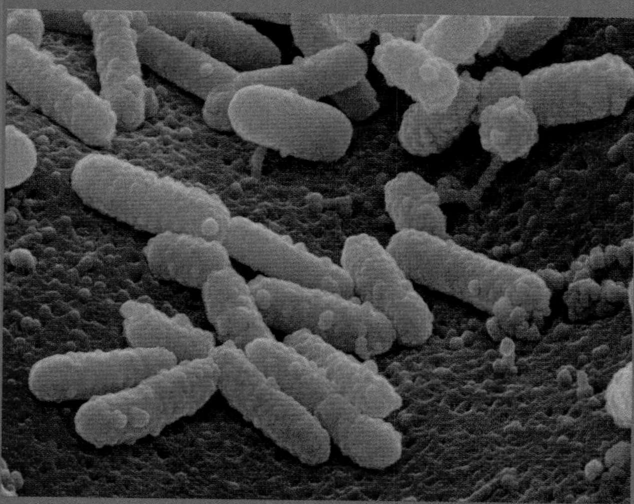

Figure 6.1 False colour scanning electron micrograph of *Escherichia coli* (×1000).

If you were asked to design a molecule which could act as the genetic material in living things, where would you start?

One of the features of the 'genetic molecule' would have to be the ability to **carry instructions** – a sort of blueprint – for the construction and behaviour of cells and the way in which they grow together to form a complete living organism. Another would be the **ability to be copied** perfectly, over and over again, so that whenever the nucleus of a cell divides it can pass on an exact copy of each 'genetic molecule' to the nuclei of each of its daughter cells.

Until the mid 1940s, biologists assumed that such a molecule must be a protein. Only proteins were thought to be complex enough to be able to carry the huge number of instructions which would be necessary to make such a complicated structure as a living organism. But during the 1940s and 1950s, a variety of evidence came to light that proved beyond doubt that the genetic molecule was not a protein at all, but DNA.

The structure of DNA and RNA

DNA stands for **deoxyribonucleic acid**, and RNA for **ribonucleic acid**. As we saw in Chapter 2, nucleic acids such as DNA and RNA, like proteins and polysaccharides, are **macromolecules** (page 29). They are also **polymers**, made up of many similar, smaller molecules joined into a long chain. The smaller molecules from which DNA and RNA molecules are made are **nucleotides**. DNA and RNA are therefore **polynucleotides**. They are often referred to simply as nucleic acids.

Nucleotides

Figure 6.2 shows the structure of nucleotides. Nucleotides are made up of three smaller components. These are:

- a nitrogen-containing base
- a pentose sugar
- a phosphate group.

There are just five different nitrogen-containing bases found in DNA and RNA. In a DNA molecule, there are four: **adenine**, **thymine**, **guanine** and **cytosine**. An RNA molecule also contains four bases, but never the base thymine. Instead, RNA molecules contain a base called **uracil**. These bases are often referred to by their first letters: **A**, **T**, **C**, **G** and **U**.

The pentose (5-carbon) sugar can be either **ribose** (in RNA) or **deoxyribose** (in DNA). As their names suggest, deoxyribose is almost the same as ribose, except that it has one fewer oxygen atoms in its molecule.

Figure 6.2 shows the five different nucleotides from which DNA and RNA molecules can be built up. Figure 6.3 shows the structure of their components in more detail; you do not need to remember these structures, but if you enjoy biochemistry you may find them interesting.

Do not confuse adenine with adenosine, which is part of the name of ATP (adenosine triphosphate) – adenosine is adenine with a sugar joined to it. And don't confuse thymine with thiamine, which is a vitamin.

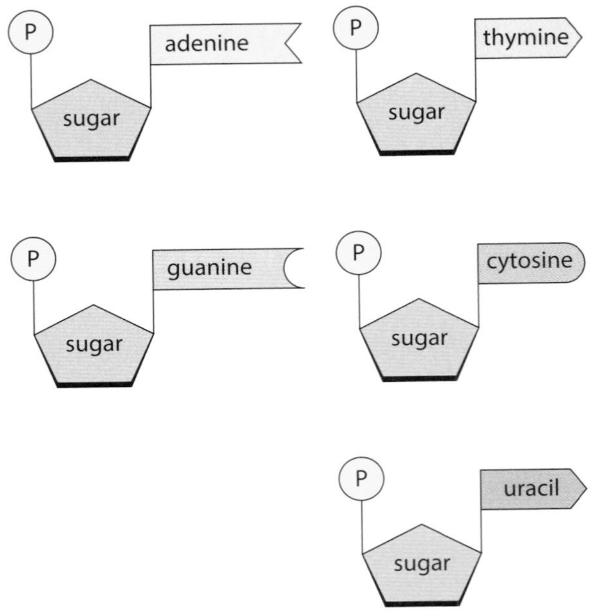

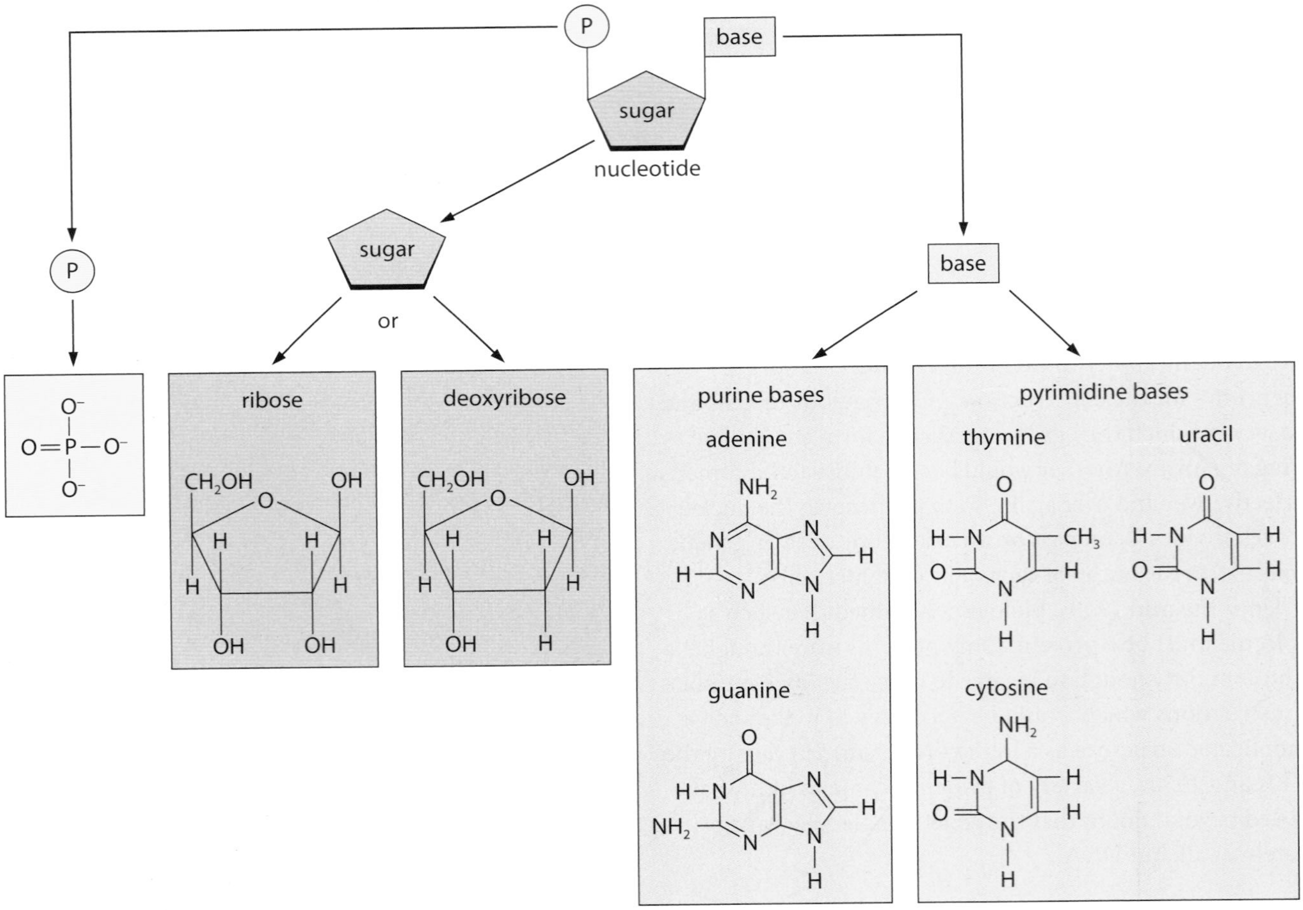

Figure 6.2 Nucleotides. A nucleotide is made of a nitrogen-containing base, a pentose sugar and a phosphate group Ⓟ.

Figure 6.3 The components of nucleotides. Note that you do not need to learn these structural formulae.

ATP

Although ATP is not part of DNA or RNA, we will look at its structure here because it is very similiar to a nucleotide.

Its structure is shown in Figure 6.4. Adenosine can be combined with one, two or three phosphate groups to give, in turn, adenosine monophosphate (AMP), adenosine diphosphate (ADP) or adenosine triphosphate (ATP).

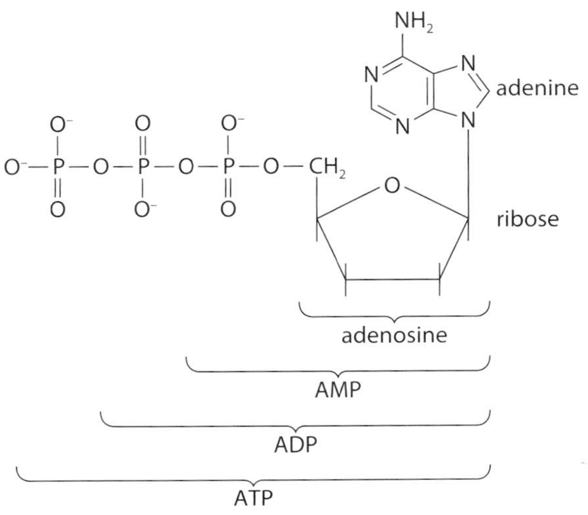

Figure 6.4 Structure of ATP.

Polynucleotides

To form the polynucleotides DNA and RNA, many nucleotides are linked together into a long chain. This takes place inside the nucleus, during interphase of the cell cycle (page 97).

Figure 6.5a shows the structure of part of a polynucleotide strand. In both DNA and RNA it is formed of alternating sugars and phosphates linked together, with the bases projecting sideways.

The covalent sugar–phosphate bonds (phosphodiester bonds) link the 5-carbon of one sugar molecule and the 3-carbon of the next. (See Chapter 2 for the numbering of carbon atoms in a sugar.) The polynucleotide strand is said to have 3′ and 5′ ends.

DNA molecules are made of **two** polynucleotide strands lying side by side, running in opposite directions. The strands are said to be antiparallel. The two strands are held together by **hydrogen bonds** between the bases (Figure 6.5b and c). The way the two strands line up is very precise.

The bases can be purines or pyrimidines. From Figure 6.3, you will see that the two purine bases, adenine and guanine, are larger molecules than the two pyrimidines, cytosine and thymine. In a DNA molecule, there is just enough room between the two sugar–phosphate backbones for one purine and one pyrimidine molecule, so a purine in one strand must always be opposite a pyrimidine in the other. In fact, the pairing of the bases is even more precise than this. Adenine always pairs with thymine, while cytosine always pairs with guanine: **A** with **T**, **C** with **G**. This complementary base pairing is a very important feature of polynucleotides, as you will see later.

DNA is often described as a double helix. This refers to the three-dimensional shape that DNA molecules form (Figure 6.5d). The hydrogen bonds linking the bases, and therefore holding the two strands together, can be broken relatively easily. This happens during DNA replication (DNA copying) and also during protein synthesis (protein manufacture). As we shall see, the breaking of the hydrogen bonds is a very important feature of the DNA molecule that enables it to perform its role in the cell.

RNA molecules, unlike DNA, remain as **single** strands of polynucleotide and can form very different three-dimensional structures. We will look at this later in the chapter when we consider protein synthesis.

DNA replication

We said at the beginning of this chapter that one of the features of a 'genetic molecule' would have to be the **ability to be copied** perfectly many times over.

It was not until 1953 that James Watson and Francis Crick (Figure 6.6) used the results of work by Rosalind Franklin (Figure 6.7) and others to work out the basic structure of the DNA molecule that we have just been looking at (Figure 6.8). To them, it was immediately obvious how this molecule could be copied perfectly, time and time again.

Watson and Crick suggested that the two strands of the DNA molecule could split apart. New nucleotides could then line up along each strand, opposite their appropriate partners, and join up to form complementary strands along each half of the original molecule. The new DNA molecules would be just like the old ones, because each base would only pair with its complementary one. Each pair of strands could then wind up again into a double helix, exactly like the original one.

This idea proved to be correct. The process is shown in Figure 6.9 on page 116. This method of copying is called semi-conservative replication, because **half** of the original molecule is **kept** (conserved) in each of the new molecules.

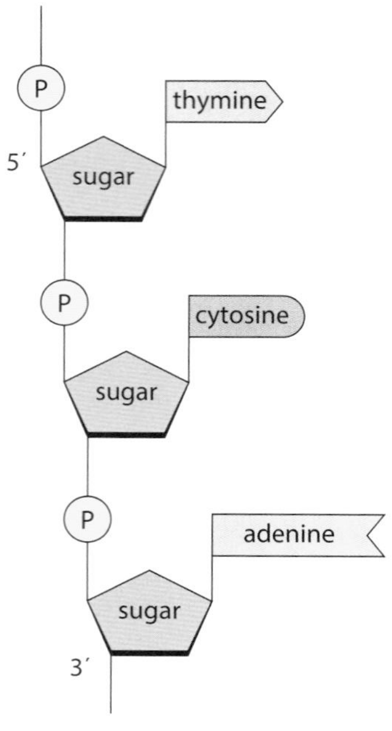

a Part of a polynucleotide. Each nucleotide is linked to the next by covalent bonds between the phosphates and sugars.

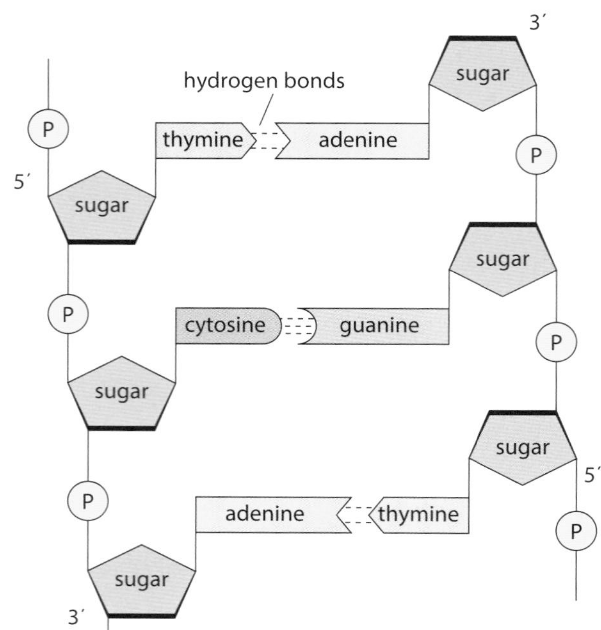

b Part of a DNA molecule. Two polynucleotides, running in opposite directions, are held together by hydrogen bonds between the bases. A links with T by two hydrogen bonds; C links with G by three hydrogen bonds. This is complementary base pairing.

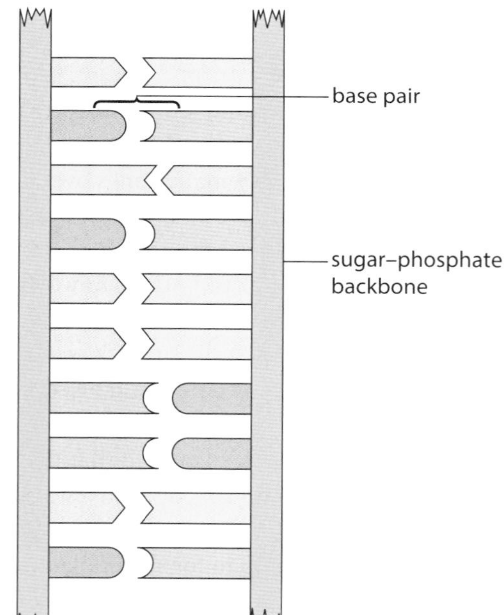

c A simplified diagram of a DNA molecule showing its backbone of alternating sugar–phosphate units, with the bases projecting into the centre creating base pairs.

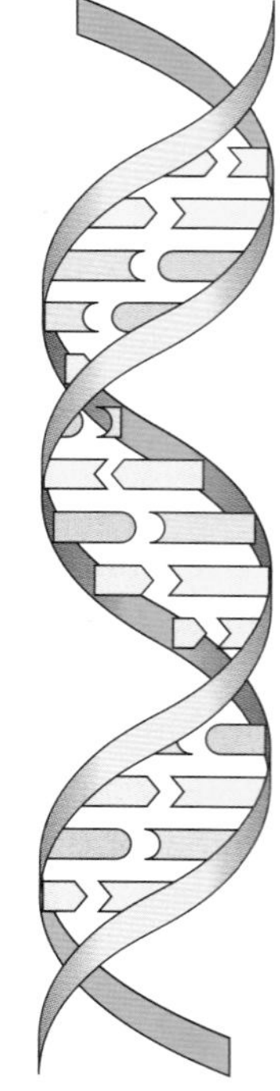

d The DNA double helix.

Figure 6.5 The structure of DNA.

The experimental evidence for the process of semi-conservative replication is described in Box 6.1 on page 117.

DNA replication takes place when a cell is not dividing. This is in interphase in eukaryotic cells (Chapter 5).

Figure 6.6 James Watson (left) and Francis Crick with their model of DNA.

Figure 6.7 Rosalind Franklin, whose X-ray diffraction images of DNA gave important clues to its structure.

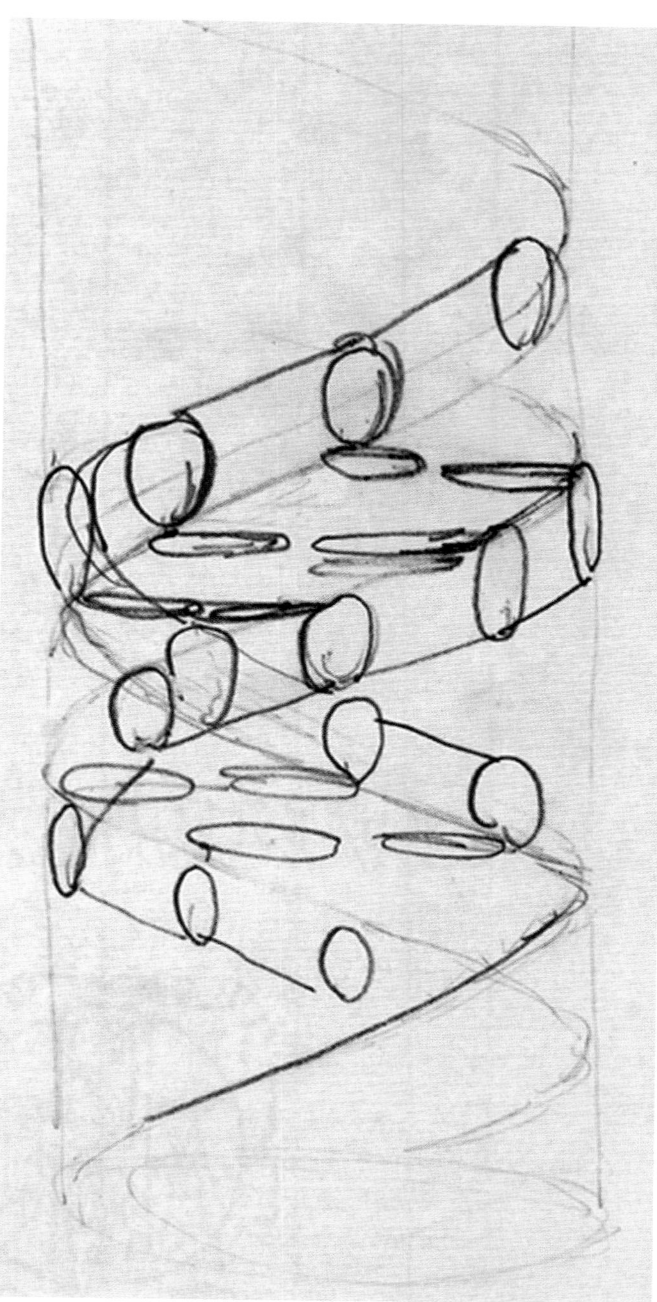

Figure 6.8 Crick's original sketch of the structure of DNA, made in 1953.

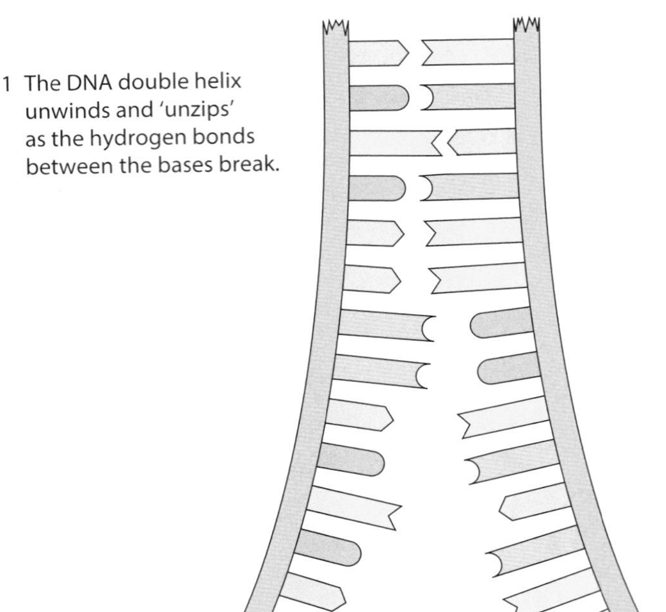

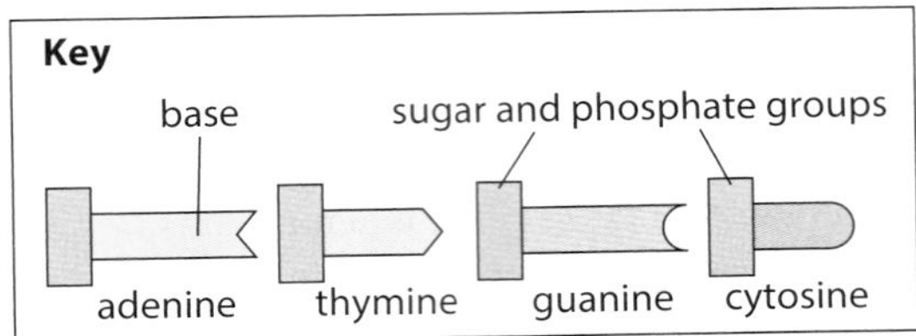

2 In the nucleus, there are nucleotides to which two extra phosphates have been added. The extra phosphates activate the nucleotides, enabling them to take part in the following reactions.

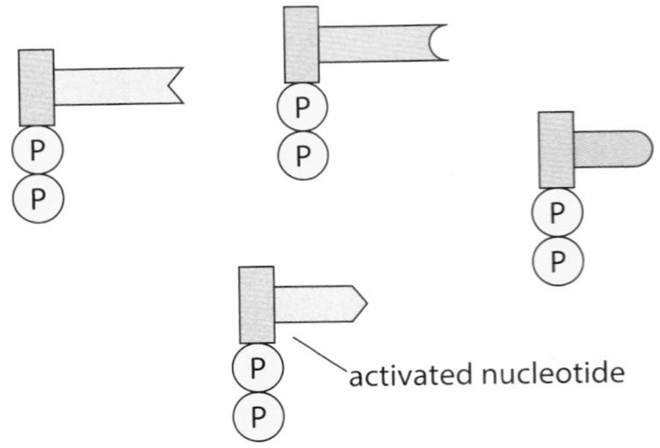

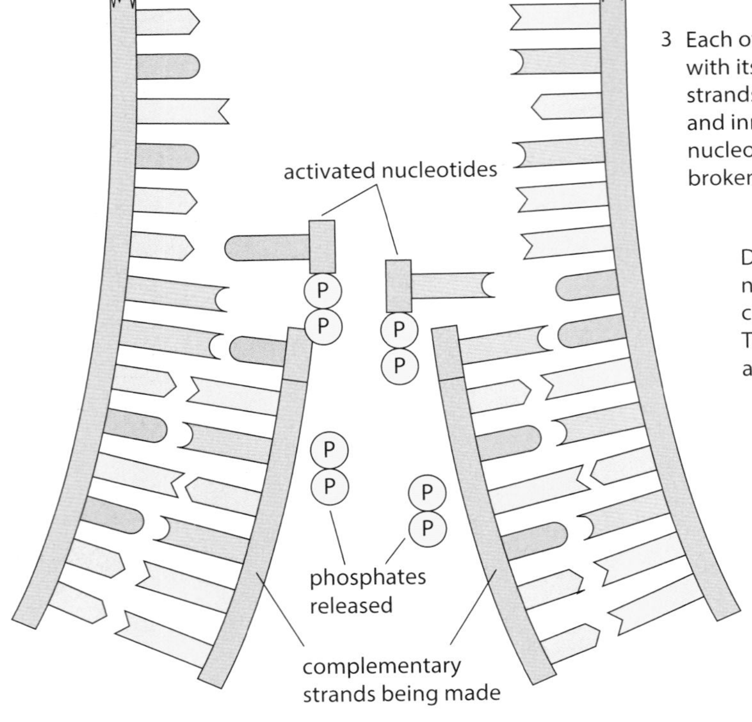

3 Each of the bases of the activated nucleotides pairs up with its complementary base on each of the old DNA strands. An enzyme, DNA polymerase, links the sugar and innermost phosphate groups of next-door nucleotides together. The two extra phosphates are broken off and released into the nucleus.

DNA polymerase will only link an incoming nucleotide to the growing new chain if it is complementary to the base on the old strand. Thus very few mistakes are made, perhaps around one in every 10^8 base pairs.

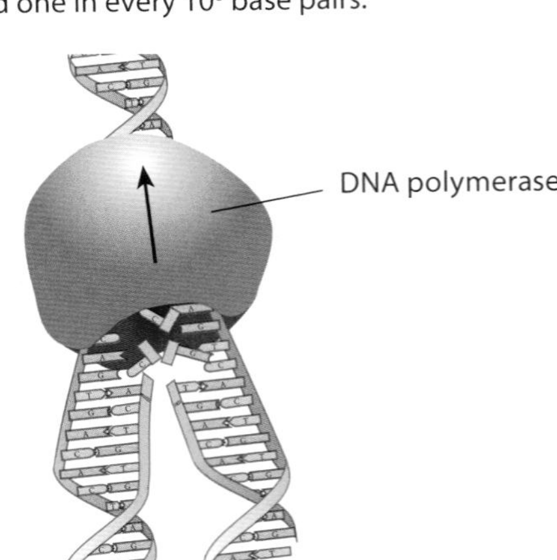

Figure 6.9 DNA replication.

QUESTION

6.1 a What other types of molecules, apart from nucleotides, are needed for DNA replication to take place? (Use Figure 6.9 to help you to answer this.) What does each of these molecules do?

b In what part of a eukaryotic cell does DNA replication take place?

BOX 6.1: Experimental evidence for the semi-conservative replication of DNA

In the 1950s, no-one knew exactly how DNA replicated. Three possibilities were suggested:

- conservative replication, in which one completely new double helix would be made from the old one (Figure 6.10a)
- semi-conservative replication, in which each new molecule would contain one old strand and one new one (Figure 6.10b)
- dispersive replication, in which each new molecule would be made of old bits and new bits scattered randomly through the molecules (Figure 6.10c).

Most people thought that semi-conservative replication was most likely, because they could see how it might work. However, to make sure, experiments were carried out in 1958 by Matthew Meselson and Franklin Stahl in America.

They used the bacterium *Escherichia coli* (*E. coli* for short), a common, usually harmless, bacterium which lives in the human alimentary canal. They grew populations of the bacterium in a food source that contained ammonium chloride as a source of nitrogen.

The experiment relied on the variation in structure of nitrogen atoms. All nitrogen atoms contain seven protons, but the number of neutrons can vary. Most nitrogen atoms have seven neutrons, so their relative atomic mass (the total mass of all the protons and neutrons in an atom) is 14. Some nitrogen atoms have eight neutrons, so their relative atomic mass is 15. These two forms are said to be **isotopes** of nitrogen.

The nitrogen atoms in the ammonium chloride that Meselson and Stahl supplied to the bacteria were the heavy isotope, nitrogen-15 (^{15}N). The bacteria used the ^{15}N to make their DNA. They were left in ammonium chloride long enough for them to divide many times, so that nearly all of their DNA contained only ^{15}N atoms, not nitrogen-14 (^{14}N). This DNA would be heavier than DNA containing ^{14}N. Some of these bacteria were then transferred to a food source in which the nitrogen atoms were all ^{14}N. Some were left there just long enough for their DNA to replicate once – about 50 minutes. Others were left long enough for their DNA to replicate two, three or more times.

DNA was then extracted from each group of bacteria. The samples were placed into a solution of caesium chloride and spun in a centrifuge. The heavier the DNA was, the closer to the bottom of the tube it came to rest.

Figure 6.11 (page 118) shows the results of Meselson and Stahl's experiment.

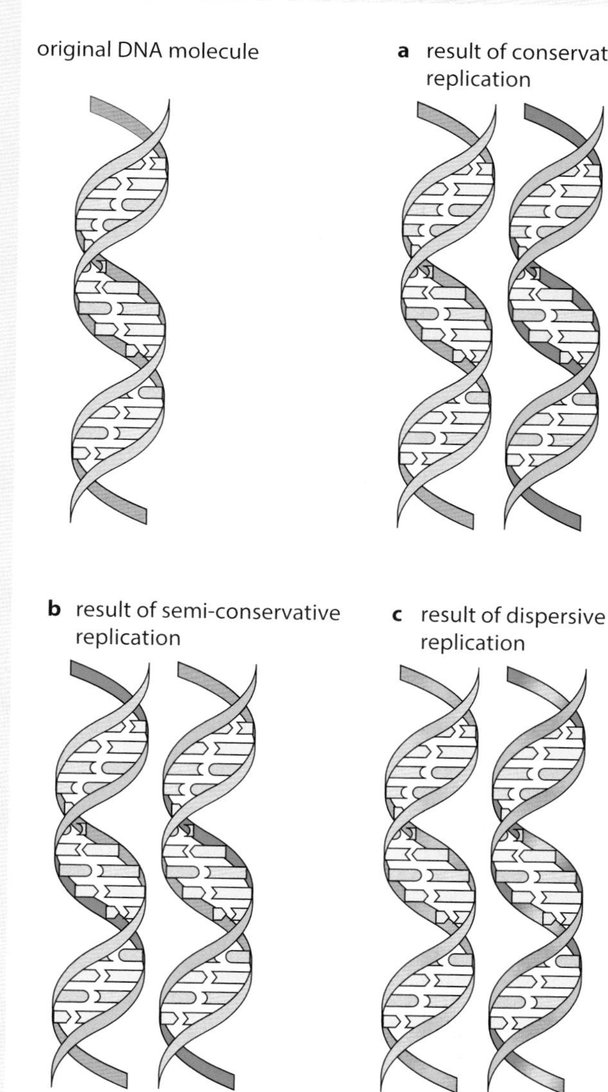

Figure 6.10 Three suggestions for the method of DNA replication.

continued ...

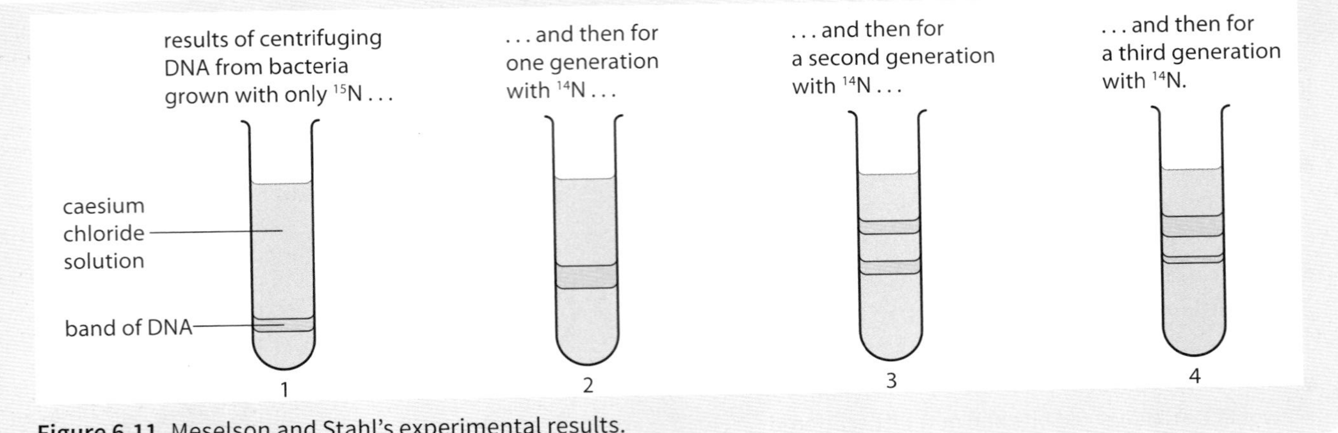

Figure 6.11 Meselson and Stahl's experimental results.

QUESTION

6.2 Look at Figure 6.11.

- **a** Assuming that the DNA has reproduced semi-conservatively, explain why the band of DNA in tube 2 is higher than that in tube 1.
- **b** What would you expect to see in tube 2 if the DNA had replicated conservatively?
- **c** What would you expect to see in tube 2 if the DNA had replicated dispersively?
- **d** Which is the first tube that provides evidence that the DNA has reproduced semi-conservatively?

Genes and mutations

DNA molecules can be enormous. The bacterium *E. coli* has just one DNA molecule, which is four million base pairs long. There is enough information to code for several thousand proteins. The total DNA of a human cell is estimated to be about 3×10^9 base pairs long. However, it is thought that only 3% of this DNA actually codes for protein. The function of much of the remainder is uncertain.

A part of a DNA molecule, where the nucleotide sequence codes for just one polypeptide, is called a **gene**, and one DNA molecule contains many genes. A change in the nucleotide sequence of a gene, which may then result in an altered polypeptide, is called a **mutation**. Most genes have several different variants called **alleles**, which originally arose by the process of mutation.

DNA, RNA and protein synthesis

DNA controls protein synthesis

How can a single type of molecule like DNA control all the activities of a cell? The answer is very logical. All chemical reactions in cells, and therefore all the cells' activities, are controlled by enzymes. Enzymes are proteins. DNA is a code for proteins, controlling which proteins are made. Thus, DNA controls the cell's activities.

Protein molecules are made up of strings of amino acids. The shape and behaviour of a protein molecule depends on the exact sequence of these amino acids – that is, its primary structure (page 40). DNA controls protein structure by determining the exact order in which the amino acids join together when proteins are made in a cell.

The triplet code

The sequence of nucleotide bases in a DNA molecule is a code for the sequence of amino acids in a polypeptide. Figure 6.12 shows a very short length of a DNA molecule, just enough to code for four amino acids.

The code is a three-letter, or triplet, code. Each sequence of three bases stands for one amino acid. The sequence is always read in the same direction and from only one of the two strands of the DNA molecule (the so-called sense strand). In this case, assume that this is the lower strand in the diagram. The complementary strand is referred to as the anti-sense strand.

Reading from the the left-hand end of the lower strand in Figure 6.12, the code is:

CAA which codes for the amino acid valine
TTT which codes for the amino acid lysine
GAA which codes for the amino acid leucine
CCC which codes for the amino acid glycine

So this short piece of DNA carries the instruction to the cell: 'Make a chain of amino acids in the sequence valine, lysine, leucine and glycine'. The complete set of triplet codes is shown in Appendix 2.

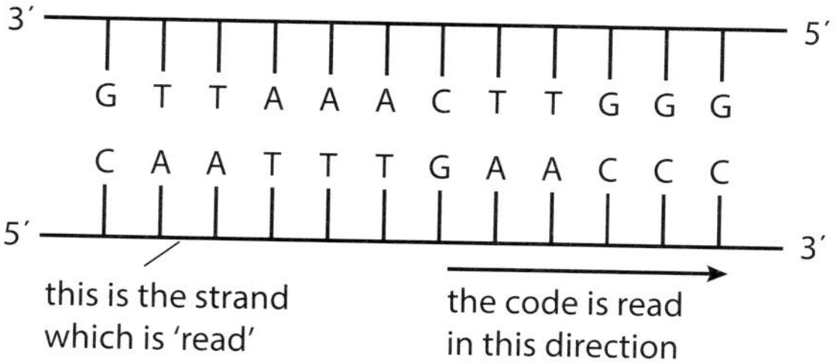

Figure 6.12 A length of DNA coding for four amino acids.

QUESTION

6.3 There are 20 different amino acids which cells use for making proteins.

- **a** How many different amino acids could be coded for by the triplet code? (Remember that there are four possible bases, and that the code is always read in just one direction on the DNA strand.)
- **b** Suggest how the 'spare' triplets might be used.
- **c** Explain why the code could not be a two-letter code.

An example of mutation: sickle cell anaemia

One mutation that has a significant effect is the one involved in the inherited blood disorder sickle cell anaemia.

Haemoglobin is the red pigment in red blood cells which carries oxygen around the body. A haemoglobin molecule is made up of four polypeptide chains, each with one iron-containing haem group in the centre. Two of these polypeptide chains are called α chains, and the other two β chains. (The structure of haemoglobin is described on pages 43 – 44.)

The gene which codes for the amino acid sequence in the β polypeptides is not the same in everyone. In most people, the β polypeptides begin with the amino acid sequence:

Val-His-Leu-Thr-Pro-Glu-Glu-Lys-

This is coded from the **Hb**^{A} (normal) allele of the gene.

But in some people, the base sequence CTT is replaced by CAT and the amino acid sequence becomes:

Val-His-Leu-Thr-Pro-Val-Glu-Lys-

This is coded from the **Hb**^{S} (sickle cell) allele of the gene.

This type of mutation is called a **substitution**. In this case, the small difference in the amino acid sequence results in the genetic disease sickle cell anaemia in individuals with two copies of the **Hb**^{S} allele. You can read more about sickle cell anaemia on pages 407–408.

Protein synthesis

The code on the DNA molecule is used to determine the sequence of amino acids in the polypeptide. Figure 6.13 on pages 120–121 describes the process in detail, but briefly the process is as follows.

The first stage is called transcription. In the nucleus, a complementary copy of the code from a gene is made by building a molecule of a different type of nucleic acid, called messenger RNA (mRNA), using one strand (the sense strand) as a template. Three RNA nucleotides are joined together by the enzyme RNA polymerase. This process copies the DNA code onto an mRNA molecule. Transcription of a gene begins when RNA polymerase binds to a control region of the DNA called a promoter and ends when the enzyme has reached a terminator sequence. At this point, the enzyme stops adding nucleotides to the growing mRNA. The hydrogen bonds holding the DNA and RNA together are broken and double-stranded DNA reforms. The last triplet transcribed onto mRNA is one of the DNA triplets coding for 'stop' (ATT, ATC or ACT – see Appendix 2).

The next stage of protein synthesis is called translation because this is when the DNA code is translated into an amino acid sequence. The mRNA leaves the nucleus and attaches to a **ribosome** in the cytoplasm (page 15).

In the cytoplasm, there are molecules of transfer RNA (**tRNA**). These have a triplet of bases at one end and a region where an amino acid can attach at the other. There are at least 20 different sorts of tRNA molecules, each with a particular triplet of bases at one end and able to attach to a specific amino acid at the other (Figure 6.14 on page 122).

The tRNA molecules pick up their specific amino acids from the cytoplasm and bring them to the mRNA on the ribosome. The triplet of bases (an anticodon) of each tRNA links up with a complementary triplet (a codon) on the mRNA molecule. Two tRNA molecules fit onto the ribosome at any one time. This brings two amino acids side by side, and a peptide bond is formed between them (page 39). Usually, several ribosomes work on the same mRNA strand at the same time. They are visible, using an electron microscope, as polyribosomes (Figure 6.15 on page 122).

So the base sequence on the DNA molecule determines the base sequence on the mRNA, which determines which tRNA molecules can link up with them. Since each type of tRNA molecule is specific for just one amino acid, this determines the sequence in which the amino acids are linked together as the polypeptide molecule is made.

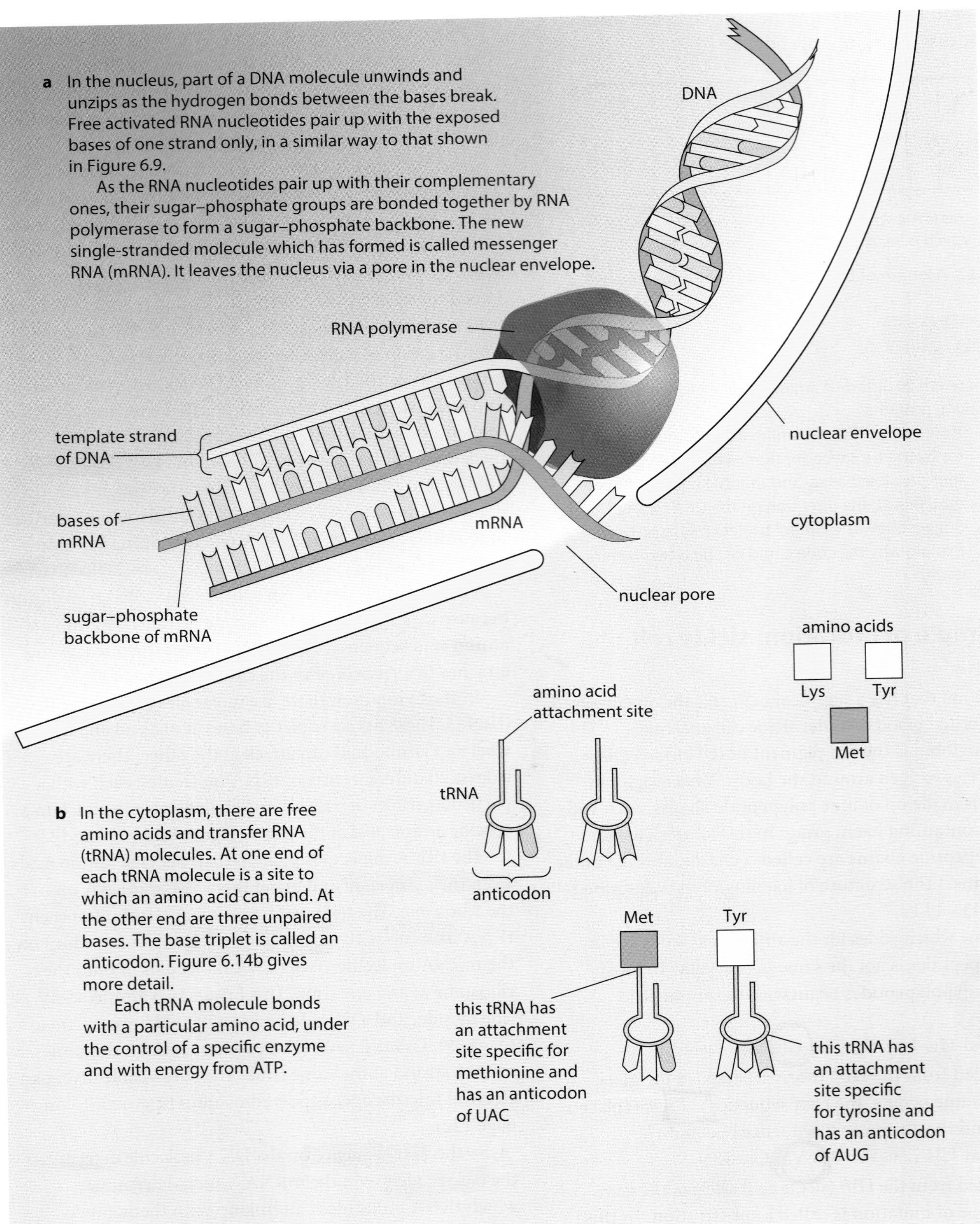

Figure 6.13 Protein synthesis – transcription.

continued ...

c Meanwhile, also in the cytoplasm, the mRNA molecule attaches to a ribosome. Ribosomes are made of ribosomal RNA (rRNA) and protein and contain a small and a large subunit. The mRNA binds to the small subunit. Six bases at a time are exposed to the large subunit.

The first three exposed bases, or codon, are always AUG. A tRNA molecule with the complementary anticodon, UAC, forms hydrogen bonds with this codon. This tRNA molecule has the amino acid methionine attached to it.

d A second tRNA molecule bonds with the next three exposed bases. This one brings a different amino acid. The two amino acids are held closely together, and a peptide bond is formed between them. This reaction is catalysed by the enzyme peptidyl transferase, which is found in the small subunit of the ribosome.

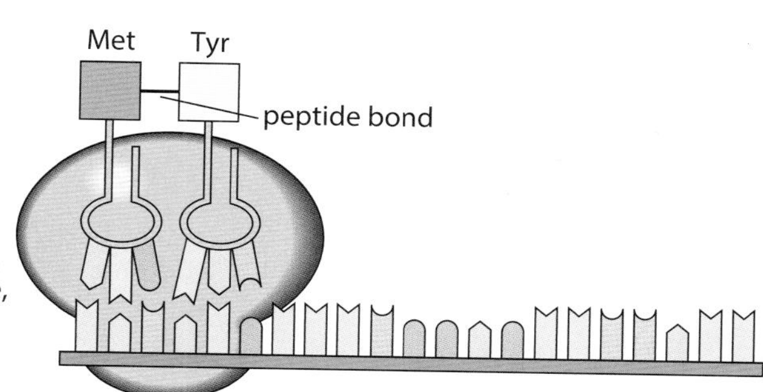

e The ribosome now moves along the mRNA, 'reading' the next three bases on the ribosome. A third tRNA molecule brings a third amino acid, which joins to the second one. The first tRNA leaves.

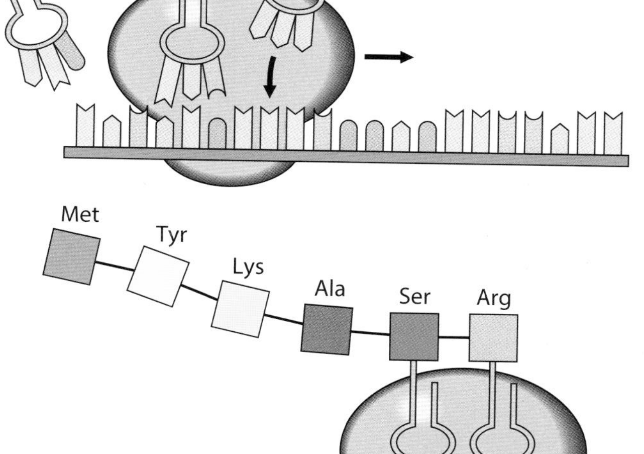

f The polypeptide chain continues to grow until a 'stop' codon is exposed on the ribosome. This is UAA, UAC or UGA.

Figure 6.13 continued Protein synthesis – translation.

a

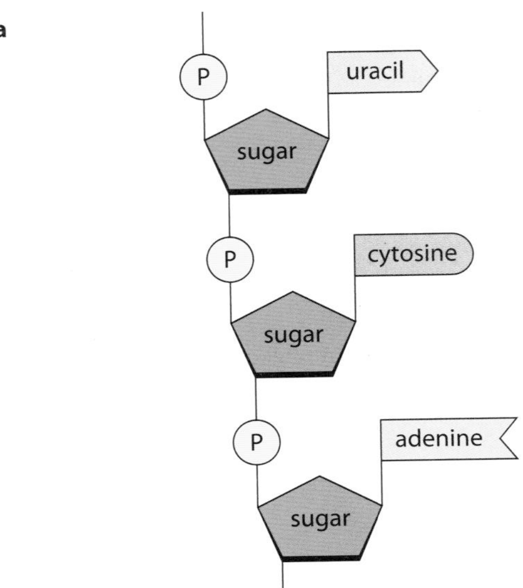

b

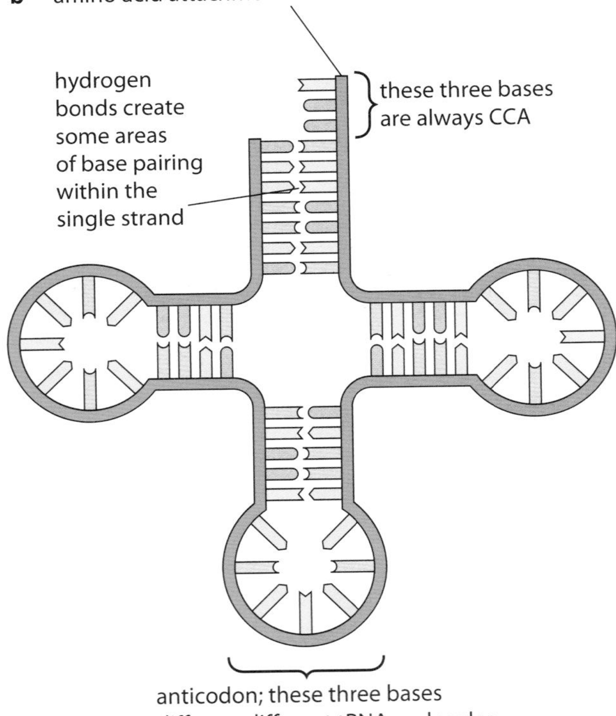

Figure 6.14 RNA **a** Part of a messenger RNA (mRNA) molecule. **b** Transfer RNA (tRNA). The molecule is a single-stranded polynucleotide, folded into a clover-leaf shape. Transfer RNA molecules with different anticodons are recognised by different enzymes, which load them with their appropriate amino acid.

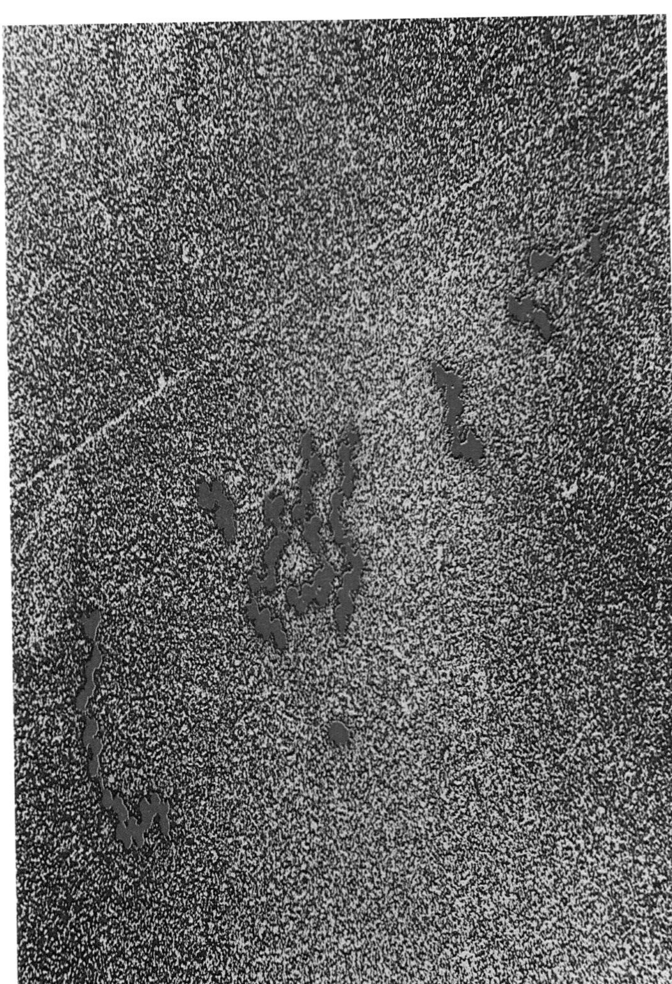

Figure 6.15 Protein synthesis in a bacterium. In bacteria, there is no nucleus, so protein synthesis can begin as soon as some mRNA has been made. Here, the long thread running from left to right is DNA. Nine mRNA molecules are being made, using this DNA as a template. Each mRNA molecule is immediately being read by ribosomes, which you can see as polyribosomes (the red blobs) attached along the mRNAs. The mRNA strand at the left-hand end is much longer than the one at the right, indicating that the mRNA is being synthesised working along the DNA molecule from right to left (× 54 000).

QUESTIONS

6.4 Summarise the differences between the structures of DNA and RNA.

6.5 Draw a simple flow diagram to illustrate the important stages in protein synthesis.

Summary

- DNA and RNA are polynucleotides, made up of long chains of nucleotides. A nucleotide contains a pentose sugar, a phosphate group and a nitrogen-containing base. In RNA, the sugar is ribose; in DNA, it is deoxyribose.
- A DNA molecule consists of two polynucleotide chains, linked by hydrogen bonds between bases. In DNA there are four bases – adenine always pairs with thymine, and cytosine always pairs with guanine.
- RNA, which comes in several different forms, has only one polynucleotide chain, although this may be twisted back on itself, as in tRNA. In RNA, the base thymine is replaced by uracil.
- DNA molecules replicate during interphase by semi-conservative replication. In DNA replication, the hydrogen bonds between the bases break, allowing free nucleotides to fall into position opposite their complementary ones on each strand of the original DNA molecule. Adjacent nucleotides are then linked, through their phosphates and sugars, to form new strands. Two complete new molecules are thus formed from one old one, each new molecule containing one old strand and one new strand.
- The sequence of nucleotide bases on a DNA molecule codes for the sequence of amino acids in a polypeptide. Each amino acid is coded for by three bases. A length of DNA coding for just one polypeptide is a gene. A change in the nucleotide sequence of DNA is a mutation, producing a new allele of the gene.
- The DNA sequences for the **Hb**$^{\mathbf{A}}$ (normal) and **Hb**$^{\mathbf{S}}$ (sickle cell) alleles of the gene for the β-globin polypeptide differ by only one base. The triplet CTT in **Hb**$^{\mathbf{A}}$ is replaced by CAT in **Hb**$^{\mathbf{S}}$, swapping the amino acid from glutamic acid to valine. This single difference in the polypeptide results in sickle cell anaemia in individuals with two **Hb**$^{\mathbf{S}}$ alleles.
- During protein synthesis, a complementary copy of the base sequence on a gene is made, by building a molecule of messenger RNA (mRNA) against one DNA strand. This stage is called transcription. After transcription, the next stage is called translation. During translation the mRNA moves to a ribosome in the cytoplasm.
- Transfer RNA (tRNA) molecules with complementary triplets of bases temporarily pair with base triplets on the mRNA, bringing appropriate amino acids. When two amino acids are held side by side, a peptide bond forms between them. The ribosome moves along the mRNA molecule, so that appropriate amino acids are gradually linked together, following the sequence laid down by the base sequence on the mRNA.

End-of-chapter questions

1 What can be found in both DNA and messenger RNA (mRNA)?

A double helix structure

B sugar–phosphate chain

C ribose

D thymine [1]

2 Which statement about base pairing in nucleic acids is **not** correct?

A Adenine can pair with either thymine or uracil.

B Guanine only pairs with cytosine.

C Thymine can pair with either adenine or uracil.

D Uracil only pairs with adenine. [1]

3 How many different arrangements of four bases into triplets can be made?

A $3 + 4$

B 3×4

C 3^4

D 4^3 [1]

4 Look at the structures of nucleotides in Figure 6.2 (page 112). Draw a nucleotide that could be found:

a in either DNA or RNA [1]

b only in DNA [1]

c only in RNA. [1]

[Total: 3]

5 Distinguish between a **nucleotide** and a **nucleic acid**. [2]

6 Copy the drawing and annotate it to explain the replication of DNA.

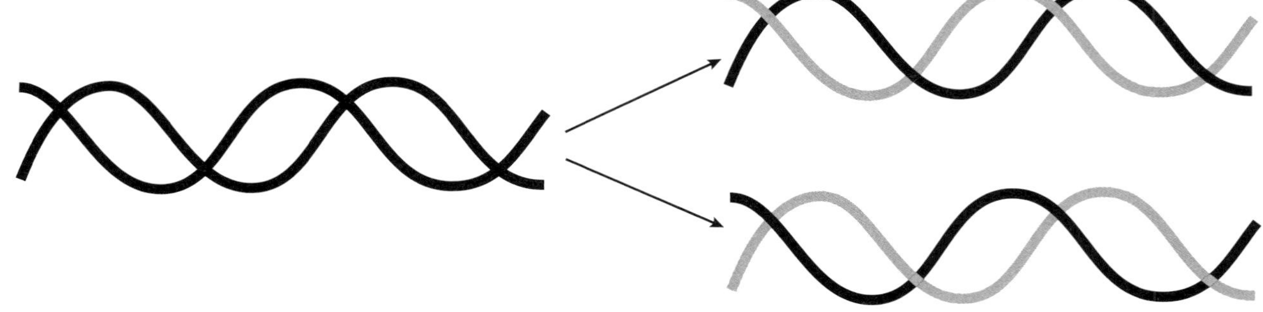

[4]

7 Use Appendix 2 (page 513) to find the sequence of amino acids that is coded by the following length of messenger RNA (mRNA):

AUGUUUCUUGAUUAA [2]

8 The table shows all the messenger RNA (mRNA) codons for the amino acid leucine. Copy the table and write in, for each codon, the transfer RNA (tRNA) anticodon that would bind with it and the DNA triplet from which it was transcribed.

mRNA codon	tRNA anticodon	DNA triplet from which mRNA was transcribed
UUA		
UUG		
CUU		
CUC		
CUA		
CUG		

[6]

9 In most people, the first six amino acids in their β-globin polypeptide chains are:

1 2 3 4 5 6
Val–His–Leu–Thr–Pro–Glu–rest of chain

The DNA triplet coding for the sixth amino acid (Glu) in most people is CTT. In some people this DNA triplet is CAT.

a What type of mutation is the change from CTT to CAT? [1]

b Use Appendix 2 (page 513) to identify the amino acid in the β-globin polypeptide chains of people with this mutation. [1]

c State the consequences for a person of having two copies of the mutated gene. [1]

[Total: 3]

10 Suggest why:

a a mutation in which one nucleotide of a triplet code is altered often makes no difference to the protein molecule coded by the DNA [2]

b the addition or deletion of three nucleotides in the DNA sequence of a gene often has less effect on the encoded protein than the addition or deletion of a single nucleotide. [4]

[Total: 6]

11 Copy and complete the following table to distinguish between the processes of transcription and translation.

	Transcription	Translation
Site in cell where process occurs		
Molecule used as a template in process		
Molecule produced by process		
Component molecules (monomers) used in process		
One other molecule that is essential for the process to occur		

[5]

12 The drawing shows a polyribosome.

a Name **X**, **Y** and **Z**. [3]

b In which direction are the ribosomes moving? Explain how you were able to decide on their direction of movement. [2]

[Total: 5]

13 In the 1940s, Chargaff and his co-workers analysed the base composition of the DNA of various organisms. The relative numbers of the bases adenine (A), cytosine (C), guanine (G) and thymine (T) of three of these organisms are shown in the table.

Organism (tissue)	Relative numbers of bases			
	A	C	G	T
ox (spleen)	27.9	20.8	22.7	27.3
ox (thymus)	28.2	21.2	21.5	27.8
yeast	31.3	17.1	18.7	32.9
virus with single-stranded DNA	24.3	18.2	24.5	32.3

Explain why:

a the relative numbers of each base in ox spleen and thymus are the same, within experimental error [2]

b the relative numbers of each base in yeast are different from those in ox spleen or thymus [2]

c the relative numbers of the bases A and T, or of C and G, are similar in ox and yeast [2]

d in the virus, the relative numbers of A and T, and of C and G, are **not** similar. [2]

[Total: 8]

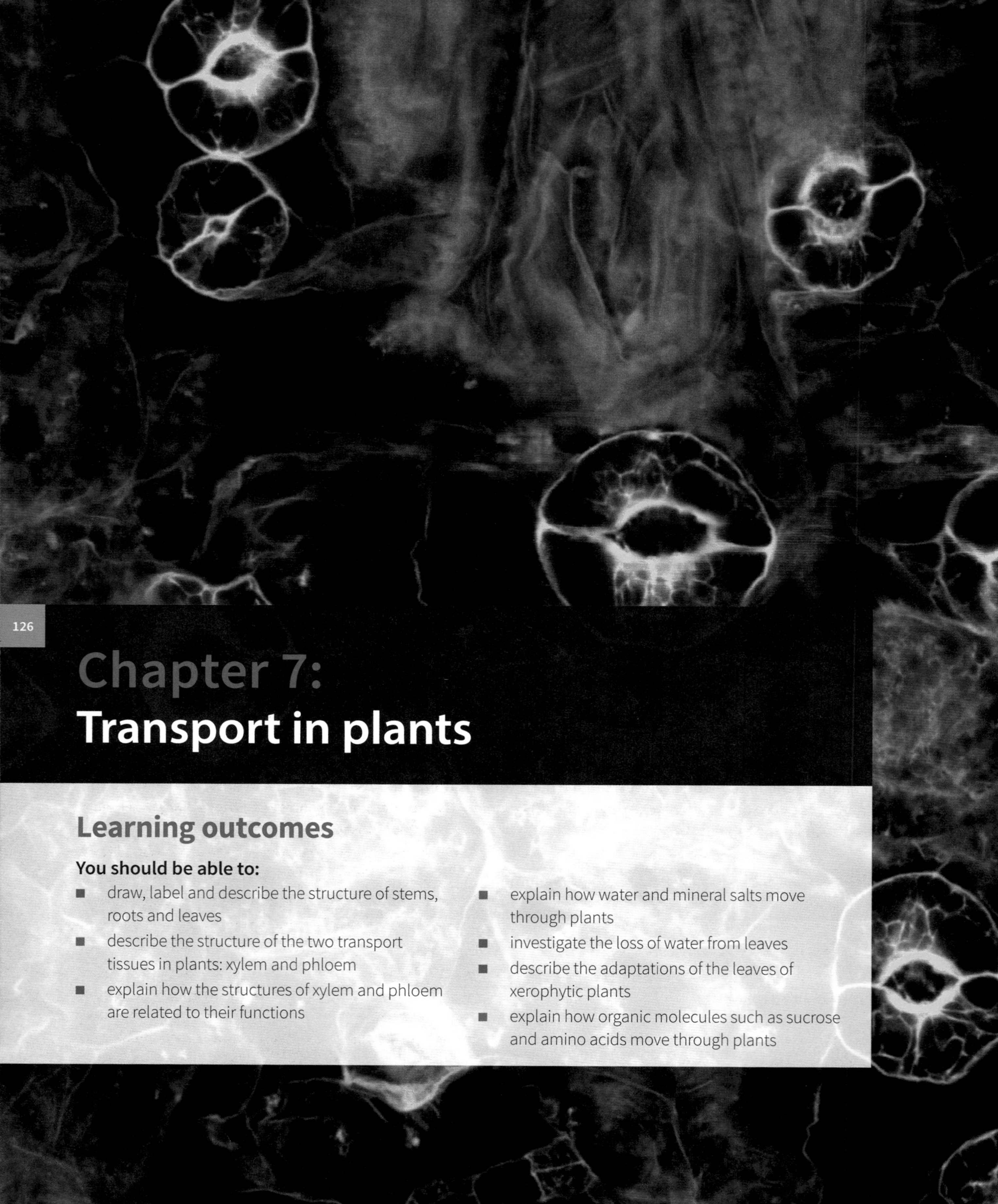

Chapter 7: Transport in plants

Learning outcomes

You should be able to:

- draw, label and describe the structure of stems, roots and leaves
- describe the structure of the two transport tissues in plants: xylem and phloem
- explain how the structures of xylem and phloem are related to their functions
- explain how water and mineral salts move through plants
- investigate the loss of water from leaves
- describe the adaptations of the leaves of xerophytic plants
- explain how organic molecules such as sucrose and amino acids move through plants

Does gold grow on trees?

Scientists studying how plants transport materials over large distances have always found eucalyptus trees interesting because the leaves tend to be concentrated at the tops of the tall trunks. Studying transport in eucalyptus trees recently provided an unexpected bonus. Scientists from Australia's national scientific agency, the CSIRO, have discovered that the roots of eucalyptus trees growing over rocks containing gold take up tiny quantities of gold and transport them throughout the plant. Gold is toxic to plant cells, but the trees get round this problem by depositing the gold in insoluble calcium oxalate crystals in the leaves and bark (Figure 7.1).

The systematic sampling of eucalyptus leaves for gold opens up a new way of finding deposits of gold in the remote Australian outback. This could save mining companies the expense of drilling in rugged terrain where exploration is difficult. It is likely that the method could also be applied to the search for other minerals such as copper and zinc.

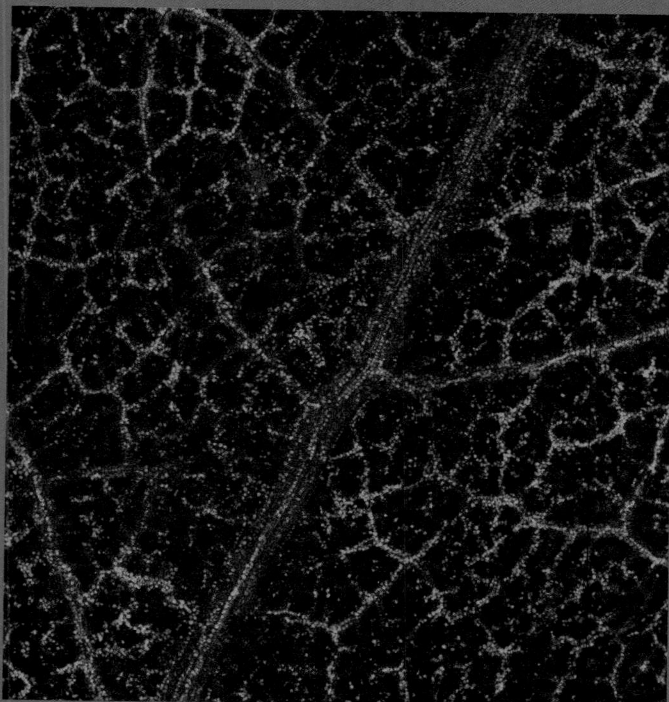

Figure 7.1 Particles of gold in a eucalyptus leaf.

The transport needs of plants

Unlike animals, plants make their own organic molecules, using the process of photosynthesis. Carbon dioxide gas is the source of carbon and light is the source of energy. The main photosynthetic organs are the leaves, which have evolved a large surface area : volume ratio for efficient capture of carbon dioxide and light. As a result, most plants do not have compact bodies like animals, but have extensive branching bodies with leaves above ground. In order to obtain the water and mineral salts also needed for nutrition, plants have extensive root systems below ground. The plant body therefore spreads out to obtain the carbon dioxide, light energy, water and inorganic mineral ions it needs from its environment to make organic molecules like sugars and amino acids.

Transport systems are therefore needed for the following reasons.

- To move substances from where they are absorbed to where they are needed – for example, water and mineral ions are absorbed by roots and transported in the xylem to other parts of the plant.
- To move substances from where they are produced to where they are needed for metabolism. For example, sugars are produced in leaves, but glucose is needed by all parts of the plant for respiration and for converting to cellulose for making cell walls in areas of growth. Glucose can be moved in phloem as part of the sucrose molecule.
- To move substances to a different part of the plant for storage – for example, to move sugars into a potato tuber for storage in the form of starch.

Unlike animals, plants do not have systems for transporting carbon dioxide and oxygen. Instead, these gases diffuse through air spaces within stems, roots and leaves.

- **Carbon dioxide.** Photosynthetic plant cells require a supply of carbon dioxide during daylight. Most photosynthetic tissue is in leaves, and most plants have evolved thin, flat leaves with a large surface area ideal for absorbing as much carbon dioxide as possible. They obtain this by diffusion from the air.
- **Oxygen.** All plant cells require a supply of oxygen for respiration, but cells which are actively photosynthesising produce more than enough oxygen for their own needs because oxygen is a waste product of photosynthesis. Cells which are not photosynthesising need to take in oxygen from their environment. Plants have much lower energy demands than animals, so they respire at much lower rates. They therefore do not need such a rapid supply of oxygen. The branching shape of plants and a network of air spaces in the plant body provide a large enough surface area for effective absorption of oxygen by diffusion.

To summarise, it is relatively easy for carbon dioxide and oxygen to diffuse into and out of the bodies of plants, reaching and leaving every cell quickly enough down diffusion gradients. Consequently, there is no need for a transport system for these gases. However, transport systems are needed for distribution of water, inorganic and organic nutrients, as well as other substances such as plant hormones.

Two systems: xylem and phloem

The design of a plant's transport system is quite different from that of a mammal. In fact, plants have two transport systems, **xylem** and **phloem**. Xylem carries mainly water and inorganic ions (mineral salts) from roots to the parts above ground. The xylem sap contained in the xylem can move in only one direction, from roots to the rest of the plant. The second system is phloem. This carries substances made by photosynthesis from the leaves to other areas of the plant. At any one time, phloem sap can be moving in different directions in different parts of the phloem.

In neither of these systems do fluids move as rapidly as blood does in a mammal, nor is there an obvious pump such as the heart. Neither plant transport system carries oxygen or carbon dioxide, which travel to and from cells and their environment by diffusion alone.

Structure of stems, roots and leaves

Stems, roots and leaves are the main organs involved in transport within plants. Organs are composed of more than one tissue. Tissues are collections of cells specialised for a particular function. The cells may be of the same type, such as parenchyma, or of different types, as in xylem and phloem.

The tissues found in stems, roots and leaves are most easily studied using prepared slides of transverse sections of these organs. Drawing low power plans of the organs and representative groups of cells of the individual tissues as seen at high power with a microscope is a useful way of understanding the structure of the organs. Structure is closely linked with function.

When making drawings using a microscope, you will need to follow the advice given in Box 7.1.

Using an eyepiece graticule and a stage micrometer will enable you to make measurements of cells, tissues and organs and will help you to show tissues in their correct proportions. See Worked example 1, page 7, for guidance on making measurements.

Monocotyledons and dicotyledons

Flowering plants (angiosperms) may be **monocotyledons** (monocots) or **dicotyledons** (dicots). Each type has its own characteristics. For example, monocotyledonous plants, such as grasses, typically have long, narrow leaves. Dicotyledonous plants typically have leaves with blades and stalks (petioles). The mechanisms of transport through both types of plant are the same, but there are differences in the distribution of xylem and phloem in their roots, stems and leaves. Only dicotyledonous plants are described in this book.

Low-power plan diagrams

Transverse sections of a typical dicotyledonous stem, root and leaf are shown in Figures 7.2, 7.3 and 7.5–7.8. In each case a labelled photomicrograph from a prepared slide is shown, followed by a low-power, labelled drawing of the same organ.

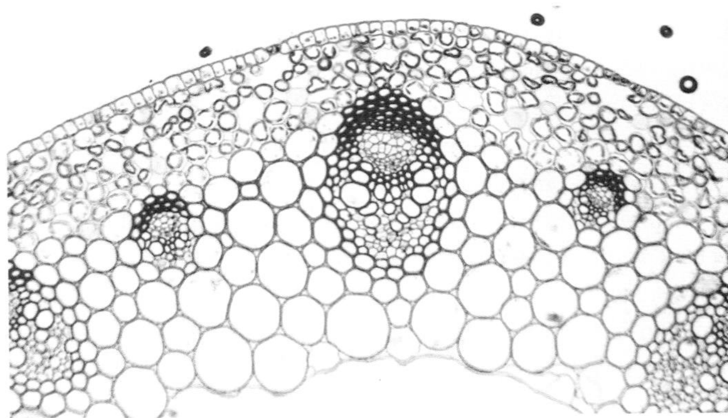

Figure 7.2 Light micrograph of part of a transverse section of a young *Ranunculus* (buttercup) stem (×60).

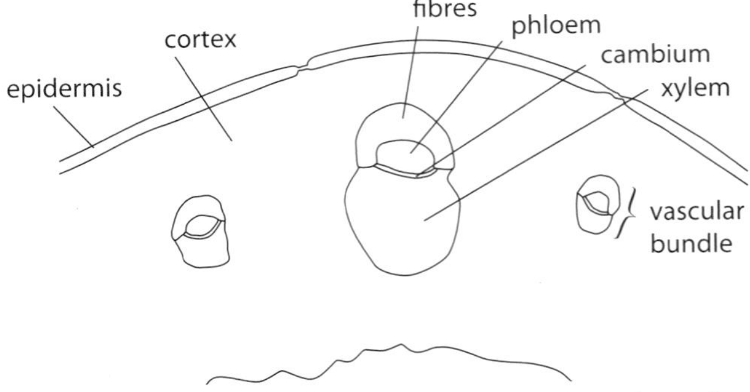

Figure 7.3 Low-power plan of the *Ranunculus* stem shown in Figure 7.2.

BOX 7.1: Biological drawing

You need the following equipment:

- pencil (HB)
- pencil sharpener
- eraser
- ruler
- plain paper.

Here are some guidelines for the quality of your drawing:

- Always use a sharp pencil, not a pen; no coloured pencils.
- Don't use shading.
- Use clear, continuous lines.
- Use accurate proportions and observation – not a textbook version.
- Make the drawing large enough – it should normally occupy more than half the available space on the page, if it is a whole organ or a tissue. Individual cells drawn at high power should be one to several centimetres in diameter.
- If you make a mistake, use a good eraser to rub out the lines completely, then redraw clearly.

For a low-power drawing (Figure 7.4):

- Don't draw individual cells.
- Draw all tissues completely enclosed by lines.
- Draw a correct interpretation of the distribution of tissues.
- A representative portion may be drawn (e.g. half a transverse section).

For a high-power drawing:

- Draw only a few representative cells.
- Draw the cell wall of all plant cells.
- Don't draw the nucleus as a solid blob.

Some guidelines for the quality of your labelling:

- Label all tissues and relevant structures.
- Identify parts correctly.
- Use a ruler for label lines and scale line.
- Label lines should stop exactly at the structure being labelled; do not use arrowheads.
- Arrange label lines neatly and ensure they don't cross over each other.
- Annotate your drawing if necessary (i.e. provide short notes with one or more of the labels in order to describe or explain features of biological interest).
- Add a scale line at the bottom of the drawing if appropriate.
- Use a sharp pencil, not a pen.

An example of a drawing of a section through the stem of *Helianthus* is shown below. Biological drawing is also covered in Chapter P1, page 262.

QUESTION

7.1 Look at Figure 7.4. List the errors in drawing technique that you can spot in the left-hand half of the drawing.

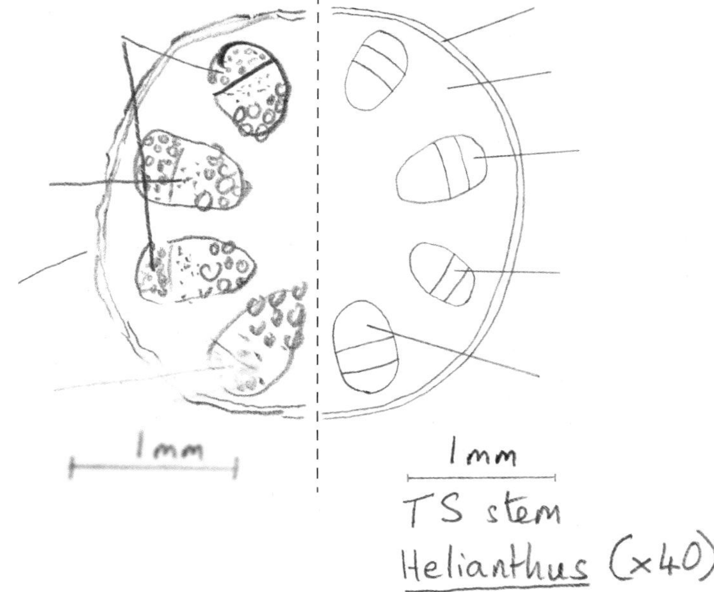

Figure 7.4 The right side of this low-power drawing shows examples of good technique, while the left side shows many of the pitfalls you should avoid.

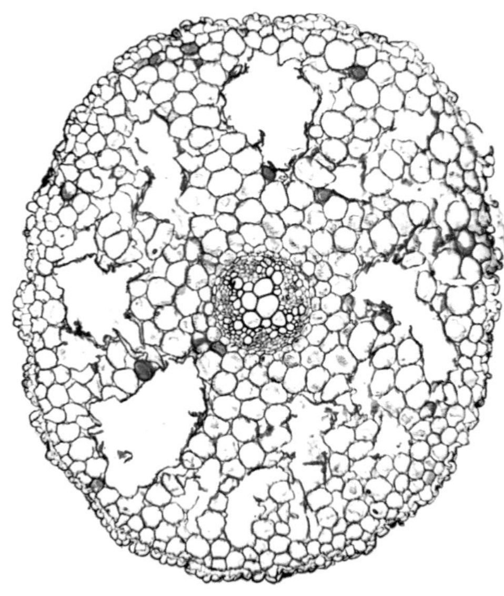

Figure 7.5 Light micrograph of a transverse section of *Ranunculus* (buttercup) root (×35).

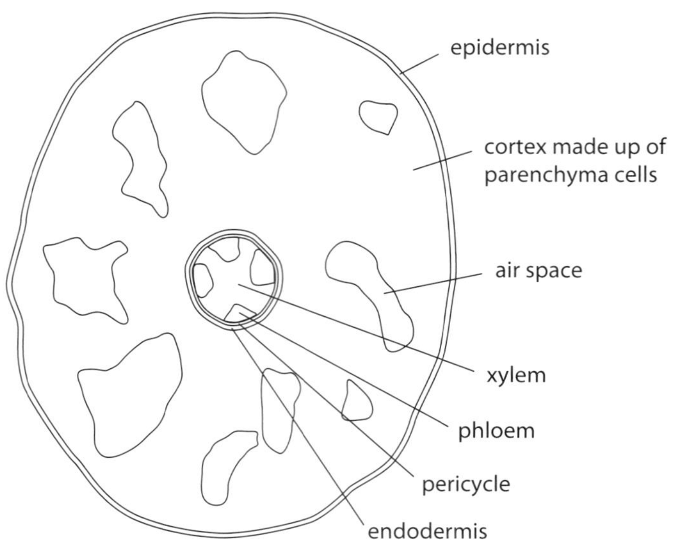

Figure 7.6 Low-power plan of the *Ranunculus* root shown in Figure 7.5.

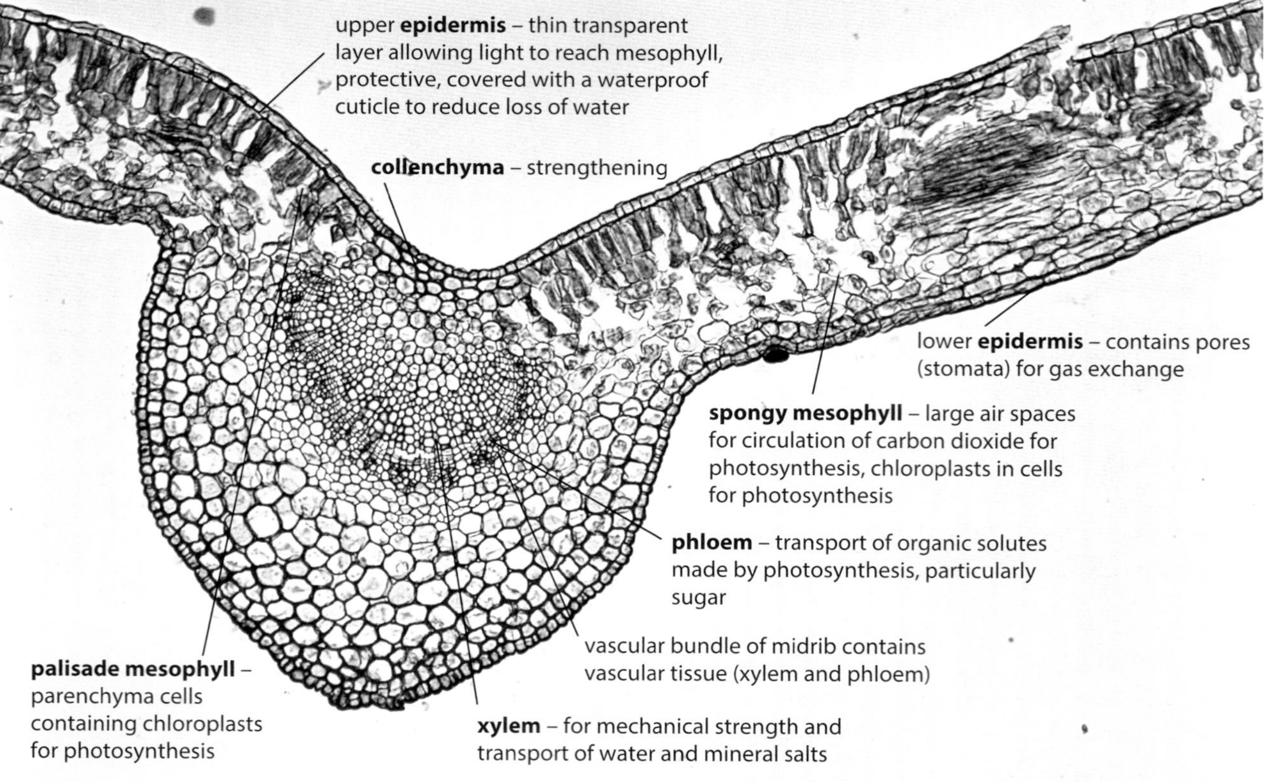

Figure 7.7 Transverse section through the midrib of a dicotyledonous leaf, *Ligustrum* (privet) (×50). Tissues are indicated in bold type.

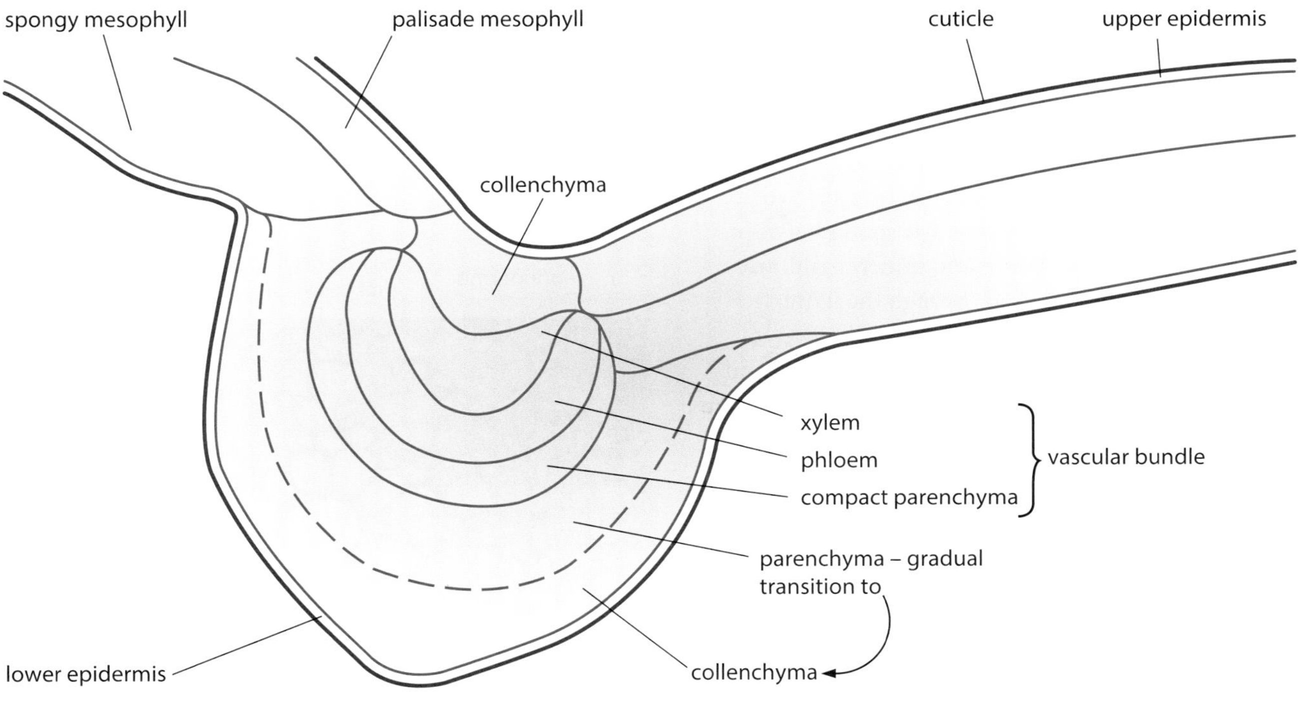

Figure 7.8 A plan diagram of the transverse section through a privet leaf shown in Figure 7.7. Parenchyma is a tissue made up of unspecialised cells. Collenchyma is made up of cells in which the walls are thickened with extra cellulose, especially at the corners, providing extra strength for support.

High-power detail diagrams

When drawing cells at high power, remember the rules given in Box 7.1 (page 129). The emphasis should be on drawing two or three representative cells of each tissue accurately, rather than trying to draw a lot of cells to make it look like the specimen on the slide. Your low-power plan shows where the cells are located.

Photomicrographs and diagrams of representative cells from the tissues of dicotyledonous stems, roots and leaves are shown in Figures 7.9 to 7.13.

A brief description of some of the tissues you will be examining is given below to help you to understand their structure. The structure and function of xylem and phloem is explained on pages 141–152.

Epidermis

This is a continuous layer on the outside of the plant, one cell thick, that provides protection (Figure 7.9). In stems and leaves it is covered with a waxy **cuticle** which is waterproof and helps to protect the organ from drying out and from infection. In leaves, it also has pores called stomata which allow entry of carbon dioxide for photosynthesis. In roots, it may have extensions called **root hairs** to increase the surface area for absorption of water and mineral salts.

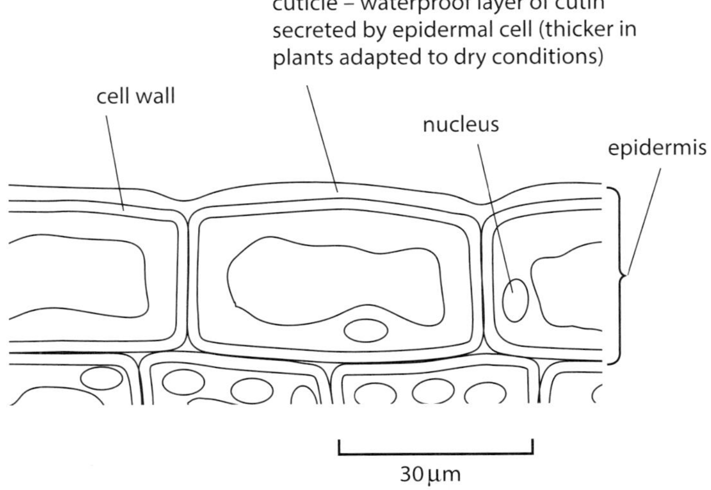

Figure 7.9 High-power detail of a transverse section of leaf epidermis.

Parenchyma

Parenchyma is made up of thin-walled cells used as packing tissue (Figure 7.10). The cells are very metabolically active and may be used for many functions. They may be used for storage of foods like starch. When they are turgid (fully inflated with water) they help to support the plant, preventing wilting. Air spaces between the cells allow gas exchange. Water and mineral salts are transported through the walls and through the living contents of the cells.

Parenchyma forms the cortex in roots and stems, and the pith in stems. The cortex is an outer region of cells. The pith is made up of similar cells but is the name given to the central region of stems.

Parenchyma contains chloroplasts in leaves, where it is modified to form the palisade and spongy mesophyll.

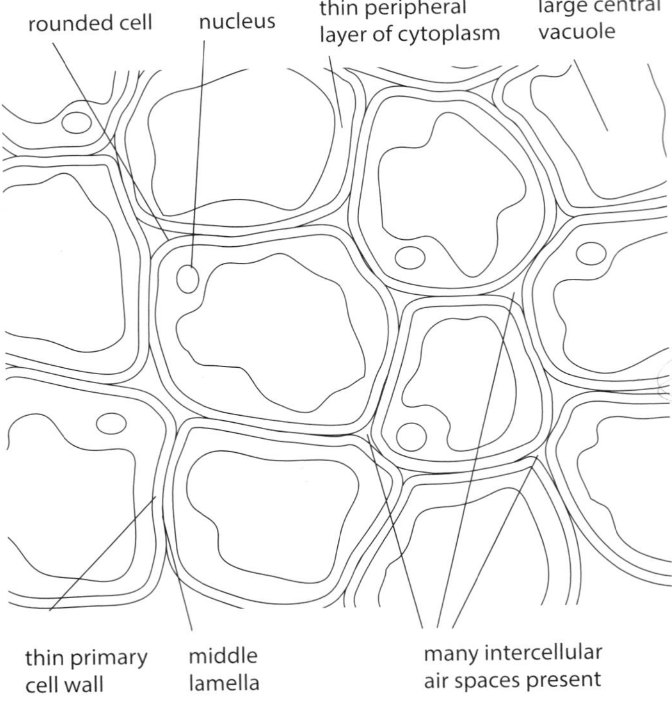

Figure 7.10 High-power detail of a transverse section of parenchyma. Cells are usually roughly spherical, though may be elongated. The average cell diameter is about 25 μm.

Collenchyma

Collenchyma is a modified form of parenchyma with extra cellulose deposited at the corners of the cells. This provides extra strength. The midrib of leaves contains collenchyma (Figures 7.7 and 7.8).

found here

Endodermis

The endodermis, like the epidermis, is one cell thick (Figure 7.11). It surrounds the vascular tissue in stems and roots. Its function in roots is explained later in this chapter.

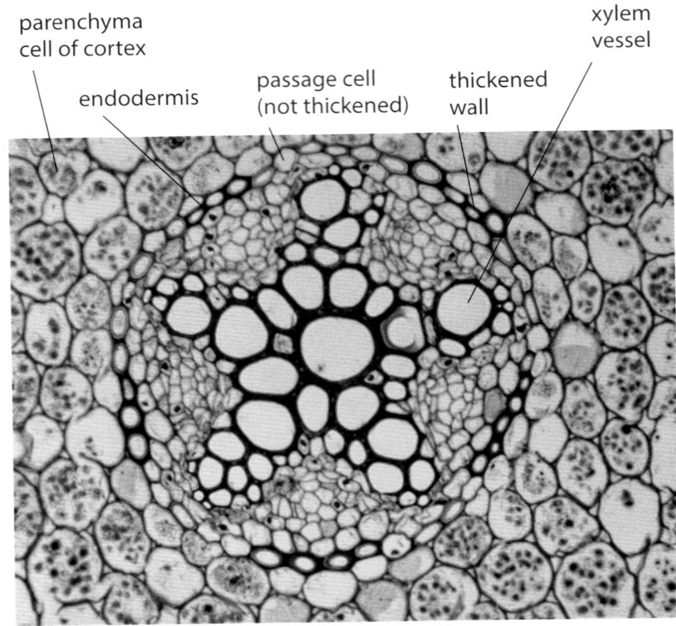

Figure 7.11 Light micrograph of part of a transverse section of a young dicotyledonous root. The endodermis with its thickened walls is shown. Note also the passage cell which allows the passage of water (×250).

Mesophyll

'Meso' means middle and 'phyll' means leaf. The mesophyll is made up of specialised parenchyma cells found between the lower and upper epidermis of the leaf (Figure 7.12). They are specialised for photosynthesis and therefore contain chloroplasts. They are of two types, palisade mesophyll and spongy mesophyll. 'Palisade' means 'column-shaped'. Spongy mesophyll is so-called because in three dimensions it is spongy in appearance, because it has many large air spaces between the cells. Palisade mesophyll cells are near the upper surface of the leaf where they receive more sunlight. They therefore contain more chloroplasts than spongy mesophyll cells.

Pericycle

This is a layer of cells, one to several cells thick, just inside the endodermis and next to the vascular tissue. In roots, it is one cell thick and new roots can grow from this layer. In stems, it is formed from a tissue called sclerenchyma (Figure 7.13). This has dead, lignified cells for extra strength.

nucleus
thin cellulose cell wall
cell of upper epidermis
palisade mesophyll
spongy mesophyll
chloroplasts
large air space

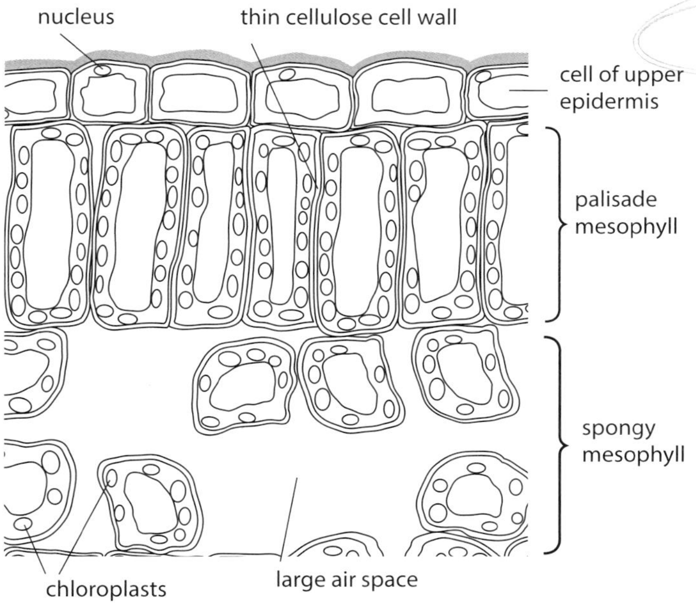

Figure 7.12 High-power detail of palisade and spongy mesophyll cells seen in transverse section.

Vascular tissue

Xylem and phloem both contain more than one type of cell and together they make the vascular tissue. 'Vascular' means having tubes for transporting fluids.

Xylem contains tubes called vessels made from dead cells called xylem vessel elements. The walls of the cells are reinforced with a strong, waterproof material called **lignin**. Xylem allows long distance transport of water and mineral salts. It also provides mechanical support and strength. In roots, it is at the centre and has a series of 'arms' between which the phloem is found (Figures 7.5 and 7.6). In stems, the xylem and phloem are found in bundles called **vascular bundles** (Figures 7.2 and 7.3). The outsides of these bundles have caps made of sclerenchyma fibres which provide extra support for the stem. Sclerenchyma fibres, like xylem vessel elements, are long, dead, empty cells with lignified walls (Figure 7.13). Unlike xylem, however, they have only a mechanical function and do not transport water.

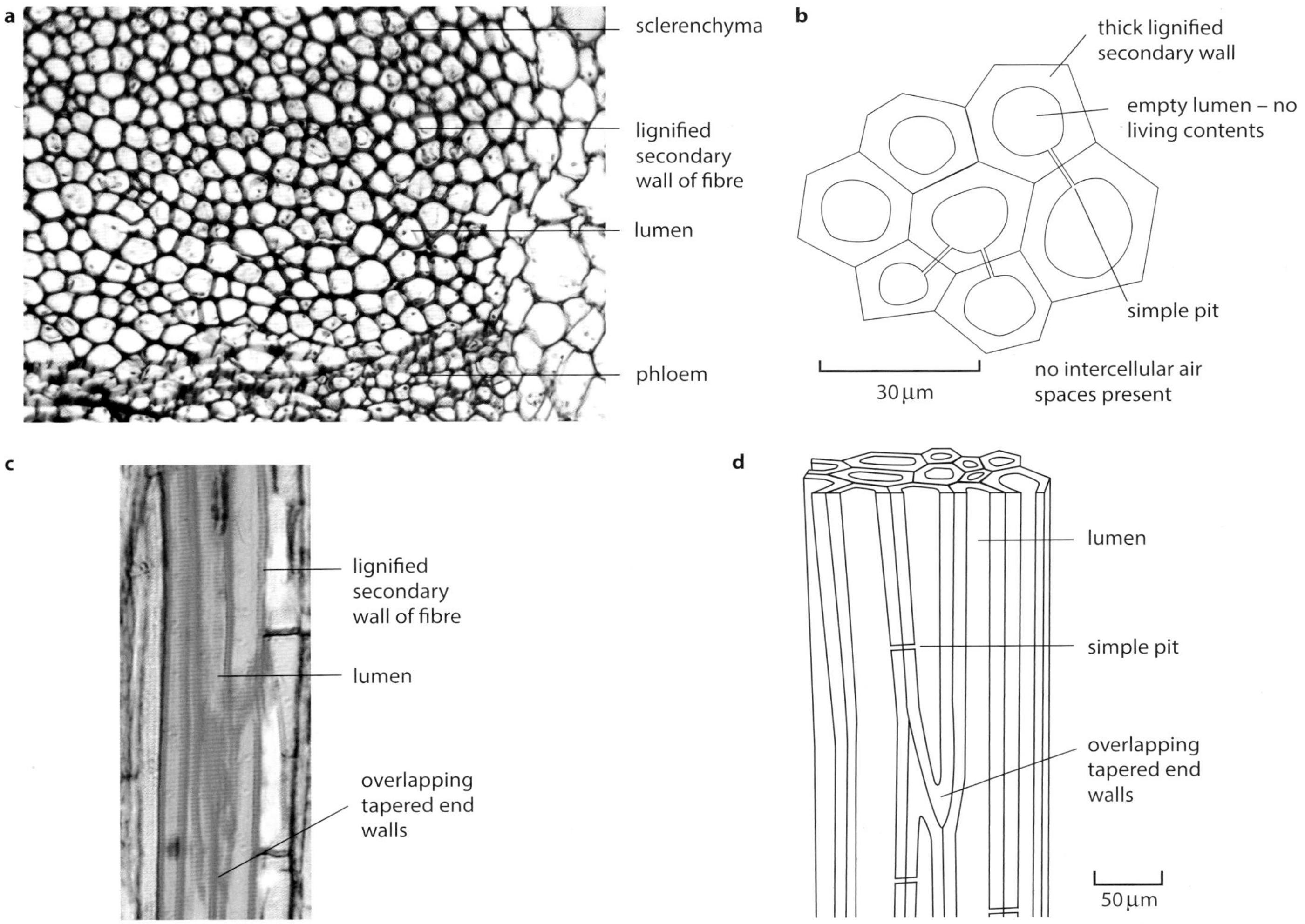

a
sclerenchyma
lignified secondary wall of fibre
lumen
phloem
b
thick lignified secondary wall
empty lumen – no living contents
simple pit
30 µm
no intercellular air spaces present
c
lignified secondary wall of fibre
lumen
overlapping tapered end walls
d
lumen
simple pit
overlapping tapered end walls
50 µm

Figure 7.13 Structure of sclerenchyma cells. **a** light micrograph in TS (×100), **b** drawing in TS, **c** light micrograph in LS (×200) and **d** drawing in LS.

Note that the distribution of the strengthening tissues, xylem and sclerenchyma, is different in roots and stems. This reflects the fact that these organs are subjected to different stresses and strains. Stems, for example, need to be supported in air, whereas roots are usually spreading through soil and are subjected to pulling strains from the parts above ground. In the roots and stems of trees and shrubs, extra xylem is made, forming wood.

The way in which the structure of xylem is related to its function is described on pages 141 to 143.

Phloem contains tubes called sieve tubes made from living cells called **sieve tube elements**. These allow long distance transport of organic compounds, particularly the sugar sucrose. The structure and function of phloem is described on pages 147 to 151.

The transport of water

Figure 7.14 outlines the pathway taken by water as it is transported through a plant. In order to understand the transport mechanism, you will need to understand that water moves from a region of higher to lower water potential, as explained in Chapter 4. The movement of water is passive as it is driven by evaporation from the leaves.

The process starts in the leaves. The energy of the Sun causes water to evaporate from the leaves, a process called **transpiration**. This reduces the water potential in the leaves and sets up a water potential gradient throughout the plant. Water moves down this gradient from the soil into the plant – for example, through its root hairs. Water then moves across the root into the xylem tissue in the

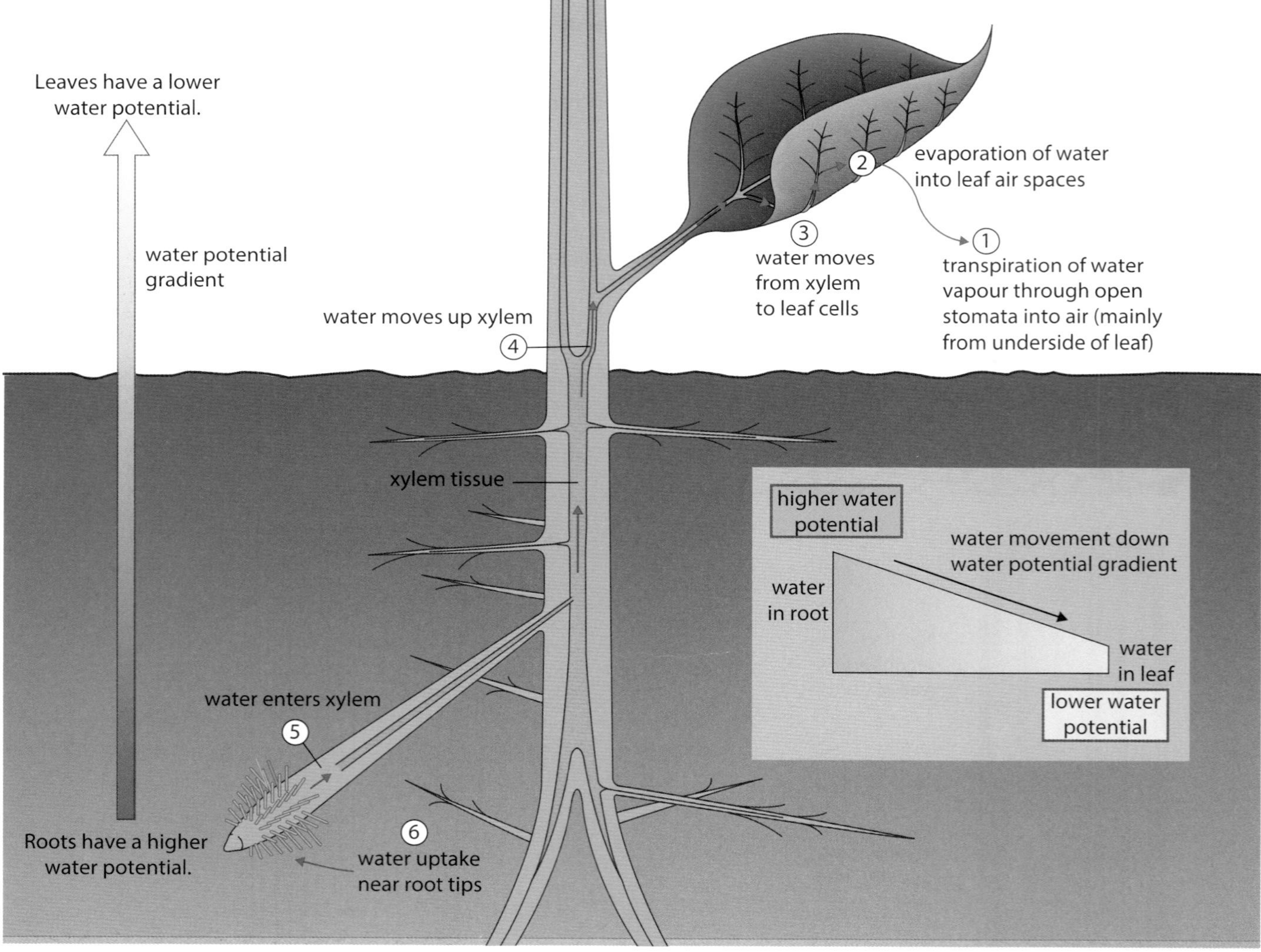

Figure 7.14 An overview of the movement of water through a plant. Water moves down a water potential gradient from the soil to the air. The process starts with loss of water vapour from the leaves (transpiration) and follows the sequence from **1** to **6** in the diagram.

centre. Once inside the xylem vessels, the water moves upwards through the root to the stem and from there into the leaves.

From leaf to atmosphere – transpiration

Figure 7.15 shows the internal structure of a dicotyledonous leaf. The cells in the **mesophyll** ('middle leaf') layers are not tightly packed, and have many spaces around them filled with air. The walls of the mesophyll cells are wet, and some of this water evaporates into the air spaces (Figure 7.16), so that the air inside the leaf is usually saturated with water vapour.

The air in the internal spaces of the leaf has direct contact with the air outside the leaf, through small pores called **stomata**. If there is a water potential gradient between the air inside the leaf (higher water potential) and the air outside (lower water potential), then water vapour will diffuse out of the leaf down this gradient.

Although some of the water in the leaf will be used, for example, in photosynthesis, most eventually evaporates and diffuses out of the leaf by the process of transpiration.

The distribution of stomata in the epidermis of leaves can be seen by examining 'epidermal peels' or 'epidermal impressions' (Box 7.2).

Transpiration is the loss of water vapour from a plant to its environment, by diffusion down a water potential gradient; most transpiration takes place through the stomata in the leaves.

QUESTION

7.2 Most stomata are usually found in the lower epidermis of leaves. Suggest why this is the case.

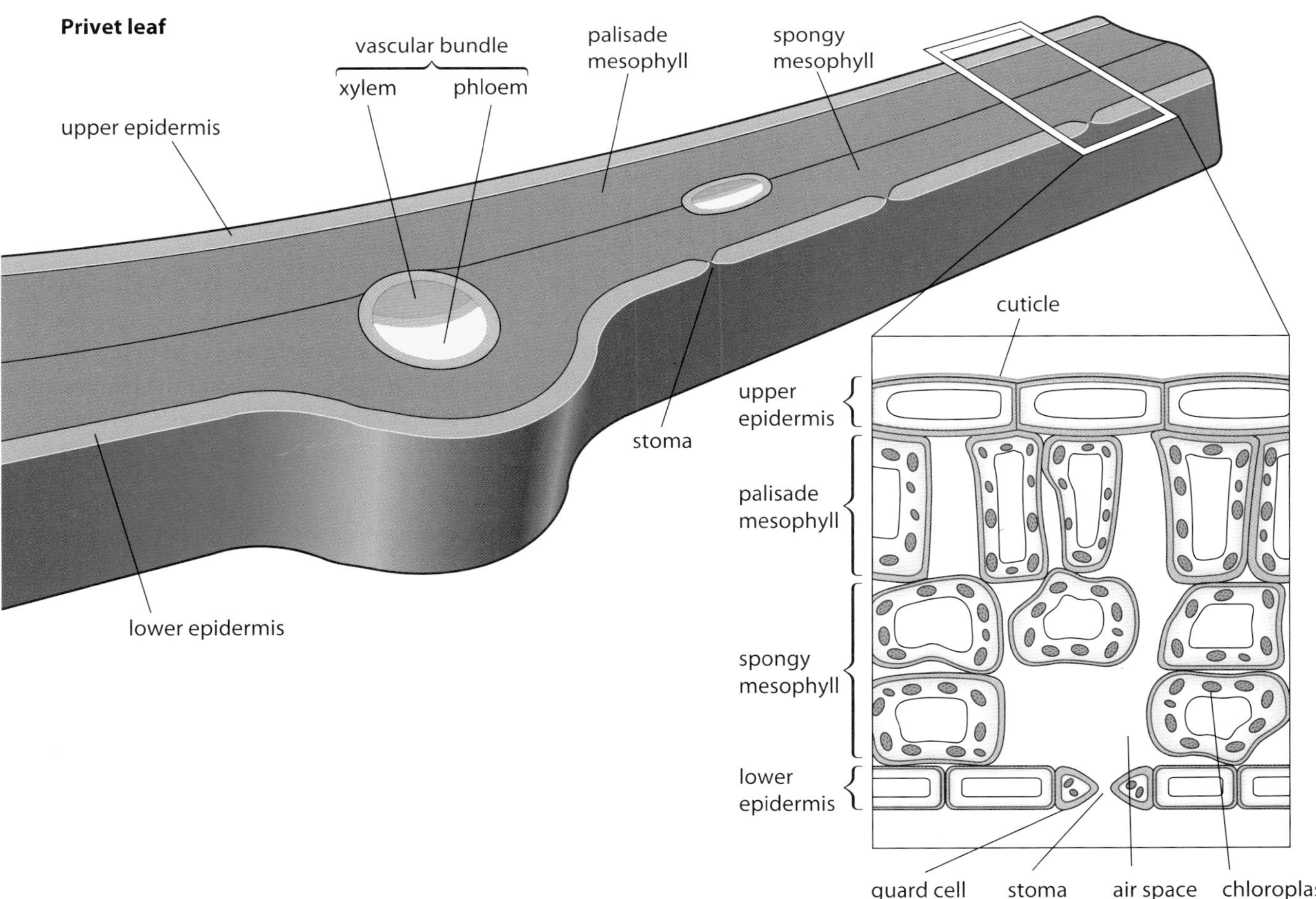

Figure 7.15 The structure of a dicotyledonous leaf. Water enters the leaf as liquid water in the xylem vessels and diffuses out as water vapour through the stomata.

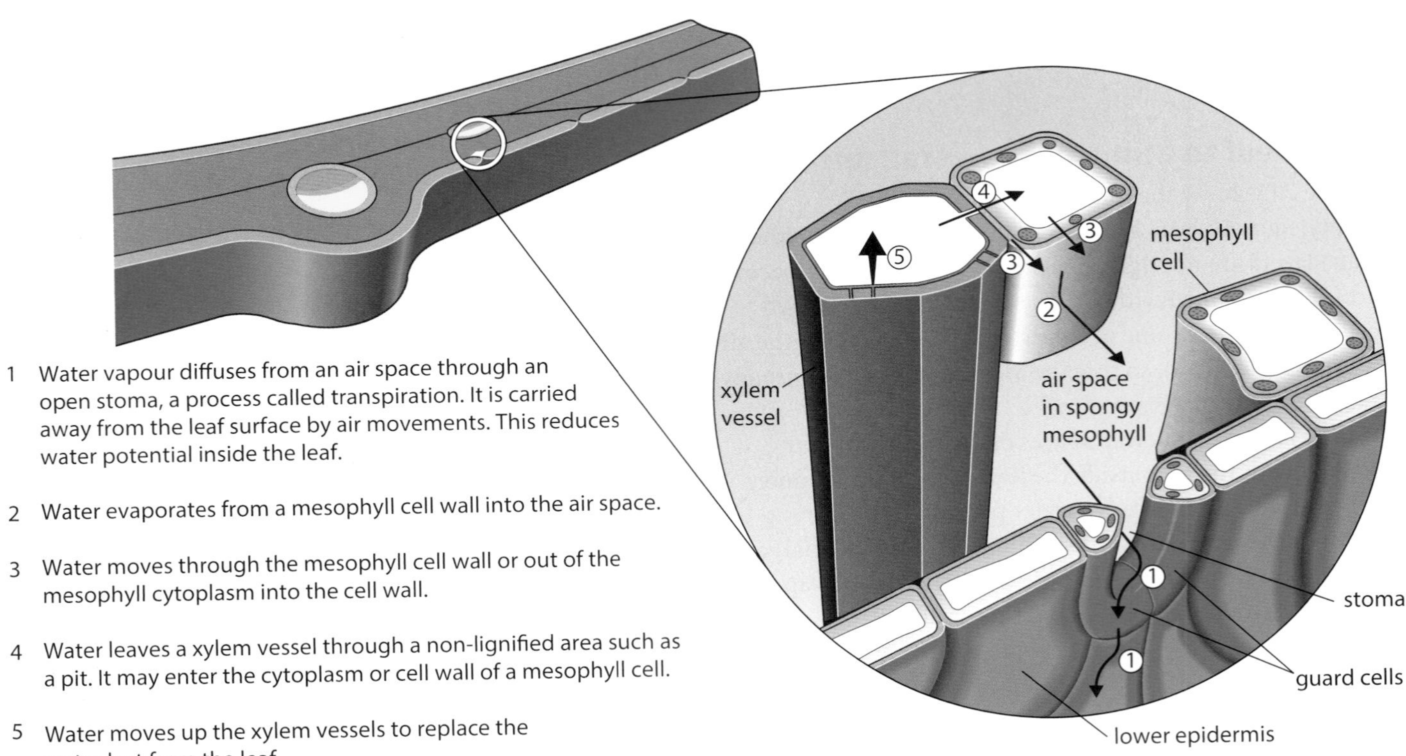

Figure 7.16 Water movement through a leaf. Water is, in effect, being pulled through the plant as a result of transpiration and evaporation. Movement of water through the plant is therefore known as the transpiration stream.

Factors affecting transpiration

- **Humidity**. If the water potential gradient between the air spaces in the leaf and the air outside becomes steeper, the rate of transpiration will increase. In conditions of low humidity, the gradient is steep, so transpiration takes place more quickly than in high humidity.
- **Wind speed and temperature**. Transpiration may also be increased by an increase in wind speed or rise in temperature.
- **Light intensity**. In most plants, stomata open during the day and close at night. Most transpiration takes place through the stomata (although a little water vapour can escape through the epidermis if the cuticle is thin), so the rate of transpiration is almost zero at night. Stomata must be open during the day to allow carbon dioxide to diffuse into the leaf for photosynthesis. This inevitably increases the rate of transpiration. Closing at night, when photosynthesis is impossible, reduces unnecessary water loss.
- **Very dry conditions**. In especially dry conditions, when the water potential gradient between the internal air spaces and the external air is steep, a plant may have to compromise by partially or completely closing its stomata to prevent its leaves drying out, even if this means reducing the rate of photosynthesis.

In hot conditions, transpiration plays an important role in cooling the leaves. As water evaporates from the cell walls inside the leaf, it absorbs heat energy from these cells, thus reducing their temperature.

If the rate at which water vapour is lost by transpiration exceeds the rate at which a plant can take up water from the soil, then the amount of water in its cells decreases. The cells become less turgid (page 85) and the plant wilts as the soft parts such as leaves lose the support provided by turgid cells (Figure 7.17). In this situation the plant will also close its stomata.

QUESTIONS

7.3 Suggest how the following factors may cause the rate of transpiration to increase:

- **a** an increase in wind speed
- **b** a rise in temperature.

7.4 How does the transpiration cooling mechanism compare with the main cooling mechanism of mammals?

7.5 Transpiration has sometimes been described as a 'necessary evil'. Explain how this statement might be justified.

Figure 7.17 The *Fatsia* plant on the left has plenty of water and remains turgid. The plant on the right is wilted because it has lost more water by transpiration than has been taken up by its roots, and so does not have sufficient water to maintain the turgidity of its cells.

BOX 7.2: Epidermal peels and impressions

Epidermal peels

It is often easy to strip the epidermis from the underlying tissues of leaves by grasping the epidermis with a pair of fine forceps and peeling a strip of epidermis away from the leaf. The epidermis is a single layer of cells, so can be clearly seen if placed on a slide and viewed with a microscope.

A suitable leaf is provided by an onion bulb. This is mainly made up of layers of fleshy storage leaves. It is very easy to remove the epidermis from one of these leaves by separating a leaf, making slits in its concave surface in the shape of a small square using a scalpel, and gently peeling off the square of epidermis using fine forceps. This can be mounted in water on a slide for viewing.

The onion is a monocot, so its epidermal cells are long and narrow and lined up in neat rows. The epidermal cells in dicot leaves tend to be irregularly distributed and rounder in shape, but with wavy edges which interlock with each other rather like jigsaw puzzle pieces, as shown in Figure 7.18. Many dicot plants may be used to make peels; lettuce and geraniums are suitable. Use the lower epidermis. Apart from the stomata, note the pair of guard cells either side of each stoma. These contain chloroplasts and control the opening and closing of the stomata.

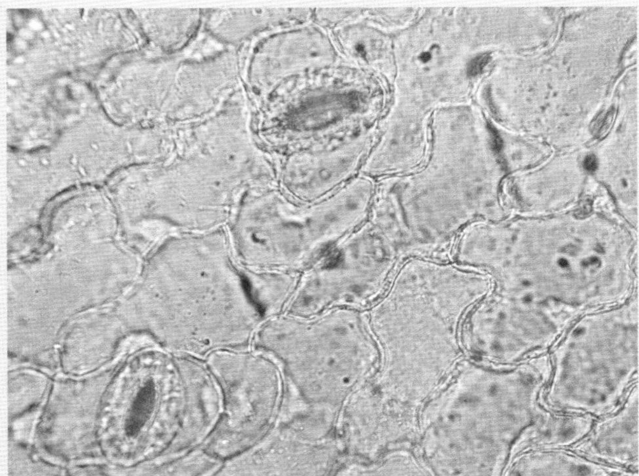

Figure 7.18 Epidermal peel of a dicot leaf (×150).

This practical lends itself to further investigative work. For example, you can investigate whether stomata are open or closed under different conditions, or estimate stomatal density (number of stomata per unit area). The latter can be done by counting the number of stomata in one field of view and then finding the area of the field of view (and hence the surface area of epidermis viewed) using the formula πr^2. The radius is found by using a calibrated eyepiece graticule (see Worked example 1 on page 7) to measure the diameter and dividing by two. Factors affecting density could be investigated, for example:

- compare upper and lower epidermis
- compare monocots and dicots (upper and lower epidermises)
- different parts of the same leaf (are stomata more frequent near veins?)
- age of leaf, particularly growing compared with fully grown
- habitat – wet or dry
- variation within a species or within one plant
- variation between species
- succulent plants and cacti.

Epidermal impressions

Impressions (or replicas) of the epidermal surface can be made with clear nail varnish. Very hairy epidermises should be avoided. The surface of the leaf is coated with a thin layer of nail varnish which is allowed to dry. It can then be carefully peeled off using fine forceps and mounted dry or in water on a slide. Alternatively, if this is too difficult, the nail varnish can be peeled off using transparent sticky tape which is then stuck on the slide.

You will be able to see whether stomata are open or closed at the point in time the peel was made. Stomatal densities can be estimated as with epidermal peels.

BOX 7.3: Comparing rates of transpiration – using a potometer

The amount of water vapour lost by transpiration from the leaves of a plant can be very great. Even in the relatively cool and moist conditions of a temperate country, a leaf may lose the volume of water contained in its leaves every 20 minutes. Thus, water must move into the leaves equally rapidly to replace this lost water.

It is not easy to measure the rate at which water vapour is leaving a plant's leaves. This makes it very difficult to investigate directly how different factors such as humidity, temperature, wind speed, or light intensity affect the rate of transpiration. However, it is relatively easy to measure the rate at which a plant stem **takes up** water. As a very high proportion of the water taken up by a stem is lost in transpiration, and as the rate at which transpiration is happening directly affects the rate of water uptake, this measurement can give a very good approximation of the rate of transpiration.

The apparatus used to measure the rate at which water is taken up by a plant is called a potometer (Figure 7.19). Everything must be completely water-tight and airtight, so that no leakage of water occurs, and so that no air bubbles break the continuous water column. To achieve this, it helps if you can insert the plant stem into the apparatus with everything submerged in water. It also helps to cut the end of the stem underwater and with a slanting cut before placing it in the potometer, as air bubbles are less likely to get trapped against it. Petroleum jelly applied around the joints helps to keep the apparatus airtight. Potometers can be simpler than this one. You can manage without the reservoir (although this does mean it takes more time and effort to refill the potometer), and the tubing can be straight rather than bent. In other words, you can manage with a straight piece of glass tubing!

As water evaporates from the plant's leaves, it is drawn into the xylem vessels that are exposed at the cut end of the stem. Water is therefore drawn along the capillary tubing. If you record the position of the meniscus at set time intervals, you can plot a graph of distance moved against time. If you expose the plant to different conditions, you can compare the rates of water uptake. The rates of water uptake will have a close relationship to the rates of transpiration under different conditions.

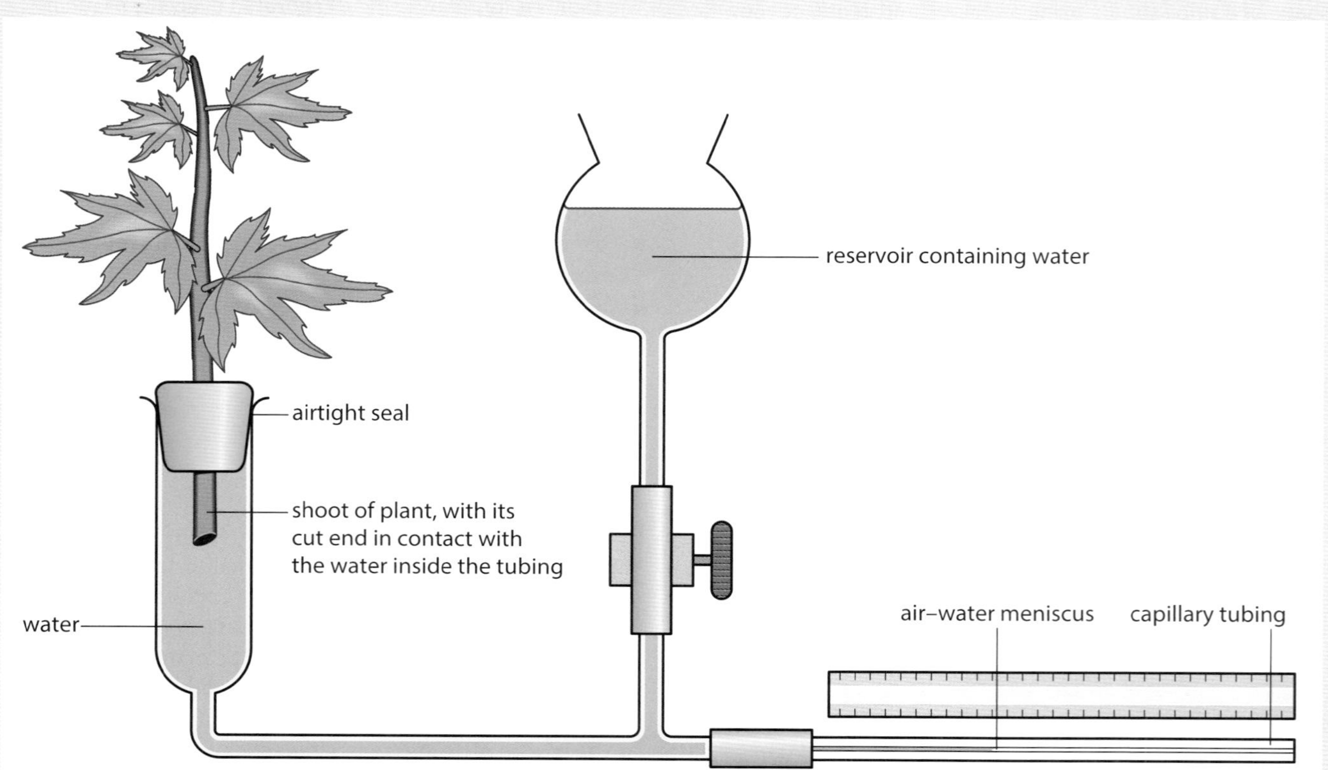

Figure 7.19 A potometer.

continued ...

Measuring the surface area of the whole leaf using grids

It is useful to be able to measure the rate of transpiration per unit leaf area. This allows valid comparisons to be made between different species, different plants of the same species or different leaves of the same plant. In order to do this, leaf area must be measured. This can be done after the experiment. A method for doing this is described below.

Grid paper is required (or graph paper) with 2 cm, 1.5 cm, 1.0 cm or 0.5 cm spacing. The smaller the grid size, the more accurate the determination of the area, but the more time consuming. Think about the best size to use before choosing.

The outline of the leaf is drawn on the grid (Figure 7.20a). Whole squares are identified first (Figure 7.20b) and their combined area calculated. Squares only partly occupied by the leaf can be treated in various ways. For example, only count it if the leaf occupies more than half the square, or find how many squares contain part of the leaf and divide this by two.

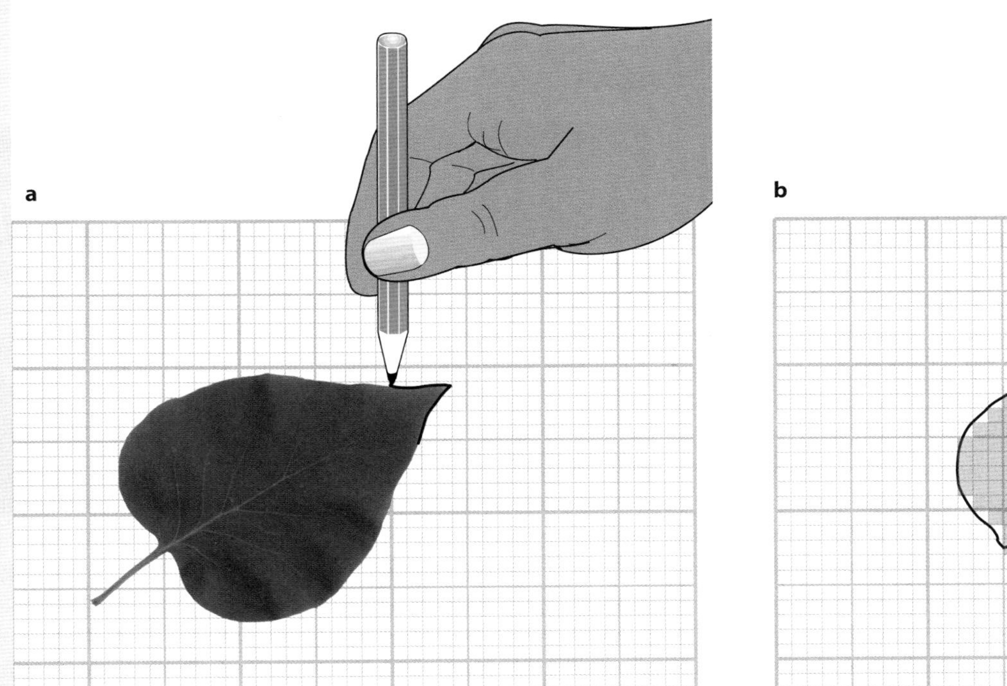

Figure 7.20 Measuring the surface area of a leaf.

QUESTION

7.6 a State three variables that must be controlled as far as possible if comparing transpiration rates from two different plants.

b Suggest how you could measure the volume of water lost from the leaves during an experiment to measure the rate of transpiration.

Xerophytes

Xerophytes (or xerophytic plants) are plants that live in places where water is in short supply. Many xerophytes have evolved special adaptations of their leaves that keep water loss down to a minimum. Some examples are shown in Figure 7.21.

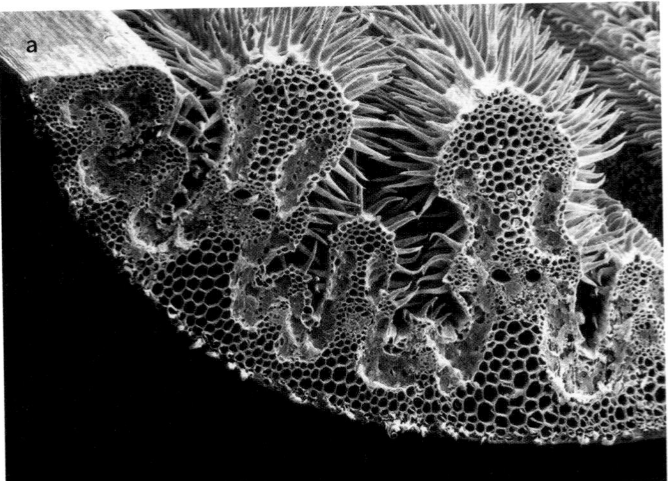

a A scanning electron micrograph of a TS through part of a rolled leaf of marram grass, *Ammophila arenaria*. This grass grows on sand dunes, where conditions are very dry. The leaves can roll up due to shrinkage of special 'hinge cells', exposing a thick, waterproof cuticle to the air outside the leaf. The cuticle contains a fatty, relatively waterproof substance called cutin. The stomata are found only in the upper epidermis and therefore open into the enclosed, humid space in the middle of the 'roll'. Hairs help to trap a layer of moist air close to the leaf surface, reducing the steepness of the diffusion gradient for water vapour.

b *Opuntia* is a cactus with flattened, photosynthetic stems that store water. The leaves are reduced to spines, which lessens the surface area from which transpiration can take place and protects the plant from being eaten by animals.

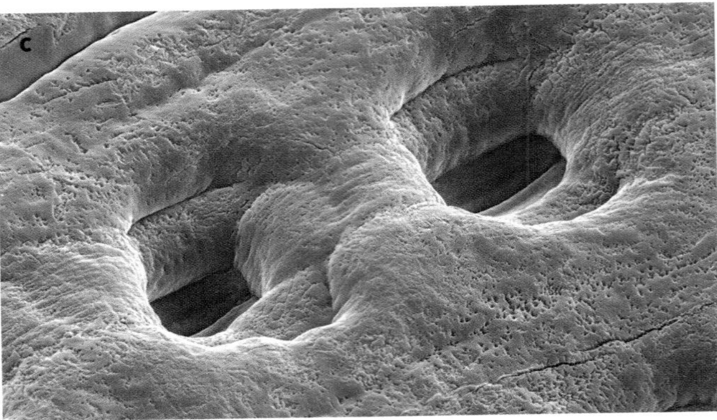

c False colour scanning electron micrograph of a needle from a Sitka spruce (×1265), a large tree native to Canada and Alaska. Its leaves are in the form of needles, greatly reducing the surface area available for water loss. In addition, they are covered in a layer of waterproof wax and have sunken stomata, as shown here.

d Scanning electron micrograph of a TS through a *Phlomis italica* leaf showing its trichomes (×20). These are tiny hair-like structures that act as a physical barrier to the loss of water, like the marram grass hairs. *Phlomis* is a small shrub that lives in dry habitats in the Mediterranean regions of Europe and North Africa.

e The cardon *Euphorbia canariensis* grows in dry areas of Tenerife. It has swollen, succulent stems that store water and photosynthesise. The stems are coated with wax, which cuts down water loss. The leaves are extremely small.

Figure 7.21 Some adaptations shown by xerophytes.

It is useful to observe the xerophytic features of marram grass (Figure 7.21a) for yourself by examining a transverse section with a light microscope and making an annotated drawing using Figure 7.22 to help you. Note the sunken stomata on the inner surface (upper epidermis) – they are at the bottoms of grooves in the leaf. Note also that the outer surface (lower epidermis) has no stomata.

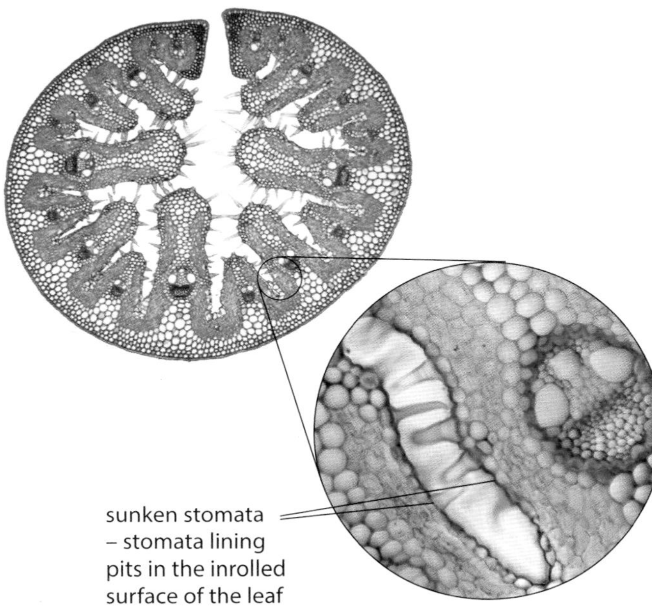

Figure 7.22 Light micrograph of a transverse section of a rolled marram grass leaf.

QUESTION

7.7 Identify six xerophytic features of leaves visible in Figures 7.21 and 7.22 and explain how each feature helps the plant to conserve water. Summarise your answer in a table using the following headings:

Xerophytic feature of leaves	How it helps to conserve water	Example (name of plant)

From xylem across the leaf

As water evaporates from the cell walls of mesophyll cells, more water is drawn into the walls to replace it. This water comes from the xylem vessels in the leaf. Water constantly moves out of these vessels through the unlignified parts of the xylem vessel walls (structure of xylem is described in the next section). The water then moves down a water potential gradient from cell to cell in the leaf along two possible pathways. In one pathway, known as the symplastic pathway, water moves from cell to cell via the plasmodesmata. In the other pathway, known as the apoplastic pathway, water moves through the cell walls. Symplastic and apoplastic pathways are both described in more detail on pages 144–145.

Xylem tissue

In order to understand how water moves into the xylem from the root, up the stem and into the leaves, scientists first had to understand the structure of xylem tissue. Of particular interest were the xylem vessels, which have certain unusual characteristics:

- they are made from cells joined end to end to form tubes
- the cells are dead
- the walls of the cells are thickened with a hard, strong material called lignin.

Xylem tissue (Figure 7.23) has two functions, namely support and transport. It contains several different types of cell.

In flowering plants, xylem tissue contains vessel elements, tracheids, fibres and parenchyma cells.

- Vessel elements and **tracheids** are the cells that are involved with the transport of water. Unlike other plants, flowering plants rely mostly on the vessel elements for their water transport, so only these are described in detail below.
- **Sclerenchyma fibres** are elongated cells with lignified walls that help to support the plant. They are dead cells; they have no living contents at all.
- **Parenchyma cells** are described on page 132.

Xylem vessels and vessel elements

Figure 7.23 shows the structure of typical xylem vessels. Vessels are made up of many elongated cells called **vessel elements**, arranged end to end.

Each vessel element begins life as a normal plant cell in whose wall **lignin** is laid down. Lignin is a very hard, strong substance, which is impermeable to water. (Wood is made of xylem and gets its strength from lignin.) As lignin builds up around the cell, the contents of the cell die, leaving a completely empty space, or **lumen**, inside.

However, in those parts of the original cell walls where groups of plasmodesmata (page 5) are found, no lignin is laid down. These non-lignified areas can be seen as 'gaps' in the thick walls of the xylem vessel, and are called **pits**. Pits are not open pores; they are crossed by permeable, unthickened cellulose cell wall. The pits in one cell link with those in the neighbouring cells, so water can pass freely from one cell to the next (Figure 7.23c).

The end walls of neighbouring vessel elements break down completely, to form a continuous tube rather like a drainpipe running through the plant. This long, non-living tube is a **xylem vessel**. It may be up to several metres long. The structural features of xylem vessels are closely related to their function, as explained in the next section.

From root to stem and leaf in the xylem

The removal of water from xylem vessels in the leaf reduces the hydrostatic pressure in the xylem vessels. (Hydrostatic pressure is pressure exerted by a liquid.) The hydrostatic pressure at the top of the xylem vessel becomes lower than the pressure at the bottom. This pressure difference causes water to move up the xylem vessels in

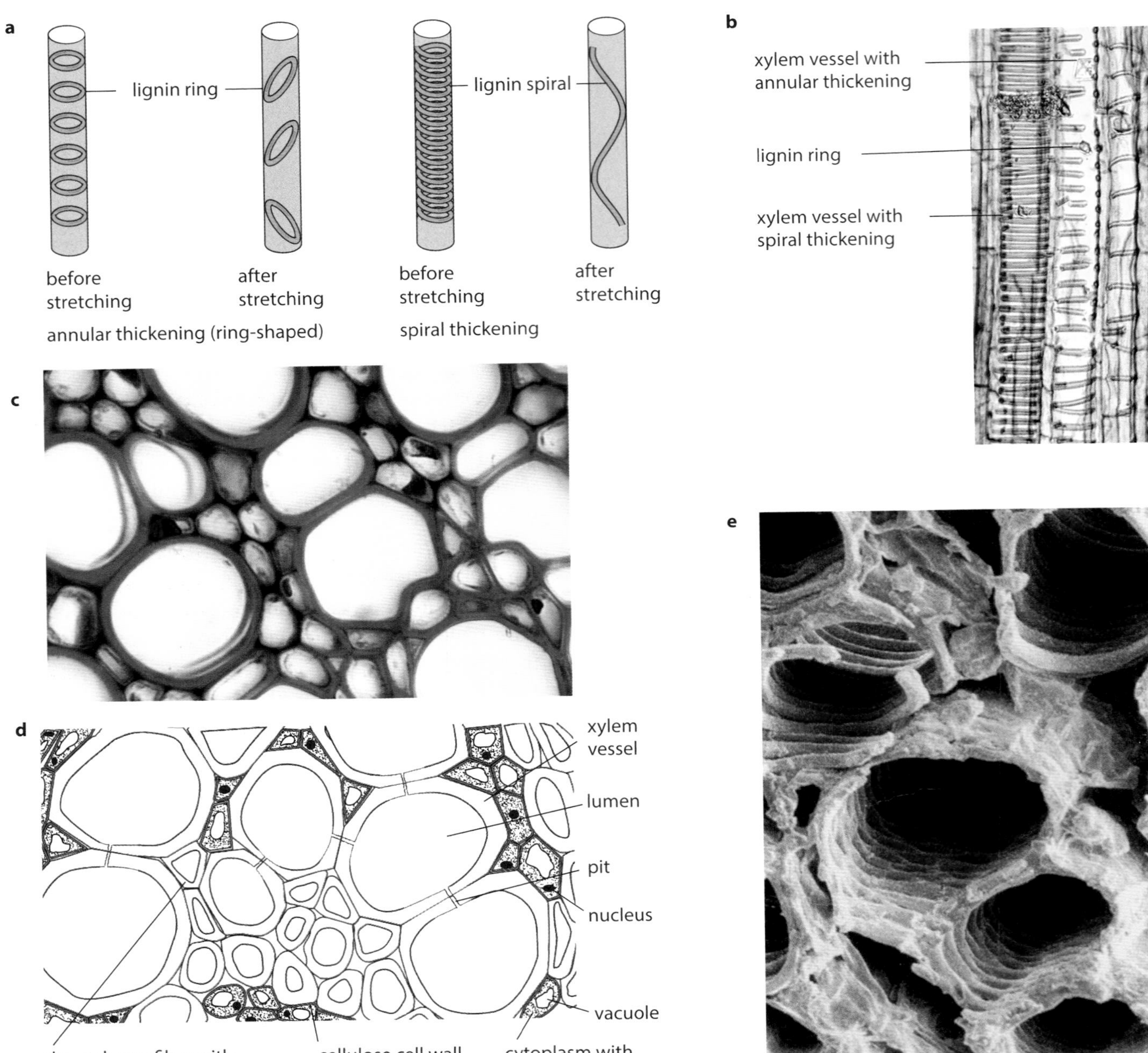

Figure 7.23 Structure of xylem. **a** Diagrams to show different types of thickening. **b** Micrograph of xylem as seen in longitudinal section. Lignin is stained red (×100). **c** and **d** Micrograph and diagram of xylem as seen in transverse section; lignin is stained red. Small parenchyma cells are also visible between the xylem vessels (×120). **e** Scanning electron micrograph of mature xylem vessels, showing reticulate (net-like) pattern of lignification (×130).

continuous columns. It is just like sucking water up a straw. When you suck a straw, you reduce the pressure at the top of the straw, causing a pressure difference between the top and bottom. The higher pressure at the bottom pushes water up the straw. Note that the lower the hydrostatic pressure, the lower the water potential, so a hydrostatic pressure gradient is also a water potential gradient.

The water in the xylem vessels, like the liquid in a 'sucked' straw, is under tension. If you suck hard on a straw, its walls may collapse inwards as a result of the pressure differences you are creating. Xylem vessels have strong, lignified walls to stop them from collapsing in this way.

The movement of water up through xylem vessels is by **mass flow**. This means that all the water molecules (and any dissolved solutes) move together, as a body of liquid, like water in a river. This is helped by the fact that water molecules are attracted to each other by hydrogen bonding (page 36); this attraction is called **cohesion**. They are also attracted to the cellulose and lignin in the walls of the xylem vessels, and this attraction is called **adhesion**. Cohesion and adhesion help to keep the water in a xylem vessel moving as a continuous column. The vessels are full of water. The fact that the cells are dead is an advantage, because it means there is no protoplasm to get in the way of transport.

If an air bubble forms in the column, the column of water breaks and the difference in pressure between the water at the top and the water at the bottom cannot be transmitted through the vessel. We say there is an air lock. The water stops moving upwards. The small diameter of xylem vessels helps to prevent such breaks from occurring. Also, the pits in the vessel walls allow water to move out into neighbouring vessels and so bypass such an air lock. Air bubbles cannot pass through pits. Pits are also important because they allow water to move out of xylem vessels to surrounding living cells.

The xylem tissue in dicotyledonous stems is arranged in a series of rods around the centre of the stem, as shown in Figure 7.24. These strong rods help to support the stem.

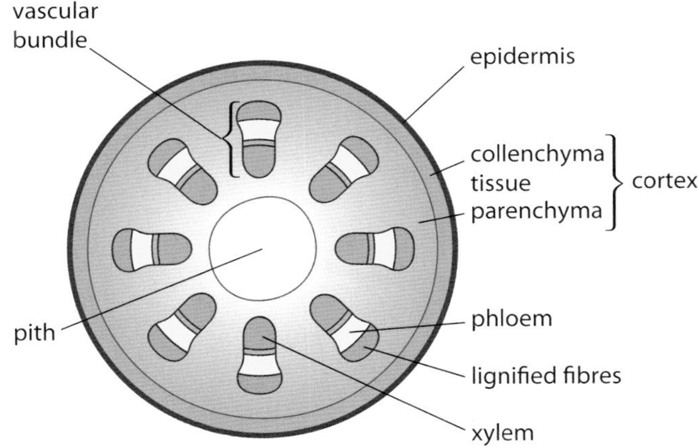

Figure 7.24 TS of a young sunflower (*Helianthus*) stem to show the distribution of tissues. The sunflower is a dicotyledonous plant.

QUESTIONS

7.8 In plants, transport of water from the environment to cells occurs in several ways. State the stages in the following types of transport:

a osmosis

b mass flow.

7.9 Explain how each of the following features adapts xylem vessels for their function of transporting water from roots to leaves.

a total lack of cell contents

b no end walls in individual xylem elements

c a diameter of between 0.01 mm and 0.2 mm

d lignified walls

e pits

Root pressure

You have seen how transpiration **reduces** the water (hydrostatic) pressure at the top of a xylem vessel compared with the pressure at the base, so causing the water to flow up the vessels. Plants may also increase the pressure difference between the top and bottom by **raising** the water pressure at the **base** of the vessels.

The pressure is raised by the active secretion of solutes, for example mineral ions, into the water in the xylem vessels in the root. Cells surrounding the xylem vessels use energy to pump solutes across their membranes and into the xylem by active transport (page 86). The presence of the solutes lowers the water potential of the solution in the xylem, thus drawing in water from the surrounding root cells. This influx of water increases the water pressure at the base of the xylem vessel.

Although root pressure may help in moving water up xylem vessels, it is not essential and is probably not significant in causing water to move up xylem in most plants. Water can continue to move up through xylem even if the plant is dead. Water transport in plants is largely a **passive** process, driven by transpiration from the leaves. The water simply moves down a continuous water potential gradient from the soil to the air.

From root hair to xylem

Figures 7.25 and 7.26 show transverse sections of a young root. The xylem vessels are in the centre of the root, unlike the arrangement in stems, where they are arranged in a ring and are nearer to the outside. Water taken up by root hairs crosses the cortex of the root and enters the xylem in the centre of the root. It does this because the water potential inside the xylem vessels is lower than the water potential in the root hairs. Therefore, the water moves down this water potential gradient across the root.

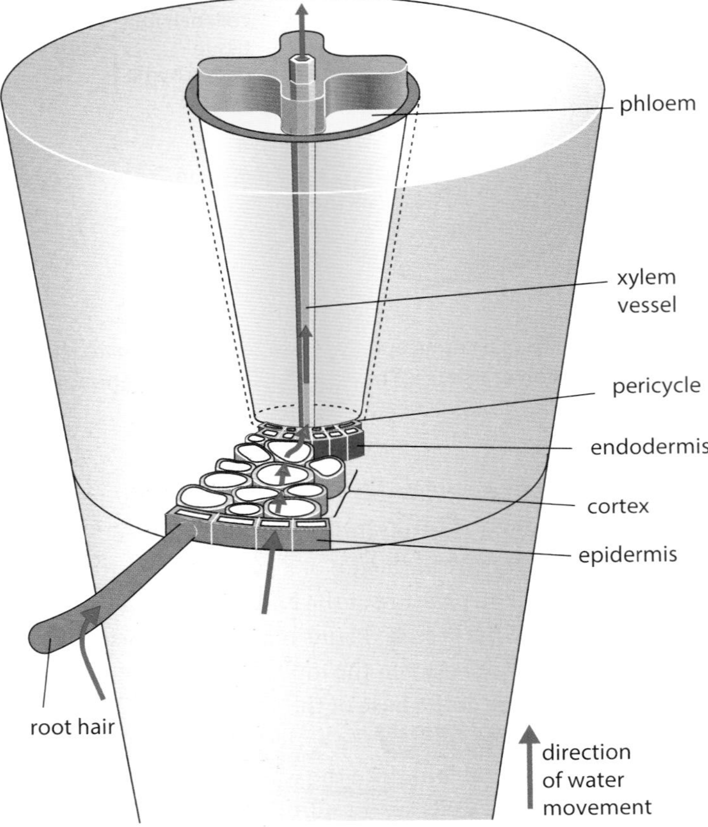

Figure 7.25 The pathway of water movement from root hair to xylem.

The water takes two routes through the cortex. Individual molecules can switch from one route to the other at any time.

The cells of the cortex, like all plant cells, are surrounded by cell walls made of several layers of cellulose fibres criss-crossing one another. Water can soak into these walls, rather as it would soak into blotting paper, and can seep across the root from cell wall to cell wall without ever entering the cytoplasm of the cortical cells. This is called the **apoplastic** pathway (Figure 7.27a).

Another possibility is for the water to move into the cytoplasm or vacuole of a cortical cell by osmosis, and then into adjacent cells through the interconnecting plasmodesmata. This is the **symplastic** pathway (Figure 7.27b).

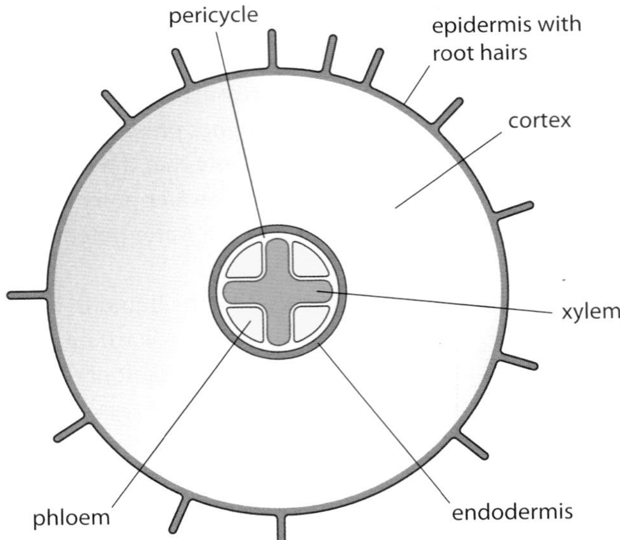

Figure 7.26 TS of a young dicotyledonous root to show the distribution of tissues.

The relative importance of these two pathways varies from plant to plant, and in different conditions. Normally, it is probable that the symplastic pathway is more important but, when transpiration rates are especially high, more water travels by the apoplastic pathway.

Once the water reaches the **endodermis** (Figures 7.25 and 7.26), the apoplastic pathway is abruptly blocked. The cells in the endodermis have a thick, waterproof, waxy band of **suberin** in their cell walls (Figure 7.28). This band, called the **Casparian strip**, forms an impenetrable barrier to water in the walls of the endodermis cells. The only way for water to cross the endodermis is through the cytoplasm of the endodermal cells. As the endodermal cells get older, the suberin deposits become more extensive, except in certain cells called **passage cells**, through which water can continue to pass freely. It is thought that this arrangement gives a plant control over what mineral ions pass into its xylem vessels, as everything has to cross cell surface membranes. It may also help with the generation of root pressure.

Once across the endodermis, water continues to move down the water potential gradient across the pericycle and towards the xylem vessels. Water moves into the xylem vessels through the pits in their walls. It then moves up the vessels towards the leaves as previously described.

a

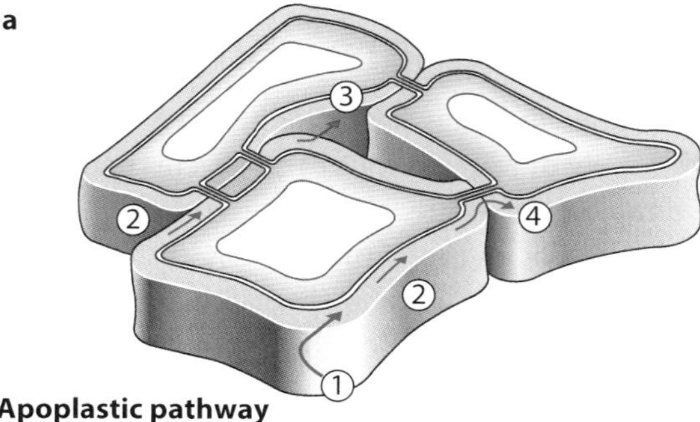

Apoplastic pathway

1 Water enters the cell wall.
2 Water moves through the cell wall.
3 Water may move from cell wall to cell wall through the intercellular spaces.
4 Water may move directly from cell wall to cell wall.

b

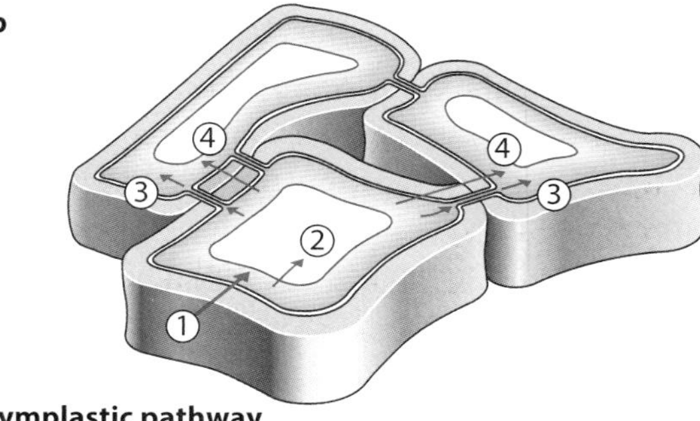

Symplastic pathway

1 Water enters the cytoplasm by osmosis through the partially permeable cell surface membrane.
2 Water moves into the sap in the vacuole, through the tonoplast by osmosis.
3 Water may move from cell to cell through the plasmodesmata.
4 Water may move from cell to cell through adjacent cell surface membranes and cell walls.

Figure 7.27 **a** Apoplastic and **b** symplastic pathways for movement of water from root hairs to xylem.

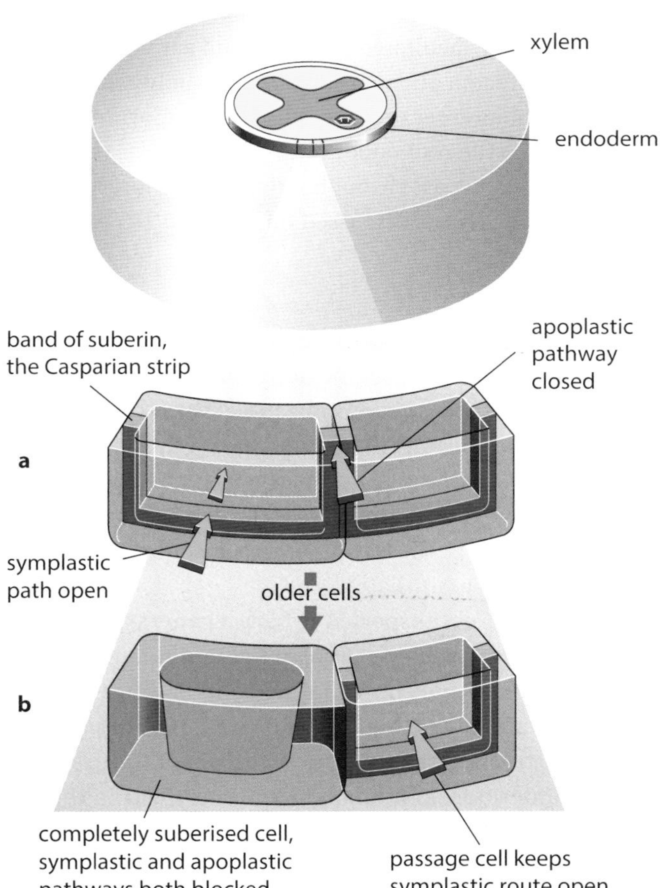

Figure 7.28 Suberin deposits close the apoplastic pathway for water in the endodermis. **a** In a young root, the suberin deposits form bands in the cell walls called Casparian strips. The symplastic path remains open. **b** In an older root, entire cells become suberised, closing the symplastic path too. Only passage cells are then permeable to water.

From soil into root hair

Figure 7.29a shows a young root. The tip is covered by a tough, protective root cap and is not permeable to water. However, just behind the tip some of the cells in the outer layer, or **epidermis**, are drawn out into long, thin extensions called **root hairs**. These reach into spaces between the soil particles, from where they absorb water.

Water moves into the root hairs by osmosis down a water potential gradient (Figure 7.29b). Although soil water contains some inorganic ions in solution, it is a relatively dilute solution and so has a relatively high water potential. However, the cytoplasm and cell sap inside the root hairs have considerable quantities of inorganic ions and organic substances such as proteins and sugars dissolved in them, and so have a relatively low water potential. Water, therefore, diffuses down this water potential gradient, through the partially permeable cell surface membrane and into the cytoplasm and vacuole of the root hair cell.

The large number of very fine root hairs provides a large surface area in contact with the soil surrounding the root, thus increasing the rate at which water can be absorbed. However, these root hairs are very delicate and often only function for a few days before being replaced by new ones as the root grows. Root hairs are also important for the absorption of mineral ions such as nitrate and magnesium.

Many plants, especially trees, have fungi located in or on their roots, forming associations called **mycorrhizas**, which serve a similar function to root hairs. The mycorrhizas act like a mass of fine roots which absorb water and nutrients, especially phosphate, from the soil and transport them into the plant. Some trees, if growing on poor soils, are unable to survive without these fungi. In return, the fungi receive organic nutrients from the plant. The name given to a relationship such as this, in which two organisms of different species both benefit, is mutualism.

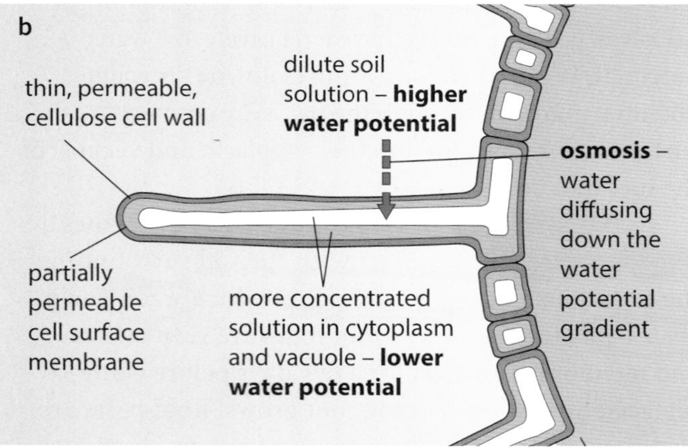

Figure 7.29 **a** A root of a young radish (*Raphanus*) plant showing the root cap and the root hairs. **b** Water uptake by a root hair cell. Mineral ions are also taken up, mostly by active transport against the concentration gradient via carrier proteins, but also by diffusion.

Transport of mineral ions

Apart from the carbohydrates made in photosynthesis, plants need a supply of mineral ions to complete their nutrition. Examples are nitrate, phosphate, sulfate, potassium, magnesium and calcium.

Mineral ions in solution are absorbed along with water by the roots, particularly by the root hairs. Their route through the plant is the same as that for water, crossing the root by apoplastic and symplastic pathways before moving in the mass flow of xylem sap up the xylem to the rest of the plant. From the xylem they enter the apoplastic and symplastic pathways again.

As well as moving by mass flow through the apoplastic pathway and xylem, mineral ions can also move by diffusion and active transport. For example, they can diffuse into the apoplastic pathway of the root from the soil and once in the apoplastic pathway can diffuse in any direction according to concentration gradients. They can also enter cells by the methods described in Chapter 4, namely diffusion, facilitated diffusion and active transport. Facilitated diffusion and active transport allow cells to control what ions enter or leave cells. One important control point is the root endodermis, where the Casparian strip forces ions to pass through living cells before they can enter the xylem, as discussed on page 145.

Translocation

The term translocation can be applied to transport in both xylem and phloem – it means literally moving from one place to another. It tends to be used more commonly to describe the transport of soluble organic substances within a plant. These are substances which the plant itself has made – for example, sugars which are made by photosynthesis in the leaves. These substances are sometimes called assimilates.

Assimilates are transported in sieve elements. Sieve elements are found in phloem tissue, along with several other types of cells including companion cells, parenchyma and fibres (Figures 7.30 and 7.31). For phloem, as with xylem, understanding the structure of the tissue is essential for understanding how transport within the tissue occurs, so we will study its structure first.

QUESTION

7.10 Give an example of an organic molecule containing:

a nitrogen
b phosphorus
c sulfur.

Sieve tubes and sieve elements

Phloem contains unique tube-like structures called sieve tubes. Unlike xylem vessels, sieve tubes are made of **living** cells. Figure 7.30 shows the structure of a **sieve tube** and its accompanying companion cells. A sieve tube is made up of many elongated sieve elements (also known as sieve tube elements), joined end to end vertically to form a continuous tube. Each sieve element is a living cell. Like a 'normal' plant cell, a sieve element has a cellulose cell wall, a cell surface membrane and cytoplasm containing endoplasmic reticulum and mitochondria. However, the amount of cytoplasm is very small and only forms a thin layer lining the inside of the wall of the cell. There is no nucleus, nor are there any ribosomes.

Perhaps the most striking feature of sieve elements is their end walls. Where the end walls of two sieve elements meet, a **sieve plate** is formed. This is made up of the walls of both elements, perforated by large pores. These pores are easily visible with a good light microscope. In living phloem, the pores are open, presenting little barrier to the free flow of liquids through them.

Companion cells

Each sieve element has at least one companion cell lying close beside it. Companion cells have the structure of a 'normal' plant cell, with a cellulose cell wall, a cell surface membrane, cytoplasm, a small vacuole and a nucleus. However, the number of mitochondria and ribosomes is rather larger than normal, and the cells are metabolically very active.

Companion cells are very closely associated with their neighbouring sieve elements. In fact, they are regarded as a single functional unit. Numerous plasmodesmata pass through their cell walls, making direct contact between the cytoplasm of the companion cell and that of the sieve element.

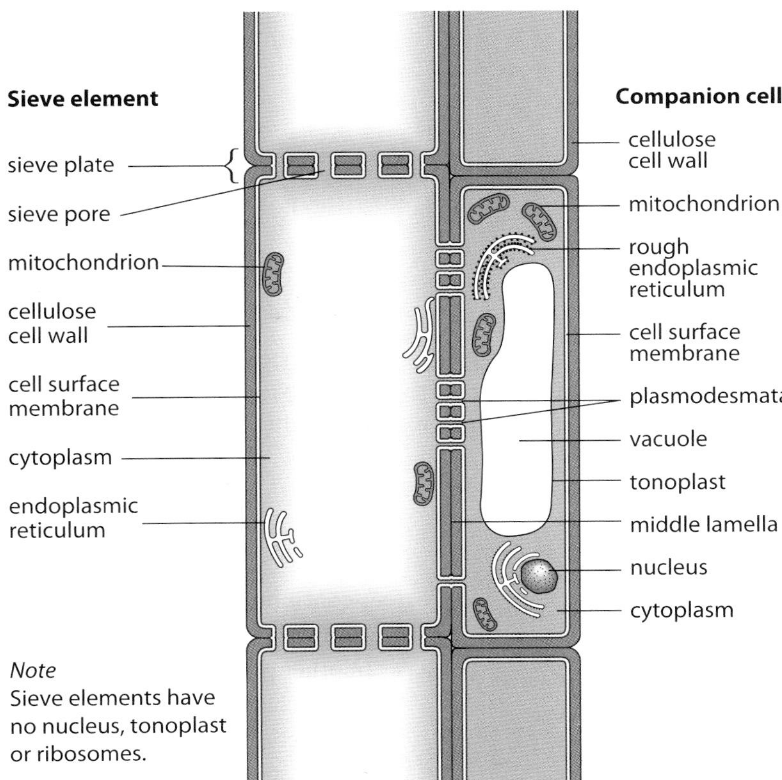

Figure 7.30 A phloem sieve tube element and its companion cell.

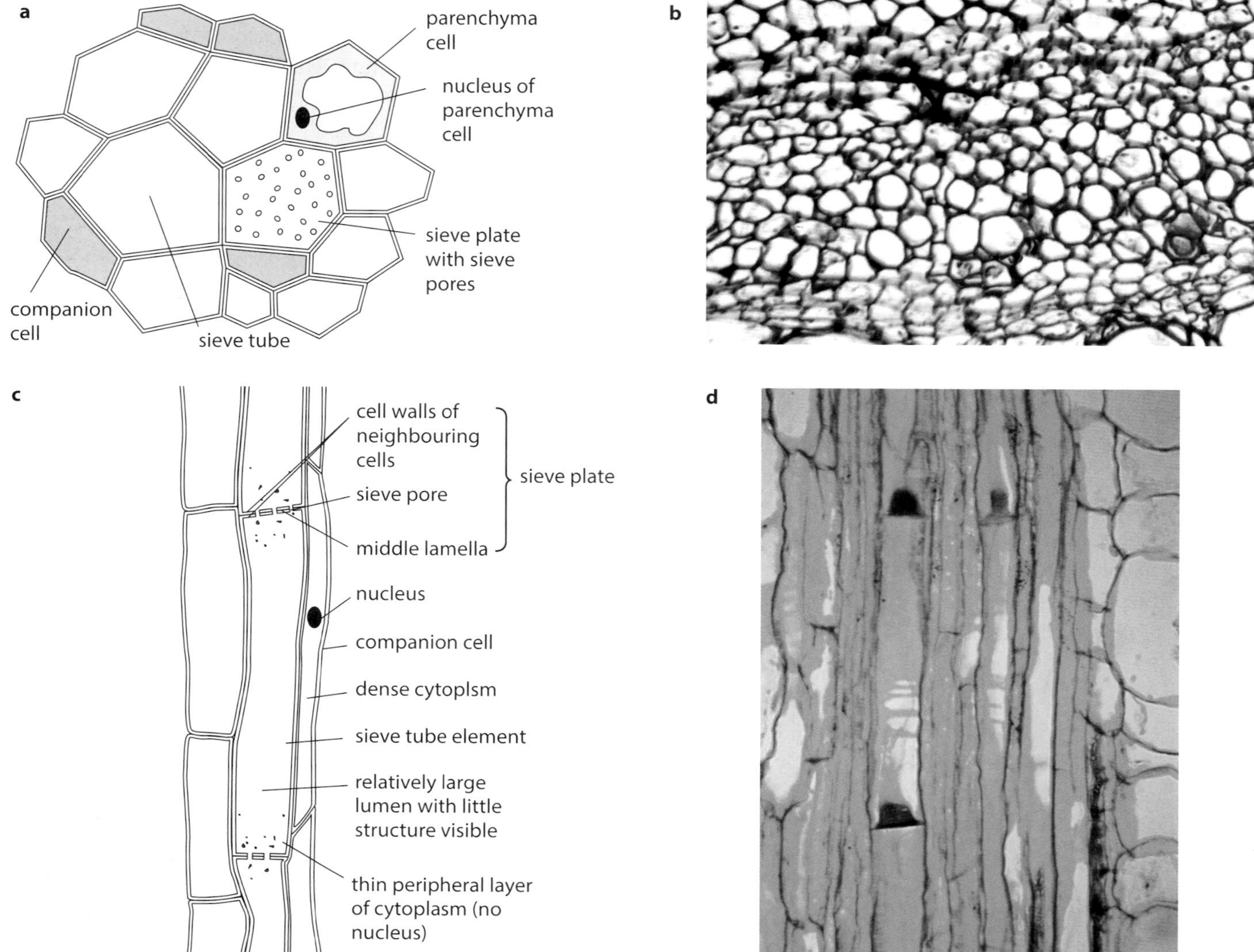

Figure 7.31 Structure of phloem. **a** Diagram in TS, **b** light micrograph in TS (×300), **c** diagram in LS, **d** light micrograph in LS (×200). The red triangles are patches of callose that formed at the sieve plates between the sieve tube elements in response to the damage done as the section was being cut. You can see companion cells, with their denser cytoplasm, lying alongside the sieve tube elements. On the far right are some parenchyma cells.

The contents of phloem sieve tubes

The liquid inside phloem sieve tubes is called **phloem sap**, or just sap. Table 7.3 shows the composition of the sap of the castor oil plant, *Ricinus communis*.

It is not easy to collect enough phloem sap to analyse its contents. The contents of sieve tubes are under very high pressure. When a sieve tube is cut, the release of pressure inside the tube causes a surge of its contents towards the cut. When the contents come up against a sieve plate, they may block it. This helps to prevent escape of the contents of the sieve tube. Then, within minutes, the sieve plate is properly sealed with a carbohydrate called **callose**, a process sometimes called 'clotting'. This can be seen in Figure 7.31. However, castor oil plants (used for the data in Table 7.3) are unusual in that their phloem sap does continue to flow from a cut for some time, making it relatively easy to collect.

In other plants, aphids may be used to sample sap. Aphids such as greenfly feed using tubular mouthparts called stylets. They insert these through the surface of the plant's stem or leaves, into the phloem (Figure 7.32). Phloem sap flows through the stylet into the aphid. If the stylet is cut near the aphid's head, the sap continues to flow; it seems that the small diameter of the stylet does not allow sap to flow out rapidly enough to trigger the plant's phloem 'clotting' mechanism.

Solute	Concentration / $mol\,dm^{-3}$
sucrose	250
potassium ions	80
amino acids	40
chloride ions	15
phosphate ions	10
magnesium ions	5
sodium ions	2
ATP	0.5
nitrate ions	0
plant growth substances (e.g. auxin, cytokinin)	small traces

Table 7.3 Composition of phloem sap.

QUESTION

7.11 Which of the substances listed in Table 7.3 have been synthesised by the plant?

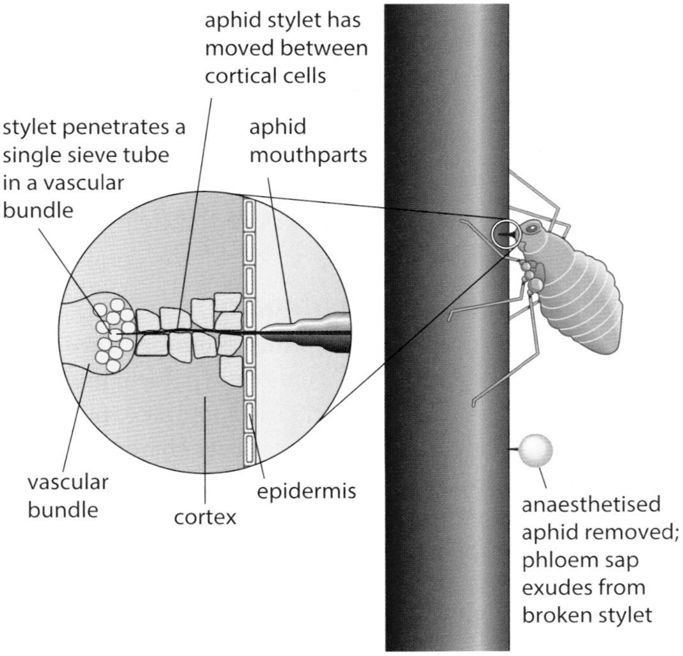

Figure 7.32 Using an aphid to collect phloem sap.

How translocation occurs

Phloem sap, like the contents of xylem vessels, moves by **mass flow**, as the aphid experiment shows. Mass flow moves organic solutes about $1\,m\,h^{-1}$ on average, about 10 000 times faster than diffusion would. Whereas in xylem vessels, the difference in pressure that causes mass flow is produced by a water potential gradient between soil and air, requiring no energy input from the plant, this is not so in phloem transport. To create the pressure differences needed for mass flow in phloem, the plant has to use energy. Phloem transport is therefore an **active** process, in contrast to the **passive** transport in xylem.

The pressure difference is produced by **active loading** of sucrose into the sieve elements at the place from which sucrose is to be transported. Any area of a plant in which sucrose is loaded into the phloem is called a **source**. This is usually a photosynthesising leaf or a storage organ. Any area where sucrose is taken out of the phloem is called a **sink** – for example, the roots.

Loading a high concentration of sucrose into a sieve element greatly decreases the water potential in the sap inside it. Therefore, water enters the sieve element, moving down a water potential gradient by osmosis. This causes a correspondingly high build up in pressure (equivalent to about six times atmospheric pressure). The pressure is referred to as hydrostatic pressure, turgor pressure or pressure potential. A pressure difference is therefore created between the source and the sink. This pressure difference causes a mass flow of water and dissolved solutes through the sieve tubes, from the high pressure area to the low pressure area (Figure 7.33). At the sink, sucrose may be

Figure 7.33 The phloem sap of the sugar maple (*Acer saccharum*) contains a high concentration of sugar and can be harvested to make maple syrup. Taps are inserted into each tree and the sap runs out under its own pressure through the plastic pipelines.

removed and used, causing the water to follow by osmosis, and thus maintaining the pressure gradient. Mass flow from source to sink is summarised in Figure 7.34.

Sinks can be anywhere in the plant, both above and below the photosynthesising leaves. Thus, sap flows both upwards and downwards in phloem (in contrast to xylem, in which flow is always upwards). Within any vascular bundle, phloem sap may be flowing upwards in some sieve tubes and downwards in others, but it can only flow one way in any particular sieve tube at any one time.

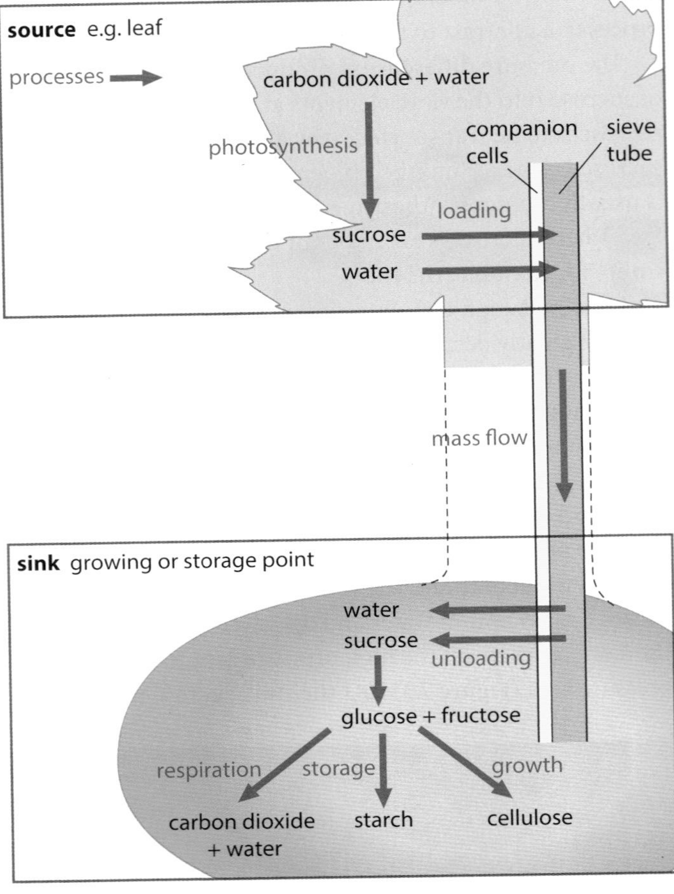

The sink is a growing point, e.g. young leaf, bud, flower or root, or a storage point, e.g. seed, fruit or tuber.

Figure 7.34 Sources, sinks and mass flow in phloem.

QUESTION

7.12 Which of the following are sources, and which are sinks?

a a nectary in a flower
b a developing fruit
c the storage tissue of a potato tuber when the buds are beginning to sprout
d a developing potato tuber

Loading sucrose into phloem

In leaf mesophyll cells, photosynthesis in chloroplasts produces triose sugars, some of which are converted into sucrose.

The sucrose, in solution, then moves from the mesophyll cell, across the leaf to the phloem tissue. It may move by the symplastic pathway (page 130), moving from cell to cell via plasmodesmata. Alternatively, it may move by the apoplastic pathway, travelling along cell walls. At the moment, little is known about how the sucrose crosses the cell surface membrane of the mesophyll cell to enter the apoplastic pathway. Which of these routes – the symplastic route or the apoplastic route – is more important depends on the species.

It is now known that the companion cells and sieve elements work together. Sucrose is loaded into a companion cell or directly into a sieve element by active transport (page 86). Figure 7.35 shows how this may be done. Hydrogen ions (H^+) are pumped out of the companion cell into its cell wall, using ATP as an energy source. This creates a large excess of hydrogen ions in the apoplastic pathway outside the companion cell. The hydrogen ions can move back into the cell down their concentration gradient, through a protein which acts as a carrier for both hydrogen ions and sucrose at the same time. The sucrose molecules are carried through this **co-transporter** molecule into the companion cell, against the concentration gradient for sucrose. The sucrose molecules can then move from the companion cell into the sieve tube, through the plasmodesmata which connect them (the symplastic pathway).

Unloading sucrose from phloem

Unloading occurs into any tissue which requires sucrose. It is probable that sucrose moves out of the phloem into these tissues using both symplastic and apoplastic routes, as with loading. Phloem unloading requires energy, and similar methods to those used for loading are probably used. Once in the tissue, the sucrose is converted into something else by enzymes, so decreasing its concentration and maintaining a concentration gradient. One such enzyme is invertase, which hydrolyses sucrose to glucose and fructose.

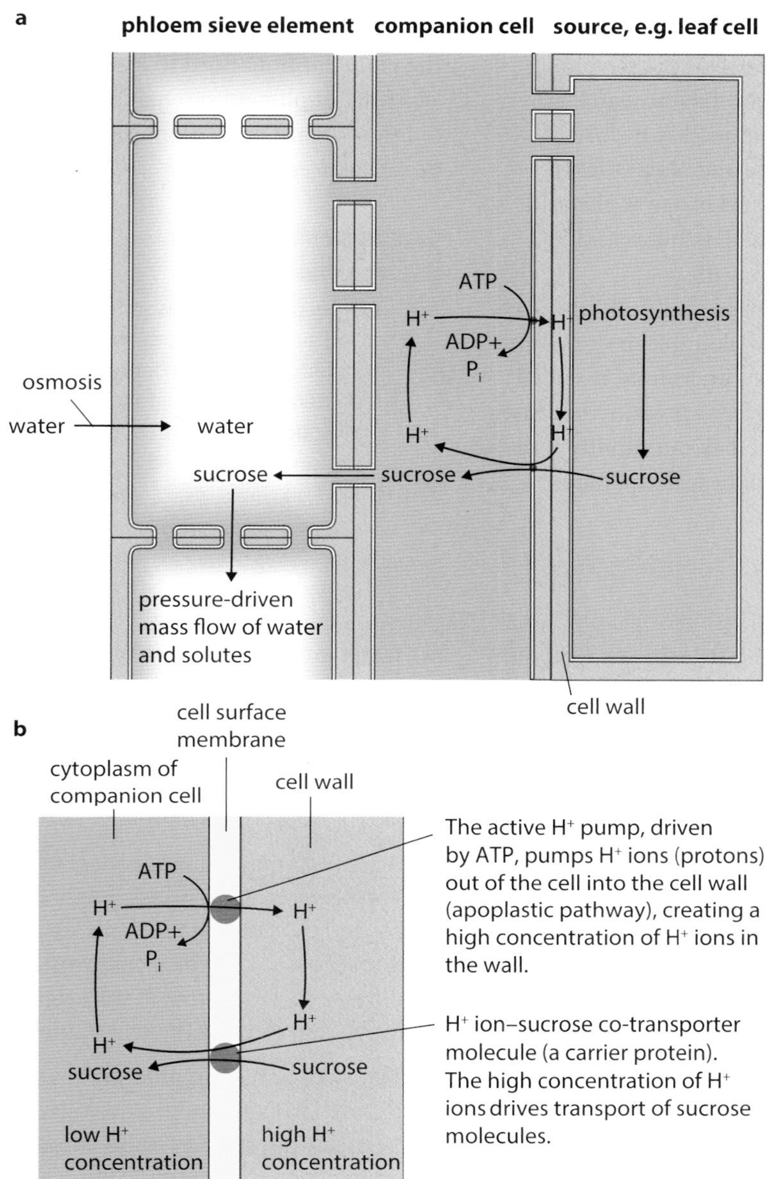

Figure 7.35 Loading phloem. **a** One of the possible methods by which sucrose is loaded and a pressure gradient generated. **b** Detail of the H^+ ion-sucrose co-transporter system.

Differences between sieve tubes and xylem vessels

From this account of translocation in the phloem, several similarities with the translocation of water in the xylem emerge. In each case, liquid moves by mass flow down a pressure gradient, through tubes formed by cells stacked end to end. So why are phloem sieve tubes so different in structure from xylem vessels?

Unlike water transport through xylem, which occurs through dead xylem vessels, translocation through phloem sieve tubes involves active loading of sucrose at sources and unloading at sinks, thus requiring living cells. Sucrose would leak out of xylem vessels, because they have no living membranes – membranes can control entry or loss of solutes.

Xylem vessels have lignified cell walls, whereas phloem sieve tubes do not. This is an advantage, because it means the cells can be dead and therefore entirely empty. Water can therefore flow unimpeded, and the dead xylem vessels with their strong walls also support the plant.

Sieve tubes also reduce resistance to flow by having only a thin layer of cytoplasm and no nuclei, but whereas the end walls of xylem elements disappear completely, those of phloem sieve elements form sieve plates. It is calculated that without the resistance of the sieve plates along the pathway, the steep positive pressure gradient inside the sieve tubes would quickly be lost, with the different pressures at source and sink quickly equilibrating. Xylem on the other hand has to withstand high **negative** pressure (tension) inside its tubes and buckling is prevented by its lignified walls.

Sieve plates also allow the phloem to seal itself up rapidly with callose if damaged – for example, by a grazing herbivore – rather as a blood vessel in an animal is sealed by clotting. Phloem sap has a high turgor pressure because of its high solute content, and would leak out rapidly if the holes in the sieve plate were not quickly sealed. Phloem sap contains valuable substances such as sucrose, which the plant cannot afford to lose in large quantity. The 'clotting' of phloem sap may also help to prevent the entry of microorganisms which might feed on the nutritious sap or cause disease.

QUESTION

7.13 Draw up a comparison table between xylem vessels and sieve tubes. Some features which you could include are: cell structure (walls, diameter, cell contents, etc.), substances transported, methods and direction of transport. Include a column giving a brief explanation for the differences in structure.

Summary

- Multicellular organisms with small surface area : volume ratios need transport systems. Flowering plants do not have compact bodies like those of animals. They spread and branch above and below ground to obtain the carbon dioxide, light energy, water and mineral ions needed for nutrition. Plants do not need systems for transporting carbon dioxide or oxygen – diffusion is sufficient.
- Water and mineral salts are transported through a plant in xylem vessels. Movement of water through a plant is a passive process in which the water moves down a water potential gradient from soil to air. The energy for this process comes from the Sun, which causes evaporation of water from the wet walls of mesophyll cells in leaves. Water vapour in the air spaces of the leaf diffuses out of the leaf through stomata, in a process called transpiration. This loss of water sets up a water potential gradient throughout the plant.
- Transpiration is an inevitable consequence of gaseous exchange in plants. Plants need stomata so that carbon dioxide and oxygen can be exchanged with the environment. The rate of transpiration is affected by several environmental factors, in particular temperature, light intensity, wind speed and humidity. It is diffcult to measure rate of transpiration directly, but water uptake can be measured using a potometer.
- Plants that are adapted to live in places where the environmental conditions are likely to cause high rates of transpiration, and where soil water is in short supply, are called xerophytes. Xerophytes have evolved adaptations that help to reduce the rate of loss of water vapour from their leaves.
- Water enters the plant through root hairs by osmosis. Water crosses the root either through the cytoplasm of cells (the symplastic pathway) or via their cell walls (the apoplastic pathway), and enters the dead, empty xylem vessels. Water also moves across the leaf by symplast and apoplast pathways.
- Water and mineral ions (xylem sap) move up xylem vessels by mass flow, as a result of pressure differences caused by loss of water from leaves by transpiration. Root pressure can also contribute to this pressure difference. Movement in the xylem is in one direction only, from roots to the rest of the plant.
- Translocation of organic solutes such as sucrose occurs through living phloem sieve tubes. Phloem sap moves by mass flow from a region known as the source to a region known as the sink.
- Sucrose is produced at the source (e.g. photosynthesising leaves) and used at the sink (e.g. a flower or a storage organ). Mass flow occurs as a result of pressure differences between the source and the sink. Active loading of sucrose into the sieve tubes at the source results in the entry of water by osmosis, thus creating a high hydrostatic pressure in the sieve tubes. Phloem sap can move in different directions in different sieve tubes.
- Both xylem vessels and phloem sieve tubes show unique structural features and distributions which are adaptations to their roles in transport.

End-of-chapter questions

1 If sucrose is actively loaded into a sieve tube, which combination of changes takes place in the sieve tube?

	Solute potential	Hydrogen ion concentration
A	decreases (becomes more negative)	decreases
B	decreases (becomes more negative)	increases
C	increases (becomes less negative)	decreases
D	increases (becomes less negative)	increases

[1]

2 Which of the following rows correctly describes the hydrostatic pressure of the two types of elements?

	Hydrostatic pressure	
	Xylem vessel element	Phloem sieve element
A	negative	negative
B	negative	positive
C	positive	negative
D	positive	positive

[1]

3 The diagram shows the effect of light intensity on the rate of transpiration from the upper and lower epidermis of a leaf. Other environmental factors were kept constant. What could explain the differences in transpiration rates from the two surfaces?

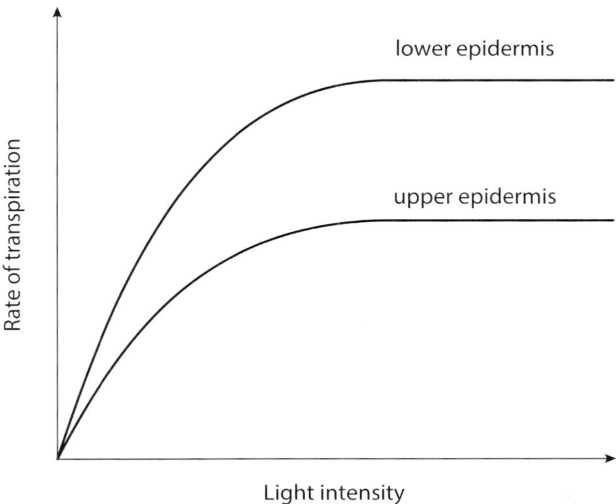

A Higher light intensities are associated with higher temperatures.
B The palisade mesophyll cells have fewer air spaces than the spongy mesophyll cells.
C The upper epidermis has fewer stomata.
D The upper epidermis is more exposed to light. [1]

4 Explain how water moves from:

a the soil into a root hair cell [3]
b one root cortex cell to another [4]
c a xylem vessel into a leaf mesophyll cell. [3]

[Total: 10]

5 a Name **three** cell types found in:

i xylem [3]

ii phloem. [3]

b State the functions of the cell types you have named. [4]

[Total: 10]

6 Arrange the following in order of water potential. Use the symbol > to mean 'greater than'.

dry atmospheric air; mesophyll cell; root hair cell; soil solution; xylem vessel contents [1]

7 Figure **a** shows changes in the relative humidity of the atmosphere during the daylight hours of one day. Figure **b** shows changes in the tension in the xylem of a tree during the same period.

a

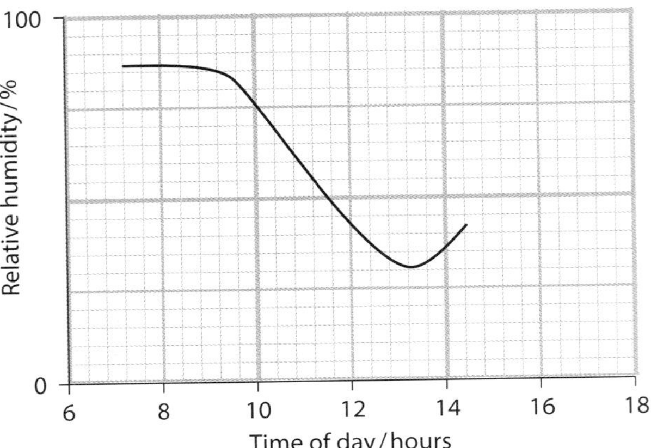

b

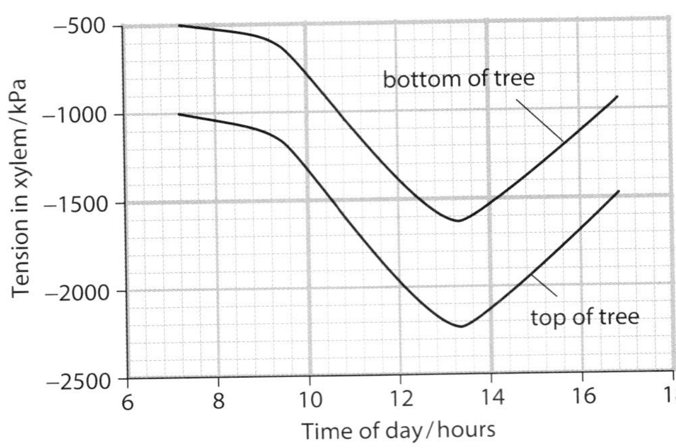

a Describe and explain the relationship between relative humidity and xylem tension. [4]

b Describe and explain the differences observed in xylem tension between the top of the tree and the bottom of the tree. [3]

[Total: 7]

8 An instrument called a dendrogram can be used to measure small changes in the diameter of a tree trunk. Typically, the instrument reveals daily changes, with the diameter at its lowest during daylight hours and at its greatest at night.

Suggest an explanation for these observations. [3]

9 The graph below shows the relationship between rate of transpiration and rate of water uptake for a particular plant.

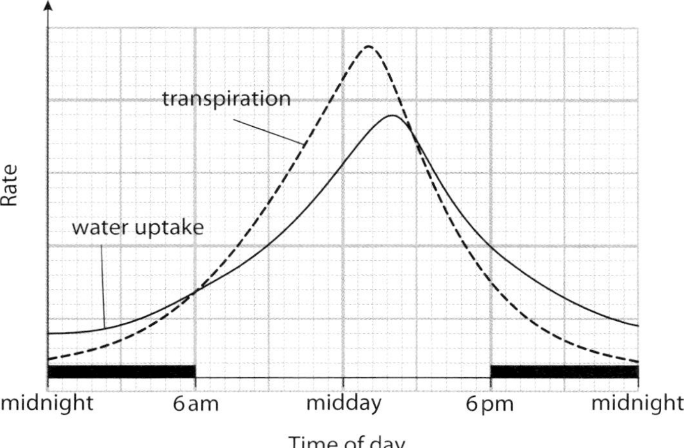

a Define the term **transpiration**. [2]

b State the **two** environmental factors which are most likely to be responsible for the changes in transpiration rate shown in the graph above. [2]

c Describe the relationship between rate of transpiration and rate of water uptake shown in the graph above. [2]

d Explain the relationship. [4]

[Total: 10]

10 The graph below shows the results of two experiments, **A** and **B**. In both cases the uptake of potassium ions into the roots of young plants was measured. Roots were thoroughly washed in pure water before the plants were added to solutions containing potassium ions. Experiment **A** was carried out at 25 °C and experiment **B** at 0 °C. In experiment **A**, potassium cyanide was added to the solution surrounding the roots after 90 minutes (see arrow in figure). Potassium cyanide is an inhibitor of respiration.

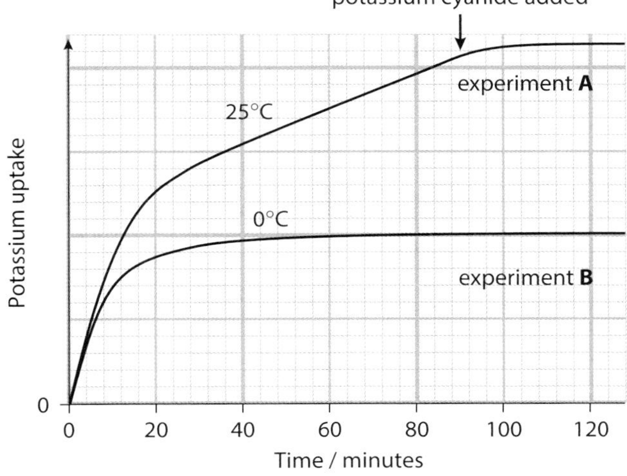

a Suggest why the roots were thoroughly washed in pure water before the experiments. [2]

b Describe the results of experiment **A** over the first 80 minutes of the experiment. [3]

c Describe the results of experiment **B**. [3]

d Explain the results described in **b** and **c** above. [max 5]

e Describe the effect of adding potassium cyanide in experiment **A**. [1]

f Explain the effect of adding potassium cyanide in experiment **A**. [2]

[Total: 16]

11 Explain how active loading of sucrose into sieve elements accounts for the following observations:

- **a** The phloem sap has a relatively high pH of about pH 8. [1]
- **b** The inside of sieve element/companion cell units is negatively charged relative to the outside. (There is a difference in electrical potential across the cell surface membrane, with a potential of about –150 mV on the inside.) [2]
- **c** ATP is present in relatively large amounts inside sieve tubes. [1]

[Total: 4]

12 Figure 7.31d on page 148 shows a sieve element with red-stained 'triangles' of callose at each end. These triangles indicate the positions of the sieve plates.

- **a** Assuming the magnification of the micrograph is $\times 100$, calculate the length of the sieve element. Show your working. (You may find it useful to refer to Chapter 1, Worked example 4, page 9, for the method of calculation.) [3]
- **b** Scientists were puzzled for many years by the fact that sieve plates were present in sieve elements, because sieve plates increase the resistance to flow. This contrasts with xylem vessel elements, which have open ends, reducing resistance to flow.
  - **i** Calculate how many sieve plates per metre a sucrose molecule would have to cross if it were travelling in the sieve tube identified in **a** above. Show your working. (Assume all the sieve elements are the same size as the one measured in Figure 7.31d.) [2]
  - **ii** What is the function of the sieve plates? [1]
  - **iii** What feature of the sieve plates allows materials to cross them? [1]
- **c** Flow rates in sieve tubes range from 0.3 to 1.5 m h^{-1} and average about 1 m h^{-1}. If the flow rate in the sieve element shown in Figure 7.31d were 1 m h^{-1}, how long would it take a sucrose molecule to travel through it? Show your working. [3]

[Total: 10]

13 Translocation of organic solutes takes place between sources and sinks.

- **a** Briefly explain under what circumstances:
  - **i** a seed could be a sink [1]
  - **ii** a seed could be a source [1]
  - **iii** a leaf could be a sink [1]
  - **iv** a leaf could be a source [1]
  - **v** a storage organ could be a sink [1]
  - **vi** a storage organ could be a source. [1]
- **b** Suggest **two** possible roles for glucose in each of the following sinks:
  - **i** a storage organ [2]
  - **ii** a growing bud. [2]

[Total: 10]

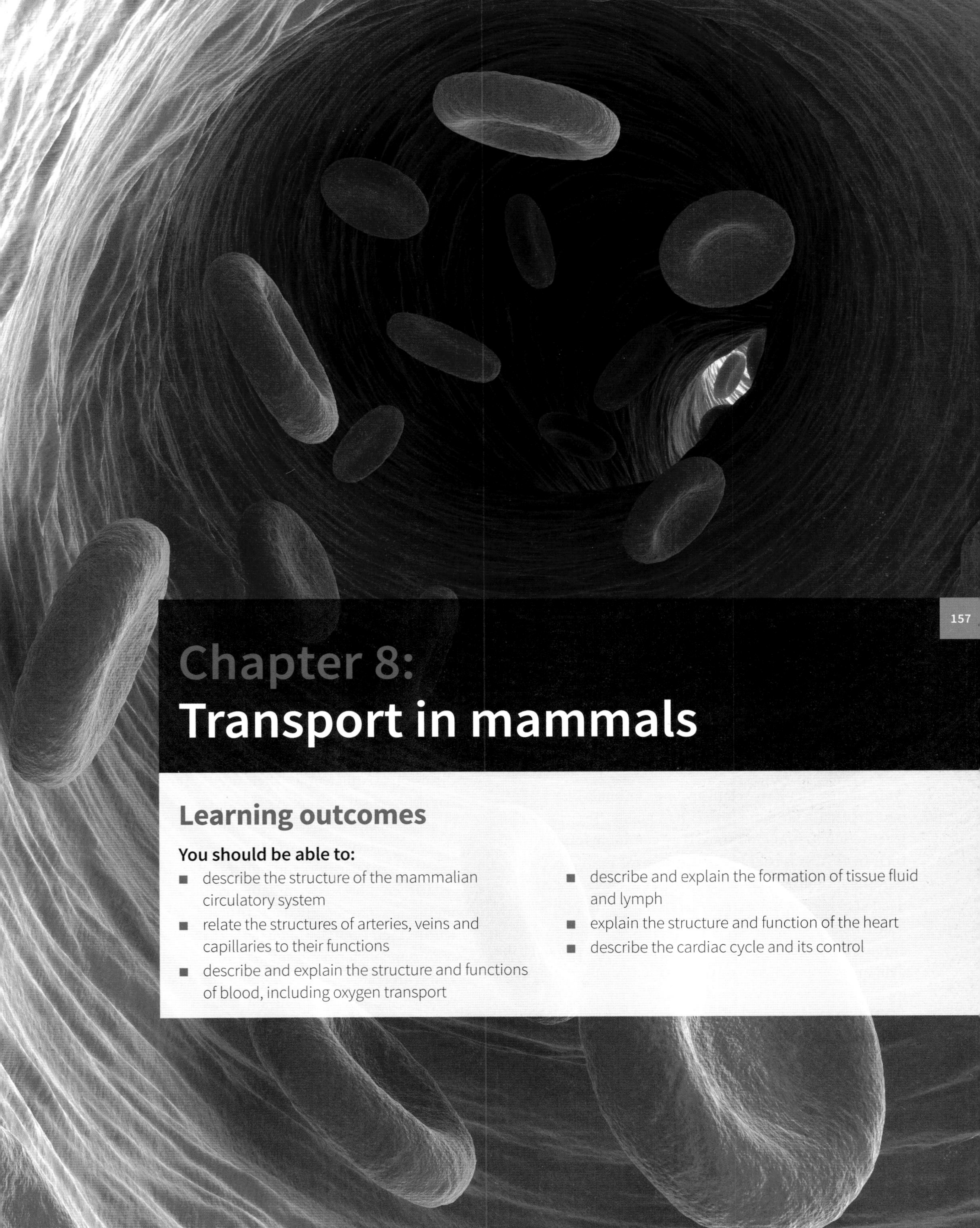

Chapter 8: Transport in mammals

Learning outcomes

You should be able to:

- describe the structure of the mammalian circulatory system
- relate the structures of arteries, veins and capillaries to their functions
- describe and explain the structure and functions of blood, including oxygen transport
- describe and explain the formation of tissue fluid and lymph
- explain the structure and function of the heart
- describe the cardiac cycle and its control

Artificial hearts

Each year, over seven million people woldwide die from heart disease, more than from any other condition. In many countries, medical help is available for people with heart disease, ranging from treatment with drugs to major heart surgery. But, until recently, the only hope for some heart patients was a heart transplant. However, the number of people needing a new heart far outweighs the number of hearts available. Many people wait years for a heart transplant, and many die from their heart disease before they get a new heart.

Petar Bilic (not his real name) thought that he was going to add to that statistic. The muscle in both of his ventricles had deteriorated so much that his heart was only just keeping him alive. No suitable heart could be found for a transplant.

Petar was very lucky. In recent years, biomedical engineers have developed a pumping device called a 'total artificial heart' (Figure 8.1). Petar's heart was completely removed, and an artificial heart put in its place. Petar was able to go home within a few weeks of his operation. The plan is that the artificial heart will keep him alive until a real heart is available for transplant.

Recently, biomedical engineers have made progress in developing a new type of heart that should work for much longer – perhaps long enough for its owner to live out a long life without any need for a heart transplant. This heart works in a different way, causing blood to flow smoothly rather than in rhythmic pulses. It would be strange to be alive, but not to have a pulse.

Figure 8.1 Moments after this photograph was taken, this artificial heart was successfully inserted into the body of a 62-year-old man, completely replacing his own heart.

Transport systems in animals

Animals, particularly relatively large animals like mammals, are far more active than plants. They rely on locomotion to find food since they cannot manufacture their own. This is energy consuming; for example, it may require the contraction of muscles. A nervous system is needed to coordinate activity and sending nerve impulses is also energy consuming. Large brains consume large amounts of energy. The greater the activity, the greater the demand for oxygen for respiration, so transport systems for oxygen have evolved, unlike in plants.

The larger and more complex animals become, the greater the demands on the transport systems. All animal cells require a supply of nutrients, such as glucose. Most living cells also need a constant supply of oxygen. There will also be waste products, such as carbon dioxide, to be disposed of. Very small organisms, such as *Paramecium*, can meet their requirements for the supply of nutrients and oxygen, and the removal of waste products, by means of diffusion. The very small distances that substances have to travel means that the speed of supply or removal by diffusion is sufficient for their needs. These tiny organisms have a large surface area compared to their total volume (a large surface area : volume ratio – see Chapter 4), so there is a relativity large area of membrane across which gases can diffuse into and out of their bodies.

Larger, more active organisms, such as insects, fish and mammals, cannot rely on diffusion alone. Cells, often deep within their bodies, are metabolically very active, with requirements for rapid supplies of nutrients and oxygen, and with relatively large amounts of waste products to be removed. These organisms have well-organised transport systems, with pumps to keep fluid moving through them.

The mammalian cardiovascular system

Figure 8.2 shows the general layout of the main transport system of mammals – the blood system or **cardiovascular system**. It is made up of a pump, the **heart**, and a system of interconnecting tubes, the **blood vessels**. The blood always

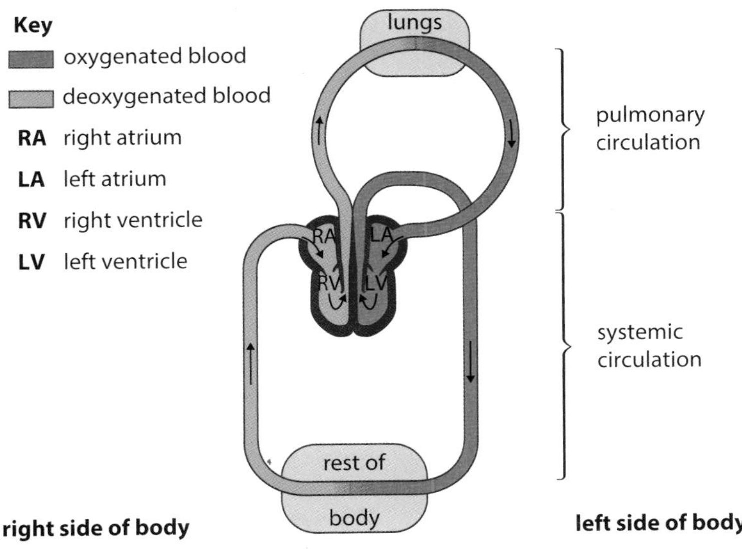

Figure 8.2 The general plan of the mammalian transport system, viewed as though looking at someone facing you. It is a closed double circulatory system.

remains within these vessels, and so the system is known as a **closed** blood system.

If you trace the journey of the blood around the body, beginning in the left ventricle of the heart, you will find that the blood travels twice through the heart on one complete 'circuit'. Blood is pumped out of the left ventricle into the **aorta** (Figure 8.3), and travels from there to all parts of the body except the lungs. It returns to the right side of the heart in the **vena cava**. This is called the **systemic circulation**.

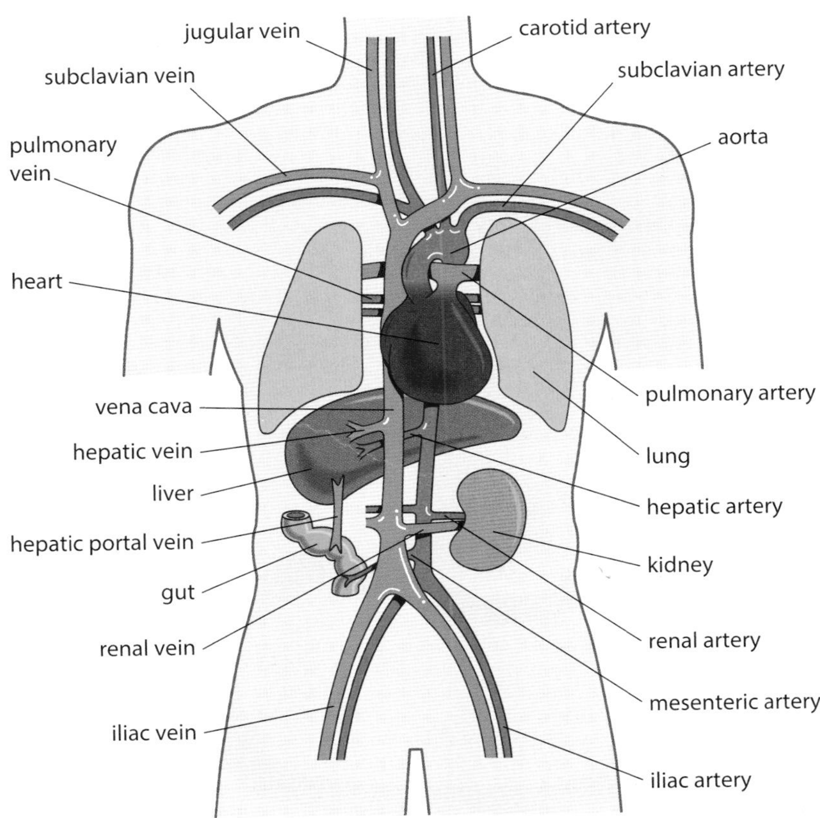

Figure 8.3 The positions of some of the main blood vessels in the human body.

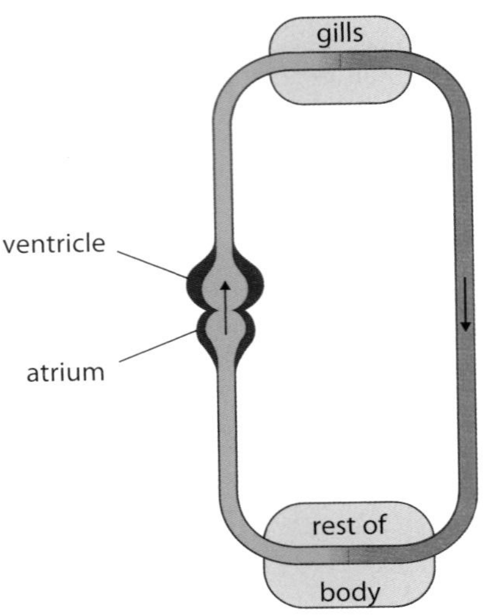

Figure 8.4 The general plan of the transport system of a fish.

The blood is then pumped out of the right ventricle into the **pulmonary arteries**, which carry it to the lungs. The final part of the journey is along the **pulmonary veins**, which return it to the left side of the heart. This is called the **pulmonary circulation**.

QUESTION

8.1 Figure 8.4 shows the general layout of the circulatory system of a fish.

- **a** How does this differ from the circulatory system of a mammal?
- **b** Suggest the possible advantages of the design of the mammalian circulatory system over that of a fish.

This combination of pulmonary circulation and systemic circulation makes a **double circulatory system**. The mammalian circulatory system is therefore a closed double circulation.

Blood vessels

The vessels making up the blood system are of three main types. Figure 8.5 shows these vessels in transverse section. Vessels carrying blood **away** from the heart are known as arteries, while those carrying blood **towards** the heart are veins. Linking arteries and veins, taking blood close to almost every cell in the body, are tiny vessels called capillaries.

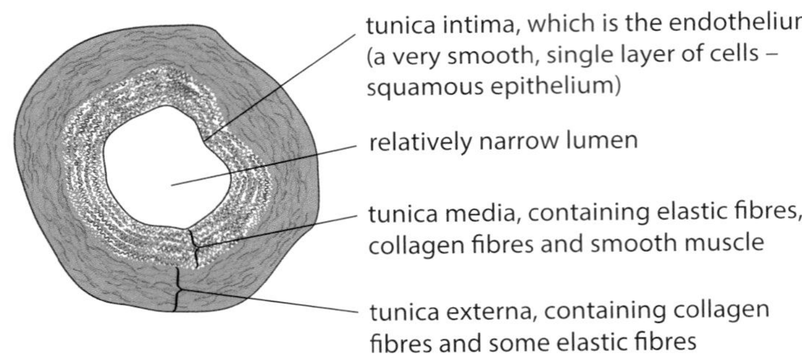

Arteries in different parts of the body vary in their structure. Arteries near the heart have especially large numbers of elastic fibres in the tunica media, as shown here. In other parts of the body, the tunica media contains less elastic tissue and more smooth muscle.

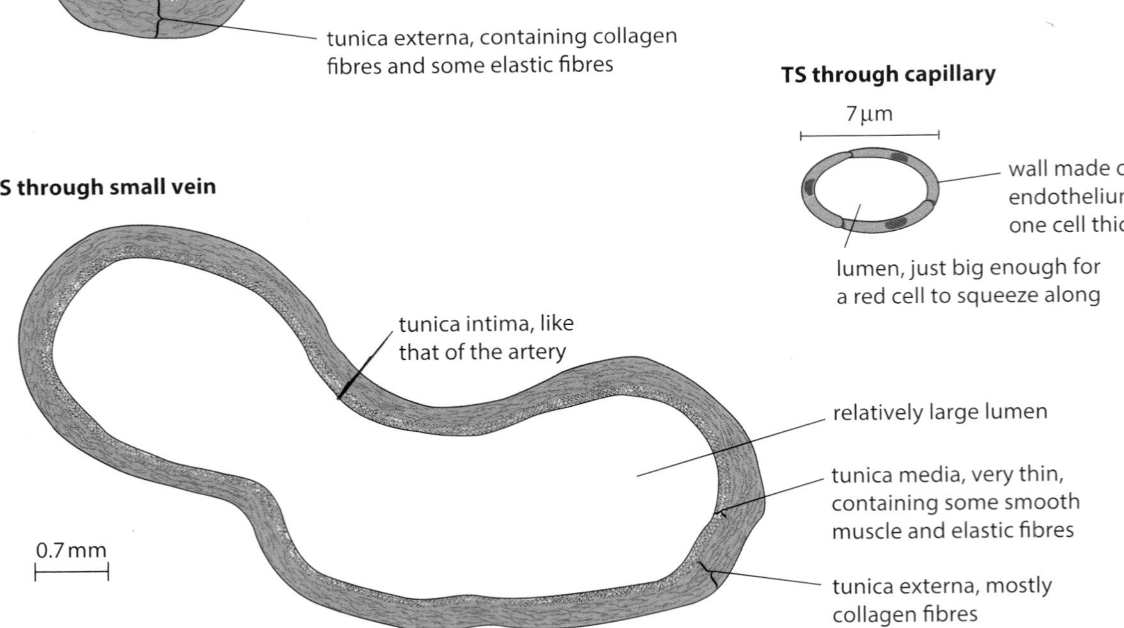

Figure 8.5 The tissues making up the walls of arteries, capillaries and veins.

Arteries

The function of arteries is to transport blood, swiftly and at high pressure, to the tissues.

The structure of the wall of an artery enables it to perform this function efficiently. Arteries and veins both have walls made up of three layers:

- an inner endothelium (lining tissue), called the **tunica intima**, made up of a layer of flat cells (squamous epithelium) fitting together like jigsaw pieces; this layer is very smooth, minimising friction with the moving blood
- a middle layer called the **tunica media** ('middle coat'), containing smooth muscle, collagen and elastic fibres
- an outer layer called the **tunica externa** ('outer coat'), containing elastic fibres and collagen fibres.

The distinctive characteristics of an artery wall are its strength and elasticity. Blood leaving the heart is at a very high pressure. Blood pressure in the human aorta may be around 120 mm Hg, or 16 kPa. To withstand such pressure, artery walls must be extremely strong. This is achieved by the thickness and composition of the artery wall.

Arteries have the thickest walls of any blood vessel (Figure 8.6). The aorta, the largest artery, has an overall diameter of 2.5 cm close to the heart, and a wall thickness of about 2 mm. Although this may not seem very great, the composition of the wall provides great strength and resilience. The tunica media, which is by far the thickest part of the wall, contains large amounts of elastic fibres. These allow the wall to stretch as pulses of blood surge through at high pressure. Arteries further away from the heart have fewer elastic fibres in the tunica media, but have more muscle fibres.

The elasticity of artery walls is important in allowing them to 'give', which reduces the likelihood that they will burst. This elasticity also has another very important function. Blood is pumped out of the heart in pulses, rushing out at high pressure as the ventricles contract, and slowing as the ventricles relax. The artery walls stretch as the high-pressure blood surges into them, and then recoil inwards as the pressure drops. Therefore, as blood at high pressure enters an artery, the artery becomes wider, reducing the pressure a little. As blood at lower pressure enters an artery, the artery wall recoils inwards, giving the blood a small 'push' and raising the pressure a little. The overall effect is to 'even out' the flow of blood. However, the arteries are not entirely effective in achieving this: if you feel your pulse in your wrist, you can feel the artery, even at this distance from your heart, being stretched outwards with each surge of blood from the heart.

Blood pressure is still measured in the old units of mm Hg even though kPa is the SI unit. mm Hg stands for 'millimetres of mercury', and refers to the distance which mercury is pushed up the arm of a U-tube. 1 mm Hg is equivalent to about 0.13 kPa.

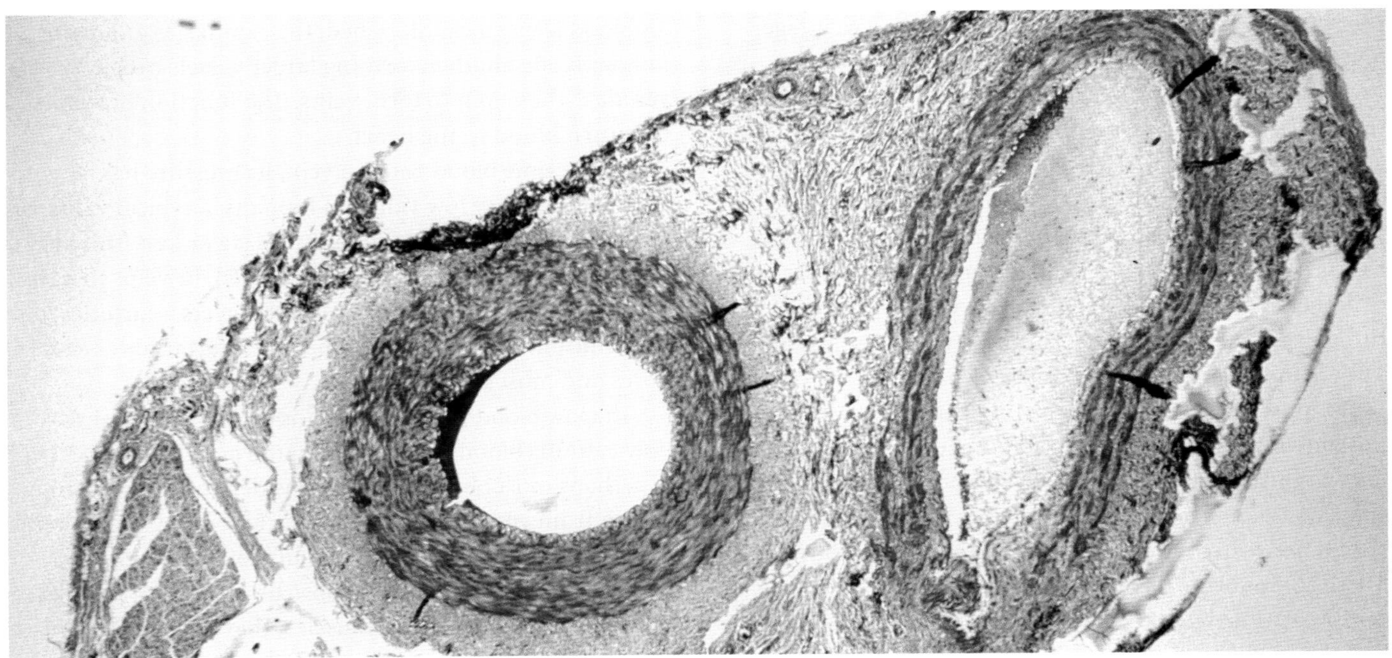

Figure 8.6 Photomicrograph of an artery (left) and a vein (right) (×110).

As arteries reach the tissue to which they are transporting blood, they branch into smaller and smaller vessels, called arterioles. The walls of arterioles are similar to those of arteries, but they have a greater proportion of smooth muscle. This muscle can contract, narrowing the diameter of the arteriole and so reducing blood flow. This helps to control the volume of blood flowing into a tissue at different times. For example, during exercise, arterioles that supply blood to muscles in your legs would be wide (dilated) as their walls relax, while those carrying blood to the gut wall would be narrow (constricted).

Blood pressure drops significantly as blood flows through the arteries. As blood enters an arteriole it may have a pressure of 85 mm Hg (11.3 kPa) but as it leaves and flows into a capillary, the pressure will have dropped to about 35 mm Hg (4.7 kPa).

QUESTION

8.2 Suggest why arteries close to the heart have more elastic fibres in their walls than arteries further away from the heart.

Capillaries

The arterioles themselves continue to branch, eventually forming the tiniest of all blood vessels, **capillaries**. The function of capillaries is to take blood as close as possible to all cells, allowing rapid transfer of substances between cells and blood. Capillaries form a network throughout every tissue in the body except the cornea and cartilage. Such networks are sometimes called **capillary beds**.

The small size of capillaries is obviously of great importance in allowing them to bring blood as close as possible to each group of cells in the body. A human capillary is approximately 7 μm in diameter, about the same size as a red blood cell (Figure 8.7). Moreover, the walls of capillaries are extremely thin, made up of a single layer of endothelial cells. As red blood cells carrying oxygen squeeze through a capillary, they are brought to within as little as 1 μm of the cells outside the capillary which need the oxygen.

In most capillaries, there are tiny gaps between the individual cells that form the endothelium. As we shall see later in this chapter, these gaps are important in allowing some components of the blood to seep through into the spaces between the cells in all the tissues of the body. These components form tissue fluid.

By the time blood reaches the capillaries, it has already lost a great deal of the pressure originally supplied to it by the contraction of the ventricles. As blood enters a capillary from an arteriole, it may have a pressure of around 35 mm Hg or 4.7 kPa; by the time it reaches the far end of the capillary, the pressure will have dropped to around 10 mm Hg or 1.3 kPa.

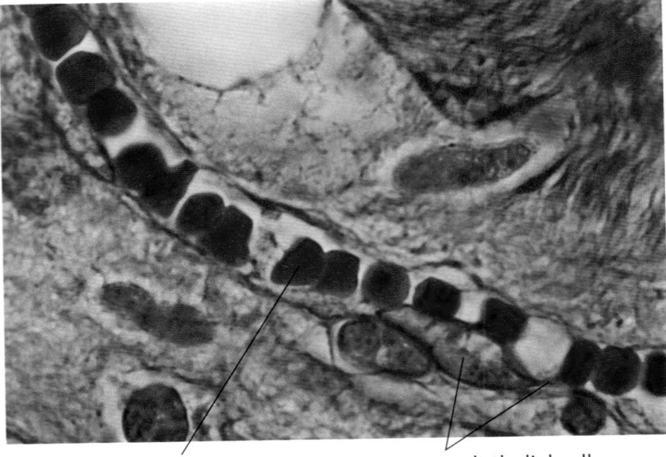

Figure 8.7 Photomicrograph of a blood capillary containing red blood cells (dark red) (×900). The cells of the endothelium are very thin, except where there is a nucleus (red).

QUESTION

8.3 Suggest why there are no blood capillaries in the cornea of the eye. How might the cornea be supplied with its requirements?

Veins

As blood leaves a capillary bed, the capillaries gradually join with one another, forming larger vessels called **venules**. These join to form **veins**. The function of veins is **to return blood to the heart**.

By the time blood enters a vein, its pressure has dropped to a very low value. In humans, a typical value for venous blood pressure is about 5 mm Hg or less. This very low pressure means that there is no need for veins to have thick walls. They have the same three layers as arteries, but the tunica media is much thinner and has far fewer elastic fibres and muscle fibres.

The low blood pressure in veins creates a problem: how can this blood be returned to the heart? The problem is perhaps most obvious if you consider how blood can return from your legs. Unaided, the blood in your leg veins would sink and accumulate in your feet. However, many of the veins run within, or very close to, several leg muscles. Whenever you tense these muscles, they squeeze inwards on the veins in your legs, temporarily raising the pressure within them.

This squeezing, in itself, would not help to push the blood back towards the heart; blood would just squidge up and down as you walked. To keep the blood flowing in the right direction, veins contain half-moon valves, or **semilunar valves**, formed from their endothelium (Figure 8.8). These valves allow blood to move towards the heart, but not away from it. Thus, when you contract your leg muscles, the blood in the veins is squeezed **up** through these valves, but cannot pass **down** through them. Note that blood samples are normally taken from veins, rather than from arteries, because of the lower pressure in them (Figure 8.9).

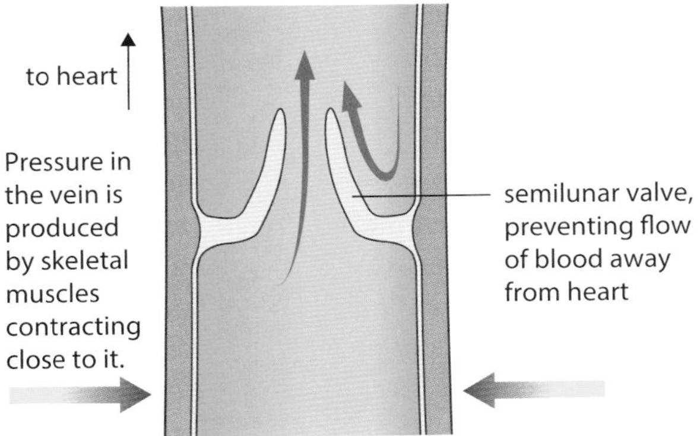

Figure 8.8 Longitudinal section through a small vein and a valve.

Figure 8.10 shows how blood pressure changes as the blood travels on one complete journey from the heart, through the systemic circulatory system, back to the heart and then through the pulmonary circulatory system.

QUESTION

8.4 Suggest reasons for each of the following.

a Normal venous pressure in the feet is about 25 mm Hg. When a soldier stands motionless at attention, the blood pressure in his feet rises very quickly to about 90 mm Hg.

b When you breathe in – that is, when the volume of the thorax increases – blood moves through the veins towards the heart.

8.5 Using Figure 8.10, describe and explain how blood pressure varies in different parts of the circulatory system.

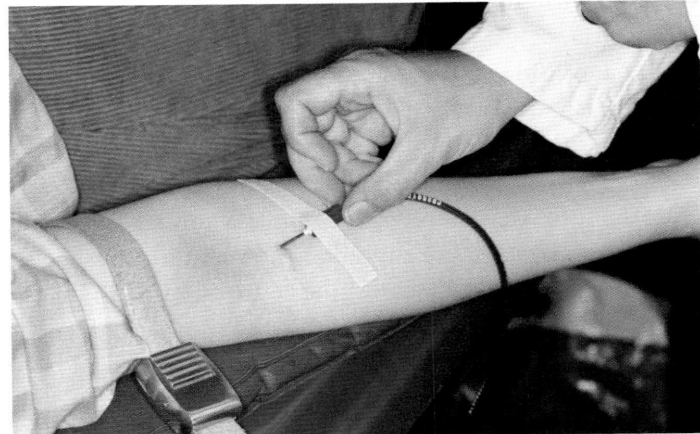

Figure 8.9 Veins are closer to the skin surface than arteries, and blood flows more gently and smoothly in them. They are therefore used when a blood sample is being taken, or to donate blood for transfusions.

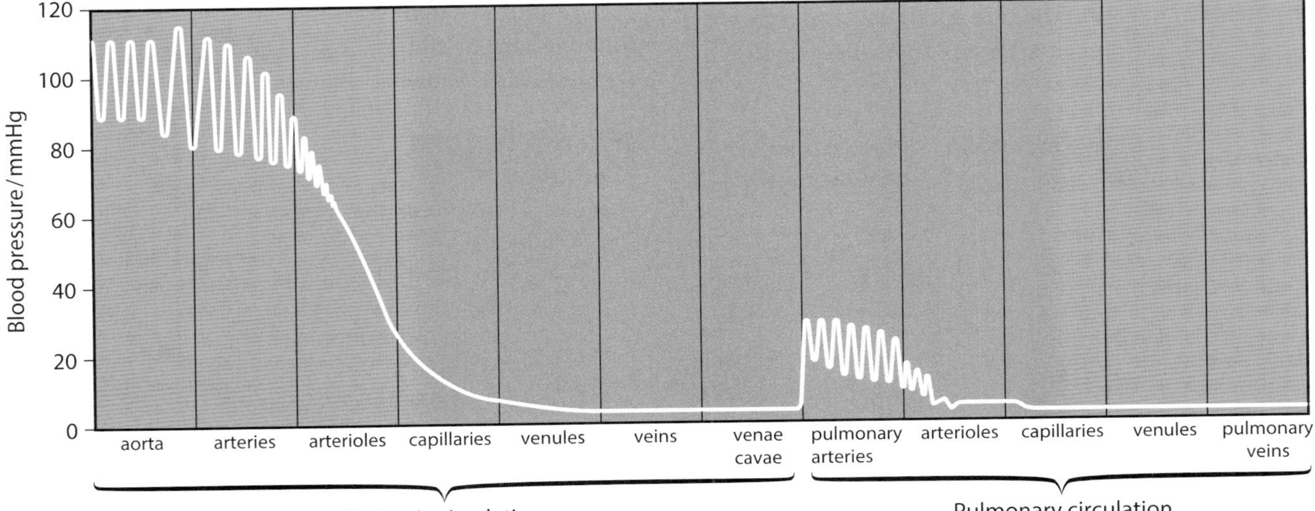

Figure 8.10 Blood pressure in different regions of the human circulatory system.

Blood plasma and tissue fluid

Blood is composed of cells floating in a pale yellow liquid called **plasma**. Blood plasma is mostly water, with a variety of substances dissolved in it. These solutes include nutrients such as glucose and waste products such as urea that are being transported from one place to another in the body. Solutes also include protein molecules, called **plasma proteins**, that remain in the blood all the time.

As blood flows through capillaries within tissues, some of the plasma leaks out through the gaps between the cells in the walls of the capillary, and seeps into the spaces between the cells of the tissues. Almost one-sixth of your body consists of spaces between your cells. These spaces are filled with this leaked plasma, which is known as **tissue fluid**.

Tissue fluid is almost identical in composition to blood plasma. However, it contains far fewer protein molecules than blood plasma, as these are too large to escape easily through the tiny holes in the capillary endothelium. Red blood cells are much too large to pass through, so tissue fluid does not contain these, but some white blood cells can squeeze through and move freely around in tissue fluid. Table 8.1 shows the sizes of the molecules of some of the substances in blood plasma, and the relative ease with which they pass from capillaries into tissue fluid.

The volume of fluid which leaves the capillary to form tissue fluid is the result of two opposing pressures. Particularly at the arterial end of a capillary bed, the blood pressure inside the capillary is enough to push fluid out into the tissue. However, we have seen that water moves by osmosis from regions of low solute concentration (high water potential) to regions of high solute concentration (low water potential) (page 83). Since tissue fluid lacks the high concentrations of proteins that exist in plasma, the imbalance leads to osmotic movement of water back into capillaries from tissue fluid. The net result of these competing processes is that fluid tends to flow **out** of capillaries into tissue fluid at the **arterial** end of a capillary bed and **into** capillaries from tissue fluid near the **venous** end of a capillary bed. Overall, however, rather more fluid flows out of capillaries than into them, so that there is a net loss of fluid from the blood as it flows through a capillary bed.

If blood pressure is too high, too much fluid is forced out of the capillaries, and may accumulate in the tissues. This build-up of fluid is called **oedema**. One of the roles of arterioles is to reduce the pressure of the blood that enters the arterioles, in order to avoid this.

Tissue fluid forms the immediate environment of each individual body cell. It is through tissue fluid that exchanges of materials between cells and the blood occur. Within our bodies, many processes take place to maintain the composition of tissue fluid at a constant level, to provide an optimum environment in which cells can work. These processes contribute to the overall process of **homeostasis** – that is, the maintenance of a constant internal environment – and include the regulation of glucose concentration, water, pH, metabolic wastes and temperature.

Substance	Relative molecular mass	Permeability of capillary walls
water	18	1.00
sodium ions	23	0.96
urea	60	0.8
glucose	180	0.6
haemoglobin	68 000	0.01
albumin	69 000	0.000 01

Table 8.1 Relative permeability of capillaries in a muscle to different substances. The permeability to water is given a value of 1. The other values are given in proportion to that of water.

Lymph

About 90% of the fluid that leaks from capillaries eventually seeps back into them. The remaining 10% is collected up and returned to the blood system by means of a series of tubes known as **lymph vessels** or **lymphatics**.

Lymphatics are tiny, blind-ending vessels, which are found in almost all tissues of the body. The end of one of these vessels is shown in Figure 8.11. Lymphatics contain

QUESTION

8.6 Use the information in Table 8.1 to answer the following.

- **a** Describe the relationship between the relative molecular mass of a substance and the permeability of capillary walls to this substance.
- **b** In a respiring muscle, would you expect the net diffusion of glucose to be **from** the blood plasma to the muscle cells or vice versa? Explain your answer.
- **c** Albumin is the most abundant plasma protein. Suggest why it is important that capillary walls should not be permeable to albumin.

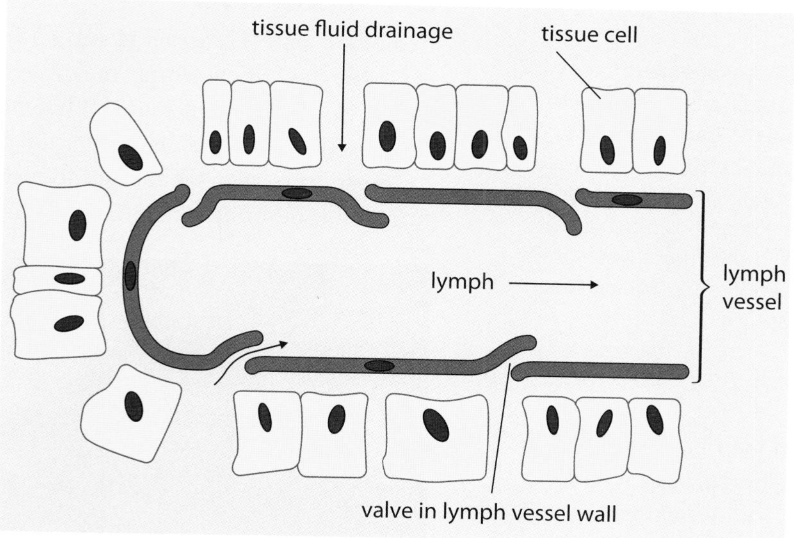

Figure 8.11 Drainage of tissue fluid into a lymph vessel.

tiny valves, which allow the tissue fluid to flow in but stop it from leaking out.

The valves in the lymph vessel walls are wide enough to allow large protein molecules to pass through. This is very important because such molecules are too big to get into blood capillaries, and so cannot be taken away by the blood. If your lymphatics did not take away the protein in the tissue fluid between your cells, you could die within 24 hours. If the protein concentration and rate of loss from plasma are not in balance with the concentration and rate of loss from tissue fluid, oedema may result.

The fluid inside lymphatics is called **lymph**. Lymph is virtually identical to tissue fluid; it has a different name more because it is in a different place than because it is different in composition.

In some tissues, the tissue fluid, and therefore the lymph, is rather different from that in other tissues. For example, the tissue fluid and lymph in the liver have particularly high concentrations of protein. High concentrations of lipids are found in lymph in the walls of the small intestine shortly after a meal. Here, lymphatics are found in each villus, where they absorb lipids from digested food.

Lymphatics join up to form larger lymph vessels, that gradually transport the lymph back to the large veins that run just beneath the collarbone, the **subclavian veins** (Figure 8.12). As in veins, the movement of fluid along the lymphatics is largely caused by the contraction of muscles around the vessels, and kept going in the right direction by valves. Lymph vessels also have smooth muscle in their walls, which can contract to push the lymph along. Lymph flow is very slow, and only about 100 $cm^3 h^{-1}$ flows through the largest lymph vessel, the thoracic duct, in a resting human. This contrasts with the flow rate of blood, of around 80 $cm^3 s^{-1}$.

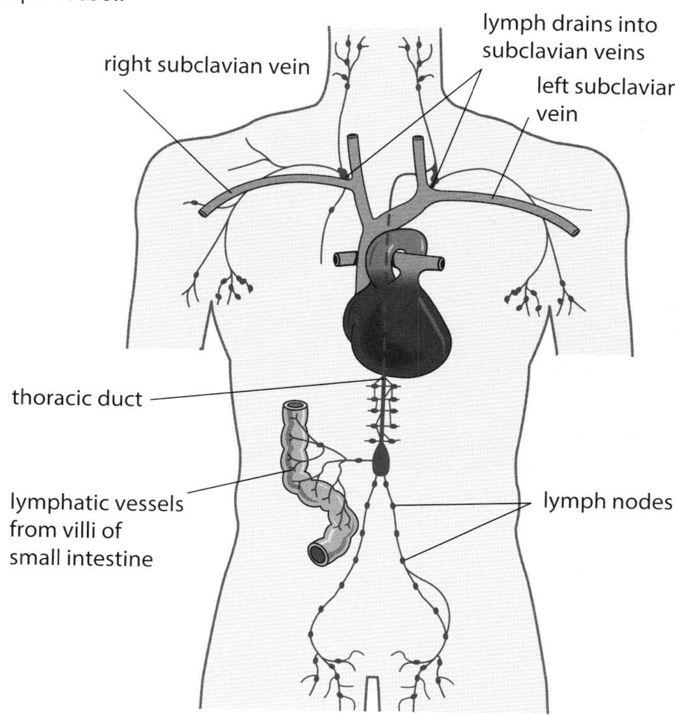

Figure 8.12 Outline of the human lymphatic system.

At intervals along lymph vessels, there are **lymph nodes**. These are involved in protection against disease. Bacteria and other unwanted particles are removed from lymph by some types of white blood cells as the lymph passes through a node, while other white blood cells within the nodes secrete **antibodies**. For more detail, see Chapter 11.

QUESTION

8.7 a We have seen that capillary walls are not very permeable to plasma proteins. Suggest where the protein in tissue fluid has come from.

b The disease kwashiorkor is caused by a diet which is very low in protein. The concentration of proteins in blood plasma becomes much lower than usual. One of the symptoms of kwashiorkor is oedema. Suggest why this is so. (You will need to think about water potential.)

Blood

You have about 5 dm^3 of blood in your body, with a mass of about 5 kg. Suspended in the blood plasma, you have around 2.5×10^{13} red blood cells, 5×10^{11} white blood cells and 6×10^{12} platelets (small cell fragments with no nucleus) (Figures 8.13 and 8.14).

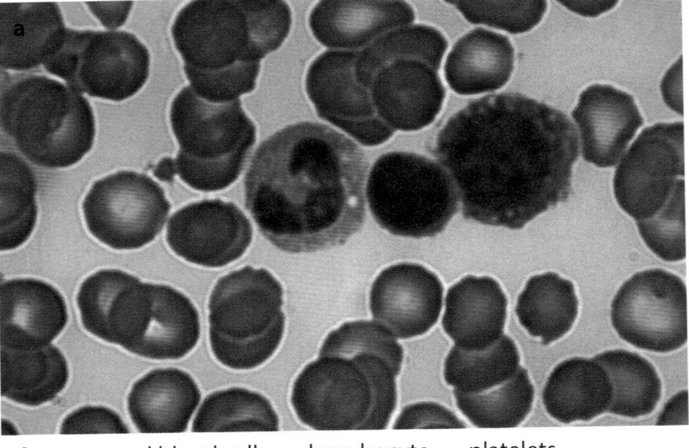

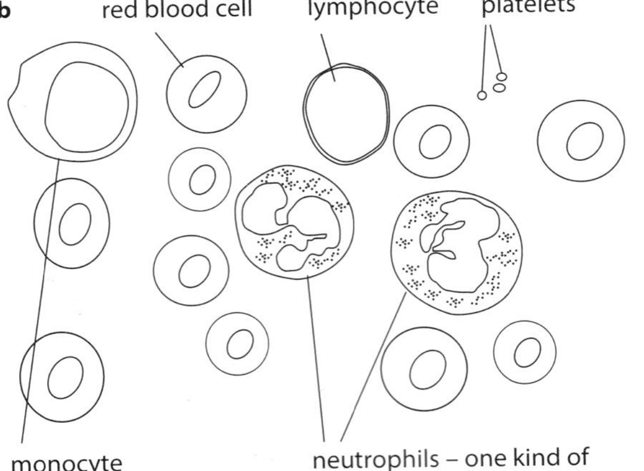

Figure 8.13 **a** Photomicrograph of human blood. It has been stained so that the nuclei of the cells are dark purple (× 1600). **b** Diagram of the types of cells seen in a stained blood film.

Red blood cells

Red blood cells (Figures 8.14 and 8.15) are also called **erythrocytes**, which simply means 'red cells'. Their red colour is caused by the pigment **haemoglobin**, a globular protein (page 43). The main function of haemoglobin is to transport oxygen from lungs to respiring tissues. This function is described in detail on pages 168–172.

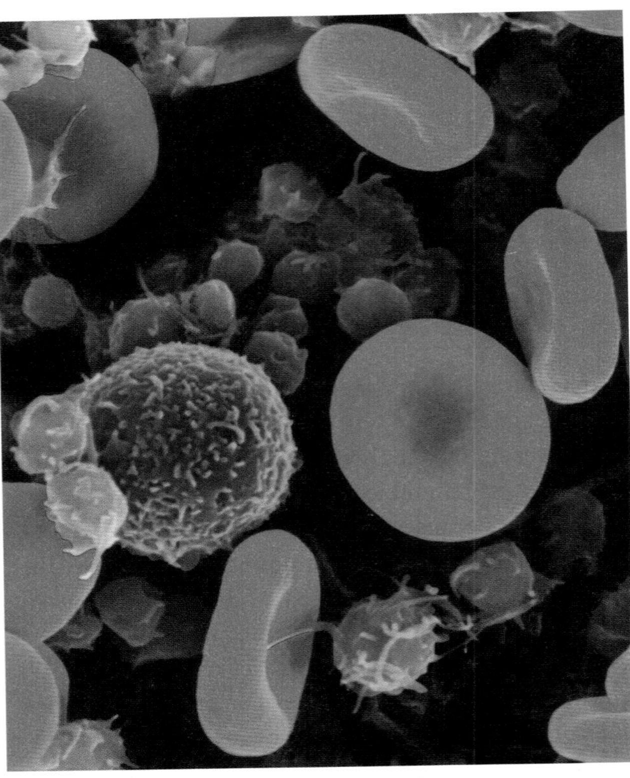

Figure 8.14 False-colour scanning electron micrograph of human blood. Red blood cells are coloured red. The blue sphere is a white blood cell. Platelets are coloured yellow (× 4000).

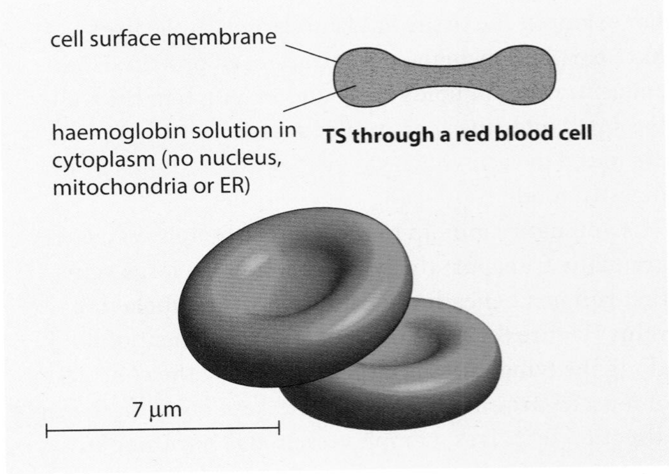

Figure 8.15 Red blood cells.

A person's first red blood cells are formed in the liver, while still a fetus inside the uterus. By the time a baby is born, the liver has stopped manufacturing red blood cells. This function has been taken over by the bone marrow. This continues, at first in the long bones such as the humerus and femur, and then increasingly in the skull, ribs, pelvis and vertebrae, throughout life. Red blood cells do not live long; their membranes become more and more fragile and eventually rupture within some 'tight spot' in the circulatory system, often inside the spleen.

The structure of a red blood cell is unusual in several ways.

- **Red blood cells are shaped like a biconcave disc.** The dent in each side of a red blood cell increases the amount of surface area in relation to the volume of the cell, giving it a large surface area : volume ratio. This large surface area means that oxygen can diffuse quickly into or out of the cell.
- **Red blood cells are very small.** The diameter of a human red blood cell is about 7 μm, compared with the diameter of an average liver cell of 40 μm. This small size means that no haemoglobin molecule within the cell is very far from the cell surface membrane, and the haemoglobin molecule can therefore quickly exchange oxygen with the fluid outside the cell. It also means that capillaries can be only 7 μm wide and still allow red blood cells to squeeze through them, so bringing oxygen as close as possible to cells which require it.
- **Red blood cells are very flexible.** Some capillaries are even narrower than the diameter of a red blood cell. The cells are able to deform so that they can pass through these vessels. This is possible because the cells have a specialised cytoskeleton (page 17), made up of a mesh-like network of protein fibres that allows them to be squashed into different shapes, but then springs back to produce the normal biconcave shape.
- **Red blood cells have no nucleus, no mitochondria and no endoplasmic reticulum.** The lack of these organelles means that there is more room for haemoglobin, so maximising the amount of oxygen which can be carried by each red blood cell.

White blood cells

White blood cells are sometimes known as **leucocytes**, which just means 'white cells'. They, too, are made in the bone marrow, but are easy to distinguish from red blood cells in a blood sample because:

- white blood cells all have a nucleus, although the shape of this varies in different types of white cell
- most white blood cells are larger than red blood cells, although one type, lymphocytes, may be slightly smaller
- white blood cells are either spherical or irregular in shape, never looking like a biconcave disc (Figures 8.14 and 8.16).

There are many different kinds of white blood cell, with a wide variety of functions, although all are concerned with fighting disease. They can be divided into two main groups.

Phagocytes are cells that destroy invading microorganisms by phagocytosis (page 87). The commonest type of phagocyte (neutrophils) can be recognised by the lobed nuclei and granular cytoplasm. Monocytes (Figure 8.13) are also phagocytes.

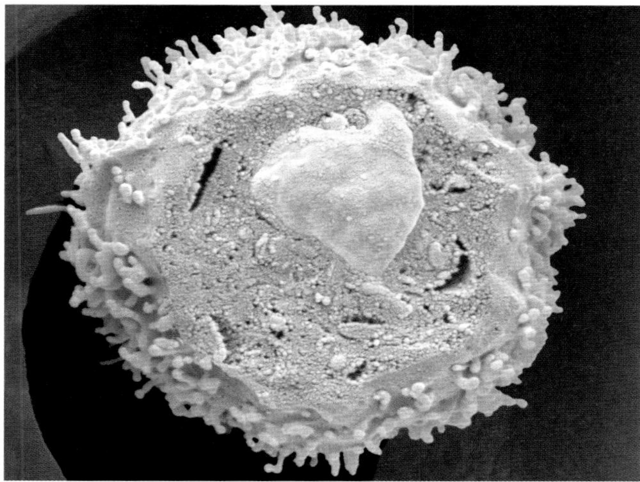

Figure 8.16 False-colour scanning electron micrograph of a section through a white blood cell (×6000). This is a lymphocyte.

QUESTIONS

8.8 Assuming that you have 2.5×10^{13} red blood cells in your body, that the average life of a red blood cell is 120 days, and that the total number of red blood cells remains constant, calculate how many new red blood cells must be made, on average, in your bone marrow each day.

8.9 Which of these functions could, or could not, be carried out by a red blood cell? In each case, briefly justify your answer.

- **a** protein synthesis
- **b** cell division
- **c** lipid synthesis
- **d** active transport

Lymphocytes also destroy microorganisms, but not by phagocytosis. Some of them secrete chemicals called antibodies, which attach to and destroy the invading cells. There are different types of lymphocytes, which act in different ways, though they all look the same. Their activities are described in Chapter 11. Lymphocytes are smaller than most phagocytes, and they have a large round nucleus and only a small amount of cytoplasm.

BOX 8.1: Observing and drawing blood vessels and cells

You should practise using prepared microscope slides to identify sections of arteries and veins. Plan diagrams can be used to show the different tissue layers in the walls of blood vessels. Look back at Box 7.1 on page 129 to remind yourself how to make plan diagrams. Figure 8.5 will help you to interpret what you see. You could also try making a plan diagram from Figure 8.6.

You can also use prepared slides to observe and draw blood cells – red cells, monocytes, neutrophils and lymphocytes. For this, you will need to use high power, and your drawings will be high-power details, showing the structures of individual cells. Look back at pages 129–133 for advice on high-power detail drawings, and some examples. Figure 8.13 will help you to interpret what you see.

Haemoglobin

A major role of the cardiovascular system is to transport oxygen from the gas exchange surfaces of the alveoli in the lungs (page 189) to tissues all over the body. Body cells need a constant supply of oxygen in order to be able to carry out aerobic respiration. Oxygen is transported around the body inside red blood cells in combination with the protein **haemoglobin** (Figure 2.23, page 42).

As we saw in Chapter 2, each haemoglobin molecule is made up of four polypeptides, each containing one haem group. Each haem group can combine with one oxygen molecule, O_2. Overall, then, each haemoglobin molecule can combine with four oxygen molecules (eight oxygen atoms).

$$\underset{\text{haemoglobin}}{\text{Hb}} + \underset{\text{oxygen}}{4O_2} \rightleftharpoons \underset{\text{oxyhaemoglobin}}{\text{HbO}_8}$$

QUESTION

8.10 In a healthy adult human, the amount of haemoglobin in $1\,dm^3$ of blood is about 150 g.

a Given that 1 g of pure haemoglobin can combine with $1.3\,cm^3$ of oxygen at body temperature, how much oxygen can be carried in $1\,dm^3$ of blood?

b At body temperature, the solubility of oxygen in water is approximately $0.025\,cm^3$ of oxygen per cm^3 of water. Assuming that blood plasma is mostly water, how much oxygen could be carried in $1\,dm^3$ of blood if there was no haemoglobin?

The haemoglobin dissociation curve

A molecule whose function is to transport oxygen from one part of the body to another must be able not only to **pick up** oxygen at the lungs, but also to **release** oxygen within respiring tissues. Haemoglobin performs this task superbly.

To investigate how haemoglobin behaves, samples are extracted from blood and exposed to different concentrations, or **partial pressures**, of oxygen. The amount of oxygen which combines with each sample of haemoglobin is then measured. The maximum amount of oxygen with which a sample can possibly combine is given a value of 100%. A sample of haemoglobin which has combined with this maximum amount of oxygen is said to be **saturated**. The amounts of oxygen with which identical samples of haemoglobin combine at lower oxygen partial pressures are then expressed as a percentage of this maximum value. Table 8.2 shows a series of results from such an investigation.

The percentage saturation of each sample can be plotted against the partial pressure of oxygen to obtain the curve shown in Figure 8.17. This is known as a **dissociation curve**.

The curve shows that at low partial pressures of oxygen, the percentage saturation of haemoglobin is very low – that is, the haemoglobin is combined with only a very little oxygen. At high partial pressures of oxygen, the percentage saturation of haemoglobin is very high; it is combined with large amounts of oxygen.

Consider the haemoglobin within a red blood cell in a capillary in the lungs. Here, where the partial pressure of oxygen is high, this haemoglobin will be 95–97% saturated with oxygen – that is, almost every haemoglobin molecule will be combined with its full complement of eight oxygen atoms. In an actively respiring muscle, on the other hand, where the partial pressure of oxygen is low, the haemoglobin will be about 20–25% saturated

Partial pressure of oxygen / kPa	1	2	3	4	5	6	7	8	9	10	11	12	13	14
Percentage saturation of haemoglobin	8.5	24.0	43.0	57.5	71.5	80.0	85.5	88.0	92.0	94.0	95.5	96.5	97.5	98.0

Table 8.2 The varying ability of haemoglobin to carry oxygen.

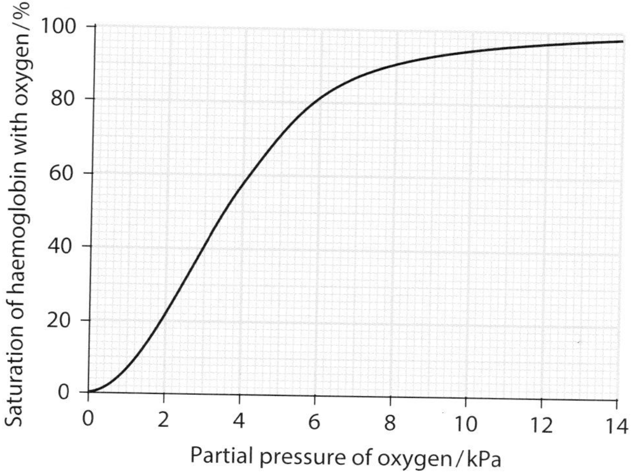

Figure 8.17 The haemoglobin dissociation curve.

with oxygen – that is, the haemoglobin is carrying only a quarter of the oxygen which it is capable of carrying. This means that haemoglobin coming from the lungs carries a lot of oxygen; as it reaches a muscle, it releases around three-quarters of it. This released oxygen diffuses out of the red blood cell and into the muscle where it can be used in respiration.

The S-shaped curve

The shape of the haemoglobin dissociation curve can be explained by the behaviour of a haemoglobin molecule as it combines with or loses oxygen molecules.

Oxygen molecules combine with the iron atoms in the haem groups of a haemoglobin molecule. You will remember that each haemoglobin molecule has four haem groups. When an oxygen molecule combines with one haem group, the whole haemoglobin molecule is slightly distorted. The distortion makes it easier for a second oxygen molecule to combine with a second haem group. This in turn makes it easier for a third oxygen molecule to combine with a third haem group. It is then still easier for the fourth and final oxygen molecule to combine.

The shape of the haemoglobin dissociation curve reflects the way that oxygen atoms combine with haemoglobin molecules. Up to an oxygen partial pressure of around 2 kPa, on average only one oxygen molecule is combined with each haemoglobin molecule. Once this oxygen molecule is combined, however, it becomes successively easier for the second and third oxygen molecules to combine, so the curve rises very steeply. Over this part of the curve, a **small** change in the partial pressure of oxygen causes a **very large** change in the amount of oxygen which is carried by the haemoglobin.

QUESTION

8.11 Use the dissociation curve in Figure 8.17 to answer these questions.

a i The partial pressure of oxygen in the alveoli of the lungs is about 12 kPa. What is the percentage saturation of haemoglobin in the capillaries in the lungs?

ii If 1 g of fully saturated haemoglobin is combined with 1.3 cm^3 of oxygen, how much oxygen will 1 g of haemoglobin in the capillaries in the lungs be combined with?

b i The partial pressure of oxygen in an actively respiring muscle is about 2 kPa. What is the percentage saturation of haemoglobin in the capillaries of such a muscle?

ii How much oxygen will 1 g of haemoglobin in the capillaries of this muscle be combined with?

The Bohr shift

The behaviour of haemoglobin in picking up oxygen at the lungs, and readily releasing it when in conditions of low oxygen partial pressure, is exactly what is needed. But, in fact, it is even better at this than is shown by the dissociation curve in Figure 8.17. This is because the amount of oxygen the haemoglobin carries is affected not only by the partial pressure of **oxygen**, but also by the partial pressure of **carbon dioxide**.

Carbon dioxide is continually produced by respiring cells. It diffuses from the cells and into blood plasma, from where some of it diffuses into the red blood cells.

In the cytoplasm of red blood cells there is an enzyme, **carbonic anhydrase**, that catalyses the following reaction:

$$\underset{\text{carbon dioxide}}{CO_2} + \underset{\text{water}}{H_2O} \overset{\text{carbonic anhydrase}}{\rightleftharpoons} \underset{\text{carbonic acid}}{H_2CO_3}$$

The carbonic acid dissociates:

$$\underset{\text{carbonic acid}}{H_2CO_3} \rightleftharpoons \underset{\text{hydrogen ion}}{H^+} + \underset{\text{hydrogencarbonate ion}}{HCO_3^-}$$

Haemoglobin readily combines with the hydrogen ions, forming **haemoglobinic acid, HHb**. In so doing, it releases the oxygen which it is carrying.

The net result of this reaction is two-fold.

- The haemoglobin 'mops up' the hydrogen ions which are formed when carbon dioxide dissolves and dissociates. A high concentration of hydrogen ions means a low pH; if the hydrogen ions were left in solution, the blood would be very acidic. By removing the hydrogen ions from solution, haemoglobin helps to maintain the pH of the blood close to neutral. It is acting as a **buffer**.
- The presence of a high partial pressure of carbon dioxide causes haemoglobin to release oxygen. This is called the Bohr effect, after Christian Bohr who discovered it in 1904. It is exactly what is needed. High concentrations of carbon dioxide are found in actively respiring tissues, which need oxygen; these high carbon dioxide concentrations cause haemoglobin to release its oxygen even more readily than it would otherwise do.

If a dissociation curve is drawn for haemoglobin at a high partial pressure of carbon dioxide, it looks like the lower curve shown on both graphs in Figure 8.18. At each partial pressure of oxygen, the haemoglobin is less saturated than it would be at a low partial pressure of carbon dioxide. The curve therefore lies below, and to the right of, the 'normal' curve.

Carbon dioxide transport

The description of the Bohr effect above explains one way in which carbon dioxide is carried in the blood. One product of the dissociation of dissolved carbon dioxide is hydrogencarbonate ions, HCO_3^-. These are initially formed in the cytoplasm of the red blood cell, because this is where the enzyme carbonic anhydrase is found. Most of the hydrogencarbonate ions then diffuse out of the red blood cell into the blood plasma, where they are carried in solution. About 85% of the carbon dioxide transported by the blood is carried in this way.

Some carbon dioxide, however, does not dissociate, but remains as carbon dioxide molecules. Some of these simply dissolve in the blood plasma; about 5% of the total is carried in this form. Other carbon dioxide molecules diffuse into the red blood cells, but instead of undergoing the reaction catalysed by carbonic anhydrase, combine directly with the terminal amine groups ($-NH_2$) of some of the haemoglobin molecules. The compound formed is called **carbaminohaemoglobin**. About 10% of the carbon dioxide is carried in this way (Figure 8.19).

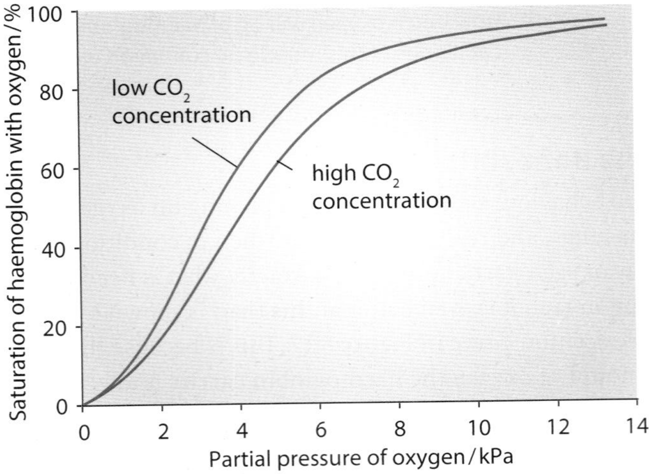

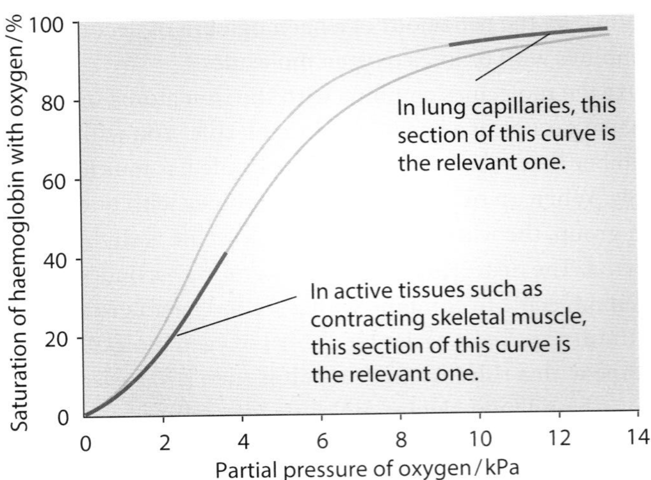

Figure 8.18 Dissociation curves for haemoglobin at two different partial pressures of carbon dioxide. The shift of the curve to the right when the haemoglobin is exposed to higher carbon dioxide concentration is called the Bohr effect.

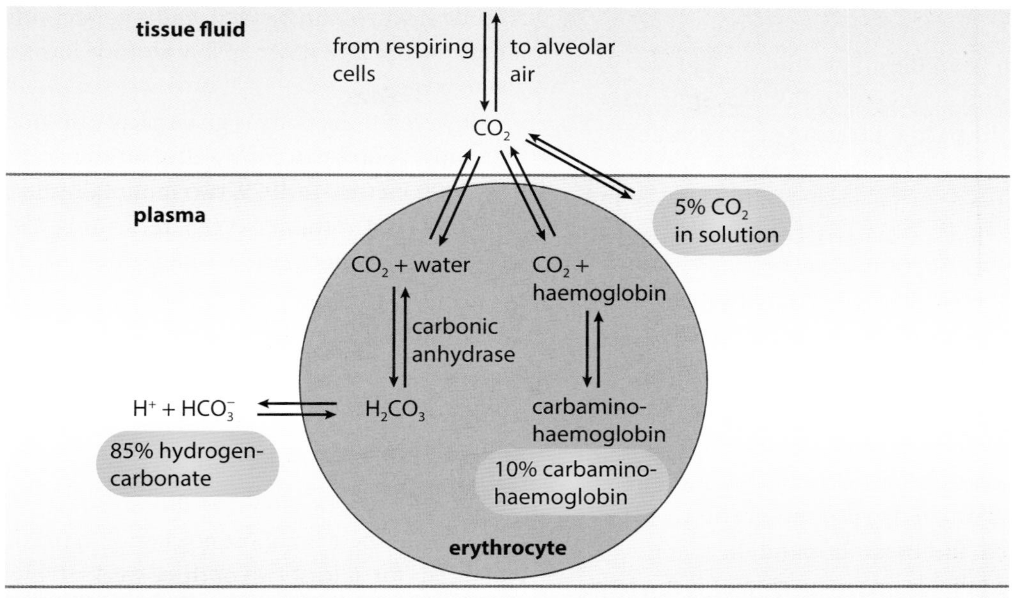

Figure 8.19 Carbon dioxide transport in the blood. The blood carries carbon dioxide partly as undissociated carbon dioxide in solution in the plasma, partly as hydrogencarbonate ions in solution in the plasma, and partly combined with haemoglobin in the red blood cells.

When blood reaches the lungs, the reactions described above go into reverse. The relatively low concentration of carbon dioxide in the alveoli compared with that in the blood causes carbon dioxide to diffuse from the blood into the air in the alveoli, stimulating the carbon dioxide of carbaminohaemoglobin to leave the red blood cell, and hydrogencarbonate and hydrogen ions to recombine to form carbon dioxide molecules once more. This leaves the haemoglobin molecules free to combine with oxygen, ready to begin another circuit of the body.

Problems with oxygen transport

The efficient transport of oxygen around the body can be impaired by many different factors.

Carbon monoxide

Despite its almost perfect structure as an oxygen-transporting molecule, haemoglobin does have one property which can prove very dangerous. It combines very readily, and almost irreversibly, with carbon monoxide.

Carbon monoxide, CO, is formed when a carbon-containing compound burns incompletely. Exhaust fumes from cars contain significant amounts of carbon monoxide, as does cigarette smoke. When such fumes are inhaled, the carbon monoxide readily diffuses across the walls of the alveoli, into blood, and into red blood cells. Here it combines with the haem groups in the haemoglobin molecules, forming **carboxyhaemoglobin**.

Haemoglobin combines with carbon monoxide 250 times more readily than it does with oxygen. Thus, even if the concentration of carbon monoxide in air is much lower than the concentration of oxygen, a high proportion of haemoglobin will combine with carbon monoxide. Moreover, carboxyhaemoglobin is a very stable compound; the carbon monoxide remains combined with the haemoglobin for a long time.

The result of this is that even relatively low concentrations of carbon monoxide, as low as 0.1% of the air, can cause death by asphyxiation. Treatment of carbon monoxide poisoning involves administration of a mixture of pure oxygen and carbon dioxide: high concentrations of oxygen to favour the combination of haemoglobin with oxygen rather than carbon monoxide, and carbon dioxide to stimulate an increase in the breathing rate.

Cigarette smoke contains up to 5% carbon monoxide (pages 190). If you breathed in 'pure' cigarette smoke for any length of time, you would die of asphyxiation. As it is, even smokers who inhale also breathe in some normal air, diluting the carbon monoxide levels in their lungs. Nevertheless, around 5% of the haemoglobin in a regular smoker's blood is permanently combined with carbon monoxide. This considerably reduces its oxygen-carrying capacity.

High altitude

We obtain our oxygen from the air around us. At sea level, the partial pressure of oxygen in the atmosphere is just over 20 kPa, and the partial pressure of oxygen in an alveolus in the lungs is about 13 kPa. If you look at the oxygen dissociation curve for haemoglobin in Figure 8.17 (page 169), you can see that at this partial pressure of oxygen, haemoglobin is almost completely saturated with oxygen.

If, however, a person climbs up a mountain to a height of 6500 metres, then the air pressure is much less (Figure 8.20). The partial pressure of oxygen in the air is only about 10 kPa, and in the lungs about 5.3 kPa. You can see from Figure 8.17 that this will mean that the haemoglobin will become only about 70% saturated in the lungs. Less oxygen will be carried around the body, and the person may begin to feel breathless and ill.

If someone travels quickly, over a period of just a few days, from sea level to a very high altitude, the body does not have enough time to adjust to this drop in oxygen availability, and the person may suffer from **altitude sickness.**

The symptoms of altitude sickness frequently begin with an increase in the rate and depth of breathing, and a general feeling of dizziness and weakness. These symptoms can be easily reversed by going down to a lower altitude.

Some people, however, can quickly become very ill indeed. The arterioles in their brain dilate, increasing the amount of blood flowing into the capillaries, so that fluid begins to leak from the capillaries into the brain tissues. This can cause disorientation. Fluid may also leak into the lungs, preventing them from functioning properly. Acute altitude sickness can be fatal, and a person suffering from it must be brought down to low altitude immediately, or given oxygen.

Figure 8.20 A climber rests on the summit of Mount Everest. He is breathing oxygen through a mask.

However, if the body is given plenty of time to adapt, then most people can cope well at altitudes up to at least 5000 metres. In 1979, two mountaineers climbed Mount Everest without oxygen, returning safely despite experiencing hallucinations and feelings of euphoria at the summit.

As the body gradually acclimatises to high altitude, a number of changes take place. Perhaps the most significant of these is that the number of red blood cells increases. Whereas red blood cells normally make up about 40–50% of the blood, after a few months at high altitude this rises to as much as 50–70%. However, this does take a long time to happen, and there is almost no change in the number of red blood cells for at least two or three weeks at high altitude.

People who live permanently at high altitude, such as in the Andes or Himalayas, show a number of adaptations to their low-oxygen environment. It seems that they are not genetically different from people who live at low altitudes, but rather that their exposure to low oxygen partial pressures from birth encourages the development of these adaptations from an early age. They often have especially broad chests, providing larger lung capacities than normal. The heart is often larger than in a person who lives at low altitude, especially the right side, which pumps blood to the lungs. They also have more haemoglobin in their blood than usual, so increasing the efficiency of oxygen transport from lungs to tissues.

QUESTIONS

8.12 Mount Everest is nearly 9000 m high. The partial pressure of oxygen in the alveoli at this height is only about 2.5 kPa. Explain what effect this would have on the supply of oxygen to body cells if a person climbed to the top of Mount Everest without a supplementary oxygen supply.

8.13 Explain how an increase in the number of red blood cells can help to compensate for the lack of oxygen in the air at high altitude.

8.14 Athletes often prepare themselves for important competitions by spending several months training at high altitude. Explain how this could improve their performance.

The heart

The heart of an adult human has a mass of around 300 g, and is about the size of your fist (Figure 8.21). It is a bag made of muscle and filled with blood. Figure 8.22 shows the appearance of a human heart, looking at it from the front of the body.

The muscle of which the heart is made is called **cardiac muscle**. Although you do not need to know the structure of cardiac muscle, you may find it interesting, and it is shown in Figure 8.23. It is made of interconnecting cells, whose cell surface membranes are very tightly joined together. This close contact between the muscle cells allows waves of electrical excitation to pass easily between them, which is a very important feature of cardiac muscle, as you will see later.

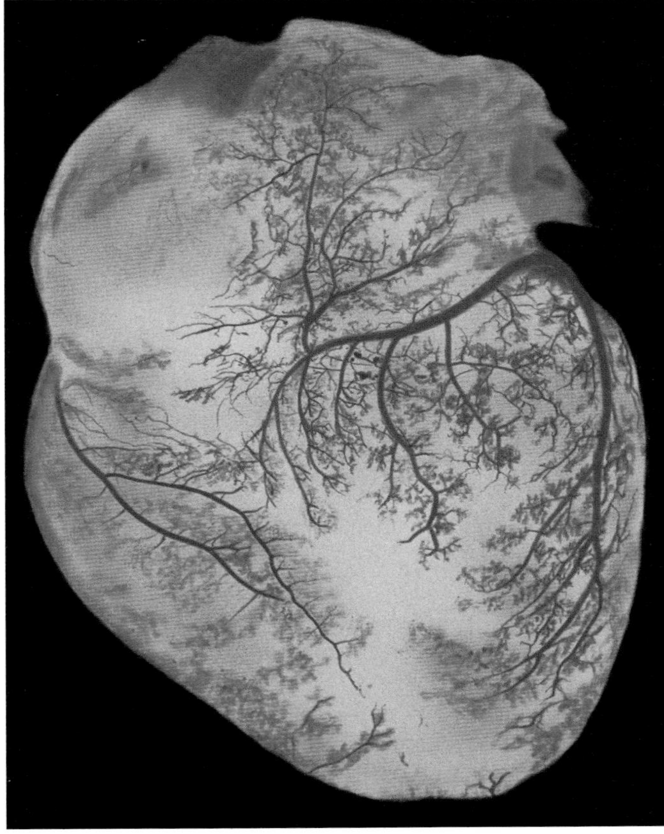

Figure 8.21 A human heart. The blood vessels in the photograph lie immediately below the surface of the heart and have been injected with gelatine containing a dye. The cardiac muscle was treated to make it transparent to a depth of 2 millimetres to allow the blood vessels to seen.

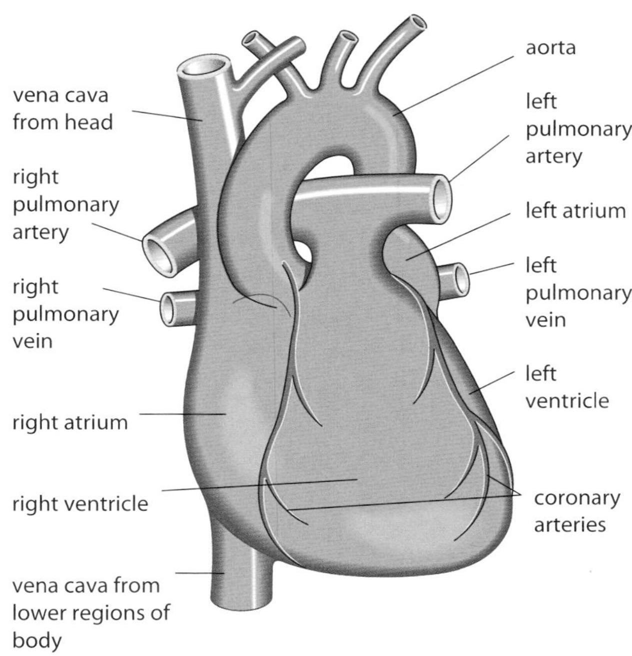

Figure 8.22 Diagram of the external structure of a human heart, seen from the front.

Figure 8.22 also shows the blood vessels that carry blood into and out of the heart. The large, arching blood vessel is the largest artery, the **aorta**, with branches leading upwards towards the head, and the main flow doubling back downwards to the rest of the body. The other blood vessel leaving the heart is the **pulmonary artery**. This, too, branches very quickly after leaving the heart, into two arteries taking blood to the right and left lungs. Running vertically on the right-hand side of the heart are the two large veins, the **venae cavae**, one bringing blood downwards from the head and the

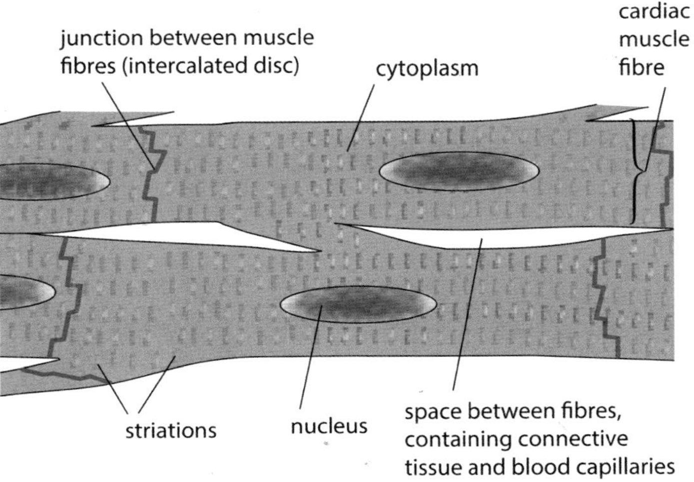

Figure 8.23 You do not need to know this structure, but you may like to compare it with striated muscle, shown in Figures 15.25 and 15.26.

other bringing it upwards from the rest of the body. The **pulmonary veins** bring blood back to the heart from the left and right lungs.

On the surface of the heart, the **coronary arteries** can be seen (Figure 8.21). These branch from the aorta, and deliver oxygenated blood to the walls of the heart itself.

If the heart is cut open vertically (Figures 8.24 and 8.25) it can be seen to contain four chambers. The two chambers on the left of the heart are completely separated from those on the right by a wall of muscle called the **septum**. Blood cannot pass through this septum; the only way for blood to get from one side of the heart to the other is for it to leave the heart, circulate around either the lungs or the rest of the body, and then return to the heart.

The upper chamber on each side of the heart is called an **atrium** (plural: **atria**), or sometimes an **auricle**. The two atria receive blood from the veins. You can see from Figure 8.25 that blood from the venae cavae flows into the right atrium, while blood from the pulmonary veins flows into the left atrium.

The lower chambers are **ventricles**. Blood flows into the ventricles from the atria, and is then squeezed out into the arteries. Blood from the left ventricle flows into the aorta, while blood from the right ventricle flows into the pulmonary arteries.

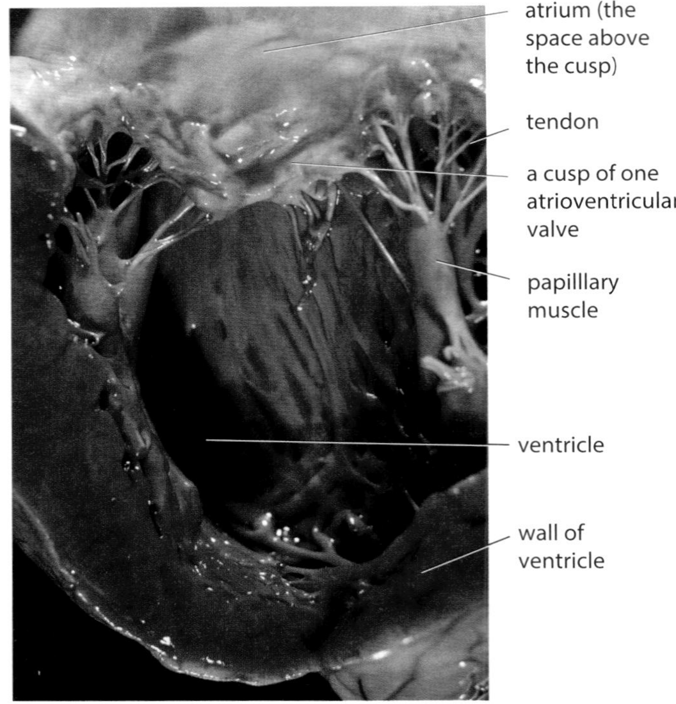

Figure 8.24 Section through part of the left side of the heart.

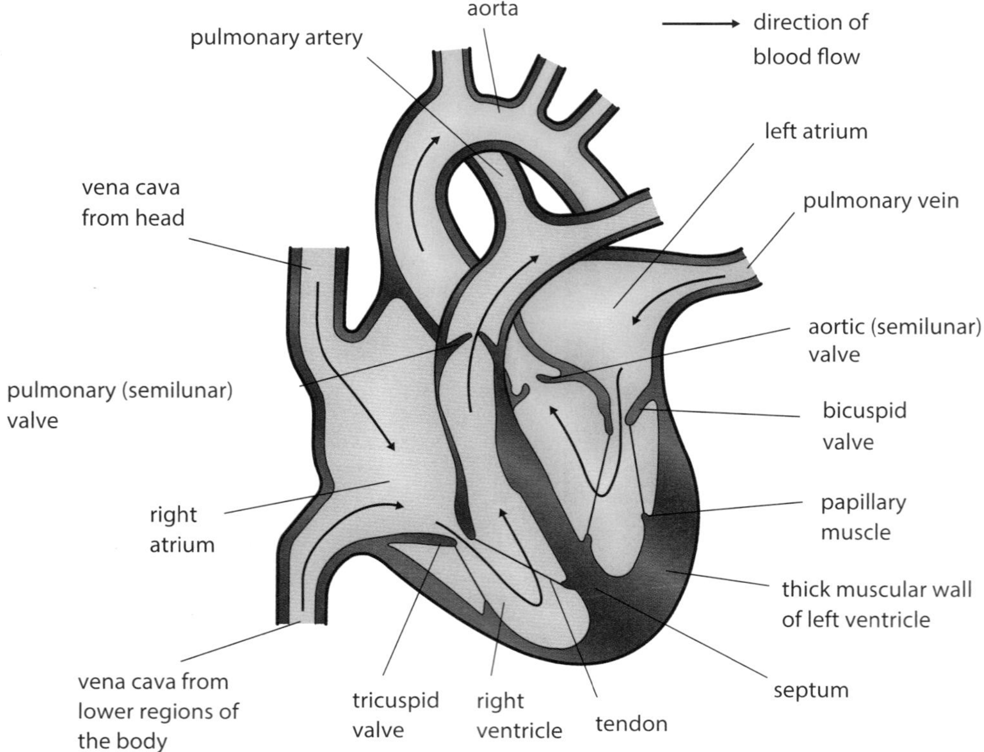

Figure 8.25 Diagrammatic section through a heart.

The atria and ventricles have valves between them, which are known as the **atrioventricular valves**. The one on the left is the **mitral** or **bicuspid** valve, and the one on the right is the **tricuspid valve**. We will now consider how all of these components work together so that the heart can be an efficient pump for the blood.

The cardiac cycle

Your heart beats around 70 times a minute. The cardiac cycle is the sequence of events which makes up one heart beat. Figure 8.26 shows three stages in this cycle.

As the cycle is continuous, a description of it could begin anywhere. We will begin with the time when the heart is filled with blood and the muscle in the atrial walls contracts. This stage is called **atrial systole**. The pressure developed by this contraction is not very great, because the muscular walls of the atria are only thin, but it is enough to force the blood in the atria down through the atrioventricular valves into the ventricles. The blood from the atria does not go back into the pulmonary veins or the venae cavae, because these have semilunar valves to prevent backflow.

About 0.1 seconds after the atria contract, the ventricles contract. This is called **ventricular systole**. The thick, muscular walls of the ventricles squeeze inwards on the blood, increasing its pressure and pushing it out of the heart. As soon as the pressure in the ventricles becomes greater than the pressure in the atria, this pressure difference pushes the atrioventricular valves shut, preventing blood from going back into the atria. Instead, the blood rushes upwards into the aorta and the pulmonary artery, pushing open the semilunar valves in these vessels as it does so.

Ventricular systole lasts for about 0.3 seconds. The muscle then relaxes, and the stage called **ventricular diastole** begins. As the muscle relaxes, the pressure in the ventricles drops. The high-pressure blood which has just been pushed into the arteries would flow back into the ventricles but for the presence of the semilunar valves, which snap shut as the blood fills their cusps.

During diastole, as the whole of the heart muscle relaxes, blood from the veins flows into the two atria. The blood is at a very low pressure, but the thin walls of the atria are easily distended, providing very little resistance to the blood flow. Some of the blood trickles downwards into the ventricles, through the atrioventricular valves. The atrial muscle then contracts, to push blood forcefully down into the ventricles, and the whole cycle begins again.

Figure 8.27 shows how the atrioventricular and semilunar valves work.

The walls of the ventricles are much thicker than the walls of the atria, because the ventricles need to develop much more force when they contract. Their contraction has to push the blood out of the heart and around the body. For the right ventricle, the force produced must be relatively small, because the blood goes only to the lungs, which are very close to the heart. If too high a pressure was developed, tissue fluid would accumulate in the lungs, hampering gas exchange.

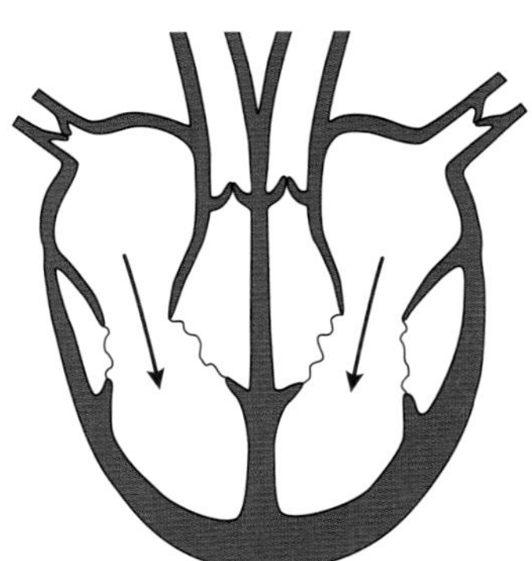

1 **Atrial systole.** Both atria contract. Blood flows from the atria into the ventricles. Backflow of blood into the veins is prevented by closure of the valves in the veins.

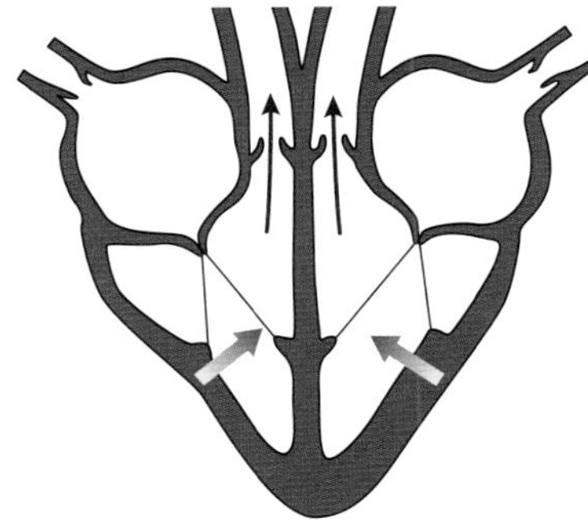

2 **Ventricular systole.** Both ventricles contract. The atrioventricular valves are pushed shut by the pressurised blood in the ventricles. The semilunar valves in the aorta and pulmonary artery are pushed open. Blood flows from the ventricles into the arteries.

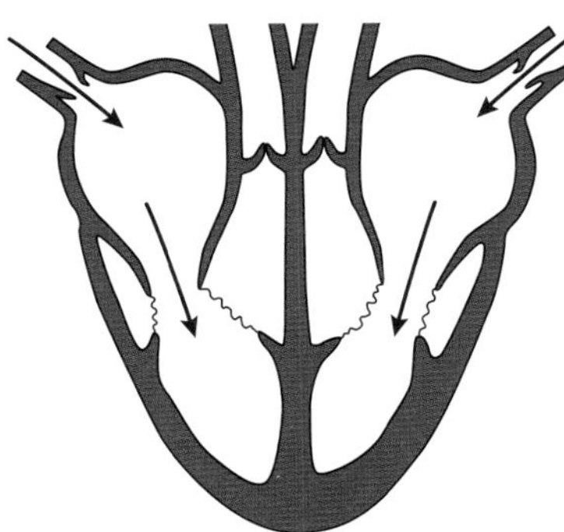

3 **Ventricular diastole.** Atria and ventricles relax. The semilunar valves in the aorta and pulmonary artery are pushed shut. Blood flows from the veins through the atria and into the ventricles.

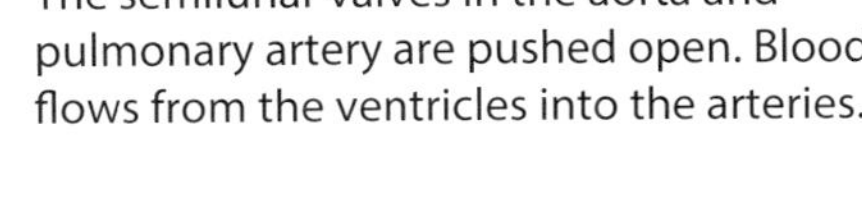

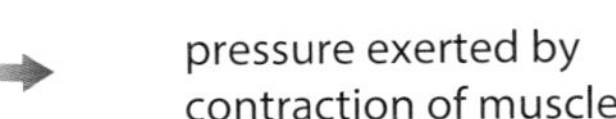

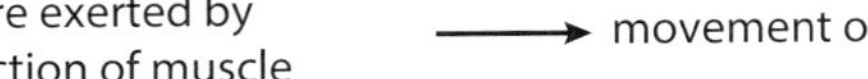

Figure 8.26 The cardiac cycle. Only three stages in this continuous process are shown.

Atrioventricular valve

During **atrial systole**, the pressure of the blood is higher in the atrium than in the ventricle, and so forces the atrioventricular valve open.
During **ventricular systole**, the pressure of the blood is higher in the ventricle than in the atrium. The pressure of the blood pushes up against the cusps of the atrioventricular valve, pushing it shut. Contraction of the papillary muscles, attached to the valve by tendons, prevents the atrioventricular valve from being forced inside-out.

atrial systole – valve open
ventricular systole – valve shut
cusp of valve
tendon
papillary muscle

Semilunar valve

During **ventricular systole**, the pressure of the blood forces the semilunar valves open.
During **ventricular diastole**, the pressure of the blood in the arteries is higher than in the ventricles. The pressure of the blood pushes into the cusps of the semilunar valves, squeezing them shut.

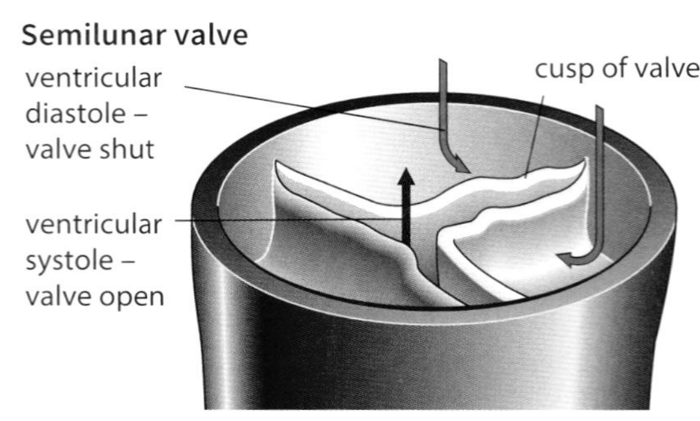

Figure 8.27 How the heart valves function.

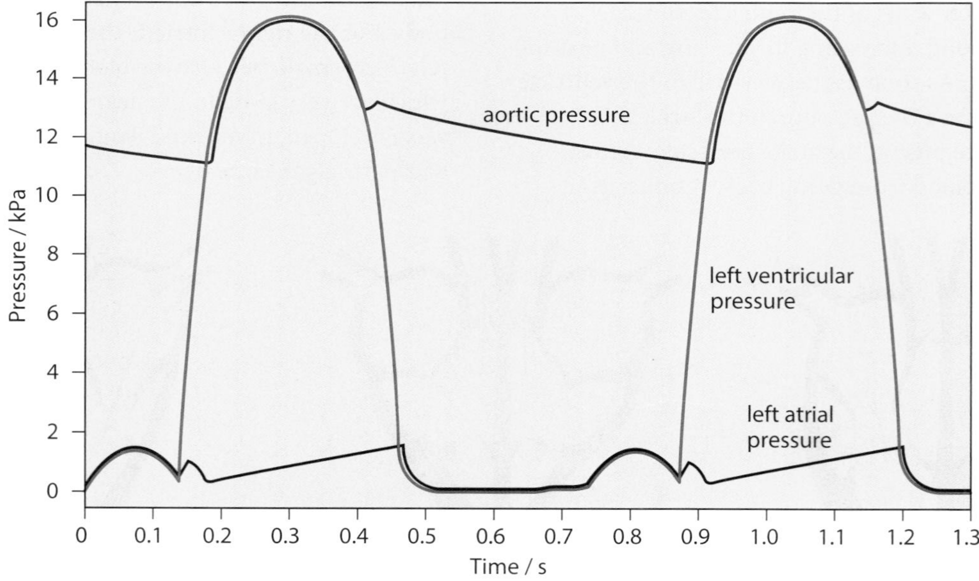

Figure 8.28 Pressure changes in the left side of the heart during the cardiac cycle.

QUESTIONS

8.15 From Figure 8.28, identify the time at which each stage shown in Figure 8.26 (page 175) is occurring.

8.16 Heart valves can become weakened and fail to close effectively. Suggest how this would affect the function of the heart, and the health of a person.

The left ventricle, however, has to develop sufficient force to supply blood to all the rest of the body organs. For most organs, most of the time, the high pressures that the left ventricle is capable of producing would be too great and, as we have seen, arterioles play an important role in reducing this pressure before blood flows into the capillaries. However, during vigorous exercise, when muscles are working hard, the arterioles supplying blood to them dilate, increasing blood flow to them. The left ventricle must be able to develop enough force to ensure that there is still sufficient blood reaching other organs, especially the brain. Kidneys require high-pressure blood all the time, to carry out ultrafiltration (page 307). Therefore, the thickness of the muscular wall of the left ventricle is much greater than that of the right.

Figure 8.28 shows the pressure changes in the left side of the heart and the aorta during two consecutive cardiac cycles. You can see that the pressure developed in the left ventricle is much greater than that in the left atrium.

Control of heart beat

Cardiac muscle differs from the muscle in all other areas of the body in that it is **myogenic**. This means that it **naturally** contracts and relaxes; it does not need to receive impulses from a nerve to make it contract. If cardiac muscle cells are cultured in a warm, oxygenated solution containing nutrients, they contract and relax rhythmically, all by themselves.

However, the individual heart muscle cells cannot be allowed to contract at their own natural rhythms. If they did, parts of the heart would contract out of sequence with other parts. The cardiac cycle would become disordered, and the heart would stop working as a pump. The heart has its own built-in controlling and coordinating system which prevents this happening.

The cardiac cycle is initiated in a specialised patch of muscle in the wall of the right atrium, called the **sinoatrial node**. It is often called the **SAN** for short, or **pacemaker**. The muscle cells of the SAN set the rhythm for all the other cardiac muscle cells. Their natural rhythm of contraction is slightly faster than that of the rest of the heart muscle. Each time the muscles of the SAN contract, they set up a wave of electrical activity which spreads out rapidly over the whole of the atrial walls. The cardiac muscle in the atrial walls responds to this excitation wave by contracting, at the same rhythm as the SAN. Thus, all the muscle in both atria contracts almost simultaneously.

As we have seen, the muscles of the ventricles do not contract until **after** the muscles of the atria. (You can imagine what would happen if they all contracted at once.) This delay is caused by a feature of the heart that briefly delays the excitation wave in its passage from the atria to the ventricles. There is a band of fibres between the atria and ventricles which does not conduct the excitation wave. Thus, as the wave spreads out from the SAN over the atrial walls, it cannot pass into the ventricle walls. The only route through is via a patch of conducting fibres, situated in the septum, known as the **atrioventricular node**, or **AVN** (Figure 8.29). The AVN picks up the excitation wave as it spreads across the atria and, after a delay of about

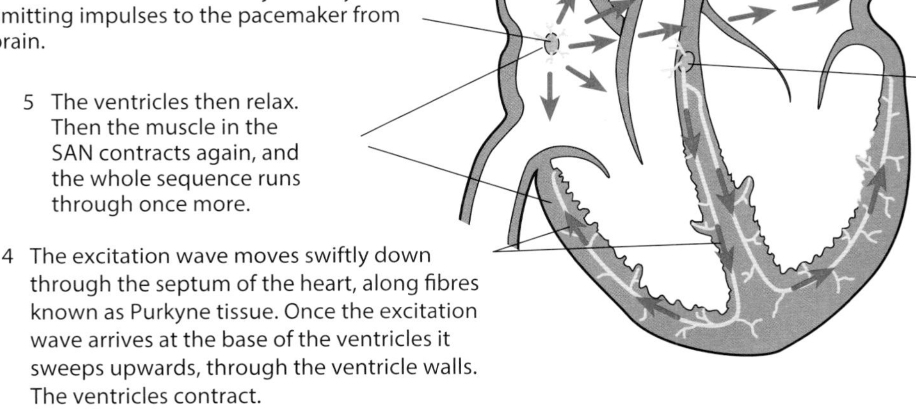

Figure 8.29 How electrical excitation waves move through the heart.

0.1 seconds, passes it on to a bunch of conducting fibres called the **Purkyne tissue**, which runs down the septum between the ventricles. This transmits the excitation wave very rapidly down to the base of the septum, from where it spreads outwards and upwards through the ventricle walls. As it does so, it causes the cardiac muscle in these walls to contract, from the bottom up, so squeezing blood upwards and into the arteries.

In a healthy heart, therefore, the atria contract and then the ventricles contract from the bottom upwards. Sometimes, this coordination of contraction goes wrong. The excitation wave becomes chaotic, passing through the ventricular muscle in all directions, feeding back on itself and re-stimulating areas it has just left. Small sections of the cardiac muscle contract while other sections are relaxing. The result is **fibrillation**, in which the heart wall simply flutters rather than contracting as a whole and then relaxing as a whole. Fibrillation is almost always fatal, unless treated instantly. Fibrillation may be started by an electric shock or by damage to large areas of muscle in the walls of the heart.

Electrocardiograms (ECGs)

It is relatively easy to detect and record the waves of excitation flowing through heart muscle. Electrodes can be placed on the skin over opposite sides of the heart, and the electrical potentials generated recorded with time. The result, which is essentially a graph of voltage against time, is an **electrocardiogram** (ECG) (Figure 8.30). You do not need to know about electrocardiograms, but you may find it interesting to relate what is shown in Figure 8.30 to what you know about the spread of electrical activity through the heart during a heart beat.

The part labelled **P** represents the wave of excitation sweeping over the atrial walls. The parts labelled **Q**, **R** and **S** represent the wave of excitation in the ventricle walls. The **T** section indicates the recovery of the ventricle walls.

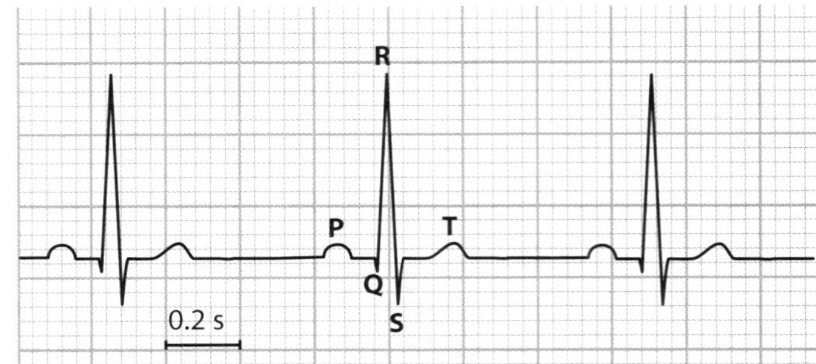

Figure 8.30 A normal ECG.

Summary

- Blood is carried away from the heart in arteries, passes through tissues in capillaries, and is returned to the heart in veins. Blood pressure drops gradually as it passes along this system.
- Arteries have thick, elastic walls, to allow them to withstand high blood pressures and to smooth out the pulsed blood flow. Arterioles are small arteries that help to reduce blood pressure and control the amount of blood flow to different tissues. Capillaries are only just wide enough to allow the passage of red blood cells, and have very thin walls to allow efficient and rapid transfer of materials between blood and cells. Veins have thinner walls than arteries and possess valves to help blood at low pressure flow back to the heart.
- Blood plasma leaks from capillaries to form tissue fluid. This is collected into lymphatics as lymph, and returned to the blood in the subclavian veins. Tissue fluid and lymph are almost identical in composition; both of them contain fewer plasma protein molecules than blood plasma, as these are too large to pass through the pores in the capillary walls.
- Red blood cells are relatively small cells. They have a biconcave shape and no nucleus. Their cytoplasm is full of haemoglobin. White blood cells include phagocytes and lymphocytes. They all have nuclei, and are either spherical or irregular in shape. Red blood cells carry oxygen in combination with haemoglobin.
- Haemoglobin picks up oxygen at high partial pressures of oxygen in the lungs, and releases it at low partial pressures of oxygen in respiring tissues. A graph showing the percentage saturation of haemoglobin at different partial pressures (concentrations) of oxygen is known as a dissociation curve.
- At high carbon dioxide concentrations, the dissociation curve shifts downwards and to the right, showing that haemoglobin releases oxygen more easily when carbon dioxide concentration is high. This is known as the Bohr effect.

- The mammalian heart has four chambers: right and left atria and right and left ventricles. The right side of the heart is divided from the left by a wall of muscle tissue called the septum. The atrial muscular walls are thin and do not exert much pressure when they contract. The ventricular walls are much more muscular and exert a sufficient pressure to drive blood to the lungs from the right ventricle and around the rest of the body from the left ventricle. The left ventricular wall is therefore much thicker and more muscular than the right ventricular wall.
- The cardiac cycle is a continuous process but can be considered in three stages. (a) Atrial systole (contraction of the atria) allows blood to flow into the ventricles from the atria. Closure of valves in the veins prevents backflow of blood into the veins. (b) Ventricular systole (contraction of the ventricles) pushes blood into the arteries by forcing open the semilunar valves. Blood is prevented from flowing back into the atria by pressure closing the atrioventricular valves. (c) In diastole (relaxation of heart muscle), the semilunar valves are pushed shut preventing backflow of blood from the arteries into the ventricles. Blood flows into the atria and ventricles from the veins.
- Beating of the heart is initiated by the sinoatrial node (SAN) or pacemaker which has its own myogenic rhythm. A wave of excitation spreads across the atria so all the heart muscle cells in the atria contract together. The wave of excitation cannot spread to the ventricles directly because of a band of non-conducting tissue. However, the atrioventricular node in the septum passes the wave to the Purkyne tissue which then causes the ventricles to contract from the bottom up shortly after the atria. This is important because it pushes the blood upwards out of the ventricles into the arteries.

End-of-chapter questions

1 The diagram shows the changes in blood pressure as blood flows through the blood vessels in the human systemic circulatory system.

Which correctly identifies the vessels labelled **P** to **S**?

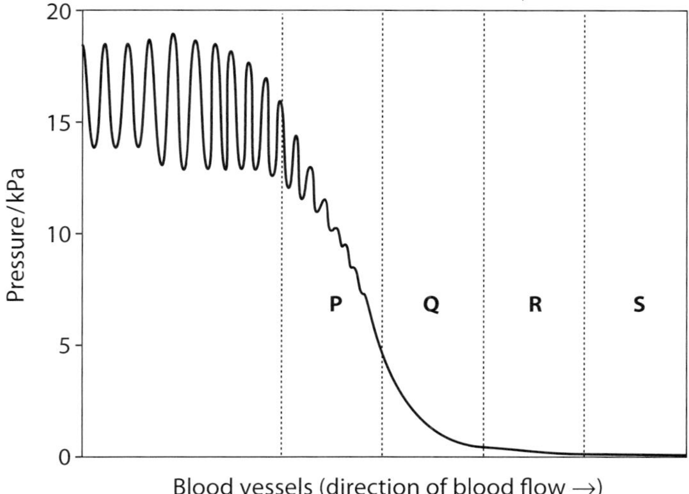

	P	Q	R	S
A	artery	capillary	arteriole	venule
B	arteriole	artery	venule	capillary
C	artery	arteriole	capillary	venule
D	venule	capillary	arteriole	artery

[1]

2 The micrograph shows an artery and a vein.

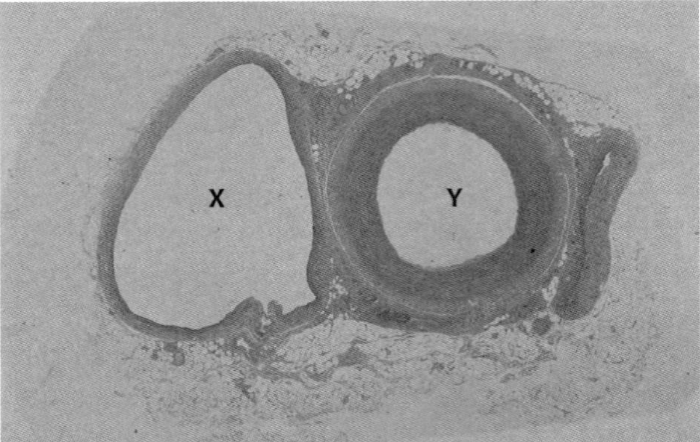

Which row correctly identifies and describes the artery and the vein?

	X	Y	Description
A	artery	vein	The artery has thick walls and the vein has thin walls.
B	artery	vein	The artery has a thin tunica media while the vein has a thick tunica media.
C	vein	artery	The artery has a thick tunica media while the vein has a thin tunica media.
D	vein	artery	The artery has thin walls and the vein has thick walls.

[1]

3 Where is the mammalian heart beat initiated?

A atrioventricular node
B left atrium
C Purkyne tissue
D sinoatrial node [1]

4 What causes the bicuspid valve to close during ventricular systole?

A a greater blood pressure in the left atrium than in the left ventricle
B a greater blood pressure in the left ventricle than in the left atrium
C contraction of muscles in the septum
D contraction of muscles in the valve [1]

5 Construct a table comparing the structure of arteries, veins and capillaries. Include both similarities and differences, and give reasons for the differences which you describe. [6]

6 Construct a table comparing blood plasma, tissue fluid and lymph. [6]

7 Explain how the structure of haemoglobin enables it to carry out its functions. (You may wish to look back at Chapter 2 to remind you about the various levels of structure of a protein molecule such as haemoglobin.) [6]

8 The following statements were all made by candidates in examination answers. Explain what is wrong with each statement.

a Oxyhaemoglobin gradually releases its oxygen as it passes from the lungs to a muscle. [2]
b The strong walls of arteries enable them to pump blood around the body. [2]
c Each red blood cell can combine with eight oxygen atoms. [2]
d Red blood cells have a large surface area so that many oxygen molecules can be attached. [2]

[Total: 8]

9 Carbon dioxide is transported in the blood in various forms.

a Describe how carbon dioxide molecules reach red blood cells from respiring cells. [2]

The diagram below shows part of a capillary network and some cells of the surrounding tissue.

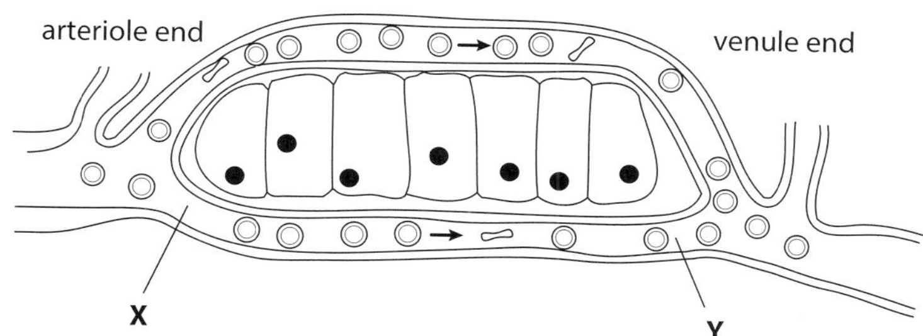

b State three ways in which the blood at **Y** differs from the blood at **X** **other than** in the concentration of carbon dioxide. [3]

An enzyme in red blood cells catalyses the reaction between carbon dioxide and water as blood flows through respiring tissues.

$$CO_2 + H_2O \xrightarrow{\text{enzyme}} H_2CO_3 \longrightarrow H^+ + HCO_3^-$$

c i Name the enzyme that catalyses this reaction. [1]

ii Explain the significance of this reaction in the transport of carbon dioxide. [3]

d The graph below shows the effect of increasing the carbon dioxide concentration on the oxygen dissociation curve for haemoglobin.

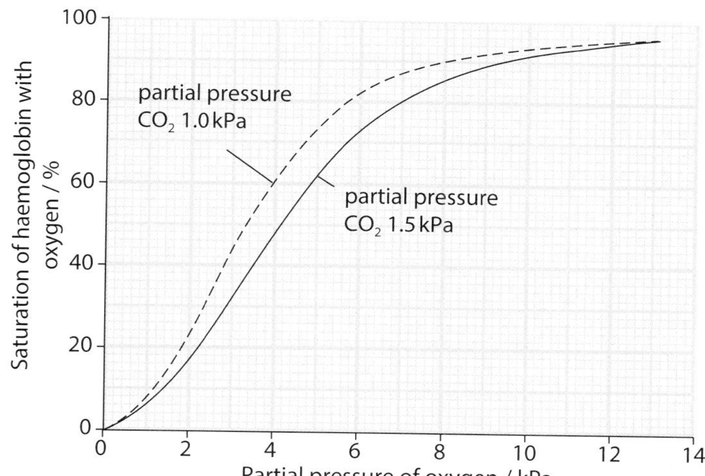

i State the percentage saturation of haemoglobin with oxygen at a partial pressure of 5 kPa of oxygen when the partial pressure of carbon dioxide is:

1.0 kPa

1.5 kPa [1]

ii The percentage saturation of haemoglobin with oxygen **decreases** as the partial pressure of carbon dioxide increases. Explain how this happens. [2]

iii Name the effect of increasing carbon dioxide concentration on the oxygen dissociation curve. [1]

iv Explain the importance of the effect of carbon dioxide on haemoglobin as shown in the graph above. [3]

[Total: 16]

Cambridge International AS and A Level Biology 9700 Paper 21, Question 2, June 2011

10 Mammals have a closed, double circulation.

a State what is meant by the term double circulation. [1]

The figure below shows part of the circulation in a mammalian tissue. The central part is enlarged to show a capillary, a cell supplied by the capillary, and vessel **Z**.

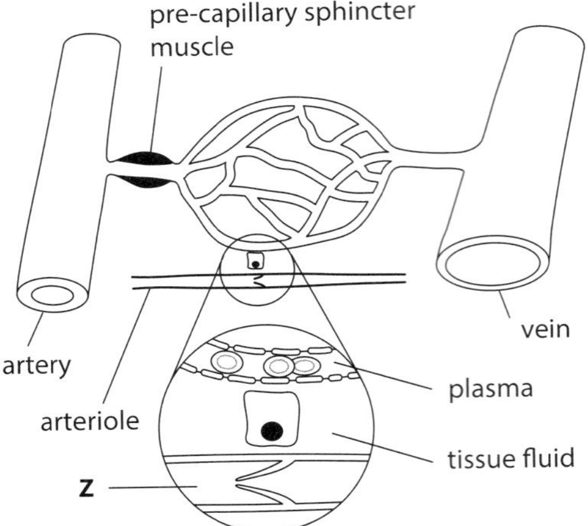

b Explain why the wall of the artery is thicker than the wall of the vein. [2]

c Suggest one role for the pre-capillary sphincter muscle shown in the figure. [1]

d With reference to the figure, describe the role of capillaries in forming tissue fluid. [3]

e **i** Describe three ways in which plasma differs from tissue fluid. [3]

ii Name the fluid in vessel **Z**. [1]

[Total: 11]

Cambridge International AS and A Level Biology 9700 Paper 2, Question 4, November 2008

11 Figure 8.28, page 176, shows the pressure changes in the left atrium, left ventricle and aorta throughout two cardiac cycles. Make a copy of this diagram.

a **i** How long does one heart beat (one cardiac cycle) last? [1]

ii What is the heart rate represented on this graph, in beats per minute? [1]

b The contraction of muscles in the ventricle wall causes the pressure inside the ventricle to rise. When the muscles relax, the pressure drops again. On your copy of the diagram, mark the following periods:

i the time when the ventricle is contracting (ventricular systole) [1]

ii the time when the ventricle is relaxing (ventricular diastole). [1]

c The contraction of muscles in the wall of the atrium raises the pressure inside it. This pressure is also raised when blood flows into the atrium from the veins, while the atrial walls are relaxed. On your copy of the diagram, mark the following periods:

i the time when the atrium is contracting (atrial systole) [1]

ii the time when the atrium is relaxing (atrial diastole). [1]

d The atrioventricular valves open when the pressure of the blood in the atria is greater than that in the ventricles. They snap shut when the pressure of the blood in the ventricles is greater than that in the atria. On your diagram, mark the points at which these valves will open and close. [1]

e The opening and closing of the semilunar valves in the aorta depends in a similar way on the relative pressures in the aorta and ventricles. On your diagram, mark the points at which these valves will open and close. [1]

f The right ventricle has much less muscle in its walls than the left ventricle, and only develops about one-quarter of the pressure developed on the left side of the heart. On your diagram, draw a line to represent the probable pressure inside the right ventricle over the 1.3 seconds shown. [1]

[Total: 9]

12 The diagram below shows a cross-section of the heart at the level of the valves.

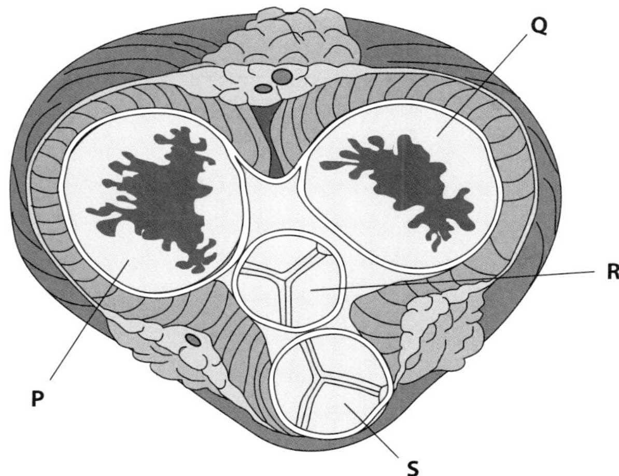

a i Copy and complete the following flow chart to show the pathway of blood through the heart. [2]

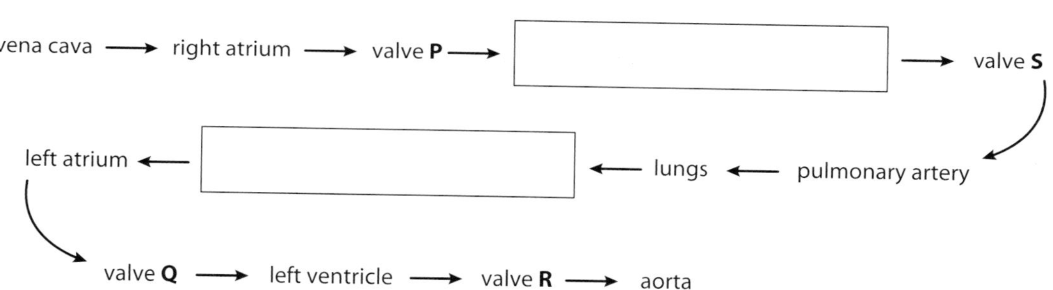

ii Explain how the valves **P** and **Q** ensure one-way flow of blood through the heart. [2]

b The cardiac cycle describes the events that occur during one heart beat. The following figure shows the changes in blood pressure that occur within the left atrium, left ventricle and aorta during one heart beat.

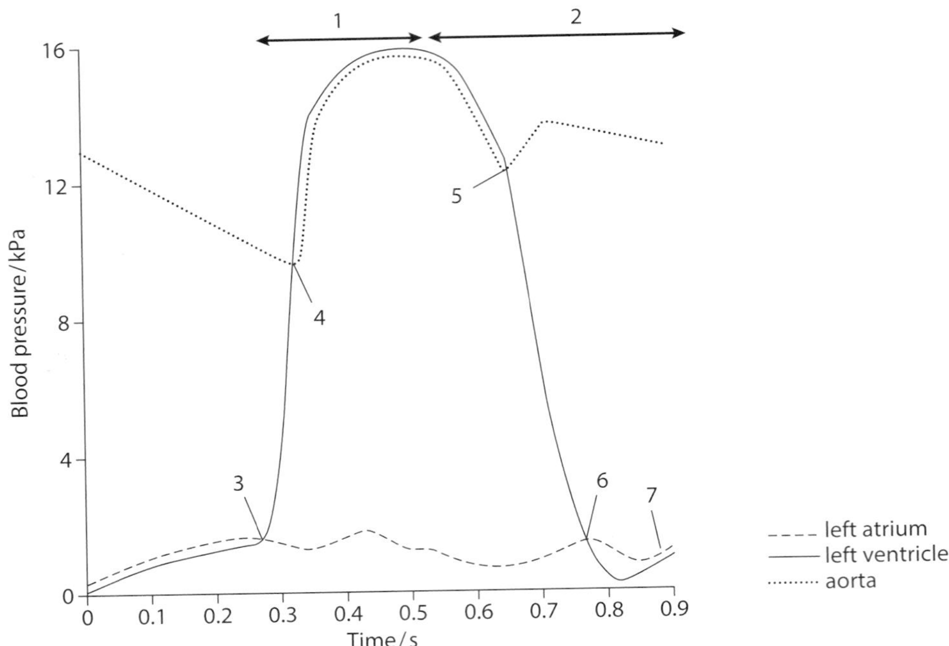

Copy and complete the table below. Match up each event during the cardiac cycle with an appropriate number from 1 to 7 on the diagram. **You should put only one number in each box. You may use each number once, more than once or not at all**.

The first answer has been completed for you.

Event during the cardiac cycle	Number
atrioventricular (bicuspid) valve opens	6
ventricular systole	
semilunar (aortic) valve closes	
left ventricle and left atrium both relaxing	
semilunar (aortic) valve opens	

[4]

c Explain the roles of the sinoatrial node (SAN), atrioventricular node (AVN) and the Purkyne tissue during one heart beat. [5]

[Total: 13]

Cambridge International AS and A Level Biology 9700 Paper 21, Question 3, May/June 2010

Chapter 9:
Gas exchange and smoking

Learning outcomes

You should be able to:

- describe the structure of the human gas exchange system
- describe and explain the distribution of tissues and cells within the gas exchange system
- describe the functions of these tissues and cells
- explain how gases are exchanged in the lungs
- explain how tobacco smoke affects the gas exchange system
- describe the effects of nicotine and carbon monoxide on the cardiovascular system

Examining the airways

The surgeon in Figure 9.1a is using an endoscope to examine the airways looking for any blockages or for possible signs of lung cancer. Endoscopes are flexible tubes with a light and camera at one end. They are used for many routine examinations of body cavities. The type of endoscope used to examine the airways and lungs is a bronchoscope.

The surgeon inserts the bronchoscope through the nose or mouth. It passes through the vocal cords in the larynx and then goes down the trachea (windpipe) and into one of the bronchi. If the bronchoscope has a video camera it is possible for others, including the patient, to view any blockages or damage to the airways on a monitor screen.

In this case, the white area at the base of the trachea is inflamed (Figure 9.1b). If the patient is a smoker, the surgeon may suspect lung cancer. He can use the bronchoscope to carry out a biopsy by removing a small amount of tissue from this area and sending it to the lab for analysis.

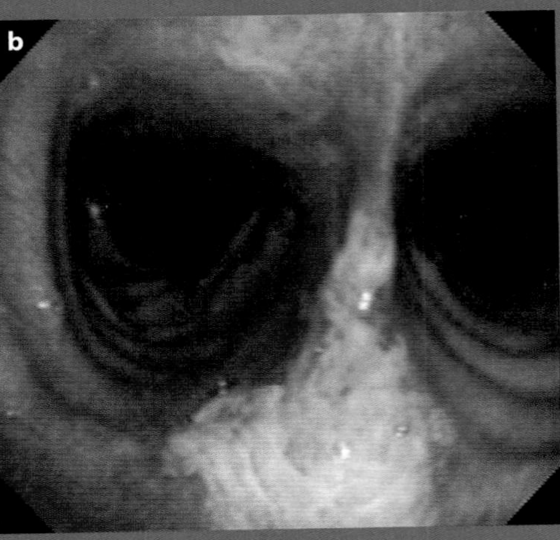

Figure 9.1 **a** A surgeon using a bronchoscope to view the interior of the airways. **b** A view through a bronchoscope of the base of the trachea where it branches into the two bronchi. The white area is inflammation of the tissues that line the trachea.

Gas exchange

The human gas exchange system links the circulatory system (Chapter 8) with the atmosphere. It is adapted to:

- clean and warm the air that enters during breathing
- maximise the surface area for diffusion of oxygen and carbon dioxide between the blood and atmosphere
- minimise the distance for this diffusion
- maintain adequate gradients for this diffusion.

Most organisms need a supply of oxygen for respiration. In single-celled organisms, the oxygen simply diffuses from the fluid outside the cell, through the cell surface membrane and into the cytoplasm. In a multicellular organism such as a human, however, most of the cells are a considerable distance away from the external environment from which the oxygen is obtained. Multicellular organisms therefore usually have a specialised **gas exchange surface** where oxygen from the external environment can diffuse into the body, and carbon dioxide can diffuse out.

In humans, the gas exchange surface is the **alveoli** (singular: **alveolus**) in the lungs. Figure 9.2 shows the distribution of alveoli in the lungs and their structure. Although each individual alveolus is tiny, the alveoli collectively have a huge surface area, probably totalling around 70 m^2 in an adult. This means that a large number of oxygen and carbon dioxide molecules can diffuse through the surface at any one moment to give us a high rate of gas exchange.

Lungs

The lungs are in the thoracic (chest) cavity surrounded by the pleural membranes, which enclose an airtight space. This space contains a small quantity of fluid to allow friction-free movement as the lungs are ventilated by the movement of the diaphragm and ribs.

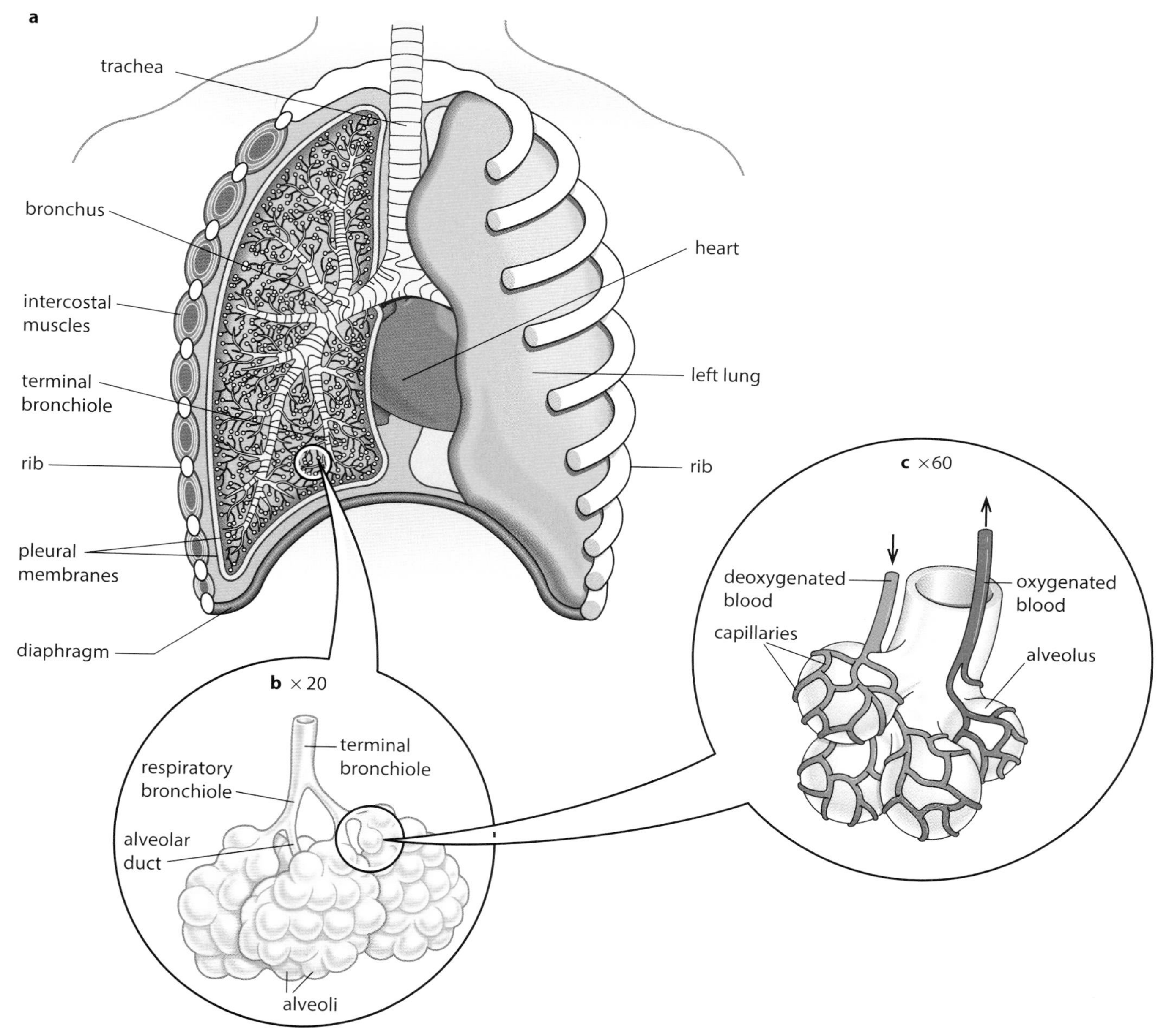

Figure 9.2 The human lungs. Air passes through **a** the trachea and bronchi to supply many branching bronchioles **b** which terminate in alveoli **c** where gas exchange occurs. A gas exchange surface of around 70 m^2 fits into the thoracic cavity, which has a capacity of about 5 dm^3.

Trachea, bronchi and bronchioles

The lungs are ventilated with air that passes through a branching system of airways (Table 9.1). Leading from the throat to the lungs is the **trachea**. At the base of the trachea are two **bronchi** (singular: **bronchus**), which subdivide and branch extensively forming a bronchial 'tree' in each lung. Each bronchus divides many times to form smaller **bronchioles**. Terminal bronchioles divide to form even narrower respiratory bronchioles that supply the alveoli with air.

Cartilage in the trachea and bronchi keeps these airways open and air resistance low, and prevents them from collapsing or bursting as the air pressure changes during breathing. In the trachea, there is a regular arrangement of C-shaped rings of cartilage; in the bronchi, there are irregular blocks of cartilage instead (Figure 9.3).

Airway	Number	Approximate diameter	Cartilage	Goblet cells	Smooth muscle	Cilia	Site of gas exchange
trachea	1	1.8 cm	yes	yes	yes	yes	no
bronchus	2	1.2 cm	yes	yes	yes	yes	no
terminal bronchiole	48 000	1.0 mm	no	no	yes	yes	no
respiratory bronchiole	300 000	0.5 mm	no	no	no	a few	no
alveolar duct	9×10^6	400 μm	no	no	no	no	yes
alveoli	3×10^9	250 μm	no	no	no	no	yes

Table 9.1 The structure of the airways from the trachea to the alveoli. The various airways are shown in Figure 9.2.

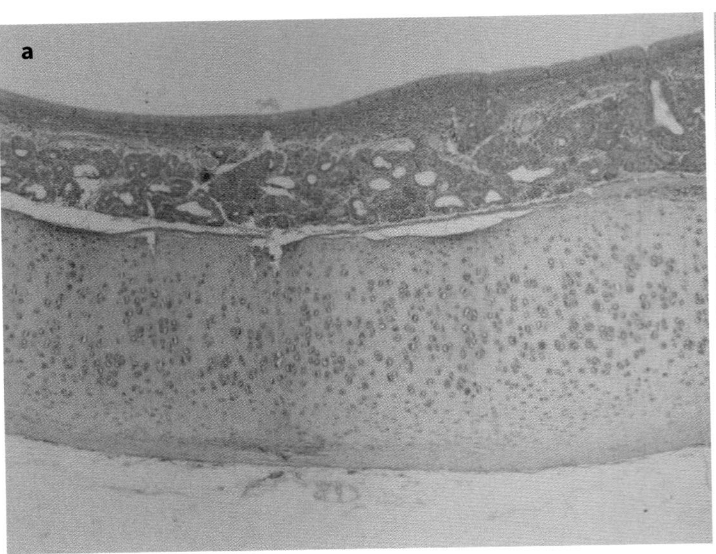

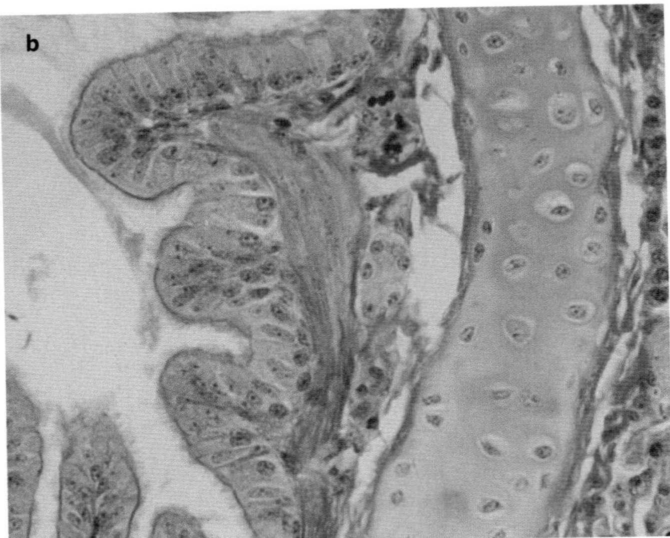

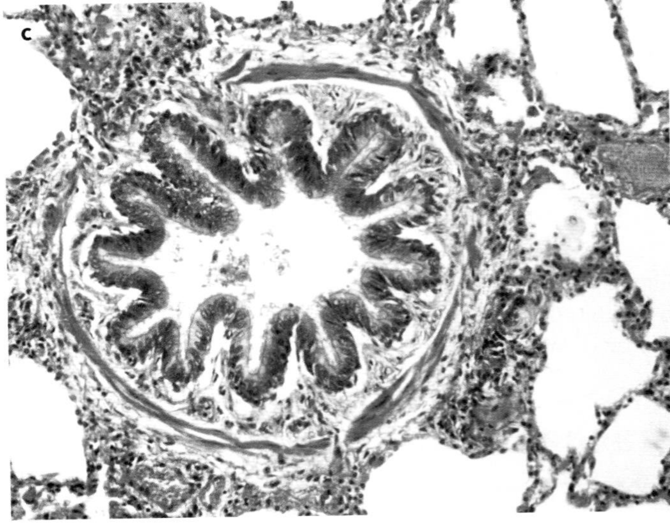

Figure 9.3

a A light micrograph of part of the trachea in transverse section (TS) (×65). The lining is comprised of ciliated epithelium which rests on a basement membrane made of protein fibres. In between the ciliated cells are goblet cells (here stained blue). Beneath the epithelium is an area of loose tissue with blood vessels and mucous glands. The trachea as a whole is supported by C-shaped rings of cartilage, a portion of which appears as the thick layer running across the bottom of the picture.

b A light micrograph of part of a bronchus in TS (×300). Between the ciliated epithelial cells, the goblet cells are stained pink. There are fewer goblet cells per cm^2 than in the trachea, and the epithelial cells are not as tall. Beneath the epithelium there are elastic fibres. Blocks of cartilage, not rings, support the bronchus, and part of one can be seen, also stained pink, stretching from the top to the bottom of the picture.

c A light micrograph of a small bronchiole in TS (×135). Surrounding the epithelium is smooth muscle. There is no cartilage. Around the bronchiole are some alveoli.

QUESTION

9.1 Use Figure 9.3 to draw plan diagrams of the walls of **a** the trachea, **b** a bronchus, **c** a bronchiole and alveoli. Label your diagrams to identify the tissues.

Bronchioles are surrounded by smooth muscle, which can contract or relax to adjust the diameter of these tiny airways. During exercise, the muscles relax to allow a greater flow of air to the alveoli. The absence of cartilage makes these adjustments possible.

Warming and cleaning the air

As air flows through the nose and the trachea, it is warmed to body temperature and moistened by evaporation from the lining, so protecting the delicate surfaces inside the lungs from desiccation (drying out). Protection is also needed against the suspended matter carried in the air, which may include dust, sand, pollen, fungal spores, bacteria and viruses. All are potential threats to the proper functioning of the lungs. Particles larger than about 5–10 μm are caught on the hairs inside the nose and the **mucus** lining the nasal passages and other airways.

In the trachea and bronchi, the mucus is produced by the **goblet cells** of the ciliated epithelium. The upper part of each goblet cell is swollen with **mucin** droplets which have been secreted by the cell. Mucus is a slimy solution of mucin, which is composed of glycoproteins with many carbohydrate chains that make them sticky and able to trap inhaled particles. The rest of the goblet cell, which contains the nucleus, is quite slender like the stem of a goblet. Mucus is also made by mucous glands beneath the epithelium. Some chemical pollutants, such as sulfur dioxide and nitrogen dioxide, can dissolve in mucus to form an acidic solution that irritates the lining of the airways.

Between the goblet cells are the ciliated cells (Figure 9.4). The continual beating of their cilia carries the carpet of mucus upwards towards the larynx at a speed of about 1 cm min^{-1}. When mucus reaches the top of the trachea it is usually swallowed so that pathogens are destroyed by the acid in the stomach.

Phagocytic white blood cells known as **macrophages** (Chapter 11) patrol the surfaces of the airways scavenging small particles such as bacteria and fine dust particles. During an infection, the macrophages are joined by other phagocytic cells which leave the capillaries to help remove pathogens.

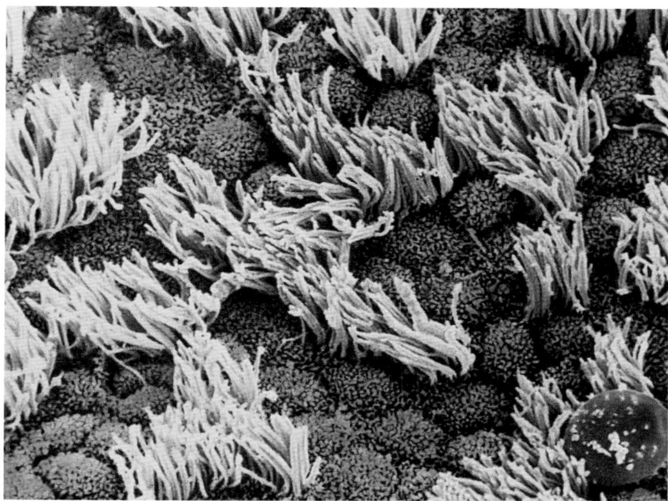

Figure 9.4 False-colour scanning electron micrograph of the surface of the trachea, showing large numbers of cilia (yellow) and some mucus-secreting goblet cells (red) (×2600).

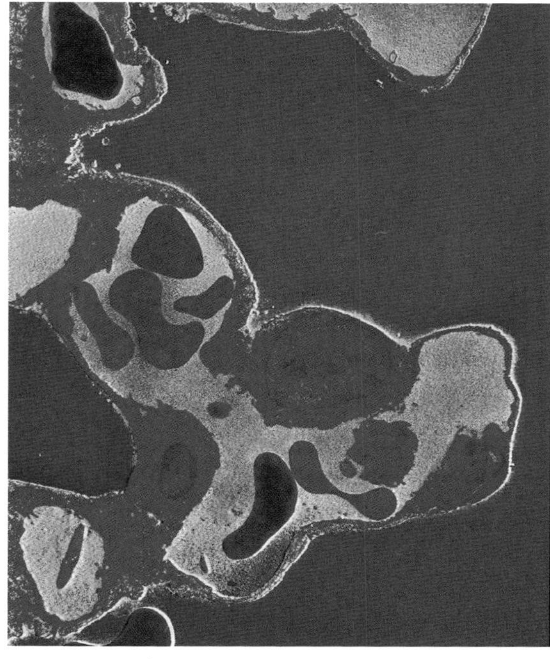

Figure 9.5 False-colour transmission electron micrograph of the lining of an alveolus. Red blood cells fill the blood capillaries (yellow), which are separated from the air (blue) by a thin layer of cells (pink) ×2100.

Alveoli

At the end of the pathway between the atmosphere and the bloodstream are the **alveoli** (Figures 9.3c and 9.5). Alveolar walls contain **elastic fibres** which stretch during inspiration and recoil during expiration to help force out air. This elasticity allows alveoli to expand according to the volume of air breathed in. When the alveoli are fully expanded during exercise, the surface area available for diffusion increases, and the air is expelled efficiently when the elastic fibres recoil.

The alveoli have extremely thin walls, each consisting of a single layer of squamous epithelial cells no more than 0.5 μm thick. Pressed closely against the alveoli walls are blood capillaries, also with very thin single-celled walls. Oxygen and carbon dioxide molecules diffuse quickly between the air and the blood because the distance is very small.

You will remember that diffusion is the net movement of molecules or ions down a concentration gradient. So, for gas exchange to take place rapidly, a steep concentration gradient must be maintained. This is done by breathing and by the movement of the blood. Breathing brings supplies of fresh air into the lungs, with a relatively high oxygen concentration and a relatively low carbon dioxide concentration. Blood is brought to the lungs with a lower concentration of oxygen and a higher concentration of carbon dioxide than the air in the alveoli. Oxygen therefore diffuses down its concentration gradient from the air in the alveoli to the blood, and carbon dioxide diffuses down its concentration gradient in the opposite direction. The blood is constantly flowing through and out of the lungs, so, as the oxygenated blood leaves, more deoxygenated blood enters to maintain the concentration gradient with each new breath.

QUESTIONS

9.2 a Describe the pathway taken by a molecule of oxygen as it passes from the atmosphere to the blood in the lungs.

b Explain how alveoli are adapted for gas exchange.

9.3 a Explain the advantage of being able to adjust the diameter of bronchioles.

b How many times does an oxygen molecule cross a cell surface membrane as it moves from the air into a red blood cell?

Smoking

The World Health Organization (WHO) considers smoking to be a disease. Until the end of the 19th century, tobacco was smoked almost exclusively by men and in pipes and cigars, involving little inhalation. Then the manufacture of cigarettes began. Smoking cigarettes became fashionable for European men during the First World War and in the 1940s women started smoking in large numbers too. The numbers of smokers in many countries, such as the UK and the USA, has decreased sharply in recent decades. Meanwhile, there has been a huge rise in the number of people smoking in countries such as China and Pakistan. It is estimated that in Pakistan as many as 40% of men and 8% of women are regular smokers. The smoking of a flavoured tobacco called shisha has become fashionable among young people across the world, with many thinking that as it is smoked through a water pipe it is safer than smoking cigarettes.

Tobacco smoke

The tobacco companies do not declare the ingredients in their products, but it is known by analysis that there are over 4000 different chemicals in cigarette smoke, many of which are toxic. Tobacco smoke is composed of 'mainstream' smoke (from the filter or mouth end) and 'sidestream' smoke (from the burning tip). When a person smokes, about 85% of the smoke that is released is sidestream smoke. Many of the toxic ingredients are in a higher concentration in sidestream than in mainstream smoke, and any other people in the vicinity are also exposed to them. Breathing in someone else's cigarette smoke is called **passive smoking**.

The main components of cigarette smoke pose a threat to human health. These are:

- tar, which contains carcinogens (cancer-causing compounds)
- carbon monoxide
- nicotine.

In general, tar and carcinogens damage the gas exchange system; carbon monoxide and nicotine damage the cardiovascular system (page 171).

- **Tar** is a mixture of compounds that settles on the lining of the airways in the lungs and stimulates a series of changes that may lead to obstructive lung diseases and lung cancer.
- **Carcinogens** are cancer-causing compounds. These cause mutations in the genes that control cell division.

Lung diseases

Lung diseases are a major cause of illness and death worldwide. Air pollution, smoking and allergic reactions are the causes of almost all cases.

The gas exchange system is naturally efficient and adaptable. Healthy people breathe with little conscious effort; for people with lung disease, every breath may be a struggle.

The lungs' large surface area of delicate tissue is constantly exposed to moving streams of air that may carry potentially harmful gases and particles. Despite the filtering system in the airways, very small particles (less than 2 µm in diameter) can reach the alveoli and stay there. The air that flows down the trachea with each breath fills the huge volume of tiny airways. The small particles settle out easily because the air flow in the depths of the lungs is very slow. Such deposits make the lungs susceptible to airborne infections such as influenza and pneumonia and, in some people, can cause an allergic reaction leading to asthma.

Every disease has a characteristic set of **signs** and **symptoms**. Signs are the visible expression of a disease which a doctor could find by examining a patient – for example, a high temperature. Symptoms cannot be detected by an examination and can only be reported by the patient – for example, a headache or nausea. Often, however, the word 'symptom' is used more loosely to cover both signs and symptoms.

Chronic (long-term) obstructive pulmonary diseases (COPD) such as asthma, chronic **bronchitis** and **emphysema** are now common in many countries. Atmospheric pollution from vehicle and industrial emissions and tobacco smoke are linked with these diseases.

Chronic bronchitis

Tar in cigarette smoke stimulates goblet cells and mucous glands to enlarge and secrete more mucus (page 189). Tar also inhibits the cleaning action of the ciliated epithelium that lines the airways. It destroys many cilia and weakens the sweeping action of those that remain. As a result, mucus accumulates in the bronchioles, and the smallest of these may become obstructed. As mucus is not moved, or at best only moved slowly, dirt, bacteria and viruses collect and block the bronchioles. This stimulates 'smoker's cough' which is an attempt to move the mucus up the airways. With time, the damaged epithelia are replaced by scar tissue, and the smooth muscle surrounding the bronchioles and bronchi becomes thicker. This thickening of the airways causes them to narrow and makes it difficult to breathe.

Infections such as pneumonia easily develop in the accumulated mucus. When there is an infection in the lungs, the linings become inflamed and this further narrows the airways. This damage and obstruction of the airways is chronic bronchitis. Sufferers have a severe cough, producing large quantities of phlegm, which is a mixture of mucus, bacteria and some white blood cells.

Emphysema

The inflammation of the constantly infected lungs causes phagocytes to leave the blood and line the airways. Phagocytes are white blood cells that remove bacteria from the body (Chapters 4 and 11). To reach the lining of the lungs from the capillaries, phagocytes release the protein-digesting enzyme elastase. This enzyme destroys elastin in the walls of the alveoli, so making a pathway for the phagocytes to reach the surface and remove bacteria.

Elastin is responsible for the recoil of the alveoli when we breathe out. With much smaller quantities of elastin in the alveolar walls, the alveoli do not stretch and recoil when breathing in and out (Figure 9.6). As a result, the bronchioles collapse during expiration, trapping air in the alveoli, which often burst. Large spaces appear where the alveoli have burst, and this reduces the surface area for gas exchange (Figure 9.7); the number of capillaries also decreases, so less oxygen is absorbed into the blood. This condition is called emphysema.

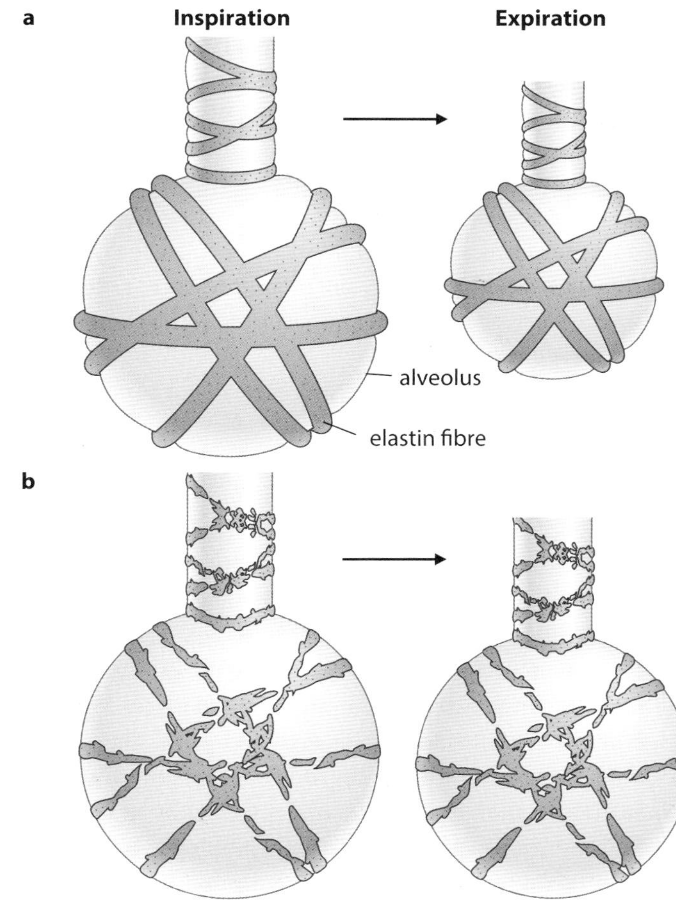

Figure 9.6 The development of emphysema. **a** Healthy alveoli partially deflate when breathing out, due to the recoil of elastin fibres. **b** Phagocytes from the blood make pathways through alveolar walls by digesting elastin and, after many years of this destruction, the alveoli do not deflate very much.

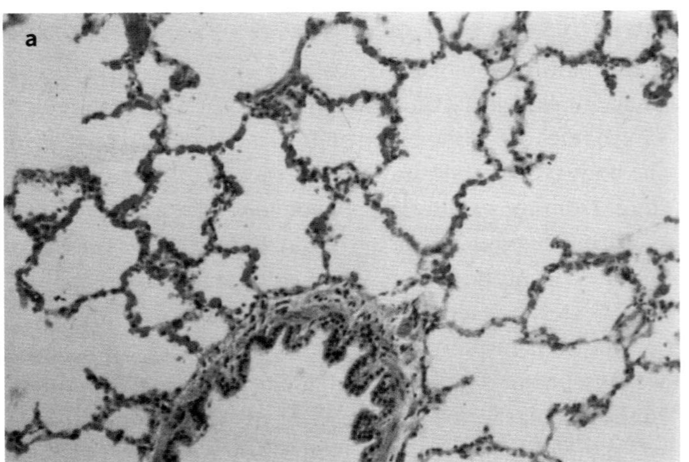

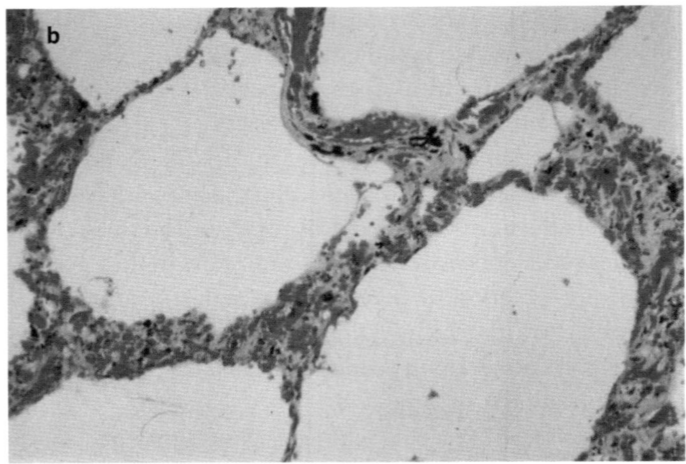

Figure 9.7 **a** Photomicrograph of normal lung tissue (×75). **b** Photomicrograph of lung tissue from a person with emphysema, showing large spaces where there should be many tiny alveoli (×75).

The loss of elastin makes it difficult to move air out of the lungs. Non-smokers can force out about 4 dm^3 of air after taking a deep breath; someone with emphysema may manage to force out only 1.3 dm^3 of air. The air remains in the lungs and is not refreshed during ventilation. Together with the reduced surface area for gas exchange, this means that many people with emphysema do not oxygenate their blood very well and have a rapid breathing rate.

As the disease progresses, the blood vessels in the lungs become more resistant to the flow of blood. To compensate for this increased resistance, the blood pressure in the pulmonary artery increases and, over time, the right side of the heart enlarges.

As lung function deteriorates, wheezing occurs and breathlessness becomes progressively worse. It may become so bad in some people that they cannot get out of bed. People with severe emphysema often need a continuous supply of oxygen through a face mask to stay alive. In Figure 9.7 you can see the destruction of alveoli that occurs in emphysema. In Figure 9.8, you can compare the appearance of diseased and relatively unaffected lung tissue in a computerised tomography (CT) scan.

Chronic bronchitis and emphysema often occur together and constitute a serious risk to health. The gradual onset of breathlessness only becomes troublesome when about half of the lungs is destroyed. Only in very rare circumstances is it reversible. If smoking is given up when still young, lung function can improve. In older people, recovery from COPD is not possible. Over 60 million people worldwide have COPD. The WHO predicts that COPD will become the third leading cause of death worldwide by 2030.

There are legal controls in many countries on emissions of pollutants from industrial, domestic and transport fuels and on conditions in work places. Many countries have also banned smoking in public places, such as shops, restaurants and on public transport.

QUESTION

9.4 Summarise the changes that occur in the lungs of people with chronic bronchitis and emphysema.

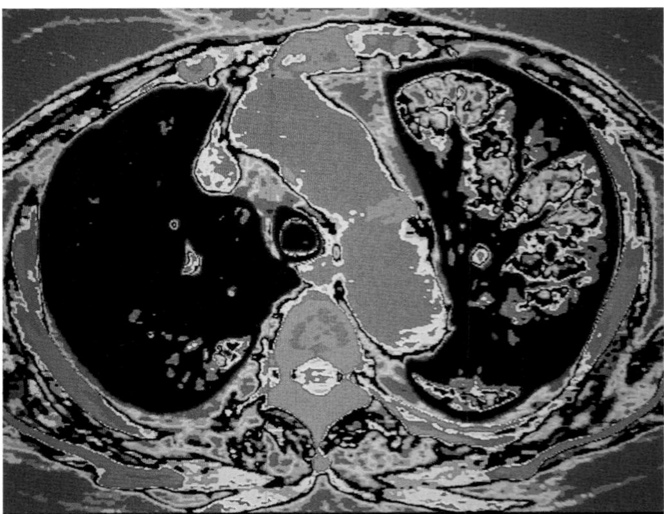

Figure 9.8 A CT scan of a horizontal section through the thorax. The two black regions are the lungs. The right lung is diseased with emphysema (blue-green areas). The left lung is relatively unaffected at the level of this scan. You can see a cross-section of a vertebra at the bottom of the scan.

Lung cancer

Tar in tobacco smoke contains several substances that are carcinogens (Chapter 5). These react, directly or via breakdown products, with DNA in epithelial cells to produce mutations, which are the first in a series of changes that lead to the development of a mass of cells, known as a tumour (Figure 5.13 on page 104 and Figure 9.9).

As the cancer develops, it spreads through the bronchial epithelium and enters the lymphatic tissues (page 164) in the lung. Cells may break away and spread to other organs (metastasis, page 105), so that secondary tumours become established. A tumour like this is known as a malignant tumour.

Lung cancer takes 20–30 years to develop. Most of the growth of a tumour occurs before there are any symptoms. The most common symptom of lung cancer is coughing up blood, as a result of tissue damage. People with lung cancer also have chest pain and find it difficult to breathe. It is rare for a cancer to be diagnosed before it reaches 1 cm in diameter.

Figure 9.9 A scanning electron micrograph of a bronchial carcinoma – a cancer in a bronchus. Cancers often develop at the base of the trachea where it divides into the bronchi as this is where most of the tar is deposited. The disorganised malignant tumour cells at the bottom right are invading the normal tissue of the ciliated epithelium (×1000).

Tumours in the lungs, such as that shown in Figure 9.9, are located by one of three methods:

- bronchoscopy, using an endoscope to allow a direct view of the lining of the bronchi (Figure 9.1)
- chest X-ray
- CT scan (similar to that shown in Figure 9.8).

By the time most lung cancers are discovered, they are well advanced; treatment involves surgery, radiotherapy and chemotherapy and is dependent on the type of lung cancer, how far it has developed and whether it has spread into other areas of the body (whether there is metastasis). If the cancer is small and in one lung, then either a part or all of the lung is removed. However, metastasis has usually happened by the time of the diagnosis so, if there are secondary tumours, surgery will not cure the disease. This is why smoking is linked to so many different cancers. Chemotherapy with anti-cancer drugs or radiotherapy with X-rays (or another form of radiation) is used.

Short-term effects on the cardiovascular system

The two components of tobacco smoke that cause short-term effects on the cardiovascular system are nicotine and carbon monoxide.

Nicotine

Nicotine is the drug in tobacco. It is absorbed very readily by the blood and travels to the brain within a few seconds. It stimulates the nervous system to reduce the diameter of the arterioles and to release the hormone adrenaline from the adrenal glands. As a result, heart rate and blood pressure increase and there is a decrease in blood supply to the extremities of the body, such as hands and feet, reducing their supply of oxygen. Nicotine also increases the risk of blood clotting.

Nicotine is a highly addictive drug that influences reward centres in the brain. It stimulates nerve endings in the brain to release the transmitter substance dopamine, which is associated with reinforcing pleasurable experiences. This makes it very hard to give up smoking.

Carbon monoxide

Carbon monoxide diffuses across the walls of the alveoli and into the blood in the lungs. It diffuses into red blood cells where it combines with haemoglobin to form the stable compound carboxyhaemoglobin (page 171). This means that haemoglobin does not become fully oxygenated. The quantity of oxygen transported in the blood may be 5–10% less in a smoker than in a non-smoker. Less oxygen is supplied to the heart muscle, putting a strain on it especially when the heart rate increases during exercise. Carbon monoxide may also damage the lining of the arteries.

These short-term effects are readily reversible in people who have not smoked for very long. Long-term smokers put the health of their cardiovascular system at risk. Damage to the walls of arteries may lead to the build-up of fatty tissue and the reduction of blood flow. **Coronary heart disease** (CHD) and **stroke** may be the result. These diseases are a major cause of death and disability. They are responsible for 20% of all deaths worldwide and up to 50% of deaths in developed countries. **Cardiovascular diseases** are multifactorial, meaning that many factors contribute to the development of these diseases. Smoking is just one among several risk factors that increase the chances of developing one of the cardiovascular diseases, such as CHD and stroke.

QUESTIONS

9.5 **a** Summarise the effects of tobacco smoke on the gas-exchange system.

b List the signs and symptoms of COPD and lung cancer.

9.6 Describe the effects of tobacco smoke on the cardiovascular system.

Summary

- Multicellular organisms often have surfaces that are specialised to allow exchange of gases to take place between their bodies and the environment. Alveoli in the lungs form the gas exchange surface in mammals.
- In the human lungs, air passes down the trachea and through a branching system of airways to reach the alveoli. The airways are lined by a ciliated epithelium with mucus-secreting goblet cells. The epithelium protects the alveoli by moving a carpet of mucus towards the throat, where it can be swallowed. There are C-shaped rings of cartilage in the trachea and irregularly shaped blocks of cartilage in the bronchi to keep the large airways open and so reduce resistance to the flow of air. Smooth muscle in the airways contracts and relaxes to adjust the diameter of the airways.
- The alveoli are lined by a squamous epithelium that gives a short diffusion distance for the exchange of oxygen and carbon dioxide. The alveoli are well supplied with blood by the many capillaries surrounding the gas exchange surface. The constant flow of blood and the continuous ventilation of the lungs maintain concentration gradients between blood and air for oxygen and carbon dioxide. Recoil of the elastic fibres surrounding the alveoli helps to move air out during expiration.
- Damage to the airways and alveoli occurs in chronic obstructive pulmonary disease (COPD). In chronic bronchitis, the airways are obstructed by mucus and infection; in emphysema, alveoli are destroyed, reducing the surface area for gas exchange. Some of the signs and symptoms of COPD are breathlessness, wheezing and constant coughing.
- Tobacco smoke contains tar, carbon monoxide and nicotine. Tar settles on the epithelium lining the bronchi and bronchioles and stimulates inflammation, an increase in the secretion of mucus and an accumulation of phagocytes from the blood. Tar contains carcinogens, which cause changes in DNA in bronchial epithelial cells, often leading to the development of a bronchial carcinoma. This is lung cancer. Two of the symptoms of lung cancer are coughing up blood and chest pains.
- Carbon monoxide combines irreversibly with haemoglobin, reducing the oxygen-carrying capacity of the blood. Nicotine stimulates the nervous system, increasing heart rate and blood pressure, and stimulating vasoconstriction, which reduces blood flow to the extremities.
- Smoking damages the cardiovascular system, increasing and increasing the risk of coronary heart disease and stroke.

End-of-chapter questions

1 The following structures are found in the walls of the gas exchange system.

1 capillaries
2 cilia
3 elastic fibres
4 goblet cells
5 smooth muscle cells

Which would be found in the lining of an alveolus?

A 1 and 3
B 1, 2 and 3
C 2 and 5
D 4 and 5 [1]

2 Cartilage is found in which structure?

A alveolus
B bronchiole
C capillary
D trachea [1]

3 Which of the following is **not** a role of elastic fibres in the gas exchange system?

A contract to decrease the volume of the alveoli during expiration
B recoil to force air out of the alveoli during expiration
C stretch to accommodate more air in the alveoli during deep breathing
D stretch to increase the surface area of the alveoli for gas exchange [1]

4 Which of the following best describes the process of gas exchange in the lungs?

A Air moves in and out of the alveoli during breathing.
B Carbon dioxide diffuses from deoxygenated blood in capillaries into the alveolar air.
C Oxygen and carbon dioxide diffuse down their concentration gradients between blood and alveolar air.
D Oxygen diffuses from alveolar air into deoxygenated blood. [1]

5 Which of the following substances in tobacco smoke damage the gas exchange system?

A carbon monoxide and carcinogens
B carbon monoxide and nicotine
C carcinogens and tar
D nicotine and tar [1]

6 Which substance in tobacco smoke decreases the oxygen-carrying capacity of haemoglobin?

A carbon dioxide
B carbon monoxide
C nicotine
D tar [1]

7 The diagram below shows an alveolus.

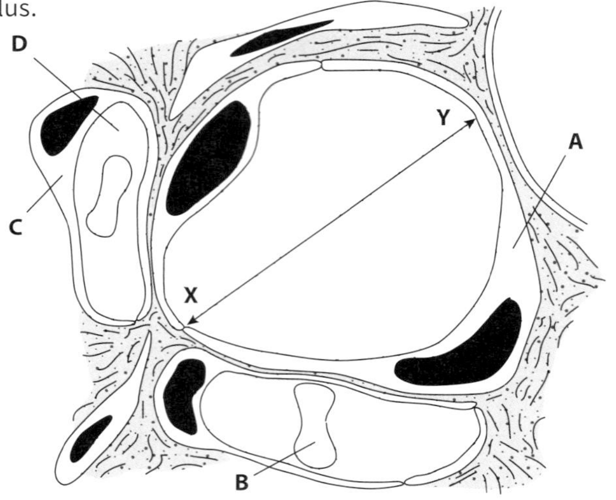

a Name:

i cells **A**, **B** and **C** [3]

ii the fluid at **D**. [1]

b Calculate the actual distance indicated by **X–Y**. Show your working. [2]

c Explain how alveoli are adapted for the exchange of gases. [4]

[Total: 10]

8 The diagram below shows two cells from the lining of the trachea.

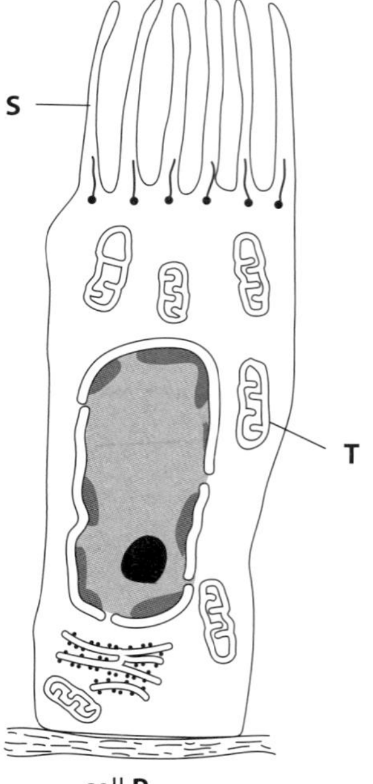

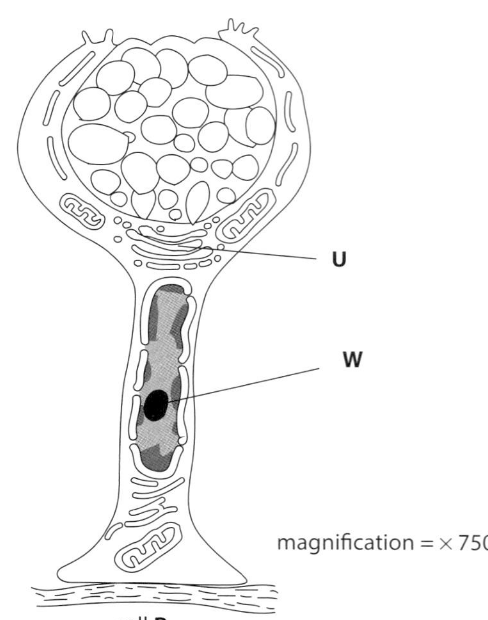

a Name:

i cells **P** and **R** [2]

ii structures **S**, **T**, **U** and **W**. [4]

b Explain:

i why cell **P** contains many of the structures labelled **T** [2]

ii the role of structure **U** in cell **R**. [2]

c Calculate the actual length of cell **P**. Show your working. [2]

d Describe the roles of cell **P** and cell **R** in the gas exchange system. [4]

[Total: 16]

9 a Copy and complete the table to compare the trachea with a respiratory bronchiole. Use a tick (✓) to indicate that the structure is present and a cross (✗) to indicate that it is not.

Structure	Trachea	Respiratory bronchiole
smooth muscle cells		
ciliated epithelial cells		
mucous glands		
cartilage		
elastic fibres		

[5]

b Describe how the alveoli are protected against infection. [5]

[Total: 10]

10 The composition of alveolar air remains fairly constant even though gases are exchanged with the blood in the capillaries that surround the alveoli.

a Describe the process of gas exchange between alveolar air and blood. [4]

b Explain why the composition of alveolar air remains fairly constant. [3]

c Suggest **three** ways in which the gas exchange system responds to the demands of exercise. [3]

[Total: 10]

11 Cigarette smoke contains tar, nicotine and carbon monoxide. Tar contains carcinogens.

a Describe the effect of tar on the lining of the bronchi in the lungs. [8]

b Describe the effects of nicotine and carbon monoxide on the cardiovascular system. [6]

[Total: 14]

12 The figure shows photomicrographs of alveoli from **1** a non-smoker (×200) and **2** a smoker (×50).

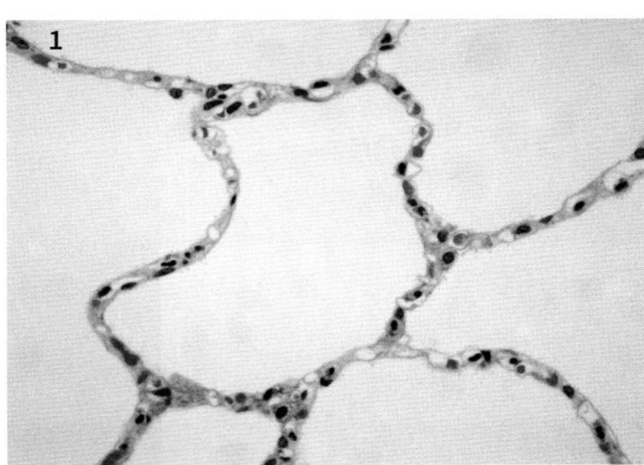

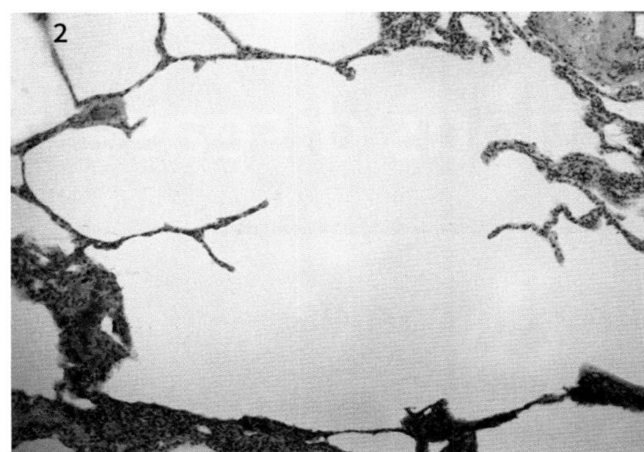

a Use the photomicrographs to describe how the lungs of this smoker differ from the lungs of the non-smoker. [4]

b Smokers whose lungs contain alveoli similar to those shown in photomicrograph 2 have poor health.

i Describe the symptoms that these people may have. [4]

ii Explain how the structure of the lungs is responsible for this poor health. [3]

[Total: 11]

Chapter 10: Infectious diseases

Learning outcomes

You should be able to:

- explain what is meant by infectious and non-infectious diseases
- name the organisms that cause some infectious diseases
- explain how cholera, measles, malaria, TB and HIV/AIDS are transmitted
- discuss the ways in which biological, social and economic factors influence the prevention and control of cholera, measles, malaria, TB and HIV/AIDS
- discuss the factors that influence global patterns of malaria, TB and HIV/AIDS
- outline how penicillin acts on bacteria and why antibiotics do not affect viruses
- explain how bacteria become resistant to antibiotics
- discuss the consequences of antibiotic resistance and how we can reduce its impact

Roll Back Malaria

Malaria is one of the most serious of all human diseases. It is endemic in about 100 countries in Latin America, Africa and South-East Asia. An infectious disease is endemic in a country if it is maintained there by transmission between people and does not occur by being brought into the country by travellers. The burden of malaria is felt most in countries in Africa: 47% of all deaths from malaria occur in Nigeria and the Democratic Republic of the Congo.

If we are to lift the burden of malaria, then we have to outsmart this disease. Many international organisations are cooperating with African governments to reduce the number of cases of malaria. For example, Roll Back Malaria is a global partnership working to implement coordinated action against malaria. Laboratory research using modern genetic methods plays an important part, but the success of new initiatives in the Roll Back Malaria campaign can only be evaluated by tracking cases of the disease and collecting accurate data. Technology is helping here too in pinpointing accurately where cases occur (Figure 10.1). Find out more by searching online for 'roll back malaria'.

Knowing when and where outbreaks occur provides key data in the fight against infectious diseases. When an outbreak of malaria occurs, the local population can be warned to take appropriate precautions to avoid getting bitten by mosquitoes. These include sleeping under bed nets and draining any bodies of water that female mosquitoes might use to lay their eggs. The local area can be sprayed to kill any mosquitoes. By tracking when and where malaria outbreaks occur, health agencies can direct their resources to control the spread of the disease and limit its effects on local populations.

Figure 10.1 Modern technology in the fight against malaria: these health workers in Kenya are using GPS to record accurate locations where there are outbreaks of malaria.

Infectious diseases are diseases that are caused by organisms known as pathogens. They are sometimes called communicable diseases as they are passed from infected to uninfected people. Some also affect animals and are passed from animals to humans.

In Chapter 9, we considered non-infectious diseases of the gas exchange and cardiovascular systems, such as lung cancer and COPD. These are long-term degenerative diseases. They are non-infectious diseases, as they are not caused by pathogens. Inherited or genetic diseases, such as cystic fibrosis and sickle cell anaemia, are another group of non-infectious diseases. There are other categories of non-infectious disease, including deficiency diseases that are caused by malnutrition and mental diseases.

The word '**disease**' implies something very serious, yet many conditions that make us feel ill, such as the common cold, are not as harmful to health as the diseases discussed in this chapter. 'Disease' is a difficult word to define satisfactorily as it covers a wide range of human conditions. A working definition is given below.

> A **disease** is an illness or disorder of the body or mind that leads to poor health; each disease is associated with a set of signs and symptoms.

As you read this chapter and Chapter 11, you can think of ways to improve this definition.

QUESTION

10.1 Use Chapter 9 and this chapter to name:

- **a** two infectious diseases and two non-infectious diseases of the gas exchange system
- **b** two non-infectious diseases of the cardiovascular system.

Many infectious diseases, such as the common cold, measles and influenza, only affect us for a short period of time. Others, such as tuberculosis (TB), may last a much longer time. Indeed, in the case of HIV/AIDS, there is as yet no cure and treatments must be taken for the whole of a person's life. Some infectious diseases can only spread from one person to another by direct contact, because the pathogen cannot survive outside the human body. Other pathogens can survive in water, human food, faeces or animals (including insects), and so are transmitted indirectly from person to person. Some people may spread a pathogen even though they do not have the disease themselves. People like this who lack symptoms are called **carriers**, and it can be very difficult to trace them as the source of an infection.

The way in which a pathogen passes from one host to another is called the **transmission** cycle. Control methods attempt to break transmission cycles by removing the conditions that favour the spread of the pathogen. Control is only possible once the cause of the disease and its method of transmission are known and understood. **Vaccination** is a major control measure for many infectious diseases; it works by making us immune so that pathogens do not live and reproduce within us and do not then spread to others (pages 233–234).

Worldwide importance of infectious diseases

Five infectious diseases of worldwide importance are discussed in detail in this chapter: cholera, malaria, HIV/AIDS, tuberculosis (TB) and measles. Smallpox is the only infectious disease that has so far been eradicated, and it is discussed in Chapter 11 (pages 235–236).

The number of people infected with cholera, malaria, HIV/AIDS, TB and measles, particularly children and young adults, remains very high, and these diseases pose serious public health problems now and for the foreseeable future. This is particularly true in parts of the world without the efficient health services available in affluent countries.

Table 10.1 shows the **causative agents** of the six chosen diseases. To control a disease, we must first know what causes it.

A new strain of **cholera** appeared in 1992 but is so far restricted to South-East Asia. The disease is on the increase, and it is thought many cases go unreported for fear of disruption to travel, tourism and trade. **Malaria** has been on the increase since the 1970s and constitutes a serious risk to health in many tropical countries. **AIDS** was officially recognised in 1981, but the infective agent

Disease	Causative agent (pathogen)	Type of organism
cholera	*Vibrio cholerae*	bacterium
malaria	four species of *Plasmodium*	protoctist
HIV/AIDS	human immunodeficiency virus (HIV)	virus
tuberculosis (TB)	*Mycobacterium tuberculosis* and *M. bovis*	bacterium
measles	a species of *Morbillivirus*	virus
smallpox (eradicated)	*Variola* virus	virus

Table 10.1 The causative agents of six infectious diseases.

(**HIV**) was in human populations for many years before it was identified. The spread of HIV infection since the early 1980s has been exponential. **TB**, once thought to be nearly eradicated, has shown an increase since the 1970s and is a considerable health risk in many countries. Thanks to vaccination, **measles** is a disease that is now very rare in developed countries, but remains a serious threat to the health of children who live in poverty in many developing countries.

Diseases that are always in populations are described as **endemic**. TB is an example of a disease endemic in the whole human population. Malaria is endemic in tropical and sub-tropical regions.

The **incidence** of a disease is the number of people who are diagnosed over a certain period of time, usually a week, month or year. The **prevalence** of a disease is the number of people who have that disease at any one time. An **epidemic** occurs when there is a sudden increase in the number of people with a disease. A **pandemic** occurs when there is an increase in the number of cases throughout a continent or across the world. The death rate from different diseases is referred to as **mortality**.

Cholera

Transmission of cholera

The features of cholera are given in Table 10.2. Cholera is caused by the bacterium *Vibrio cholerae* (Figure 10.2).

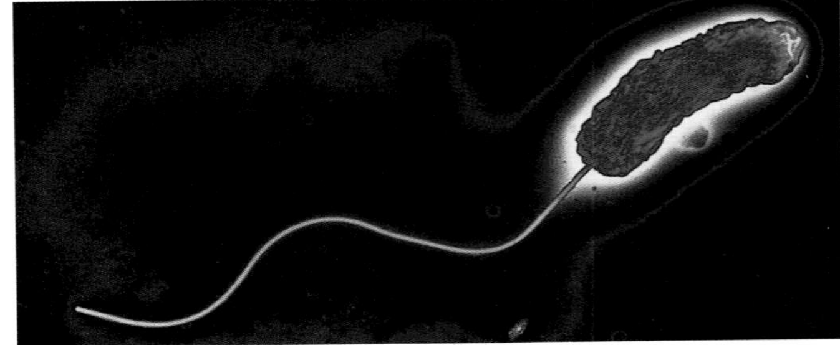

Figure 10.2 An electron micrograph of *Vibrio cholerae*. The faeces of an infected person are full of these bacteria, each with its distinctive flagellum (×13 400).

As the disease is water-borne, cholera occurs where people do not have access to proper sanitation (clean water supply) and uncontaminated food. Infected people, three-quarters of whom may be symptomless carriers, pass out large numbers of bacteria in their faeces. If these contaminate the water supply, or if infected people handle food or cooking utensils without washing their hands, then bacteria are transmitted to uninfected people.

Pathogen	*Vibrio cholerae*
Methods of transmission	food-borne, water-borne
Global distribution	Asia, Africa, Latin America
Incubation period	two hours to five days
Site of action of pathogen	wall of small intestine
Clinical features	severe diarrhoea ('rice water'), loss of water and salts, dehydration, weakness
Method of diagnosis	microscopical analysis of faeces
Annual incidence worldwide	3–5 million
Annual mortality worldwide	100 000–120 000

Table 10.2 The features of cholera.

To reach their site of action in the small intestine, the bacteria have to pass through the stomach. If the contents are sufficiently acidic (pH less than 4.5), the bacteria are unlikely to survive. However, if the bacteria do reach the small intestine, they multiply and secrete a toxin, **choleragen**, which disrupts the functions of the epithelium lining the intestine, so that salts and water leave the blood. This causes severe diarrhoea and the loss of fluid can be fatal if not treated within 24 hours.

Treating cholera

Almost all people with cholera who are treated make a quick recovery. A death from cholera is an avoidable death. The disease can be controlled quite cheaply by a solution of salts and glucose given intravenously to rehydrate the body (Figure 10.3). If people can drink, they are given **oral rehydration therapy**. Glucose is effective, because it is absorbed into the blood and takes ions (for example, sodium and potassium ions) with it. It is important to make sure that a patient's fluid intake equals fluid losses in urine and faeces, and to maintain the osmotic balance of the blood and tissue fluids (Chapter 4).

Preventing cholera

In developing countries, large cities that have grown considerably in recent years, but as yet have no sewage treatment or clean water, create perfect conditions for the spread of the disease. Increasing quantities of untreated faeces from a growing population favour cholera's survival. Countries that have huge debts do not have the financial resources to tackle large municipal projects such as providing drainage and a clean water supply to large areas of substandard housing. In many countries, the use of raw human sewage to irrigate vegetables is a common cause of the disease, as are inadequate cooking, or washing in contaminated water. Areas of the world where cholera is endemic are West and East Africa and Afghanistan.

Cholera is now almost unknown in the developed world, as a result of sewage treatment and the provision of clean piped water, which is chlorinated to kill bacteria. The transmission cycle has been broken.

Health authorities always fear outbreaks of cholera and other diarrhoeal diseases following natural disasters. In Haiti in 2010, a cholera epidemic broke out several months after the earthquake that destroyed large parts of the country.

Travellers from areas free of cholera to those where cholera is endemic used to be advised to be vaccinated, although the vaccine only provided short-term protection. This recommendation has now been dropped.

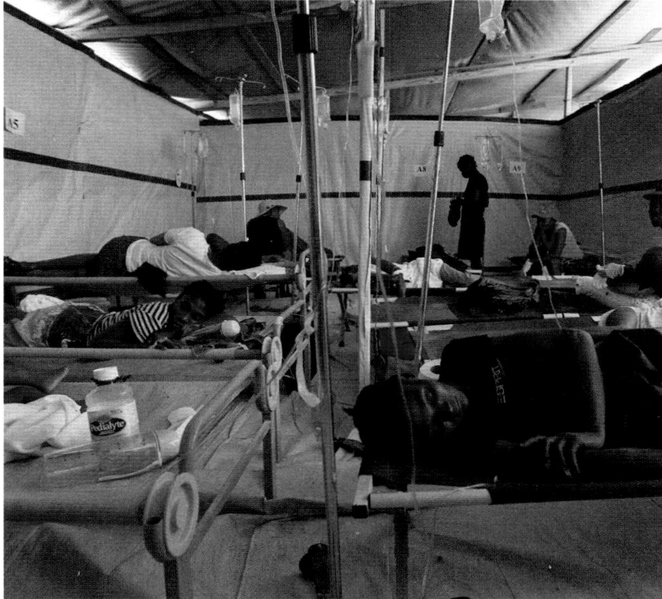

Figure 10.3 People being treated in Port-au-Prince in Haiti during the 2010 cholera epidemic. The drips contain a solution of salts to replace those lost through severe diarrhoea. Cholera causes many deaths when normal life is disrupted by war and by natural catastrophes such as earthquakes.

Strains of cholera

There are many different strains of *V. cholerae*. Until the 1990s, only the strain known as O1 caused cholera. Between 1817 and 1923, there were six pandemics of cholera. Each originated in what is now Bangladesh, and was caused by the 'classical' strain of cholera, O1. A seventh pandemic began in 1961 when a variety of O1, named 'El Tor', originated in Indonesia. El Tor soon spread to India, then to Italy in 1973, reaching South America in January 1991, where it caused an epidemic in Peru. The discharge of a ship's sewage into the sea may have been responsible. Within days of the start of the epidemic, the disease had spread 2000 km along the coast, and within four weeks had moved inland. In February and March of 1991, an average of 2550 cases a day were being reported. People in neighbouring countries were soon infected. In Peru, many sewers discharge straight onto shellfish beds. Organisms eaten as seafood, especially filter-feeders such as oysters and mussels, become contaminated because they concentrate cholera bacteria when sewage is pumped into the sea. Fish and shellfish are often eaten raw. Because the epidemic developed so rapidly in Peru, it is thought that the disease probably spread through contaminated seafood.

A new strain, known as *V. cholerae* O139, originated in Chennai (then called Madras) in October 1992 and has spread to other parts of India and Bangladesh. This strain threatens to be responsible for an eighth pandemic. It took El Tor two years to displace the 'classical' strain in India; O139 replaced El Tor in less than two months, suggesting that it may be more virulent. Many adult cases have been reported, suggesting that previous exposure to El Tor has not given immunity to O139 (Chapter 11).

QUESTIONS

10.2 List the ways in which cholera is transmitted from person to person.

10.3 One person can excrete 10^{13} cholera bacteria a day. An infective dose is 10^6. How many people could one person infect in one day?

10.4 Explain why there is such a high risk of cholera following natural disasters such as earthquakes, hurricanes, typhoons and floods.

10.5 Describe the precautions that a visitor to a country where cholera is endemic can take to avoid catching the disease.

Malaria

Transmission of malaria

The features of this disease are summarised in Table 10.3. Most cases of malaria are caused by one of four species of the protoctist *Plasmodium*, whose life cycle is shown in Figure 10.4. Genetic analysis of infections shows that some species of *Plasmodium* that cause malaria in monkeys also affect humans.

Female *Anopheles* mosquitoes feed on human blood to obtain the protein they need to develop their eggs. If the person they bite is infected with *Plasmodium*, they will take up some of the pathogen's gametes with the blood meal. Male and female gametes fuse in the mosquito's gut and develop to form infective stages, which move to the mosquito's salivary glands. When the mosquito feeds again, she injects an anticoagulant from her salivary glands that prevents the blood meal from clotting, so that it flows out of the host into the mosquito. The infective stages pass from the mosquito's salivary glands into the human's blood together with the anticoagulant in the saliva. The parasites enter the red blood cells, where they multiply (Figures 10.5 and 10.6).

The female *Anopheles* mosquito is therefore a vector of malaria and she transmits the disease when she passes the infective stages into an uninfected person. Malaria may also be transmitted during blood transfusion and when unsterile needles are re-used. *Plasmodium* can also pass across the placenta from mother to fetus.

Pathogen	*Plasmodium falciparum, P. vivax, P. ovale, P. malariae*
Method of transmission	insect vector: female *Anopheles* mosquito
Global distribution	throughout the tropics and sub-tropics (endemic in 106 countries)
Incubation period	from a week to a year
Site of action of pathogen	liver, red blood cells, brain
Clinical features	fever, anaemia, nausea, headaches, muscle pain, shivering, sweating, enlarged spleen
Method of diagnosis	microscopical examination of blood (Figure 10.5); dip stick test for malaria antigens in blood
Annual incidence worldwide	about 207 million cases of malaria in 2012 (about 80% are in Africa)
Annual mortality worldwide	about 630 000 deaths in 2012 (about 90% are in Africa)

Table 10.3 The features of malaria.

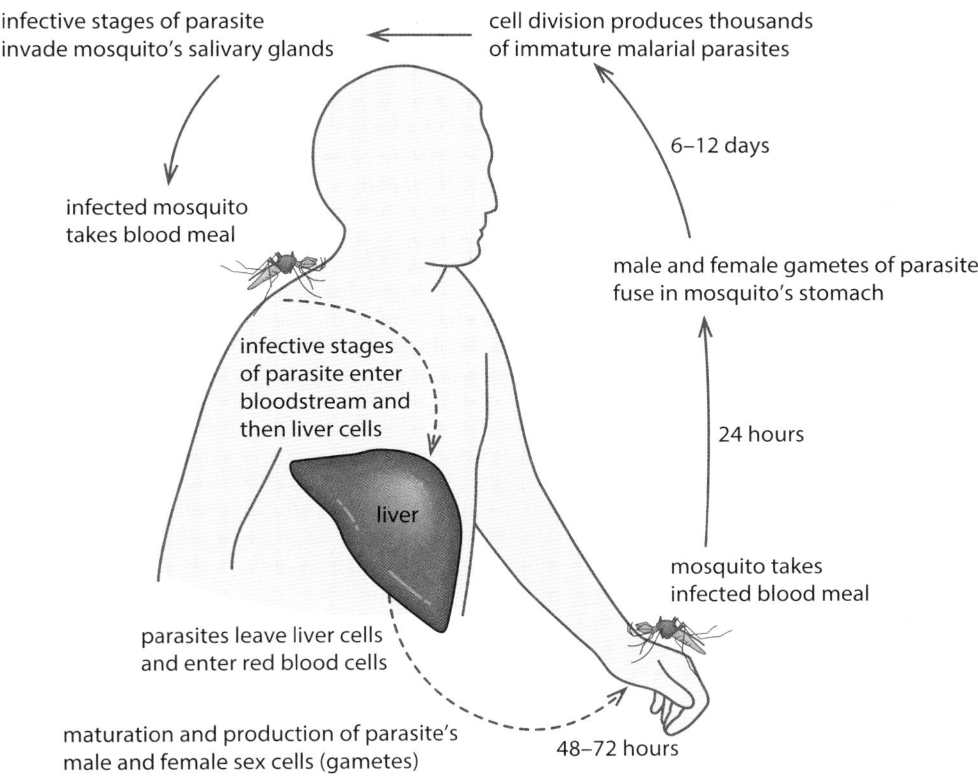

Figure 10.4 The life cycle of *Plasmodium*. The parasite has two hosts: the sexual stage occurs in mosquitoes, the asexual stage in humans. The time between infection and appearance of parasites inside red blood cells is 7–30 days in *P. falciparum*; longer in other species.

Plasmodium multiplies in both hosts, the human and the mosquito; at each stage there is a huge increase in the number of parasites, and this improves the chances of infecting another mosquito or human host.

If people are continually re-infected by different strains of malaria they become immune (Chapter 11). However, this only happens if they survive the first five years of life, when mortality from malaria is very high. The immunity only lasts as long as people are in contact with the disease. This explains why epidemics in places where malaria is not endemic can be very serious, and why malaria is more dangerous in those areas where it only occurs during and after the rainy season. This often coincides with the time of maximum agricultural activity, so the disease has a disastrous effect on the economy: people cannot cultivate the land when they are sick.

A **vector** is an organism which carries a disease from one person to another or from an animal to a human. Do not confuse it with the causative agent, which in this case is *Plasmodium*.

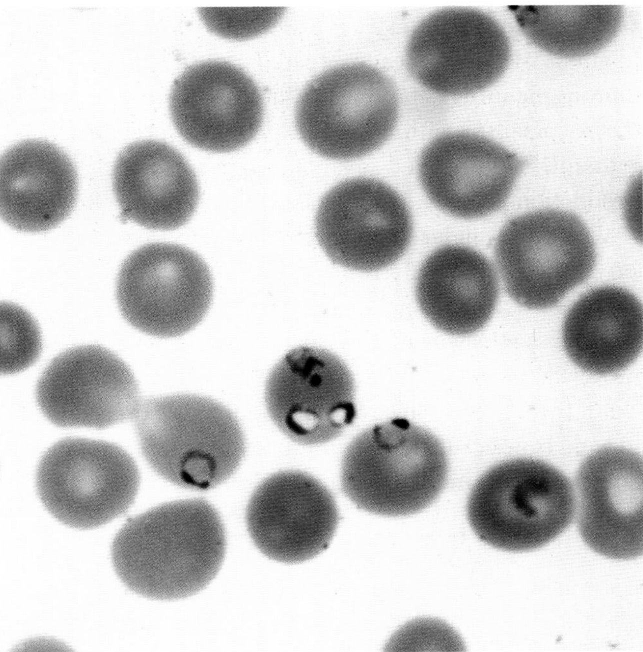

Figure 10.5 Red blood cells infected with *Plasmodium falciparum*. Notice the characteristic 'signet ring' appearance of the parasites inside the red blood cells (× 1300).

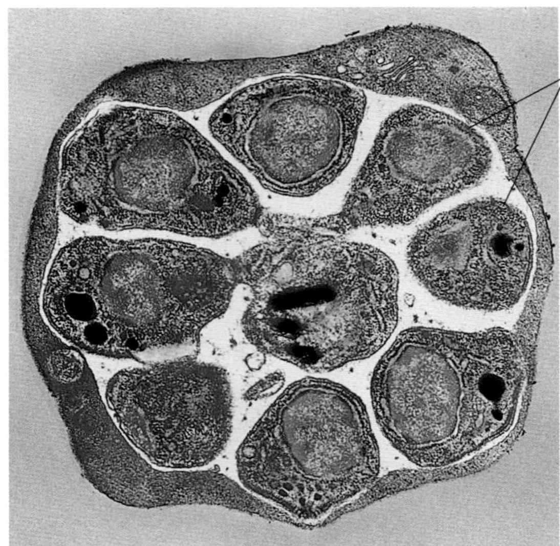

Figure 10.6 A transmission electron micrograph of a section through a red blood cell packed tightly with malarial parasites. *Plasmodium* multiplies inside red blood cells; this cell will soon burst, releasing parasites which will infect other red blood cells.

Treating malaria

Anti-malarial drugs such as quinine and chloroquine are used to treat infected people. They are also used as **prophylactic** (preventative) drugs, stopping an infection occurring if a person is bitten by an infected mosquito. Prophylactic drugs are taken before, during and after visiting an area where malaria is endemic. Chloroquine inhibits protein synthesis and prevents the parasite spreading within the body. Another prophylactic, proguanil, has the added advantage of inhibiting the sexual reproduction of *Plasmodium* inside the biting mosquito.

Where anti-malarial drugs have been used widely, there are strains of drug-resistant *Plasmodium* – the drug is no longer effective against the pathogen. Chloroquine resistance is widespread in parts of South America, Africa and New Guinea. Newer drugs such as mefloquine are used in these areas. However, mefloquine is expensive and sometimes causes unpleasant side-effects such as restlessness, dizziness, vomiting and disturbed sleep. Resistance to mefloquine has developed in some areas, notably the border regions of Thailand.

The antibiotic doxycycline is also used as a prophylactic drug. The drug artesunate, derived from the plant compound artemisin, is used in combination with mefloquine to treat infections of *P. falciparum*.

People from non-malarial countries visiting many parts of the tropics are at great risk of contracting malaria. Doctors in developed countries, who see very few cases of malaria, often misdiagnose it as influenza, since the initial symptoms are similar. Many of these cases are among settled immigrants who have been visiting relatives in Africa or India. These people do not take prophylactic drugs, because they do not realise that they have lost their immunity.

Preventing malaria

There are three main ways to control malaria:

- reduce the number of mosquitoes
- avoid being bitten by mosquitoes
- use drugs to prevent the parasite infecting people.

It is possible to kill the insect vector and break the transmission cycle. Mosquitoes lay their eggs in water. Larvae hatch and develop in water but breathe air by coming to the surface. Oil can be spread over the surfaces of water to make it impossible for mosquito larvae and pupae to breathe. Marshes can be drained and vegetation cleared.

Two biological control measures that can be used are:

- stocking ponds, irrigation and drainage ditches and other permanent bodies of water with fish which feed on mosquito larvae
- spraying a preparation containing the bacterium *Bacillus thuringiensis*, which kills mosquito larvae but is not toxic to other forms of life.

However, mosquitoes will lay their eggs in any small puddle or pool, which makes it impossible to completely eradicate breeding sites, especially in the rainy season.

The best protection against malaria is to avoid being bitten. People are advised to sleep beneath mosquito nets and use insect repellents. Soaking mosquito nets in insecticide every six months has been shown to reduce mortality from malaria. People should not expose their skin when mosquitoes are active at dusk.

Worldwide control of malaria

In the 1950s, the World Health Organization (WHO) coordinated a worldwide eradication programme. Although malaria was cleared from some countries, the programme was not generally successful. There were two main reasons for this:

- *Plasmodium* became resistant to the drugs used to control it
- mosquitoes became resistant to DDT and the other insecticides that were used at the time, such as dieldrin.

This programme was also hugely expensive and often unpopular. People living in areas where malaria was temporarily eradicated by the programme lost their immunity and suffered considerably, even dying, when the disease returned. Some villagers in South-East Asia lost the roofs of their houses because dieldrin killed a parasitic wasp that controlled the numbers of thatch-eating caterpillars. Some spray teams were set upon and killed by angry villagers in New Guinea. The programme could have been more successful if it had been tackled more sensitively, with more involvement of local people. In the 1970s, war and civil unrest destroyed much of the infrastructure throughout Africa and South-East Asia, making it impossible for mosquito control teams to work effectively.

The reasons for the worldwide concern over the spread of malaria are:

- an increase in drug-resistant forms of *Plasmodium*
- an increase in the proportion of cases caused by *P. falciparum*, the form that causes severe, often fatal malaria
- difficulties in developing vaccines against malaria
- an increase in the number of epidemics, because of climatic and environmental changes that favour the spread of mosquitoes
- the migration of people from areas where malaria is endemic, for economic and political reasons.

Malaria is still one of the world's biggest threats to health: 40% of the world's population lives in areas where there is a risk of malaria. Between 2000 and 2011, control measures have achieved a decrease in mortality rates of about 25% across the world, and 33% in the WHO's African region.

Control methods now concentrate on working within the health systems to improve diagnosis, improve the supply of effective drugs and promote appropriate methods to prevent transmission. Several recent advances give hope that malaria may one day be controlled. The introduction of simple dip stick tests for diagnosing malaria means that diagnosis can be done quickly without the need for laboratories. The whole genome of *Plasmodium* has been sequenced, and this may lead to the development of effective vaccines. Several vaccines are being trialled, but it is not likely that a successful vaccine will be available for some time. Drugs are used in combination to reduce the chances of drug resistance arising.

Three factors may lead to improvements in the control of malaria:

- use of modern techniques in gene sequencing and drug design
- development of vaccines targeted against different stages of the parasite's life cycle
- a renewed international will to remove the burden of disease from the poorest parts of the world, allied to generous donations from wealthy individuals and foundations.

QUESTIONS

10.6 Describe how malaria is transmitted.

10.7 Describe the biological factors that make malaria a difficult disease to control.

10.8 Describe the precautions that people can take to avoid catching malaria.

Acquired immune deficiency syndrome (AIDS)

Features of AIDS and HIV are listed in Table 10.4. AIDS is caused by the human immunodeficiency virus (HIV) (Figure 10.7).

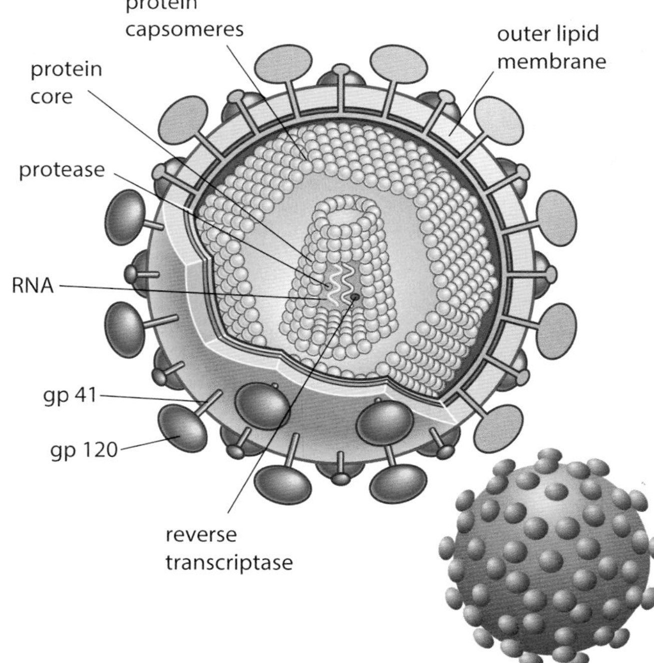

Figure 10.7 Human immunodeficiency virus (HIV). The outer envelope contains two glycoproteins: gp120 and gp41. The protein core contains genetic material (RNA) and two enzymes: a protease and reverse transcriptase. Reverse transcriptase uses the RNA as a template to produce DNA (page 466) once the virus is inside a host cell.

Pathogen	human immunodeficiency virus
Methods of transmission	in semen and vaginal fluids during sexual intercourse, infected blood or blood products, contaminated hypodermic syringes, mother to fetus across placenta, at birth, mother to infant in breast milk
Global distribution	worldwide, especially in sub-Saharan Africa and South-East Asia
Incubation period	initial incubation a few weeks, but up to ten years or more before symptoms of AIDS may develop
Site of action of pathogen	T helper lymphocytes, macrophages, brain cells
Clinical features	HIV infection – flu-like symptoms and then symptomless AIDS – opportunistic infections including pneumonia, TB and cancers; weight loss, diarrhoea, fever, sweating, dementia
Method of diagnosis	testing blood, saliva or urine for the presence of antibodies produced against HIV
Estimated total number of people infected with HIV worldwide in 2012	35.5 million (69% of these in sub-Saharan Africa)
Estimated number of new cases of HIV infection worldwide in 2012	2.3 million
Estimated number of deaths from AIDS-related diseases worldwide in 2012	1.6 million (UNAIDS estimate)

Table 10.4 The features of HIV/AIDS.

HIV is a retrovirus, which means that its genetic material is RNA, not DNA. Once inside a host cell, the viral RNA is converted 'back' to DNA (hence 'retro') to be incorporated into human chromosomes. The virus infects and destroys cells of the body's immune system (Figure 10.8) so that their numbers gradually decrease. These cells, known as **helper T cells** (page 230), control the immune system's response to infection. When the numbers of these cells are low, the body is unable to defend itself against infection, so allowing a range of pathogens to cause a variety of **opportunistic infections**. AIDS is not a disease; it is a collection of these opportunistic diseases associated with immunodeficiency caused by HIV infection.

Since HIV is an infective agent, AIDS is called an acquired immunodeficiency to distinguish it from other types – for example, an inherited form.

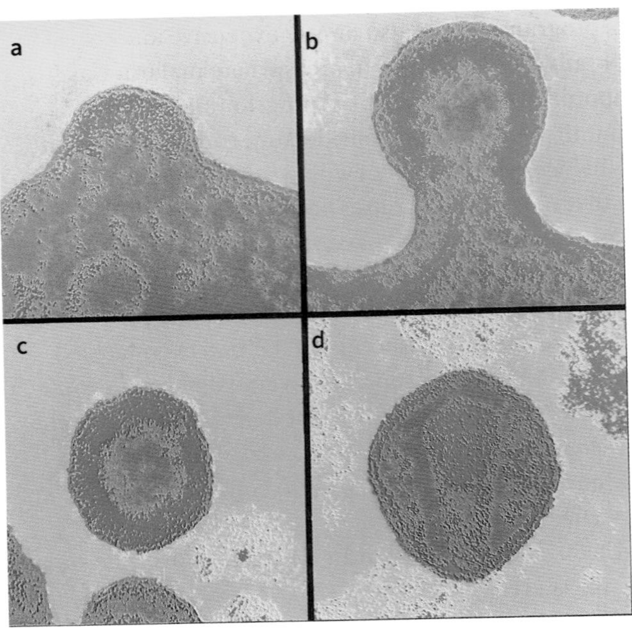

Figure 10.8 A series of transmission electron micrographs showing HIV budding from the surface of an infected lymphocyte and becoming surrounded by a membrane derived from the cell surface membrane of the host cell (×176 000). **a** The viral particle first appears as a bump, **b** which then buds out and **c** is eventually cut off. **d** The outer shell of dense material and the less dense core are visible in the released virus.

Transmission of HIV

After initial uncertainties in the early 1980s surrounding the emergence of an apparently new disease, it soon became clear that an epidemic and then a pandemic was underway. The WHO estimated that by 2010 over 25 million people had died of HIV/AIDS.

HIV is a virus that is spread by intimate human contact; there is no vector (unlike in malaria) and the virus is unable to survive outside the human body (unlike cholera or malaria pathogens). Transmission is only possible by direct exchange of body fluids. In practice, this means that HIV is spread most easily through sexual intercourse, blood donation and the sharing of needles used by intravenous drug users. HIV is also transmitted from mother to child across the placenta and, more often, through the mixing of blood during birth.

The initial epidemic in North America and Europe was amongst male homosexuals who practised anal intercourse and had many sex partners, two forms of behaviour that put them at risk. The mucous lining of the rectum is not as thick as that of the vagina, and there is less natural lubrication. As a result, the rectal lining is easily damaged during intercourse and the virus can pass from semen to blood.

Having multiple partners, both homosexual and heterosexual, allows the virus to spread more widely. Also at high risk of infection were haemophiliacs who were treated with a clotting substance (factor 8) isolated from blood pooled from many donors. Such blood products are now largely synthetic (page 475). Much of the transmission of HIV has been by heterosexual intercourse. This is particularly rapid in some African states, where equal numbers of males and females are now HIV positive (HIV+).

Figure 10.9 shows the global distribution of HIV/AIDS. The statistics below show how serious the pandemic is in sub-Saharan Africa.

- 70% of the world's deaths from AIDS occur in Africa.
- In 2007 it was estimated that 15 million people had died of HIV/AIDS in sub-Saharan Africa since the beginning of the pandemic.
- 25% of the adult population of Botswana is infected with HIV.
- Between 15% and 25% of people aged 15–49 in Botswana and Zimbabwe are infected with HIV.
- Over 16 million children are estimated to have lost one or both parents to AIDS; in some places this is 25% of the population under 15.
- The prevalence of HIV among women attending antenatal clinics in Zimbabwe was around 20% in 2012.
- A large proportion of women in Rwanda are HIV positive following the use of rape as a genocidal weapon in the civil war of the early 1990s.
- The average life expectancy in South Africa dropped from 65 to 55 during 1995–1999.

HIV is a slow virus and, after infection, there may not be any symptoms until years later. Some people who have the virus even appear not to develop any initial symptoms, although there are often flu-like symptoms for several weeks after becoming infected. At this stage, a person is HIV positive but does not have AIDS.

The infections that can opportunistically develop to create AIDS tend to be characteristic of the condition. Two of these are caused by fungi: oral thrush caused by *Candida albicans*, and a rare form of pneumonia caused by *Pneumocystis jiroveci*. During the early years of the AIDS epidemic, people in developed countries died within 12 hours of contracting this unusual pneumonia. Now this condition is managed much better and drugs are prescribed to prevent the disease developing.

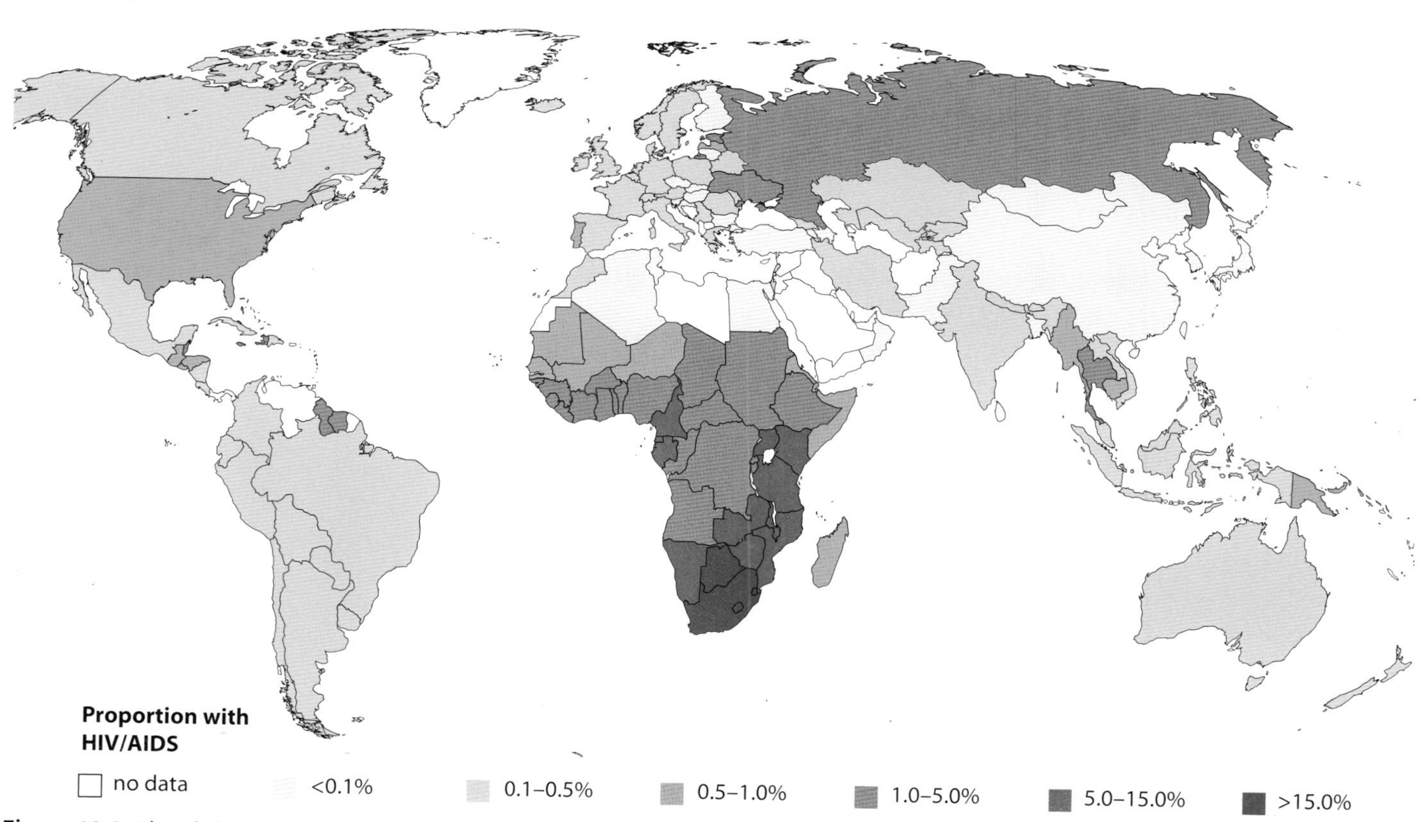

Figure 10.9 The global distribution of HIV/AIDS in 2010.

As and when the immune system collapses further, it becomes less effective in finding and destroying cancers. A rare form of skin cancer, Kaposi's sarcoma, caused by a herpes-like virus, is associated with AIDS. Kaposi's sarcoma and cancers of internal organs are now the most likely causes of death of people with AIDS in developed countries, along with degenerative diseases of the brain, such as dementias.

At about the same time that AIDS was first reported on the west coast of the USA and in Europe, doctors in Central Africa reported seeing people with similar opportunistic infections. We have seen that HIV/AIDS is now widespread throughout sub-Saharan Africa from Uganda to South Africa. It is a serious public health problem here because HIV infection makes people more vulnerable to existing diseases such as malnutrition, TB and malaria. AIDS is having an adverse effect on the economic development of countries in the region, as it affects sexually active people in their 20s and 30s who are also potentially the most economically productive, and the purchase of expensive drugs drains government funds. The World Bank estimated that AIDS had reversed 10–15 years of economic growth for some African states by the end of the 20th century.

Treating HIV/AIDS

There is as yet no cure for AIDS and no vaccine for HIV. No-one knows how many people with HIV will progress to developing full-blown AIDS. Some people think it is 100%, although a tiny minority of HIV-positive people do appear to have immunity (Chapter 11) and can live as entirely symptomless carriers.

Drug therapy can slow down the onset of AIDS quite dramatically, so much so that some HIV-positive people in developed countries are adjusting to a suddenly increased life expectancy. However, the drugs are expensive and have a variety of side-effects ranging from the mild and temporary (rashes, headaches, diarrhoea) to the severe and permanent (nerve damage, abnormal fat distribution). If used in combination, two or more drugs which prevent the replication of the virus inside host cells can prolong life, but they do not offer a cure. The drugs are similar to DNA nucleotides (for example, zidovudine is similar to the nucleotide that contains the base thymine). Zidovudine binds to the viral enzyme reverse transcriptase and blocks its action. This stops the replication of the viral genetic material and leads to an increase in some of the body's lymphocytes. A course of combination therapy (taking several drugs) can be very complicated to follow. The pattern and timing of medication through the day must be strictly followed. People who are unable to keep to such a regimen can become susceptible to strains of HIV that have developed resistance to the drugs.

Preventing HIV/AIDS

The spread of HIV/AIDS is difficult to control. The virus's long latent stage means it can be transmitted by people who are HIV positive but who show no symptoms of AIDS and do not know they are infected. The virus changes its surface proteins, which makes it hard for the body's immune system to recognise it (Chapter 11). This also makes the development of a vaccine very difficult.

For the present, public health measures are the only way to stop the spread of HIV. People can be educated about the spread of the infection and encouraged to change their behaviour so as to protect themselves and others. Condoms, femidoms and dental dams are the only effective methods of reducing the risk of infection during intercourse, as they form a barrier between body fluids, reducing the chances of transmission of the virus. Some countries have promoted the use of condoms as well as other measures. As a result, infection rates in these countries have slowed and the number of new cases reported each year has either decreased or remained the same year on year. It is estimated that the rate of HIV infection across the world decreased by 25% between 2001 and 2009.

QUESTIONS

10.9 Suggest why the true total of AIDS cases worldwide may be much higher than reported.

10.10 Suggest why condoms are not fully effective at preventing HIV infection.

10.11 Suggest the types of advice which might be offered as part of an HIV/AIDS education programme.

Contact tracing is an important part of controlling the spread of HIV. If a person who is diagnosed as HIV positive is willing and able to identify the people whom he or she has put at risk of infection by sexual intercourse or needle sharing, then these people will be offered an HIV test. This test identifies the presence of antibodies to HIV, although these only appear several weeks after the initial infection.

Injecting drug users are advised to give up their habit, stop sharing needles or take their drug in some other way. Needle-exchange schemes operate in some places to exchange used needles for sterile ones to reduce the chances of infection with HIV and other blood-borne diseases.

Blood collected from blood donors is screened for HIV and heat-treated to kill any viruses. People who think they may have been exposed to the virus are strongly discouraged from donating blood. In some low-income countries, not all donated blood is tested. Anyone concerned about becoming infected by blood transfusion during an operation may donate their own blood before the operation to be used instead of blood from a blood bank.

Widespread testing of a population to find people who are HIV positive is not expensive, but governments are reluctant to introduce such testing because of the infringement of personal freedom. In the developed world, HIV testing is promoted most strongly to people in high-risk groups, such as male homosexuals, prostitutes, injecting drug users and their sexual partners. If tested positive, they can be given the medical and psychological support they need. In Africa and South-East Asia, the epidemic is not restricted to such easily identifiable groups and widespread testing is not feasible due to the difficulty of reaching the majority of the population and organising testing. People in these regions find out that they are HIV positive when they develop the symptoms of AIDS.

Both viral particles and infected lymphocytes are found in breast milk. Mother-to-child transmission is reduced by treating HIV-positive women and their babies with drugs. However, HIV-positive women in high-income countries are advised not to breastfeed their babies, because of the risk of transmission even if they have a secure supply of drugs during this period. In contrast, HIV-positive women in low- and middle-income countries are advised to breastfeed, especially if they have a secure supply of drugs during this period, as the protection this gives against other diseases and the lack of clean water to make up formula milk may outweigh the risks of transmitting HIV.

QUESTIONS

10.12 Children in Africa with sickle cell anaemia or malaria often receive blood transfusions. Explain how this puts them at risk of HIV infection.

10.13 Explain why the early knowledge of HIV infection is important in transmission control.

Tuberculosis (TB)

Table 10.5 gives the main features of this disease. TB is caused by either of two bacteria, *Mycobacterium tuberculosis* (Figure 10.10) and *Mycobacterium bovis*. These are pathogens that live inside human cells, particularly in the lungs. This is the first site of infection, but the bacteria can spread throughout the whole body and even infect the bone tissue.

Pathogen	*Mycobacterium tuberculosis*; *Mycobacterium bovis*
Methods of transmission	airborne droplets (*M. tuberculosis*); via undercooked meat and unpasteurised milk (*M. bovis*)
Global distribution	worldwide
Incubation period	few weeks or up to several years
Site of action of pathogen	primary infection in lungs; secondary infections in lymph nodes, bones and gut
Clinical features	racking cough, coughing blood, chest pain, shortness of breath, fever, sweating, weight loss
Methods of diagnosis	microscopical examination of sputum for bacteria, chest X-ray
Annual incidence worldwide in 2012	8.6 million
Annual mortality worldwide in 2012	1.3 million (including 320 000 deaths of people who were HIV+)

Table 10.5 The features of TB.

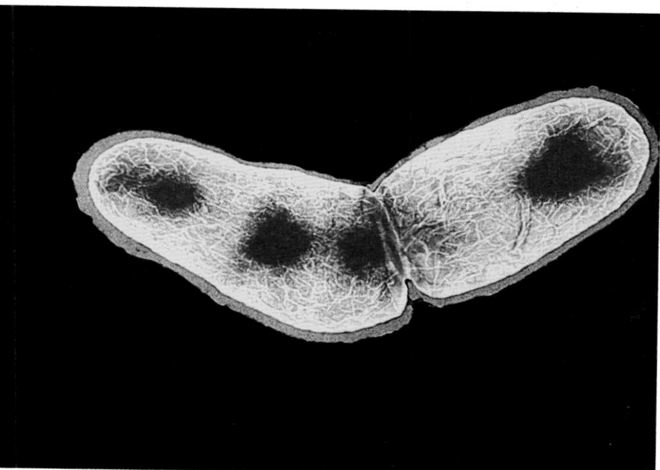

Figure 10.10 False-colour transmission electron micrograph of *Mycobacterium tuberculosis* dividing into two. It may multiply like this inside the lungs and then spread throughout the body or lie dormant, becoming active many years later.

Some people become infected and develop TB quite quickly, while in others the bacteria remain inactive for many years. It is estimated that about 30% of the world's population is infected with TB without showing any symptoms of the infection; people with this inactive infection do not spread the disease to others. However, the bacteria can later become active, and this is most likely to happen when people are weakened by other diseases, suffer from malnutrition or become infected with HIV. Those who have the active form of TB often suffer from debilitating illness for a long time. They have a persistent cough and, as part of their defence, cells release hormone-like compounds, which cause fever and suppress the appetite. As a result, people with TB lose weight and often look emaciated (Figure 10.11).

TB is often the first opportunistic infection to strike HIV-positive people. HIV infection may reactivate dormant infections of *M. tuberculosis* which may have been present from childhood or, if people are uninfected, make them susceptible to infection. TB is now the leading cause of death among HIV-positive people. The HIV pandemic has been followed very closely by a TB pandemic.

Transmission of TB

TB is spread when infected people with the active form of the illness cough or sneeze and the bacteria are carried in the air in tiny droplets of liquid. Transmission occurs when people who are uninfected inhale the droplets. TB spreads most rapidly among people living in overcrowded conditions. People who sleep close together in large numbers are particularly at risk. The disease primarily attacks the homeless and people who live in poor, substandard housing; those with low immunity, because of malnutrition or being HIV positive, are also particularly vulnerable.

The form of TB caused by *M. bovis* also occurs in cattle and is spread to humans in meat and milk. It is estimated that there were about 800 000 deaths in the UK between 1850 and 1950 as a result of TB transmitted from cattle. Very few now acquire TB in this way in developed countries for reasons explained later, although meat and milk still remain a source of infection in some developing countries.

The incidence of TB in the UK decreased steeply well before the introduction of a vaccine in the 1950s, because of improvements in housing conditions and diet. The antibiotic streptomycin was introduced in the 1940s, and this hastened the decrease in the incidence of TB. This pattern was repeated throughout the developed world.

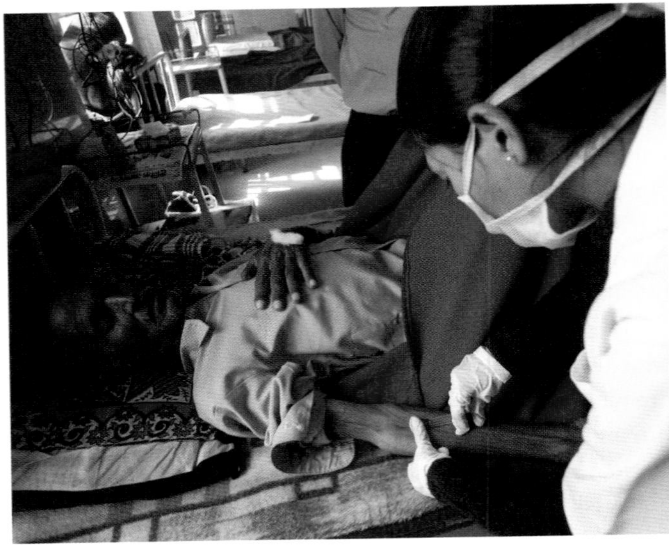

Figure 10.11 A TB patient undergoes treatment in a hospital in India.

Once thought to be practically eradicated, TB is on the increase. There are high rates of incidence all across the developing world and in Russia and surrounding countries (Figure 10.12). High rates are also found in cities with populations of migrants from countries where TB is more common. Parts of London, for example, have rates of TB much higher than the rest of the UK. The incidence in such areas is as high as in less economically developed countries. This increase is due in part to the following factors:

- some strains of TB bacteria are resistant to drugs
- the HIV/AIDS pandemic
- poor housing in inner cities and homelessness
- the breakdown of TB control programmes; partial treatment for TB increases the chance of drug resistance in *Mycobacterium*.

Treating TB

When a doctor first sees a person with the likely symptoms of TB, samples of the sputum (mucus and pus) from their lungs are collected for analysis. The identification of the TB bacteria can be done very quickly by microscopy. If TB is confirmed, then patients should be isolated while they are in the most infectious stage (which is at two to four weeks). This is particularly if they are infected with a drug-resistant strain of the bacterium. The treatment involves using several drugs to ensure that all the bacteria are killed. If not killed, drug-resistant forms remain to continue the infection. The treatment is a long one (six to nine months, or longer), because it takes a long time to kill the bacteria, which are slow growing and are not very sensitive to the drugs used. Unfortunately, many people

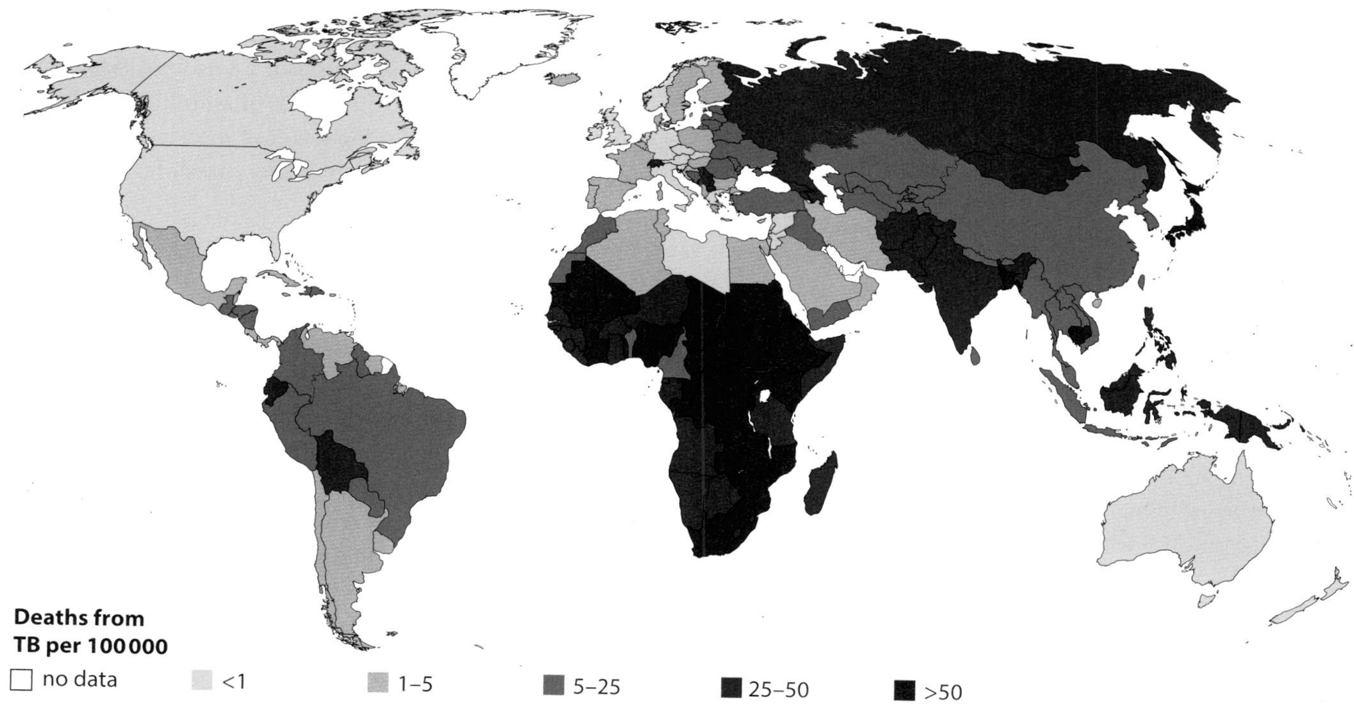

Figure 10.12 The global distribution of TB in 2010 (data from WHO).

do not complete their course of drugs, because they think that when they feel better, they are cured. People who do not complete their treatment may be harbouring drug-resistant bacteria and may spread these to others if the bacteria become active.

Drug-resistant TB

Strains of drug-resistant *M. tuberculosis* were identified when treatment with antibiotics began in the 1950s. Antibiotics act as selective agents killing drug-sensitive strains and leaving resistant ones behind. Drug resistance occurs as a result of mutation in the bacterial DNA. Mutation is a random event and occurs with a frequency of about one in every thousand bacteria. If three drugs are used in treatment, then the chance of resistance arising to all three of them by mutation is reduced to one in a thousand million. If four drugs are used, the chance is reduced to one in a billion.

If TB is not treated or the person stops the treatment before the bacteria are completely eliminated, the bacteria spread throughout the body, increasing the likelihood that mutations will arise, as the bacteria survive for a long time and multiply. Stopping treatment early can mean that *M. tuberculosis* develops resistance to all the drugs being used. People who do not complete a course of treatment are highly likely to infect others with drug-resistant forms of TB. It is 10 to 15 others, especially if the person lives in overcrowded conditions.

The WHO promotes a scheme to ensure that patients complete their course of drugs. DOTS (direct observation treatment, short course) involves health workers or responsible family members making sure that patients take their medicine regularly for six to eight months. The drugs widely used are isoniazid and rifampicin, often in combination with others. This drug therapy cures 95% of all patients, and is twice as effective as other strategies.

Multiple-drug-resistant forms of TB (MDR-TB) now exist. MDR-TB strains of TB are resistant to at least the two main drugs used to treat TB – isoniazid and rifampicin – which are known as first-line drugs. In 1995, an HIV unit in London reported an outbreak of MDR-TB with a form of *M. tuberculosis* that was resistant to five of the major drugs used to treat the disease, including isoniazid, which is the most successful drug. Extensively (or extremely) drug-resistant TB (XDR-TB) has also emerged as a very serious threat to health, especially for those people who are HIV-positive . XDR-TB strains are resistant to first-line drugs **and** to the drugs used to treat MDR-TB. These resistant strains of TB do not respond to the standard six-month treatment with first-line anti-TB drugs and can take two years or more to treat with drugs that are less potent and much more expensive. DOTS is helping to reduce the spread of MDR strains of TB.

Preventing TB

Contact tracing (page 208) and the subsequent testing of contacts for the bacterium are essential parts of controlling TB. Contacts are screened for symptoms of TB infection, but the diagnosis can take up to two weeks.

The only vaccine available for TB is the BCG vaccine, which is derived from *M. bovis* and protects up to 70–80% of people who receive it. The effectiveness of the vaccine decreases with age unless there is exposure to TB. Many countries with high numbers of people with TB use the BCG vaccine to protect children from getting the disease. Countries such as the UK and USA do not include BCG vaccination in their immunisation programmes. Instead, it may be given only to people who are at high risk of becoming infected because, for example, they live with an adult who is being treated for the disease.

TB can be transmitted between cattle and humans. To prevent people catching TB in this way, cattle are routinely tested for TB and any found to be infected are destroyed. TB bacteria are killed when milk is pasteurised. These control methods are very effective and have reduced the incidence of human TB caused by *M. bovis* considerably, so that it is virtually eliminated in countries where these controls operate. In the UK, less than 1% of the 9000 new cases of TB each year are due to *M. bovis*.

QUESTION

10.14 a Compare the global distribution of HIV/AIDS and TB as shown in Figure 10.9 (page 207) and Figure 10.12 (page 211).

b Explain why there is a high death rate from TB in countries with a high proportion of the population who are HIV-positive.

Measles

Measles is caused by a virus which enters the body and multiplies inside cells in the upper respiratory tract (nasal cavity and trachea). There are no symptoms for 8–14 days after the initial infection and then a rash appears and a fever develops. Other symptoms are a runny nose, a cough, red and watery eyes (conjunctivitis) and small white spots that may develop inside the cheeks. Treatment involves bed rest and taking medicines to lower the fever; there are no specific medicines for measles. After about ten days the disease clears up and there are rarely any complications. If complications do occur, they tend to be serious: pneumonia, ear and sinus infections, brain damage and convulsions may follow a measles infection. Some of these cases are fatal and, among malnourished children living in overcrowded conditions, measles is a serious disease and a major cause of death. About 10% of all infant deaths in developing countries are the result of measles infections. Measles is also responsible for many cases of childhood blindness.

Measles is one of the most contagious diseases. When people infected with measles sneeze or cough they release droplets containing many millions of virus particles. If these are inhaled by uninfected people who have no immunity to the disease, it is almost inevitable that they will become infected and develop symptoms. The disease rarely affects infants under eight months of age, as they have passive immunity in the form of antibodies that have crossed the placenta from their mother. However, as these antibodies are gradually destroyed the children lose their immunity (page 232). Measles used to be a common childhood disease in developed countries, but the incidence of the disease fell steeply after the introduction of a vaccine in the early 1960s.

Measles is a major disease in developing countries, particularly in cities where people live in overcrowded, insanitary conditions and where there is a high birth rate. The measles virus is transmitted easily in these conditions and it infects mainly malnourished infants suffering from vitamin A deficiency. There are estimated to be over 20 million cases of measles worldwide each year, most of which are in Africa, South-East Asia, India, Pakistan Bangladesh and some countries of the Middle East. The death rate from the disease has fallen from 630 000 in 1990 to 158 000 in 2011, largely as a result of a mass vaccination programme (Figure 11.18, page 236). There are very few cases of measles in developed countries and most outbreaks are caused by someone entering the country with the disease that they caught elsewhere. When outbreaks occur, as in Indiana in the USA in 2005 and in South Wales in the UK in 2012/2013, they can spread rapidly among those at risk. These two epidemics occurred mainly among children whose parents had not had them vaccinated. Adults who were not vaccinated and who did not have measles as a child are at risk of severe complications in epidemics like this.

QUESTION

10.15 Make a table, similar to the tables on cholera, malaria, HIV/AIDS and TB, to summarise information about measles.

Antibiotics

An antibiotic is a drug that kills or stops the growth of bacteria, without harming the cells of the infected organism. Antibiotics are derived from living organisms, although they are often made more effective by various chemical processes. There are a wide range of antibiotics to treat bacterial infections. Other antimicrobial drugs such as isoniazid, used for the treatment of TB, are synthetic (made in laboratories).

How antibiotics work

Antibiotics interfere with some aspect of growth or metabolism of the target bacterium (Figure 10.13). These include:

- synthesis of bacterial cell walls (Figure 1.30, page 21)
- activity of proteins in the cell surface membrane (Chapter 4)
- enzyme action (Chapter 3)
- DNA synthesis (replication, Chapter 6, pages 113–118)
- protein synthesis (Chapter 6, pages 119–122).

Bacterial cells have walls made of peptidoglycans (page 21). These are long molecules containing peptides (chains of amino acids) and sugars. In the bacterial cell wall, peptidoglycans are held together by cross-links that form between them. Penicillin prevents the synthesis of the cross-links between the peptidoglycan polymers in the cell walls of bacteria by inhibiting the enzymes that build these cross-links. This means that penicillin is only active against bacteria while they are growing.

When a newly formed bacterial cell is growing, it secretes enzymes called autolysins, which make little holes in its cell wall. These little holes allow the wall to stretch so that new peptidoglycan chains can link together. Penicillin prevents the peptidoglycan chains from linking up, but the autolysins keep making new holes. The cell wall therefore becomes progressively weaker. Bacteria live in watery environments and take up water by osmosis. When they are weakened, the cell walls cannot withstand the pressure potential exerted on them by the cell contents and the cells burst (Figure 10.14).

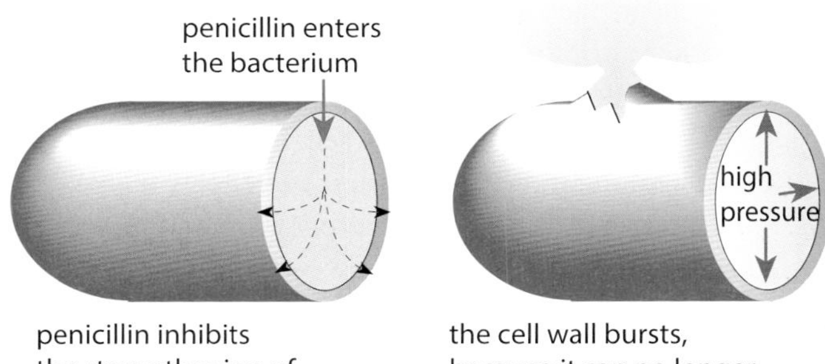

Figure 10.14 How penicillin works.

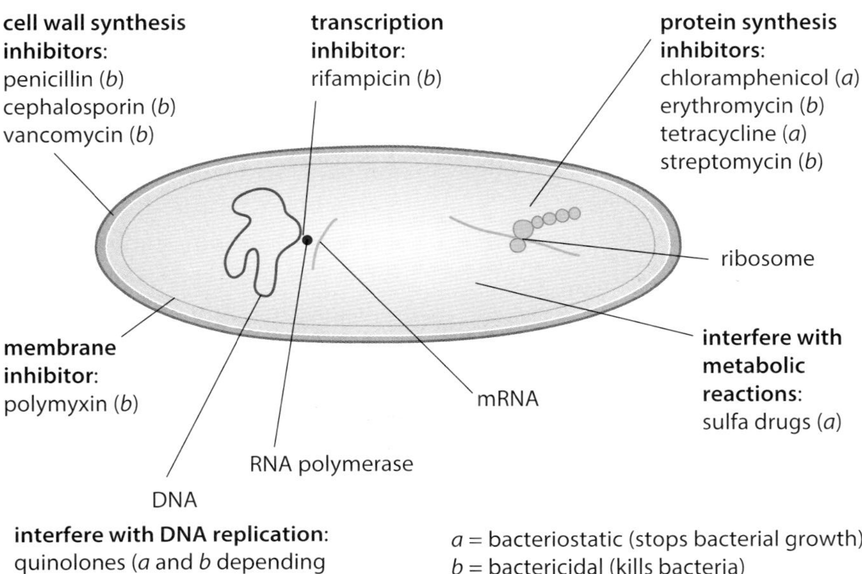

Figure 10.13 The sites of action of antibiotics in bacteria.

QUESTION

10.16 Explain why antibiotics are not effective against viruses.

This explains why penicillin does not affect human cells. Our cells do not have walls. This also explains why penicillin and other antibiotics do not affect viruses, which do not even have cells, let alone cell walls. Viruses do not have the targets shown in Figure 10.13. For example, when a virus replicates, it uses the host cell's mechanisms for transcription and translation and antibiotics do not bind to the proteins that host cells use in these processes. Eukaryotic cells have proteins that are different from those in bacteria so they are unaffected by such antibiotics. Other drugs, called antivirals, are used to control viral infections. There are fewer antivirals than there are antibiotics.

Penicillin first became available for treating disease in the 1940s. It was hailed as a wonder drug that could be used to wipe out all the diseases caused by bacteria. To begin with, this seemed to be true, but very quickly it became clear that this was not going to happen even though other antibiotics, such as streptomycin, soon became available.

Some types of bacteria are not sensitive to particular antibiotics: for example, penicillin is not effective against *M. tuberculosis.* Even among the types of bacteria that were killed by penicillin, there were certain strains that were not affected. These strains had become resistant to antibiotics. During the 70 years since the introduction of antibiotics, most pathogenic bacteria have become resistant to one or more types of antibiotic.

Antibiotic resistance

Penicillin has no effect on *M. tuberculosis* because the thick cell wall of this bacterium is not very permeable and because the bacterium has a gene that codes for an enzyme that catalyses the breakdown of penicillin. Proteins in the membranes of other species of bacteria can inactivate antibiotics so they have no effect; bacterial membranes also have proteins that pump out antibiotics if they enter the cytoplasm. In some cases, the antibiotic simply cannot bind to the intended site of action.

Bacteria that are sensitive to an antibiotic are described as being susceptible to that antibiotic. They may become resistant if they gain a gene coding for a protein that protects them from the antibiotic.

Soil bacteria have many resistance mechanisms as they grow in an environment where there are many molecules that interfere with their metabolism. These resistance mechanisms are very similar to those found in pathogenic bacteria. Before the introduction of antibiotics, enzymes known as beta-lactamases were not common among pathogenic bacteria. The genes for these enzymes have spread into many different forms of bacteria and it is believed that they have come from soil bacteria.

Penicillin has a structure that can be broken down by β-lactamase (penicillinase) enzymes. Pathogenic bacteria that have become resistant to penicillin have often done so because they have acquired the genes that code for these enzymes.

Antibiotic resistance can arise when an existing gene within the bacterial genome changes spontaneously to give rise to a nucleotide sequence that codes for a slightly different protein that is not affected by the antibiotic. This change in DNA is a mutation. When someone takes penicillin to treat a bacterial infection, bacteria that are susceptible to penicillin will die. In most cases, if the dose is followed correctly, this will be the entire population of the disease-causing bacteria. However, if the dose is not followed, perhaps because people stop taking the penicillin when they feel better as the symptoms disappear, then some susceptible bacteria survive and if any mutations occur these might confer resistance. The next time there is an infection with this strain of bacteria, penicillin may not be effective.

Bacteria have only one copy of each gene, since they only have a single loop of double-stranded DNA. This means that a mutant gene will have an immediate effect on any bacterium possessing it. These individuals have a tremendous advantage. Bacteria without this mutant gene will be killed, while those bacteria resistant to penicillin survive and reproduce. Bacteria reproduce asexually by binary fission; the DNA in the bacterial chromosome is replicated and the cell divides into two, with each daughter cell receiving a copy of the chromosome. This happens very rapidly in ideal conditions, and even if there was initially only one resistant bacterium, it might produce ten thousand million descendants within 24 hours. A large population of a penicillin-resistant strain of a bacterium would result. This method of spreading antibiotic resistance in a population of bacteria is called **vertical transmission**.

Genes for antibiotic resistance often occur on plasmids, which are small loops of double-stranded DNA (Figure 1.30, page 21). Plasmids are quite frequently transferred from one bacterium to another, even between different species. This happens during conjugation when a tube forms between two bacteria to allow the movement of DNA. During conjugation, plasmids are transferred from a donor bacterium to a recipient. Transfer of part of the DNA from the bacterial chromosome also occurs in the same way. This method of transmission is **horizontal transmission** (Figure 10.15). Thus it is possible for resistance to a particular antibiotic to arise in one species of bacterium and be passed on to another.

Vertical transmission

resistant parent cell

bacterial chromosome and plasmid replicate

cell division

Daughter cells each receive a copy of the plasmid and are resistant.

Horizontal transmission

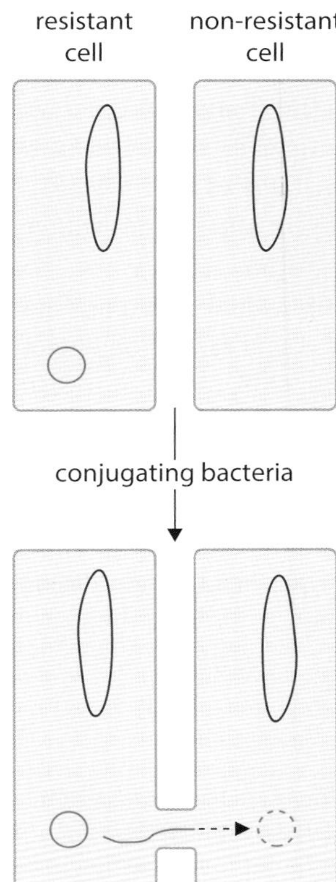

A single DNA strand of the plasmid is transferred. Each bacterium then synthesises a complementary strand. Both cells are now resistant.

Figure 10.15 Vertical and horizontal transmission of resistance in bacteria.

The more we misuse antibiotics, the greater the selection pressure we exert on bacteria to evolve resistance to them. Antibiotic-resistant strains of bacteria are continually appearing (Figure 10.16). Antibiotic-resistant infections increase the risk of death, and are often associated with long stays in hospital, and sometimes serious complications.

There is a constant race to find new antibiotics as resistant strains keep arising.

Where there is widespread use of antibiotics, such as in hospitals or on farms, resistance quickly spreads among different species of bacteria. Resistance may first appear in a non-pathogenic bacterium, but then be passed to a pathogenic species. Bacteria living where there is widespread use of antibiotics may have plasmids carrying resistance genes for several different antibiotics, giving **multiple resistance**. This presents major problems for doctors. For example, methicillin-resistant *Staphylococcus aureus* (MRSA) has become a problem in hospitals around the world and in prisons in the USA. It is now also infecting people in the general population. MRSA caused

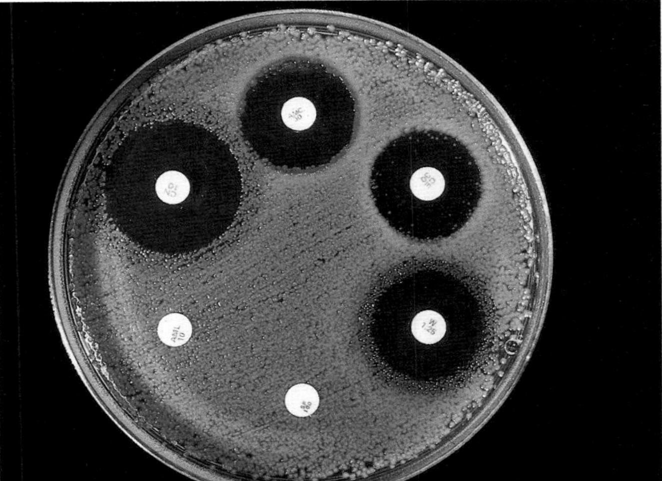

Figure 10.16 The grey areas on the agar jelly in this Petri dish are colonies of the bacterium *Escherichia coli*. The white discs are pieces of card impregnated with different antibiotics. Where there are clear areas around the discs the antibiotic has prevented the bacteria from growing. However, you can see that this strain of *E. coli* is resistant to the antibiotics on the discs at the bottom left and has been able to grow right up to the discs.

dangerous infections after surgery, which were mostly controlled by vancomycin – an antibiotic often used as a last resort for treating infections when everything else has failed, so as to lessen the chances of the development of more such resistant organisms. Then, another bacterium common in hospitals, *Enterococcus faecalis*, developed resistance to vancomycin and this resistance passed to *S. aureus*.

Recently, antibiotics called carbapens have been the antibiotics of last resort for use on bacteria with multiple resistance. In 2009, carbapen-resistant *Klebsiella pneumoniae* was found in Greece. By 2010 it was also found in Cyprus, Hungary and Italy, and in Greece the proportion of infections of *K. pneumoniae* that were carbapen-resistant had risen to over 25%. In 2010, Greece used more antibiotics per head of population than any other European country.

QUESTIONS

10.17 a Describe the ways in which bacteria can protect themselves against the effects of antibiotics.
b Explain how antibiotic resistance may arise in a bacterial population.
c Distinguish between vertical and horizontal transmission in bacteria.

10.18 Suggest why an organism resistant to many antibiotics has evolved in hospitals and is common in prisons.

Choosing effective antibiotics

Antibiotics should be chosen carefully. Testing antibiotics against the strain of the bacterium isolated from people ensures that the most effective antibiotic can be used in treatment.

As fast as we develop new antibiotics, bacteria seem to develop resistance to them. It follows from this that there is a constant search for new antibiotics, especially ones that work in a completely different way from those currently in use.

Fortunately, a bacterium resistant to a particular antibiotic may not be resistant to that antibiotic with a slightly altered chemical structure. Chemists can make such semi-synthetic antibiotics to extend the range available. However, many experts believe that we will not be able to keep up and that soon there will be no antibiotics left to treat diseases. This is fast becoming the case with gonorrhoea, a sexually transmitted infection.

Clearly we should try to reduce the number of circumstances in which bacteria develop resistance to antibiotics. Some of the ways in which we can do this include:

- using antibiotics only when appropriate and necessary; not prescribing them for viral infections
- reducing the number of countries in which antibiotics are sold without a doctor's prescription
- avoiding the use of so-called wide-spectrum antibiotics and using instead an antibiotic specific to the infection (known as narrow spectrum)
- making sure that patients complete their course of medication
- making sure that patients do not keep unused antibiotics for self-medication in the future
- changing the type of antibiotics prescribed for certain diseases so that the same antibiotic is not always prescribed for the same disease
- avoiding using antibiotics in farming to prevent, rather than cure, infections.

QUESTION

10.19 Suggest how each of the following might decrease the chances of an antibiotic-resistant strain of bacteria developing:
a limiting the use of antibiotics to cases where there is a real need
b regularly changing the type of antibiotic that is prescribed for a particular disease
c using two or more antibiotics together to treat a bacterial infection.

QUESTION

10.20 Figure 10.17 shows the results of an antibiotic sensitivity test carried out on a pathogenic strain of the human gut bacterium *Escherichia coli* O157.

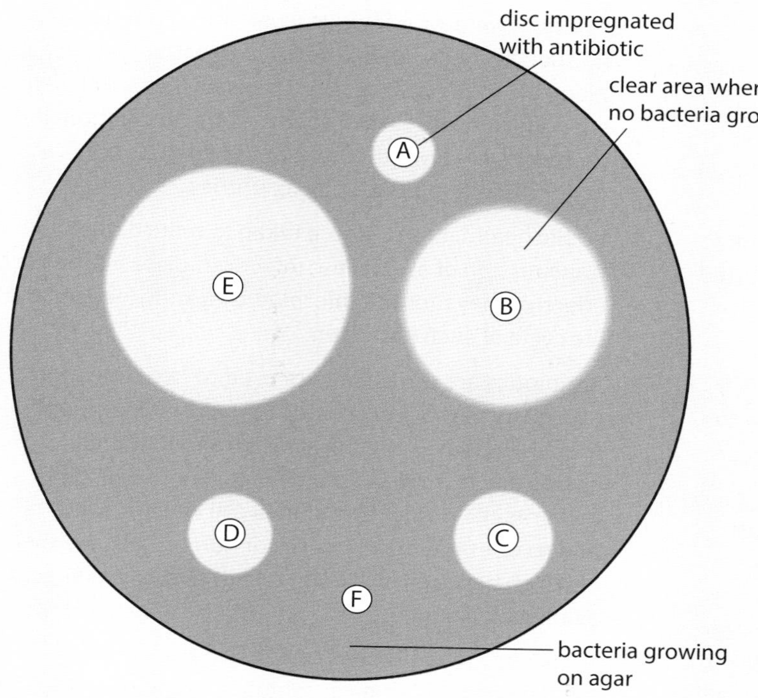

Figure 10.17 An antibiotic sensitivity test for a pathogenic strain of *E. coli*. Table 10.6 shows the inhibition zone diameters for the six antibiotics.

Bacteria are collected from faeces, food or water, and grown on an agar medium. Various antibiotics are absorbed onto discs of filter paper and placed on the agar plate. The plate is incubated, and the diameters of the inhibition zones where no bacteria are growing are measured.

If the diameter of the inhibition zone for an antibiotic is equal to or less than (≤) the figure given in the first column of Table 10.6, the bacteria are resistant to it. If the diameter is equal to or greater than (≥) the figure in the right-hand column, the bacteria are sensitive, and the antibiotic may be chosen for treatment.

Which of the antibiotics in Figure 10.17 and Table 10.6 would be chosen to treat the patient with the pathogenic strain of *E. coli* O157? Explain your answer.

Antibiotic	Inhibition zone diameter / mm	
	Resistant	Sensitive
A	≤ 11	≥ 14
B	≤ 12	≥ 18
C	≤ 9	≥ 14
D	≤ 11	≥ 22
E	≤ 12	≥ 15
F	≤ 14	≥ 19

Table 10.6 Inhibition zone diameters for the antibiotics of Figure 10.17.

Summary

- The term disease is defined as a disorder or illness that disrupts the normal functioning of the body or mind. Infectious diseases are caused by organisms known as pathogens that invade the body. Non-infectious diseases are all other diseases that are not caused by pathogens. There are many categories of non-infectious disease including genetic diseases (inherited) and deficiency diseases (caused by malnutrition).
- Cholera, malaria, HIV/AIDS, tuberculosis (TB) and measles are all examples of infectious diseases. Smallpox was an infectious disease but was eradicated in the late 20th century.
- Cholera is caused by the bacterium *Vibrio cholerae* and is transmitted in water or food contaminated by the faeces of infected people. Cholera can be controlled by treating patients with intravenous or oral rehydration therapy and making sure that human faeces do not reach the water supply. The disease is prevented by providing clean, chlorinated water and good sanitation.
- Malaria is caused by four species of *Plasmodium*. The most dangerous is *P. falciparum*. Malaria is transmitted by an insect vector: female *Anopheles* mosquitoes transfer *Plasmodium* from infected to uninfected people. Malaria is controlled in three main ways: by reducing the number of mosquitoes through insecticide spraying or draining breeding sites; by using mosquito nets (more effective if soaked in insecticide); by using drugs to prevent *Plasmodium* infecting people.

- AIDS is a set of diseases caused by the destruction of the immune system by infection with human immunodeficiency virus (HIV). HIV is transmitted in certain body fluids: blood, semen, vaginal secretions and breast milk. HIV primarily infects economically active members of populations in developing countries and has an extremely adverse effect on social and economic development.
- The transmission of HIV can be controlled by using barrier methods (e.g. condoms and femidoms) during sexual intercourse. Educating people to practise safer sex is the only control method currently available to health authorities. Contact tracing is used to find people who may have contracted HIV, so that they can be tested and counselled.
- Life expectancy can be greatly extended by using combinations of drugs which interfere with the replication of HIV. However, such treatment is expensive, is difficult to maintain and has unpleasant side-effects. There is no vaccine for HIV and no cure for AIDS.
- TB is caused by the bacteria *Mycobacterium tuberculosis* and *M. bovis*. *M. tuberculosis* is spread when people infected with the active form of the disease release bacteria in droplets of liquid when they cough or sneeze. Transmission occurs when uninfected people inhale the bacteria. This is most likely to happen where people live in overcrowded conditions, and especially where many sleep close together.
- Many people have the inactive form of TB in their lungs, but they do not have the disease and do not spread it. The inactive bacteria may become active in people who are malnourished or who become infected with HIV. *M. bovis* causes TB in cattle, but can be passed to humans. Drugs are used to treat people with the active form of TB. The treatment may take nine months or more as it is difficult to kill the bacteria. Contact tracing is used to find people who may have caught the disease. These people are tested for TB and treated if found to be infected. The BCG vaccine provides some protection against TB, but its effectiveness varies in different parts of the world.
- Measles is an extremely contagious disease caused by a virus that inhabits the cells of the nasal cavity and trachea. Infected people sneeze or cough out droplets which contain millions of virus particles. If these are inhaled by a person with no immunity, it is almost certain that they will be infected with the disease. Symptoms include fever and a rash. There is no specific medicine for measles – treatment is rest and medicine to reduce the fever. Measles is controlled in economically developed countries by vaccination and there are very few outbreaks of the disease, but is a major disease in developing countries.
- Public health measures are taken to reduce the transmission of all of these infectious diseases, but to be effective they must be informed by a knowledge of the life cycle of each pathogen.
- Antibiotics are drugs that are used to treat infections caused by pathogenic bacteria. They are compounds that are made by microorganisms and modified chemically to increase their effectiveness. Penicillin prevents the production of new cell walls in bacteria and so does not affect viruses or human cells, neither of which have cell walls. Not all antibiotics are effective against all bacteria.
- Resistance to antibiotics can arise because some bacteria may, by chance, contain a resistance gene. The bacteria survive when exposed to the antibiotic and can then reproduce to form a large population of bacteria all containing this gene. This is called vertical transmission of resistance. Resistance can also be spread between bacteria by horizontal transmission. This is when plasmids are transferred between bacteria during conjugation.
- The widespread and indiscriminate use of antibiotics has led to the growth of resistant strains of bacteria. This poses a serious challenge to the maintenance of health services in the 21st century.

End-of-chapter questions

1 Cholera, malaria, measles and tuberculosis (TB) are infectious diseases. Which row shows the type of organism that causes each of these diseases?

	Cholera	Malaria	Measles	Tuberculosis
A	bacterium	protoctist	virus	bacterium
B	bacterium	virus	bacterium	protoctist
C	protoctist	insect	bacterium	virus
D	virus	protoctist	virus	bacterium

[1]

2 Non-infectious diseases are best defined as:

A diseases caused by malnutrition

B all diseases of old age

C all diseases that are **not** caused by a pathogen

D all diseases that can be transmitted from mother to child. [1]

3 An antibiotic sensitivity test was carried out on bacteria isolated from a patient with a blood disease. Four antibiotics were tested, A, B, C and D. The results are shown in the figure.

Which antibiotic should be chosen to treat the blood disease? [1]

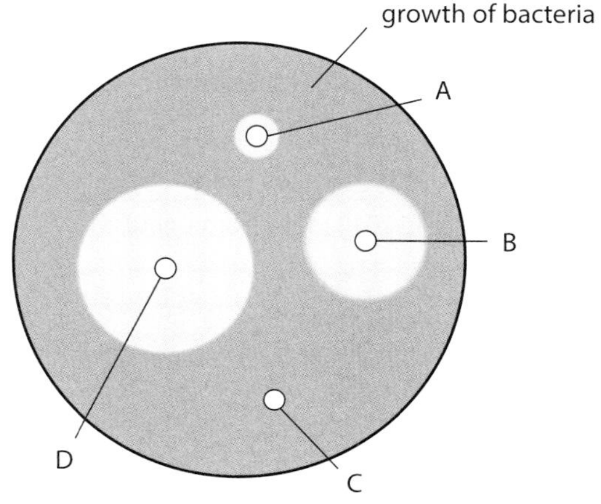

4 Which of the following diseases is transmitted by an insect vector?

A cholera

B HIV/AIDS

C malaria

D TB [1]

5 Rearrange the order of the following statements to give a flow diagram showing the evolution of resistance to the antibiotic streptomycin by the bacterium *Escherichia coli.*

1 Most of the population of *E. coli* is resistant to streptomycin.

2 A mutation in a DNA triplet of a plasmid, changing TTT to TTG, gives an *E. coli* bacterium resistance to streptomycin.

3 The resistant bacterium divides and passes copies of the plasmid to its offspring.

4 Sensitive bacteria die in the presence of streptomycin as a selective agent.

5 The frequency of the mutated gene in the population increases.

6 The resistant bacterium has a selective advantage and survives. [3]

6 a State **three** ways in which HIV is transmitted. [3]

The table shows statistics published by UNAIDS for four regions of the world and the global totals for HIV/AIDS in 2010.

Region	Number of people newly infected with HIV	Number of people living with HIV	Percentage of the adult population with HIV	Percentage of young people (15–24) with HIV	Number of people who died from AIDS
Eastern Europe and Central Asia	160 000	1.5 million	0.9	0.6 (male) 0.5 (female)	90 000
Sub-Saharan Africa	1.9 million	22.9 million	5.0	1.4 (male) 3.3 (female)	1.2 million
South and South-East Asia	270 000	4.0 million	0.3	0.1 (male) 0.1 (female)	250 000
North America	58 000	1.3 million	0.6	0.3 (male) 0.2 (female)	20 000
Global total	**2.7 million**	**34.0 million**	**0.8**	**0.3 (male) 0.6 (female)**	**1.8 million**

b Suggest **three** sources of data that UNAIDS may have used to compile the data in the table. [3]

c Explain why it is important to collect the data on the HIV/AIDS pandemic shown in the table. [3]

d i For North America, the ratio of the number of people dying from AIDS to the number of people living with HIV in 2010 was 20 000 : 1.3 million or 0.015 : 1.
Calculate the ratio for sub-Saharan Africa. [1]

ii Suggest reasons for the difference between the ratios for North America and sub-Saharan Africa. [3]

[Total: 13]

7 a Describe how malaria is transmitted. [3]

The figure shows the global distribution of malaria in 2010 (data from UNAIDS).

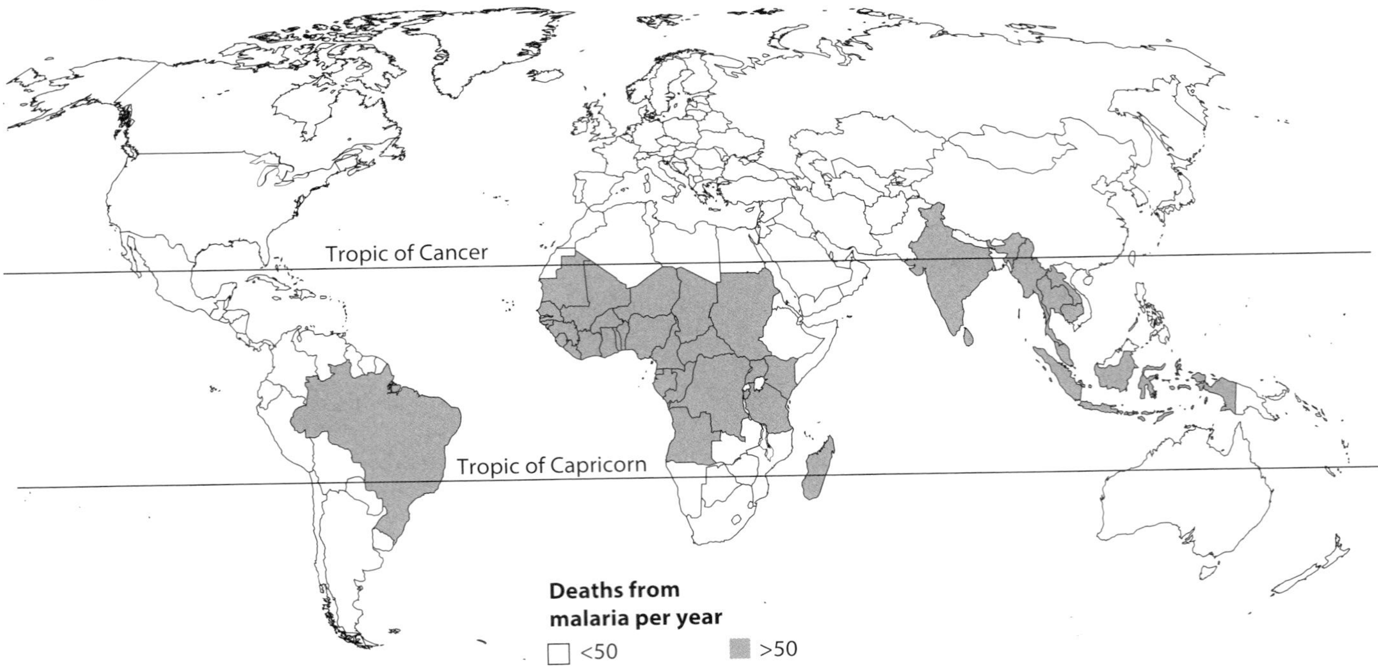

b Describe and explain the global distribution of malaria. [5]

c Outline the biological reasons for the difficulties in developing and introducing control methods for malaria. [6]

[Total: 14]

8 a Describe how cholera is transmitted. [2]

The table shows the number of cases of cholera and deaths from the disease for the five countries with the greatest outbreaks as reported to the WHO in 2010.

Country	Region	Total number of cases	Number of deaths	Case fatality rate / %
Haiti	Caribbean	179 379	3 990	
Cameroon	West Africa	10 759	657	6.10
Nigeria		44 456	1 712	3.85
Democratic Republic of the Congo		13 884	182	1.31
Papua New Guinea	Australasia	8 997	95	1.06
Total	**All regions of the world**	**317 534**	**7 543**	**2.38**

b With reference to the table:

i calculate the case fatality rate for Haiti in 2010 [1]

ii suggest why the case fatality rate varies between countries [3]

iii explain why it is important that the WHO collects data on outbreaks of cholera. [3]

c The WHO also collects data on 'imported' cases of cholera. Among countries reporting these cases in 2010 were Australia, Malaysia and the USA.

i Suggest what is meant by the term 'imported case'. [1]

ii Explain why there are no epidemics of cholera in highly economically developed countries such as Australia and the USA. [2]

[Total: 12]

9 a i Name the causative organism of TB. [1]

ii Explain how TB is transmitted. [2]

b i State the regions of the world with the highest number of cases of TB. [3]

ii Suggest reasons for the high number of cases of TB in some parts of the world. [4]

[Total: 10]

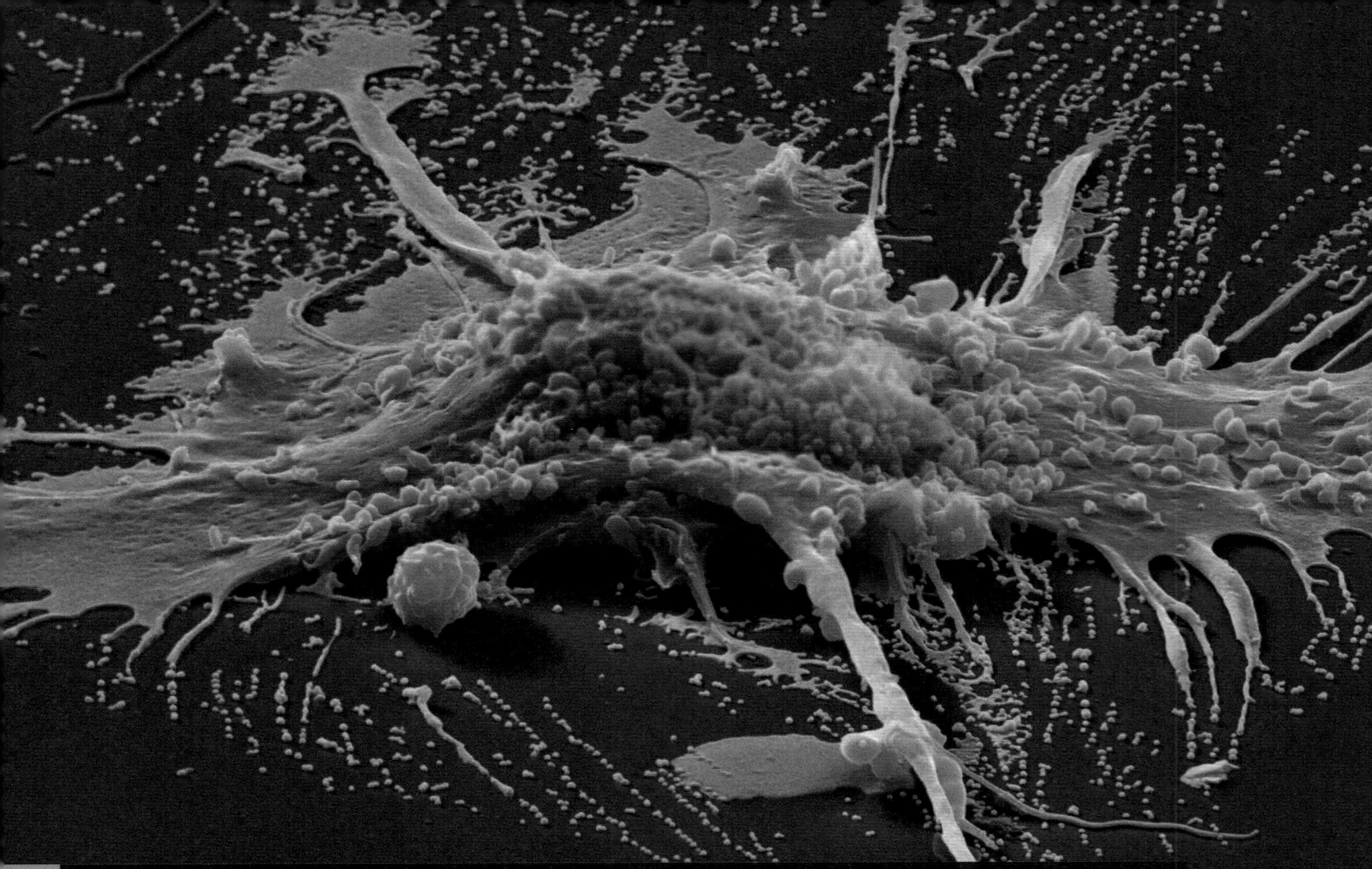

Chapter 11: Immunity

Learning outcomes

You should be able to:

- outline how and where white blood cells originate
- explain how the immune system distinguishes between the body's own cells (self) and anything foreign, such as pathogens and transplanted tissues (non-self)
- describe how phagocytes and lymphocytes respond during an infection
- describe and explain the significance of the increase in white blood cell counts in people with infections and leukaemias
- describe how autoimmune diseases occur
- explain how the structure of antibody molecules is related to their functions
- outline how monoclonal antibodies are produced and used in diagnosis and treatment
- distinguish between the four different types of immunity
- describe how vaccination is used to control some infectious diseases
- discuss why a vaccination programme was successful in eradicating smallpox, and the reasons why measles, cholera, malaria and TB have not been eradicated

Smallpox was first; polio second?

You must have been vaccinated against polio. You may wonder why people make such a fuss about having a vaccination against a disease that you are unlikely ever to come across. There may be very few cases of polio now, but in the past large numbers of people contracted the disease. In many cases, the disease was mild, but the virus that causes polio can infect the base of the nerves and cause paralysis. This happened in about 1% of all cases.

Vaccines against polio became available in the 1950s and immediately mass vaccination programmes began. These have proved very successful: the last case of polio in the Americas was in 1991. In 1994, it was declared that transmission of polio had been broken so that the disease was no longer endemic in the Western Hemisphere. Success has been slower coming in the rest of the world. In 2013, there were only 369 cases of polio throughout the world. Most of these occurred in three countries: Nigeria, Somalia and Pakistan.

India has mobilised huge numbers of medical staff and around 230 000 volunteers to vaccinate children throughout the country (Figure 11.1). This didn't just happen once, but on many occasions in the government's drive to eradicate polio. Vaccination programmes in the other countries where polio remains endemic have not been as successful as in India. One reason is that some people have resisted attempts by medical staff to vaccinate the local population. Some medical staff have been attacked, even killed, while working on polio eradication campaigns in Pakistan.

In 1980, the World Health Organization announced that the viral disease smallpox had been eradicated as there had not been a new case anywhere in the world for the previous three years. Polio is the most likely candidate for the second infectious disease to be eradicated. You can follow the progress of the campaign by searching online for 'polio eradication'.

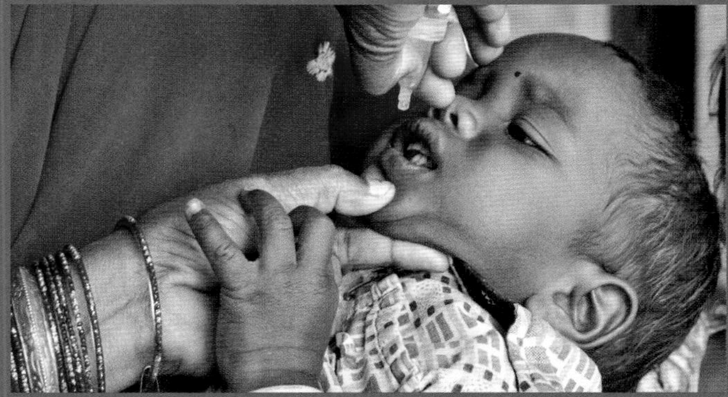

Figure 11.1 On 15 April 2012, 170 million children across India were vaccinated against polio as part of the government's programme to eradicate the disease.

We now consider in detail something that was mentioned in Chapter 10: our ability to defend ourselves against infection by pathogens (disease-causing organisms). We have seen that some people experience few or no symptoms when exposed to certain infectious diseases. Even though a person may be a carrier of disease to other people, he or she has immunity. How is this possible? The disease measles is used as an example here.

Most people have measles only once. In most cases, it is very unlikely that anyone surviving the disease will have it again. They are **immune**. Their body's internal defence system has developed a way of recognising the measles virus and preventing it from doing any harm again. Immunity is the protection against disease provided by the body's internal defence or immune system. The external cellular and chemical barriers that protect us from infection are part of that defence system.

Defence against disease

External defence system

We have a variety of mechanisms to protect ourselves against infectious diseases such as measles and those described in the previous chapter. Many pathogens do not harm us, because, if we are healthy, we have physical, chemical and cellular defences that prevent them entering. For example, the epithelia that cover the airways are an effective barrier to the entry of pathogens (pages 187–189); hydrochloric acid in the stomach kills many bacteria that we ingest with our food and drink; blood clotting is a defence mechanism that stops the loss of blood and prevents the entry of pathogens through wounds in the skin.

Internal defence system

If pathogens do successfully enter the body, white blood cells (Chapter 8) can recognise them as something foreign and destroy them.

White blood cells are part of the immune system and they recognise pathogens by the distinctive, large molecules that cover their surfaces, such as proteins, glycoproteins, lipids and polysaccharides, and the waste materials which some pathogens produce. Any molecule which the body recognises as foreign is an **antigen**.

There are two types of white blood cell, namely phagocytes and lymphocytes. Before looking at their structure and function in detail, it will be useful to look at an example of the immune response in humans. This example introduces further important features of the immune system, namely the ability to distinguish between self and non-self, and the production of antibodies. Antibodies are glycoprotein molecules that act against specific antigens (page 228).

Each of us has molecules on the surfaces of our cells that are not found in other organisms, or even in other humans. These are often called **cell surface antigens**. Although they do not stimulate production of antibodies in us, they will do if they enter someone else. The cell surface antigens of the human ABO blood group system are a good example. If you are blood group A, then you have certain glycolipids on your red blood cells that are not on the red cells of people who are blood group B. If blood of type A is given to someone who has blood of type B during a transfusion, then the recipient will recognise these blood cells as foreign and start to produce antibodies. The recipient's immune system has recognised the antigens on blood cells of type A as non-self. If blood of type B is used during the transfusion, as it should be, then the recipient's immune system recognises the antigens on the red blood cells as self and no antibodies are produced.

The response of lymphocytes to the presence of a foreign antigen is known as the immune response. In some cases lymphocytes respond by producing antibodies; in others they respond by killing cells that have become infected by pathogens.

QUESTION

11.1 a Explain the terms **antigen, antibody** and **immune response** in your own words.

b Explain why blood of type B is not given to someone with blood type A during a blood transfusion.

Cells of the immune system

The cells of the immune system originate from the bone marrow. There are two groups of these cells involved in defence:

- phagocytes (neutrophils and macrophages)
- lymphocytes.

All of these cells are visible among red blood cells when a blood smear is stained to show nuclei as shown in Figure 11.2.

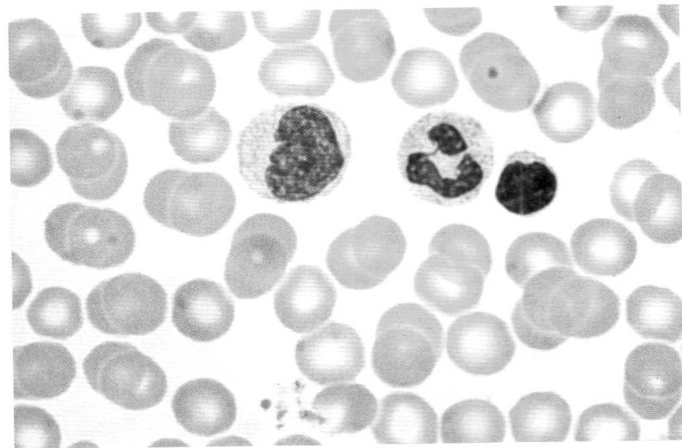

Figure 11.2 A monocyte (left), which will develop into a macrophage, a neutrophil (centre) and a lymphocyte (right), together with red blood cells in a blood smear which has been photographed through a light microscope. The cytoplasm of the neutrophil contains vacuoles full of hydrolytic enzymes (×1000).

An **antigen** is a substance that is foreign to the body and stimulates an immune response.

An **antibody** is a glycoprotein (immunoglobulin) made by plasma cells derived from B-lymphocytes, secreted in response to an antigen; the variable region of the antibody molecule is complementary in shape to its specific antigen.

The **immune response** is the complex series of responses of the body to the entry of a foreign antigen; it involves the activity of lymphocytes and phagocytes.

Non-self refers to any substance or cell that is recognised by the immune system as being foreign and will stimulate an immune response.

Self refers to substances produced by the body that the immune system does not recognise as foreign, so they do not stimulate an immune response.

Phagocytes

Phagocytes are produced throughout life in the bone marrow. They are stored there before being distributed around the body in the blood. They are scavengers, removing any dead cells as well as invasive microorganisms.

Neutrophils are a kind of phagocyte and form about 60% of the white cells in the blood (Figure 11.3). They travel throughout the body, often leaving the blood by squeezing through the walls of capillaries to 'patrol' the tissues. During an infection, neutrophils are released in large numbers from their stores, but they are short-lived cells.

Macrophages are also phagocytes but are larger than neutrophils and tend to be found in organs such as the lungs, liver, spleen, kidney and lymph nodes, rather than remaining in the blood. After they are made in the bone marrow, macrophages travel in the blood as **monocytes**, which develop into macrophages once they leave the blood and settle in the organs, removing any foreign matter found there.

Macrophages are long-lived cells and play a crucial role in initiating immune responses, since they do not destroy pathogens completely, but cut them up to display antigens that can be recognised by lymphocytes.

Phagocytosis

If pathogens invade the body and cause an infection, some of the cells under attack respond by releasing chemicals such as **histamine**. These, with any chemicals released by the pathogens themselves, attract passing neutrophils to the site. (This movement towards a chemical stimulus is called chemotaxis.) The neutrophils destroy the pathogens by phagocytosis (Figure 4.19, page 87, and Figure 11.4).

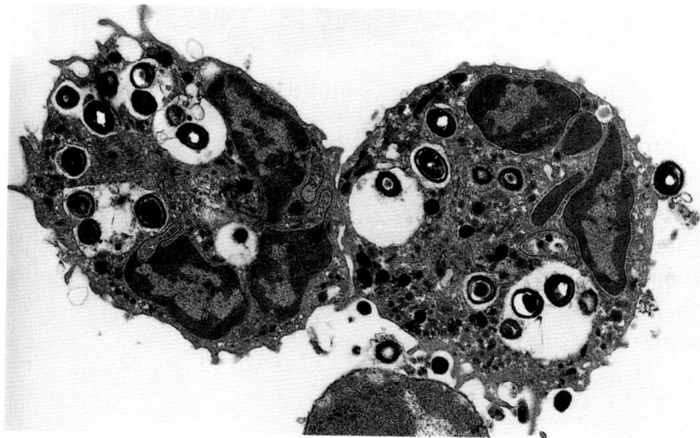

Figure 11.3 A transmission electron micrograph of two neutrophils that have ingested several *Staphylococcus* bacteria (×4000). Notice at the extreme right, one bacterium being engulfed. Compare this photograph with Figure 11.4.

The neutrophils move towards the pathogens, which may be clustered together and covered in antibodies. The antibodies further stimulate the neutrophils to attack the pathogens. This is because neutrophils have receptor proteins on their surfaces that recognise antibody molecules and attach to them. When the neutrophil attaches to the pathogen, the neutrophil's cell surface membrane engulfs the pathogen, and traps it within a phagocytic vacuole in a process called endocytosis. Digestive enzymes are secreted into the phagocytic vacuole, so destroying the pathogen.

Neutrophils have a short life: after killing and digesting some pathogens, they die. Dead neutrophils often collect at a site of infection to form pus.

QUESTION

11.2 a State the site of origin of phagocytes.

b Looking at Figure 11.2:

i describe the differences between the neutrophil and the lymphocyte

ii calculate the actual size of the neutrophil.

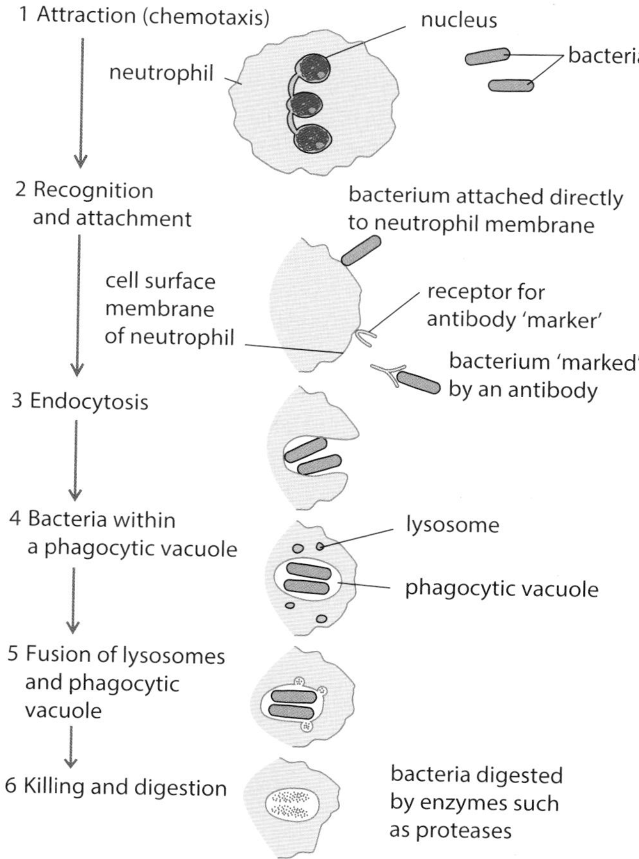

Figure 11.4 The stages of phagocytosis.

Lymphocytes

Lymphocytes are a second type of white blood cell. They play an important role in the immune response.

Lymphocytes are smaller than phagocytes. They have a large nucleus that fills most of the cell (Figure 11.2). There are two types of lymphocyte, both of which are produced before birth in bone marrow.

- B-lymphocytes (**B cells**) remain in the bone marrow until they are mature and then spread throughout the body, concentrating in lymph nodes and the spleen.
- T-lymphocytes (**T cells**) leave the bone marrow and collect in the **thymus** where they mature. The thymus is a gland that lies in the chest just beneath the sternum. It doubles in size between birth and puberty, but after puberty it shrinks.

Only mature lymphocytes can carry out immune responses. During the maturation process, many different types of B- and T-lymphocyte develop, perhaps many millions. Each type is specialised to respond to one antigen, giving the immune system as a whole the ability to respond to almost any type of pathogen that enters the body. When mature, all these B and T cells circulate between the blood and the lymph (Chapter 8). This ensures that they are distributed throughout the body so that they come into contact with any pathogens **and** with each other.

Immune responses depend on B and T cells interacting with each other to give an effective defence. We will look in detail at the roles of B and T cells and how they interact in the following section. Briefly, however, some T cells coordinate the immune response, stimulating B cells to divide and then secrete antibodies into the blood; these antibodies recognise the antigens on the pathogens and help to destroy the pathogens. Other T cells seek out and kill any of the body's own cells that are infected with pathogens. To do this they must make direct contact with infected cells.

B-lymphocytes

As each B cell matures, it gains the ability to make just one type of antibody molecule. Many different types of B cell develop in each of us, perhaps as many as 10 million. While B cells are maturing, the genes that code for antibodies are changed in a variety of ways to code for different antibodies. Each cell then divides to give a small number of cells that are able to make the same type of antibody. Each small group of identical cells is called a **clone**. At this stage, the antibody molecules do not leave the B cell but remain in the cell surface membrane. Here, part of each antibody forms a glycoprotein receptor, which can combine specifically with one type of antigen. If that antigen enters the body, there will be some mature B cells with cell surface receptors that will recognise it (Figure 11.5).

Figure 11.6 shows what happens to B cells during the immune response when an antigen enters the body on two separate occasions. When the antigen enters the body for the first time the small numbers of B cells with receptors complementary to the antigen are stimulated to divide by mitosis. This stage is known as clonal selection. The small clone of cells divides repeatedly by mitosis in the clonal expansion stage so that huge numbers of identical B cells are produced over a few weeks.

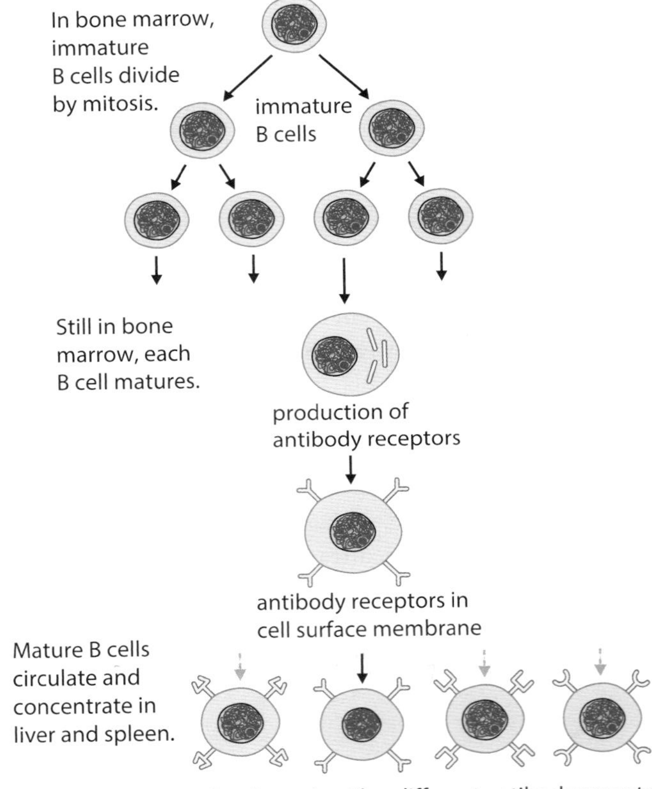

Figure 11.5 Origin and maturation of B-lymphocytes. As they mature in bone marrow, the cells become capable of secreting one type of antibody molecule with a specific shape. Some of these molecules become receptor proteins in the cell surface membrane and act like markers. By the time of a child's birth, there are millions of different B cells, each with a specific antibody receptor.

QUESTIONS

11.3 State the sites of origin and maturation of B-lymphocytes (B cells) and T-lymphocytes (T cells).

11.4 Suggest why the thymus gland becomes smaller after puberty.

Some of these activated B cells become **plasma cells** that produce antibody molecules very quickly – up to several thousand a second. Plasma cells secrete antibodies into the blood, lymph or onto the linings of the lungs and the gut (Figure 11.7). These plasma cells do not live long: after several weeks their numbers decrease. The antibody molecules they have secreted stay in the blood for longer, however, until they too eventually decrease in concentration.

Other B cells become memory cells. These cells remain circulating in the body for a long time. If the same antigen is reintroduced a few weeks or months after the first infection, memory cells divide rapidly and develop into plasma cells and more memory cells. This is repeated on every subsequent invasion by the same antigen, meaning that the infection can be destroyed and removed before any symptoms of disease develop.

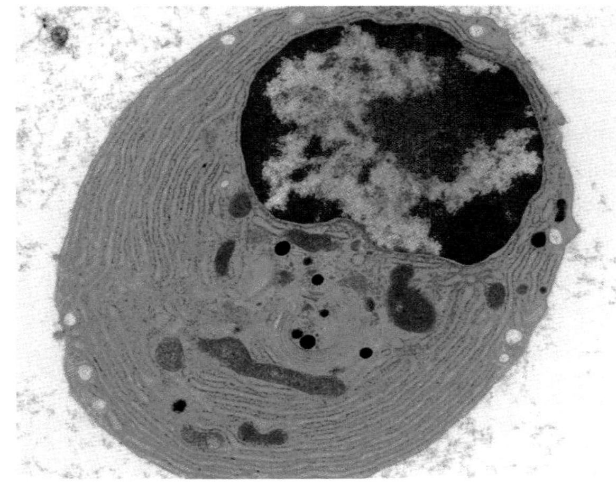

Figure 11.7 Electron micrograph of the contents of a plasma cell (×3500). There is an extensive network of rough endoplasmic reticulum in the cytoplasm (green) for the production of antibody molecules, which plasma cells secrete into blood or lymph by exocytosis (Chapter 4). The mitochondria (blue) provide ATP for protein synthesis and the movement of secretory vesicles.

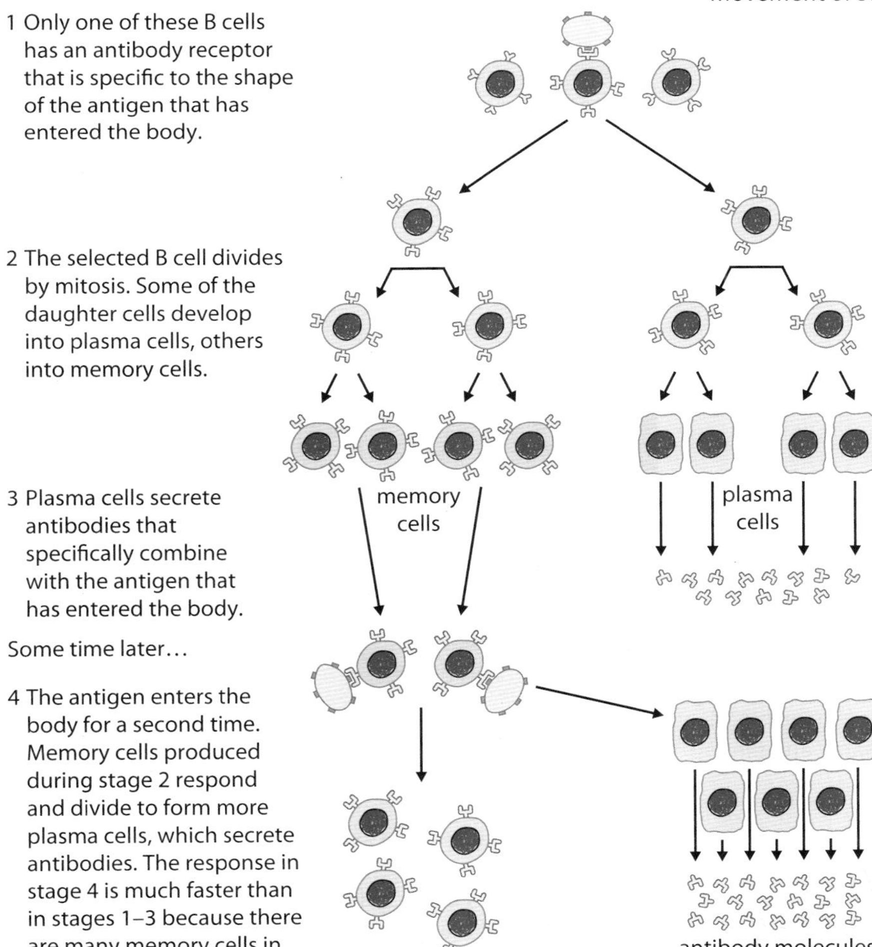

Figure 11.6 The function of B-lymphocytes during an immune response. The resulting changes in antibody concentration are shown in Figure 11.8.

Figure 11.8 shows the changes in the concentration of antibody molecules in the blood when the body encounters an antigen. The first or **primary response** is slow because, at this stage, there are very few B cells that are specific to the antigen. The **secondary response** is faster because there are now many memory cells, which quickly divide and differentiate into plasma cells. During the primary response, the number of cells in each clone of B cells that is selected has increased in size. There are many more B cells specific to the pathogen that has invaded the body. As you can see in Figure 11.8, many more antibodies are produced in the secondary response.

Memory cells are the basis of **immunological memory**; they last for many years, often a lifetime. This explains why someone is very unlikely to catch measles twice. There is only one strain of the virus that causes measles, and each time it infects the body there is a fast secondary response. However, we do suffer repeated infections of the common cold and influenza, because there are many different and new strains of the viruses that cause these diseases, each one having different antigens. Each time a pathogen with different antigens infects us, the primary response must occur before we become immune, and during that time we often become ill.

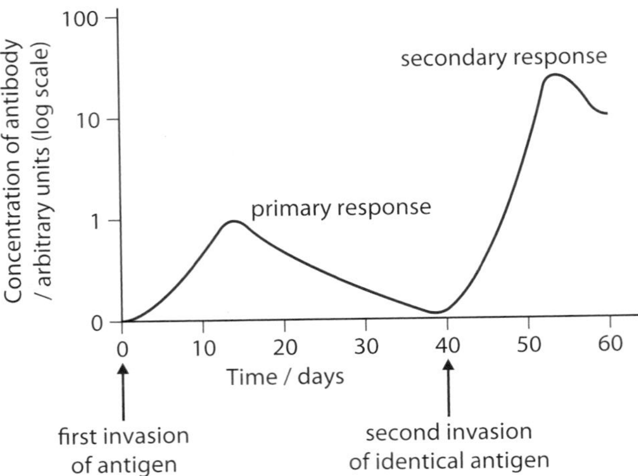

Figure 11.8 The changes in antibody concentration in the blood during a primary and secondary response to the same antigen.

QUESTIONS

11.5 a Calculate the diameter of the plasma cell in Figure 11.7 at its widest point.

b Explain how plasma cells, such as the one shown in Figure 11.7, are adapted to secrete large quantities of antibody molecules.

11.6 Explain why B cells divide by mitosis during an immune response.

11.7 Explain why people are often ill for several weeks after they catch a disease, even though they can make antibodies against the disease.

Antibodies

Antibodies are all globular glycoproteins with quaternary structure (page 42). They form the group of plasma proteins called **immunoglobulins**. The basic molecule common to all antibodies consists of four polypeptide chains: two 'long' or 'heavy' chains and two 'short' or 'light' chains (Figures 11.9 and 11.10). Disulfide bonds hold the chains together. Each molecule has two identical antigen-binding sites, which are formed by both light and heavy chains. The sequences of amino acids in these regions make the specific three-dimensional shape which binds to just one type of antigen. The antigen-binding sites form the **variable region,** which is different on each type of antibody molecule produced. The 'hinge' region gives the flexibility for the antibody molecule to bind around the antigen.

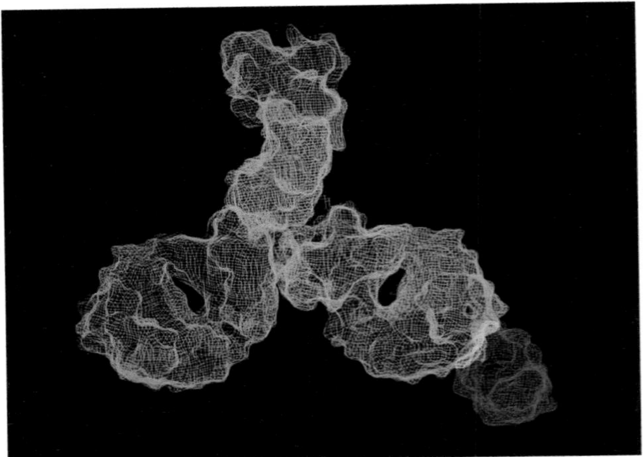

Figure 11.9 A model of an antibody made using computer graphics. The main part (green) is the antibody molecule, and the small part in the bottom right-hand corner (red) is an antigen at one of the two antigen-binding sites. Compare this with Figure 11.10. This type of antibody molecule with four polypeptides is known as immunoglobulin G, IgG for short. Larger types of antibody molecules are IgA with four antigen binding sites and IgM with ten.

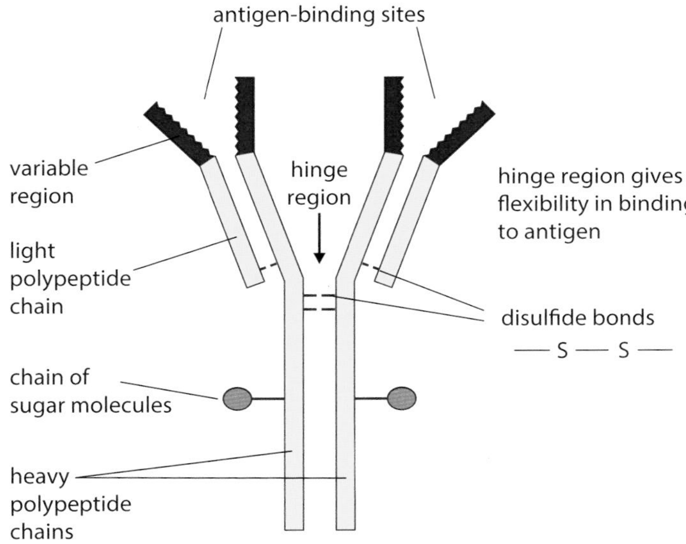

Figure 11.10 A diagram of an antibody molecule. Antigen–antibody binding occurs at the variable regions. An antigen fits into the binding site like a substrate fitting into the active site of an enzyme. The constant region of the molecule is shown in yellow, and it is identical in all antibodies like this that have four polypeptides.

Figure 11.11 shows the different ways in which antibodies work to protect the body from pathogens. As we saw earlier, some antibodies act as labels to identify antigens as appropriate targets for phagocytes to destroy. A special group of antibodies are **antitoxins** which block the toxins released by bacteria such as those that cause diphtheria and tetanus.

QUESTIONS

11.8 Explain why polysaccharides would not be suitable for making antibody molecules.

11.9 Explain why only some B cells respond during an immune response to a pathogen.

11.10 There are many different strains of the rhinovirus, which causes the common cold. Explain why people can catch several different colds in the space of a few months.

1

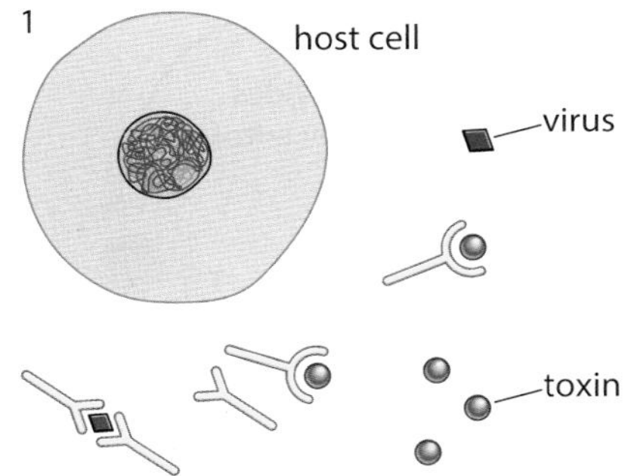

Antibodies combine with viruses and bacterial toxins preventing them entering or damaging cells.

2

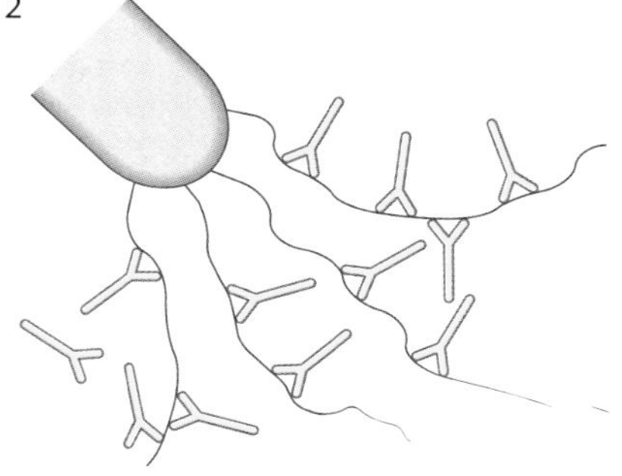

Antibodies attach to flagella of bacteria making them less active and easier for phagocytes to engulf.

3

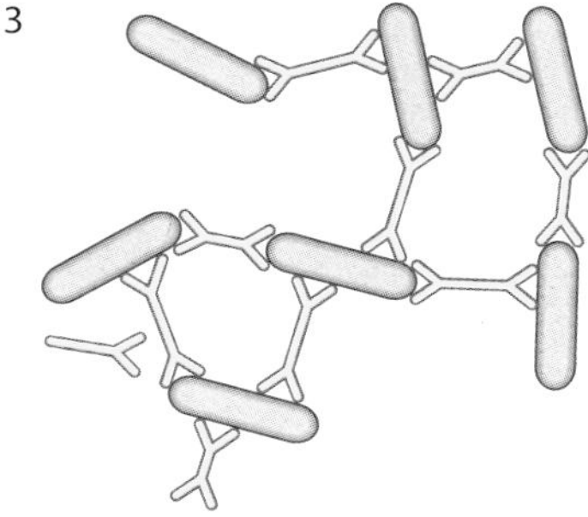

Antibodies with multiple antigen binding sites cause agglutination (clumping together) of bacteria reducing the chances of spread throughout the body.

4

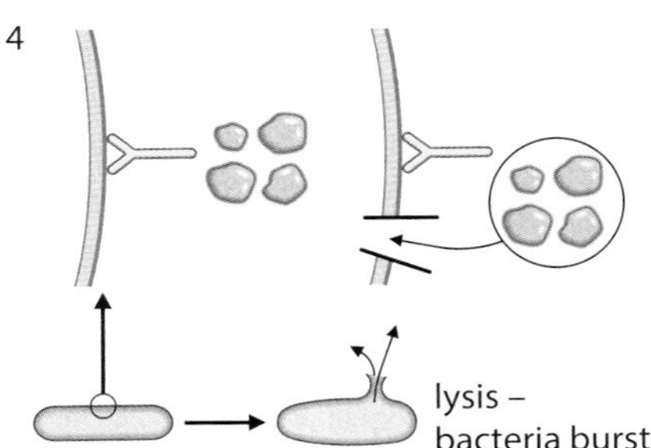

Together with other molecules, some antibodies 'punch' holes in the cell walls of bacteria, causing them to burst when they absorb water by osmosis.

5

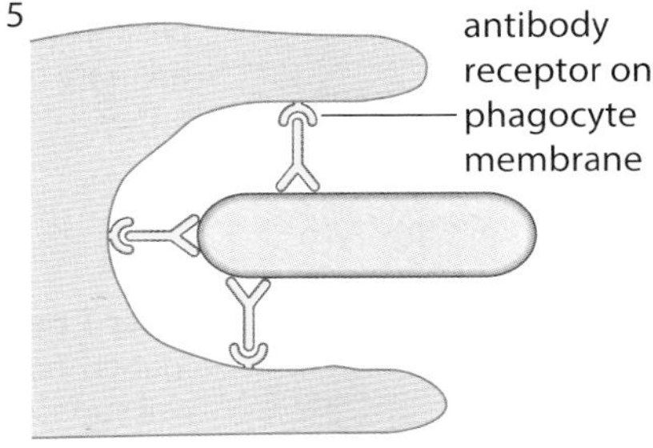

Antibodies coat bacteria, making it easier for phagocytes to ingest them; phagocytes have receptor proteins for the heavy polypeptide chains of antibodies.

6

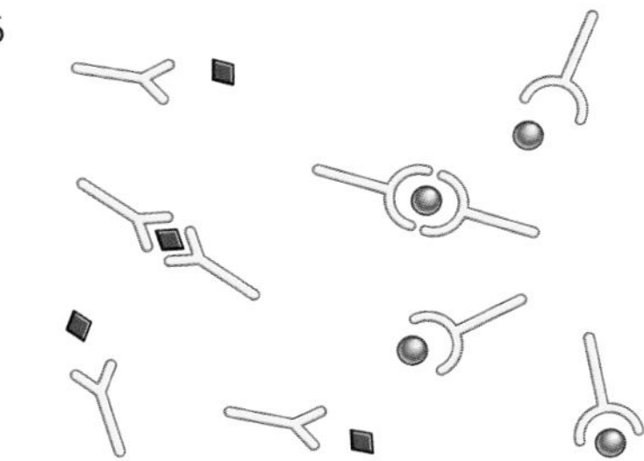

Antibodies combine with toxins, neutralising them and making them harmless; these antibodies are called antitoxins.

Figure 11.11 The functions of antibodies. Antibodies have different functions according to the type of antigen to which they bind. Note that the diagrams of antibodies are purely symbolic and do not represent their actual shapes, sizes or positions of binding sites.

T-lymphocytes

Mature T cells have specific cell surface receptors called T cell receptors (Figure 11.12). T cell receptors have a structure similar to that of antibodies, and they are each specific to one antigen. T cells are activated when they encounter this antigen on another cell of the host (that is, on the person's own cells). Sometimes this cell is a macrophage that has engulfed a pathogen and cut it up to expose the pathogen's surface molecules, or it may be a body cell that has been invaded by a pathogen and is similarly displaying the antigen on its cell surface membrane as a kind of 'help' signal. The display of antigens on the surface of cells in this way is known as **antigen presentation**. Those T cells that have receptors complementary to the antigen respond by dividing by mitosis to increase the number of cells. T cells go through the same stages of clonal selection and clonal expansion as clones of B cells.

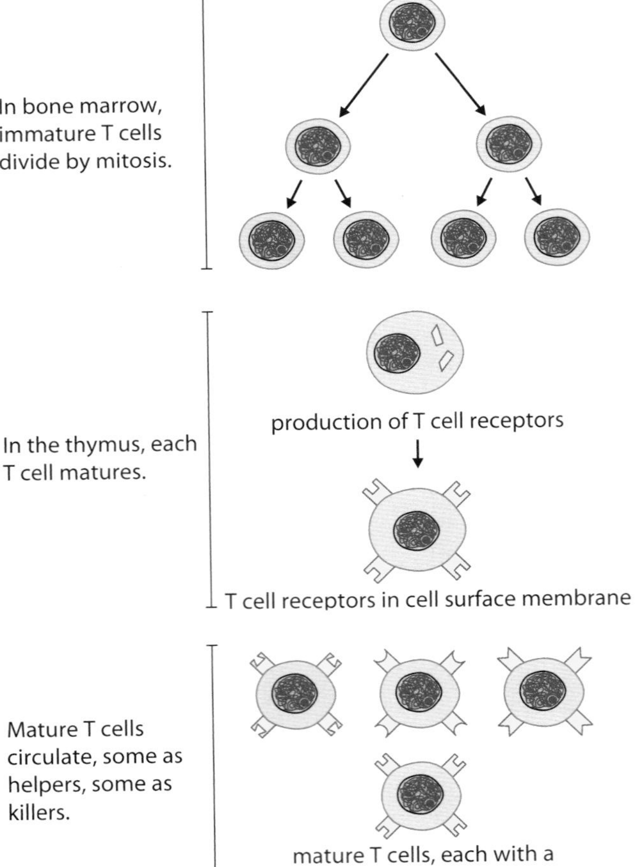

Figure 11.12 Origin and maturation of T-lymphocytes. As T cells mature in the thymus gland they produce T cell receptor proteins. Each cell has a specific receptor. Some cells become helper T cells, others become killer T cells.

There are two main types of T cell:

- helper T cells
- **killer T cells** (or cytotoxic T cells).

When helper T cells are activated, they release hormone-like **cytokines** that stimulate appropriate B cells to divide, develop into plasma cells and secrete antibodies. Some T helper cells secrete cytokines that stimulate macrophages to carry out phagocytosis more vigorously. Killer T cells search the body for cells that have become invaded by pathogens and are displaying foreign antigens from the pathogens on their cell surface membranes. Killer T cells recognise the antigens, attach themselves to the surface of infected cells, and secrete toxic substances such as hydrogen peroxide, killing the body cells and the pathogens inside (Figure 11.13). Some helper T cells

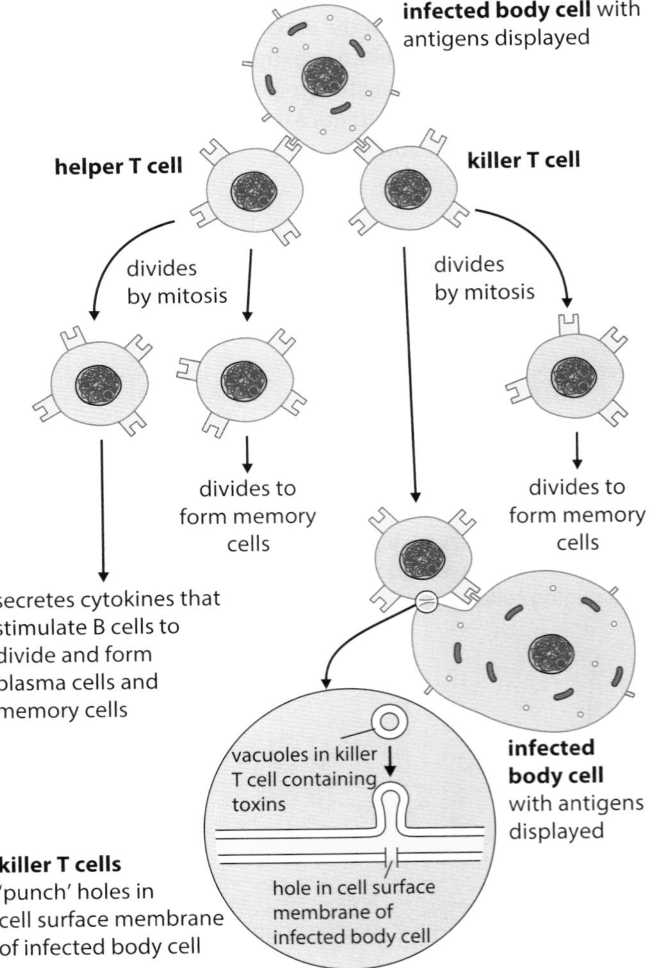

Figure 11.13 The functions of T-lymphocytes during an immune response. Helper T cells and killer T cells with T cell receptor proteins specific to the antigen respond and divide by mitosis. Activated helper T cells stimulate B cells to divide and develop into plasma cells (Figure 11.6, page 227). Killer T cells attach themselves to infected cells and kill them.

secrete cytokines that stimulate macrophages to carry out phagocytosis more vigorously, or that stimulate killer T cells to divide by mitosis and to differentiate by producing vacuoles full of toxins.

Memory helper T cells and memory killer T cells are produced, which remain in the body and become active very quickly during the secondary response to antigens.

QUESTION

11.11 Outline the functions of B-lymphocytes (B cells) and T-lymphocytes (T cells) and describe how they interact during an immune response.

Numbers of white blood cells

Blood tests are routinely carried out to help doctors diagnose diseases and to assess the success of treatments. Blood samples are taken from patients and sent for analysis in laboratories that use automated cell counters. The results usually include the numbers of red and white blood cells and platelets. Platelets are small cell fragments that do not have a nucleus; they are formed from the break-up of cells in the bone marrow (Figure 8.13, page 163). They release substances that stimulate blood clotting. The results of such blood tests are given as absolute values as in the second column of Table 11.1. The results for specific white blood cells, such as neutrophils and lymphocytes, are given as absolute numbers or as percentages of the white cell count. There is considerable variation in these numbers between people. The third column of Table 11.1 gives examples of these normal ranges.

The number of neutrophils in the blood increases during bacterial infections and whenever tissues become inflamed and die. The number of lymphocytes in the blood increases in viral infections and in TB. Most of the lymphocytes that circulate in the blood are T cells. In some blood tests, the numbers of T cells are recorded. The human immunodeficiency virus (HIV) invades helper T cells and causes their destruction, so blood tests for people who are HIV+ record the numbers of specific T cells. The normal value is between 500 and 1500 cells mm^{-3}. The specific T cell numbers provide useful information on the progress of the disease and the success of treatments, by monitoring decline in T cell number and therefore assessing the deleterious effect on the immune system.

QUESTION

11.12 Use the typical values in the second column of Table 11.1 to answer the following questions.

a Calculate the total numbers of red blood cells and white blood cells in $1\,mm^3$

i in males

ii in females.

b Calculate the white blood cell counts as percentages of the totals you calculated in **a**.

c Express the results for neutrophils and lymphocytes as percentages of the white blood cell counts for the males and females.

d Calculate the numbers of neutrophils and lymphocytes in $1\,dm^3$ of blood. Express your answers using standard form.

Cellular component of blood	Numbers mm^{-3} of blood		Percentage of cellular component within all cellular components
	Typical values	**Normal range of values**	
red blood cells	5500 000 (males) 4800 000 (females)	4600 000–6200 000	93–96
platelets	300 000	150 000–400 000	4–7
white blood cells	7500	4500–10 000	0.1–0.2
			Percentages of all white blood cells
of which: neutrophils	4500	3000–6000	30–80
B lymphocytes	400	70–600	15–40
T lymphocytes	1500	500–2500	
other white blood cells	1100	800–2000	5–14
Totals	**5807 500 (males)** **5107 500 (females)**		

Table 11.1 The results of blood tests are given as absolute numbers and compared with the normal ranges. They are often given as the numbers in $1\,mm^3$ which is the same as $1\,\mu l$ (microlitre) or $1 \times 10^{-9}\,dm^3$ of blood. As numbers of blood cells vary considerably, results are often expressed as percentages, as in the fourth column.

All the white cells in the blood originate from stem cells in the bone marrow. There are two groups of bone marrow stem cells:

- myeloid stem cells that give rise to neutrophils, monocytes and platelets
- lymphoid stem cells that give rise to lymphocytes, both B and T cells.

These stem cells divide rapidly to produce huge numbers of mature, differentiated blood cells that function in specific parts of the immune response. Leukaemias are cancers of these stem cells. The cells divide uncontrollably to give many cells which do not differentiate properly and disrupt the production of normal blood cells including red blood cells and platelets. These malignant cells fill up the bone marrow and then flow into the blood and into the lymphatic system.

In myeloid leukaemias, the stem cells responsible for producing neutrophils divide uncontrollably and the number of immature cells increases. In lymphoblastic leukaemias, the cancerous cells are those that give rise to lymphocytes.

The immature white blood cells are produced very quickly and they disrupt the normal balance of components in the blood. This means that the body does not have enough red blood cells or platelets. This causes anaemia and increases the risk of excessive bleeding. Also, the numbers of mature neutrophils and lymphocytes decrease so that people with these cancers become more susceptible to infections; they are said to be immunosuppressed.

There are acute and chronic forms of both types of leukaemia. Acute leukaemias develop very quickly, have severe effects and need to be treated immediately after they are diagnosed; chronic leukaemias may take many years to develop and changes in blood cell counts are usually monitored over time so that treatment is given when it is most likely to cure the disease. Blood tests are used to help diagnose these diseases, monitor their progress and assess the effectiveness of treatments.

Active and passive immunity

The type of immunity described so far occurs during the course of an infection. This type of immunity is called active immunity because the person makes their own antibodies. This happens when the lymphocytes are activated by antigens on the surface of pathogens that have invaded the body. As this activation occurs naturally during an infection it is called **natural active immunity**.

The immune response can also be activated artificially, either by injecting antigens into the body or – for certain diseases such as polio – taking them by mouth. This is the basis of **artificial active immunity**, more commonly known as vaccination. The immune response is similar to that following an infection, and the effect is the same – long-term immunity. In both natural and artificial active immunity, antibody concentrations in the blood follow patterns similar to those shown in Figure 11.8 (page 228).

In both forms of active immunity, it takes time for sufficient numbers of B and T cells to be produced to give an effective defence. If a person becomes infected with a potentially fatal disease such as tetanus, a more immediate defence than that provided by active immunity is needed for survival. Tetanus kills quickly, before the body's natural primary response can take place. So people who have a wound that may be infected with the bacterium that causes tetanus are given an injection of **antitoxin**. This is a preparation of human antibodies against the tetanus toxin. The antibodies are collected from blood donors who have recently been vaccinated against tetanus. Antitoxin provides immediate protection, but this is only temporary as the antibodies are not produced by the body's own B cells and are therefore regarded as foreign themselves. They are removed from the circulation by phagocytes in the liver and spleen.

This type of immunity is called passive immunity because the person has not produced the antibodies themself. B and T cells have not been activated, and plasma cells have not produced any antibodies. More specifically, antitoxins provide **artificial passive immunity**, because the antibodies have not entered the body by a natural process: they have come from another person who has encountered the antigen.

The immune system of a newborn infant is not as effective as that of a child or an adult. However, infants are not entirely unprotected against pathogens, because antibodies from their mothers cross the placenta during pregnancy and remain in the infant for several months (Figure 11.14). For example, antibodies against measles may last for four months or more in the infant's blood.

Active immunity is immunity gained when an antigen enters the body, an immune response occurs and antibodies are produced by plasma cells.

Passive immunity is immunity gained without an immune response; antibodies are injected (artificial) or pass from mother to child across the placenta or in breast milk (natural).

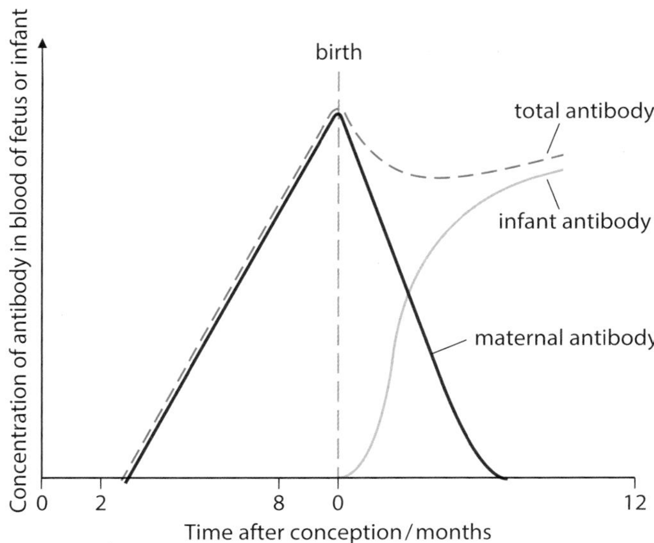

Figure 11.14 The concentrations of antibody in the blood of a fetus and an infant.

Colostrum, the thick yellowish fluid produced by a mother's breasts for the first four or five days after birth, contains a type of antibody known as IgA. Some of these antibodies remain on the surface of the infant's gut wall, while others pass into the blood undigested. IgA acts in the gut to prevent the growth of bacteria and viruses and also circulates in the blood. This is **natural passive immunity**.

The features of active and passive immunity are compared in Table 11.2 and Figure 11.15.

Vaccines

A **vaccine** is a preparation containing antigens which is used to stimulate an immune response artificially. It may contain a whole live microorganism, a dead one, a harmless version (known as an attenuated organism), a harmless form of a toxin (known as a toxoid) or a

Immunity	Features				
	Antigen encountered	**Immune response**	**Time before antibodies appear in blood**	**Production of memory cells**	**Protection**
active	yes	yes	1–2 weeks during an immune response	yes	permanent
passive	no	no	immediate	no	temporary

Table 11.2 Features of active and passive immunity.

Active immunity
Immunity developed after contacting pathogens inside the body.

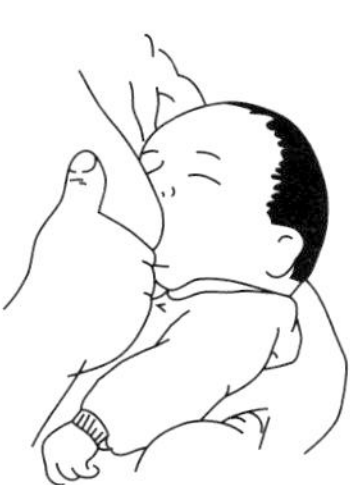

Natural
infection

Artificial
injection of live or attenuated pathogen

Passive immunity
Immunity provided by antibodies or antitoxins provided from outside the body.

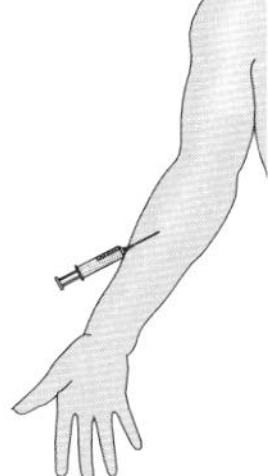

Natural
antibodies from a mother in breast milk or across the placenta

Artificial
injection of antibodies or antitoxin

Figure 11.15 Active and passive immunity.

Natural immunity is immunity gained by being infected (active) or by receiving antibodies from the mother across the placenta or in breast milk (passive).

Artificial immunity is immunity gained either by vaccination (active) or by injecting antibodies (passive).

Vaccination is giving a vaccine containing antigens for a disease, either by injection or by mouth; vaccination confers artificial active immunity.

QUESTIONS

11.13 Explain the difference between artificial active immunisation (vaccination) and artificial passive immunisation.

11.14 **a** Explain the pattern of maternal and infant antibody shown in Figure 11.14.

b Explain the advantages of natural passive immunity for newborn infants.

11.15 Explain the difference between immunity to a disease and resistance to an antibiotic.

preparation of surface antigens. Vaccines are either given by injection into a vein or muscle, or are taken orally (by mouth). Some are produced using techniques of genetic engineering (Chapter 19).

Immunity derived from a natural infection is often extremely good at providing protection, because the immune system has met living organisms which persist inside the body for some time, so the immune system has time to develop an effective response. When possible, vaccination tries to mimic this. Sometimes this works very well, when vaccines contain live microorganisms. The microorganisms reproduce, albeit rather slowly, so that the immune system is continually presented with a large dose of antigens. Less effective are those vaccines that do not mimic an infection because they are made from dead bacteria or viruses that do not replicate inside the body.

Some vaccines are highly effective, and one injection may well give a lifetime's protection. Less effective vaccines need booster injections to stimulate secondary responses that give enhanced protection (Figure 11.8, page 228). It is often a good idea to receive booster injections if you are likely to be exposed to a disease, even though you may have been vaccinated as a child.

Problems with vaccines

Poor response

Some people do not respond at all, or not very well, to vaccinations. This may be because they have a defective immune system and as a result do not develop the necessary B and T cell clones. It may also be because they suffer from malnutrition, particularly protein-energy malnutrition (an inadequate intake of protein), and do not have enough protein to make antibodies or clones of lymphocytes. These people are at a high risk of developing infectious diseases and transmitting them to people who have no immunity.

Live virus and herd immunity

People vaccinated with a live virus may pass it out in their faeces during the primary response and may infect others. This is why it is better to vaccinate a large number of people at the same time to give **herd immunity**, or to ensure that all children are vaccinated within a few months of birth. Herd immunity interrupts transmission in a population, so that those who are susceptible never encounter the infectious agents concerned.

Antigenic variation

In spite of years of research, there are no vaccines for the common cold. The type of rhinovirus that causes most colds has around 100 different strains. It may be impossible to develop a vaccine that protects against all of these.

The influenza virus mutates regularly to give different antigens. When there are only minor changes in the viral antigen, memory cells will still recognise them and start a secondary response. These minor changes are called **antigenic drift**. More serious are major changes in antigen structure – known as **antigenic shift** – when influenza viruses change their antigens considerably and the protective immunity given by vaccination against a previous strain is ineffective against the new one. The World Health Organization (WHO) recommends the type of vaccine to use according to the antigens that are common at the time. The vaccine is changed almost every year.

There are, as yet, no effective vaccines in use against the diseases that are caused by protoctists, such as malaria and sleeping sickness. This is because these pathogens are eukaryotes with many more genes than bacteria and viruses have. They can have many hundreds or even thousands of antigens on their cell surfaces. *Plasmodium*, which causes malaria, passes through three stages in its life cycle while it is in the human host. Each stage has its own specific antigens. This means that effective vaccines would have to contain antigens to all three stages or be specific to the infective stage. The latter would only work if the immune system could give an effective response in the short period of time (a few hours) between the mosquito bite and the infection of liver cells (Figure 10.4, page 203).

Several vaccines have been trialled for malaria. In one trial, a vaccine known as RTS,S reduced the risk of young children being infected with the malaria parasite by about half and reduced the chances of getting the most serious form of the disease by more than a third. The vaccine could become available for use in 2015. *Trypanosoma*, the causative agent of sleeping sickness, has about a thousand different antigens and changes them every four or five days. This makes it impossible for the immune system to respond effectively. After several weeks, the body is completely overwhelmed by the parasite, with fatal consequences.

Antigenic concealment

Some pathogens evade attack by the immune system by living inside cells. When *Plasmodium* enters liver cells or red blood cells, it is protected against antibodies in the plasma. Some parasitic worms conceal themselves by covering their bodies in host proteins, so they remain invisible to the immune system. Other pathogens suppress the immune system by parasitising cells such as macrophages and T cells. It is very difficult to develop effective vaccines against these pathogens, because there is such a short period of time for an immune response to occur before the pathogen 'hides'.

Another example is *Vibrio cholerae* (the causative agent of cholera), which remains in the intestine where it is beyond the reach of many antibodies. There are oral vaccines (taken by mouth) which provide limited protection against cholera.

QUESTIONS

11.16 Explain why malnourished children give very weak responses to vaccines.

11.17 Explain why humans cannot produce an effective immune response to an infection by *Trypanosoma*.

11.18 Name one pathogen that parasitises:

a macrophages

b helper T cells.

The eradication of smallpox

Smallpox was an acute, highly infectious disease caused by the variola virus and transmitted by direct contact. It was a terrible disease. Red spots containing a transparent fluid would appear all over the body (Figure 11.16). These then filled with thick pus. Eyelids became swollen and could become 'glued' together. Sufferers often had to be prevented from tearing at their flesh. Many people who recovered were permanently blind and disfigured by scabs left when the pustules dried out. Smallpox killed 12–30% of its victims.

The WHO started an eradication programme in 1956; in 1967 it stated its intention to rid the world of the disease within ten years. There were two main aspects of the programme: vaccination and surveillance. Successful attempts were made across the world to vaccinate in excess of 80% of populations at risk of the disease. When a case of smallpox was reported, everyone in the household and the 30 surrounding households, as well as other relatives and possible contacts in the area, was vaccinated. This **ring vaccination** protected everyone who could possibly have come into contact with a person with the disease, reduced the chances of transmission and contained the disease. The last places with cases of smallpox were in East Africa, Afghanistan and the Indian subcontinent. Eradication was most difficult in Ethiopia and Somalia, where many people lived in remote districts well away from main roads which were no more than dirt tracks. In the late 1970s, the two countries went to war and, even though large parts of Ethiopia were overrun by the Somalis, the eradication programme continued. The last case of smallpox was reported in Somalia in 1977. The WHO finally declared the world free of smallpox in 1980.

How the eradication programme succeeded

The eradication programme was successful for a number of reasons.

- The variola virus was stable; it did not mutate and change its surface antigens. This meant that the same vaccine could be used everywhere in the world throughout the campaign. It was therefore cheap to produce.
- The vaccine was made from a harmless strain of a similar virus (vaccinia) and was effective because it was a 'live' vaccine.
- The vaccine was freeze-dried and could be kept at high temperatures for as long as six months. This made it suitable for use in the tropics.
- Infected people were easy to identify.
- The vaccine was easy to administer and was even more effective after the development of a stainless steel, re-usable needle for its delivery. This 'bifurcated needle' had two prongs, which were used to push the vaccine into the skin.
- The smallpox virus did not linger in the body after an infection to become active later and form a reservoir of infection.
- The virus did not infect animals, which made it easier to break the transmission cycle.
- Many 16- to 17-year-olds became enthusiastic vaccinators and suppliers of information about cases; this was especially valuable in remote areas.

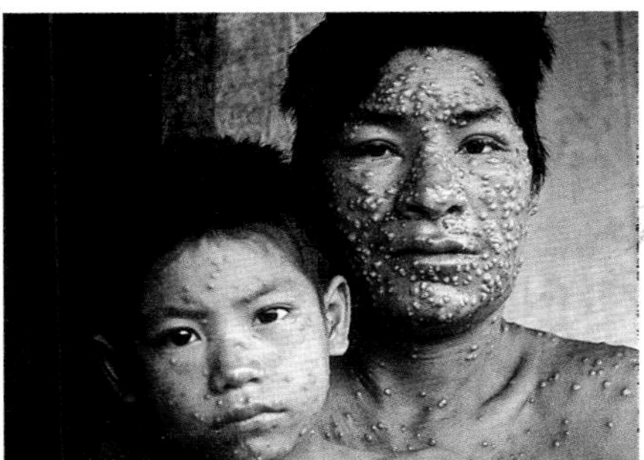

Figure 11.16 Among the last cases of smallpox, this parent and child of the Kampa people of the Amazon region, South America, show the characteristic pustules of smallpox.

The eradication of smallpox is a medical success story. It has been more difficult to repeat this success with other infectious diseases. This is partly because of the more unstable political situation since the late 1970s and 1980s, particularly in Africa, Latin America and parts of Asia such as Pakistan. Public health facilities are difficult to organise in developing countries with poor infrastructure, few trained personnel and limited financial resources. They are almost impossible to maintain during periods of civil unrest or during a war.

Preventing measles

Measles is a preventable disease and one that could be eradicated by a worldwide surveillance and vaccination programme. However, a programme of one-dose-vaccination has not eliminated the disease in any country, despite high coverage of the population. This is explained by the poor response to the vaccine shown by some children who need several boosters to develop full immunity. In large cities with high birth rates and shifting populations, it can be difficult to give boosters, follow up cases of measles and trace contacts. Migrants and refugees can form reservoirs of infection, experiencing epidemics within their communities and spreading the disease to surrounding populations. This makes measles a very difficult disease to eradicate, even with high vaccination coverage.

Measles is highly infectious and it is estimated that herd immunity of 93–95% is required to prevent transmission in a population. As the currently available vaccine has a success rate of 95%, this means that the whole population needs to be vaccinated and infants must be vaccinated within about eight months of birth.

Many countries achieve up to 80% or more coverage with measles vaccination (Figures 11.17 and 11.18).

The Americas have been free of endemic measles since 2002 with any cases being the result of someone bringing in the disease from somewhere else in the world. With an estimated coverage of only about 75% in Africa, India and South-East Asia, it is likely that the disease will still persist for many years to come.

Figure 11.17 The success of immunisation programmes relies on people, such as these Red Cross workers in Nairobi, Kenya, ensuring that all families know when and where vaccinations are available.

QUESTION

11.19 a Distinguish between herd immunity and ring immunity.

b Explain the biological reasons for the difficulty in developing successful vaccines for cholera, malaria and TB.

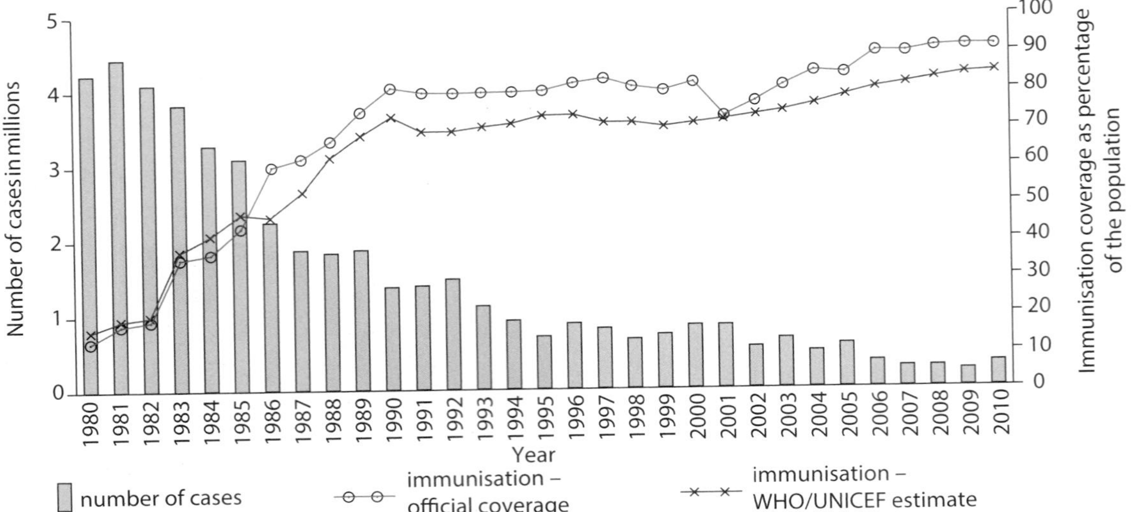

Figure 11.18 The global measles immunisation programme coordinated by the WHO, which has caused the number of cases reported each year to decrease significantly.

Autoimmune diseases – a case of mistaken identity

Not only does the body mount immune responses against pathogens and harmful substances from outside the body, but it can also attack itself leading, in some cases, to severe symptoms. Diseases of this type are **autoimmune diseases**. They occur because the immune system attacks one or more self-antigens, usually proteins. During the maturation of T cells in the thymus, many millions of cells are destroyed because they have T cell receptors that are complementary to self-antigens. However, some of these T cells evade destruction and under certain circumstances are activated to stimulate an immune response against the body's own proteins. This starts an attack often involving antibodies and killer T cells against parts of the body. In some autoimmune diseases, the attack is localised and directed against one organ; in others it is directed against the whole body (Table 11.3).

Myasthenia gravis (MG), which means grave (serious) muscle weakness, is an autoimmune disease that targets the neuromuscular junctions between motor neurones (nerve cells) and skeletal muscle cells. Motor neurones transmit nerve impulses to muscle cells. Where the motor neurones terminate they release acetylcholine, a cell signalling molecule, into the small gaps between neurones and muscle fibres. Acetylcholine binds with receptor proteins on cell surface membranes of the muscle fibres (Figure 11.19). The interaction of acetylcholine and its receptor stimulates channels to open allowing sodium ions to move through the membranes. The influx of sodium ions begins a series of events that result in muscle contraction (pages 348–349, Chapter 15).

Autoimmune disease	Area of body affected	Main effects of the disease
myasthenia gravis	neuromuscular junctions	progressive muscle weakness
multiple sclerosis	central nervous system	progressive paralysis
rheumatoid arthritis	joints	progressive destruction of the joints
type 1 diabetes	islets of Langerhans – endocrine tissue in the pancreas	destruction of cells that secrete insulin
systemic lupus erythromatosus	skin, kidneys and joints	progressive deformity

Table 11.3 Five autoimmune diseases.

a Normal

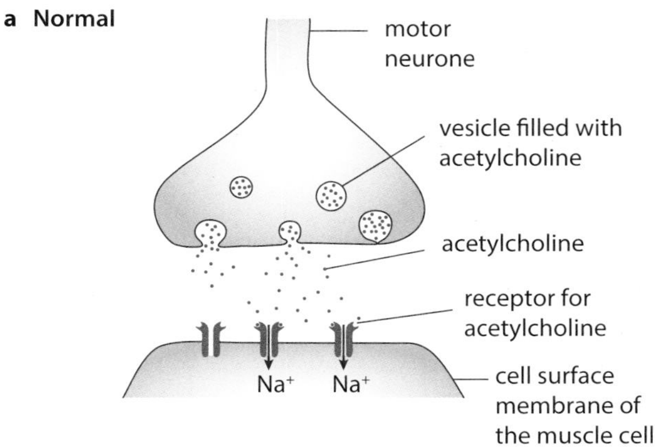

b Myasthenia gravis

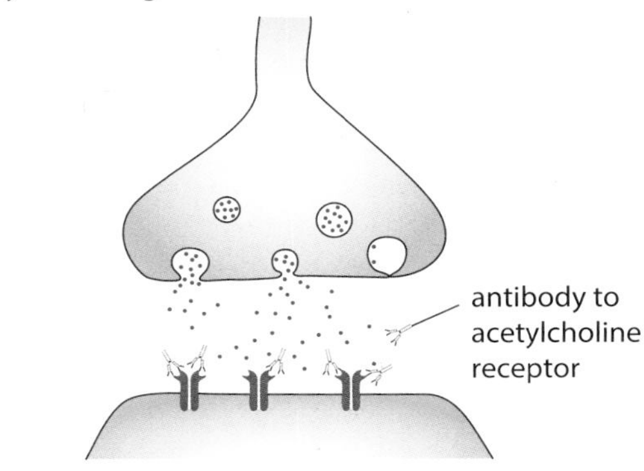

Figure 11.19 **a** Acetylcholine is released by motor neurones. It stimulates muscle cells to contract by combining with receptor proteins, which allow sodium ions into the muscle cell. **b** In myasthenia gravis, antibodies are secreted that block the receptor proteins and then cause their destruction so they do not allow the movement of sodium ions.

People with MG have helper T cells that are specific for these cell surface receptors for acetylcholine. Under certain conditions, these cells stimulate a clone of B cells to differentiate into plasma cells and secrete antibodies that bind to the receptor so blocking the transmission of impulses from motor neurones. Muscle fibres absorb the receptor–antibody complexes and destroy them. Without acetylcholine receptors, muscle cells do not receive any stimulation, and because the muscles are not stimulated muscle tissue starts to break down.

The symptoms of MG vary greatly between people with this autoimmune disease. The typical symptom is muscle weakness that gets worse with activity and improves with rest. Affected muscles become fatigued very easily (Figure 11.20). This means that symptoms are usually worse at the end of the day and after exercise.

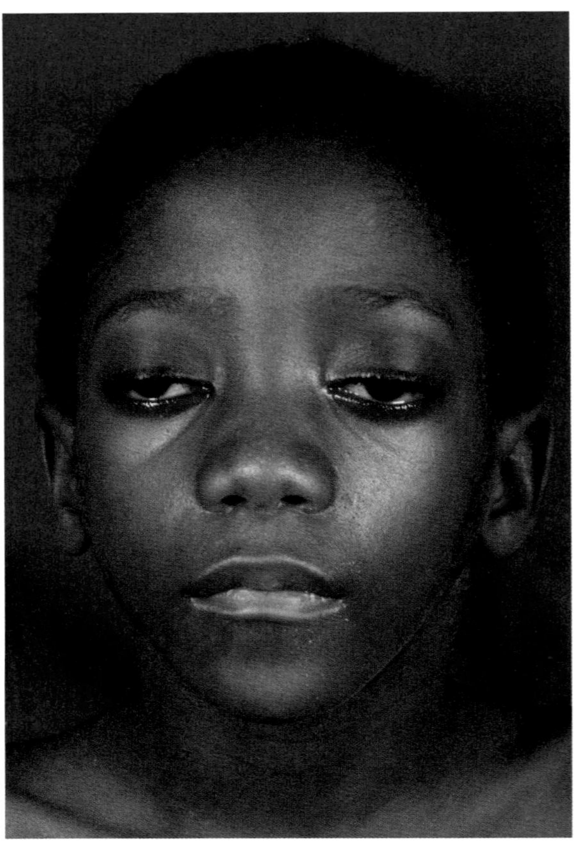

Figure 11.20 Drooping eyelids are a common early symptom of myasthenia gravis. These muscles are in constant use and tire quickly. If after two years, this is the only symptom, then MG is not likely to progress to other muscles.

There are effective treatments for MG, such as a drug that inhibits the enzyme in synapses that breaks down acetylcholine. The effect of this drug is to increase the concentration of acetylcholine in synapses so its action in stimulating muscle fibres to contract lasts for longer. Surgical removal of the thymus gland is also an effective treatment, possibly because it is the site of maturation of the helper T cells that stimulate B cells to produce antibodies to the acetylcholine receptors.

Other autoimmune diseases include multiple sclerosis, rheumatoid arthritis and one form of diabetes. Multiple sclerosis develops when nerve cells in the brain and spinal cord lose the insulating myelin sheaths that surround them. This can happen anywhere in the central nervous system (CNS) and appears to be quite random. The degenerating areas, known as plaques, can be detected using MRI scans. With the loss of the protective myelin, the neurones stop conducting impulses and there is a loss of the functions controlled by the areas of the CNS concerned. Symptoms are muscle weakness, loss of sensory input from the skin and other areas, poor vision and mental problems.

Rheumatoid arthritis is another long-term destructive process, this time occurring in the joints. It starts with the finger and hand joints and then spreads to the shoulders and other joints. Tendons become inflamed and there is constant muscle spasm and pain. People with rheumatoid arthritis find it hard to keep mobile.

Type 1 insulin-dependent diabetes is thought to be caused partly by a virus infection that makes the cells that secrete insulin in the pancreas unrecognisable as self. Killer T cells enter the islets of Langerhans and destroy the cells that produce insulin (Chapter 14).

The causes of autoimmune diseases are not well known and are the subject of much research. Although MG often runs in families it is not an inherited condition. However, people with certain alleles of genes involved in cell recognition are at a higher risk of developing MG than those without these alleles.

Genetic factors are involved as it has been shown that susceptibility to the diseases is inherited. However, environmental factors are also very important as the increase in prevalence of these diseases in the developed world over the last 50 years suggests. Also people who have moved from an area where these diseases are rare, such as Japan, to places where they are more common, such as the United States, have an increased chance of developing one of these diseases.

QUESTION

11.20 With reference to myasthenia gravis, explain the meaning of the term autoimmune disease.

Monoclonal antibodies

Figure 11.6 (page 227) shows that, during an immune response, B cells become plasma cells that secrete antibodies in response to the presence of a non-self antigen. As you have seen, antibodies bind to pathogens and kill them or mark them for destruction by phagocytes. Antibodies have high degrees of specificity. This specificity of antibodies has made them very desirable for use in the diagnosis and treatment of diseases.

For some time, though, no-one could see how to manufacture antibodies on a large scale. This requires a very large number of cells of a particular B cell clone, all secreting identical or **monoclonal antibodies** (Mabs). There is a major problem in achieving this. B cells that divide by mitosis do not produce antibodies, and plasma cells that secrete antibodies do not divide.

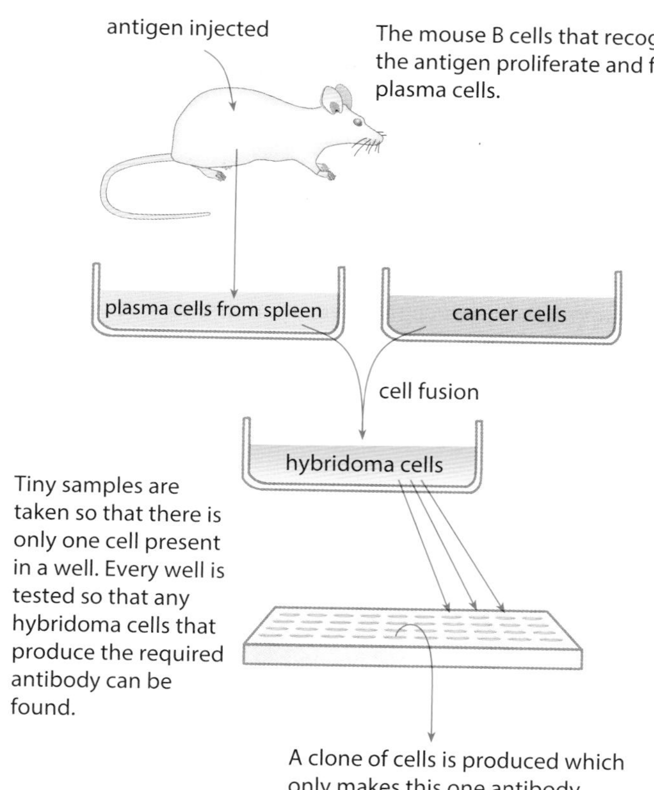

Figure 11.21 How monoclonal antibodies are produced. Monoclonal antibodies have many different uses in research and in medicine, both in diagnosis and in treatment.

The breakthrough came in the 1970s using the technique of cell fusion. A small number of plasma cells producing a particular antibody were fused with cancer cells. Cancer cells, unlike other cells, go on dividing indefinitely. The cell produced by this fusion of a plasma cell and a cancer cell is called a **hybridoma** (Figure 11.21). The hybridoma cells divide by mitosis **and** secrete antibodies.

Using monoclonal antibodies in diagnosis

Mabs have many different uses in medicine and new applications for them are the subject of research. They are used both for diagnosis and treatment. For example, Mabs can be used to locate the position of blood clots in the body of a person thought to have a deep vein thrombosis. The antibodies are produced by injecting a mouse with human fibrin, the main protein found in blood clots. The mouse makes many plasma cells that secrete the antibody against fibrin; these plasma cells are collected from its spleen. The plasma cells are fused with cancer cells to form hybridomas that secrete the antifibrin antibody. A radioactive chemical that produces gamma radiation is attached to each antibody molecule to make radioactively labelled antibodies.

The labelled antibodies are then introduced into the patient's blood. As the Mabs are carried around the body in the bloodstream, they bind to any fibrin molecules with which they come into contact. The radioactivity emitted by these labelled antibodies is used to detect where they are in the body. A gamma-ray camera is used to detect the exact position of the antibodies in the person's body. The position of the labelled Mabs indicates the position of any blood clots.

There are now many Mabs available to diagnose hundreds of different medical conditions. For example, they can be used to locate cancer cells, which have proteins in their cell surface membranes that differ from the protiens on normal body cells and can therefore be detected by antibodies (Figure 11.22). They can also be used to identify the exact strain of a virus or bacterium that is causing an infection; this speeds up the choice of the most appropriate treatment for the patient. Mabs are also used routinely in blood typing before transfusion, and tissue typing before transplants.

Using monoclonal antibodies in treatment

When monoclonal antibodies are used in diagnosis, they are normally administered on just one occasion. Monoclonal antibodies used as a treatment need to be administered more than once, and that presents problems.

Figure 11.22 These scientists are checking tissue samples to see whether the cells are cancerous. They do this with monoclonal antibodies specific to cancer cells.

The antibodies are produced by mice, rabbits or other laboratory animals. When introduced into humans, they trigger an immune response because they are foreign (non-self) and act as antigens. This problem has now been largely overcome by humanising Mabs in two ways:

- altering the genes that code for the heavy and light polypeptide chains of the antibodies so that they code for human sequences of amino acids, rather than mouse or rabbit sequences (Figure 11.10, page 229)
- changing the type and position of the sugar groups that are attached to the heavy chains to the arrangement found in human antibodies.

Some monoclonal antibodies have had significant success in treatments that involve modifying immune responses. These monoclonals are trastuzumab, ipilimumab, infliximab and rituximab.

Trastuzumab (also known as Herceptin™) is used in the treatment of some breast cancers. It is a humanised mouse monoclonal antibody. It binds to a receptor protein that is produced in abnormal quantities in the cell surface membranes of some breast cancers.

The receptor protein is not unique to cancer cells, but cells with between 10 and 100 times the usual number of these receptor molecules in their cell surface membranes can only be cancer cells. Trastuzumab binds to these cells, and this marks them out for destruction by the immune system.

Ipilimumab is a more recent cancer therapy for melanoma, a type of skin cancer. It also works by activating the immune system, but in a different way to trastuzumab. Ipilimumab binds to a protein produced by T cells (page 230), the role of which is to reduce the immune response. By blocking the action of the protein, an immune response can be maintained against the cancer cells.

Infliximab is used to treat rheumatoid arthritis, which is an autoimmune disease (Table 11.3, page 237). This Mab binds to a protein secreted by T cells that causes damage to the cartilage in joints and blocks its action. Most of those treated for rheumatoid arthritis receive this therapy at monthly or two-monthly intervals, so it is important that the monoclonal antibody is humanised and does not itself trigger an immune response.

Rituximab is used to control B-lymphocytes. This Mab binds to a cell surface receptor protein found on the surface of B cells, but not on the surface of plasma cells. Following binding to B cells, rituximab causes a variety of changes that lead to the death of the cell. Rituximab is used to treat diseases in which there is an overproduction or inappropriate production of B cells, such as leukaemias and autoimmune diseases. Reducing the numbers of B cells appears to reduce the severity of diseases, such as multiple sclerosis and rheumatoid arthritis. Rituximab has also been trialled successfully on myasthenia gravis.

QUESTION

11.21 Mabs anti-A and anti-B are used in blood typing to distinguish between the A, B, AB and O blood groups. The Mab anti-D is used to detect the Rhesus blood group antigen. This antigen, known as D, is present on the red blood cells of people who have Rhesus positive blood, but absent from those who are Rhesus negative. If the red blood cells agglutinate and clump together when tested with a Mab, then the antigen is present. If there is no agglutination then the antigen is absent. For example, agglutination with anti-A indicates the presence of antigen A on the red blood cells; if no agglutination occurs, antigen A is absent.

a Make a table to show the expected results when blood of the following types is tested separately with anti-A, anti-B and anti-D:
A positive; B positive; AB negative; AB positive; O positive.

b Explain why it is important to carry out blood tests on people who need blood transfusions or who are about to receive a transplanted organ, such as a kidney.

11.22 a What are the advantages of using monoclonal antibodies in diagnosis?

b Suggest how treating cancers using monoclonal antibodies could cause fewer side-effects than treating them using conventional anti-cancer drugs or radiotherapy.

Summary

- Phagocytes and lymphocytes are the cells of the immune system. Phagocytes originate in the bone marrow and are produced there throughout life. There are two types of phagocyte. Neutrophils circulate in the blood and enter infected tissues; macrophages remain inside tissues. Neutrophils and macrophages destroy bacteria and viruses by phagocytosis.
- Antigens are 'foreign' (non-self) molecules that stimulate the immune system.
- Lymphocytes also originate in bone marrow. There are two types: B-lymphocytes (B cells) and T-lymphocytes (T cells). B cells and T cells gain glycoprotein receptors that are specific to each cell as they mature. Each lymphocyte divides to form a small clone of cells that spreads throughout the body in the blood and in the lymphoid tissue (e.g. lymph nodes and spleen). B cells mature in bone marrow. T cells mature in the thymus gland. During maturation, many T cells are destroyed, as they express receptors that interact with self-antigens. If left to circulate in the body, they would destroy cells and tissues. The T cells that are not destroyed recognise non-self antigens, such as those on the surfaces of pathogens.
- During an immune response, those B and T cells that have receptors specific to the antigen are activated. When B cells are activated, they form plasma cells which secrete antibodies. T cells do not secrete antibodies; their surface receptors are similar to antibodies and identify antigens. T cells develop into either helper T cells or killer T cells (cytotoxic T cells). Helper T cells secrete cytokines that control the immune system, activating B cells and killer T cells, which kill infected host cells.
- During an immune response, memory cells are formed which retain the ability to divide rapidly and develop into active B or T cells on a second exposure to the same antigen (immunological memory).
- Antibodies are globular glycoproteins. They all have one or more pairs of identical heavy polypeptides and of identical light polypeptides. Each type of antibody interacts with one antigen via the specific shape of its variable region. Each molecule of the simplest antibody (IgG) can bind to two antigen molecules. Larger antibodies (IgM and IgA) have more than two antigen-binding sites. Antibodies agglutinate bacteria, prevent viruses infecting cells, coat bacteria and viruses to aid phagocytosis, act with plasma proteins to burst bacteria, and neutralise toxins.
- Blood cell counts give useful information in diagnosis and in monitoring the success of treatments for certain medical conditions. During infectious diseases, the white blood cell count often increases as neutrophils are released from stores in the bone marrow (neutrophils travel in the blood to the site(s) of infection to destroy the pathogens by phagocytosis).
- Leukaemias are cancers of the stem cells in the bone marrow that give rise to blood cells. In myeloid leukaemias, the stem cells responsible for producing neutrophils divide uncontrollably and the number of immature cells increases. In lymphoblastic leukaemias, the cancerous cells are those that give rise to lymphocytes.
- Active immunity is the production of antibodies and active T cells during a primary immune response to an antigen acquired either naturally by infection or artificially by vaccination. This gives long-term immunity. Passive immunity is the introduction of antibodies either naturally across the placenta or in breast milk, or artificially by injection. This gives temporary immunity.
- Vaccination confers artificial active immunity by introducing a small quantity of an antigen by injection or by mouth. A vaccine may be a whole living organism, a dead one, a harmless version of a toxin (toxoid) or a preparation of antigens. It is difficult to develop successful vaccines against diseases caused by organisms that have many different strains, express different antigens during their life cycle within humans (antigenic variation), or infect parts of the body beyond the reach of antibodies (antigenic concealment).
- Smallpox was eradicated by a programme of surveillance, contact tracing and 'ring' vaccination, using a 'live' vaccine against the only strain of the smallpox virus.
- Measles is a common cause of death among infants in poor communities. It is difficult to eradicate because a wide coverage of vaccination has not been achieved and malnourished children do not respond well to just one dose of the vaccine. Vaccines for TB and cholera are not very effective. Vaccines for malaria are being trialled and may become available for widespread use. Antibodies are not very effective against *V. cholerae* and its toxin, as the bacterium and its toxin remain in the alimentary canal. Antibodies cannot cross cell surface membranes to reach *M. tuberculosis*, which infects host cells, especially macrophages.

- *Plasmodium*, the malaria parasite, spends most of its time within human cells and is likewise out of reach of antibodies. It is also a eukaryote with a much larger genome and the ability to produce many different antigens.
- Autoimmune diseases occur when the immune system mistakenly identifies self-antigens as foreign and mounts an immune response against them. Myasthenia gravis is an example of an autoimmune disease in which antibodies are produced against receptors on muscle fibres for acetylcholine which is released by the ends of motor neurones to stimulate muscle contraction. The antibodies block and trigger the destruction of the receptors so that muscle fibres do not respond to stimulation by motor neurones. Muscle weakness is a main symptom.
- Monoclonal antibodies are antibodies that are all identical to each other. Monoclonal antibodies are produced by fusing a plasma cell with a cancer cell to produce a hybridoma, which divides repeatedly to form many genetically identical cells that all produce the same antibody. Monoclonal antibodies are used in diagnosis – for example, in locating blood clots in veins, and in the treatment of diseases such as breast cancer.

End-of-chapter questions

1 A student made drawings of four blood cells as shown in the figure.

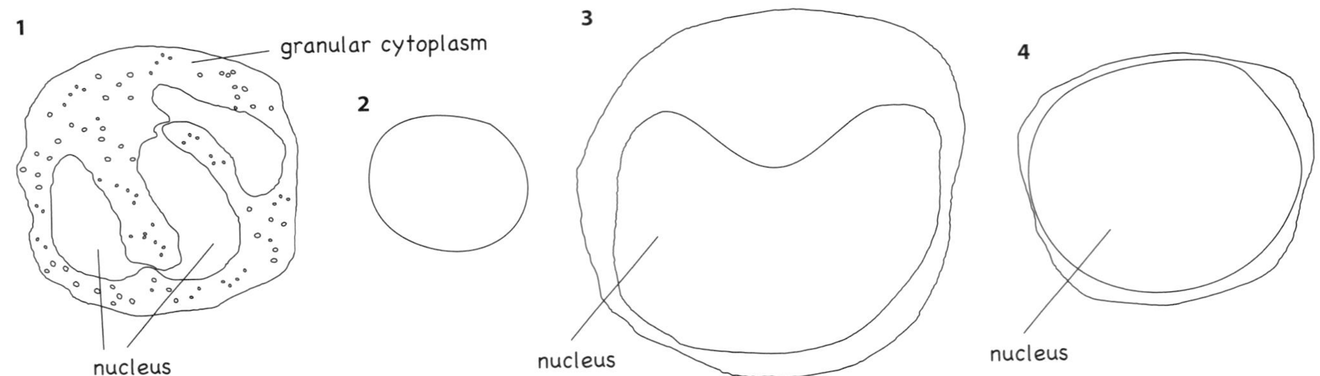

The correct identification of the cells is:

	1	2	3	4
A	lymphocyte	monocyte	red blood cell	neutrophil
B	monocyte	lymphocyte	neutrophil	red blood cell
C	neutrophil	red blood cell	monocyte	lymphocyte
D	red blood cell	neutrophil	lymphocyte	monocyte

[1]

2 The following occur during the response to infection.

1 attachment of bacteria to cell surface membrane of phagocyte
2 movement of phagocyte to site of infection by bacteria
3 formation of a phagocytic vacuole
4 fusion of lysosomes to the phagocytic vacuole
5 infolding of cell surface membrane
6 release of enzymes into the phagocytic vacuole

In which order do these events occur?

A 1, 2, 3, 4, 6, 5
B 1, 2, 3, 5, 4, 6
C 2, 1, 3, 6, 5, 4
D 2, 1, 5, 3, 4, 6

[1]

3 Which of the following explains why antibody molecules have quaternary structure?

- **A** antibodies have a variable region
- **B** antibodies have complex 3D shapes
- **C** antibodies have four polypeptides
- **D** antibodies have more than one polypeptide [1]

4 Which type of immunity is provided by vaccination?

- **A** artificial active
- **B** artificial passive
- **C** natural active
- **D** natural passive [1]

5 What describes the cells producing monoclonal antibodies?

- **A** a clone of B-lymphocytes, all of which secrete different antibodies
- **B** a clone of hybridoma cells, all of which secrete identical proteins
- **C** a clone of T-lymphocytes, all of which secrete the same cytokine
- **D** a clone of T-lymphocytes, all of which secrete identical antibodies [1]

6 Tetanus is a bacterial disease that may be acquired during accidents in which a wound is exposed to the soil.

B-lymphocytes originate from stem cells, mature and circulate around the body. Following infection by tetanus bacteria, some B-lymphocytes will become activated as shown in the figure.

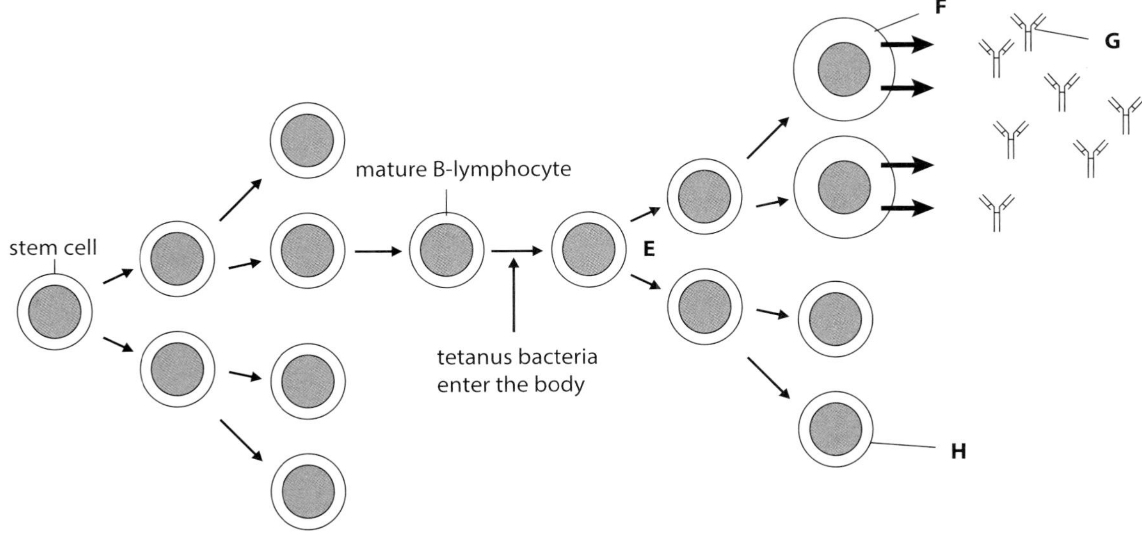

- **a** With reference to the figure, name:
  - **i** the place where the stem cells divide to form B-lymphocytes [1]
  - **ii** the type of division that occurs at **E** [1]
  - **iii** the activated B-lymphocyte, **F** [1]
  - **iv** the molecule **G**. [1]
- **b** Use the information in the figure to explain the differences between the following pairs of terms:
  - **i** antigen and antibody [3]
  - **ii** self and non-self. [2]
- **c** Explain how cell **H** is responsible for long-term immunity to tetanus. [3]

[Total: 12]

7 Phagocytes and lymphocytes are both present in samples of blood.

a Describe how the structure of a phagocyte differs from the structure of a lymphocyte. [3]

T-lymphocytes are involved in immune responses to pathogens that invade the body. Immune responses involve the following:

- antigen presentation
- clonal selection
- clonal expansion.

Certain groups of T-lymphocytes are activated when the body is infected with the measles virus.

b Using the information above, describe what happens to T-lymphocytes during an immune response to measles. [6]

c State how the response of B-lymphocytes during an immune response is different to the response of T-lymphocytes. [2]

[Total: 11]

8 Measles is a common viral infection. Babies gain passive immunity to measles.

a Explain:

i the term passive immunity [2]

ii how babies gain passive immunity. [2]

A vaccine for measles has been available since the 1960s. Global vaccination programmes include providing vaccination for measles, but it is important that the vaccine is not given to babies too early.

b Explain why:

i the vaccine for measles should not be given too early [3]

ii measles has not been eradicated, even though a vaccine has existed since the 1960s. [3]

c Smallpox was an infectious disease that was finally eradicated in the late 1970s. Explain how vaccination was used in the eradication of smallpox. [6]

[Total: 16]

9 The diagram shows an antibody molecule.

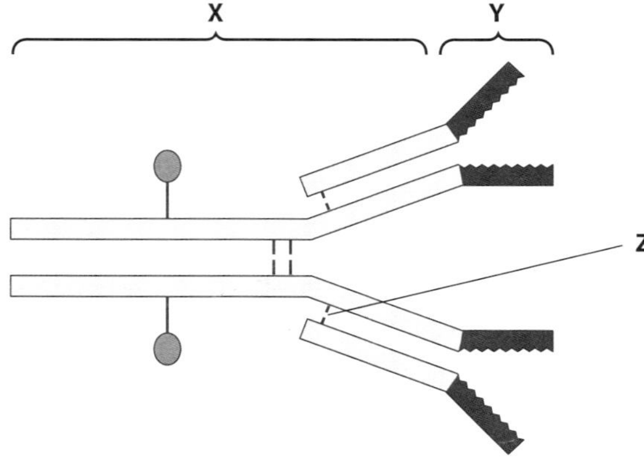

a Describe briefly how antibody molecules are produced and secreted. [4]

b Name:

i the regions **X** and **Y** [2]

ii the bond labelled **Z**. [1]

c Explain how the structure of an antibody is related to its function. [4]

[Total: 11]

10 a Rearrange the following statements to produce a flow diagram to show how monoclonal antibodies are produced.

1 hybridoma cells cultured
2 mouse injected with antigen
3 mouse plasma cells fused with cancer cells
4 hybridoma cells making appropriate antibody cloned
5 mouse B-lymphocytes that recognise the antigen become plasma cells [5]

b A monoclonal antibody has been produced that binds to a glycoprotein called tumour-associated antigen (TAG). TAG is found in ovarian cancer and colon cancer cells.

The antibody is labelled with the radioactive isotope indium-111, (^{111}In). This isotope emits low-energy gamma rays that can pass through soft tissues of the body. The half-life of indium-111 is 2.8 days.

i Explain how this monoclonal antibody can be used in the diagnosis of cancer. [3]

ii Suggest why indium-111 is a suitable radioactive label for this diagnostic antibody. [3]

iii Suggest how the antibody could be modified in order to treat a cancer. [2]

[Total: 13]

11 The table below shows the blood cell counts for four young female patients at a hospital. Each patient has a different medical condition.

Patient	Medical condition	Red blood cell count / millions cells mm^{-3}	White blood cell counts / cells mm^{-3}				
			All white blood cells	Neutrophils	Lymphocytes		
					All	B cells	T cells
A	an unknown bacterial infection	5.7	11 000	7500	3500	2000	1500
B	HIV/AIDS	3.6	3885	2085	1800	130	1670 (of which 150 are helper T cells)
C	acute myeloid leukaemia	1.1	47 850	4800 (mature) 40 800 (immature)	2250	350	1900
D	acute lymphoblastic leukaemia	3.2	156 700	450	1250 (mature) 155 000 (immature)	400	850

Use the data in the table above and Table 11.1 on page 231 to answer the following questions.

a Make a table to show how the blood cell counts of each of the four female patients differ from the typical values and the ranges given in Table 11.1. [3]

b Explain the differences that you have described in **a**. [8]

c Patient A had a blood test before she fell ill. The red blood cell count was 4.6×10^6 mm^{-3} and the white blood cell count was 7000 cells mm^{-3}.

i Calculate the changes in cell counts between this test result and the result shown in the table above. [2]

ii Use your results for **i** to calculate the percentage change in the numbers of red and white blood cells in 1 mm^3 of blood. [2]

d Blood cell counting is an automated process and results can be available very quickly. Suggest why the results for a single sample of blood taken from a patient on admission to hospital should be interpreted with care. [3]

[Total: 18]

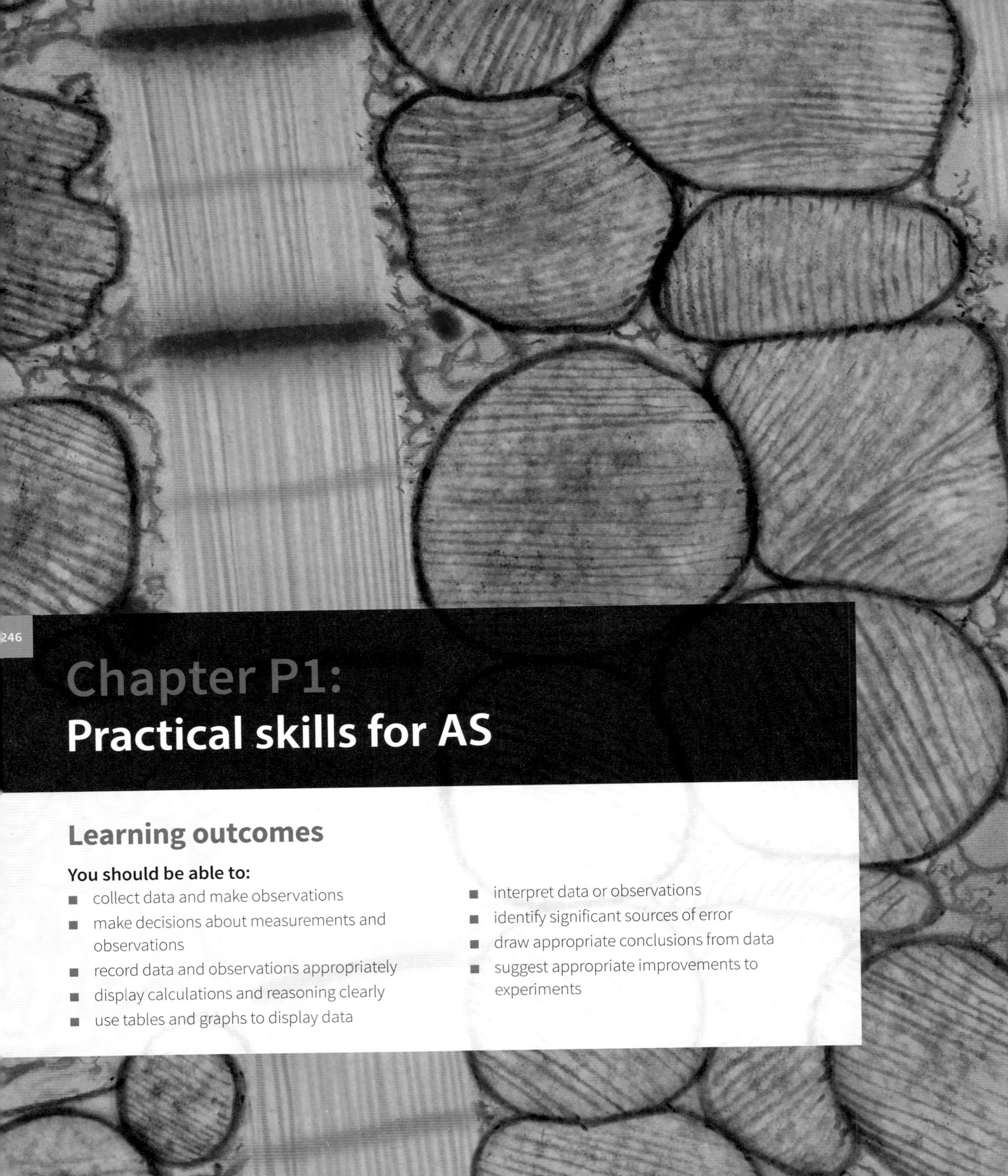

Chapter P1: Practical skills for AS

Learning outcomes

You should be able to:

- collect data and make observations
- make decisions about measurements and observations
- record data and observations appropriately
- display calculations and reasoning clearly
- use tables and graphs to display data
- interpret data or observations
- identify significant sources of error
- draw appropriate conclusions from data
- suggest appropriate improvements to experiments

The practical skills that you will develop during your AS level course are tested by a practical examination. In this examination, you will work in a laboratory, answering questions on an examination paper. The questions will test a range of skills, which can be classified into three groups:

- manipulation, measurement and observation
- presentation of data and observations
- analysis, conclusions and evaluations.

During your AS level course, each time you carry out an experiment, you will use a variety of these skills. In this chapter, we will look at the different components of the skills in detail, and consider what you must be able to do to work to the best of your abilities.

The practical examination usually includes two questions, sometimes three. One of these questions will probably be a 'wet practical' – an experiment that will involve you in manipulating apparatus, perhaps making up and measuring out solutions, making measurements and observations, recording them and drawing conclusions from a set of results. One of the questions will involve observing a biological structure, perhaps using a microscope, and recording your observations in the form of a diagram.

Experiments

Many of the experiments that you will do during your course involve investigating how one thing affects another. For example:

- investigating how enzyme concentration affects the rate of activity of rennin
- investigating how temperature affects the rate of activity of catalase
- investigating how surface area affects the rate of diffusion
- investigating how the concentration of a solution affects the percentage of onion cells that become plasmolysed.

We will concentrate on the first of these experiments – the effect of enzyme concentration on the rate of activity of rennin – to illustrate how you should approach an experiment, and how you should answer questions about it.

Rennin is an enzyme that clots milk. It is found in the stomachs of young mammals, which are fed on milk. Rennin, also known as chymosin, is used commercially in cheese-making. Its substrate is a protein called casein. In fresh milk, the casein molecules are dispersed in the milk as little **micelles** (groups of molecules organised rather like a cell membrane), which spread evenly through the milk to form a homogeneous emulsion. Rennin splits the casein molecules into smaller molecules, which breaks up the micelles and causes the protein to clump together into small lumps, a process called clotting (Figure P1.1). These lumps separate out from the liquid milk, producing the curd that can be made into cheese.

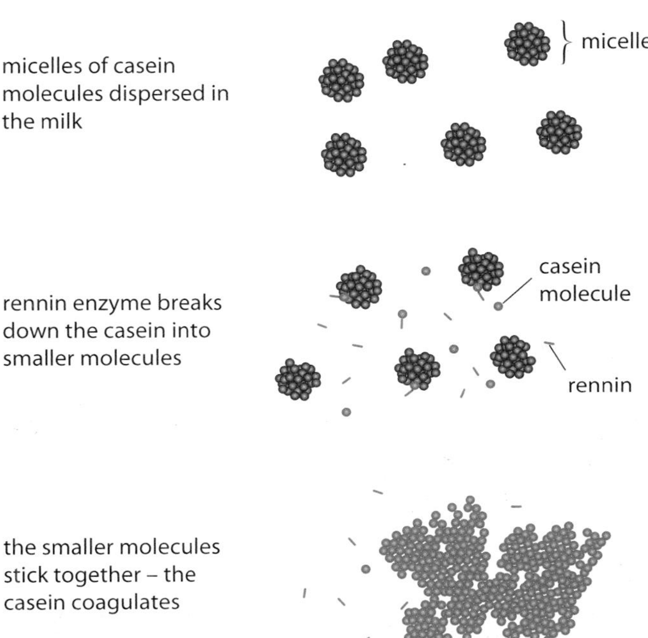

Figure P1.1 The effect of rennin on milk.

Variables and making measurements

In this experiment, you would be investigating the effect of the concentration of rennin on the rate at which it causes milk to clot.

The concentration of rennin is the **independent variable**. This is the factor whose values you decide on, and which you change.

The rate at which the rennin causes the milk to clot is the **dependent variable**. This is the variable which is not under your control; you do not know what its values will be until you collect your results.

In an experiment such as this, it is important that all other variables that might affect the results are kept constant. These are sometimes called **standardised variables** or **controlled variables**. In this experiment, important standardised variables would include:

- temperature
- the type of milk used
- pH.

If you allowed any of these to change during the experiment, then they could affect the results and would make it difficult – if not impossible – to know what effect the enzyme concentration was having on the rate of reaction.

It is important, when doing any experiment, to decide which variables you should keep constant – you should know which are the important ones to keep constant, and which do not really matter. In this case, anything that could affect the rate of enzyme activity – other than the independent variable, enzyme concentration – must be kept constant. Other variables, such as the amount of light, the time of day, or the kind of glassware you use, are unlikely to have any significant effects, so you do not need to worry about them.

Changing the independent variable

You may be asked to decide what values of the independent variable to use in your experiment. You will need to make decisions about the range and the intervals.

The **range** of the independent variable is the spread of values from lowest to highest. In this case, you might use concentrations of rennin ranging from 0 to 1%. If you are asked to do this in an examination, you will usually be given some clues that will help you to decide the range. For example, if you are given a solution with a concentration of 1% to work with, then that will be your highest concentration, because you cannot make a more concentrated solution from it, only more dilute ones.

The **interval** is the 'gap' between the values that you choose within the range. In this case, you could use concentrations of rennin of 0, 0.2, 0.4, 0.6, 0.8 and 1%. The interval would then be 0.2%. Another possibility would be to use a series of values that are each one tenth of each other – 0.0001, 0.001, 0.01, 0.1 and 1.0. In either case, you can produce this range of concentrations by diluting the original solution. Figure P1.2 explains how to do this.

If you are planning to display your results as a graph, then you should have at least five values for the independent variable. You cannot really see any trends or patterns in a line graph unless you have at least five plotted points.

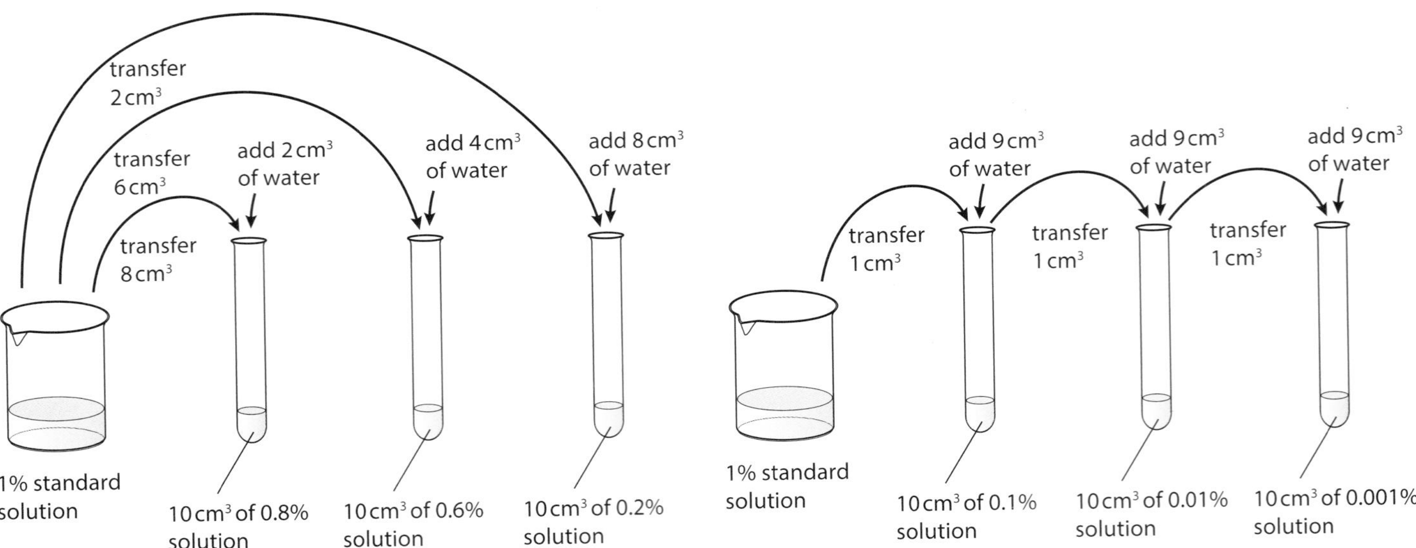

Figure P1.2 Producing a range of concentrations from a standard solution.

Measuring the dependent variable

You may be told exactly how to measure the dependent variable. Sometimes, however, you may have to decide the best way to do this.

In an enzyme experiment such as the rennin one, there are three possible methods for taking measurements. (You can remind yourself about ways of measuring reaction rate by looking back at pages 57–58 in Chapter 3.)

- You could determine the initial rate of reaction – taking measurements very quickly to find how much product has been formed, or how much substrate has disappeared, in the first minute or so of the reaction.
- You could leave the reaction to take place until it has completely finished – that is, all the substrate has been converted to product – and record the time taken to do this.
- You could time how long it takes to reach a clearly identifiable stage of the reaction, called an end-point.

Let us say that, for the rennin reaction, you will use the last of these three methods. You will measure the time taken for the rennin to produce clots of milk that stick to the sides of the test tube in which the reaction is taking place. You add the rennin to the milk in the tube, and twist the tube gently. The end-point is the moment when you first see little clots remaining on the sides of the tube (Figure P1.3).

The following boxes describe apparatus and techniques which you might not use but should be aware of. Box P1.1 describes using a colorimeter to measure colour changes as a dependent variable, while Box P1.2 describes how to use a haemocytometer to count cells. You won't be expected to use these in an examination but you may be shown these apparatus in class or in a video or animation.

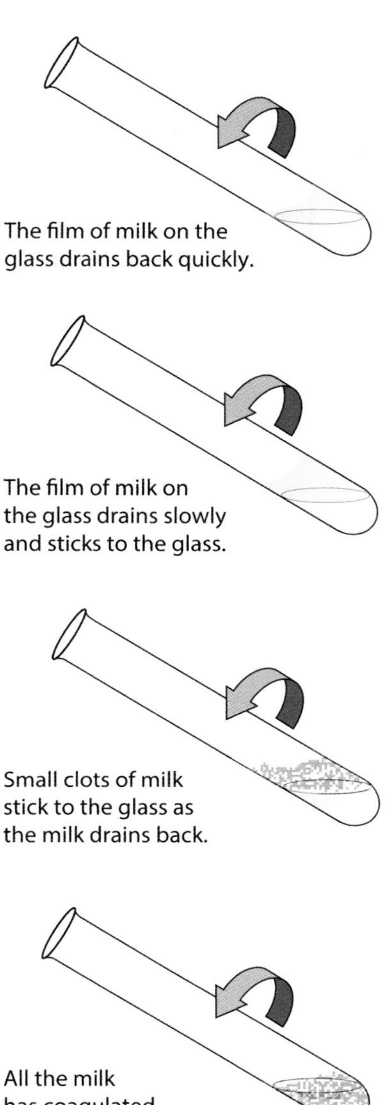

Figure P1.3 Determining the end-point of the clotting of milk by rennin.

QUESTION

P1.1 A student is investigating the effect of temperature on the activity of amylase on starch.

a Identify the independent variable and the dependent variable in this investigation.

b Suggest a suitable range for the independent variable. Explain why you have chosen this range.

c Suggest a suitable interval for the independent variable. Explain why you have chosen this interval.

QUESTION

P1.2 Look back at the experiment described in Question P1.1. Suggest **two** ways in which you could measure the dependent variable.

BOX P1.1: Observing and measuring colour changes and intensities

In some experiments, the dependent variable is colour. It is important to communicate clearly when describing colour changes. Use simple words to describe colours – for example, red, purple, green. You can qualify these by using simple terms such as 'pale' or 'dark'. You could also use a scale such as + for the palest colour, ++ for the next darkest and so on – if you do that, be sure to include a key.

Always state the actual colour that you observe. For example, if you are doing a Benedict's test and get a negative result, do not write 'no change', but say that the colour is blue.

It is sometimes useful to use **colour standards**, against which you can compare a colour you have obtained in your results. For example, if you are doing a Benedict's test, you could first carry out the test using a set volume of a series of solutions with known concentrations of glucose, using excess Benedict's solution. Stand these in a rack, and use them as colour comparison for the results you obtain with an unknown solution.

You can use a **colorimeter** to provide you with quantitative measurements of colour intensity in a solution. You will not be expected to use a colorimeter in the practical examination, but it might be a sensible suggestion if you are asked how you could improve the reliability of the results collected in an experiment.

A colorimeter is an instrument that measures the amount of light that is absorbed by a tube containing a coloured liquid. The deeper the colour, the more light is absorbed.

It is important to choose a suitable colour of light to shine through your coloured liquid. For example, if you want to measure how much red pigment is in a sample, you should use green light. (Red things look red because they reflect red light but absorb blue and green light.) You can change the colour of the light by using different coloured filters, which are supplied with the colorimeter (Figure P1.4).

Special tubes called cuvettes are used to contain the liquid. To measure the concentration of an unknown solution, a colorimeter has to be calibrated using standards of known concentration.

Calibration:

Step 1 Put a set volume of liquid that is identical to your samples of known concentrations, but which does not contain any red pigment, into a cuvette. This is called the blank. Put the blank into the colorimeter, and set the absorbance reading on the colorimeter to 0.

Step 2 Put the same volume of one of your samples of known concentrations (standards) into an identical cuvette. Put this into the colorimeter, and read the absorbance.

Step 3 Put the blank back into the colorimeter, and check that the absorbance still reads 0. If it does not, start from step 1 again.

Step 4 Repeat steps 2 and 3 with each standard.

Step 5 Use your readings to draw a calibration curve (absorbance against concentration).

Measuring unknown samples:

Step 5 Measure the absorbance of each of your samples containing unknown concentrations of red pigment. Use your calibration curve to determine concentrations.

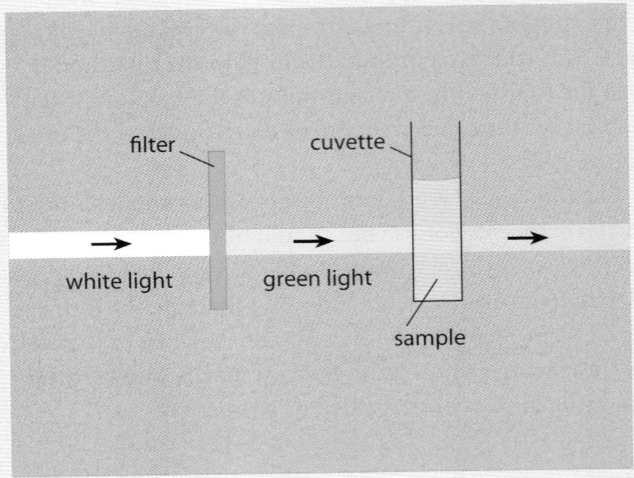

Figure P1.4 The path of light through the filter and cuvette in a colorimeter.

BOX P1.2: Using a grid or haemocytometer to count cells

You may need to count the number of cells in a given **area** or **volume** of a sample. To count the number in a given **area**, you could use a grid in the eyepiece of your microscope or marked on a special slide. You can then count the number of cells in a known area of the grid.

To count the number of cells in a known **volume** of liquid, you can use a haemocytometer. You could use it, for example, to count the number of red blood cells in a known volume of a blood sample, or the number of yeast cells in a known volume of a sample from a culture.

The haemocytometer is a special microscope slide with two sets of ruled grids in the centre, with deep grooves either side. Between the grooves, the surface of the slide is 0.1 mm lower than elsewhere. This produces a platform, called the counting chamber (Figure P1.5).

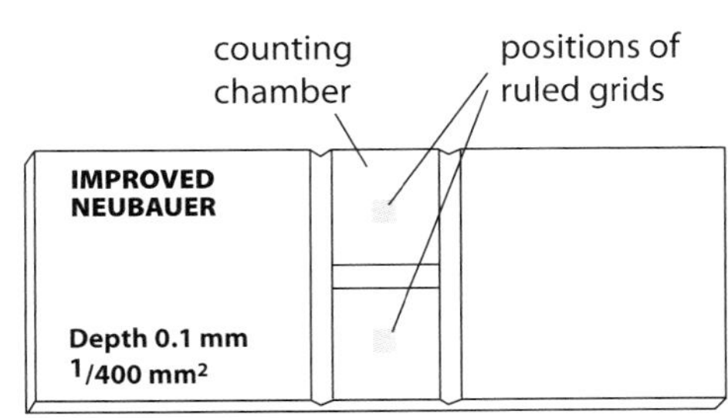

Figure P1.5 Haemocytometer.

To use the haemocytometer, a special coverslip is placed over the counting chamber. It must be slid into place and pressed down firmly until you can see 'interference' colours where the coverslip overlaps the slide. (These are called Newton's rings.) When you can see these, the distance between the rulings on the slide and the coverslip is exactly 0.1 mm (Figure P1.6).

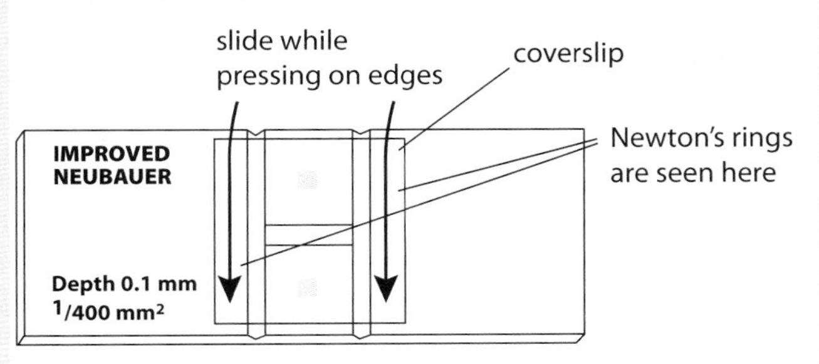

Figure P1.6 Preparing the haemocytometer.

Now use a pipette to carefully introduce your sample into the space between the coverslip and the platform. The liquid should cover the platform completely, but not spill over into the grooves (Figure P1.7).

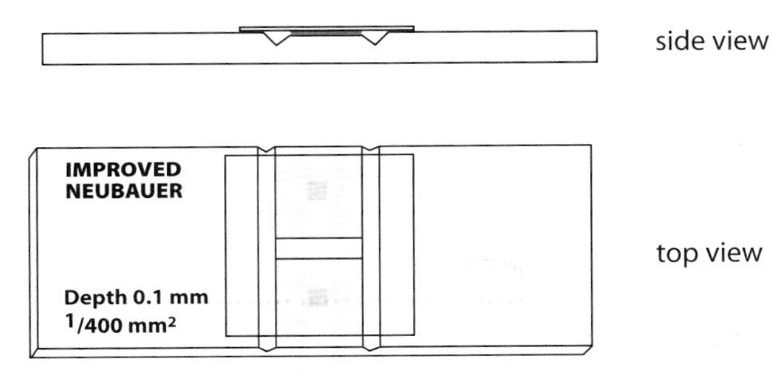

Figure P1.7 Haemocytometer containing the sample.

Place the haemocytometer on a microscope stage, and use the low-power objective to locate one of the ruled grids. Move up to the ×40 objective, and focus on the block of squares at the centre of the ruled grids (containing the numbers 1 to 5 in Figure P1.8). This is made up of 25 squares, each containing 16 smaller squares. Count the number of cells you can see in any five blocks of 16 small squares. If a cell is on the boundary line at the top or on the left, it is counted. If it is on the boundary line at the bottom or right, it is not counted (Figure P1.8).

continued ...

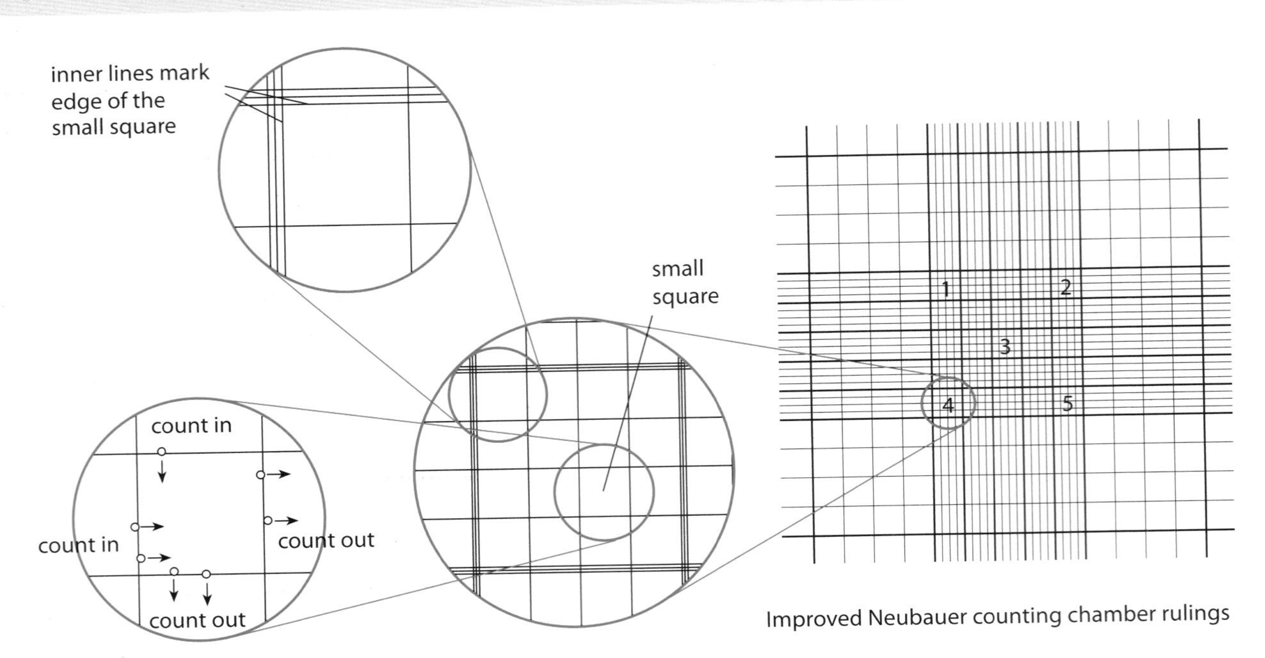

Figure P1.8 Counting cells in the haemocytometer.

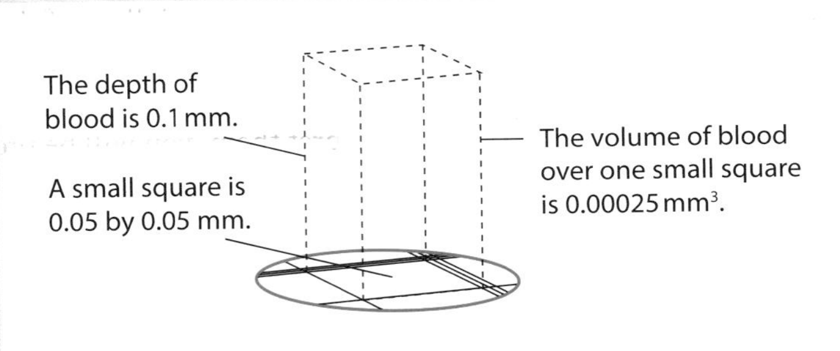

Figure P1.9 Calculating the number of cells in a unit volume of the sample.

Now you can calculate the number of cells in a unit volume of your sample. Look at Figures P1.8 and P1.9. Each small square has sides of 0.05 mm. The depth of the sample is 0.1 mm.

Therefore the volume of liquid over one small square is $0.05 \times 0.05 \times 0.1 = 0.00025\ \text{mm}^3$.

You counted cells in $5 \times 16 = 80$ of these small squares. This was a total volume of $80 \times 0.00025 = 0.02\ \text{mm}^3$.

Let's say you counted *N* cells. If there are *N* cells in $0.02\ \text{mm}^3$, then there are $N \times 10^6 \div 0.02$ cells in $1\ \text{dm}^3$.

Controlling or changing other variables

We have seen that it is very important to try to keep all significant variables other than the independent and dependent variable as constant as possible throughout your experiment. In an examination question, you might be expected not only to identify which variables you should keep constant, but also how you would do this.

Two variables which often crop up in experiments, either as independent variables or standardised variables, are **temperature** and pH. You may also need to explain how to change or control light intensity, windspeed or humidity.

Temperature

To control temperature, you can use a **water bath**. If you are lucky, you might be provided with an electronically controlled water bath. It is more likely that you will have to make your own water bath, using a large beaker of water. Figure P1.10 shows how you can do this.

Whichever method you use, it is important to measure the temperature carefully.

- Don't assume that, just because an electric water bath says the temperature is 30 °C, it really is that temperature. Use a thermometer, held in the water and not touching the sides or base of the container, to measure the temperature.
- If possible, read the thermometer while its bulb is still in the water – if you take it out to read it, then you will be measuring the temperature of the air.
- If you are standing tubes of liquid in the water bath, you should allow time for their temperatures to reach the same temperature as the water. This can take a surprisingly long time. In an exam, you may only be able to allow about five minutes for this. It is a good idea to measure the temperature of the actual liquids in the tubes, rather than assuming that they are at the same temperature as the liquid in the water bath.
- If you are doing an enzyme experiment, you may need to bring both the enzyme solution and the substrate solution to the same temperature before you add one to the other. Stand them, in two separate tubes, in the same water bath and leave them to reach the desired temperature. Remember that, as soon as you add the enzyme and substrate to one another, the reaction will start, so don't do that until you are ready to begin taking measurements.

pH

To control pH, you can use **buffer** solutions. These are solutions that have a particular pH, and that keep that pH even if the reaction taking place produces an acidic or alkaline substance that would otherwise cause pH to change. You simply add a measured volume of the buffer to your reacting mixture.

You can measure pH using an indicator. Universal Indicator is especially useful, as it produces a range of colours over the whole pH range, from 0 to 14. You will not be expected to remember these colours – if you need to interpret them, you will be provided with a colour chart.

Alternatively, you may be able to measure pH using a pH meter.

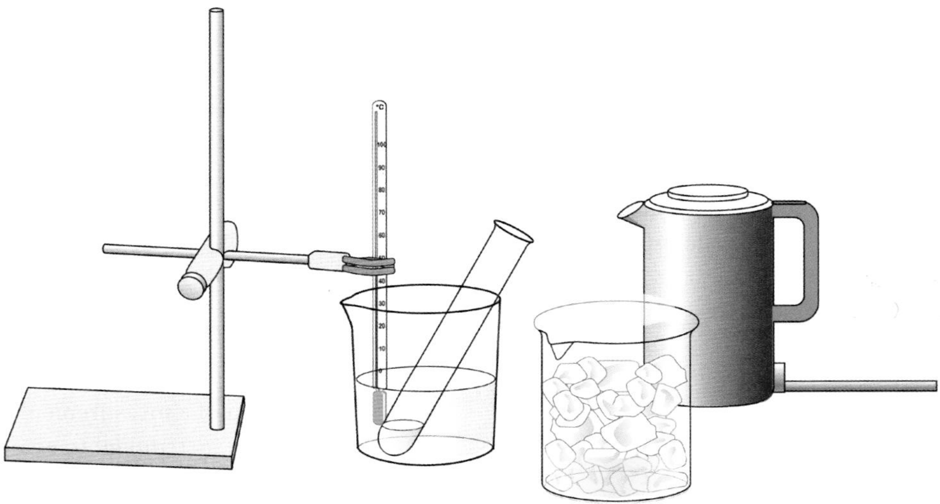

Figure P1.10 Controlling temperature using a water bath. You can use this method for producing a range of temperatures if temperature is your independent variable, or for keeping temperature constant if it is a standardised variable.

Light intensity

The simplest way to control light intensity is to vary the distance of a light source. In general:

$$\text{light intensity is proportional to } \frac{1}{\text{distance}^2}$$

For example, if you were to do an experiment investigating the effect of light intensity on the rate of photosynthesis, you could use a lamp at a measured distance from a photosynthesising piece of pondweed to provide light. It is obviously important to ensure that this is the **only** source of light for the weed.

A problem with this method of varying light intensity is that the lamp will also produce a significant amount of heat, and this introduces another variable into your experiment: the temperature of the water surrounding the pondweed. Placing a transparent piece of plastic between the lamp and weed will allow light to pass through, but will greatly reduce the transmission of longer wavelength radiation, which will reduce the heating effect from the lamp.

Windspeed

You may need to vary windspeed if you are investigating rate of transpiration. This can be done using a fan, which you can place at different distances from the plant.

You will probably not be able to measure the actual windspeed.

Humidity

Humidity is a measure of the water content of the air. You are most likely to want to vary or control humidity when you are investigating rates of transpiration.

Humidity can be increased by placing a container of water close to the plant, or by covering the plant with a plastic bag. The bag will prevent the water vapour lost from the plant's leaves from escaping into the atmosphere, and will therefore eventually produce an environment in which the air contains as much water vapour as it possibly can.

Humidity can be reduced by placing a container of calcium chloride close to the plant. Calcium chloride absorbs water vapour.

You will probably not be able to measure the actual humidity of the air.

Biological material

It is often very difficult to make sure that samples of biological material used in an experiment are all identical. Depending on the kind of material you are using, you should try to keep as many features as possible the same. These could include: age, storage conditions, genotype (including sex), mass, volume, position in the organism from which the sample was taken.

QUESTION

P1.3 Look back at the experiment described in Question P1.1.

- **a** Describe how you would change and measure the independent variable.
- **b** Suggest two important variables that you should try to keep constant.
- **c** Describe how you would keep these variables constant.

Controls

Think back to the rennin experiment. How can we be sure that it is the rennin that is making the milk clot, and not some other factor? To check this, we need to use a **control**, where the factor that we are investigating is absent.

In this experiment, the control is a tube that has no rennin in it. Everything else must be the same, so the same volume of water is added to match the volume of enzyme solution that is added to all the other tubes.

Another possible control could be a tube containing boiled rennin solution. Boiling denatures the rennin enzyme so it is inactive.

More about measurements – accuracy, precision and reliability

No measuring instrument is perfect. We can never be completely certain that a measurement we make gives us an absolutely 'correct' value.

The **accuracy** of a measurement is how 'true' it is. If you are measuring a temperature, then the accuracy of your measurement will depend on whether or not the thermometer is perfectly calibrated. If the thermometer is accurate, then when the meniscus is at 31 °C, the temperature really is exactly 31 °C.

The **precision** of a measurement depends on the ability of the measuring instrument to give you the same reading every time it measures the same thing. This doesn't have to be the 'true' value. So, if your thermometer **always** reads 32 °C when the temperature is really 31 °C, it obviously is not accurate but it **is** precise.

The **reliability** of a measurement is the degree of trust that you can have in it. If your measurements are reliable, then you would expect to get the same ones if you repeated them on other occasions.

Reliability is affected by both the accuracy and precision of your measuring instruments, and also by the kind of measurements that you are making. For example, in the rennin experiment, you have to make a decision about exactly when the end-point is reached. This is really difficult: there is no precise moment at which you can say the clots definitely form, so your measurement of the time at which this happens will be very unreliable.

One of the best ways of dealing with poor reliability is to **repeat** the readings several times. For example, with the rennin experiment, you could set up three tubes of each concentration of the enzyme and measure the time to the end-point for each of them. You would then have three results for this particular enzyme concentration, which you could use to calculate a mean.

Estimating uncertainty in measurement

When you use a measuring instrument with a scale, such as a thermometer or syringe, you will often find that your reading doesn't lie exactly on one of the lines on the scale. For example, if a thermometer has a scale marked off in intervals of 1 °C, you could probably read it to the nearest 0.5 °C. We can assume that each reading you make could be inaccurate by 0.5 °C. This is the possible error in your measurement. So, if your temperature reading was 31.5 °C, we would show this as 31.5 °C ± 0.5 °C. In general, the potential error in a measurement is half of the value of the smallest division on the scale you are reading from. Figure P1.11 explains why this is so.

In some situations, you might make two measurements, and your result is the difference between them. For example, you might want to measure a change in temperature. You measure the temperature at the start as 28.5 °C (error = ±0.5 °C) and at the end as 39.0 °C (error once again = ±0.5 °C). You then calculate the rise in temperature as 39.0 − 28.5 °C = 10.5 °C. But you have to assume that both of the measurements could be out by 0.5 °C in the same direction. Your total error is therefore 1 °C, and you should write the answer as 10.5 °C ±1 °C.

You might be asked to express the size of the error as a percentage. To do this, you divide the error in the measurement by the measurement itself, and multiply by 100.

Here, the temperature rise was measured as 10.5 °C, and the error was ±1 °C. Therefore:

$$\text{percentage error} = (1 \div 10.5) \times 100 = 9.5\%$$

Errors in measurement are not the only important sources of error in biology experiments, and you can read more about this in the section on identifying significant sources of error on page 261.

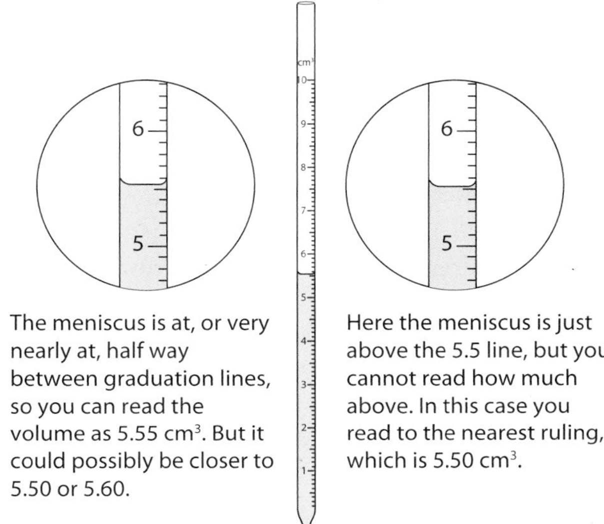

Figure P1.11 Estimating uncertainty in measurement when using a scale.

Recording quantitative results

Most of the experiments that you will do, either during your course or in the practical examination, will involve the collection and display of quantitative (numerical) results. You may be given a results table to complete, but often you will have to design and draw your own results table.

Table P1.1 shows a results table that you could use for your results from the experiment investigating the effect of enzyme concentration on the rate of activity of rennin. Three repeat readings were made for each enzyme concentration, and a mean has been calculated.

There are several important points to note about this results table, which you should always bear in mind whenever you construct and complete one.

Rennin concentration / %	Time to reach end-point / s			
	1st reading	2nd reading	3rd reading	Mean
0.0	did not clot	did not clot	did not clot	did not clot
0.2	67.2	68.9	67.8	68.0
0.4	48.1	46.9	47.3	47.4
0.6	30.1	31.9	30.1	30.7
0.8	20.3	19.2	19.9	19.8
1.0	13.1	***18.9***	12.7	12.9

Table P1.1 Results for an experiment to investigate the effect of enzyme concentration on the rate of activity of rennin. The reading in ***bold italics*** is an anomalous result and has been excluded from the calculation of the mean.

- The table is drawn with ruled columns, rows and a border. The purpose of a results table is to record your results clearly, so that you and others can easily see what they are, and so that you can use them easily to draw a graph or to make calculations. Drawing neat, clear lines makes it much easier to see the results at a glance.
- The columns are clearly headed with the quantity and its unit. (Use SI units.) Sometimes, you might want to arrange the table the other way round, so that it is the rows that are headed. Sometimes, both rows and columns might need to include units. The important thing to remember is that the units go in the heading, not with the numerical entries in the table.
- The results are organised in a sensible sequence. The values for the rennin concentration go up from the lowest to the highest.
- The independent variable (rennin concentration) comes first, followed by the readings of the dependent variable (time taken to reach end-point).
- Each measurement of the dependent variable is taken to the same number of decimal places. You would have used a stopwatch to take these readings, and it probably gave a reading to one hundredth, or even one thousandth, of a second. So the first reading on the watch could have been 67.207. However, as we have seen, it is very difficult to judge this end-point, so to suggest that you can time it to the nearest thousandth of a second is not sensible. You can perhaps justify, however, recording the values to the nearest one tenth of a second, rounding up or down the reading on the watch.

- The values calculated for the mean are given to the same number of decimal places as the individual readings. This is very important to remember. If you have only recorded the individual readings to the nearest one tenth of a second, then it is wrong to suggest you can calculate the mean to one hundredth or one thousandth of a second.
- In the last row, the readings for the rennin at a concentration of 1% contain an anomalous result. The second reading (shown in ***bold italics***) is clearly out of line with the other two, and looks much too close to the readings for the 0.8% rennin solution. You can't know what went wrong here, but something clearly did. If you are in a position to do so, the best thing to do about an anomalous result is to measure it again. However, if you can't do that, then you should ignore it. Do not include it in your calculation of the mean.
  The mean for this row is therefore calculated as $(13.1 + 12.7) \div 2 = 12.9$.
- The first row of the table records that the milk 'did not clot'. An alternative way of recording this would be to record the time as infinite (symbol: ∞). This can then be converted to a rate like all the other results by calculating $\frac{1}{\text{time}}$. Note that $\frac{1}{\infty} = 0$ (zero rate).

QUESTION

P1.4 Look back at the experiment described in Question P1.1, and your answers to Questions P1.2 and P1.3. Construct a results table, with full headings, in which you could record your results.

Constructing a line graph

You will generally want to display the results in a table as a graph. Figure P1.12 shows a line graph constructed using the results in Table P1.1.

Once again, there are several important points to note about this graph, which you should always bear in mind whenever you construct and complete a graph.

- The independent variable goes on the *x*-axis (horizontal axis), and the dependent variable on the *y*-axis (vertical axis).
- Each axis is fully labelled, including the units. Usually, you can simply copy the headings that you have used in the results table.
- The scale on each axis goes up in equal intervals, such as 1 s, 2 s, 5 s or 10 s intervals. You would not therefore, have an axis that read 20 °C, 30 °C, 50 °C, 60 °C, 80 °C.

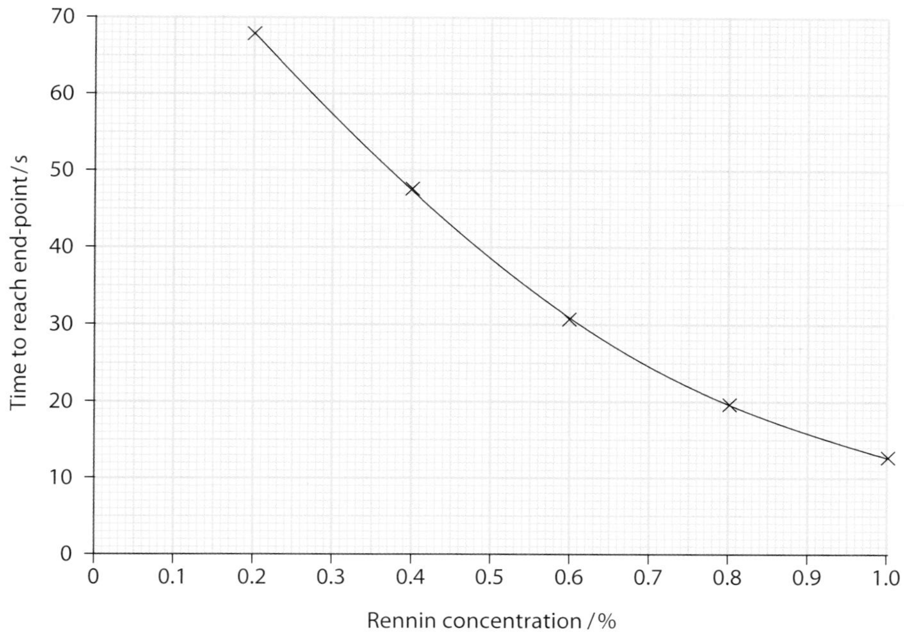

Figure P1.12 Line graph displaying the results in Table P1.1.

- The intervals chosen make it easy to read intermediate values.
- The scales cover the entire range of the values to be plotted, but don't go too far above and below them. This makes best use of the graph paper. The more spread out the scale is, the easier it is to see any trends and patterns in the relationship between the independent variable and the dependent variable. (Note that there is not always a need to begin your scale at 0.)
- The points are plotted as neat, carefully placed crosses. An acceptable alternative is a dot with a circle around it. Do not use a simple dot, as this may be hidden when the line is drawn.
- A best-fit line has been drawn. This is a smooth line which shows the trend that the points seem to fit. There is not a single perfect place to put a best-fit line, but you should ensure that approximately the same number of points, roughly the same distances from the line, lie above and below it. There is no need for your line to go through either the first or last point – these points are no more 'special' than any of the others, so should not get special treatment.
- An alternative way to draw the line would be to join each point to the next one with a ruled, straight line. Generally, you should use a best-fit line when told to do so, or when you can see a clear trend in which you have confidence. If you are not sure of the trend, then draw straight lines between points.
- It is almost always incorrect to extend the line beyond the plotted points (extrapolate). However, it can sometimes be allowable to do this – for example, if you have not made a measurement when the independent variable (x-axis value) is 0, and when you are absolutely certain that the dependent variable (y-axis value) would also be 0 at that point. You could then extend your line back to meet the origin, at point 0,0.

Constructing bar charts and histograms

Not all experiments generate results that can be displayed as a line graph. Some results are best shown in bar charts or histograms.

A **bar chart** is drawn when you have a discontinuous variable on the *x*-axis and a continuous variable on the *y*-axis. A **discontinuous variable** is one where there is no continuous relationship between the items listed on the scale. Each category is discrete. Figure P1.13 shows an example. The *x*-axis lists five species of tree; each type of tree is separate from the others, and there is no continuous relationship between them. The bars are therefore drawn with gaps between them.

A **continuous variable** is one where there is a smooth, numerical relationship between the values. (Line graphs always have a continuous variable on both the *x*-axis and *y*-axis, as in Figure P1.12.) Sometimes, you will want to draw a graph where there is a continuous range of categories on the *x*-axis, and the frequency with which each of these categories occurs is shown on the *y*-axis. In this case, the bars are drawn so that they touch. This kind of graph is a **histogram**, or a **frequency diagram**. Figure P1.14 shows an example.

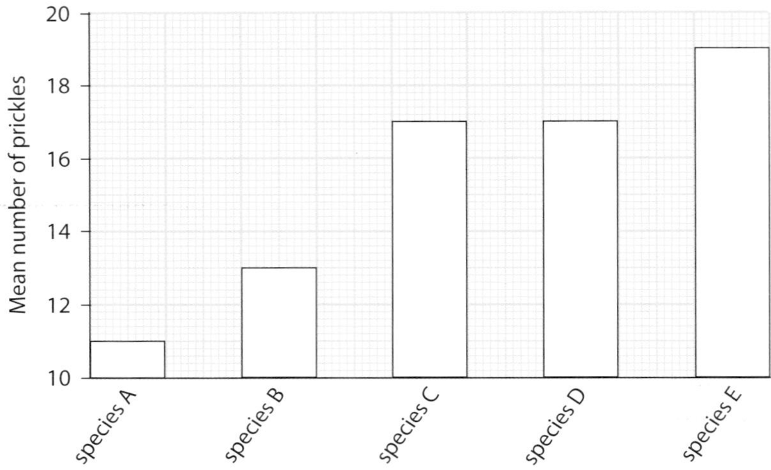

Figure P1.13 Bar chart showing the mean number of prickles on leaves from five different species of tree.

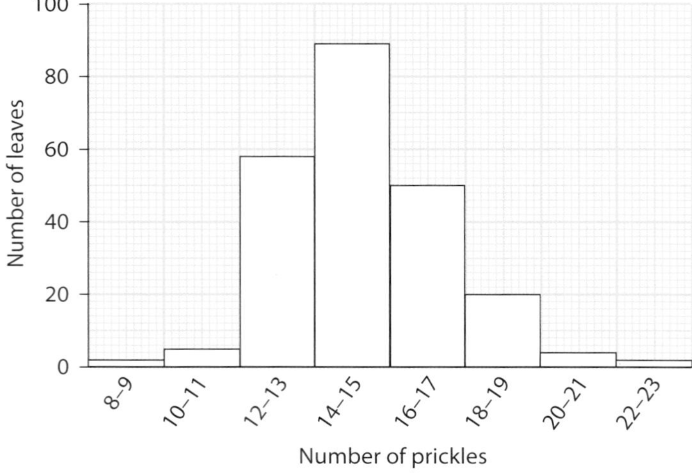

Figure P1.14 Frequency diagram (histogram) showing the numbers of leaves with different numbers of prickles on a holly tree.

Making conclusions

A conclusion is a simple, well-focused and clear statement describing what you can deduce from the results of your experiment. The conclusion should relate to the initial question you were investigating or the hypothesis you were testing.

For example, the results shown in Table P1.1 could lead you to write a conclusion like this:

The greater the concentration of rennin, the shorter the time taken to reach the end-point. An increase in rennin concentration increases the rate of reaction.

Describing data

You may be asked to describe your results in detail. You can do this from a table of results, but it is often best done using the graph that you have drawn. The graph is likely to show more clearly any trends and patterns in the relationship between your independent and dependent variables.

For example, you could describe the results shown in Figure P1.12 (page 257) like this:

When no rennin was present, no end-point was reached (time taken = ∞), indicating that no reaction was taking place. At a concentration of 0.2% rennin, the end-point was reached in a mean time of 68.0 seconds. As the concentration of rennin increased, the mean time to reach the end-point decreased, with the shortest mean time (12.9 s) occurring at a concentration of 1% rennin. This indicates that the rate of reaction increases as the concentration of rennin increases.

The line on the graph is a curve with decreasing gradient, not a straight line, so the relationship between concentration of rennin and the rate of reaction is not proportional (linear). The curve is steepest for the lower concentrations of rennin, gradually flattening out for the higher concentrations. This shows that a 0.2% increase in rennin concentration has a greater effect on reaction rate at low rennin concentrations than at high rennin concentrations.

There are several points to bear in mind when you are describing results shown on a graph. (These same points are important when you are dealing with graphs on structured question papers, as well as on practical examination papers.)

- Begin by describing the overall trend – the overall relationship between what is shown on the x-axis and on the y-axis.
- Look for any changes in gradient on the graph, and describe these. In this case, the change in gradient is a steady one (the gradient gets gradually less and less as the rennin concentration increases). Sometimes, there are sharp changes in gradient at particular key points, and you should focus on those and describe the gradient changes and precisely where they occur.
- Quote figures from the graph. You will need to pick on points of particular interest (for example, where gradient changes occur), and quote the coordinates of those points – that is, you should state both the x-axis value **and** the y-axis value.
- Take great care not to use phrases that suggest something is happening over time, if time is not shown on the x-axis. For example, it would be totally wrong to say that the gradient of the graph in Figure P1.12 (page 257) 'is steep at first and gradually gets less'. 'At first' suggests time – but the x-axis shows concentration, not time. Words such as 'more quickly', 'slower' and 'rapidly' should all be avoided unless the x-axis shows time.

Making calculations from data

You may be asked to carry out a calculation from a set of results – either the results that you have collected, or a set of results that is presented to you.

It is very important to show every single step in any calculation that you make. For example, you might be given a set of five measurements and asked to find the mean value. You should set out your calculation clearly, like this:

measurements: 12.5 μm, 18.6 μm, 13.2 μm, 10.8 μm, 11.3 μm

$$\text{mean} = \frac{(12.5 + 18.6 + 13.2 + 10.8 + 11.3)}{5}$$

$$= \frac{66.4}{5}$$

$$= 13.3\,\mu\text{m}$$

Remember that, even though your calculator will show an answer of 13.28, you must give your answer to only one decimal place, the same as for the original measurements.

You could also be asked to calculate a gradient at one or more points on your graph. Figure P1.15 explains how to do this when there is a straight line, and also when the line is curved.

QUESTION

P1.5 a Choose two different points on the graph in Figure P1.12 (page 257), and calculate the gradient at each point. Remember to show all the steps in your calculation fully and clearly.

b Use your calculated values to add to the description of the results given on page 259.

A third type of calculation you could be asked to do is to find the percentage change. To do this:

1 find the difference between the first reading and the second reading, by subtracting one from the other

2 divide this value by the first reading, and multiply by 100; this figure gives you the percentage change – remember to state whether the change is an increase or a decrease.

For example, imagine that the mass of a plant on day 1 was 250 g. On day 5, after it had lost a lot of water by transpiration, its mass was 221 g.

$$\text{change in mass} = 250 - 221 = 29\,\text{g}$$

$$\text{percentage change in mass} = \frac{\text{change in mass}}{\text{original mass}} \times 100$$

$$= \frac{29}{250} \times 100$$

$$= 11.6\%\ \text{decrease}$$

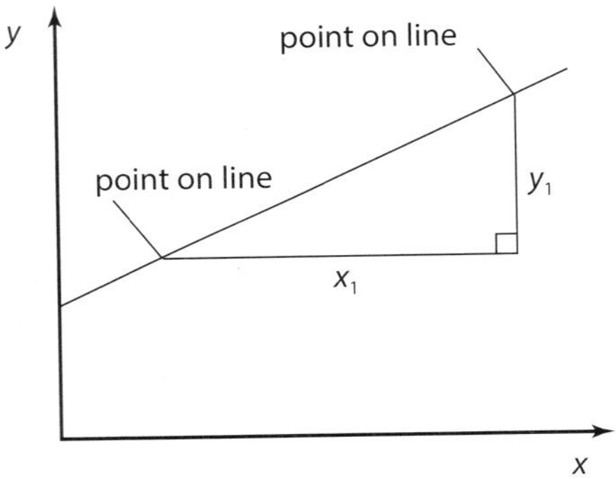

To determine the gradient of a straight line graph:

1 Select two points which are at least half as far apart as the length of the line of the graph.
2 Draw a right-angle triangle between these points.
3 Calculate the gradient using the lengths of the triangle sides x_1 and y_1:

$$\text{gradient} = \frac{y_1}{x_1}$$

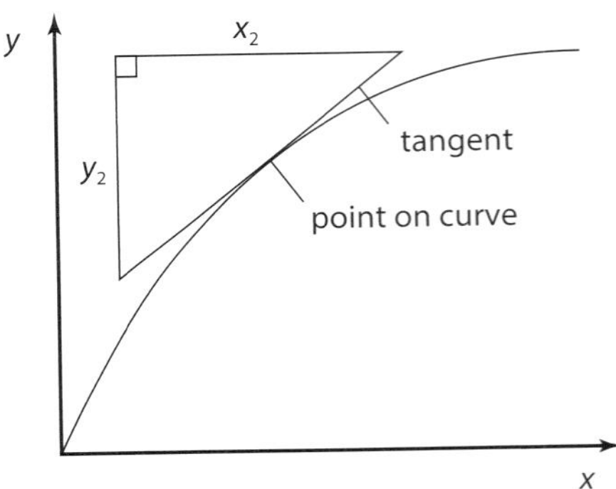

To determine the gradient at a point on a curved graph:

1 Draw a tangent to the curve at that point, making sure it is at least half as long as the line of the graph.
2 Draw a right-angle triangle on the tangent.
3 Calculate the gradient using the lengths of the triangle sides x_2 and y_2:

$$\text{gradient} = \frac{y_2}{x_2}$$

Figure P1.15 Calculating the gradients of a straight line and a curve at a point.

Explaining your results

You may be asked to explain your results. This requires you to use your scientific knowledge to explain **why** the relationship you have found between your independent variable and your dependent variable exists.

QUESTION

P1.6 Use your knowledge and understanding of enzyme activity to explain the results shown in Figure P1.12 (page 257).

Identifying sources of error and suggesting improvements

You will often be asked to identify important sources of error in your experiment. It is very important to realise that you are not being asked about mistakes that you might have made – for example, not reading the thermometer correctly, or not measuring out the right volume of a solution, or taking a reading at the wrong time. These are all avoidable human mistakes and you should not be making them!

Sources of error are unavoidable limitations of your apparatus, measuring instruments, experimental technique or experimental design that prevent your results from being totally reliable (page 255). They generally fall into the three major categories given below.

- **Uncertainty in measurements** resulting from lack of accuracy or precision in the measuring instruments that you were using, and from the limitations in reading the scale. These are described on page 255. These errors are likely to be the same all through your experiment. They will be about the same size, and act in the same direction, on all of your readings and results. They are **systematic errors**.
- **Difficulties in controlling the standardised variables.** For example, if you were using a water bath to maintain a constant temperature in the rennin experiment, it may have been impossible to keep the temperature absolutely constant. Variations in temperature could have affected the rate of activity of the rennin, making it impossible to be sure that all changes in rate of activity were due to differences in your independent variable – the concentration of the rennin. These errors are likely to be different for different stages of your investigation. They are **random errors.**
- **Difficulties in measuring the dependent variable**, due to human limitations. For example, in the rennin experiment, you needed to judge the end-point, which is impossible to do precisely using just the human eye. You may sometimes have chosen a moment for the end-point that was relatively early, and sometimes chosen a moment that was too late. So these types of errors are also random errors.

It is very important to learn to spot the really important sources of error when you are doing an experiment. Be aware of when you are having difficulties, and don't assume that this is just because you are not very good at doing practical work! If you carry out the rennin experiment, you will quickly realise how difficult it is to keep your water bath at exactly the correct temperature and to judge the end-point of the reaction precisely. These are the really important errors in this experiment, and they outweigh any others such as the error in measuring a volume or in measuring temperature.

If you are asked to suggest improvements to an experiment, your suggestions should be focused on reducing these sources of error. Improvements could include:

- using measuring instruments that are likely to be more precise, accurate or reliable – for example, measuring volumes with a graduated pipette rather than a syringe
- using techniques for measuring the dependent variable that are likely to be more reliable – for example, using a colorimeter to measure colour changes, rather than the naked eye
- using techniques or apparatus that are better able to keep standardised variables constant, such as using a thermostatically controlled water bath rather than a beaker of water
- controlling important variables that were not controlled in the original experiment (note that it is also important to say **how** you would control these variables)
- doing repeats so that you have several readings of your dependent variable for each value of your independent variable, and then calculating a mean value of the dependent variable.

Drawings

One of the questions in a practical examination will usually require you to make observations of a photograph or specimen – which will often be on a microscope slide – and to record your observations as a diagram or drawing.

You may be asked to calculate a magnification, which is described on pages 6–9 in Chapter 1. You could be asked to use a graticule to measure an object using a microscope, and perhaps to calculate its real size by calibrating the graticule against a stage micrometer. This is described on page 7 in Chapter 1. Remember, in every case, to show every step of your working clearly and fully.

You do not have to be a good artist to be good at doing biological drawings. A good biological drawing looks simple and uncomplicated. It should:

- be drawn with clear, single lines
  – do not have several 'goes' at a line so that it ends up being fuzzy
  – use an HB pencil and a good eraser, so that when you make a mistake (which you almost certainly will) you can rub it out completely
- show the overall shape, and the proportions of the different components of the structure you are drawing, accurately
- not include shading or colouring
- be large, using most of the space available but not going outside that space (for example, it should not go over any of the words printed on the page).

See also Box 7.1, page 129, Chapter 7.

It is very important to draw what you can see, and not what you think you should see. The microscope slide that you are given might be something that is different from anything you have seen before. During your course, you should become confident in using a microscope and learn to look carefully at what you can see through the eyepiece.

You may be asked to draw a **low-power plan** (a **plan diagram**). This is a diagram in which only the outlines of the different **tissues** are shown. Figure 7.8 on page 131 is a plan diagram of the transverse section through a privet leaf shown in Figure 7.7 on page 130. Plan diagrams should never show individual cells. You will need to look carefully to determine where one tissue ends and another one begins, and it may be helpful to move up to a high-power objective lens to help with this. You can then go back down to a lower power lens to enable you to see the whole area that you will show in your drawing.

You may be asked to draw a more detailed drawing, using high power. This is sometimes called a **high-power detail**, and it generally does show individual cells. Figure P1.16 shows a drawing of some plant cells, made from what is seen using the high-power objective on a light microscope.

You may be asked to label your drawings. The label lines should be drawn with a ruler and pencil, and the end of the line should precisely touch the part of the diagram you are labelling. Do not use arrowheads. The label lines should not cross over one another. The labels themselves should be written horizontally (no matter what angle the label line is at), and should not be written on the drawing.

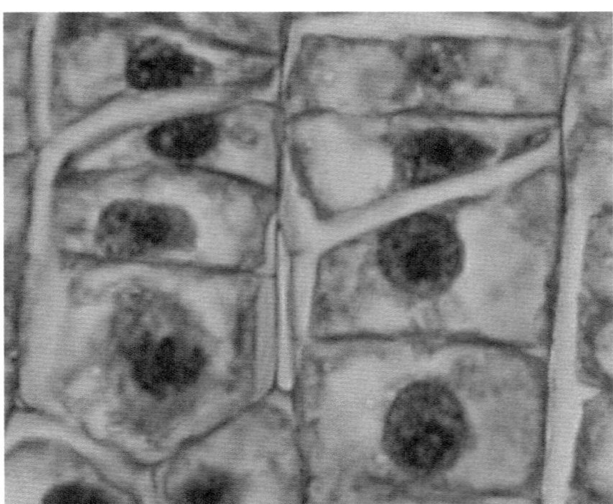

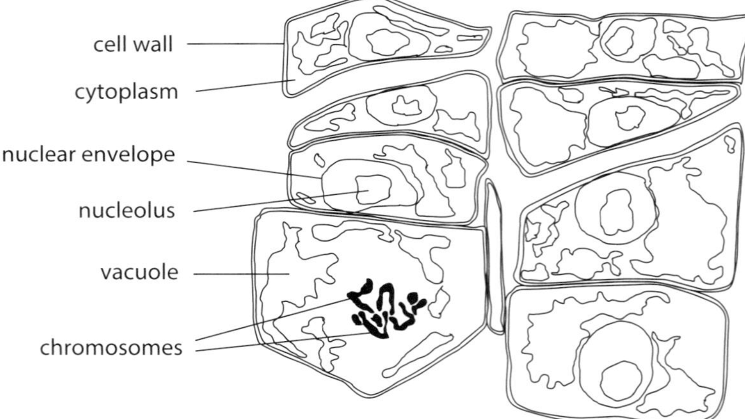

Figure P1.16 A high-power drawing of a group of plant cells, showing the detail visible using a high-power objective lens of a light microscope.

Summary

- In an experiment investigating the effect of one variable on another, the independent variable is the one that you change and the dependent variable is the one that you measure. All other variables should be controlled (kept constant). The range of the independent variable is the spread from lowest to highest value. The interval is the distance between each value in the range. Temperature can be kept constant or varied using a water bath. pH can be kept constant or varied using buffer solutions.
- The accuracy of a measurement is how true it is. For example, an accurate measuring cylinder reads exactly 50 cm^3 when it contains 50 cm^3 of liquid. The precision of a measuring instrument is how consistent it is in giving exactly the same reading for the same value. The reliability of a set of measurements is the degree of trust that you can have in them. A reliable set of measurements are likely to be very similar if you are able to do the same experiment again. If you are concerned about reliability, then do at least three repeat measurements for each value of your independent variable, and calculate a mean.
- In general, the error in any measurement is half the value of the smallest division on the scale. For example, on a measuring cylinder marked in 2 cm^3 divisions, the error in any reading will be ± 1 cm^3. If you are taking two readings and calculating the difference between them, then the error is ± 1 cm^3 for each reading, making a total error of ± 2 cm^3.
- Results tables should be constructed with the independent variable in the first column and the readings for the dependent variable(s) in the next column(s). Units go in the headings, not in the body of the table. Each value should be recorded to the same number of decimal places. This is also the case for any calculated values.
- In a line graph, the independent variable goes on the *x*-axis and the dependent variable on the *y*-axis. Headings must include units. Scales must go up in even and sensible steps. Points should be plotted as small crosses or as encircled dots. Lines should be best-fit or ruled between successive points. Do not extrapolate.
- Bar charts are drawn when there is a discontinuous variable on the *x*-axis. Bars in a bar chart do not touch. Frequency diagrams or histograms are drawn when there is a continuous variable on the *x*-axis. Bars touch.
- Conclusions should be short and to the point. They should use the results to answer the question posed by the investigation. They should not go beyond what is shown by the results. Do not confuse conclusion with discussion.
- When describing data displayed on a graph, begin by stating the general trend and then describe any points at which the gradient of the curve changes. Quote figures from from the *x*-axis and *y*-axis coordinates for these points. Do not use language suggesting time (e.g. 'faster') if time is not shown on the *x*-axis or *y*-axis.
- Show every small step whenever you are asked to do a calculation.
- Do not confuse mistakes with experimental errors. Mistakes should not happen. Experimental errors are often unavoidable, unless you have the opportunity to use a better technique or better apparatus. Systematic errors are those which have the same magnitude and direction throughout the experiment, and are usually caused by limitations in the measuring instruments. Random errors are those which vary in magnitude and direction during the experiment, and may be caused by difficulty in controlling variables or in making judgements. When asked to suggest improvements in an experiment, concentrate on the main sources of error and suggest ways of reducing them.
- When making drawings from a microscope, a low-power plan should show only the outlines of tissues and no individual cells. Be prepared to go up to high power to get more information about where one tissue ends and another begins. High-power drawings should show as much detail as possible, including details of individual cells.

End-of-chapter questions

1 An investigation is carried out into the effect of substrate concentration on the activity of catalase. What could be the dependent variable?

A the concentration of catalase

B the pH of the enzyme solution

C the rate of production of oxygen

D the temperature of the substrate [1]

2 An investigation is carried out into the effect of temperature on the activity of lipase. Separate tubes of substrate solution and enzyme solution are left in temperature-controlled water baths for ten minutes before mixing. Why is this done?

A to activate the enzyme

B to allow time for the enzyme and substrate to react

C to control the independent variable

D to keep a standardised variable constant [1]

3 Copy and complete the table.

Investigation	Independent variable	Dependent variable	Two important control variables
The effect of sucrose concentration on plasmolysis of onion cells			
The effect of pH on the rate of activity of amylase			
The effect of temperature on the percentage of open stomata in a leaf			

[9]

4 For this question you need two sheets of graph paper.

The light micrographs below are **a** a cross section of a young root, and **b** a representative part of a young stem of *Ranunculus* (buttercup).

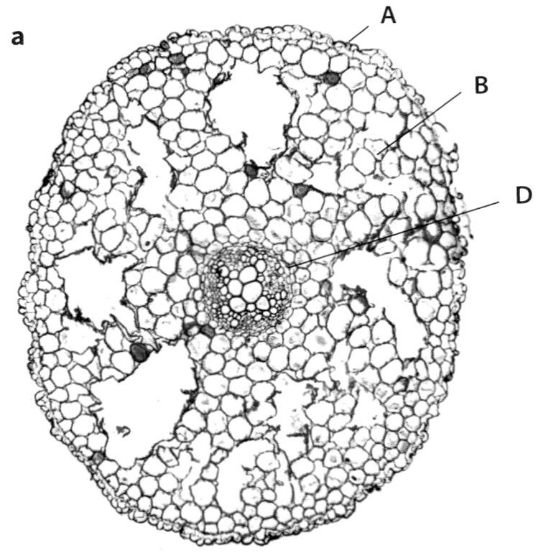

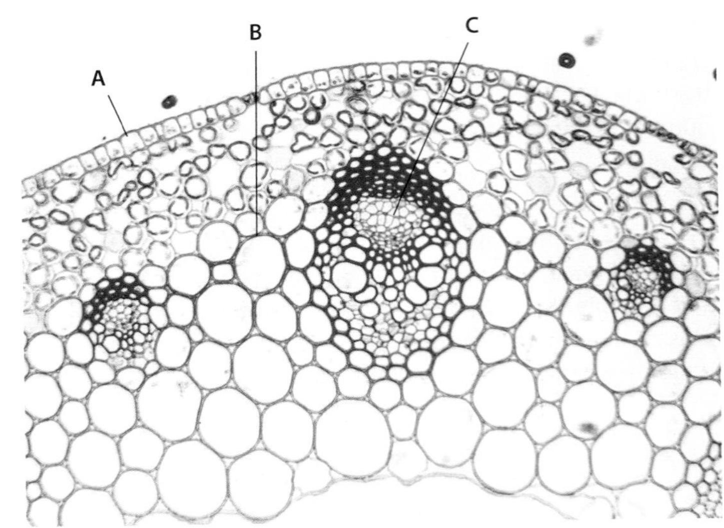

a Name the tissues **A**, **B**, **C** and **D**. [4]

b i On one of the sheets of graph paper, draw the outline of the root. Use at least half the width of the graph paper when making your drawing.

Now draw inside your outline a low-power plan of the xylem only. Be as accurate as you can in drawing the correct proportions compared with the overall size of the root – you may find it useful to make some measurements with a ruler. [4]

ii Now take the second sheet of graph paper and draw the outline of the stem. It does not have to be exactly the same size as your drawing of the root.

Carefully make a low-power plan to show the vascular bundles **only**. Draw in outline the lignified tissues sclerenchyma and xylem, and the tissue labelled **C** between them. [2]

iii Sclerenchyma and xylem are tissues which contain dead cells whose walls are thickened with a mechanically strong substance called lignin. Lignin is used for strength and support. Count the number of squares of graph paper covered by lignified tissue (xylem) in the root. Count the squares that are more than half included in the drawing as whole squares, and do not count squares that are less than half included. [1]

iv Count the number of squares covered by the whole root section (including the lignified tissue). [1]

v Calculate the percentage of squares occupied by lignified tissue in the root as follows:

$$\frac{\text{number of squares occupied by lignified tissue}}{\text{number of squares occupied by whole root}} \times 100$$

[1]

vi Repeat steps **iii** to **v** for the stem (remember lignified tissue in the stem is sclerenchyma plus xylem). [3]

vii Assuming the results you have obtained are typical of the whole stem, suggest an explanation for the difference in percentage of lignified tissue in the root and the stem. [2]

c If you try to imagine these structures in three dimensions, the lignified tissue in the root is a central rod, but in the stem it is a circle of separate rods. Suggest the reasons for the different distribution of lignified tissues in the root and the stem. [2]

[Total: 20]

5 A student decided to investigate the effect of temperature on the activity of enzymes in yeast. The student measured the activity of the enzymes by counting the number of bubbles of carbon dioxide which were released in three minutes.

The results of the student's investigation are shown in the table.

Temperature / °C	Enzyme activity / mean number of carbon dioxide bubbles released per minute
15	5
20	7
30	11
35	15
40	18

a i Plot a graph of the data shown in the table. [4]

ii From the graph, estimate the enzyme activity at 25 °C. [1]

iii Suggest how the student should make sure that the results of this investigation are as accurate as possible and as reliable as possible. [3]

b In carrying out this investigation, the student made the hypothesis that 'The activity of the enzymes in yeast increases as temperature increases.' State whether you think this hypothesis is supported by the student's results. Explain your answer. [2]

[Total: 10]

Cambridge International AS and A Level Biology 9700 Paper 31, Question 1c and d, June 2009

6 A student investigated the time taken for the complete digestion of starch by amylase found in the saliva of 25 individuals of a species of mammal.

A sample of saliva was collected from each individual and mixed with 5 cm^3 of starch suspension. Samples of the mixture were tested for the presence of starch.

The student recorded the time taken for the complete digestion of starch.

The investigation was repeated with the same individuals on the following day.

The results of the student's investigation are shown in the table.

Time taken for complete digestion of starch / min	Number of individuals	
	day 1	day 2
35	2	8
40	6	10
45	9	4
50	5	2
55	3	1

a Plot a graph to display these data. [4]

b Describe the patterns in the results. [3]

c Suggest a reason for the differences between the results for day 1 and day 2. [1]

d Suggest how you might control the variables in this investigation to compare a different species of mammal with the mammal studied. [3]

[Total: 11]

Cambridge International AS and A Level Biology 9700 Paper 33, Question 1b, November 2009

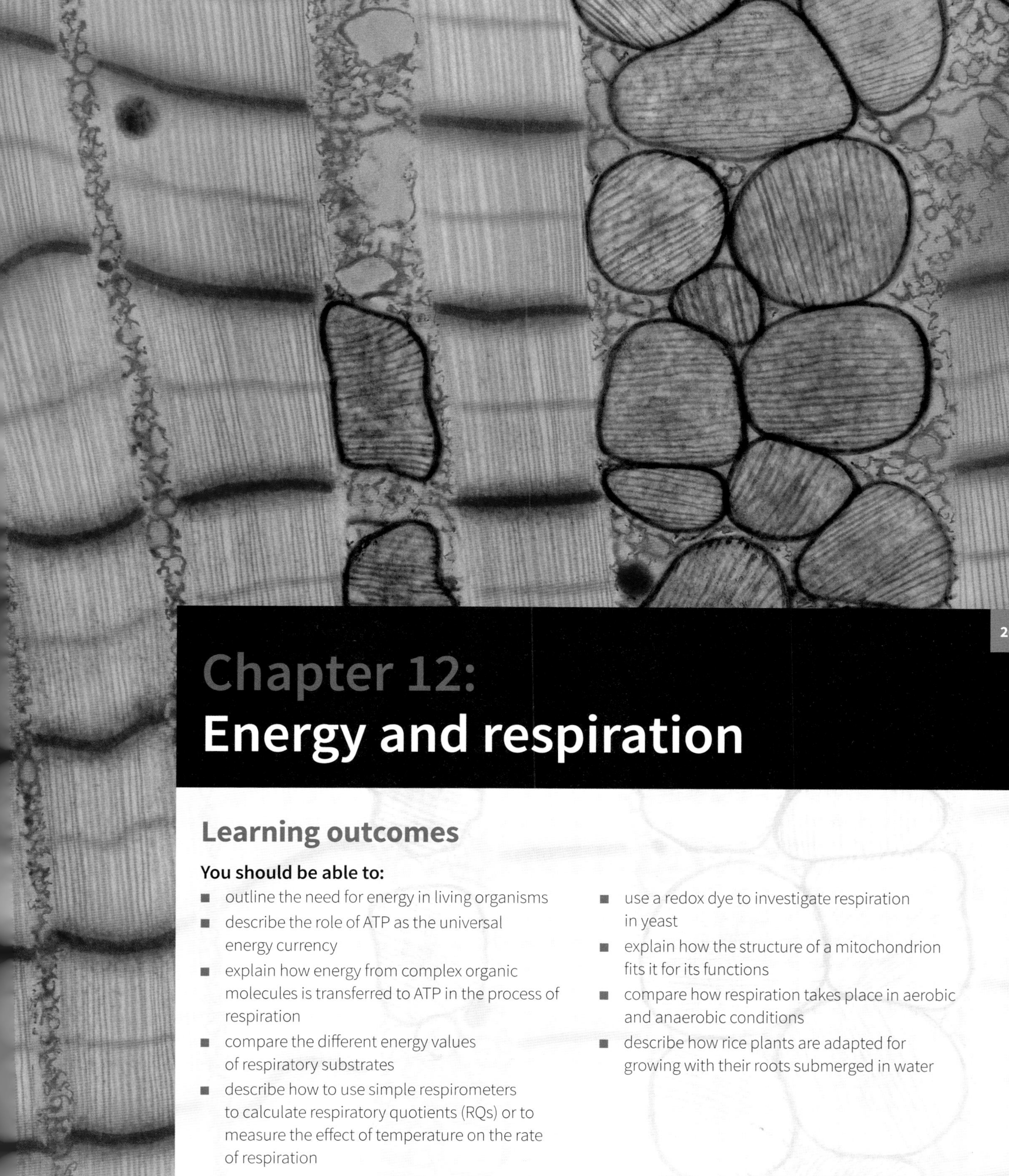

Chapter 12: Energy and respiration

Learning outcomes

You should be able to:

- outline the need for energy in living organisms
- describe the role of ATP as the universal energy currency
- explain how energy from complex organic molecules is transferred to ATP in the process of respiration
- compare the different energy values of respiratory substrates
- describe how to use simple respirometers to calculate respiratory quotients (RQs) or to measure the effect of temperature on the rate of respiration
- use a redox dye to investigate respiration in yeast
- explain how the structure of a mitochondrion fits it for its functions
- compare how respiration takes place in aerobic and anaerobic conditions
- describe how rice plants are adapted for growing with their roots submerged in water

A baby with three parents

The small circles of DNA in a mitochondrion code for the ribosomal RNA and transfer RNAs that the organelle needs to be able to synthesise proteins, and for 13 proteins involved in its function of respiration. Mutations of these genes result in serious metabolic disorders, particularly affecting tissues that have a high energy demand, such as nerve tissue, cardiac muscle and liver cells. These mutations are inherited from the mother, since only she contributes mitochondria to a fertilised egg. Some form of mitochondrial disorder is found in about 1 in 200 babies born worldwide each year. Of these, only around ten show very severe symptoms.

Women in danger of passing on a potentially fatal mitochondrial disorder to their babies can be helped by an in-vitro fertilisation treatment involving three biological parents: mother, father and a female donor with healthy mitochondria.

The nucleus is removed from an egg from the mother and inserted into a donor egg, from which the nucleus has been removed. Then the donor egg, with its healthy mitochondria and a nucleus from the mother, is placed in a dish and about 100 000 motile sperm cells from the father are added. Alternatively, the DNA of a single sperm cell from the father can be injected into the egg (Figure 12.1).

The fertilised egg is inserted into the mother's uterus, where a healthy embryo develops. The baby's chromosomal genes have been inherited from its mother and father. Only the mitochondrial genes have come from the donor; less than 0.1% of the baby's genes.

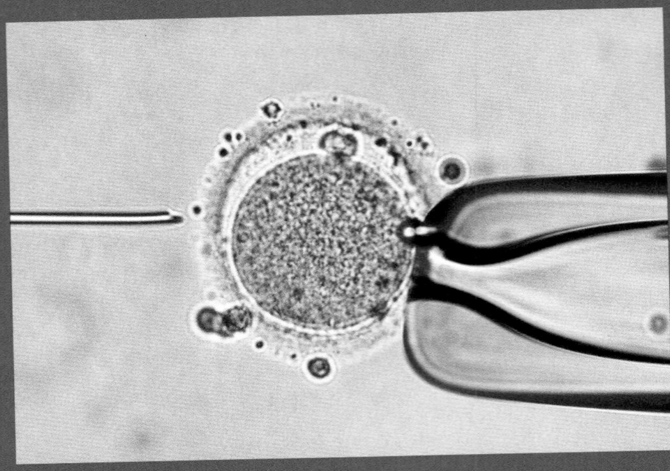

Figure 12.1 A microneedle (left) about to penetrate an egg to inject the DNA of a sperm cell. On the right, a flat-nosed pipette is used to hold the egg steady.

The need for energy in living organisms

All living organisms require a continuous supply of energy to stay alive, either from the absorption of light energy or from chemical potential energy (energy stored in nutrient molecules). The process of photosynthesis transfers light energy to chemical potential energy, and so almost all life on Earth depends on photosynthesis, either directly or indirectly. Photosynthesis supplies living organisms with two essential requirements: an energy supply and usable carbon compounds.

All biological macromolecules such as carbohydrates, lipids, proteins and nucleic acids contain carbon. All living organisms therefore need a source of carbon. Organisms that can use an inorganic carbon source in the form of carbon dioxide are called **autotrophs**. Those needing a ready-made organic supply of carbon are **heterotrophs**.

An organic molecule is a compound including carbon and hydrogen. The term originally meant a molecule derived from an organism, but now includes all compounds of carbon and hydrogen even if they do not occur naturally.

Organic molecules can be used by living organisms in two ways. They can serve as 'building bricks' for making other organic molecules that are essential to the organism, and they can represent chemical potential energy that can be released by breaking down the molecules in respiration (page 272). This energy can then be used for all forms of work. Heterotrophs depend on autotrophs for both materials and energy (Figure 12.2).

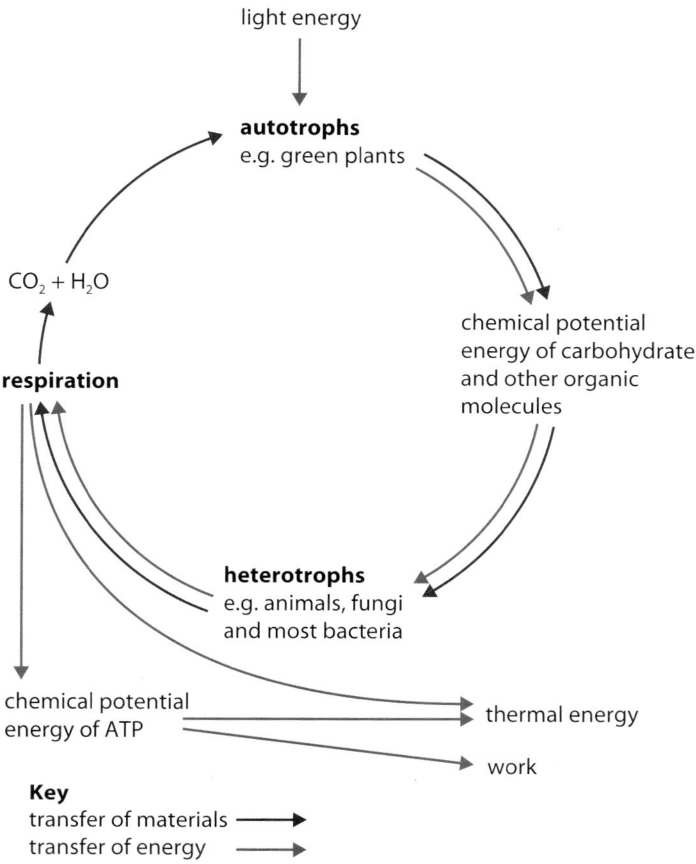

Figure 12.2 Transfer of materials and energy in an ecosystem.

Work

Work in a living organism includes:

- the synthesis of complex substances from simpler ones (anabolic reactions), such as the synthesis of polysaccharides from monosaccharides, lipids from glycerol and fatty acids, polypeptides from amino acids, and nucleic acids from nucleotides
- the active transport of substances against a diffusion gradient, such as the activity of the sodium–potassium pump (Figure 4.18, page 87)
- mechanical work such as muscle contraction (page 344) and other cellular movements; for example, the movement of cilia and flagella (page 189), amoeboid movement and the movement of vesicles through cytoplasm
- in a few organisms, bioluminescence and electrical discharge.

Mammals and birds use thermal energy (heat) that is released from metabolic reactions to maintain a constant body temperature. Most animals are ectotherms. The thermal energy that warms them comes from outside their bodies. Mammals and birds are endotherms, releasing enough thermal energy within their bodies to maintain them above the temperature of their surroundings when necessary. They also maintain a constant body temperature through negative feedback loops (page 301).

For a living organism to do work, energy-requiring reactions must be linked to those that yield energy. In the complete oxidation of glucose ($C_6H_{12}O_6$) in aerobic conditions, a large quantity of energy is made available:

$$C_6H_{12}O_6 + 6O_2 \rightarrow 6CO_2 + 6H_2O + 2870\,\text{kJ}$$

Reactions such as this take place in a series of small steps, each releasing a small quantity of the total available energy. Multi-step reactions allow precise control via feedback mechanisms (Chapter 3). Moreover, the cell could not usefully harness the total available energy if all of it were made available at one instant.

Although the complete oxidation of glucose to carbon dioxide and water has a very high energy yield, the reaction does not happen easily. Glucose is actually quite stable, because of the **activation energy** that has to be added before any reaction takes place (Figure 12.3). In living organisms, the activation energy is overcome by lowering it using enzymes (page 56), and also by raising the energy level of the glucose by phosphorylation (page 272).

Theoretically, the energy released from each step of respiration could be harnessed directly to some form of work in the cell. However, a much more flexible system actually occurs in which energy-yielding reactions in all organisms are used to make an intermediary molecule, **ATP**.

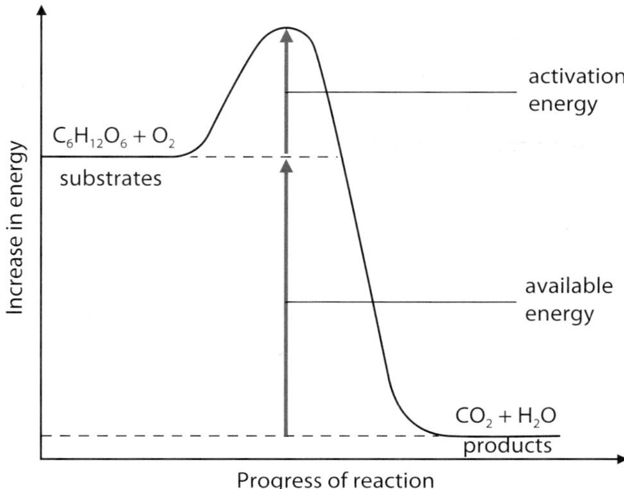

Figure 12.3 Oxidation of glucose.

ATP

ATP as energy 'currency'

Look back at the structure of ATP, shown in Figure 6.4, page 113.

When a phosphate group is removed from ATP, adenosine diphosphate (ADP) is formed and 30.5 kJ mol^{-1} of energy is released. Removal of a second phosphate produces adenosine monophosphate (AMP), and 30.5 kJ mol^{-1} of energy is again released. Removal of the last phosphate, leaving adenosine, releases only 14.2 kJ mol^{-1} (Figure 12.4). In the past, the bonds attaching the two outer phosphate groups have been called high-energy bonds, because more energy is released when they are broken than when the last phosphate is removed. This description is misleading and should be avoided, since the energy does not come simply from breaking those bonds, but rather from changes in chemical potential energy of all parts of the system.

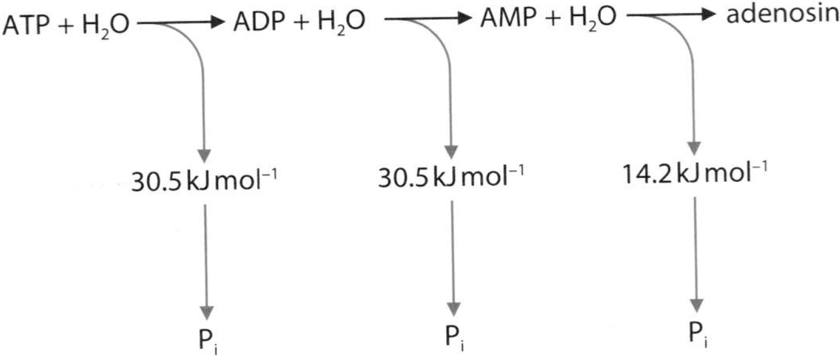

Figure 12.4 Hydrolysis of ATP (P_i is inorganic phosphate, H_3PO_4).

These reactions are all reversible. It is the interconversion of ATP and ADP that is all-important in providing energy for the cell:

$$ATP + H_2O \rightleftharpoons ADP + H_3PO_4 \pm 30.5\,kJ$$

The rate of interconversion, or turnover, is enormous. It is estimated that a resting human uses about 40 kg of ATP in 24 hours, but at any one time contains only about 5 g of ATP. During strenuous exercise, ATP breakdown may be as much as 0.5 kg per minute.

The cell's energy-yielding reactions are linked to ATP synthesis. The ATP is then used by the cell in all forms of work. ATP is the universal intermediary molecule between energy-yielding and energy-requiring reactions used in a cell, whatever the type of cell. In other words, ATP is the 'energy currency' of the cell. The cell 'trades' in ATP, rather than making use of a number of different intermediates. ATP is a highly suitable molecule for this role. Not only is it readily hydrolysed to release energy, it is also small and water-soluble. This allows it to be easily transported around the cell.

Energy transfers are inefficient. Some energy is converted to thermal energy whenever energy is transferred. At the different stages in a multi-step reaction such as respiration, the energy made available may not perfectly correspond with the energy needed to synthesise ATP. Any excess energy is converted to thermal energy. Also, many energy-requiring reactions in cells use less energy than that released by hydrolysis of ATP to ADP. Again, any extra energy will be released as thermal energy.

Be careful to distinguish between molecules used as energy currency and as energy storage. An energy currency molecule acts as the immediate donor of energy to the cell's energy-requiring reactions. An energy storage molecule is a short-term (glucose or sucrose) or long-term (glycogen, starch or triglyceride) store of chemical potential energy.

Synthesis of ATP

Energy for ATP synthesis can become available in two ways. In respiration, energy released by reorganising chemical bonds (chemical potential energy) during glycolysis and the Krebs cycle (pages 272–273) is used to make some ATP. However, most ATP in cells is generated using electrical potential energy. This energy is from the transfer of electrons by electron carriers in mitochondria and chloroplasts. It is stored as a difference in proton (hydrogen ion) concentration across some phospholipid membranes in mitochondria and chloroplasts, which are essentially impermeable to protons. Protons are then allowed to flow down their concentration gradient (by facilitated diffusion) through a protein that spans the phospholipid bilayer. Part of this protein acts as an enzyme that synthesises ATP and is called ATP synthase. The transfer of three protons allows the production of one ATP molecule, provided that ADP and an inorganic phosphate group (P_i) are available inside the organelle. This process occurs in both mitochondria (page 275) and chloroplasts (page 288) and is summarised in Figure 12.5. The process was first proposed by Peter Mitchell in 1961 and is called chemiosmosis.

Note that a hydrogen atom consists of one proton and one electron. The loss of an electron forms a hydrogen ion, which is a single proton.

$$H \rightleftharpoons H^+ + e^-$$

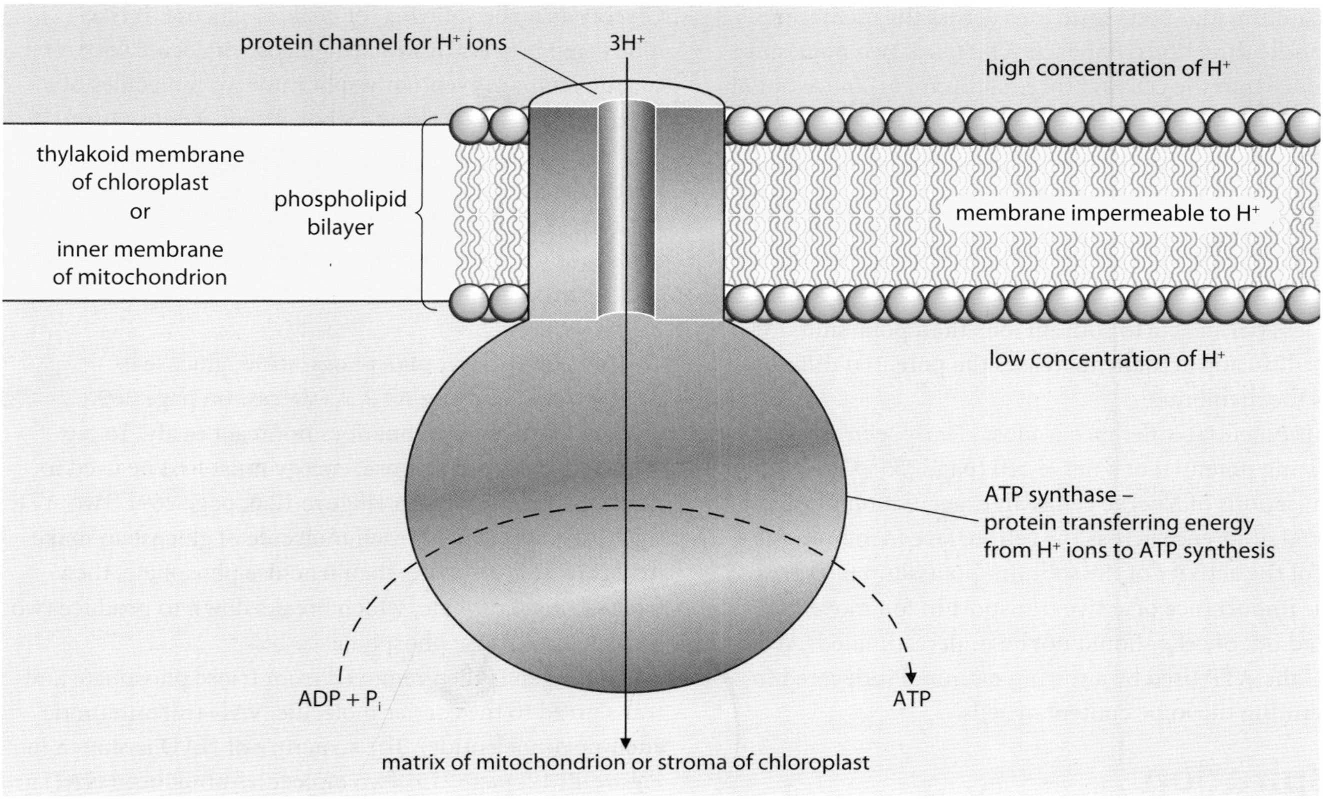

Figure 12.5 ATP synthesis (chemiosmosis).

ATP synthase has three binding sites (Figure 12.6) and a part of the molecule (γ) that rotates as hydrogen ions (H^+) pass. This produces structural changes in the binding sites and allows them to pass sequentially through three phases:

- binding ADP and P_i
- forming tightly bound ATP
- releasing ATP.

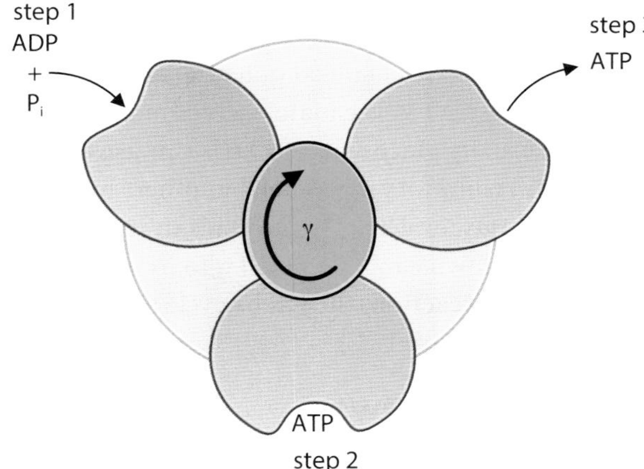

Figure 12.6 Transverse section (TS) of ATP synthase showing its activity.

QUESTION

12.1 Write the equation for the reaction catalysed by ATP synthase.

The role of ATP in active transport

Active transport is the movement of molecules or ions across a partially permeable membrane against a concentration gradient. Energy is needed, from ATP, to counteract the tendency of these particles to move by diffusion down the gradient.

All cells show differences in the concentration of ions, in particular sodium and potassium ions, inside the cell with respect to the surrounding solution. Most cells seem to have sodium pumps in the cell surface membrane that pump sodium ions out of the cell. This is usually coupled with the ability to pump potassium ions from the surrounding solution into the cell.

The sodium–potassium pump is a protein that spans the cell surface membrane (Figure 4.18, page 87). It has binding sites for sodium ions (Na^+) and for ATP on the inner side, and for potassium ions (K^+) on the outer side. The protein acts as an ATPase and catalyses the hydrolysis of ATP to ADP and inorganic phosphate, releasing energy to drive the pump. Changes in the shape of the protein move sodium and potassium ions across the membrane in opposite directions. For each ATP used, two potassium ions move into the cell and three sodium ions move out of the cell. As only two potassium ions are added to the cell contents for every three sodium ions removed, a potential difference is created across the membrane that is negative inside with respect to the outside. Both sodium and potassium ions leak back across the membrane, down their diffusion gradients. However, cell surface membranes are much less permeable to sodium ions than potassium ions, so this diffusion actually increases the potential difference across the membrane.

This potential difference is most clearly seen as the resting potential of a nerve cell (page 333). One of the specialisations of a nerve cell is an exaggeration of the potential difference across the cell surface membrane as a result of the activity of the sodium–potassium pump.

The importance of active transport in ion movement into and out of cells should not be underestimated. About 50% of the ATP used by a resting mammal is devoted to maintaining the ionic content of cells.

Respiration

Respiration is a process in which organic molecules act as a fuel. The organic molecules are broken down in a series of stages to release chemical potential energy, which is used to synthesise ATP. The main fuel for most cells is carbohydrate, usually glucose. Many cells can use only glucose as their respiratory substrate, but others break down fatty acids, glycerol and amino acids in respiration.

Glucose breakdown can be divided into four stages: **glycolysis**, the **link reaction**, the **Krebs cycle** and **oxidative phosphorylation** (Figure 12.7).

The glycolytic pathway

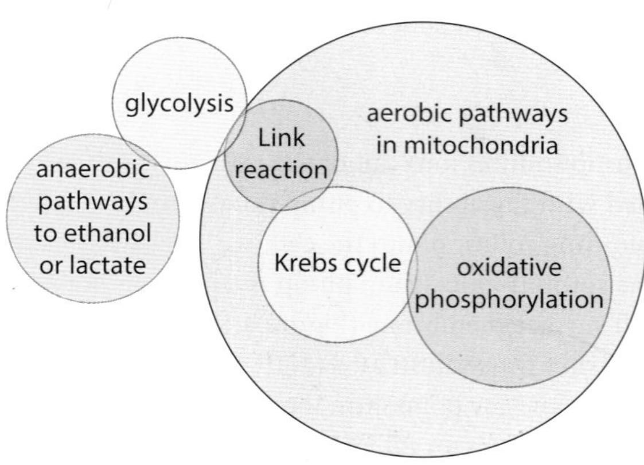

Figure 12.7 The sequence of events in respiration.

Glycolysis is the splitting, or lysis, of glucose. It is a multi-step process in which a glucose molecule with six carbon atoms is eventually split into two molecules of pyruvate, each with three carbon atoms. Energy from ATP is needed in the first steps, but energy is released in later steps, when it can be used to make ATP. There is a net gain of two ATP molecules per molecule of glucose broken down. Glycolysis takes place in the cytoplasm of a cell. A simplified flow diagram of the pathway is shown in Figure 12.8.

In the first stage, phosphorylation, glucose is phosphorylated using ATP. As we saw on page 269, glucose is energy-rich but does not react easily. To tap the bond energy of glucose, energy must first be used to make the reaction easier (Figure 12.3, page 269). Two ATP molecules are used for each molecule of glucose to make first glucose phosphate, then fructose phosphate, then fructose bisphosphate, which breaks down to produce two molecules of triose phosphate.

Hydrogen is then removed from triose phosphate and transferred to the carrier molecule NAD (nicotinamide adenine dinucleotide). The structure of NAD is shown in Figure 12.12, page 275. Two molecules of reduced NAD are produced for each molecule of glucose entering glycolysis. The hydrogens carried by reduced NAD can easily be transferred to other molecules and are used in oxidative phosphorylation to generate ATP (page 273).

The end-product of glycolysis, pyruvate, still contains a

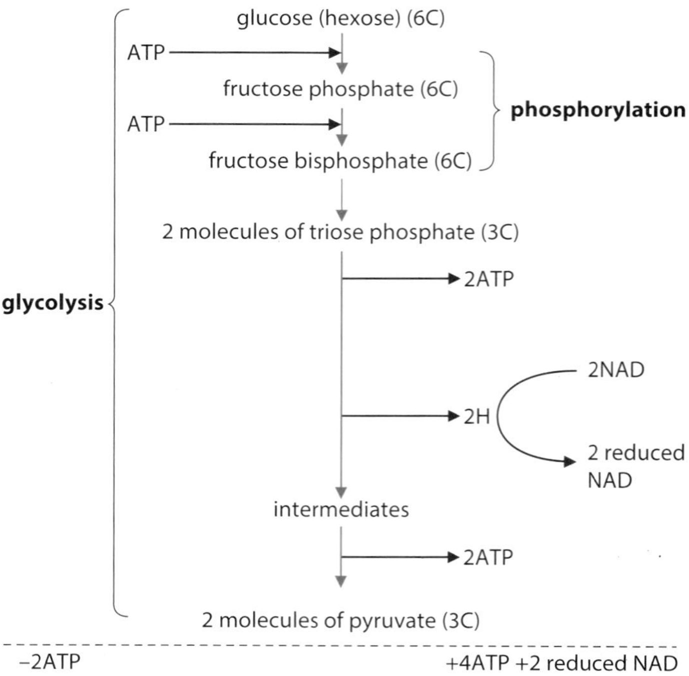

Figure 12.8 The glycolytic pathway.

great deal of chemical potential energy. When free oxygen is available, some of this energy can be released via the Krebs cycle and oxidative phosphorylation. However, the pyruvate first enters the link reaction, which takes place in the mitochondria (page 275).

The link reaction

Pyruvate passes by active transport from the cytoplasm, through the outer and inner membranes of a mitochondrion and into the mitochondrial matrix. Here it is decarboxylated (this means that carbon dioxide is removed), dehydrogenated (hydrogen is removed) and combined with coenzyme A (CoA) to give acetyl coenzyme A. This is known as the **link reaction** (Figure 12.9). Coenzyme A is a complex molecule composed of a nucleoside (adenine plus ribose) with a vitamin (pantothenic acid), and acts as a carrier of acetyl groups to the Krebs cycle. The hydrogen removed from pyruvate is transferred to NAD.

$$\text{pyruvate} + \text{CoA} + \text{NAD} \rightleftharpoons \text{acetyl CoA} + CO_2 + \text{reduced NAD}$$

Fatty acids from fat metabolism may also be used to produce acetyl coenzyme A. Fatty acids are broken down in the mitochondrion in a cycle of reactions in which each turn of the cycle shortens the fatty acid chain by a two-carbon acetyl unit. Each of these can react with coenzyme A to produce acetyl coenzyme A, which, like that produced from pyruvate, now enters the Krebs cycle.

The Krebs cycle

The **Krebs cycle** (also known as the citric acid cycle or tricarboxylic acid cycle) was discovered in 1937 by Hans Krebs. It is shown in Figure 12.9.

The Krebs cycle is a closed pathway of enzyme-controlled reactions.

- Acetyl coenzyme A combines with a four-carbon compound (oxaloacetate) to form a six-carbon compound (citrate).
- The citrate is decarboxylated and dehydrogenated in a series of steps, to yield carbon dioxide, which is given off as a waste gas, and hydrogens which are accepted by the carriers NAD and FAD (page 275).
- Oxaloacetate is regenerated to combine with another acetyl coenzyme A.

For each turn of the cycle, two carbon dioxide molecules are produced, one FAD and three NAD molecules are reduced, and one ATP molecule is generated via an intermediate compound.

Although part of aerobic respiration, the reactions

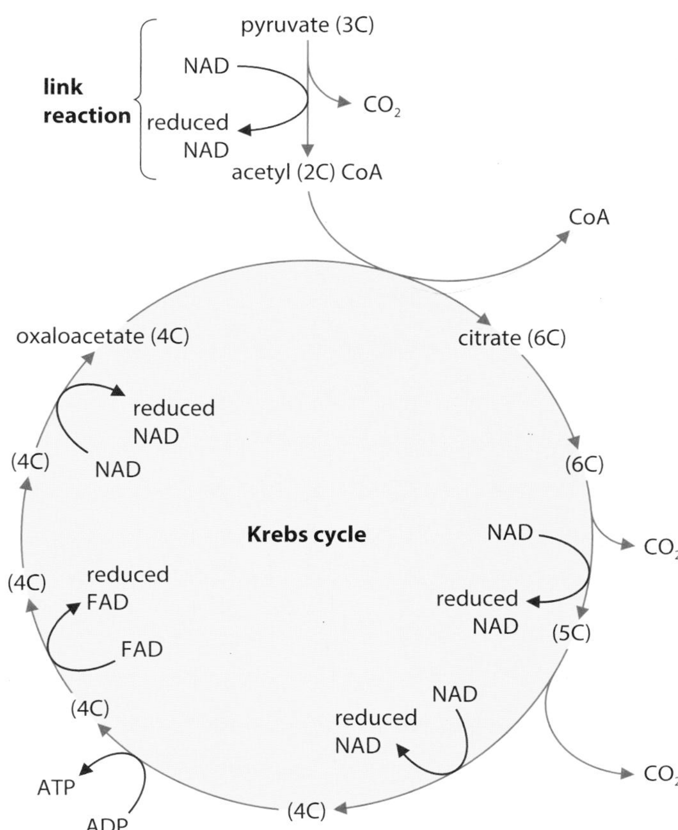

Figure 12.9 The link reaction and the Krebs cycle.

of the Krebs cycle make no use of molecular oxygen. However, oxygen is necessary for the final stage of aerobic respiration, which is called oxidative phosphorylation.

The most important contribution of the Krebs cycle to the cell's energetics is the release of hydrogens, which can be used in oxidative phosphorylation to provide energy to make ATP.

QUESTION

12.2 Explain how the events of the Krebs cycle can be cyclical.

Oxidative phosphorylation and the electron transport chain

In the final stage of aerobic respiration, **oxidative phosphorylation**, the energy for the phosphorylation of ADP to ATP comes from the activity of the **electron transport chain**. Oxidative phosphorylation takes place in the inner mitochondrial membrane (Figure 12.10).

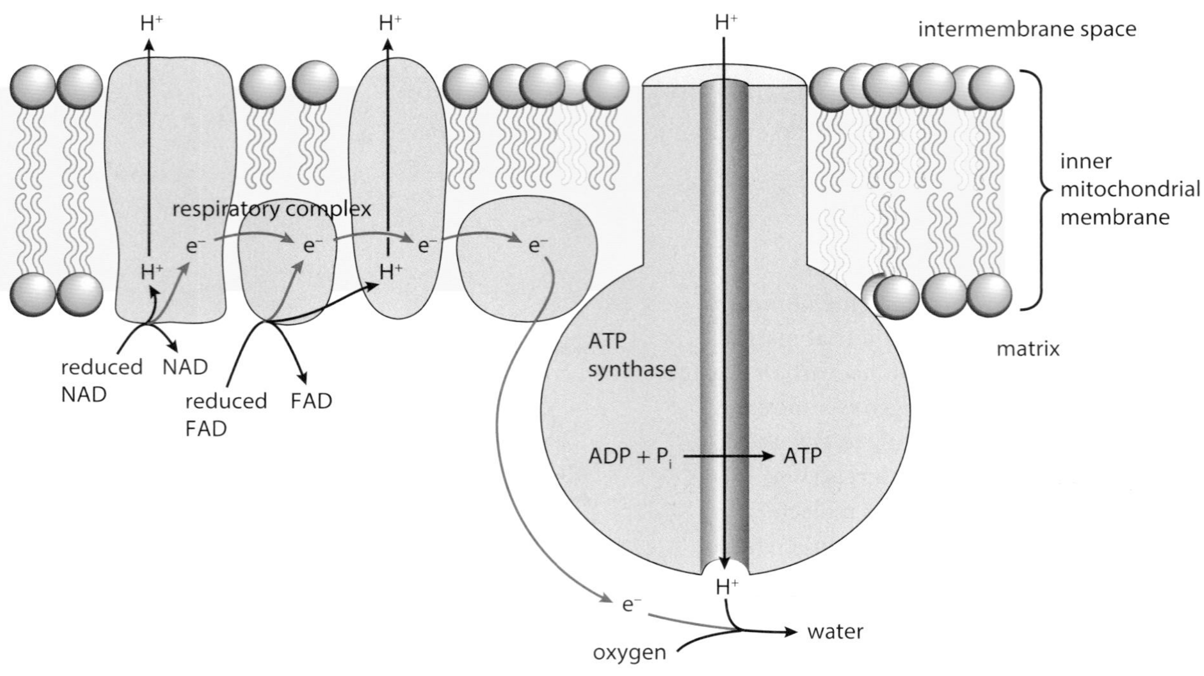

Figure 12.10 Oxidative phosphorylation: the electron transport chain.

Reduced NAD and reduced FAD are passed to the electron transport chain. Here, the hydrogens are removed from the two hydrogen carriers and each is split into its constituent proton (H^+) and electron (e^-). The energetic electron is transferred to the first of a series of electron carriers.

Most of the carriers are associated with membrane proteins, of which there are four types. A functional unit, called a respiratory complex, consists of one of each of these proteins, arranged in such a way that electrons can be passed from one to another down an energy gradient.

As an electron moves from one carrier at a higher energy level to another one at a lower level, energy is released. Some of this energy is used to move protons from the matrix of the mitochondrion (Figure 12.11) into the space between the inner and outer membranes of the mitochondrial envelope. This produces a higher concentration of protons in the intermembrane space than in the matrix, setting up a concentration gradient.

Now, protons pass back into the mitochondrial matrix through protein channels in the inner membrane, moving down their concentration gradient. Associated with each channel is the enzyme ATP synthase. As the protons pass through the channel, their electrical potential energy is used to synthesise ATP in the process called chemiosmosis (Figure 12.5, page 271).

Finally, oxygen has a role to play as the final electron acceptor. In the mitochondrial matrix, an electron and a proton are transferred to oxygen, reducing it to water. The process of aerobic respiration is complete.

The sequence of events in respiration and their sites are shown in Figure 12.11. The balance sheet of ATP used and synthesised for each molecule of glucose entering the respiration pathway is shown in Table 12.1.

Theoretically, three molecules of ATP can be produced from each molecule of reduced NAD, and two molecules of ATP from each molecule of reduced FAD. However, this yield cannot be achieved unless ADP and P_i are available inside the mitochondrion. About 25% of the total energy yield of electron transfer is used to transport ADP into the mitochondrion and ATP into the cytoplasm. Hence, each reduced NAD

	ATP used	**ATP made**	**Net gain in ATP**
glycolysis	−2	4	+2
link reaction	0	0	0
Krebs cycle	0	2	+2
oxidative phosphorylation	0	28	+28
Total	−2	**34**	**+32**

Table 12.1 Balance sheet of ATP use and synthesis for each molecule of glucose entering respiration.

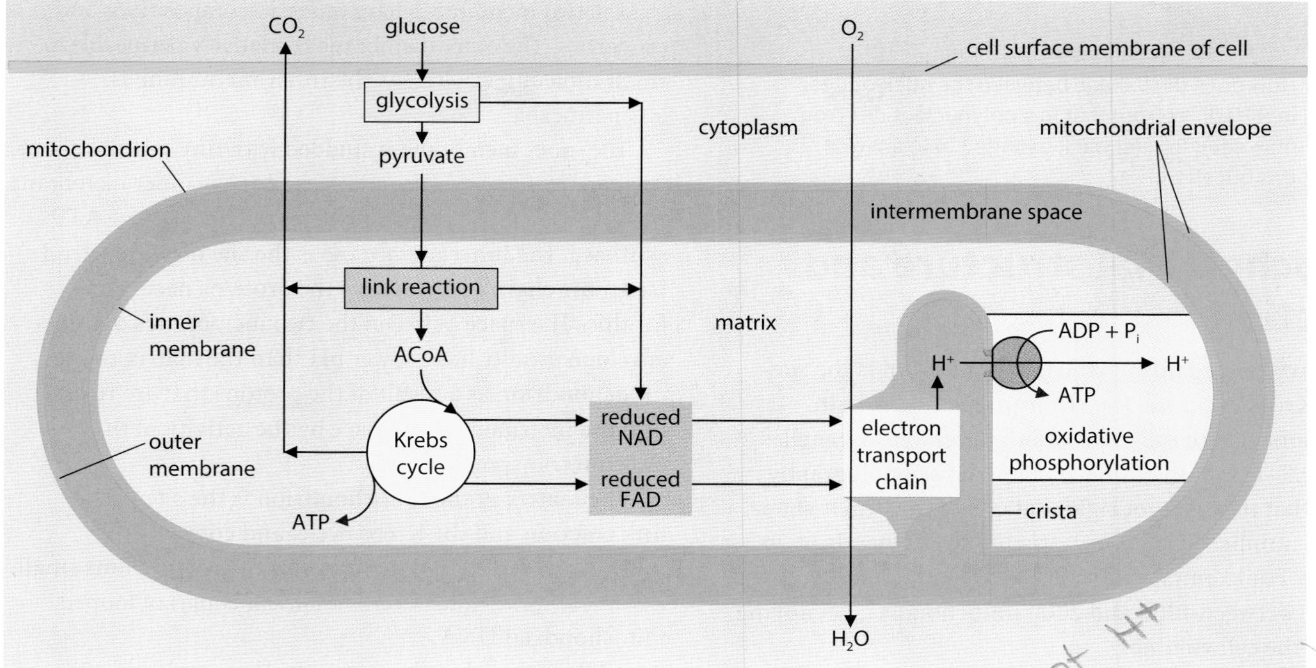

Figure 12.11 The sites of the events of respiration in a cell. ACoA = acetyl coenzyme A.

molecule entering the chain produces on average two and a half molecules of ATP, and each reduced FAD produces one and a half molecules of ATP. The number of ATP molecules actually produced varies in different tissues and different circumstances, largely dependent on how much energy is used to move substances into and out of the mitochondria.

Hydrogen carrier molecules

NAD is made of two linked nucleotides (Figure 12.12). Both nucleotides contain ribose. One nucleotide contains the nitrogenous base adenine. The other has a nicotinamide ring, which can accept a hydrogen ion and two electrons, thereby becoming reduced.

$$\text{NAD} + 2\text{H} \rightleftharpoons \text{reduced NAD}$$
$$\text{NAD}^+ + 2\text{H} \rightleftharpoons \text{NADH}^+ + \text{H}^+$$

A slightly different form of NAD has a phosphate group instead of the hydrogen on carbon 1 in one of the ribose rings. This molecule is called NADP (nicotinamide adenine dinucleotide phosphate) and is used as a hydrogen carrier molecule in photosynthesis.

FAD (flavin adenine dinucleotide) is similar in function to NAD and is used in respiration in the Krebs cycle. FAD is made of one nucleotide containing ribose and adenine and one with an unusual structure involving a linear molecule, ribitol, instead of ribose.

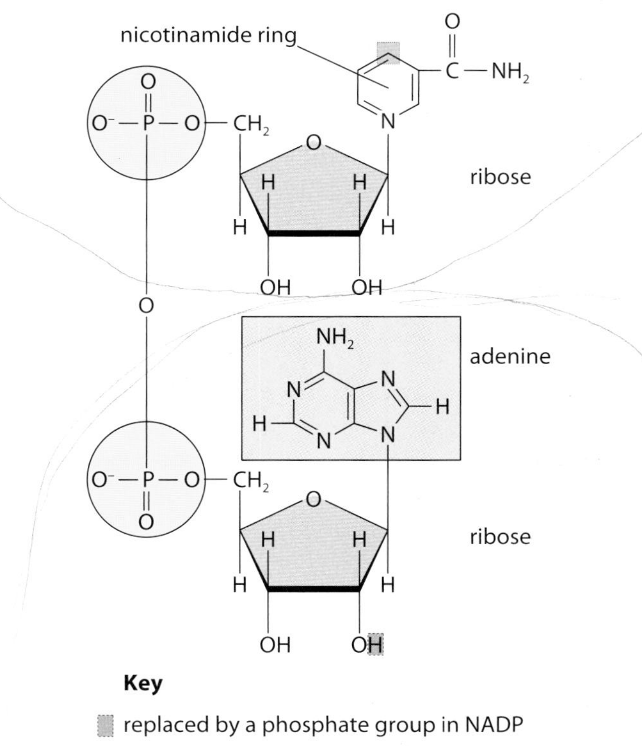

Figure 12.12 NAD (nicotinamide adenine dinucleotide). You do not need to learn the structure of this molecule, but may like to compare it with the units that make up DNA and RNA.

QUESTION

12.3 How does the linkage between the nucleotides in NAD differ from that in a polynucleotide? (You may need to refer back to page 114 to answer this question.)

Mitochondrial structure and function

In eukaryotic organisms, the mitochondrion is the site of the Krebs cycle and the electron transport chain. Mitochondria are rod-shaped or filamentous organelles about 0.5–1.0 μm in diameter. Time-lapse photography shows that they are not rigid, but can change their shape.

The number of mitochondria in a cell depends on its activity. For example, highly active mammalian liver cells contain between 1000 and 2000 mitochondria, occupying 20% of the cell volume.

The structure of a mitochondrion is shown in Figures 1.22 (page 16) and 12.13. Like a chloroplast, each mitochondrion is surrounded by an envelope of two phospholipid membranes (page 74). The outer membrane is smooth, but the inner is much folded inwards to form cristae (singular: crista). These cristae give the inner membrane a large total surface area. Cristae in mitochondria from different types of cell show considerable variation, but, in general, mitochondria from active cells have longer, more densely packed cristae than mitochondria from less active cells.

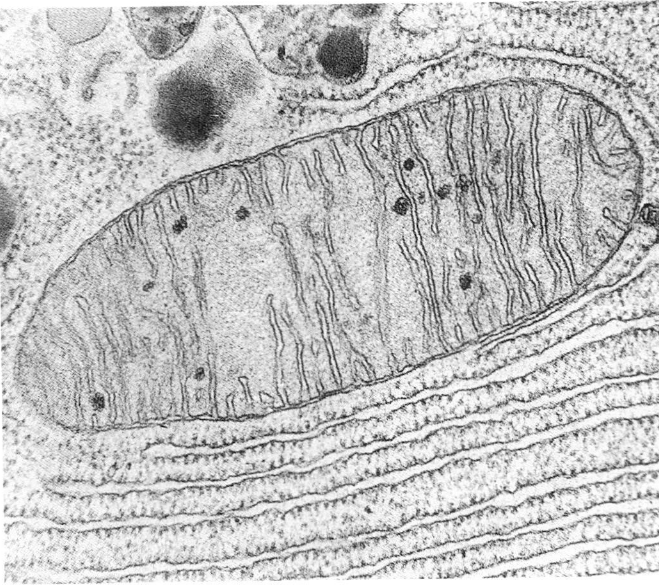

Figure 12.13 Transmission electron micrograph of a mitochondrion from a pancreas cell (×15 000).

The two membranes have different compositions and properties. The outer membrane is relatively permeable to small molecules, whereas the inner membrane is less permeable.

The inner membrane is studded with tiny spheres, about 9 nm in diameter, which are attached to the inner membrane by stalks (Figure 12.14). The spheres are the enzyme **ATP synthase**. The inner membrane is the site of the electron transport chain and contains the proteins necessary for this. The space between the two membranes of the envelope usually has a lower pH than the matrix of the mitochondrion as a result of the protons that are released into the intermembrane space by the activity of the electron transport chain.

The matrix of the mitochondrion is the site of the link reaction and the Krebs cycle, and contains the enzymes needed for these reactions. It also contains small (70 S) ribosomes and several identical copies of looped mitochondrial DNA.

ATP is formed in the matrix by the activity of ATP synthase on the cristae. The energy for the production of ATP comes from the proton gradient between the intermembrane space and the matrix. The ATP can be used for all the energy-requiring reactions of the cell, both inside and outside the mitochondrion.

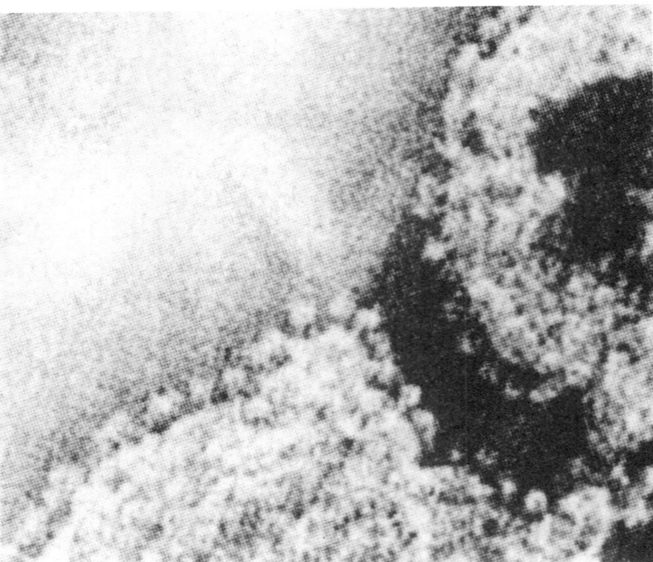

Figure 12.14 Transmission electron micrograph of ATP synthase particles on the inner membrane of a mitochondrion (×400 000).

QUESTIONS

12.4 Calculate the number of reduced NAD and reduced FAD molecules produced for each molecule of glucose entering the respiration pathway when oxygen is available.

12.5 Using your answer to Question **12.4**, calculate the number of ATP molecules produced for each molecule of glucose in oxidative phosphorylation.

12.6 Explain why the important contribution of the Krebs cycle to cellular energetics is the release of hydrogens and not the direct production of ATP.

12.7 Explain how the structure of a mitochondrion is adapted for its functions in aerobic respiration.

Respiration without oxygen

When free oxygen is not present, hydrogen cannot be disposed of by combination with oxygen. The electron transfer chain therefore stops working and no further ATP is formed by oxidative phosphorylation. If a cell is to gain even the two ATP molecules for each glucose yielded by glycolysis, it is essential to pass on the hydrogens from the molecules of reduced NAD that are made in glycolysis. There are two different anaerobic pathways that solve the problem of 'dumping' this hydrogen. Both pathways take place in the cytoplasm of the cell.

In various microorganisms such as yeast, and in some plant tissues, the hydrogen from reduced NAD is passed to ethanal (CH_3CHO). This releases the NAD and allows glycolysis to continue. The pathway is shown in Figure 12.15. First, pyruvate is decarboxylated to ethanal; then the ethanal is reduced to ethanol (C_2H_5OH) by the enzyme alcohol dehydrogenase. The conversion of glucose to ethanol is referred to as **alcoholic fermentation**.

In other microorganisms, and in mammalian muscles when deprived of oxygen, pyruvate acts as the hydrogen acceptor and is converted to **lactate** by the enzyme lactate dehydrogenase (named after the reverse reaction, which it also catalyses). Again, the NAD is released and allows glycolysis to continue in anaerobic conditions. This pathway, known as **lactic fermentation**, is shown in Figure 12.16.

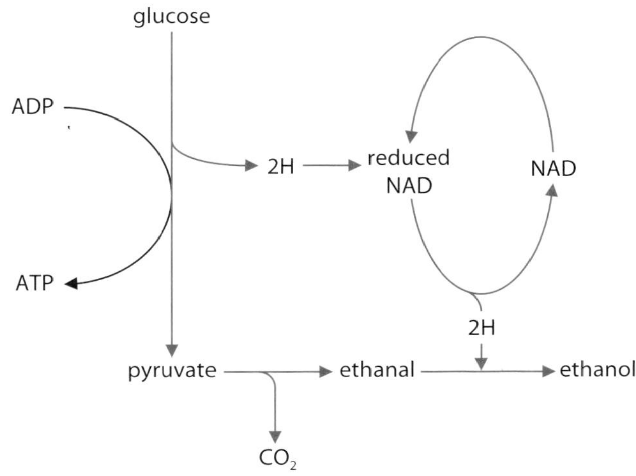

Figure 12.15 Alcoholic fermentation.

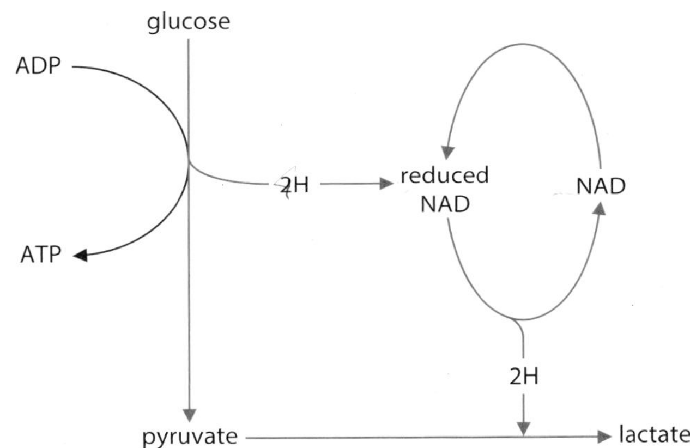

Figure 12.16 Lactic fermentation.

These reactions 'buy time'. They allow the continued production of at least some ATP even though oxygen is not available as the hydrogen acceptor. However, as the products of anaerobic reaction, ethanol or lactate, are toxic, the reactions cannot continue indefinitely. The pathway leading to ethanol cannot be reversed, and the remaining chemical potential energy of ethanol is wasted. The lactate pathway can be reversed in mammals. Lactate is carried by the blood plasma to the liver and converted back to pyruvate. The liver oxidises some (20%) of the incoming lactate to carbon dioxide and water via aerobic respiration when oxygen is available again. The remainder of the lactate is converted by the liver to glycogen.

Figure 12.17 shows what happens to the oxygen uptake of a person before, during and after taking strenuous exercise. Standing still, the person absorbs oxygen at the resting rate of $0.2\,dm^3\,min^{-1}$. (This is a measure of the person's metabolic rate.) When exercise begins, more oxygen is needed to support aerobic respiration in the person's muscles, increasing the overall demand to $2.5\,dm^3\,min^{-1}$. However, it takes four minutes for the heart and lungs to meet this demand, and during this time lactic fermentation occurs in the muscles. Thus the person builds up an oxygen deficit. For the next three minutes, enough oxygen is supplied. When exercise stops, the person continues to breathe deeply and absorb oxygen at a higher rate than when at rest. This post-exercise uptake of extra oxygen, which is 'paying back' the oxygen deficit, is called the oxygen debt. The oxygen is needed for:

- conversion of lactate to glycogen in the liver
- reoxygenation of haemoglobin in the blood
- a high metabolic rate, as many organs are operating at above resting levels.

Respiratory substrates

Although glucose is the essential respiratory substrate for some cells such as neurones in the brain, red blood cells and lymphocytes, other cells can oxidise lipids and amino acids. When lipids are respired, carbon atoms are removed in pairs, as acetyl coenzyme A, from the fatty acid chains and fed into the Krebs cycle. The carbon–hydrogen skeletons of amino acids are converted into pyruvate or into acetyl coenzyme A.

Energy values of respiratory substrates

Most of the energy liberated in aerobic respiration comes from the oxidation of hydrogen to water when reduced NAD and reduced FAD are passed to the electron transport chain. Hence, the greater the number of hydrogens in the structure of the substrate molecule, the greater the energy value. Fatty acids have more hydrogens per molecule than carbohydrates do, and so lipids have a greater energy value per unit mass, or **energy density**, than carbohydrates or proteins. The energy value of a substrate is determined by burning a known mass of the substance in oxygen in a calorimeter (Figure 12.18).

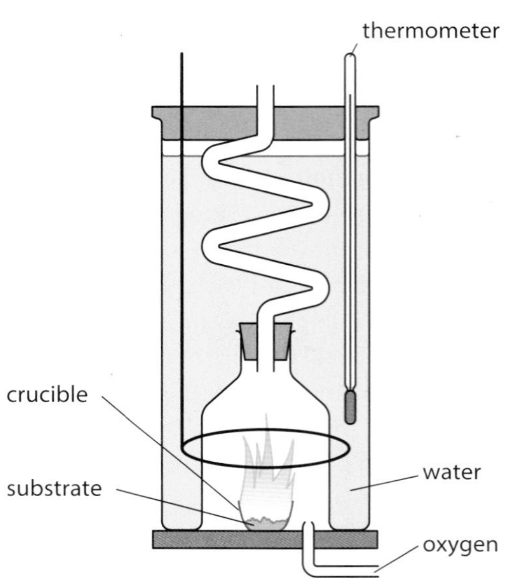

Figure 12.18 A simple calorimeter in which the energy value of a respiratory substrate can be measured.

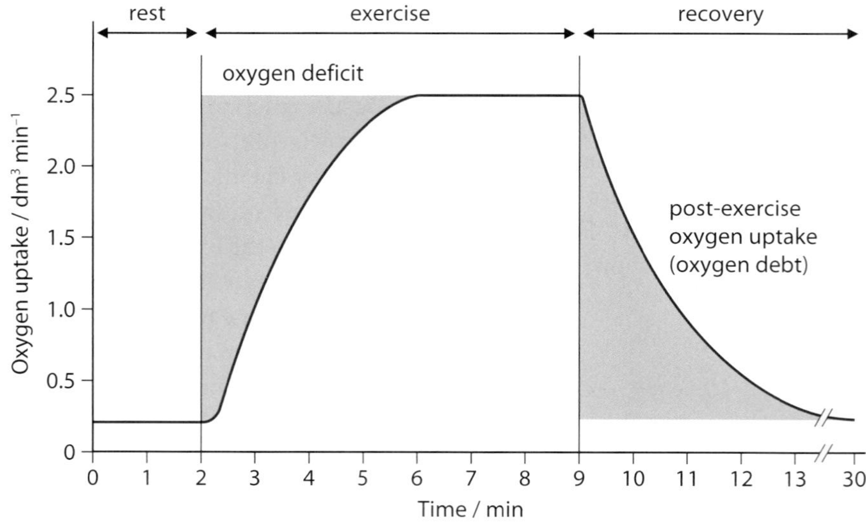

Figure 12.17 Oxygen uptake before, during and after strenuous exercise.

The energy liberated by oxidising the substrate can be determined from the rise in temperature of a known mass of water in the calorimeter. Typical energy values are shown in Table 12.2.

Respiratory substrate	Energy density / $kJ\,g^{-1}$
carbohydrate	15.8
lipid	39.4
protein	17.0

Table 12.2 Typical energy values.

Respiratory quotient (RQ)

The overall equation for the aerobic respiration of glucose shows that the number of molecules, and hence the volumes, of oxygen used and carbon dioxide produced are the same:

$$C_6H_{12}O_6 + 6O_2 \rightarrow 6CO_2 + 6H_2O + \text{energy}$$

So the **ratio** of oxygen taken in and carbon dioxide released is 1 : 1. However, when other substrates are respired, the ratio of the volumes of oxygen used and carbon dioxide given off differ. It follows that measuring this ratio, called the respiratory quotient (RQ), shows what substrate is being used in respiration. It can also show whether or not anaerobic respiration is occurring.

$$RQ = \frac{\text{volume of carbon dioxide given out in unit time}}{\text{volume of oxygen taken in in unit time}}$$

Or, from an equation,

$$RQ = \frac{\text{moles or molecules of carbon dioxide given out}}{\text{moles or molecules of oxygen taken in}}$$

For the aerobic respiration of glucose:

$$RQ = \frac{CO_2}{O_2} = \frac{6}{6} = 1.0$$

When the fatty acid oleic acid (from olive oil) is respired aerobically, the equation is:

$$C_{18}H_{34}O_2 + 25.5O_2 \rightarrow 18CO_2 + 17H_2O + \text{energy}$$

For the aerobic respiration of oleic acid:

$$RQ = \frac{CO_2}{O_2} = \frac{18}{25.5} = 0.7$$

Typical RQs for the aerobic respiration of different substrates are shown in Table 12.3.

Respiratory substrate	Respiratory quotient (RQ)
carbohydrate	1.0
lipid	0.7
protein	0.9

Table 12.3 Respiratory quotients of different substrates.

QUESTION

12.8 Calculate the RQ for the aerobic respiration of the fatty acid stearic acid ($C_{18}H_{36}O_2$).

What happens when respiration is not aerobic? The equation for the alcoholic fermentation of glucose in a yeast cell is:

$$C_6H_{12}O_6 \rightarrow 2C_2H_5OH + 2CO_2 + \text{energy}$$

$$RQ = \frac{CO_2}{O_2} = \frac{2}{0} = \infty$$

In reality, some respiration in the yeast cell will be aerobic, and so a small volume of oxygen will be taken up and the RQ will be less than infinity. High values of RQ indicate that alcoholic fermentation is occurring. Note that no RQ can be calculated for muscle cells using the lactate pathway, as no carbon dioxide is produced:

$$\text{glucose } (C_6H_{12}O_6) \rightarrow 2 \text{ lactic acid } (C_3H_6O_3) + \text{energy}$$

BOX 12.1: Oxygen uptake

Oxygen uptake during respiration can be measured using a **respirometer**. A respirometer suitable for measuring the rate of oxygen consumption of seeds or small terrestrial invertebrates at different temperatures is shown in Figure 12.19.

Carbon dioxide produced in respiration is absorbed by a suitable chemical such as soda-lime or a concentrated solution of potassium hydroxide (KOH) or sodium hydroxide (NaOH). Any decrease in the volume of air surrounding the organisms results from their oxygen consumption. Oxygen consumption in unit time can be measured by reading the level of the manometer fluid against the scale.

Changes in temperature and pressure alter the volume of air in the apparatus, and so the temperature of the surroundings must be kept constant while readings are taken – for example, by using a thermostatically controlled water bath. The presence of a control tube containing an equal volume of inert material to the volume of the organisms used helps to compensate for changes in atmospheric pressure.

Once measurements have been taken at a series of temperatures, a graph can be plotted of oxygen consumption against temperature.

The same apparatus can be used to measure the RQ of an organism. First, oxygen consumption at a particular temperature is found ($x\,\text{cm}^3\,\text{min}^{-1}$). Then the respirometer is set up with the same organism at the same temperature, but with no chemical to absorb carbon dioxide. The manometer scale will show whether the volumes of oxygen absorbed and carbon dioxide produced are the same. When the volumes **are** the same, the level of the manometer fluid will not change and the RQ = 1. When more carbon dioxide is produced than oxygen absorbed, the scale will show an increase in the volume of air in the respirometer (by $y\,\text{cm}^3\,\text{min}^{-1}$). The RQ can then be calculated:

$$RQ = \frac{CO_2}{O_2} = \frac{x+y}{x}$$

Conversely, when less carbon dioxide is produced than oxygen absorbed, the volume of air in the respirometer will decrease (by $z\,\text{cm}^3\,\text{min}^{-1}$) and the calculation will be:

$$RQ = \frac{CO_2}{O_2} = \frac{x-z}{x}$$

Another way of investigating the rate of respiration of yeast is to use a redox dye such as a solution of dichlorophenolindophenol (DCPIP) (Box 13.1 and page 289) or of methylene blue. These dyes do not damage cells and so can be added to a suspension of yeast cells. When reduced, these blue dyes become colourless. The rate of change from blue to colourless is a measure of the rate of respiration of the yeast.

This technique can be used to investigate the effect of various factors on yeast respiration, such as temperature, substrate concentration or different substrates.

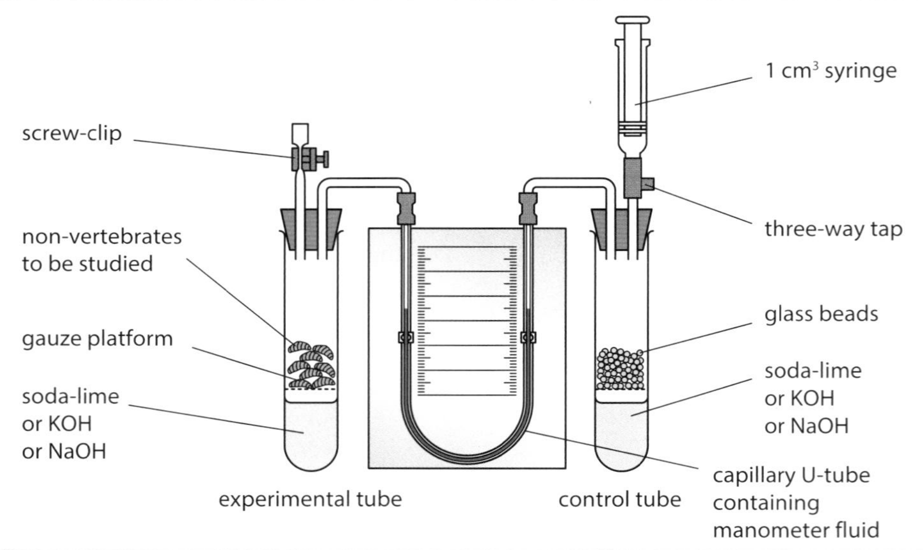

Figure 12.19 A respirometer.

QUESTION

12.9 Outline the steps you would take to investigate the effect of temperature on respiration rate.

Adaptations of rice for wet environments

Although rice can grow in dry conditions, it is often grown in 'paddies' – fields where the ground is intentionally flooded. Rice can tolerate growing in water, whereas most of the weeds that might compete with it are not able to do so (Figure 12.20).

Most plants cannot grow in deep water because their roots do not get enough oxygen. Oxygen is required for aerobic respiration, which provides ATP as an energy source for active transport and other energy-consuming processes such as cell division. Nor, if the leaves are submerged, can photosynthesis take place, because there is not enough carbon dioxide available. This happens because gases diffuse much more slowly in water than they do in air. Moreover, the concentrations of dissolved oxygen and dissolved carbon dioxide in water are much less than they are in air. This is especially true in rice paddies, where the rich mud in which the rice roots are planted contains large populations of microorganisms, many of which are aerobic and take oxygen from the water.

Some varieties of rice respond to flooding by growing taller. As the water rises around them, they keep growing upwards so that the top parts of their leaves and flower spikes are always held above the water. This allows oxygen and carbon dioxide to be exchanged through the stomata on the leaves.

The stems of the rice plants contain loosely packed cells forming a tissue known as aerenchyma (Figure 12.21). Gases are able to diffuse through the aerenchyma to other parts of the plant, including those under the water. This is supplemented by air that is trapped in between the ridges of the underwater leaves. These leaves have a hydrophobic, corrugated surface that holds a thin layer of air in contact with the leaf surface.

Nevertheless, the cells in the submerged roots do still have to use alcoholic fermentation at least some of the time. Ethanol can therefore build up in the tissues. Ethanol is toxic, but the cells in rice roots can tolerate much higher levels than most plants. They also produce more alcohol dehydrogenase, which breaks down ethanol. This allows the plants to grow actively even when oxygen is scarce, using ATP produced by alcoholic fermentation.

QUESTION

12.10 List the features that make rice adapted to grow when partly submerged in water.

Figure 12.20 Rice growing in Madagascar. The blocks of rice were planted at different times and are at different stages of growth.

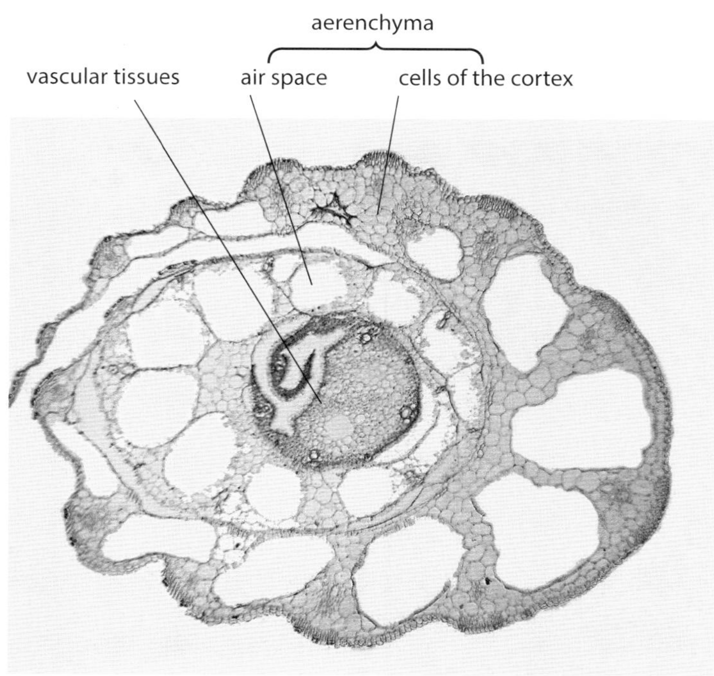

Figure 12.21 Photomicrograph of a cross-section of a rice stem near its tip, with a leaf base around it. Lower down, the stem is completely hollow (×140).

Summary

- Organisms must do work to stay alive. The energy input necessary for this work is either light, for photosynthesis, or the chemical potential energy of organic molecules. Work includes anabolic reactions, active transport and movement.
- Some organisms, such as mammals and birds, use thermal energy released from metabolic reactions to maintain their body temperature. Reactions that release energy must be harnessed to energy-requiring reactions. This involves an intermediary molecule, ATP. ATP can be synthesised from ADP and phosphate using energy, and hydrolysed to ADP and phosphate to release energy. ATP therefore acts as an energy currency in all living organisms.
- Respiration is the sequence of enzyme-controlled steps by which an organic molecule, usually glucose, is broken down so that its chemical potential energy can be used to make the energy currency, ATP. In aerobic respiration, the sequence involves four main stages: glycolysis, the link reaction, the Krebs cycle and oxidative phosphorylation.
- In glycolysis, glucose is first phosphorylated and then split into two triose phosphate molecules. These are further oxidised to pyruvate, giving a small yield of ATP and reduced NAD. Glycolysis occurs in the cell cytoplasm. When oxygen is available (aerobic respiration), the pyruvate passes to the matrix of a mitochondrion. In a mitochondrion, in the link reaction, pyruvate is decarboxylated and dehydrogenated and the remaining 2C acetyl unit combined with coenzyme A to give acetyl coenzyme A.
- The acetyl coenzyme A enters the Krebs cycle in the mitochondrial matrix and donates the acetyl unit to oxaloacetate (4C) to make citrate (6C). The Krebs cycle decarboxylates and dehydrogenates citrate to oxaloacetate in a series of small steps. The oxaloacetate can then react with another acetyl coenzyme A from the link reaction.
- Dehydrogenation provides hydrogen atoms, which are accepted by the carriers NAD and FAD. These pass to the inner membrane of the mitochondrial envelope, where they are split into protons and electrons.
- In the process of oxidative phosphorylation, the electrons are passed along a series of carriers. Some of the energy released in oxidative phosphorylation is used to move protons from the mitochondrial matrix to the intermembrane space. The movement of electrons sets up a gradient of protons across the inner membrane of the mitochondrial envelope. Protons pass back into the matrix, moving down their concentration gradient through protein channels in the inner membrane. An enzyme, ATP synthase, is associated with each of the proton channels. ATP synthase uses the electrical potential energy of the proton gradient to phosphorylate ADP to ATP. At the end of the carrier chain, electrons and protons are recombined and reduce oxygen to water.
- In the absence of oxygen as a hydrogen acceptor (in alcoholic and lactic fermentations), a small yield of ATP is made through glycolysis, then dumping hydrogen into other pathways in the cytoplasm which produce ethanol or lactate. The lactate pathway can be reversed in mammals when oxygen becomes available. The oxygen needed to remove the lactate produced during lactic fermentation is called the oxygen debt.
- The energy values of respiratory substrates depend on the number of hydrogen atoms per molecule. Lipids have a higher energy density than carbohydrates or proteins. The respiratory quotient (RQ) is the ratio of the volume of oxygen absorbed and the volume of carbon dioxide given off in respiration. The RQ reveals the nature of the substrate being respired. Carbohydrate has an RQ of 1.0, lipid 0.7 and protein 0.9. Oxygen uptake, and hence RQ, can be measured using a respirometer.

End-of-chapter questions

1 What does **not** occur in the conversion of glucose to two molecules of pyruvate?

A hydrolysis of ATP
B phosphorylation of ATP
C phosphorylation of triose (3C) sugar
D reduction of NAD [1]

2 Where does each stage of aerobic respiration occur in a eukaryotic cell?

	Link reaction	Krebs cycle	Oxidative phosphorylation
A	cytoplasm	mitochondrial matrix	mitochondrial cristae
B	mitochondrial cristae	cytoplasm	mitochondrial matrix
C	cytoplasm	mitochondrial cristae	mitochondrial matrix
D	mitochondrial matrix	mitochondrial matrix	mitochondrial cristae

[1]

3 The diagram summarises how glucose can be used to produce ATP, without the use of oxygen.

glucose ⟶ X ⟶ Y in mammals
↓
Z in yeast

Which compounds are represented by the letters X, Y and Z?

	X	Y	Z
A	ethanol	pyruvate	lactate
B	lactate	ethanol	pyruvate
C	pyruvate	ethanol	lactate
D	pyruvate	lactate	ethanol

[1]

4 Distinguish between:

a an energy currency molecule and an energy storage molecule [2]
b decarboxylation and dehydrogenation. [2]
[Total: 4]

5 State the roles in respiration of:

a NAD [1]
b coenzyme A [1]
c oxygen. [1]
[Total: 3]

6 Copy and complete the table to show how much ATP is used and produced for each molecule of glucose respired in the various stages of respiration.

	ATP used	ATP produced	Net gain in ATP
glycolysis			
link reaction			
Krebs cycle			
oxidative phosphorylation			
Total			

[5]

7 a Explain why the energy value of lipid is more than twice that of carbohydrate. [2]

b Explain what is meant by **respiratory quotient** (RQ). [2]

c Copy and complete the table to show the respiratory substrates with each of the given RQs.

Respiratory substrate	RQ
	1.0
	0.7
	0.9

[3]

d Measurements of oxygen uptake and carbon dioxide production by germinating seeds in a respirometer showed that 25 cm^3 of oxygen was used and 17.5 cm^3 of carbon dioxide was produced over the same time period.

i Calculate the RQ for these seeds. [2]

ii Identify the respiratory substrate used by the seeds. [1]

e Dahlia plants store a compound called inulin, which is a polymer of fructose. The structure of fructose is shown in the diagram.

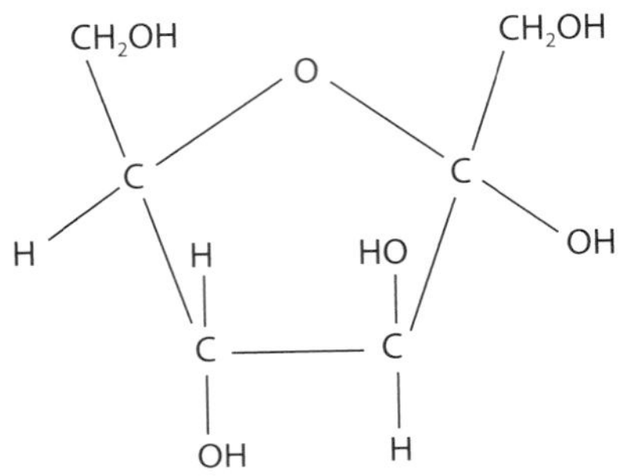

Calculate the RQ when inulin is hydrolysed and then respired aerobically. [2]

[Total: 12]

8 Copy and complete the following passage describing the adaptations of rice for growing with its roots submerged in water.

The stems and leaves of rice plants have very large .... in tissue called ...., which allow oxygen to pass from the air to the .... The roots are very shallow, giving them access to the higher concentration of .... in surface water. When oxygen concentrations fall, the roots can oxidise glucose through .... This produces ...., which is toxic. However, the root cells are tolerant of higher concentrations of this than are most cells and they also contain high concentrations of the enzyme .... to break it down. [7]

9 In aerobic respiration, the Krebs cycle is regarded as a series of small steps. One of these steps is the conversion of succinate to fumarate by an enzyme, succinate dehydrogenase.

a State the role played by dehydrogenase enzymes in the Krebs cycle **and** explain briefly the importance of this role in the production of ATP. [3]

b An investigation was carried out into the effect of different concentrations of aluminium ions on the activity of succinate dehydrogenase. The enzyme concentration and all other conditions were kept constant. The graph below shows the results of this investigation.

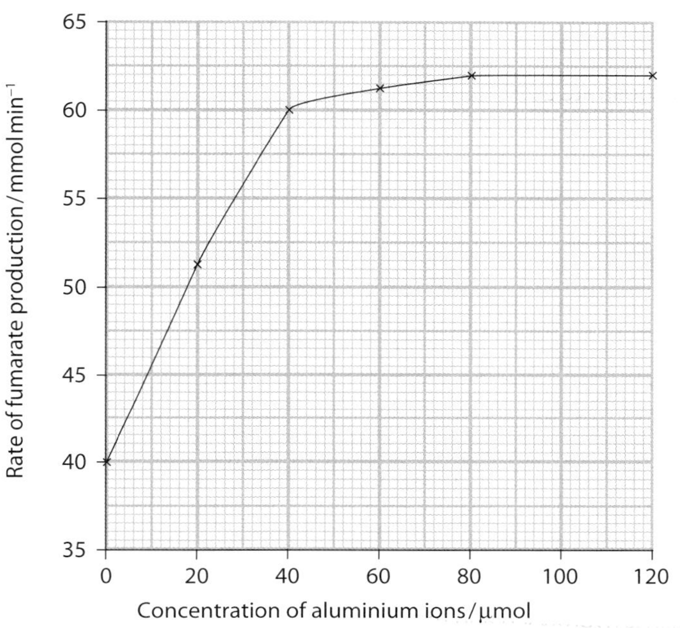

With reference to the graph:

i describe the effect of the concentration of aluminium ions on the rate of production of fumarate [2]

ii suggest an explanation for this effect. [2]

[Total: 7]

Cambridge International AS and A Level Biology 9700/04, Question 7, October/November 2007

Chapter 13: Photosynthesis

Learning outcomes

You should be able to:

- describe the absorption of light energy in the light dependent stage of photosynthesis
- explain the transfer of this energy to the light independent stage of photosynthesis and its use in the production of complex organic molecules
- describe the role of chloroplast pigments in the absorption of light energy
- discuss how the structure of a chloroplast fits it for its functions
- explain how environmental factors influence the rate of photosynthesis
- describe how C4 plants are adapted for high rates of carbon fixation at high temperatures

Fuel from algae

Despite millions of hours of research, we still have not managed to set up a chemical manufacturing system that can harvest light energy and use it to make complex chemicals, in the way that plants and some protoctists do. So, why not just let the cells do it for us?

Figure 13.1 shows a photobioreactor – a series of tubes containing the single-celled photosynthetic organism *Chlorella*. Provide light, carbon dioxide and minerals, and the cells photosynthesise. Bioreactors like this are being used around the world to produce biomass for animal feed, and chemicals that can be used as food additives or in the manufacture of cosmetics. They can also be used to convert energy from the Sun into ethanol or biodiesel but, so far, the bioreactors cannot produce biomass cheaply enough to compete with the use of fossil fuels.

Figure 13.1 A photobioreactor.

An energy transfer process

As you have seen at the beginning of Chapter 12, the process of photosynthesis transfers light energy into chemical potential energy of organic molecules. This energy can then be released for work in respiration (Figure 12.2). Almost all the energy transferred to all the ATP molecules in all living organisms is derived from light energy used in photosynthesis by autotrophs. Such **photoautotrophs** include green plants, the photosynthetic prokaryotes and both single-celled and many-celled protoctists (including the green, red and brown algae). A few autotrophs do not depend on light energy, but use chemical energy sources. These **chemoautotrophs** include the nitrifying bacteria that are so important in the nitrogen cycle. Nitrifying bacteria obtain their energy from oxidising ammonia (NH_3) to nitrite (NO_2^-), or nitrite to nitrate (NO_3^-).

An outline of the process

Photosynthesis is the trapping (fixation) of carbon dioxide and its subsequent reduction to carbohydrate, using hydrogen from water. It takes place inside chloroplasts (Figure 13.2)

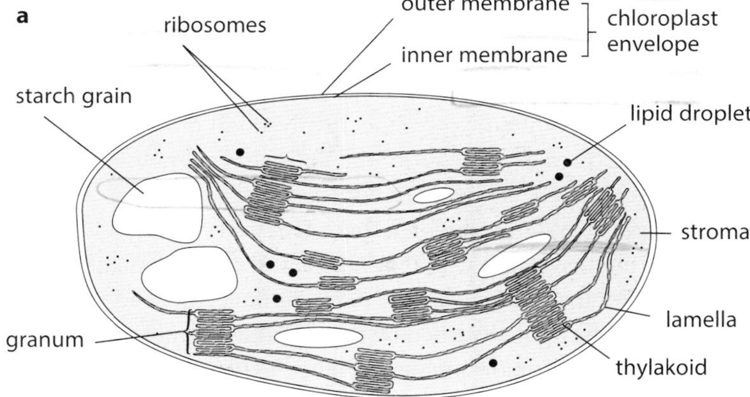

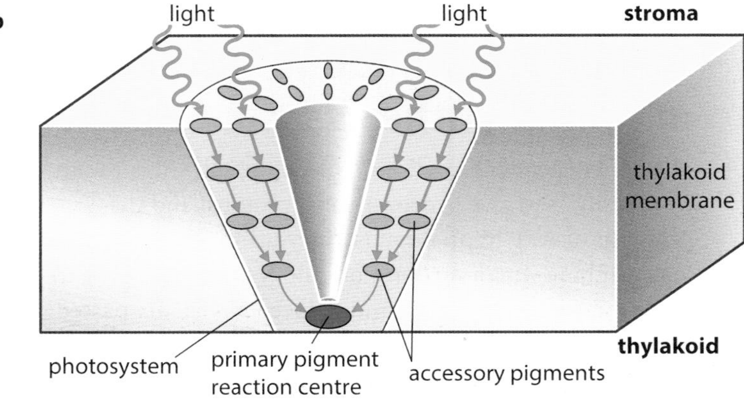

Figure 13.2 **a** A diagram of a chloroplast. **b** A photosystem: a light-harvesting cluster of photosynthetic pigments in a chloroplast thylakoid membrane. Only a few of the pigment molecules are shown.

An overall equation for photosynthesis in green plants is:

$$nCO_2 + nH_2O \xrightarrow[\text{of chlorophyll}]{\text{light energy in the presence}} (CH_2O)n + nO_2$$

carbon dioxide, water → carbohydrate, oxygen

Hexose sugars and starch are commonly formed, so the following equation is often used:

$$6CO_2 + 6H_2O \xrightarrow[\text{of chlorophyll}]{\text{light energy in the presence}} C_6H_{12}O_6 + 6O_2$$

carbon dioxide, water → carbohydrate, oxygen

Two sets of reactions are involved. These are the **light dependent reactions**, for which light energy is necessary, and the **light independent reactions**, for which light energy is not needed. The light dependent reactions only take place in the presence of suitable pigments that absorb certain wavelengths of light (pages 295–296). Light energy is necessary for the splitting (photolysis) of water into hydrogen and oxygen; oxygen is a waste product. Light energy is also needed to provide chemical energy, in the form of ATP, for the reduction of carbon dioxide to carbohydrate in the light independent reactions.

The photosynthetic pigments involved fall into two categories: **primary pigments** and **accessory pigments**. The pigments are arranged in light-harvesting clusters called **photosystems** of which there are two types, I and II. In a photosystem, several hundred accessory pigment molecules surround a primary pigment molecule, and the energy of the light absorbed by the different pigments is passed to the primary pigment (Figure 13.2b). The primary pigments are two forms of chlorophyll (pages 295–296). These primary pigments are said to act as **reaction centres**.

The light dependent reactions of photosynthesis

The light dependent reactions include the splitting of water by **photolysis** to give hydrogen ions (protons) and the synthesis of ATP in **photophosphorylation**. The hydrogen ions combine with a carrier molecule NADP (page 275), to make reduced NADP. ATP and reduced NADP are passed from the light dependent to the light independent reactions.

Photophosphorylation of ADP to ATP can be cyclic or non-cyclic, depending on the pattern of electron flow in one or both types of photosystem.

Cyclic photophosphorylation

Cyclic photophosphorylation involves only photosystem I. Light is absorbed by photosystem I and is passed to the primary pigment. An electron in the chlorophyll molecule is excited to a higher energy level and is emitted from the chlorophyll molecule. This is called photoactivation. Instead of falling back into the photosystem and losing its energy as thermal energy or as fluorescence, the excited electron is captured by an electron acceptor and passed back to a chlorophyll molecule via a chain of electron carriers. During this process, enough energy is released to synthesise ATP from ADP and an inorganic phosphate group (P_i) by the process of chemiosmosis (page 270). The ATP then passes to the light independent reactions.

Non-cyclic photophosphorylation

Non-cyclic photophosphorylation involves both photosystems in the so-called 'Z scheme' of electron flow (Figure 13.3). Light is absorbed by both photosystems and excited electrons are emitted from the primary pigments of both reaction centres. These electrons are absorbed by electron acceptors and pass along chains of electron carriers, leaving the photosystems positively charged. The primary pigment of photosystem I absorbs electrons from photosystem II. Its primary pigment receives replacement electrons from the splitting (photolysis) of water. As in cyclic photophosphorylation, ATP is synthesised as the electrons lose energy while passing along the carrier chain.

Photolysis of water

Photosystem II includes a water-splitting enzyme that catalyses the breakdown of water:

$$H_2O \rightarrow 2H^+ + 2e^- + \tfrac{1}{2}O_2$$

Oxygen is a waste product of this process. The hydrogen ions combine with electrons from photosystem I and the carrier molecule NADP to give reduced NADP.

$$2H^+ + 2e^- + \text{NADP} \rightarrow \text{reduced NADP}$$

Reduced NADP passes to the light independent reactions and is used in the synthesis of carbohydrate.

The photolysis of water can be demonstrated by the Hill reaction.

The Hill reaction

Redox reactions are oxidation–reduction reactions and involve the transfer of electrons from an electron donor (reducing agent) to an electron acceptor (oxidising agent). Sometimes hydrogen atoms are transferred, so that dehydrogenation is equivalent to oxidation.

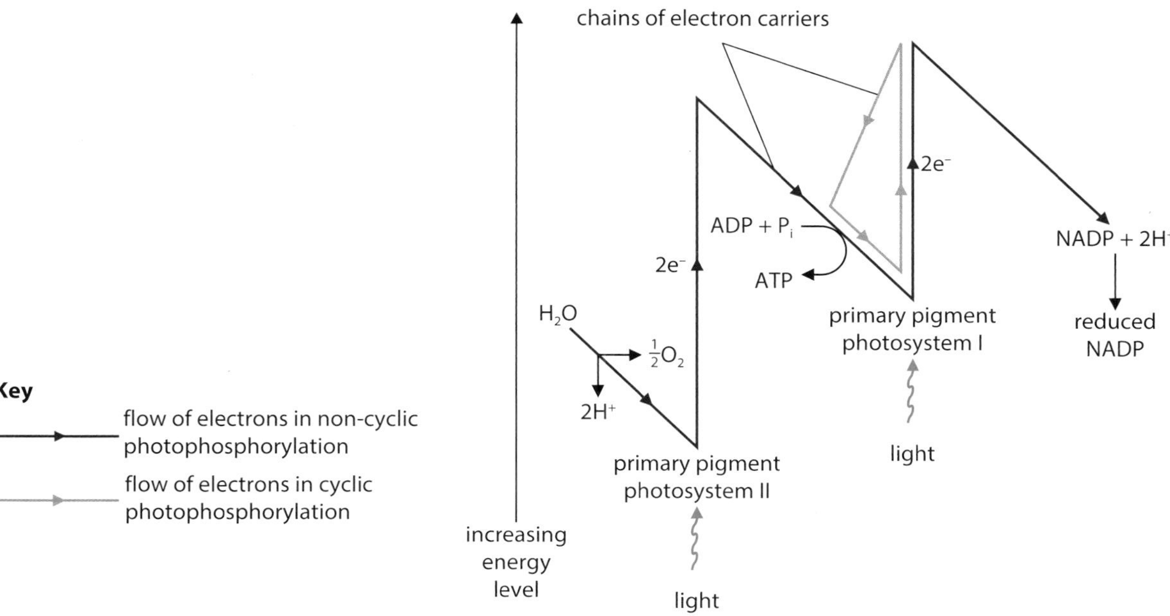

Figure 13.3 The 'Z scheme' of electron flow in photophosphorylation.

In 1939, Robert Hill showed that isolated chloroplasts had 'reducing power' and liberated oxygen from water in the presence of an oxidising agent. The 'reducing power' was demonstrated by using a redox agent that changed colour on reduction. This technique can be used to investigate the effect of light intensity or of light wavelength on the rate of photosynthesis of a suspension of chloroplasts. Hill used Fe^{3+} ions as his acceptor, but various redox agents, such as the blue dye DCPIP (dichlorophenolindophenol), can substitute for the plant's NADP in this system (Figure 13.4). DCPIP becomes colourless when reduced:

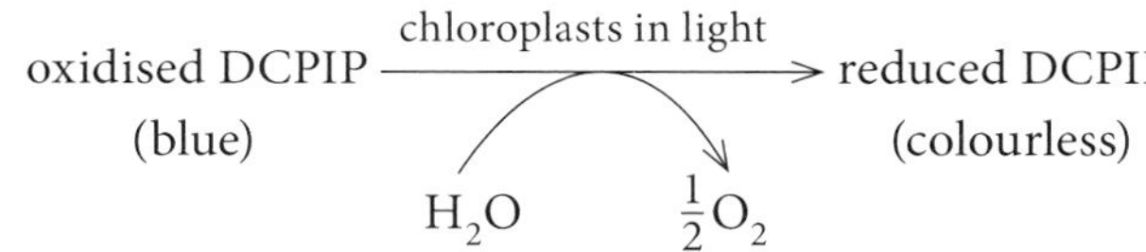

Figure 13.4 shows classroom results of this reaction.

BOX 13.1: Investigating the Hill reaction

Chloroplasts can be isolated from a leafy plant, such as lettuce or spinach, by liquidising the leaves in ice-cold buffer and then filtering or centrifuging the resulting suspension to remove unwanted debris. Working quickly and using chilled glassware, small tubes of buffered chloroplast suspension with added DCPIP solution are placed in different light intensities or in different wavelengths of light and the blue colour assessed at intervals.

The rate of loss of blue colour (as measured in a colorimeter or by matching the tubes against known concentrations of DCPIP solution) is a measure of the effect of the factor being investigated (light intensity or the wavelength of light) on chloroplast activity.

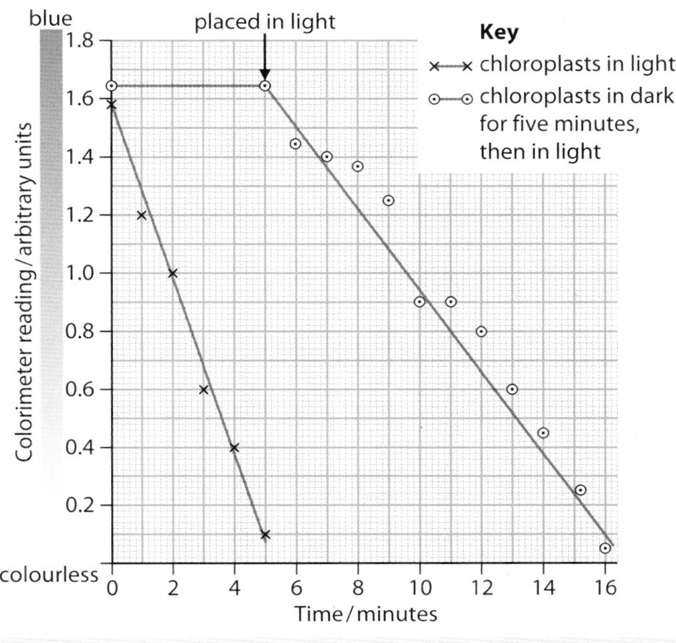

Figure 13.4 The Hill reaction. Chloroplasts were extracted from lettuce and placed in buffer solution with DCPIP. The colorimeter reading is proportional to the amount of DCPIP remaining unreduced.

QUESTIONS

13.1 Examine the two curves shown in Figure 13.4 and explain:

a the downward trend of the two curves

b the differences between the two curves.

13.2 Explain what contribution the discovery of the Hill reaction made to an understanding of the process of photosynthesis.

The light independent reactions of photosynthesis

The fixation of carbon dioxide is a light independent process in which carbon dioxide combines with a five-carbon sugar, ribulose bisphosphate (RuBP), to give two molecules of a three-carbon compound, glycerate 3-phosphate (GP). (This compound is also sometimes known as PGA.)

GP, in the presence of ATP and reduced NADP from the light dependent stages, is reduced to triose phosphate (TP) (three-carbon sugar). This is the point at which carbohydrate is produced in photosynthesis. Most (five-sixths) of the triose phosphates are used to regenerate RuBP, but the remainder (one-sixth) are used to produce other molecules needed by the plant. Some of these triose phosphates condense to become hexose phosphates which, in turn, are used to produce starch for storage, sucrose for translocation around the plant, or cellulose for making cell walls. Others are converted to glycerol and fatty acids to produce lipids for cellular membranes or to acetyl coenzyme A for use in respiration or in the production of amino acids for protein synthesis.

This cycle of events was worked out by Calvin, Benson and Bassham between 1946 and 1953, and is usually called the Calvin cycle (Figure 13.5). The enzyme ribulose bisphosphate carboxylase (rubisco), which catalyses the combination of carbon dioxide and RuBP, is the most common enzyme in the world.

Chloroplast structure and function

In eukaryotic organisms, the photosynthetic organelle is the **chloroplast**. In dicotyledons, chloroplasts can be seen with a light microscope and appear as biconvex discs about 3–10 μm in diameter. There may be only a few chloroplasts in a cell or as many as 100 in some palisade mesophyll cells.

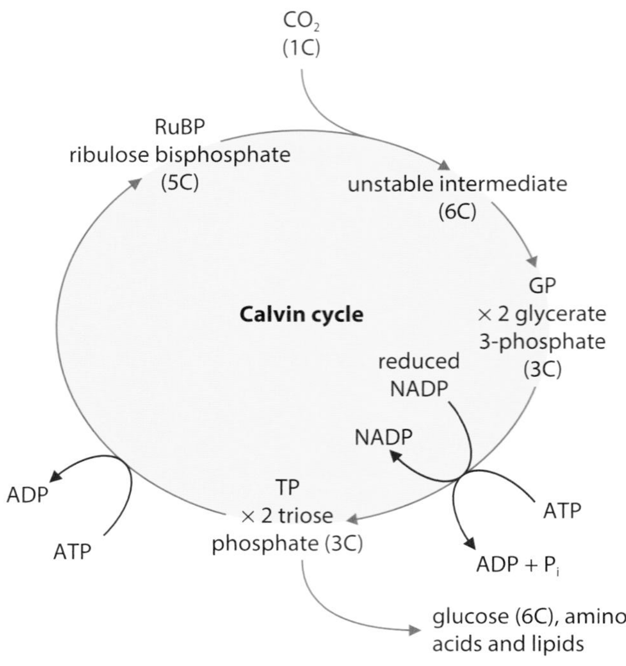

Figure 13.5 The Calvin cycle.

The structure of a chloroplast is shown in Figures 13.2a and 13.6. Each chloroplast is surrounded by an envelope of two phospholipid membranes. A system of membranes also runs through the ground substance, or stroma. The membrane system is the site of the light dependent reactions of photosynthesis. It consists of a series of flattened fluid-filled sacs, or thylakoids, which in places form stacks, called **grana**, that are joined to one another by membranes. The membranes of the grana provide a large surface area, which holds the pigments, enzymes and electron carriers needed for the light dependent reactions. The membranes make it possible for a large number of pigment molecules to be arranged so that they can absorb as much light as necessary. The pigment molecules are also arranged in particular light-harvesting clusters for efficient light absorption. In each photosystem, the different pigments are arranged in the thylakoid in funnel-like structures (Figure 13.2, page 287). Each pigment passes energy to the next member of the cluster, finally 'feeding' it to the chlorophyll *a* reaction centre (primary pigment). The membranes of the grana hold ATP synthase and are the site of ATP synthesis by chemiosmosis (page 270).

The stroma is the site of the light independent reactions. It contains the enzymes of the Calvin cycle, sugars and organic acids. It bathes the membranes of the grana and so can receive the products of the light dependent reactions. Also within the stroma are small (70 S) ribosomes, a loop of DNA, lipid droplets and

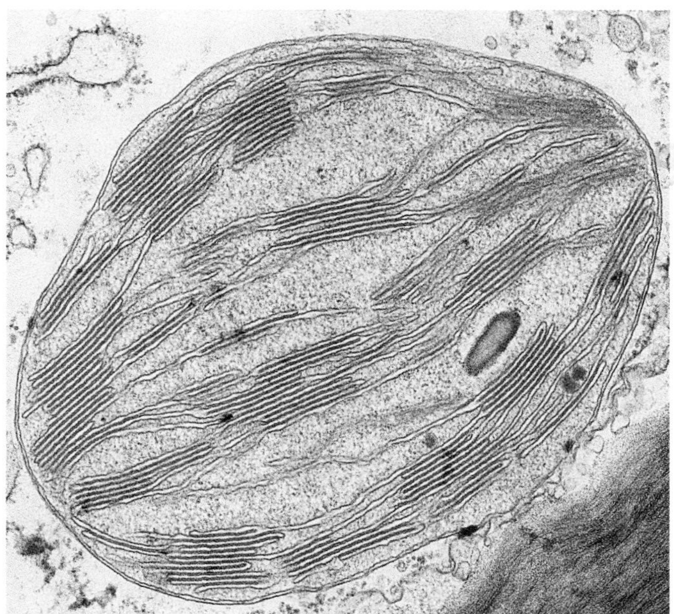

Figure 13.6 Transmission electron micrograph of a chloroplast from *Potamogeton* leaf (×27 000). See also Figure 1.29.

starch grains. The loop of DNA codes for some of the chloroplast proteins, which are made by the chloroplast's ribosomes. However, other chloroplast proteins are coded for by the DNA in the plant cell nucleus.

QUESTION

13.3 List the features of a chloroplast that aid photosynthesis.

Factors necessary for photosynthesis

You can see from the equation on page 288 that certain factors are necessary for photosynthesis to occur, namely the presence of a suitable photosynthetic pigment, a supply of carbon dioxide, water and light energy.

Factors affecting the rate of photosynthesis

The main external factors affecting the rate of photosynthesis are light intensity and wavelength, temperature and carbon dioxide concentration.

In the early 1900s, F. F. Blackman investigated the effects of light intensity and temperature on the rate of photosynthesis. At constant temperature, the rate of photosynthesis varies with the light intensity, initially increasing as the light intensity increases (Figure 13.7). However, at higher light intensities, this relationship no longer holds and the rate of photosynthesis reaches a plateau.

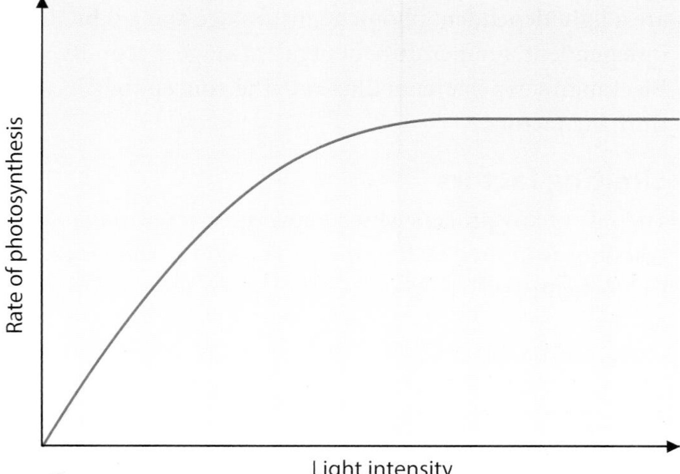

Figure 13.7 The rate of photosynthesis at different light intensities and constant temperature.

The effect on the rate of photosynthesis of varying the temperature at constant light intensities can be seen in Figure 13.8. At high light intensity the rate of photosynthesis increases as the temperature is increased over a limited range. At low light intensity, increasing the temperature has little effect on the rate of photosynthesis.

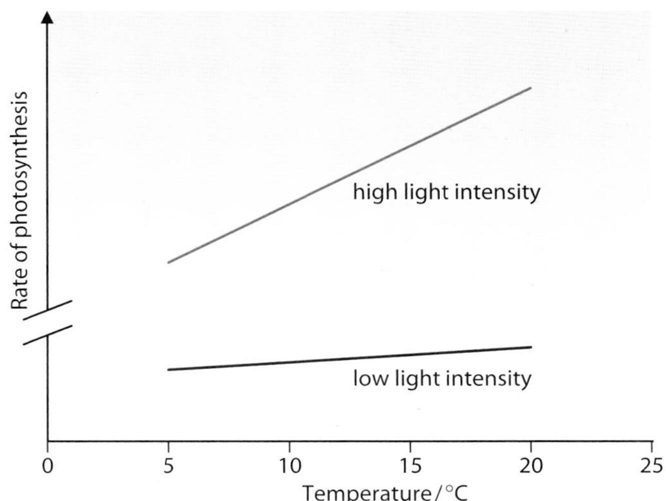

Figure 13.8 The rate of photosynthesis at different temperatures and constant light intensities.

These two experiments illustrate two important points. Firstly, from other research we know that photochemical reactions are not generally affected by temperature. However, these experiments clearly show that temperature affects the rate of photosynthesis, so there must be two sets of reactions in the full process of photosynthesis. These are a light dependent photochemical stage and a light independent, temperature dependent stage. Secondly, Blackman's experiments illustrate the concept of limiting factors.

Limiting factors

The rate of any process which depends on a series of reactions is limited by the slowest reaction in the series. In biochemistry, if a process is affected by more than one factor, the rate will be limited by the factor which is nearest its lowest value.

Look at Figure 13.9. At low light intensities, the limiting factor governing the rate of photosynthesis is the light intensity; as the intensities increase so does the rate. But at high light intensity, one or more other factors must be limiting, such as temperature or carbon dioxide supply.

As you will see in the next section of this chapter, not all wavelengths of light can be used in photosynthesis. This means that the wavelengths of light that reach a plant's leaves may limit its rate of photosynthesis (Figure 13.16b, page 295).

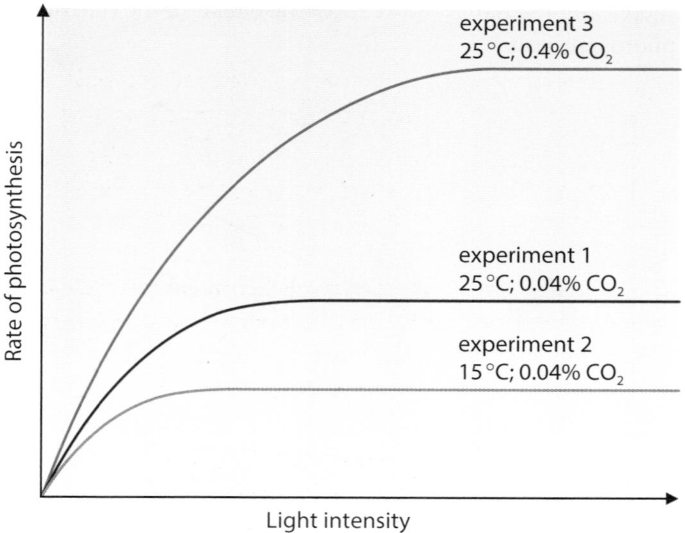

Figure 13.9 The rate of photosynthesis at different temperatures and different carbon dioxide concentrations. (0.04% CO_2 is about atmospheric concentration.)

QUESTION

13.4 Examine Figure 13.9, which shows the effect of various factors on the rate of photosynthesis, and explain the differences between the results of:

a experiments 1 and 2

b experiments 1 and 3.

At constant light intensity and temperature, the rate of photosynthesis initially increases with an increasing concentration of carbon dioxide, but again reaches a plateau at higher concentrations. A graph of the rate of photosynthesis at different concentrations of carbon dioxide has the same shape as that for different light intensities (Figure 13.9). At low concentrations of carbon dioxide, the supply of carbon dioxide is the rate-limiting factor. At higher concentrations of carbon dioxide, other factors are rate-limiting, such as light intensity or temperature.

The effects of these limiting factors on the rate of photosynthesis are easily investigated by using an aquatic plant such as *Elodea* or *Cabomba* in a simple apparatus as shown in Figure 13.10. The number of bubbles of gas (mostly oxygen) produced in unit time from a cut stem of the plant can be counted in different conditions. Alternatively, the gas can be collected and the volume produced in unit time can be measured. This procedure depends on the fact that the rate of production of oxygen is a measure of the rate of photosynthesis.

Growing plants in protected environments

An understanding of the effect of environmental factors on the rate of photosynthesis allows their management when crops are grown in protected environments, such as glasshouses. The aim is to increase the yield of the crop concerned.

For example, many hectares of tomato plants are grown in glasshouses. In the most sophisticated of these, sensors monitor the light intensity, the humidity of the atmosphere and the concentration of carbon dioxide around the plants. The plants grow hydroponically – that is, with their roots in a nutrient solution whose nutrient content can be varied at different stages of the plants' growth. All of these factors are managed by a computer to maximise the yield of the crop.

Such glasshouse-grown crops have the added advantage that insect pests and fungal diseases are more easily controlled than is possible with field-grown crops, further improving yield.

BOX 13.2: Investigating the rate of photosynthesis using an aquatic plant

Elodea, or other similar aquatic plants, can be used to investigate the effect on the rate of photosynthesis of altering the:

- **light intensity** – by altering the distance, *d*, of a small light source from the plants (light intensity is proportional to $\frac{1}{d^2}$)
- **wavelength of light** – by using different colour filters, making sure that they each transmit the same light intensity
- **concentration of carbon dioxide** – by adding different quantities of sodium hydrogencarbonate ($NaHCO_3$) to the water surrounding the plant
- **temperature** of the water surrounding the plant – using a large container, such as a beaker, to help maintain the chosen temperatures.

The aquatic plant needs to be well illuminated before use and the chosen stem needs to be cut cleanly just before putting it into a test tube (Figure 13.10).

The bubbles given off are mostly oxygen, but contain some nitrogen. To prevent these gases from dissolving in the water, rather than forming bubbles, the water needs to be well aerated (by bubbling air through it) before use.

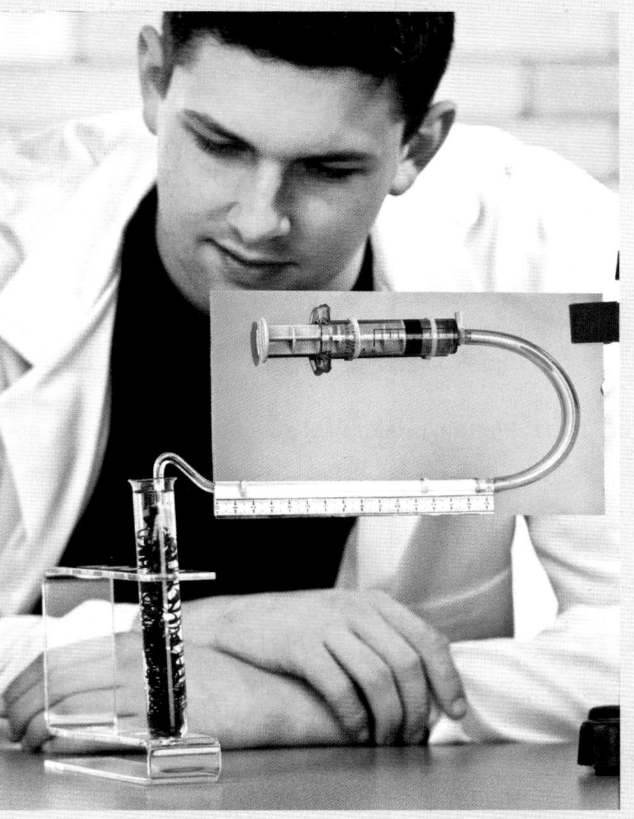

Figure 13.10 Investigating the rate of photosynthesis using an aquatic plant.

C4 plants

In the light independent stage of photosynthesis, you may remember that carbon dioxide combines with RuBP to form a six-carbon compound, which immediately splits to form two three-carbon molecules (page 290). Plants that do this are called C3 plants.

However, maize and sorghum plants – and most other tropical grasses – do something different. The first compound that is produced in the light independent reaction contains **four** carbon atoms. They are therefore called C4 plants.

Avoiding photorespiration

Why do tropical grasses need to do something different from other plants in the light independent stage of photosynthesis? The reason is a problem with the enzyme rubisco. This enzyme catalyses the reaction of carbon dioxide with RuBP. But, unfortunately, it can also catalyse the reaction of oxygen with RuBP. When this happens, less photosynthesis takes place, because some of the RuBP is being 'wasted' and less is available to combine with carbon dioxide. This unwanted reaction is known as photorespiration. It happens most readily in high temperatures and high light intensity – that is, conditions that are found at low altitudes in tropical parts of the world.

Tropical grasses such as maize, sorghum and sugar cane have evolved a method of avoiding photorespiration. They keep RuBP and rubisco well away from high oxygen concentrations. The cells that contain RuBP and rubisco are arranged around the vascular bundles, and are called bundle sheath cells (Figures 13.11, 13.12 and 13.13). They have no direct contact with the air inside the leaf.

Carbon dioxide is absorbed by another group of cells, the **mesophyll** cells, which **are** in contact with air (Figure 13.13). The mesophyll cells contain an enzyme called PEP carboxylase, which catalyses the combination of carbon dioxide from the air with a three-carbon substance called **phosphoenolpyruvate**, or PEP. The compound formed from this reaction is oxaloacetate (Figure 13.14).

Still inside the mesophyll cells, the oxaloacetate is converted to malate, and this is passed on to the bundle sheath cells. Now the carbon dioxide is removed from the malate molecules and delivered to RuBP by rubisco in the normal way. The light independent reaction then proceeds as usual.

Enzymes in C4 plants generally have higher optimum temperatures than those in C3 plants. This is an adaptation to growing in hot climates. For example, in one study it was found that in amaranth, which is a C4 plant, the optimum temperature for the activity of PEP carboxylase is around 45 °C. If the temperature drops to 15 °C, the enzyme loses around 70% of its activity. By contrast, the same enzyme in peas, which are C3 plants, was found to have an optimum temperature of around 30 °C and could continue to work at much lower temperatures than in amaranth.

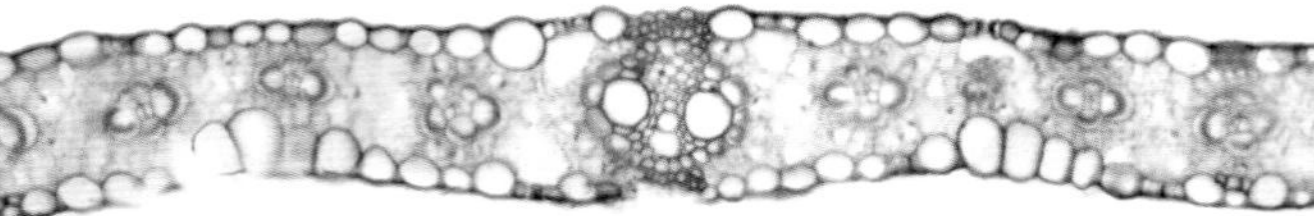

Figure 13.11 Photomicrograph of a section through a leaf of maize (×125).

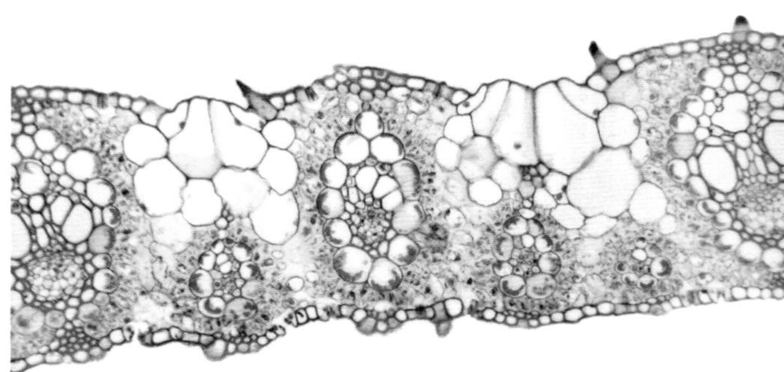

Figure 13.12 Photomicrograph of a section through a leaf of sugar cane (×120).

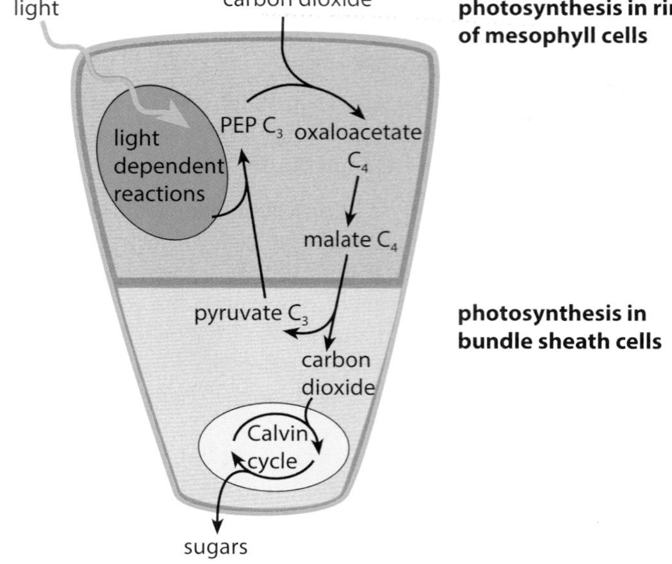

Figure 13.14 C4 photosynthesis.

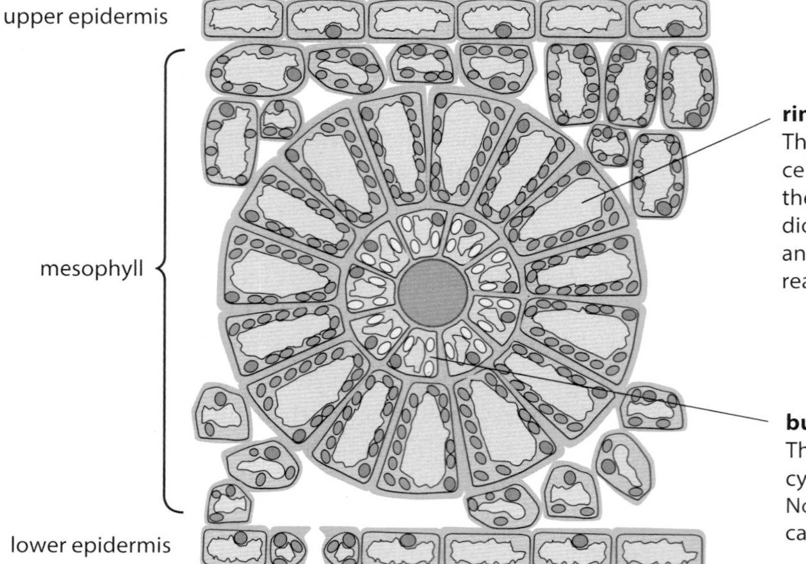

Figure 13.13 Tissues surrounding a vascular bundle of a C4 leaf.

QUESTION

13.5 Some of the most productive crop plants in the world are C4 plants. However, rice grows in tropical regions and is a C3 plant. Research is taking place into the possibility of producing genetically modified rice that uses the C4 pathway in photosynthesis. Explain how this could increase yields from rice.

Trapping light energy

Chloroplasts contain several different pigments, and these different pigments absorb different wavelengths of light. The photosynthetic pigments of higher plants form two groups: the **chlorophylls** (primary pigments) and the **carotenoids** (accessory pigments) (Table 13.1).

Group	Pigment	Colour
chlorophylls	chlorophyll *a* chlorophyll *b*	yellow-green blue-green
carotenoids	β carotene xanthophyll	orange yellow

Table 13.1 The colours of the commonly occurring photosynthetic pigments.

Chlorophylls absorb mainly in the red and blue-violet regions of the light spectrum. They reflect green light, which is why plants look green. The structure of chlorophyll *a* is shown in Figure 13.15. The carotenoids absorb mainly in the blue-violet region of the spectrum.

Figure 13.15 Structure of chlorophyll *a*. You do not need to learn this molecular structure.

An **absorption spectrum** is a graph of the absorbance of different wavelengths of light by a pigment (Figure 13.16a).

An **action spectrum** is a graph of the rate of photosynthesis at different wavelengths of light (Figure 13.16b). This shows the effectiveness of the different wavelengths, which is, of course, related to their absorption and to their energy content. The shorter the wavelength, the greater the energy it contains.

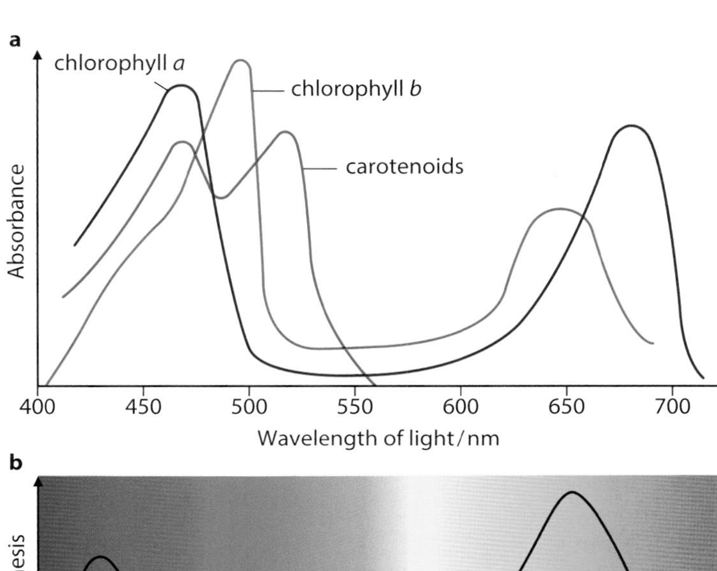

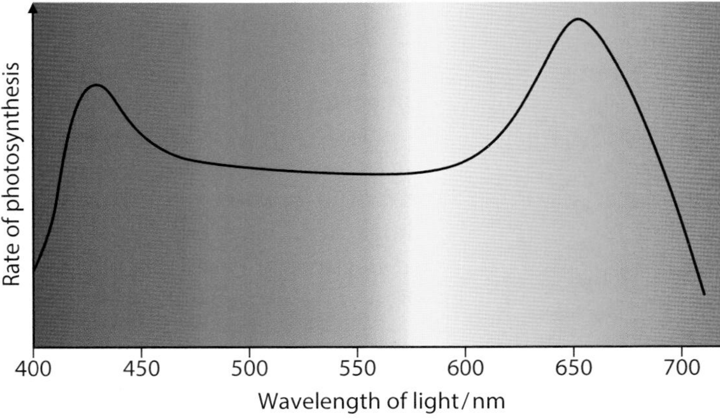

Figure 13.16 **a** Absorption spectra of chlorophylls *a* and *b*, and carotenoid pigments. **b** Photosynthetic action spectrum.

QUESTION

13.6 Compare the absorption spectra shown in Figure 13.16a with the action spectrum shown in Figure 13.16b.

a Identify and explain any similarities in the absorption and action spectra.

b Identify and explain any differences between the absorption and action spectra.

If you illuminate a solution of chlorophyll *a* or *b* with ultraviolet light, you will see a red fluorescence. (In the absence of a safe ultraviolet light, you can illuminate the pigment with a standard fluorescent tube.) The ultraviolet light is absorbed and electrons are excited but, in a solution that only contains extracted pigment, the absorbed energy cannot usefully be passed on to do work. The electrons return to their unexcited state and the absorbed energy is transferred to the surroundings as thermal energy and as light at a longer (less energetic) wavelength than that which was absorbed, and is seen as the red fluorescence. In the functioning photosynthetic system, it is this energy that drives the process of photosynthesis.

You can easily extract chloroplast pigments from a leaf to see how many pigments are present, by using paper chromatography as shown in Figure 13.17.

You can calculate the R_f value for each pigment, using this equation:

$$R_f = \frac{\text{distance travelled by pigment spot}}{\text{distance travelled by solvent}}$$

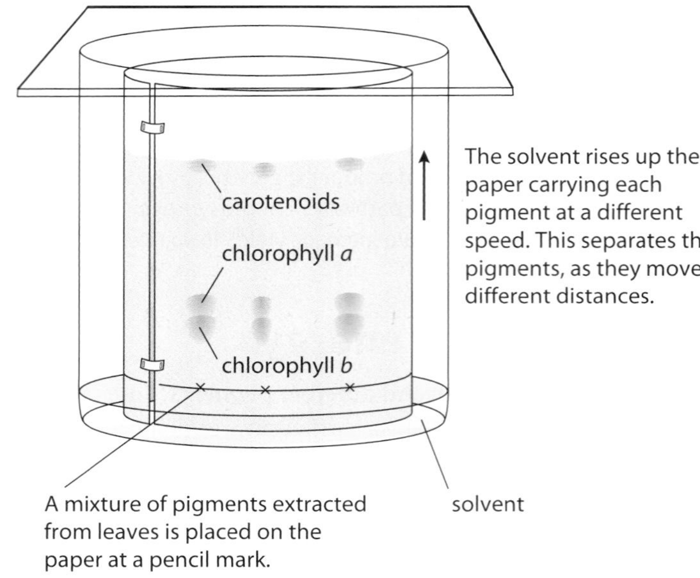

Figure 13.17 Chromatography of pigments in chloroplasts.

These will vary depending on the solvent used, but in general carotenoids have R_f values close to 1, chlorophyll *b* has a much lower R_f value and chlorophyll *a* has an R_f value between those of carotenoids and chlorophyll *b*.

Summary

- In photosynthesis, light energy is absorbed by chlorophyll pigments and converted to chemical energy, which is used to produce complex organic molecules. In the light dependent reactions, water is split by photolysis to give hydrogen ions, electrons and oxygen. The hydrogen ions and electrons are used to reduce the carrier molecule, NADP, and the oxygen is given off as a waste product.
- ATP is synthesised in the light dependent reactions of cyclic and non-cyclic photophosphorylation. During these reactions, the photosynthetic pigments of the chloroplast absorb light energy and give out excited electrons. Energy from the electrons is used to synthesise ATP. ATP and reduced NADP are the two main products of the light dependent reactions of photosynthesis, and they then pass to the light independent reactions.
- In the light independent reactions, carbon dioxide is trapped by combination with a 5C compound, RuBP, which acts as an acceptor molecule. This reaction is catalysed by the enzyme ribulose bisphosphate carboxylase (rubisco), which is the most common enzyme in the world. The resulting 6C compound splits to give two molecules of a 3C compound, GP (also known as PGA). GP is reduced to carbohydrate, using ATP and reduced NADP from the light dependent reactions. This carbohydrate can be converted into other carbohydrates, amino acids and lipids or used to regenerate RuBP. This sequence of light independent events is called the Calvin cycle.
- Chloroplasts are adapted for the efficient absorption of light for the process of photosynthesis. When a process is affected by more than one factor, the rate of the process will be limited by the factor closest to its lowest value. The rate of photosynthesis is subject to various such limiting factors, including light intensity and wavelength, carbon dioxide concentration and temperature.
- Some tropical crops are adapted for high rates of carbon fixation at high temperatures by having a leaf structure that separates initial carbon fixation from the light independent stage, and by the high optimum temperatures of the enzymes concerned.
- A graph of the particular wavelengths of light that are absorbed by a photosynthetic pigment is called an absorption spectrum. A graph of the rate of photosynthesis at different wavelengths of light is called an action spectrum.
- The different pigments present in a chloroplast can be separated by paper chromatography.

End-of-chapter questions

1 What are the products of the light dependent reactions of photosynthesis?

A ATP, RuBP and reduced NAD

B ATP, oxygen and reduced NADP

C GP, oxygen and reduced NAD

D GP, reduced NADP and RuBP [1]

2 Where in the chloroplast are the products of photophosphorylation used?

A envelope

B granum

C stroma

D thylakoid [1]

3 In separate experiments, an actively photosynthesising plant was supplied with one of two labelled reactants:

water containing the ^{18}O isotope of oxygen
carbon dioxide containing the ^{17}O isotope of oxygen.

In which products of photosynthesis would these isotopes be found?

	^{18}O	^{17}O
A	oxygen produced by chloroplast grana	carbohydrate produced by the chloroplast stroma
B	oxygen produced by the chloroplast stroma	carbohydrate produced by chloroplast grana
C	carbohydrate produced by chloroplast grana	oxygen produced by the chloroplast stroma
D	carbohydrate produced by the chloroplast stroma	oxygen produced by chloroplast grana

[1]

4 a Explain how the inner membrane system of a chloroplast makes it well adapted for photosynthesis. [5]

b Copy the table below and insert ticks or crosses to show which structural features are shared by a plant chloroplast and a typical prokaryotic cell.

✓= structural feature shared; ✗ = structural feature not shared.

Structural feature	Structural feature shared by chloroplast and typical prokaryotic cell
circular DNA	
DNA combined with structural protein to form chromosomes	
ribosomes about 18 nm in diameter	
complex arrangement of internal membranes	
peptidoglycan wall	
size ranges overlap	

[6]

[Total: 11]

5 a When isolated chloroplasts are placed in buffer solution with a blue dye such as DCPIP or methylene blue and illuminated, the blue colour disappears. Explain this observation. [4]

b Name the compound, normally present in photosynthesis, that is replaced by the blue dye in this investigation. [1]

[Total: 5]

6 Distinguish between:

a cyclic and non-cyclic photophosphorylation [2]

b photophosphorylation and oxidative phosphorylation [2]

c the roles of NAD and NADP in a plant. [2]

[Total: 6]

7 a Draw a simple flow diagram of the Calvin cycle to show the relative positions in the cycle of the following molecules:

CO_2 (1C)

GP/PGA (3C)

triose phosphate (3C)

RuBP (5C). [4]

b Show the point in the cycle at which the enzyme rubisco is active. [1]

[Total: 5]

8 a Explain what is meant by a **limiting factor**. [1]

b List **four** factors that may be rate-limiting in photosynthesis. [4]

c At low light intensities, increasing the temperature has little effect on the rate of photosynthesis.
At high light intensities, increasing the temperature increases the rate of photosynthesis.
Explain these observations. [5]

[Total: 10]

9 a Copy and complete the table to show the differences between mesophyll and bundle sheath cells in C4 plants. Insert a tick (✓) when an item is present in the cell and a cross (✗) when it is not. [7]

Item	Mesophyll cell	Bundle sheath cell
PEP carboxylase		
rubisco		
RuBP		
enzymes of Calvin cycle		
high concentration of oxygen		
light dependent reactions		
contact with air spaces		

b Explain what is meant by **photorespiration**. [2]

[Total: 9]

10 a Distinguish between an **absorption spectrum** and an **action spectrum**. [4]

b Pondweed was exposed to each of three different wavelengths of light for the same length of time.
For each wavelength, the number of bubbles produced from the cut ends of the pondweed were counted and are shown in the table.

Wavelength of light / nm	Mean number of bubbles produced in unit time
450	22
550	3
650	18

Explain these results. [4]

[Total: 8]

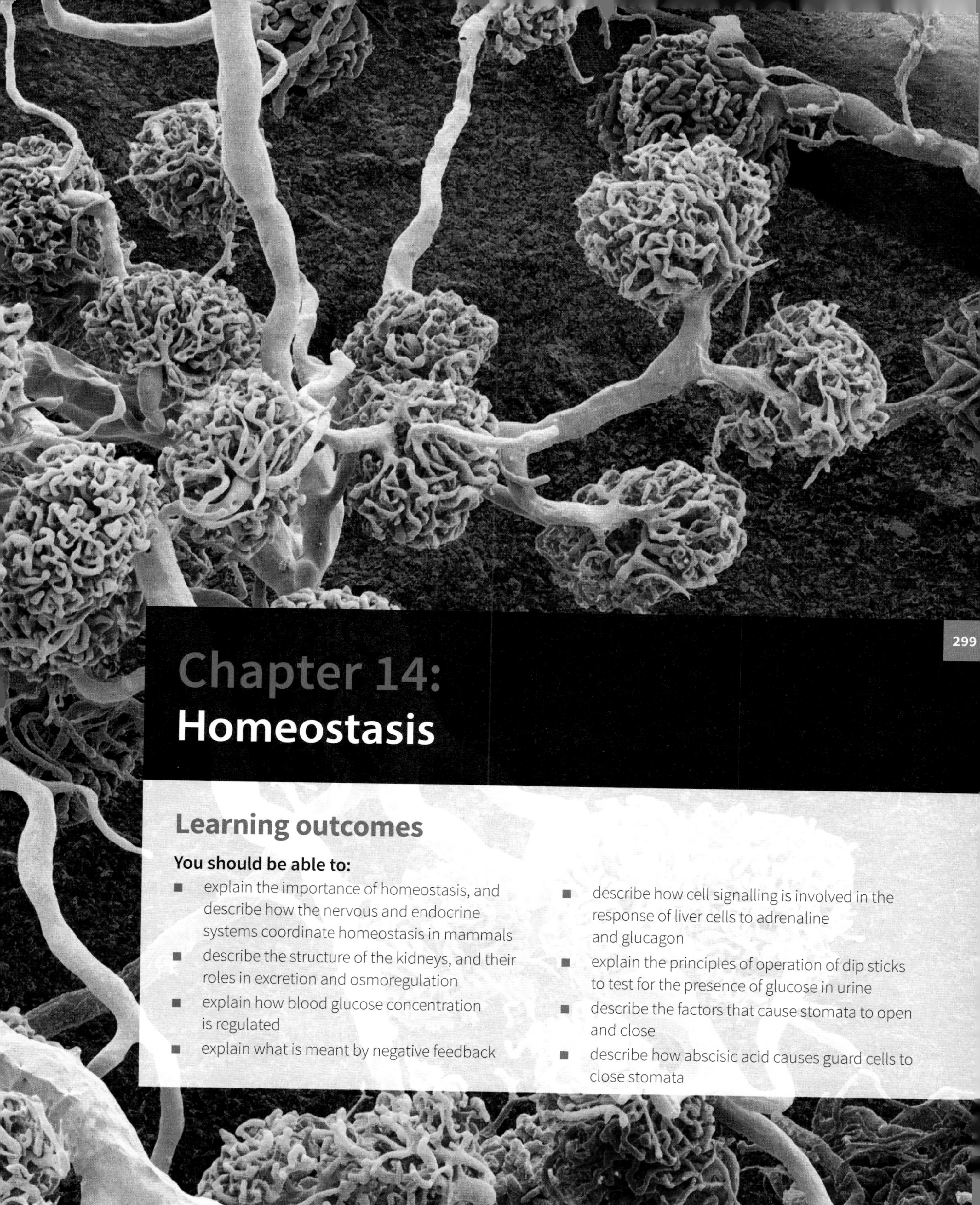

Chapter 14: Homeostasis

Learning outcomes

You should be able to:

- explain the importance of homeostasis, and describe how the nervous and endocrine systems coordinate homeostasis in mammals
- describe the structure of the kidneys, and their roles in excretion and osmoregulation
- explain how blood glucose concentration is regulated
- explain what is meant by negative feedback
- describe how cell signalling is involved in the response of liver cells to adrenaline and glucagon
- explain the principles of operation of dip sticks to test for the presence of glucose in urine
- describe the factors that cause stomata to open and close
- describe how abscisic acid causes guard cells to close stomata

The black bear's big sleep

It is metabolically expensive for a mammal to maintain a constant, warm body temperature in long winters when it is very cold and food is hard to find. Black bears feed well during the summer to build up stores of energy-rich fat (Figure 14.1). In the autumn and early winter, the bears dig dens for themselves, or find a ready-made one in somewhere like a cave, curl up and sleep until the weather improves. Their metabolism adjusts for this lengthy period of inactivity when they do not eat, drink, urinate or defecate. Their stores of fat and some muscle protein provide energy. The waste product of protein breakdown is urea, which is filtered from the blood by the kidneys. The kidneys continue to produce urine, but it is all reabsorbed by the bladder. The urea cannot be stored; instead it is recycled by the bear's gut bacteria. These break it down to ammonia and carbon dioxide, which are absorbed into the blood. Carbon dioxide is breathed out and ammonia combined with glycerol from the breakdown of fat to make amino acids. The amino acids are used to synthesise the enzymes that are needed in larger quantities for the increased hydrolysis of fat during the bear's hibernation.

Figure 14.1 During the summer, black bears build up stores of fat for survival during the seven months or so when they do not eat.

To function efficiently, organisms have control systems to keep internal conditions near constant, a feature known as homeostasis. This requires information about conditions inside the body and the surroundings, which are detected by sensory cells. Some of the physiological factors controlled in homeostasis in mammals are:

- core body temperature
- metabolic wastes, particularly carbon dioxide and urea
- blood pH
- blood glucose concentration
- water potential of the blood
- the concentrations in the blood of the respiratory gases, oxygen and carbon dioxide.

First, we will look at the need for mammals to maintain a stable internal environment, and then consider how they maintain a constant core body temperature.

Internal environment

The internal environment of an organism refers to all the conditions inside the body. These are the conditions in which the cells function. For a cell, its immediate environment is the tissue fluid that surrounds it.

Many features of the tissue fluid influence how well the cell functions. Three features of tissue fluid that influence cell activities are:

- **temperature** – low temperatures slow down metabolic reactions; at high temperatures proteins, including enzymes, are denatured and cannot function
- **water potential** – if the water potential decreases, water may move out of cells by osmosis, causing metabolic reactions in the cell to slow or stop; if the water potential increases, water may enter the cell causing it to swell and maybe burst
- **concentration of glucose** – glucose is the fuel for respiration, so lack of it causes respiration to slow or stop, depriving the cell of an energy source; too much glucose may cause water to move out of the cell by osmosis, again disturbing the metabolism of the cell.

In general, homeostatic mechanisms work by controlling the composition of blood, which therefore controls the composition of tissue fluid. See page 164 to remind yourself how this happens. There are control mechanisms for the different aspects of the blood and tissue fluid. These include the three physiological factors listed above.

Homeostatic control

Most control mechanisms in living organisms use a **negative feedback** control loop (Figure 14.2) to maintain homeostatic balance. This involves a **receptor** (or **sensor**) and an **effector**. Effectors include muscles and glands. The receptor detects **stimuli** that are involved with the condition (or physiological factor) being regulated. A stimulus is any change in a factor, such as a change in blood temperature or the water content of the blood. The body has receptors which detect external stimuli and other receptors that detect internal stimuli. These receptors send information about the changes they detect through the nervous system to a **central control** in the brain or spinal cord. This sensory information is known as the **input**. The central control instructs an effector to carry out an action, which is called the **output**. These actions are sometimes called **corrective actions** as their effect is to correct (or reverse) the change. Continuous monitoring of the factor by receptors produces a steady stream of information to the control centre that makes continuous adjustments to the output. As a result, the factor fluctuates around a particular 'ideal' value, or **set point**. This mechanism to keep changes in the factor within narrow limits is known as **negative feedback**. In these systems, an increase in the factor results in something happening that makes the factor decrease. Similarly, if there is a decrease in the factor, then something happens to make it increase. Homeostatic mechanisms involve negative feedback as it minimises the difference between the actual value of the factor and the ideal value or set point. The factor never stays exactly constant, but fluctuates a little above and a little below the set point.

The homeostatic mechanisms in mammals require information to be transferred between different parts of the body. There are two coordination systems in mammals that do this: the nervous system and the endocrine system.

- In the nervous system, information in the form of electrical impulses is transmitted along nerve cells (neurones).
- The endocrine system uses chemical messengers called **hormones** that travel in the blood, in a form of long-distance cell signalling.

QUESTION

14.1 a Describe the immediate environment of a typical cell within the body of a mammal.

b Explain why it is important that the internal environment of a mammal is carefully regulated.

c Explain how the following are involved in maintaining the internal environment of a mammal: stimuli, receptors, central control, coordination systems and effectors.

d i Explain the meaning of the terms **homeostasis** and **negative feedback**.

ii Distinguish between the input and the output in a homeostatic control mechanism.

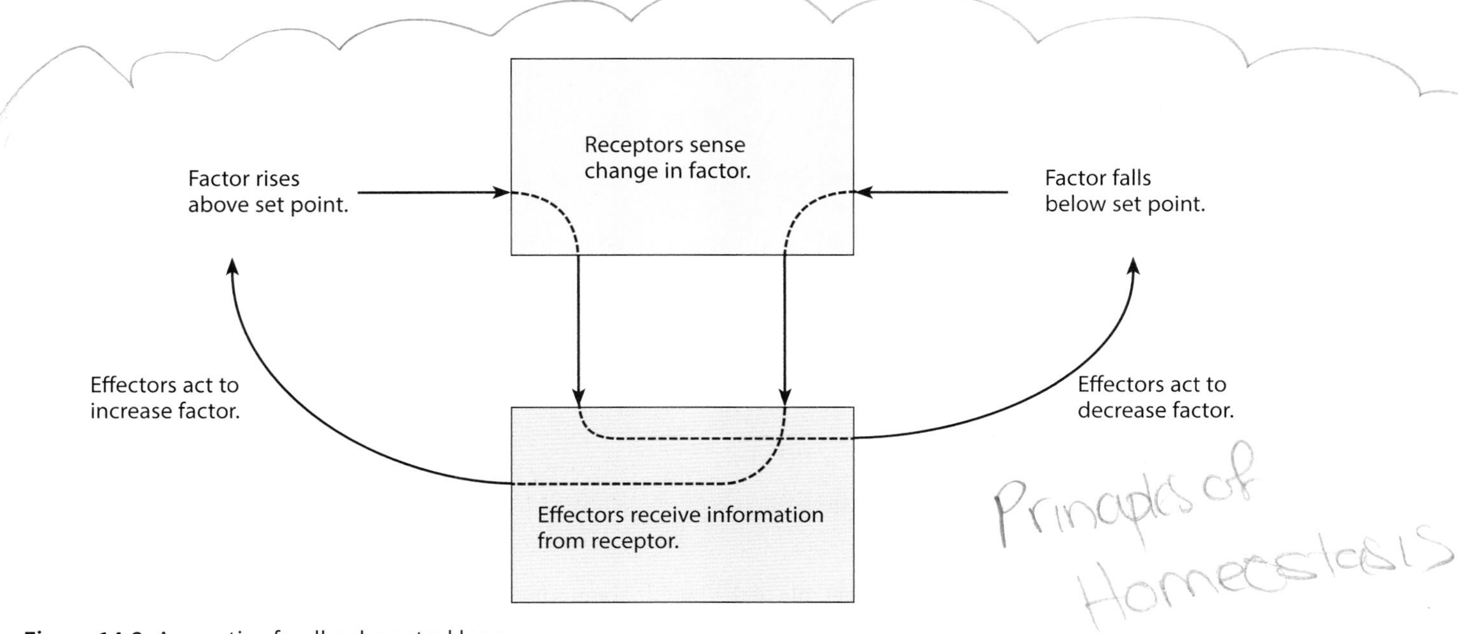

Figure 14.2 A negative feedback control loop.

The control of body temperature

Thermoregulation is the control of body temperature. This involves both coordination systems – nervous and endocrine. All mammals generate heat and have ways to retain it within their bodies. They also have physiological methods to balance heat gain, retention of body heat and heat loss so that they can maintain a constant body temperature. As a result, they are not dependent on absorbing heat from their surroundings and can be active at any time of day or night, whatever the external temperature. Most other animals, with the exception of birds, rely on external sources of heat and are often relatively inactive when it is cold.

The heat that mammals generate is released during respiration (page 272). Much of the heat is produced by liver cells that have a huge requirement for energy. The heat they produce is absorbed by the blood flowing through the liver and distributed around the rest of the body.

The hypothalamus (Figure 14.3) in the brain is the central control for body temperature; it is the body's thermostat. This region of the brain receives a constant input of sensory information about the temperature of the blood and about the temperature of the surroundings. The hypothalamus has thermoreceptor cells that continually monitor the temperature of the blood flowing through it. The temperature it monitors is the **core temperature** – the temperature inside the body that remains very close to the set point, which is 37 °C in humans. This temperature fluctuates a little, but is kept within very narrow limits by the hypothalamus.

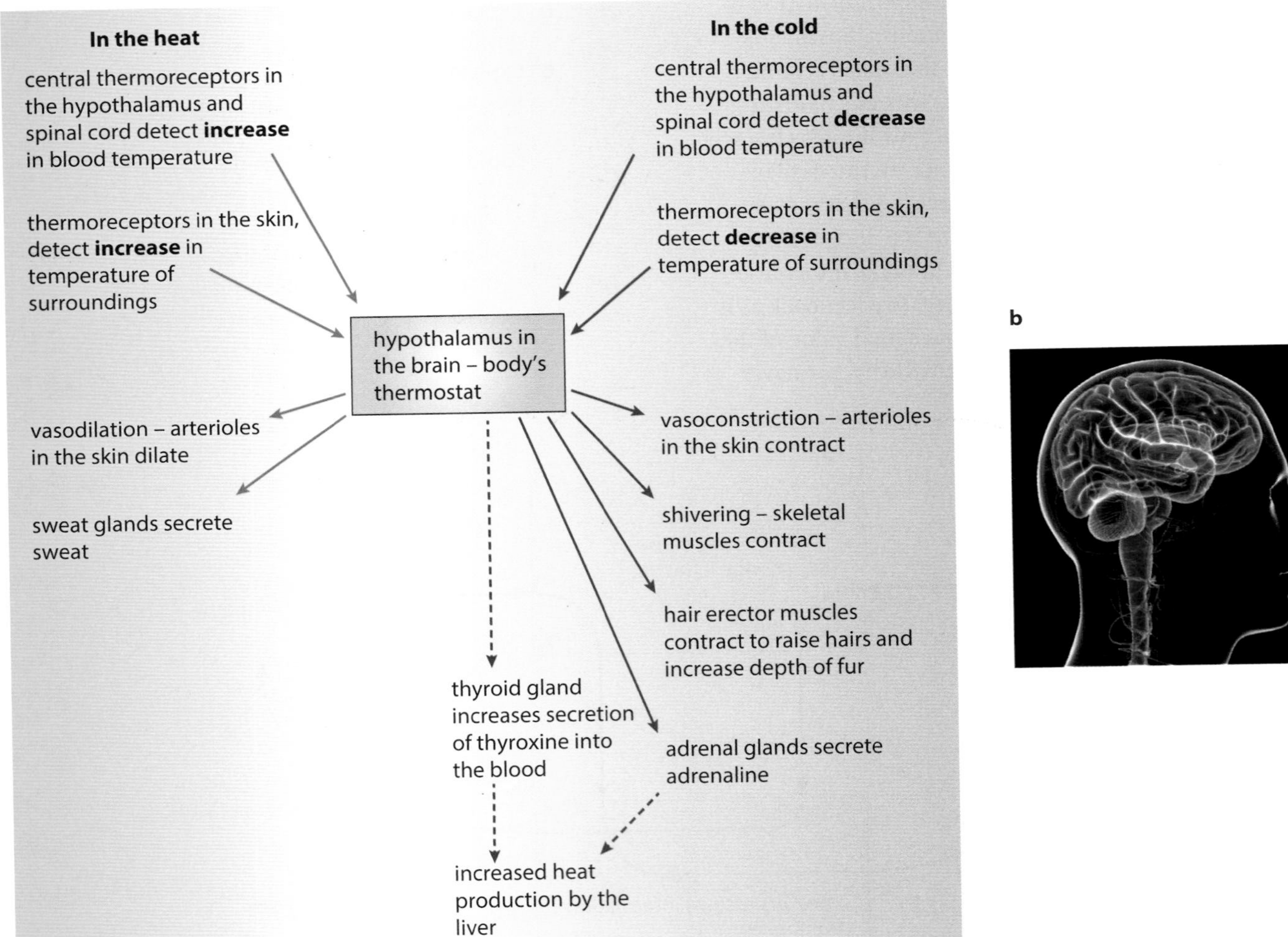

Figure 14.3 **a** A summary diagram to show the central role of the hypothalamus in thermoregulation when it is hot and when it is cold. The hypothalamus communicates with other regions of the body by using nerves (solid lines) and hormones (dashed lines). **b** The position of the hypothalamus, shown in red, in the brain.

The hypothalamus receives information about temperature from other sources as well. The skin contains receptors that monitor changes in skin temperature. The skin temperature is the first to change if there is a change in the temperature of the surroundings. These skin receptors give an 'early warning' about a possible change in core temperature. If the core temperature decreases, or if the temperature receptors in the skin detect a decrease in the temperature of the surroundings, the hypothalamus sends impulses that activate the following physiological responses.

- **Vasoconstriction** – muscles in the walls of arterioles that supply blood to capillaries near the skin surface contract. This narrows the lumens of the arterioles and reduces the supply of blood to the capillaries so that less heat is lost from the blood.
- **Shivering** – the involuntary contraction of skeletal muscles generates heat which is absorbed by the blood and carried around the rest of the body.
- **Raising body hairs** – muscles at the base of hairs in the skin contract to increase the depth of fur so trapping air close to the skin. Air is a poor conductor of heat and therefore a good insulator. This is not much use in humans, but is highly effective for most mammals.
- **Decreasing the production of sweat** – this reduces the loss of heat by evaporation from the skin surface.
- **Increasing the secretion of adrenaline** – this hormone from the adrenal gland increases the rate of heat production in the liver.

The hypothalamus also stimulates higher centres in the brain to bring about some behavioural responses. Some animals respond by curling up to reduce the surface area exposed to the air and by huddling together. We respond by finding a source of warmth and putting on warm clothing.

When an increase in environmental temperature is detected by skin receptors or the central thermoreceptors, the hypothalamus increases the loss of heat from the body and reduces heat production.

- **Vasodilation** – the muscles in the arterioles in the skin relax, allowing more blood to flow through the capillaries so that heat is lost to the surroundings.
- **Lowering body hairs** – muscles attached to the hairs relax so they lie flat, reducing the depth of fur and the layer of insulation.
- **Increasing sweat production** – sweat glands increase the production of sweat which evaporates on the surface of the skin so removing heat from the body.

The behavioural responses of animals to heat include resting or lying down with the limbs spread out to increase the body surface exposed to the air. We respond by wearing loose fitting clothing, turning on fans or air conditioning and taking cold drinks.

When the environmental temperature decreases gradually, as it does with the approach of winter in temperate climates, the hypothalamus releases a hormone which activates the anterior pituitary gland (page 312) to release thyroid stimulating hormone (TSH). TSH stimulates the thyroid gland to secrete the hormone thyroxine into the blood. Thyroxine increases metabolic rate, which increases heat production especially in the liver. When temperatures start to increase again, the hypothalamus responds by reducing the release of TSH by the anterior pituitary gland so less thyroxine is released from the thyroid gland.

There are two other examples of the role of negative feedback in homeostasis later in this chapter: osmoregulation and blood glucose control. Sometimes control mechanisms do not respond in the way described so far. If a person breathes air that has very high carbon dioxide content, this produces a high concentration of carbon dioxide in the blood. This is sensed by carbon dioxide receptors, which cause the breathing rate to increase. So the person breathes faster, taking in even more carbon dioxide, which stimulates the receptors even more, so the person breathes faster and faster. This is an example of a **positive feedback**. You can see that positive feedback cannot play any role in keeping conditions in the body constant! However, this method of control is involved in several biological processes including the transmission of nerve impulses (page 335)

QUESTION

14.2 Use Figure 14.2 on page 301 to make a flow diagram to show the negative feedback loop that keeps temperature constant in a mammal. Your diagram should include the names of the receptors and effectors, and the actions that the effectors take.

Excretion

Many of the metabolic reactions occurring within the body produce unwanted substances. Some of these are toxic (poisonous). The removal of these unwanted products of metabolism is known as excretion.

Many excretory products are formed in humans, but two are made in much greater quantities than others. These are **carbon dioxide** and urea. Carbon dioxide is produced continuously by cells that are respiring aerobically. The waste carbon dioxide is transported from the respiring cells to the lungs, in the bloodstream (page 170). Gas exchange occurs within the lungs, and carbon dioxide diffuses from the blood into the alveoli; it is then excreted in the air we breathe out.

Urea is produced in the **liver**. It is produced from excess amino acids and is transported from the liver to the kidneys, in solution in blood plasma. The kidneys remove urea from the blood and excrete it, dissolved in water, as **urine**. Here, we will look more fully at the production and excretion of urea.

Deamination

If more protein is eaten than is needed, the excess cannot be stored in the body. It would be wasteful, however, simply to get rid of all the excess, because the amino acids provide useful energy. To make use of this energy, the liver removes the amino groups in a process known as deamination.

Figure 14.4a shows how deamination takes place. In the liver cells, the amino group ($-NH_2$) of an amino acid is removed, together with an extra hydrogen atom. These combine to produce ammonia (NH_3). The keto acid that remains may enter the Krebs cycle and be respired, or it may be converted to glucose, or converted to glycogen or fat for storage.

Ammonia is a very soluble and highly toxic compound. In many aquatic animals, such as fish that live in fresh water, ammonia diffuses from the blood and dissolves in the water around the animal. However, in terrestrial animals, such as humans, ammonia would rapidly build up in the blood and cause immense damage. Damage is prevented by converting ammonia immediately to urea, which is less soluble and less toxic. Several reactions, known as the urea cycle, are involved in combining ammonia and carbon dioxide to form urea. These are simplified as shown in Figure 14.4b. An adult human produces around 25–30 g of urea per day.

Urea is the main nitrogenous excretory product of humans. We also produce small quantities of other nitrogenous excretory products, mainly creatinine and uric acid. A substance called **creatine** is made in the liver, from certain amino acids. Much of this creatine is used in the muscles, in the form of creatine phosphate, where it acts as an energy store (Chapter 15). However, some is converted to creatinine and excreted. Uric acid is made from the breakdown of purines from nucleotides, not from amino acids.

Urea diffuses from liver cells into the blood plasma. All of the urea made each day must be excreted, or its concentration in the blood would build up and become dangerous. As the blood passes through the kidneys, the urea is filtered out and excreted. To explain how this happens, we must first look at the structure of a kidney.

QUESTION

14.3 a Name the nitrogenous waste substances excreted by mammals.

b Explain why it is important that carbon dioxide and nitrogenous wastes are excreted and not allowed to accumulate in the body.

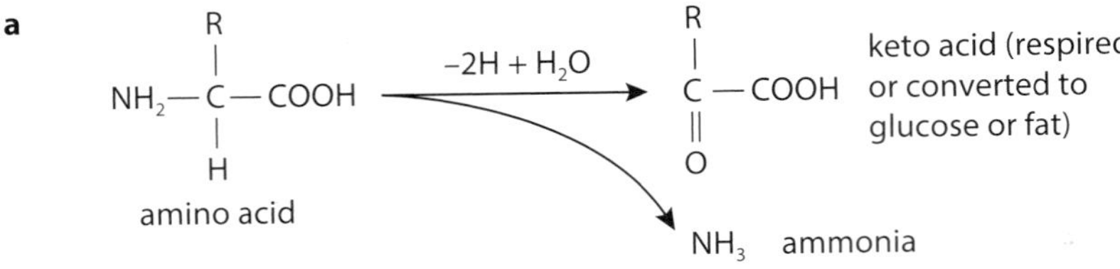

Figure 14.4 **a** Deamination and **b** urea formation.

The structure of the kidney

Figure 14.5 shows the position of the kidneys in the body, together with their associated structures. Each kidney receives blood from a **renal artery**, and returns blood via a **renal vein**. A narrow tube, called the **ureter**, carries urine from the kidney to the bladder. From the bladder a single tube, the **urethra**, carries urine to the outside of the body.

A longitudinal section through a kidney (Figure 14.6) shows that it has three main areas. The whole kidney is covered by a fairly tough **capsule**, beneath which lies the **cortex**. The central area is made up of the **medulla**. Where the ureter joins, there is an area called the **pelvis**.

A section through a kidney, seen through a microscope (Figure 14.7), shows it to be made up of

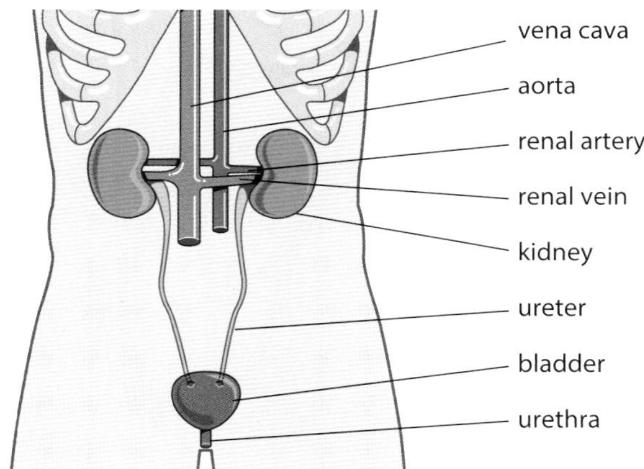

Figure 14.5 Position of the kidneys and associated structures in the human body.

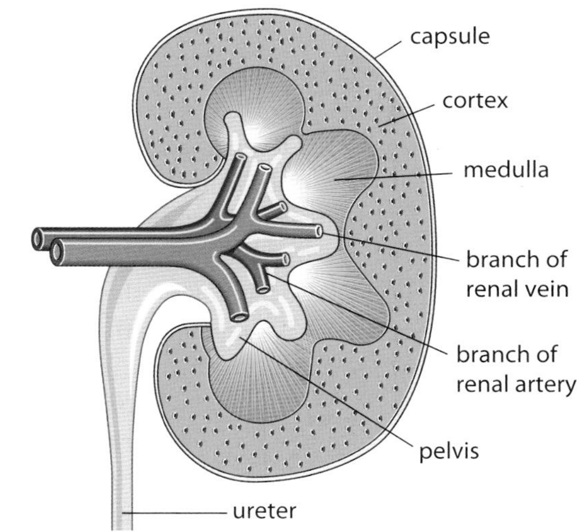

Figure 14.6 A kidney cut in half vertically.

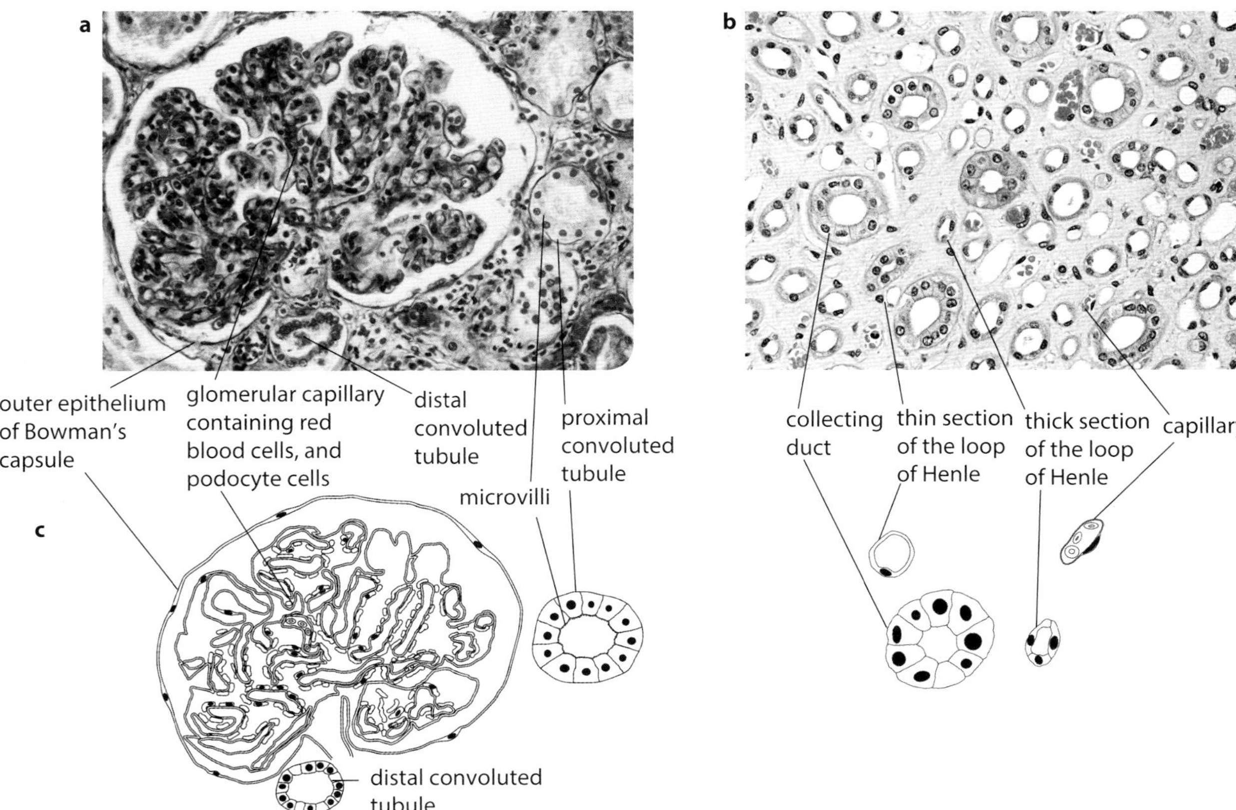

Figure 14.7 **a** Photomicrograph of a section through the cortex of the kidney showing a glomerulus and Bowman's capsule surrounded by proximal and distal convoluted tubules (×150); **b** photomicrograph of a section through the medulla of a kidney (×300); **c** interpretive drawings.

thousands of tiny tubes, called nephrons, and many blood vessels. Figure 14.8a shows the position of a single nephron, and Figure 14.8b shows its structure. One end of the tube forms a cup-shaped structure called a Bowman's capsule, which surrounds a tight network of capillaries called a glomerulus. The glomeruli and capsules of all the nephrons are in the cortex of the kidney. From the capsule, the tube runs towards the centre of the kidney, first forming a twisted region called the **proximal convoluted tubule**, and then a long hairpin loop in the medulla, the loop of Henle. The tubule then runs back upwards into the cortex, where it forms another twisted region called the **distal convoluted tubule**, before finally joining a collecting duct that leads down through the medulla and into the pelvis of the kidney.

Blood vessels are closely associated with the nephrons (Figure 14.9). Each glomerulus is supplied with blood by a branch of the renal artery called an afferent **arteriole**. The capillaries of the glomerulus rejoin to form an efferent **arteriole**. The efferent arteriole leads off to form a network of capillaries running closely alongside the rest of the nephron. Blood from these capillaries flows into a branch of the renal vein.

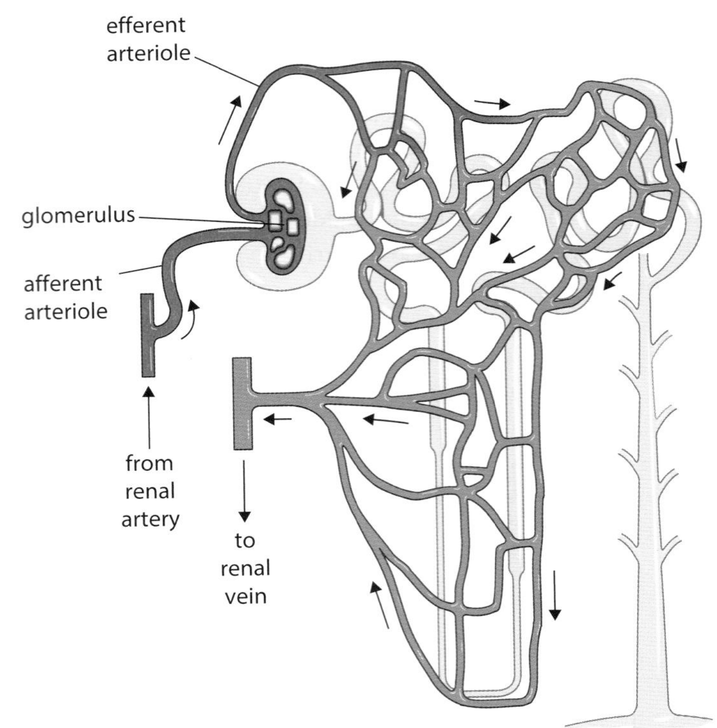

Figure 14.9 The blood supply associated with a nephron.

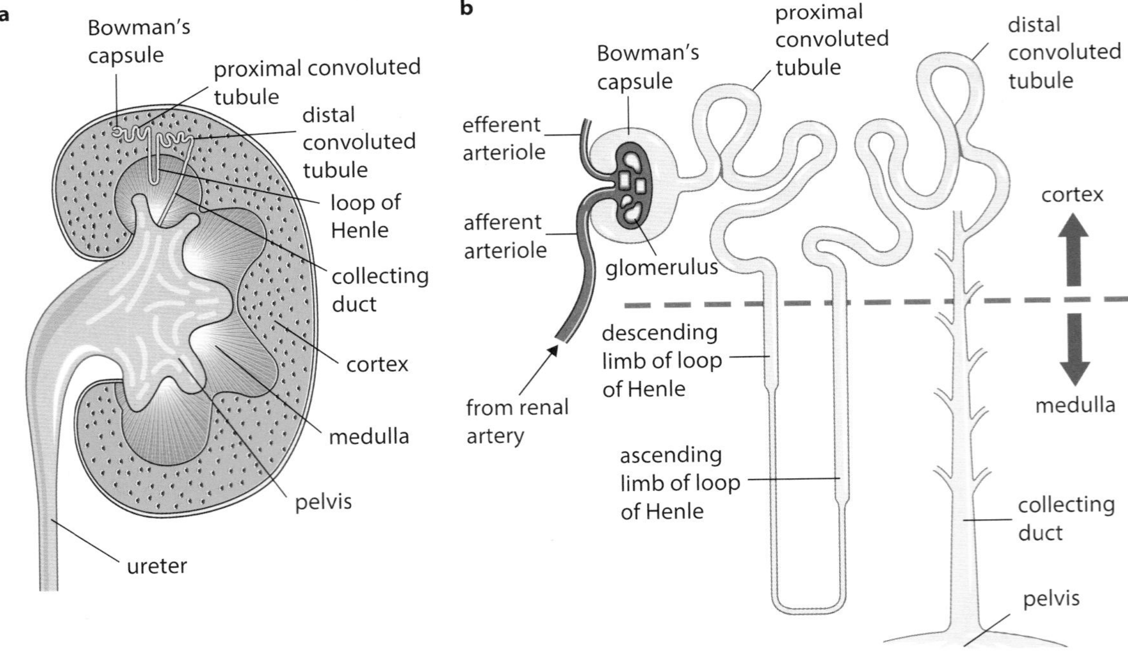

Figure 14.8 a Section through the kidney to show the position of a nephron; **b** a nephron.

The kidney makes urine in a two-stage process. The first stage, ultrafiltration, involves filtering small molecules, including urea, out of the blood and into the Bowman's capsule. From the Bowman's capsule the molecules flow along the nephron towards the ureter. The second stage, selective reabsorption, involves taking back any useful molecules from the fluid in the nephron as it flows along.

Ultrafiltration

Figure 14.10 shows a section through part of a glomerulus and Bowman's capsule. The blood in the glomerular capillaries is separated from the lumen of the Bowman's capsule by two cell layers and a basement membrane. The first cell layer is the lining, or endothelium, of the capillary. Like the endothelium of most capillaries, this has gaps in it, but there are far more gaps than in other capillaries: each endothelial cell has thousands of tiny holes in it. Next comes the **basement membrane**, which is made up of a network of collagen and glycoproteins. The second cell layer is formed from **epithelial cells**, which make up the inner lining of the Bowman's capsule. These cells have many tiny finger-like projections with gaps in between them, and are called podocytes (Figure 14.11).

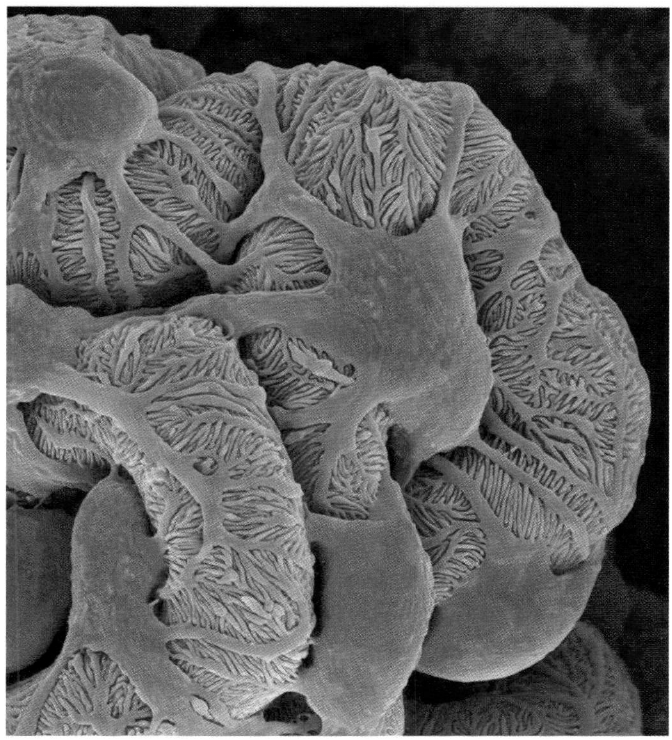

Figure 14.11 A false-colour scanning electron micrograph of podocytes (×3900). The podocytes are the blue-green cells with their extensions wrapped around the blood capillary, which is purple.

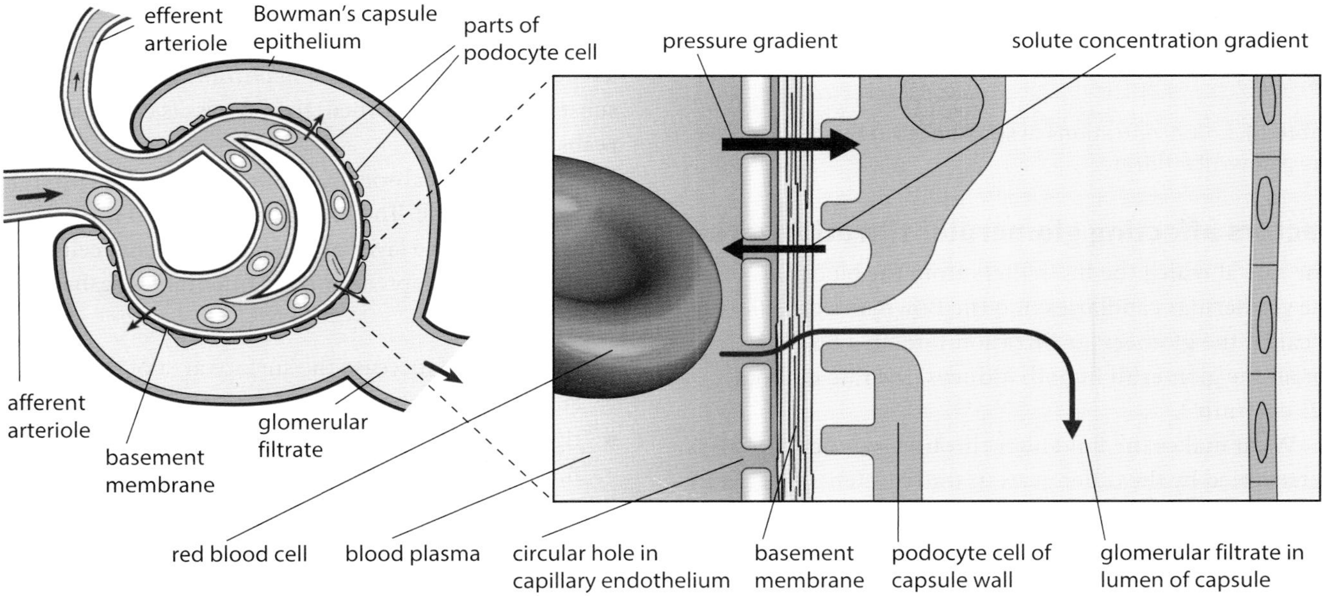

Figure 14.10 Detail of the endothelium of a glomerular capillary and Bowman's capsule. The arrows show how the net effect of higher pressure in the capillary and lower solute concentration in the Bowman's capsule is that fluid moves out of the capillary and into the lumen of the capsule. The basement membrane acts as a molecular filter.

The holes in the capillary endothelium and the gaps between the podocytes are quite large, and make it easy for substances dissolved in the blood plasma to get through from the blood into the capsule. However, the basement membrane stops large protein molecules from getting through. Any protein molecule with a relative molecular mass of around 69 000 or more cannot pass through the basement membrane, and so cannot escape from the glomerular capillaries. This basement membrane therefore acts as a filter. Blood cells, both red and white, are also too large to pass through this barrier, and so remain in the blood. Table 14.1 shows the relative concentrations of substances in the blood and in the glomerular filtrate. You will see that glomerular filtrate is identical to blood plasma except that there are almost no plasma proteins in it.

Substance	Concentration in blood plasma / $g\,dm^{-3}$	Concentration in glomerular filtrate / $g\,dm^{3}$
water	900	900
proteins	80.0	0.05
amino acids	0.5	0.5
glucose	1.0	1.0
urea	0.3	0.3
uric acid	0.04	0.04
creatinine	0.01	0.01
inorganic ions (mainly Na^+, K^+ and Cl^-)	7.2	7.2

Table 14.1 Concentrations of substances in the blood and in the glomerular filtrate.

Factors affecting glomerular filtration rate

The rate at which the fluid filters from the blood in the glomerular capillaries into the Bowman's capsule is called the glomerular filtration rate. In a human, for all the glomeruli in both kidneys, the rate is about $125\,cm^3\,min^{-1}$.

What makes the fluid filter through so quickly? This is determined by the differences in water potential between the plasma in glomerular capillaries and the filtrate in the Bowman's capsule. You will remember that water moves from a region of higher water potential to a region of lower water potential, down a water potential gradient (page 83). Water potential is lowered by the presence of solutes, and raised by high pressures.

Inside the capillaries in the glomerulus, the blood pressure is relatively high, because the diameter of the afferent arteriole is wider than that of the efferent arteriole, causing a head of pressure inside the glomerulus. This tends to raise the water potential of the blood plasma above the water potential of the contents of the Bowman's capsule (Figure 14.9).

However, the concentration of solutes in the blood plasma in the capillaries is **higher** than the concentration of solutes in the filtrate in the Bowman's capsule. This is because, while most of the contents of the blood plasma filter through the basement membrane and into the capsule, the plasma protein molecules are too big to get through, and so stay in the blood. This difference in solute concentration tends to make the water potential in the blood capillaries **lower** than that of the filtrate in the Bowman's capsule.

Overall, though, the effect of differences in pressure outweighs the effect of the differences in solute concentration. Overall, the water potential of the blood plasma in the glomerulus is higher than the water potential of the filtrate in the capsule. So water continues to move down the water potential gradient from the blood into the capsule.

Reabsorption in the proximal convoluted tubule

Many of the substances in the glomerular filtrate need to be kept in the body, so they are reabsorbed into the blood as the fluid passes along the nephron. As only certain substances are reabsorbed, the process is called **selective reabsorption**.

Most of the reabsorption takes place in the proximal convoluted tubule. The lining of this part of the nephron is made of a single layer of cuboidal epithelial cells. These cells are adapted for their function of reabsorption by having:

- microvilli to increase the surface area of the inner surface facing the lumen
- tight junctions that hold adjacent cells together so that fluid cannot pass **between** the cells (all substances that are reabsorbed must go through the cells)
- many mitochondria to provide energy for sodium–potassium (Na^+–K^+) pump proteins in the outer membranes of the cells
- co-transporter proteins in the membrane facing the lumen.

Blood capillaries are very close to the outer surface of the tubule. The blood in these capillaries has come directly from the glomerulus, so it has much less plasma in it than usual and has lost much of its water and many of the ions and other small solutes.

The basal membranes of the cells lining the proximal convoluted tubule are those nearest the blood capillaries. Sodium–potassium pumps in these membranes move sodium ions out of the cells (Figure 14.12). The sodium ions are carried away in the blood. This lowers the concentration of sodium ions inside the cell, so that they passively diffuse into it, down their concentration gradient, from the fluid in the lumen of the tubule. However, sodium ions do not diffuse freely through the membrane: they can only enter through special co-transporter proteins in the membrane. There are several different kinds of co-transporter protein, each of which transports something else, such as a glucose molecule or an amino acid, at the same time as the sodium ion.

The passive movement of sodium ions **into** the cell down their concentration gradient provides the energy to move glucose molecules, even against a concentration gradient. This movement of glucose, and of other solutes, is an example of indirect or secondary active transport, since the energy (as ATP) is used in the pumping of sodium ions, not in moving these solutes. Once inside the cell, glucose diffuses down its concentration gradient, through a transport protein in the basal membrane, into the blood.

All of the **glucose** in the glomerular filtrate is transported out of the proximal convoluted tubule and into the blood. Normally, no glucose is left in the filtrate, so no glucose is present in urine. Similarly, **amino acids**, **vitamins**, and many **sodium** and **chloride** ions (Cl^-) are reabsorbed in the proximal convoluted tubule.

The removal of these solutes from the filtrate greatly increases its water potential. The movement of solutes into the cells and then into the blood decreases the water potential there, so a water potential gradient exists between filtrate and blood. Water moves down this gradient through the cells and into the blood. The water and reabsorbed solutes are carried away, back into the circulation.

Surprisingly, quite a lot of urea is reabsorbed too. Urea is a small molecule which passes easily through cell membranes. Its concentration in the filtrate is considerably higher than that in the capillaries, so it diffuses passively through the cells of the proximal convoluted tubule and into the blood. About half of the urea in the filtrate is reabsorbed in this way.

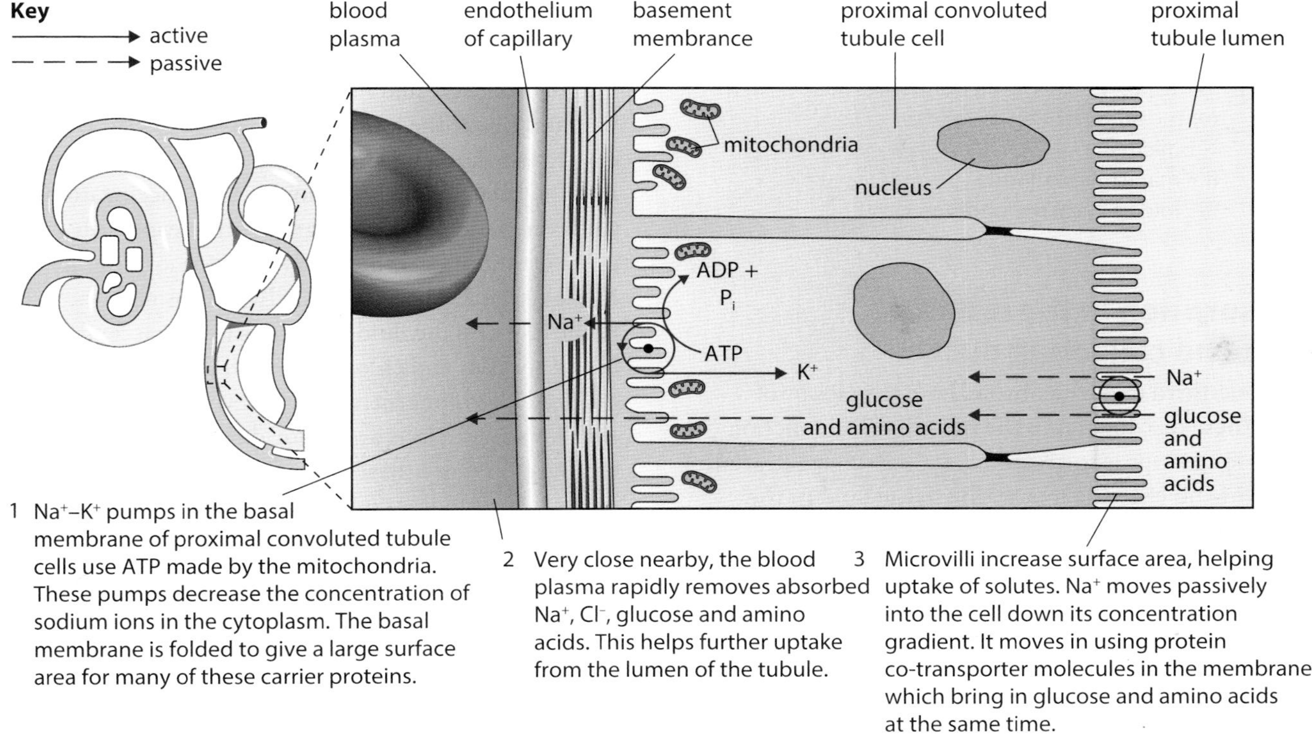

Figure 14.12 Reabsorption in the proximal convoluted tubule.

The other two nitrogenous excretory products, uric acid and creatinine, are not reabsorbed. Indeed, creatinine is actively **secreted** by the cells of the proximal convoluted tubule into its lumen.

The reabsorption of so much water and solutes from the filtrate in the proximal convoluted tubule greatly reduces the volume of liquid remaining. In an adult human, around 125 cm^3 of fluid enter the proximal tubules every minute, but only about 64% of this passes on to the next region of each nephron, the loop of Henle.

QUESTIONS

14.4 a Where has the blood in the capillaries surrounding the proximal convoluted tubule come from?

b What solutes will this blood contain that are **not** present in the glomerular filtrate?

c How might this help in the reabsorption of water from the proximal convoluted tubule?

d State the name of the process by which water is reabsorbed.

14.5 a Calculate the volume of filtrate that enters the loops of Henle from the proximal convoluted tubules each minute.

b Although almost half of the urea in the glomerular filtrate is reabsorbed from the proximal convoluted tubule, the **concentration** of urea in the fluid in the nephron actually increases as it passes along the proximal convoluted tubule. Explain why this is so.

c Explain how each of these features of the cells in the proximal convoluted tubules adapts them for the reabsorption of solutes:

i microvilli

ii many mitochondria

iii folded basal membranes.

Reabsorption in the loop of Henle and collecting duct

About one-third of our nephrons have long loops of Henle. These dip down into the medulla. The function of these long loops is to create a very high concentration of sodium and chloride ions in the tissue fluid in the medulla. As you will see, this enables a lot of water to be reabsorbed from the fluid in the collecting duct, as it flows through the medulla. This allows the production of very concentrated urine, which means that water is conserved in the body, rather than lost in urine, helping to prevent dehydration.

Figure 14.13a shows the loop of Henle. The hairpin loop runs deep into the medulla of the kidney, before turning back towards the cortex again. The first part of the loop is the **descending limb**, and the second part is the **ascending limb**. These differ in their permeabilities to water. The descending limb is permeable to water, whereas the ascending limb is not.

To explain how the loop of Henle works, it is best to start by describing what happens in the ascending limb. The cells that line this region of the loop actively transport sodium and chloride ions out of the fluid in the loop, into the tissue fluid. This **decreases** the water potential in the tissue fluid and **increases** the water potential of the fluid inside the ascending limb.

The cells lining the descending limb are **permeable** to water and also to sodium and chloride ions. As the fluid flows down this loop, water from the filtrate moves down a water potential gradient into the tissue fluid by osmosis. At the same time, sodium and chloride ions diffuse **into** the loop, down their concentration gradient. So, by the time the fluid has reached the very bottom of the hairpin, it contains much less water and many more sodium and chloride ions than it did when it entered from the proximal convoluted tubule. The fluid becomes more concentrated towards the bottom of the loop. The longer the loop, the more concentrated the fluid can become. The concentration in human kidneys can be as much as four times the concentration of blood plasma.

This concentrated fluid flows up the ascending limb. As the fluid inside the loop is so concentrated, it is relatively easy for sodium and chloride ions to leave it and pass into the tissue fluid, even though the concentration in the tissue fluid is also very great. Thus, especially high concentrations of sodium and chloride ions can be built up in the tissue fluid between the two limbs near the bottom of the loop. As the fluid continues up the ascending limb, losing sodium and chloride ions all the time, it becomes gradually less concentrated. However, it is still relatively easy for sodium ions and chloride ions to be actively removed, because these higher parts of the ascending loop are next to less concentrated regions of tissue fluid. All the way up, the concentration of sodium and chloride ions inside the tubule is never very different from the concentration in the tissue fluid, so it is never too difficult to pump sodium and chloride ions out of the tubule into the tissue fluid.

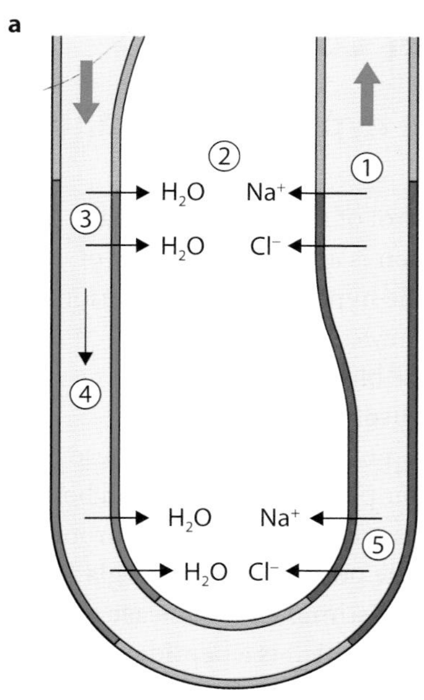

1 Na^+ and Cl^- are actively transported out of the ascending limb.

2 This raises the concentration of Na^+ and Cl^- in the tissue fluid.

3 This in turn causes the loss of water from the descending limb.

4 The loss of water concentrates Na^+ and Cl^- in the descending limb.

5 Na^+ and Cl^- ions diffuse out of this concentrated solution in the lower part of the ascending limb.

Key

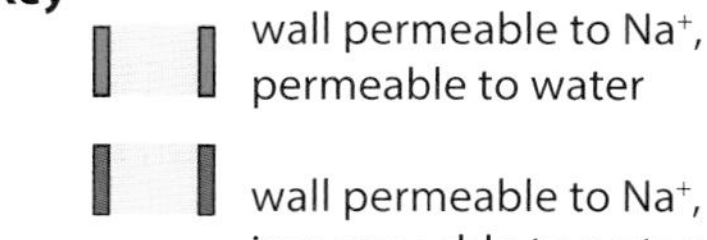

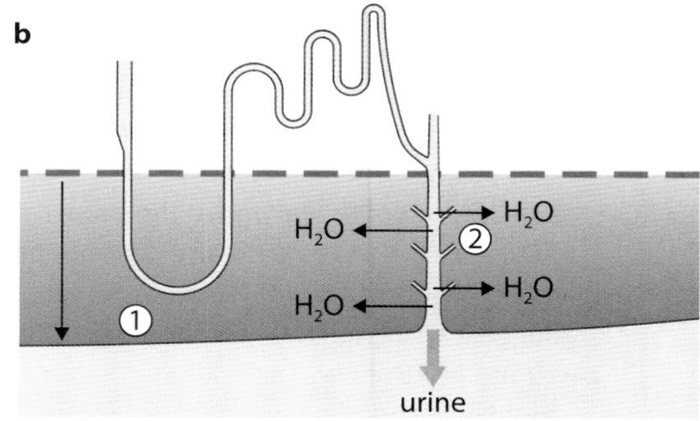

1 The tissue in the deeper layers of the medulla contains a very concentrated solution of Na^+, Cl^- and urea.
2 As urine passes down the collecting duct, water can pass out of it by osmosis. The reabsorbed water is carried away by the blood in the capillaries.

Figure 14.13 How the loop of Henle allows the production of concentrated urine. **a** The counter-current mechanism in the loop of Henle builds up high concentrations of sodium ions and chloride ions in the tissue fluid of the medulla. **b** Water can pass out of the fluid in the collecting duct by osmosis, as the surrounding tissue fluid has a lower water potential.

Having the two limbs of the loop running side by side like this, with the fluid flowing down in one and up in the other, enables the maximum concentration of solutes to be built up both inside and outside the tube at the bottom of the loop. This mechanism is called a counter-current multiplier.

But the story is not yet complete. You have seen that the fluid flowing up the ascending limb of the loop of Henle loses sodium and chloride ions as it goes, so becoming more dilute and having a higher water potential. The cells of the ascending limb of the loop of Henle and the cells lining the collecting ducts are permeable to urea, which diffuses into the tissue fluid. As a result, urea is also concentrated in the tissue fluid in the medulla. In Figure 14.13b you can see that the fluid continues round through the distal convoluted tubule into the **collecting duct**, which runs down into the medulla again. It therefore passes once again through the regions where the solute concentration of the tissue fluid is very high and the water potential very low. Water therefore can move out of the collecting duct, by osmosis, until the water potential of urine is the same as the water potential of the tissue fluid in the medulla, which may be much greater than the water potential of the blood. The degree to which this happens is controlled by antidiuretic hormone (ADH).

The ability of some small mammals, such as rodents, to produce a very concentrated urine is related to the relative thickness of the medulla in their kidneys. The maximum concentration of urine that we can produce is four times that of our blood plasma. Desert rodents, such as gerbils and kangaroo rats, can produce a urine that is about 20 times the concentration of their blood plasma. This is possible because the medulla is relatively large and the cells that line the ascending limb of their loops have deep infolds with many Na^+–K^+ pumps and cytoplasm filled with many mitochondria, each with many cristae that allow the production of much ATP to provide the energy for the pumping of sodium ions into the tissue fluid.

Reabsorption in the distal convoluted tubule and collecting duct

The first part of the distal convoluted tubule functions in the same way as the ascending limb of the loop of Henle. The second part functions in the same way as the collecting duct, so the functions of this part of the distal convoluted tubule and the collecting duct will be described together.

In the distal convoluted tubule and collecting duct, sodium ions are actively pumped from the fluid in the tubule into the tissue fluid, from where they pass into the blood. Potassium ions, however, are actively transported **into** the tubule. The rate at which these two ions are moved into and out of the fluid in the nephron can be varied, and helps to regulate the concentration of these ions in the blood.

QUESTION

14.6 a Figure 14.14 shows the relative rate at which fluid flows through each part of a nephron. If water flows into an impermeable tube such as a hosepipe, it will flow **out** of the far end at the same rate that it flows **in**. However, this clearly does not happen in a nephron. Consider what happens in each region, and suggest an explanation for the shape of the graph.

b Figure 14.15 shows the relative concentrations of four substances in each part of a nephron. Explain the shapes of the curves for: **i** glucose, **ii** urea, **iii** sodium ions **iv** potassium ions.

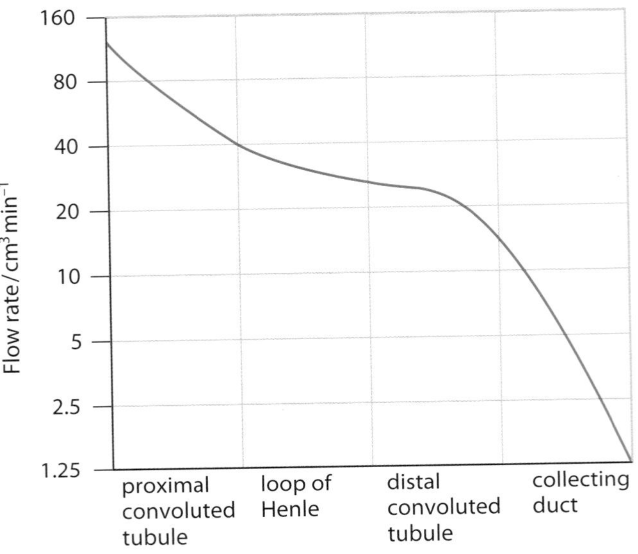

Figure 14.14 Flow rates in different parts of a nephron.

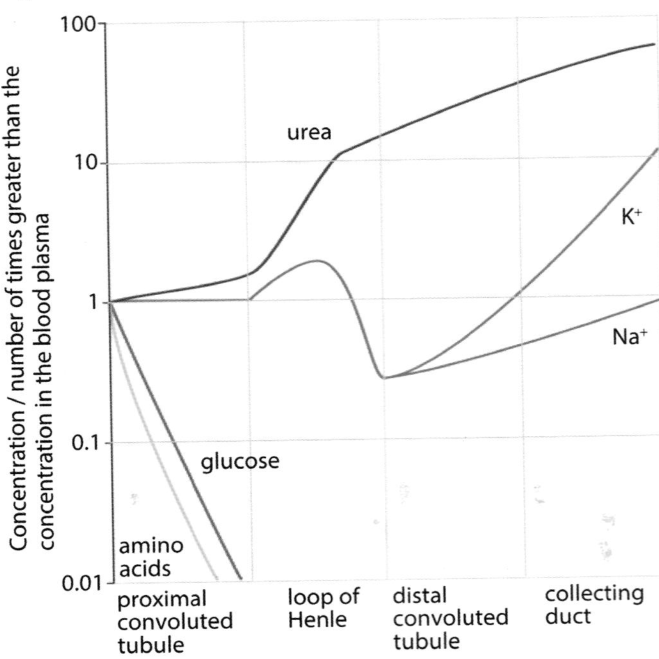

Figure 14.15 Relative concentrations of five substances in different parts of a nephron.

Control of water content

Osmoreceptors, the hypothalamus and ADH

Osmoregulation is the control of the water potential of body fluids. This regulation is an important part of **homeostasis** and involves the hypothalamus, posterior pituitary gland and the kidneys.

The water potential of the blood is constantly monitored by specialised sensory neurones in the hypothalamus, known as **osmoreceptors**. When these cells detect a **decrease** in the water potential of the blood below the set point, nerve impulses are sent along the neurones to where they terminate in the posterior pituitary gland (Figure 14.16). These impulses stimulate the release of **antidiuretic hormone (ADH)**, which is a peptide hormone made of nine amino acids. Molecules of ADH enter the blood in capillaries and are carried all over the body. The effect of ADH is to reduce the loss of water in the urine by making the kidney reabsorb as much water as possible. The word 'diuresis' means the production of dilute urine. Antidiuretic hormone gets its name because it stops dilute urine being produced, by stimulating the reabsorption of water.

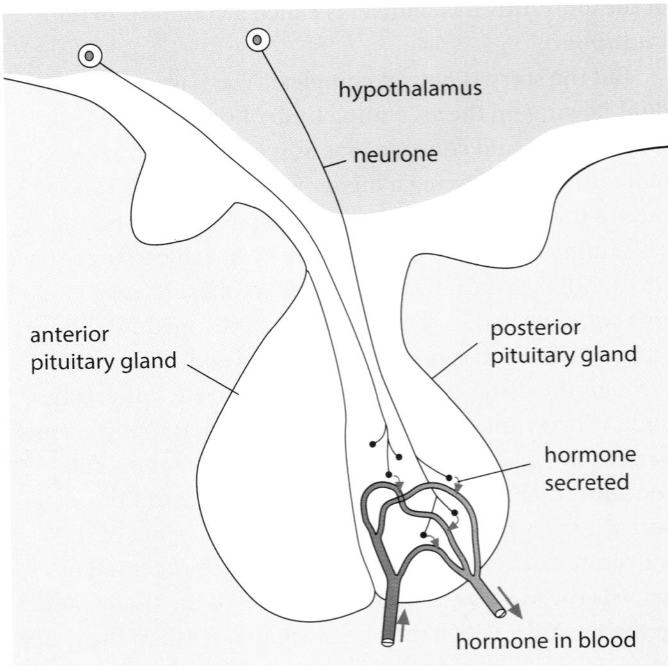

Figure 14.16 ADH is produced by neurones in the hypothalamus and is released into the blood where the neurones terminate in the posterior pituitary gland.

How ADH affects the kidneys

You have seen that water is reabsorbed by osmosis from the fluid in the nephron as the fluid passes through the collecting ducts. The cells of the collecting duct are the target cells for ADH. This hormone acts on the cell surface membranes of the collecting ducts cells, making them more permeable to water than usual (Figure 14.17).

This change in permeability is brought about by increasing the number of the water-permeable channels known as aquaporins in the cell surface membrane of the collecting duct cells (Figure 14.18). ADH molecules bind to receptor proteins on the cell surface membranes, which in turn activate enzymes inside the cells. The cells contain ready-made vesicles that have many aquaporins in their

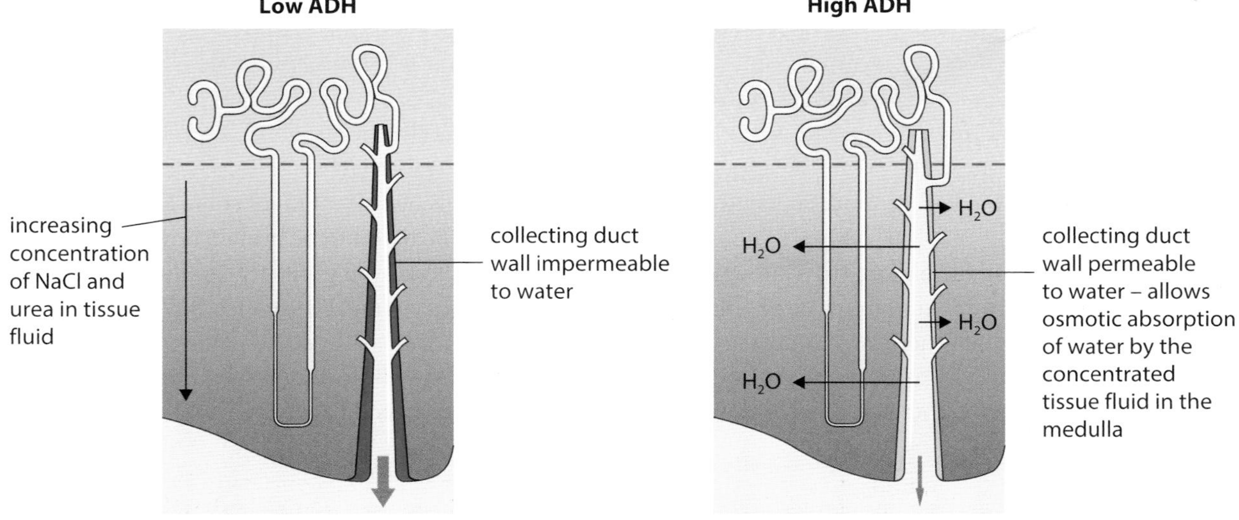

Figure 14.17 The effects of ADH on water reabsorption from the collecting duct.

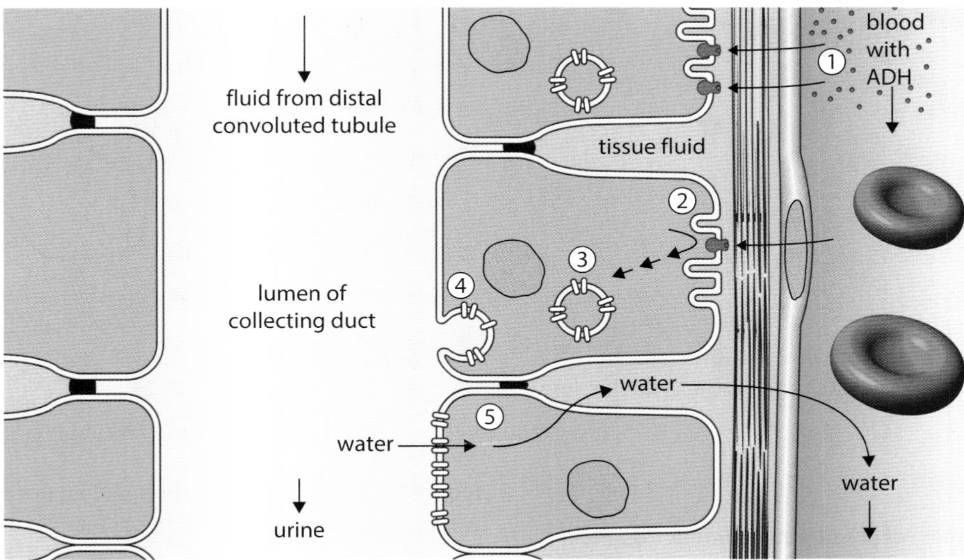

1 ADH binds to receptors in the cell surface membrane of the cells lining the collecting duct.
2 This activates a series of enzyme-controlled reactions, ending with the production of an active phosphorylase enzyme.
3 The phosphorylase causes vesicles, surrounded by membrane containing water-permeable channels (aquaporins), to move to the cell surface membrane.
4 The vesicles fuse with the cell surface membrane.
5 Water can now move freely through the membrane, down its water potential gradient, into the concentrated tissue fluid and blood plasma in the medulla of the kidney.

Figure 14.18 How ADH increases water reabsorption in the collecting duct.

membranes. Once the enzymes in each cell are activated by the arrival of ADH, these vesicles move towards the cell surface membrane and fuse with it, so increasing the permeability of the membrane to water.

So, as the fluid flows down through the collecting duct, water molecules move through the aquaporins (Figure 14.19), out of the tubule and into the tissue fluid. This happens because the tissue fluid in the medulla has a very low water potential and the fluid in the collecting ducts has a very high water potential. The fluid in the collecting duct loses water and becomes more concentrated. The secretion of ADH has caused the increased reabsorption of water into the blood. The volume of urine which flows from the kidneys into the bladder will be smaller, and the urine will be more concentrated (Figure 14.20).

What happens when you have more than enough water in the body – for example, after enjoying a large volume of your favourite drink? When there is an **increase** in the water potential of the blood, the osmoreceptors in the hypothalamus are no longer stimulated and the neurones in the posterior pituitary gland stop secreting ADH. This affects the cells that line the collecting ducts. The aquaporins are moved out of the cell surface membrane of the collecting duct cells, back into the cytoplasm as part of the vesicles. This makes the collecting duct cells impermeable to water. The fluid flows down the collecting duct without losing any water, so a dilute urine collects in the pelvis and flows down the ureter to the bladder. Under these conditions, we tend to produce large volumes of dilute urine, losing much of the water we drank, in order to keep the water potential of the blood constant.

The collecting duct cells do not respond immediately to the reduction in ADH secretion by the posterior pituitary gland. This is because it takes some time for the ADH already in the blood to be broken down; approximately half of it is destroyed every 15–20 minutes. However, once ADH stops arriving at the collecting duct cells, it takes only 10–15 minutes for aquaporins to be removed from the cell surface membrane and taken back into the cytoplasm until they are needed again.

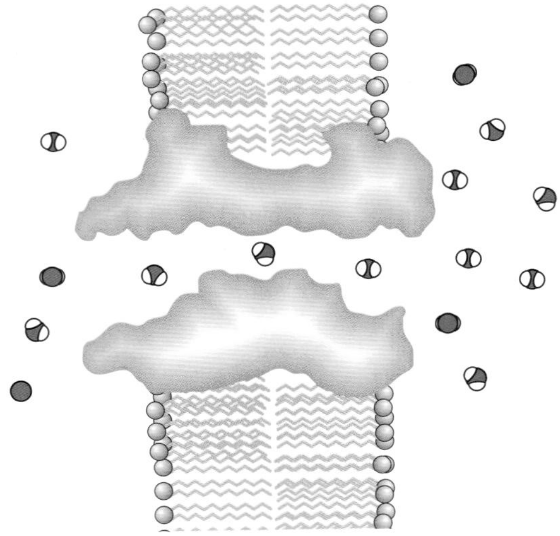

Figure 14.19 Aquaporin protein channels allow water to diffuse through membranes such as those in the cells that line collecting ducts.

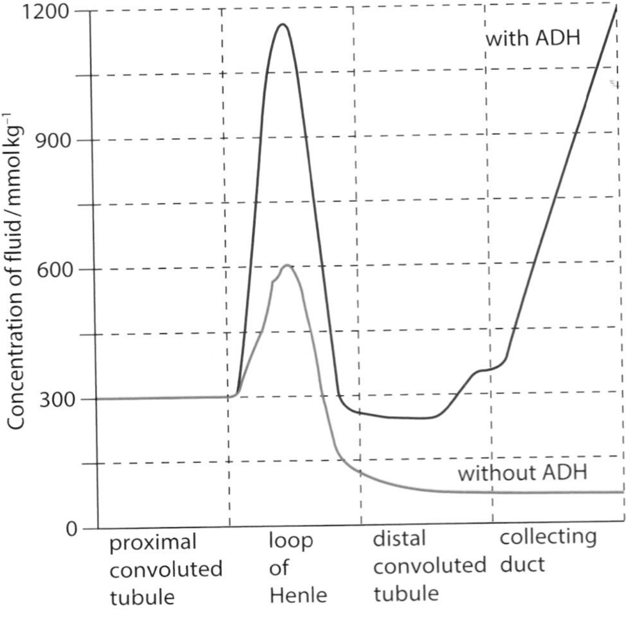

Figure 14.20 The concentration of fluid in different regions of a nephron, with and without the presence of ADH.

QUESTION

14.7 a Use the example of blood water content to explain the terms **set point** and **homeostasis**.

b Construct a flow diagram to show how the water potential of the blood is controlled. In your diagram identify the following: receptor, input, effector and output. Indicate clearly how different parts of the body are coordinated and show how negative feedback is involved.

The control of blood glucose

Carbohydrate is transported through the human bloodstream in the form of glucose in solution in the blood plasma. Glucose is converted into the polysaccharide glycogen, a large, insoluble molecule made up of many glucose units linked together by 1–4 glycosidic bonds with 1–6 branching points (page 33). Glycogen is a short-term energy store that is found in liver and muscle cells and is easily converted to glucose (Figure 14.24, page 317).

In a healthy human, each 100 cm^3 of blood normally contains between 80 and 120 mg of glucose. If the concentration decreases below this, cells may not have enough glucose for respiration, and may be unable to carry out their normal activities. This is especially important for cells that can respire only glucose, such as brain cells. Very high concentrations of glucose in the blood can also cause major problems, again upsetting the normal behaviour of cells. The homeostatic control of blood glucose concentration is carried out by two hormones secreted by endocrine tissue in the pancreas. This tissue consists of groups of cells, known as the islets of Langerhans, which are scattered throughout the pancreas. The word islet means a small island, as you might find in a river. The islets contain two types of cells:

- α cells secrete glucagon
- β cells secrete insulin.

The α and β cells act as the receptors and the central control for this homeostatic mechanism; the hormones glucagon and insulin coordinate the actions of the effectors.

Figure 14.21 shows how the blood glucose concentration fluctuates within narrow limits around the set point, which is indicated by the dashed line.

After a meal containing carbohydrate, glucose from the digested food is absorbed from the small intestine and passes into the blood. As this blood flows through the pancreas, the α and β cells detect the increase in glucose concentration. The α cells respond by stopping the secretion of glucagon, whereas the β cells respond by secreting insulin into the blood plasma. The insulin is carried to all parts of the body, in the blood.

Insulin is a signalling molecule. As it is a protein, it cannot pass through cell membranes to stimulate the mechanisms within the cell directly. Instead, insulin binds to a receptor in the cell surface membrane and affects the cell indirectly through the mediation of intracellular messengers (Figure 14.22 and pages 77–79).

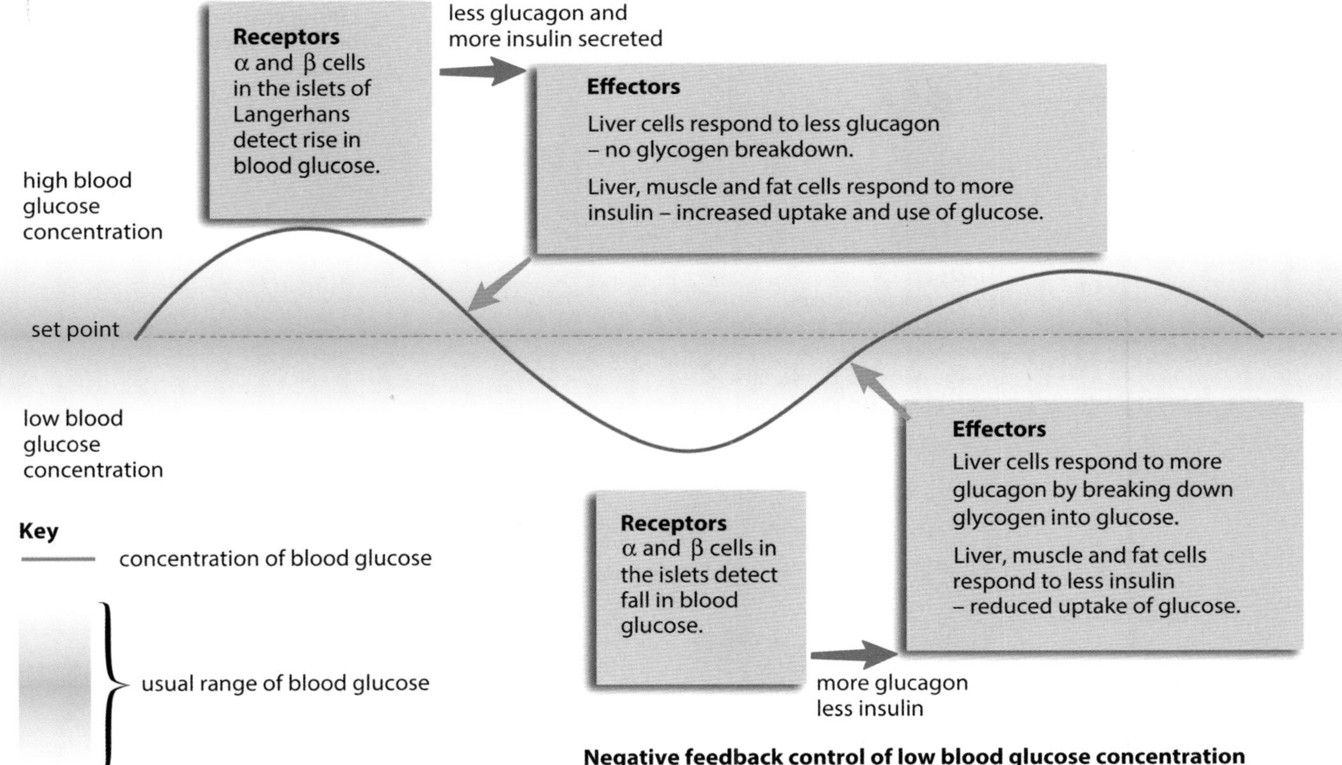

Figure 14.21 The control mechanism for the concentration of glucose in the blood.

There are insulin receptors on many cells, such as those in the liver, muscle and adipose (fat storage) tissue. Insulin stimulates cells with these receptors to increase the rate at which they absorb glucose from the blood, convert it into glycogen and use it in respiration. This results in a decrease in the concentration of glucose in the blood.

Glucose can only enter cells through transporter proteins known as GLUT. There are several different types of GLUT proteins. Muscle cells have the type called GLUT4. Normally, the GLUT proteins are kept in the cytoplasm in the same way as the aquaporins in collecting duct cells (page 314). When insulin molecules bind to receptors on muscle cells, the vesicles with GLUT4 proteins are moved to the cell surface membrane and fuse with it. GLUT4 proteins facilitate the movement of glucose into the cell (Figure 14.22). Brain cells have GLUT1 proteins and liver cells have GLUT2 proteins, which are always in the cell surface membrane, and their distribution is not altered by insulin.

Insulin also stimulates the activation of the enzyme glucokinase, which phosphorylates glucose. This traps glucose inside cells, because phosphorylated glucose cannot pass through the transporters in the cell surface membrane. Insulin also stimulates the activation of two other enzymes, phosphofructokinase and glycogen synthase, which together add glucose molecules to glycogen. This increases the size of the glycogen granules inside the cell (Figure 14.23).

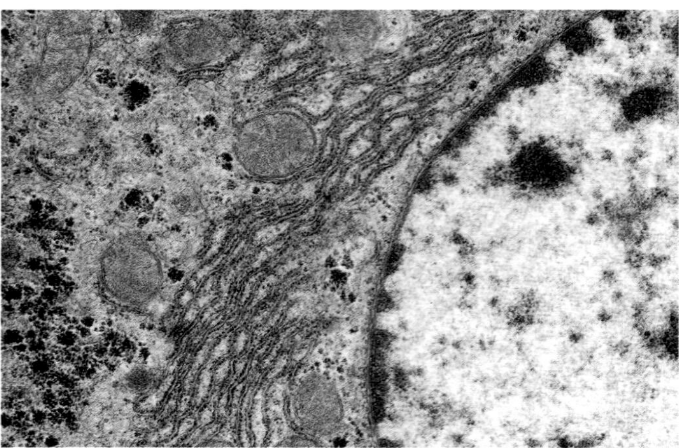

Figure 14.23 Transmission electron micrograph of part of a liver cell (×22 000). The dark spots are glycogen granules in the cytoplasm. Mitochondria can also be seen.

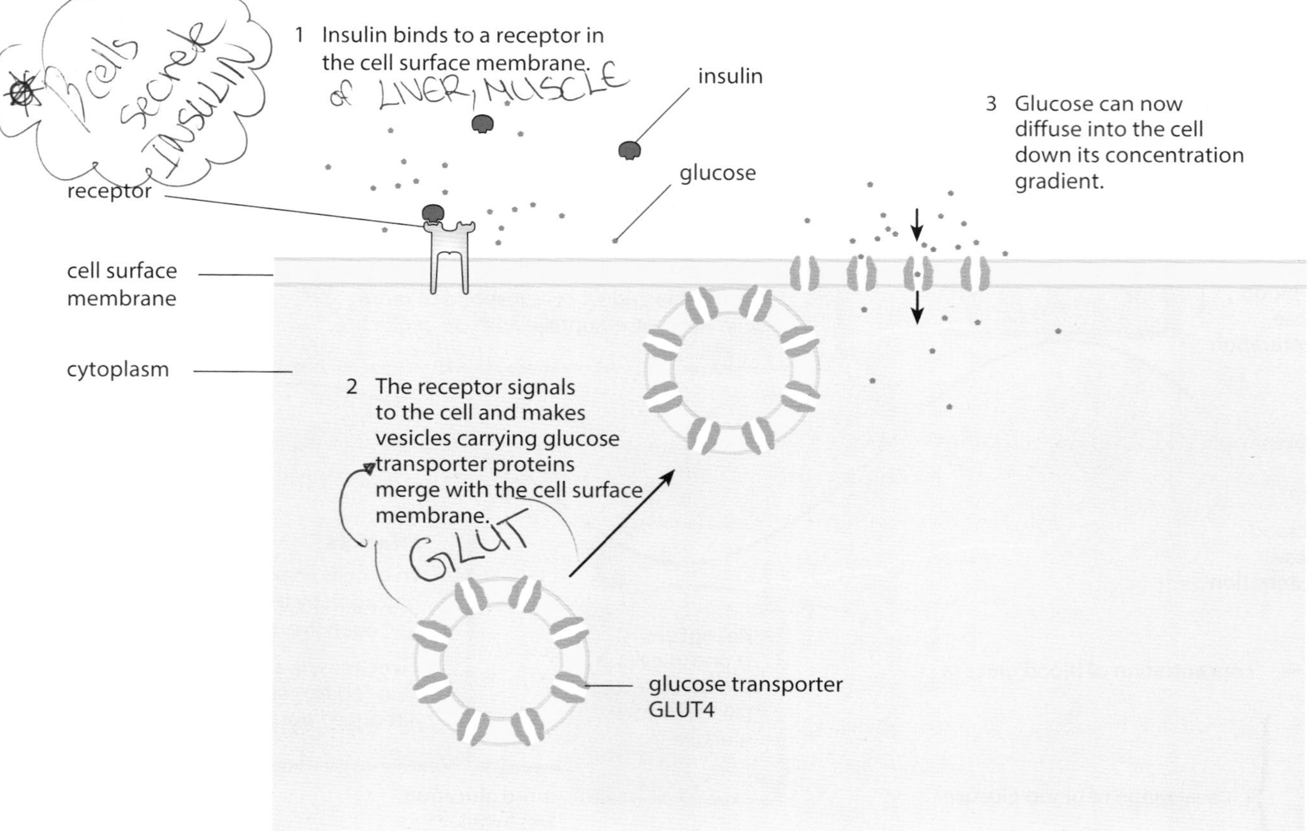

Figure 14.22 Insulin increases the permeability of muscle cells to glucose by stimulating the movement of vesicles with GLUT4 to the cell surface membrane.

A decrease in blood glucose concentration is detected by the α and β cells in the pancreas. The α cells respond by secreting glucagon, while the β cells respond by stopping the secretion of insulin.

The decrease in the concentration of insulin in the blood reduces the rates of uptake and use of glucose by liver and muscle cells. Uptake still continues, but at a lower rate. Glucagon binds to different receptor molecules in the cell surface membranes of liver cells. The method of cell signalling is the same as described on page 19 and in Figure 4.7. The binding of glucagon to a receptor activates a G protein that in turn activates an enzyme within the membrane that catalyses the conversion of ATP to cyclic AMP, which is a second messenger (Figure 14.24). Cyclic AMP binds to kinase enzymes within the cytoplasm that activate other enzymes. Kinase enzymes activate enzymes by adding phosphate groups to them in a process known as phosphorylation. This enzyme cascade amplifies the original signal from glucagon.

Glycogen phosphorylase is at the end of the enzyme cascade: when activated, it catalyses the breakdown of glycogen to glucose. It does this by removing glucose units from the numerous 'ends' of glycogen. This increases the concentration of glucose inside the cell so that it diffuses out through GLUT2 transporter proteins into the blood. Glucose is also made from amino acids and lipids in a process known as **gluconeogenesis**, which literally means the formation of 'new' glucose.

As a result of glucagon secretion, the liver releases extra glucose to increase the concentration in the blood. Muscle cells do not have receptors for glucagon and so do not respond to it.

Glucagon and insulin work together as part of the negative feedback system in which any deviation of the blood glucose concentration from the set point stimulates actions by effectors to bring it back to normal.

Blood glucose concentrations never remain constant, even in the healthiest person. One reason for this is the inevitable time delay between a change in the blood glucose concentration and the onset of actions to correct it. Time delays in control systems result in oscillation, where things do not stay absolutely constant but sometimes rise slightly above and sometimes drop slightly below the 'required' level.

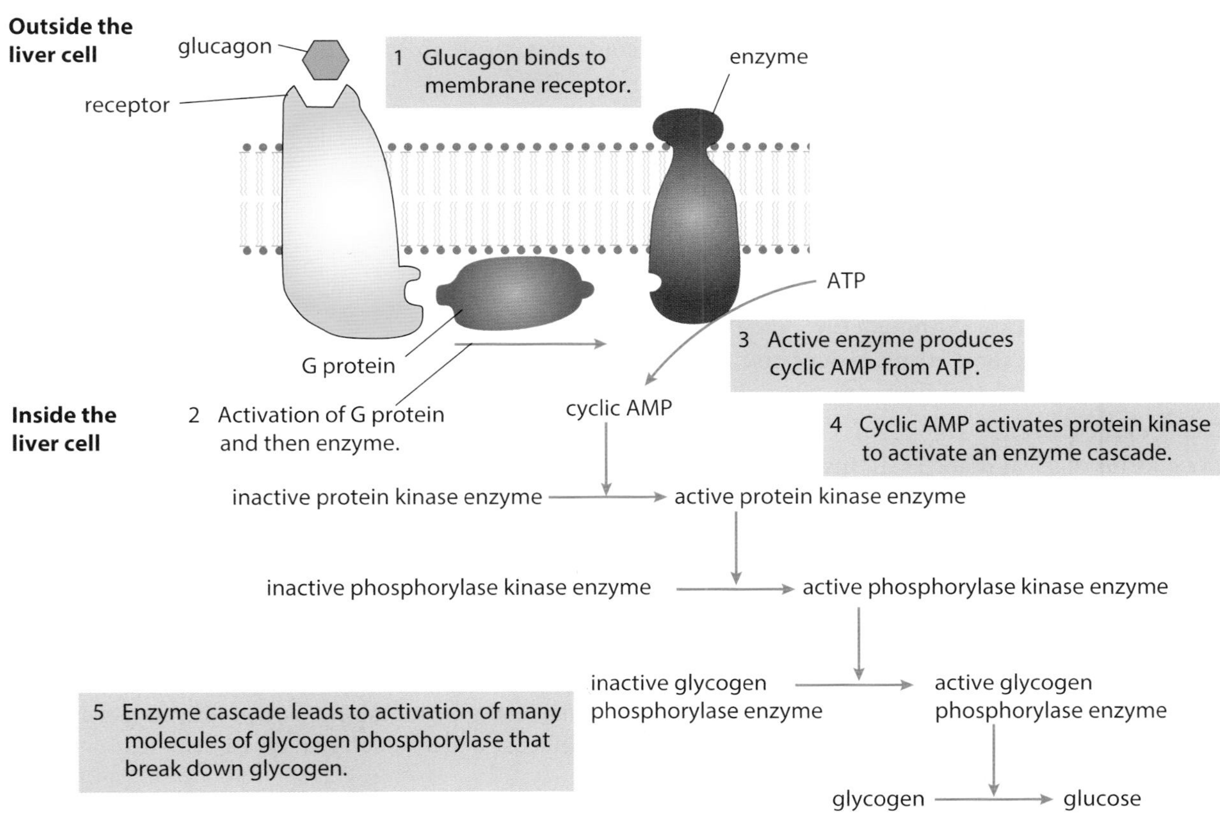

Figure 14.24 Glucagon stimulates the activation of glycogen phosphorylase enzymes in liver cells through the action of cyclic AMP.

The hormone adrenaline also increases the concentration of blood glucose. It does this by binding to different receptors on the surface of liver cells that activate the same enzyme cascade and lead to the same end result – the breakdown of glycogen by glycogen phosphorylase. Adrenaline also stimulates the breakdown of glycogen stores in muscle during exercise. The glucose produced remains in the muscle cells where it is needed for respiration.

QUESTIONS

14.8 The control of blood glucose concentration involves a negative feedback mechanism.
- **a** What are the stimuli, receptors and effectors in this control mechanism?
- **b** Explain how negative feedback is involved in this homeostatic mechanism. (You may have to look back to page 301.)

14.9 a Name the process by which glucose enters and leaves cells.
- **b** Explain why:
  - **i** muscle cells do not have receptors for glucagon
  - **ii** there are second messengers for insulin and glucagon
  - **ii** insulin and glucagon have different second messengers.

Diabetes mellitus

Sugar diabetes, or diabetes mellitus, is one of the most common metabolic diseases in humans. In 2013, the International Diabetes Federation estimated that 382 million people, or approximately 8.3% of the world's adult population, had this disease. Although the percentages are higher among some ethnic groups and in some countries than others, it is a disease that is increasing steeply everywhere.

There are two forms of sugar diabetes. In **insulin-dependent diabetes**, which is also known as **type 1 diabetes**, the pancreas seems to be incapable of secreting sufficient insulin. It is thought that this might be due to a deficiency in the gene that codes for the production of insulin, or because of an attack on the β cells by the person's own immune system. Type 1 diabetes is sometimes called juvenile-onset diabetes, because it usually begins very early in life.

The second form of diabetes is called **non-insulin-dependent diabetes** or **type 2 diabetes**. In this form of diabetes, the pancreas does secrete insulin, but the liver and muscle cells do not respond properly to it. Type 2 diabetes begins relatively late in life and is often associated with diet and obesity.

The symptoms of both types of diabetes mellitus are the same. After a carbohydrate meal, glucose is absorbed into the blood, and the concentration increases and stays high (Figure 14.25). Normally there is no glucose in urine, but if the glucose concentration in the blood becomes very high, the kidney cannot reabsorb all the glucose, so that some passes out in the urine. Extra water and salts accompany this glucose, and the person consequently feels extremely hungry and thirsty.

In a diabetic person, uptake of glucose into cells is slow, even when there is plenty of glucose in the blood. Thus cells lack glucose and metabolise fats and proteins as alternative energy sources. This can lead to a build-up of substances in the blood called keto-acids (or ketones). These are produced when the body switches to metabolising fat and they decrease the blood pH. The combination of dehydration, salt loss and low blood pH can cause coma in extreme situations.

Between meals, the blood glucose concentration of a person with untreated diabetes may decrease steeply. This is because there is no glycogen to mobilise, as it was not stored when there was plenty of glucose. Once again, coma may result, this time because of a lack of glucose for respiration.

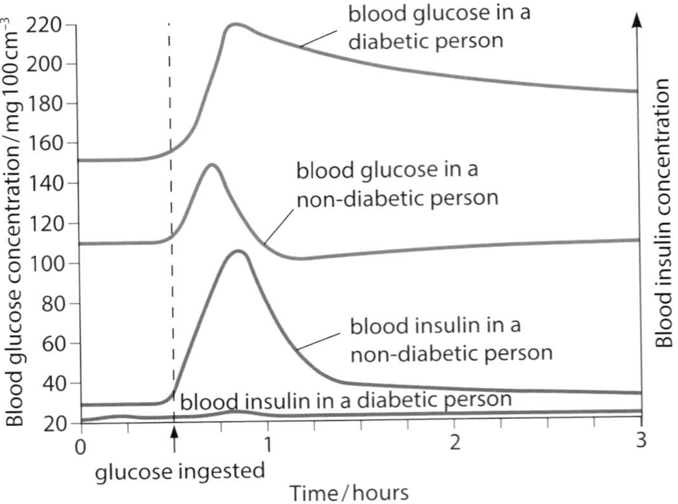

Figure 14.25 Concentrations of blood glucose and insulin following intake of glucose in a person with normal control of blood glucose and a person with type 1 diabetes.

People with type 1 diabetes receive regular injections of insulin, which they learn to do themselves. They must also take blood samples to check that the insulin is effective (Figure 14.26). Some people have mini-pumps which deliver the exact volumes of insulin that they need when they need them. A carefully controlled diet also helps to maintain a near-constant concentration of glucose in the blood.

People with type 2 diabetes rarely need to have insulin injections; instead they can use diet and regular and frequent exercise to keep their blood glucose within normal limits (Figure 14.21, page 315). Diabetics now receive insulin made by genetically engineered cells (page 466).

Figure 14.26 A nurse teaches a girl with type 1 diabetes to inject insulin. The girl may have to receive injections of insulin throughout her life.

Urine analysis

It is much easier to collect a urine sample from someone than a blood sample. Simple tests on urine can give early indications of health problems, which can then be investigated more thoroughly.

The presence of glucose and ketones in urine indicates that a person may have diabetes. If blood glucose concentration increases above a certain value, known as the renal threshold, not all of the glucose is reabsorbed from the filtrate in the proximal convoluted tubule of the kidney and some will be present in the urine.

The presence of protein in the urine indicates that there is something wrong with the kidneys. Most protein molecules are too large to be filtered. However, some protein molecules are filtered (Table 14.1), but these are reabsorbed by endocytosis in the proximal convoluted tubule, broken down and the amino acids absorbed into the blood. It is not unusual for some protein to be present in the urine for short periods of time, such as during a high fever, after vigorous exercise and during pregnancy. However, a large quantity or the long-term presence of protein in the urine indicates that there may be a disease affecting the glomeruli or there is a kidney infection. Protein in the urine is also associated with high blood pressure, which is a risk factor in heart disease.

Dip sticks and biosensors

Dip sticks (also known as test strips) can be used to test urine for a range of different factors including pH, glucose, ketones and protein. Dip sticks for detecting glucose contain the enzymes glucose oxidase and **peroxidase**. These two enzymes are immobilised on a small pad at one end of the stick. The pad is immersed in urine and if it contains glucose, glucose oxidase catalyses a chemical reaction in which glucose is oxidised into a substance called gluconolactone. Hydrogen peroxide is also produced. Peroxidase catalyses a reaction between hydrogen peroxide and a colourless chemical in the pad to form a brown compound. The resulting colour of the pad is matched against a colour chart. The chart shows the colours that indicate different concentrations of glucose. The more glucose that is present, the darker the colour (Figure 14.27).

One problem with urine tests is that they do not indicate the current blood glucose concentration, but rather whether the concentration was higher than the renal threshold in the period of time while urine was collecting in the bladder.

A biosensor like the one in Figure 14.28 allows people with diabetes to check their blood to see how well they are controlling their glucose concentration. Like the

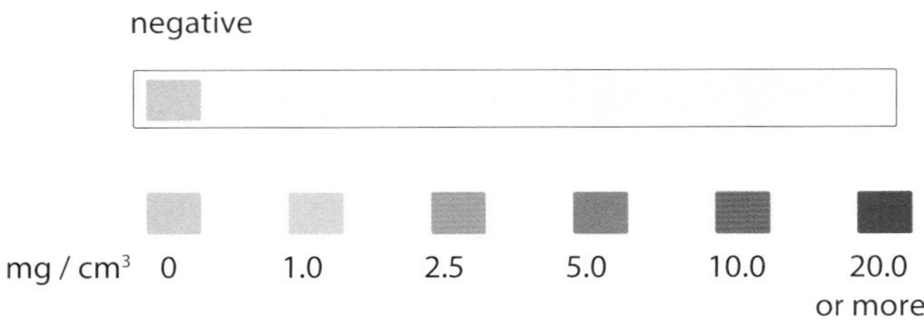

Figure 14.27 A dip stick can be used to test for the presence of glucose in urine. These dip sticks are used by people with diabetes to check whether their urine contains any glucose.

dip sticks, the biosensor uses a pad impregnated with glucose oxidase. A small sample of blood is placed on the pad which is inserted into the machine. Glucose oxidase catalyses the reaction to produce gluconolactone and at the same time a tiny electric current is generated. The current is detected by an electrode, amplified and read by the meter which produces a reading for blood glucose concentration within seconds. The more glucose that is present, the greater the current and the greater the reading from the biosensor.

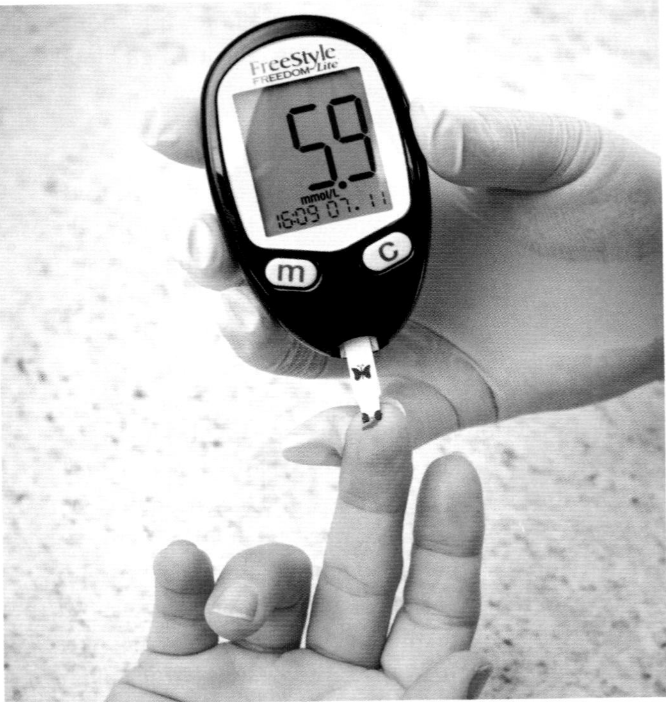

Figure 14.28 Biosensors use biological materials, such as enzymes, to measure the concentration of molecules such as glucose. This glucose biosensor is used to check the glucose concentration in a sample of blood. The meter shows a reading in the normal range.

QUESTIONS

14.10 **a** Explain why insulin cannot be taken by mouth.
b Suggest how people with type 1 diabetes can monitor the effectiveness of the insulin that they take.
c Suggest how people with type 2 diabetes can control their blood glucose.

14.11 Suggest advantages of using an electronic biosensor to measure blood glucose concentration, rather than using a dip stick to measure glucose in urine.

Homeostasis in plants

It is as important for plants to maintain a constant internal environment as it is for animals. For example, mesophyll cells in leaves require a constant supply of carbon dioxide if they are to make best use of light energy for photosynthesis. We have seen how low concentrations of carbon dioxide limit the rate of photosynthesis (page 292). Stomata control the entry of carbon dioxide into leaves (Figure 14.29). Strictly speaking, a stoma is the hole between the **guard cells**, but the term is usually used to refer to the two guard cells and the hole between them. Stomata may look very simple, but guard cells are highly specialised cells that respond to a wide range of environmental stimuli and thus control the internal atmosphere of the leaf.

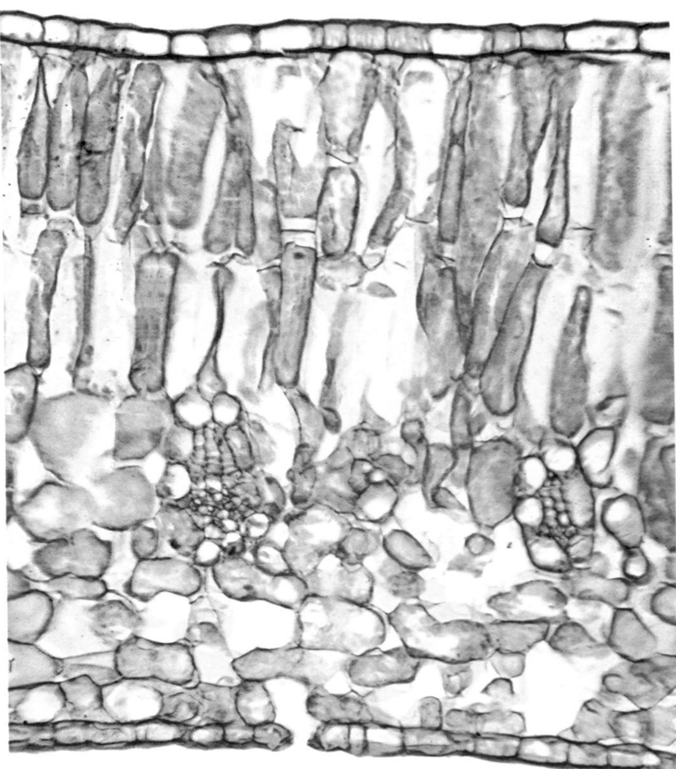

Figure 14.29 Photomicrograph of a TS of *Helianthus* leaf (×100). An open stoma is visible in the lower epidermis.

Stomata show daily rhythms of opening and closing. Even when kept in constant light or constant dark, these rhythms persist (Figure 14.30). Opening during the day maintains the inward diffusion of carbon dioxide and the outward diffusion of oxygen. However, it also allows the outward diffusion of water vapour in transpiration (pages 134–136). The closure of stomata at night when photosynthesis cannot occur reduces rates of transpiration and conserves water.

Stomata respond to changes in environmental conditions. They open in response to:

- increasing light intensity
- low carbon dioxide concentrations in the air spaces within the leaf.

When stomata are open, leaves gain carbon dioxide for photosynthesis, but tend to lose much water in transpiration.

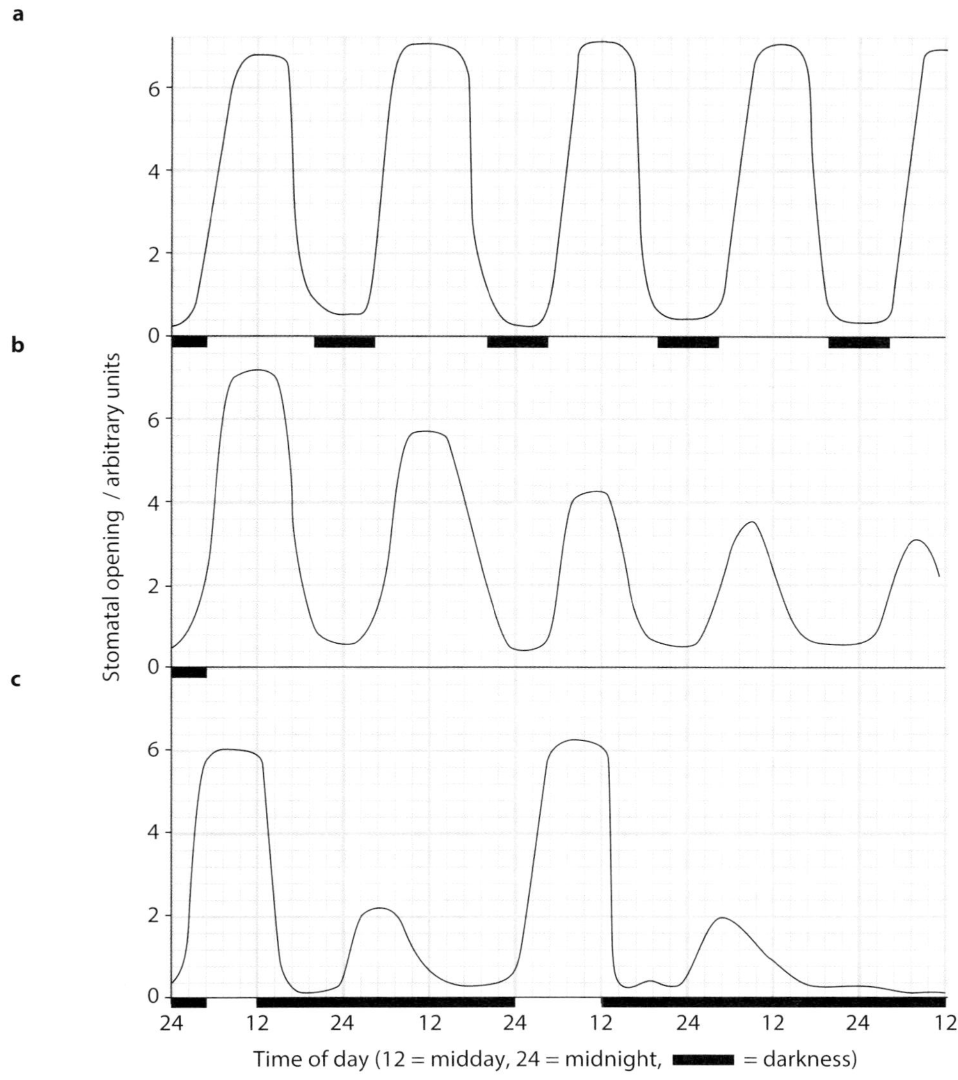

Figure 14.30 **a** The opening of stomata was measured in leaves of *Tradescantia* over several days to reveal a daily rhythm of opening and closing. This rhythm persisted even when the plants were kept in constant light **b**; and in long periods of constant dark **c**.

Stomata close in response to:

- darkness
- high carbon dioxide concentrations in the air spaces in the leaf
- low humidity
- high temperature
- water stress, when the supply of water from the roots is limited and/or there are high rates of transpiration.

The disadvantage of closing is that during daylight, the supply of carbon dioxide decreases so the rate of photosynthesis decreases. The advantage is that water is retained inside the leaf which is important in times of water stress.

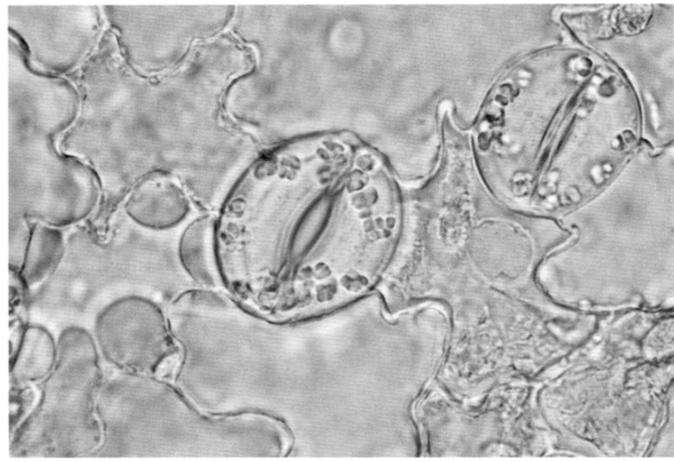

Figure 14.31 Photomicrograph of two stomata and guard cells in a lower epidermis of a leaf of *Tradescantia* (×870).

Opening and closing of stomata

Each stomatal pore is surrounded by two guard cells (Figure 14.31). Guard cells open when they gain water to become turgid and close when they lose water and become flaccid.

Guard cells gain and lose water by osmosis. A decrease in water potential is needed before water can enter the cells by osmosis. This is brought about by the activities of transporter proteins in their cell surface membranes. ATP-powered proton pumps in the membrane actively transport hydrogen ions, H^+, out of the guard cells. The decrease in the hydrogen ion concentration inside the cells causes channel proteins in the cell surface membrane to open so that potassium ions, K^+, move into the cell. They do this because the removal of hydrogen ions has left the inside of the cell negatively charged compared with the outside, and as potassium ions have a positive charge, they are drawn down an electrical gradient towards the negatively charged region. They also diffuse into the cells down a concentration gradient. Such a combined gradient is an electrochemical gradient (Figure 14.32).

The extra potassium ions inside the guard cells lower the solute potential, and therefore the water potential. Now there is a water potential gradient between the outside and the inside of the cell, so water moves in by osmosis through aquaporins in the membrane. This increases the turgor of the guard cells, and the stoma opens. Guard cells have unevenly thickened cell walls. The wall adjacent to the pore is very thick, whereas the wall furthest from the pore is thin. Bundles of cellulose microfibrils are arranged as hoops around the cells so that, as the cell becomes turgid, these hoops ensure that the cell mostly increases in length and not diameter. Since the ends of the two guard cells are joined and the thin outer walls bend more readily than the thick inner walls, the guard cells become curved. This opens the pore between the two cells.

Stoma closed

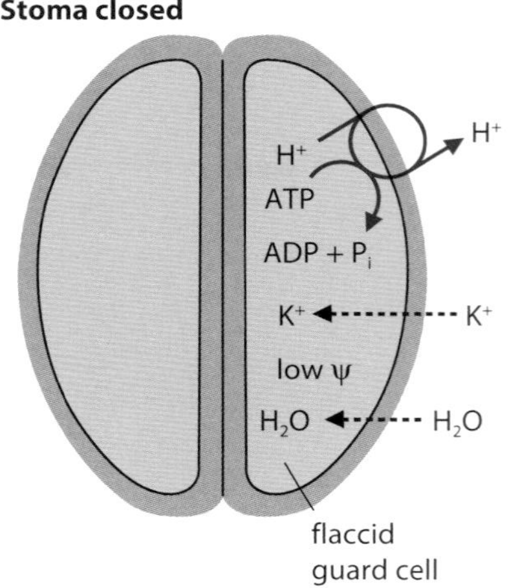

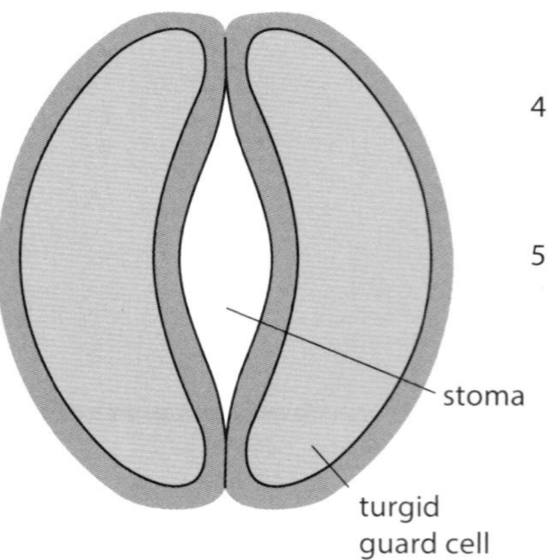

1. ATP-powered proton pumps in the cell surface membrane actively transport H^+ out of the guard cell.
2. The low H^+ concentration and negative charge inside the cell causes K^+ channels to open. K^+ diffuses into the cell down an electrochemical gradient.
3. The high concentration of K^+ inside the guard cell lowers the water potential (ψ).
4. Water moves in by osmosis, down a water potential gradient.
5. The entry of water increases the volume of the guard cells, so they expand. The thin outer wall expands most, so the cells curve apart.

Figure 14.32 How a stoma is opened. Guard cells do not have plasmodesmata, so all exchanges of water and ions must occur across the cell surface membranes through the pump and channel proteins.

Stomata close when the hydrogen ion pump proteins stop and potassium ions leave the guard cells and enter neighbouring cells. Now there is a water potential gradient in the opposite direction, so water leaves the guard cells so that they become flaccid and close the stoma. Closing of the stomata has significant effects on the plant. It reduces the uptake of carbon dioxide for photosynthesis and reduces the rate of transpiration. As transpiration is used for cooling the plant and also for maintaining the transpiration stream that supplies water and mineral ions to the leaves, stomatal closure only occurs when reducing the loss of water vapour and conserving water is the most important factor. In conditions of water stress, the hormone **abscisic acid (ABA)** is produced in plants to stimulate stomatal closure.

Abscisic acid and stomatal closure

Abscisic acid has been found in a very wide variety of plants, including ferns and mosses as well as flowering plants. ABA can be found in every part of the plant, and is synthesised in almost all cells that possess chloroplasts or amyloplasts (organelles like chloroplasts, but that contain large starch grains and no chlorophyll).

One role of ABA is to coordinate the responses to stress; hence it is known as a **stress hormone**. If a plant is subjected to difficult environmental conditions, such as very high temperatures or much reduced water supplies, then it responds by secreting ABA. In a plant in drought conditions, the concentration of ABA in the leaves can rise to 40 times that which would normally be present. This high concentration of ABA stimulates the stomata to close, reducing the loss of water vapour from the leaf.

If ABA is applied to a leaf, the stomata close within just a few minutes. Although it is not known exactly how ABA achieves the closure of stomata, it seems that guard cells have ABA receptors on their cell surface membranes, and it is possible that when ABA binds with these it inhibits the proton pumps to stop hydrogen ions being pumped out. ABA also stimulates the movement of calcium ions into the cytoplasm through the cell surface membrane and the tonoplast (membrane around the vacuole). Calcium acts as a second messenger to activate channel proteins to open that allow negatively charged ions to leave the guard cells. This, in turn, stimulates the opening of channel proteins which allows the movement of potassium ions out of the cells. At the same time, calcium ions also stimulate the closure of the channel proteins that allow potassium ions to enter. The loss of ions raises the water potential of the cells, water passes out by osmosis, the guard cells become flaccid and the stomata close.

QUESTION

14.12 a Describe the mechanism of stomatal opening.
b Explain when it is an advantage for plants to close their stomata.
c Outline how abscisic acid functions to stimulate the closure of stomata.

Summary

- Mammals keep their internal environment relatively constant, so providing steady and appropriate conditions within which cells can carry out their activities. This is known as homeostasis. Homeostatic balance requires receptors that detect changes in physiological factors such as the temperature, water potential and pH of the blood.
- Effectors are the cells, tissues and organs (including muscles and glands) that carry out the functions necessary to restore those factors to their set points. Homeostatic control systems use negative feedback in which any change in a factor stimulates actions by effectors to restore the factor to its set point.
- Thermoregulation involves maintaining a constant core body temperature. The hypothalamus is the body's thermostat monitoring the temperature of the blood and the temperature of the surroundings. It controls the processes of heat production, heat loss and heat conservation to keep the core body temperature close to its set point.
- Excretion is the removal of toxic waste products of metabolism, especially carbon dioxide and urea. The deamination of excess amino acids in the liver produces ammonia, which is converted into urea, the main nitrogenous waste product. Urea is excreted in solution in water, as urine.

- The kidneys regulate the concentration of various substances in the body fluids, by excreting appropriate amounts of them. Each kidney is made up of thousands of nephrons and their associated blood vessels. The kidneys produce urine by ultrafiltration and selective reabsorption, plus some secretion of unwanted substances. Different regions of a nephron have different functions, and this is reflected in the structure of the cells that make up their walls.
- Blood is brought to the glomerulus in an afferent arteriole. High hydrostatic pressure in the glomerulus forces substances through the capillary walls, the basement membrane and inner lining of the Bowman's capsule. The basement membrane acts as a filter, allowing only small molecules through. This filtrate collects in the Bowman's capsule and then enters the proximal convoluted tubule, where most reabsorption occurs by diffusion and active transport; substances are also reabsorbed in the distal convoluted tubule and collecting duct.
- The loop of Henle acts as a counter-current multiplier, producing high concentrations of sodium and chloride ions in the tissue fluid in the medulla. This tissue has a very low water potential. Water is reabsorbed from fluid in the collecting duct by osmosis if the body is dehydrated. The water content of the blood is controlled by changing the amount of water excreted in the urine by the kidneys. This is done by regulating the permeability of the walls of the collecting ducts to water, and hence the volume of water reabsorbed from the collecting ducts into the blood. The permeability is increased by the hormone ADH, which is secreted by the posterior pituitary gland in response to stimulation of osmoreceptors in the hypothalamus.
- The concentration of glucose in the blood is controlled by the action of insulin and glucagon, which are hormones secreted by the islets of Langerhans in the pancreas and affect liver and muscle cells. The use of negative feedback keeps the blood glucose concentration near the set point.
- Tests on urine are carried out to test for a variety of substances, including glucose, ketones and proteins. The presence of glucose and ketones suggests that a person has diabetes. The presence of proteins suggests they might have kidney damage, a kidney infection or high blood pressure.
- Immobilised glucose oxidase is used on dipsticks to detect the presence of glucose in urine. The enzyme changes glucose to gluconolactone and hydrogen peroxide, which causes a colour change in another chemical on the stick. Biosensors use the same principle but produce a small electric current instead of a colour change, which provides a direct digital readout.
- Stomata control the movement of gases between the atmosphere and the air space inside a leaf. They allow the inward diffusion of carbon dioxide to be used in photosynthesis and the outward diffusion of water vapour in transpiration. Each stoma consists of two guard cells either side of a pore. Guard cells are highly specialised cells that respond to changes in light intensity and carbon dioxide concentrations inside the leaf.
- In general, guard cells open during the day and close at night although this rhythm persists in continuous light and in continuous dark. The cell surface membranes of guard cells contain proton pumps that actively transport hydrogen ions out of the cells. This stimulates the inward movement of potassium ions down their electrochemical gradient. The potassium ions decrease the water potential of the guard cells so water enters by osmosis, the cells become turgid and open the stoma. To close the stoma, the proton pumps stop working and the potassium ions flow out of the cells. This raises the water potential inside the cells, so water passes out by osmosis. The guard cells become flaccid and this closes the stoma.
- Abscisic acid (ABA) is a plant hormone that is synthesised by any cells in a plant that contain chloroplasts or amyloplasts, especially in stress conditions. The presence of large concentrations of abscisic acid in leaves stimulates stomata to close, reduces the rate of transpiration and conserves water inside the plant.

End-of-chapter questions

1 Which of the following is an incorrect statement about the homeostatic control of the concentration of glucose in the blood?

A Negative feedback is involved in the control of blood glucose.

B The concentration of glucose in the blood fluctuates within narrow limits.

C The hypothalamus is the control centre for blood glucose.

D The liver is an effector organ in blood glucose homeostasis. [1]

2 Glucose is small enough to be filtered from the blood in glomeruli in the kidney, but is not normally found in the urine. This is because glucose is:

A reabsorbed in distal convoluted tubules

B reabsorbed in proximal convoluted tubules

C reabsorbed along the whole length of the nephrons

D respired by cells in the kidney [1]

3 Which of the following occurs in the body in response to the secretion of glucagon?

A conversion of glucose to glycogen in liver cells

B decrease in the blood glucose concentration

C increased uptake of glucose by muscle cells

D production of cyclic AMP in target cells [1]

4 In response to dehydration, ADH is secreted by the posterior pituitary gland. One of its effects is to stimulate:

A a reduction in the glomerular filtration rate

B an increase in the number of aquaporins in the cell membranes of collecting duct cells

C an increase in the uptake of water by cells in the proximal convoluted tubules of nephrons

D an increase in the volume of urine produced by the kidneys [1]

5 Rearrange the following statements to make a flow diagram of the mechanism of opening a stoma.

1 volume of guard cell increases

2 H^+ transported out of guard cells

3 water enters guard cells by osmosis

4 K^+ diffuses into guard cells

5 guard cells curve to open stoma

6 water potential of guard cells falls

7 K^+ channels open [4]

2, 7, 4, 6, 3, 1, 5

6 a Explain the meaning of the term **excretion**. [3]

b The figure is a photomicrograph of part of the kidney.

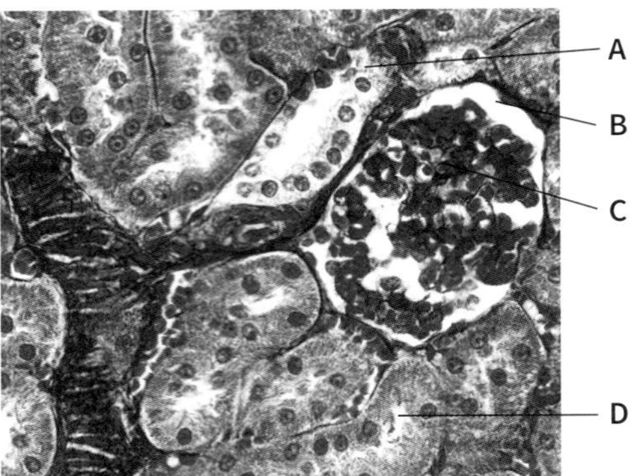

Magnification: ×180

i Name **A**, **B**, **C** and **D**. [4]

ii Identify the region of the kidney shown in the figure and give a reason for your identification. [2]

iii Calculate the actual maximum width of the structure labelled **A**. Show your working. [2]

[Total: 11]

7 The control of the water content of the blood is an example of homeostasis.

a Name the part of the body that monitors the water potential of the blood. [1]

In an investigation of the factors that influence urine production, a person drank one litre of water. The person's urine was collected at half-hourly intervals for four hours after drinking. The results are shown as line **A** on the figure. On the following day, the same person drank one litre of a dilute salt solution and the urine was collected in the same way (line **B**). Dilute salt solution has about the same water potential as blood plasma.

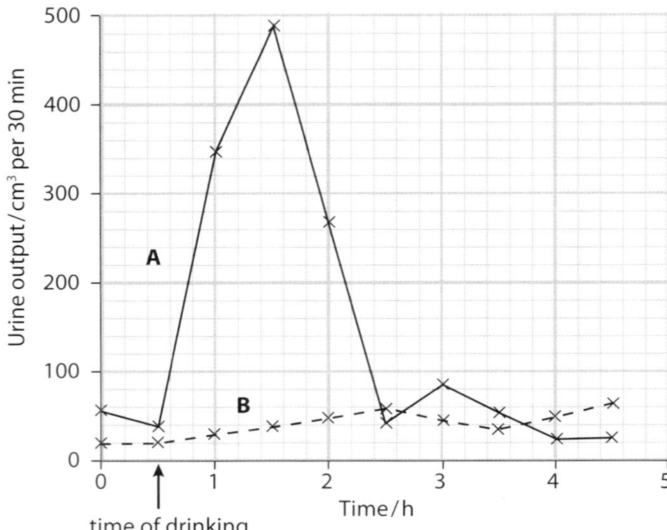

b Calculate how much urine was produced in the two hours after drinking the litre of water. [1]

c Explain why the person produced so much urine after drinking the litre of water. [4]

d Suggest why the results during the second day were so different from those on the first day. [2]

e Explain why negative feedback, and not positive feedback, is involved in homeostatic mechanisms. [5]

[Total: 13]

8 An investigation was carried out to determine the response of pancreatic cells to an increase in the glucose concentration of the blood. A person who had been told not to eat or drink anything other than water for 12 hours then took a drink of a glucose solution. Blood samples were taken from the person at one hour intervals for five hours, and the concentrations of glucose, insulin and glucagon in the blood were determined. The results are shown in the graph below.

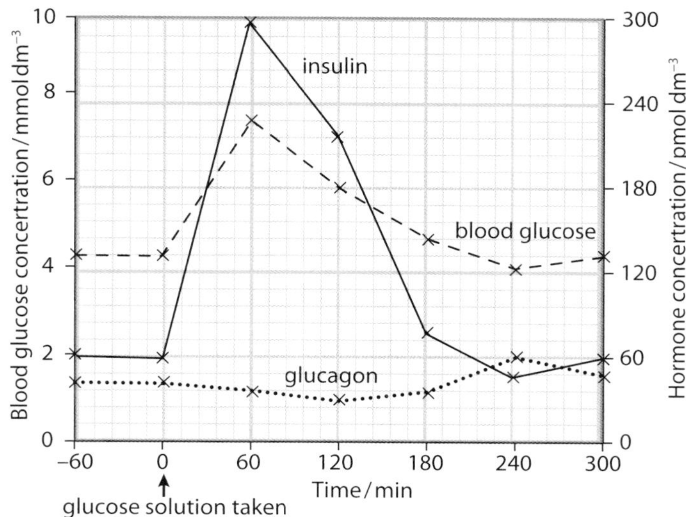

a i Explain why the person was told not to eat or drink anything other than water for 12 hours before having the glucose drink. [3]

ii Use the information in the figure to describe the response of the pancreatic cells to an increase in the glucose concentration. [4]

iii Outline the role of insulin when the glucose concentration in the blood increases. [5]

b i Suggest how the results will change if the investigation continued longer than five hours without the person taking any food. [4]

ii Outline the sequence of events that follows the binding of glucagon to its membrane receptor on a liver cell. [6]

[Total: 22]

9 a Explain what is meant by a biosensor. [2]

b Copy and complete the following passage describing the action of a biosensor.

Many people with diabetes use a biosensor to measure the concentration of glucose in their blood. The biosensor uses the enzyme ...., which is .... on a pad. This enzyme converts glucose into gluconolactone and .... An electrode in the biosensor produces a tiny .... current, the size of which is .... to the concentration of glucose in the blood. The current is read by the meter, which produces a reading for blood glucose concentration. If the reading is too high, the person needs to take .... to lower it. [6]

[Total: 8]

10 Abscisic acid (ABA) is a weak acid. Its structure can be represented as ABA-H. It dissociates into positively charged H^+ ions (protons) and negatively charged ABA^- ions as shown:

$$\text{ABA-H} \rightleftharpoons \text{ABA}^- + \text{H}^+$$

The following observations have been made by scientists:

- light stimulates proton (H^+ ion) uptake into the grana of chloroplasts
- ABA-H can diffuse into and out of chloroplasts, but ABA^- cannot.

This information is summarised in the diagram below.

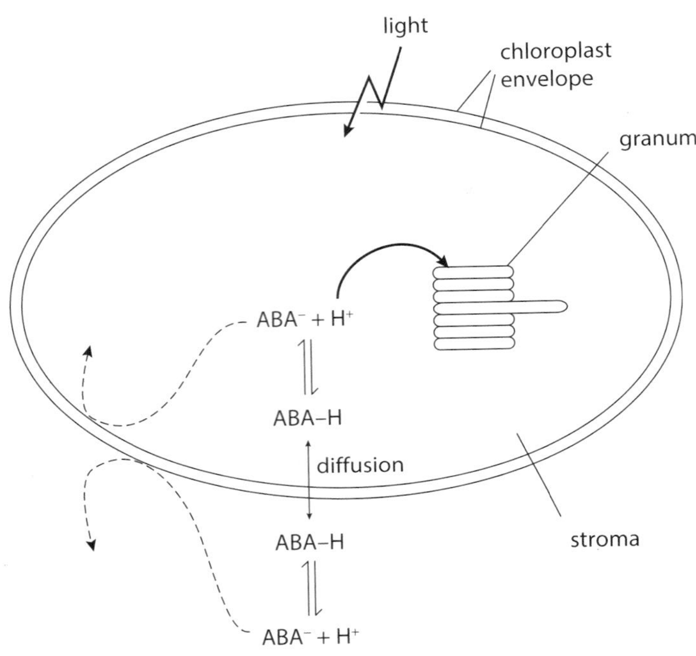

a Using all the information provided, predict what happens to the pH of the stroma in the light. [1]

b i When light shines on the chloroplast, dissociation of ABA-H is stimulated. Explain why this happens. [2]

ii Explain the effect that this will have on diffusion of ABA-H into or out of the chloroplast. [2]

When the mesophyll cells of leaves become dehydrated, some of the ABA stored in the chloroplasts is released into the transpiration stream in the apoplast.

c ABA travels in the apoplast pathway to the guard cells. Explain why this is an advantage when the leaf is dehydrated. [2]

[Total: 7]

11 a Describe the structure of a kidney, including its associated blood vessels. [6]

b Describe the mechanisms involved in reabsorption in the proximal convoluted tubule **and** describe how the epithelial cells of the proximal convoluted tubule are adapted to carry out this process. [9]

[Total: 15]

Cambridge International AS and A Level Biology 9700 Paper 41, Question 10, May/June 2012

Chapter 15: Coordination

Learning outcomes

You should be able to:

- compare the ways in which mammals coordinate responses to internal and external stimuli
- describe the structure and function of neurones, including their roles in simple reflexes
- outline the roles of sensory receptor cells
- describe and explain the transmission of nerve impulses
- describe and explain the structure, function and roles of synapses
- describe the ultrastructure of striated muscle and explain how muscles contract in response to impulses from motor neurones
- explain the roles of hormones in controlling the human menstrual cycle and outline the biological basis of contraceptive pills
- describe how responses in plants, including the rapid response of the Venus fly trap, are controlled

Where biology meets psychology

We have five senses: touch, sight, hearing, taste and smell. It's a controversial view, but some people believe in extrasensory perception (ESP), telepathy and having premonitions as a 'sixth sense'. Recent research suggests that we detect subtle changes, which we cannot put into words, so imagine it is an extra sense. Some people also have synaesthesia – a condition where stimulation of, say, hearing also produces a visual response (Figure 15.1).

But we do have a genuine sixth sense, one which we take for granted. In his essay 'The Disembodied Lady', the neurologist Oliver Sacks relates the story of a woman who woke up one day to find she had lost any sense of having a body. All the sensory neurones from the receptors in her muscles and joints had stopped sending impulses. She had no feedback from her muscles and could not coordinate her movements. The only way she could live without this sixth sense was to train herself to rely entirely on her eyesight for coordinating her muscles. A man with the same condition describes the efforts needed to do this as equivalent to running a marathon every day. Curiously, the night before Oliver Sacks's patient found she had total loss of body awareness, she dreamt about it.

Figure 15.1 Crossed wires? By studying electrical activity in the brain, researchers have found that some people do indeed hear colour and see sound.

Most animals and plants are complex organisms, made up of many millions of cells. Different parts of the organism perform different functions. It is essential that information can pass between these different parts, so that their activities are coordinated. Sometimes, the purpose of this information transfer is to coordinate the regulation of substances within the organism, such as the control of blood glucose concentrations in mammals. Sometimes, the purpose may be to change the activity of some part of the organism in response to an external stimulus, such as moving away from something that may do harm.

There are communication systems within animals that coordinate the activities of receptors and effectors. The information they receive comes from the internal and the external environment. So there are receptors that detect stimuli inside the body and receptors that detect stimuli in the surrounding environment. There are examples of these in Table 15.1 on page 338.

In animals, including mammals, there are two types of information transfer that are used to coordinate the body's activities:

- nerves that transmit information in the form of electrical impulses
- chemical messengers called hormones that travel in the blood.

In Chapter 14, you saw that aspects of homeostasis are controlled by hormones that are secreted into the blood by glands, such as the pituitary glands and the pancreas. The glands that secrete hormones make up the body's endocrine system. In this chapter, we look first at the nervous system and then, on page 349, at further aspects of the endocrine system.

Coordination in plants also involves the use of electrical impulses for fast responses and hormones (also known as plant growth regulators) for coordinating slower responses to stimuli. We look at these methods of coordination at the end of this chapter.

Nervous communication

The mammalian nervous system is made up of the brain and spinal cord, which form the central nervous system (CNS), and the cranial and spinal nerves, which form the peripheral nervous system (PNS) (Figure 15.2). Cranial nerves are attached to the brain and spinal nerves to the spinal cord. Information is transferred in the form of nerve impulses, which travel along nerve cells at very high speeds. Nerve cells are also known as neurones, and they carry information directly to their target cells. Neurones coordinate the activities of sensory receptors such as those in the eye, decision-making centres in the CNS, and effectors such as muscles and glands.

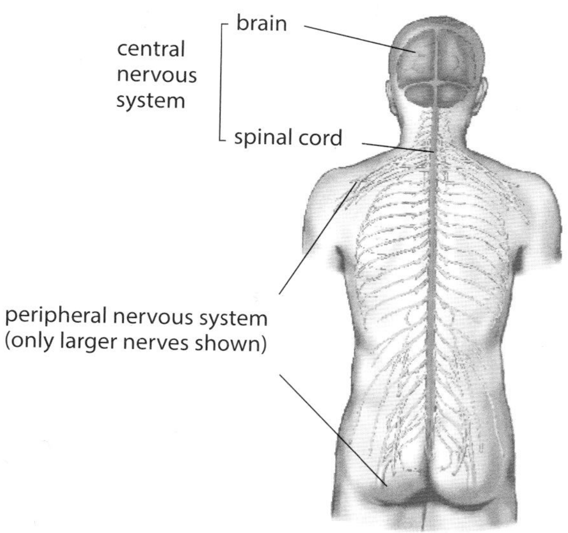

Figure 15.2 The human nervous system.

Neurones

There are three types of neurone (Figure 15.3), each with a different function:

- **sensory neurones** transmit impulses from receptors to the CNS
- **intermediate neurones** (also known as relay or connector neurones) transmit impulses from sensory neurones to motor neurones
- **motor neurones** transmit impulses from the CNS to effectors.

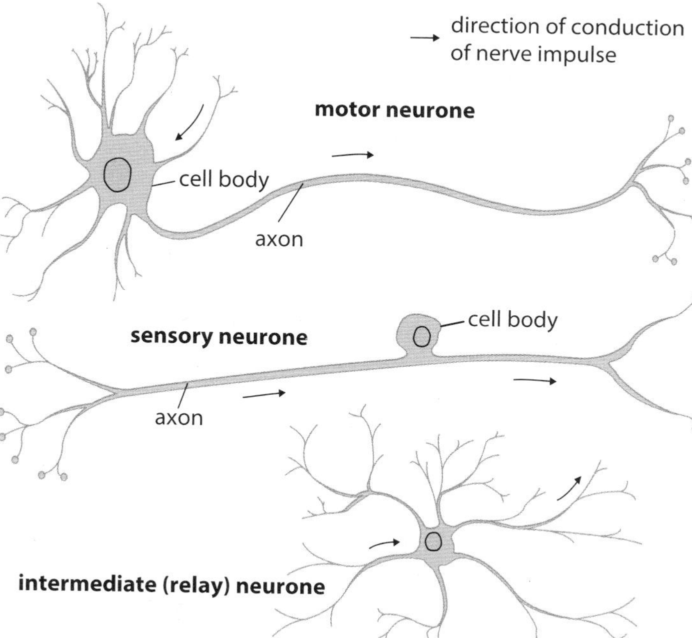

Figure 15.3 Motor, sensory and intermediate neurones.

Figure 15.4 shows the structure of a mammalian neurone. This is a motor neurone, which transmits impulses from the brain or spinal cord to a muscle or gland.

The cell body of a motor neurone lies within the spinal cord or brain (Figure 15.5). The nucleus of a neurone is always in its cell body. Often, dark specks can be seen in the cytoplasm. These are small regions of rough endoplasmic reticulum that synthesise proteins.

Thin cytoplasmic processes extend from the cell body. Some are very short and often have many branches – these are dendrites. A motor neurone has many highly branched dendrites to give a large surface area for the endings of other neurones. The axon is much longer and

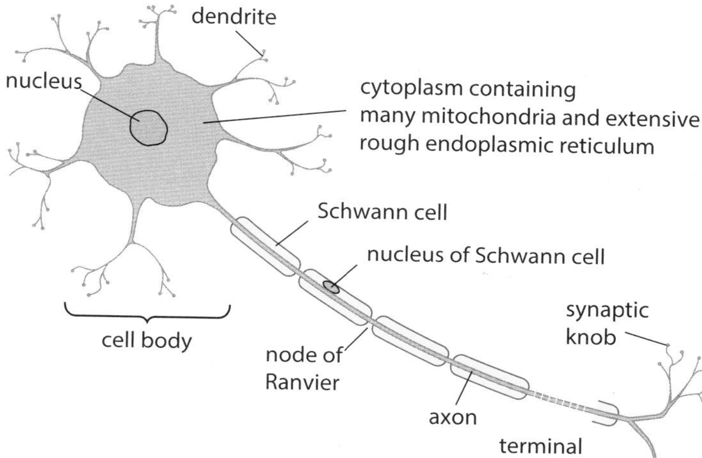

Figure 15.4 A motor neurone. The axon may be over a metre long.

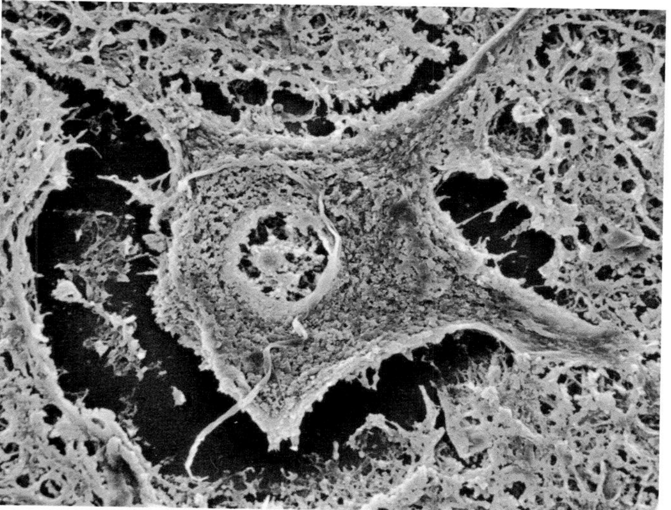

Figure 15.5 An electron micrograph of the cell body of a motor neurone within the spinal cord (×1000).

conducts impulses over long distances. A motor neurone with its cell body in your spinal cord might have its axon running all the way to one of your toes, so axons may be extremely long. Within the cytoplasm of an axon there are some organelles such as mitochondria. The ends of the branches of the axon have large numbers of mitochondria, together with many vesicles containing chemicals called transmitter substances. These vesicles are involved in passing impulses to an effector cell such as a muscle cell or a gland.

A sensory neurone has the same basic structure as a motor neurone, but it has one long axon with a cell body that may be near the source of stimuli or in a swelling of a spinal nerve known as a ganglion (Figure 15.8). (Note that the term 'dendron' is no longer used.) Relay neurones are found entirely within the central nervous system.

QUESTION

15.1 Make a table to compare the structure and function of motor and sensory neurones.

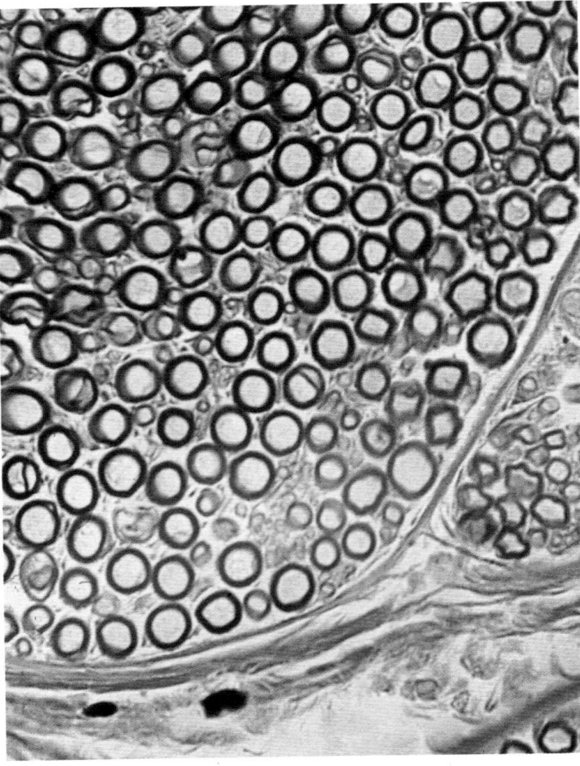

Figure 15.6 A photomicrograph of a transverse section (TS) of a nerve (×500). The circles are axons of sensory and motor neurones in cross-section. Some of these are myelinated (the ones with dark lines around) and some are not. Each group of axons is surrounded by a perineurium (red lines). Several such groups make a complete nerve.

Myelin

For most of their length, the axons of motor and sensory neurones are protected within nerves. Figure 15.6 shows a cross-section of a nerve full of neurones. You can see that some of these are surrounded by thick dark rings. This is myelin, which is made by specialised cells – Schwann cells – that surround the axons of some neurones. You can see these Schwann cells surrounding the motor neurone in Figures 15.4 and 15.7.

Not all axons are protected by myelin. You can see in Figure 15.6 that there are some neurones without dark rings; these are unmyelinated neurones. About two-thirds of our motor and sensory neurones are unmyelinated.

Myelin is made when Schwann cells wrap themselves around the axon all along its length. Figure 15.7 shows one such cell, viewed as the axon is cut transversely. The Schwann cell spirals around, enclosing the axon in many layers of its cell surface membrane. This enclosing sheath, called the **myelin sheath**, is made largely of lipid, together with some proteins. The sheath affects the speed of conduction of the nerve impulse (page 337). The small, uncovered areas of axon between Schwann cells are called nodes of Ranvier. They occur about every 1–3 mm in human neurones. The nodes themselves are very small, around 2–3 μm long.

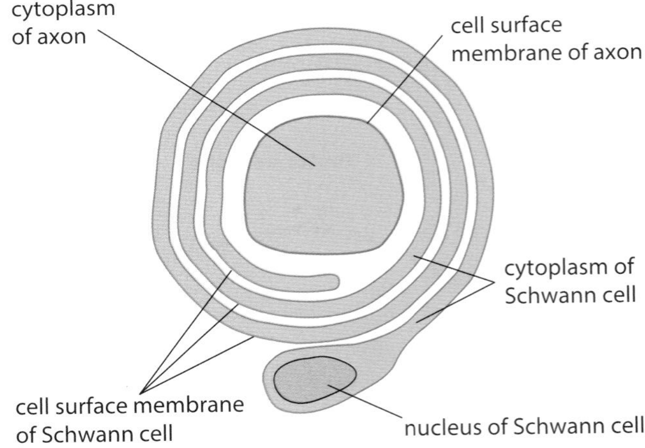

Figure 15.7 Transverse section of the axon of a myelinated neurone.

A reflex arc

Figure 15.8 shows how a sensory neurone, an intermediate neurone and a motor neurone work together to bring about a response to a stimulus. A **reflex arc** is the pathway along which impulses are transmitted from a receptor to an effector without involving 'conscious' regions of the brain. Figure 15.8 shows the structure of a spinal reflex arc in which the impulse is passed from neurone to neurone inside the spinal cord. Other reflex arcs may have no intermediate neurone, and the impulse passes directly from the sensory neurone to the motor neurone. There are also reflex arcs in the brain – for example, those controlling focusing and how much light enters the eye.

Within the spinal cord, the impulse will also be passed on to other neurones which take the impulse up the cord to the brain. This happens at the same time as impulses are travelling along the motor neurone to the effector. The effector therefore responds to the stimulus before there is any voluntary response involving the conscious regions of the brain. This type of reaction to a stimulus is called a **reflex action**. It is a fast, automatic response to a stimulus; the response to each specific stimulus is always the same. Reflex actions are a very useful way of responding to danger signals such as the touch of a very hot object on your skin or the sight of an object flying towards you.

QUESTION

15.2 Think of a reflex action other than the four already mentioned. State the precise stimulus, name the receptor that first detects this stimulus and the effector that responds to it, and describe the way in which the effector responds.

Transmission of nerve impulses

Neurones transmit electrical impulses. These impulses travel very rapidly along the cell surface membrane from one end of the cell to the other, and are **not** a flow of electrons like an electric current. Rather, the signals are very brief changes in the distribution of electrical charge **across** the cell surface membrane called **action potentials**, caused by the very rapid movement of sodium ions and potassium ions into and out of the axon.

Resting potential

Some axons in some organisms such as squids and earthworms are very wide; it is possible to insert tiny electrodes into their cytoplasm to measure the changes in electrical charge. Figure 15.9 shows just part of an unmyelinated axon. In a resting axon, it is found that the inside of the axon always has a slightly negative electrical potential compared with the outside (Figures 15.9 and 15.10). The difference between these potentials, called the **potential difference**, is often between −60 mV and −70 mV. In other words, the electrical potential of the inside of the axon is between 60 and 70 mV **lower** than the outside. This difference is the **resting potential**.

The resting potential is produced and maintained by the **sodium–potassium pumps** in the cell surface membrane (Figure 15.10b and page 271). These constantly move sodium ions, Na^+, out of the axon, and potassium ions, K^+, into the axon. The sodium–potassium pumps are membrane proteins that use energy from the hydrolysis of ATP to move both of these ions against their concentration gradients. Three sodium ions are removed from the axon for every two potassium ions brought in.

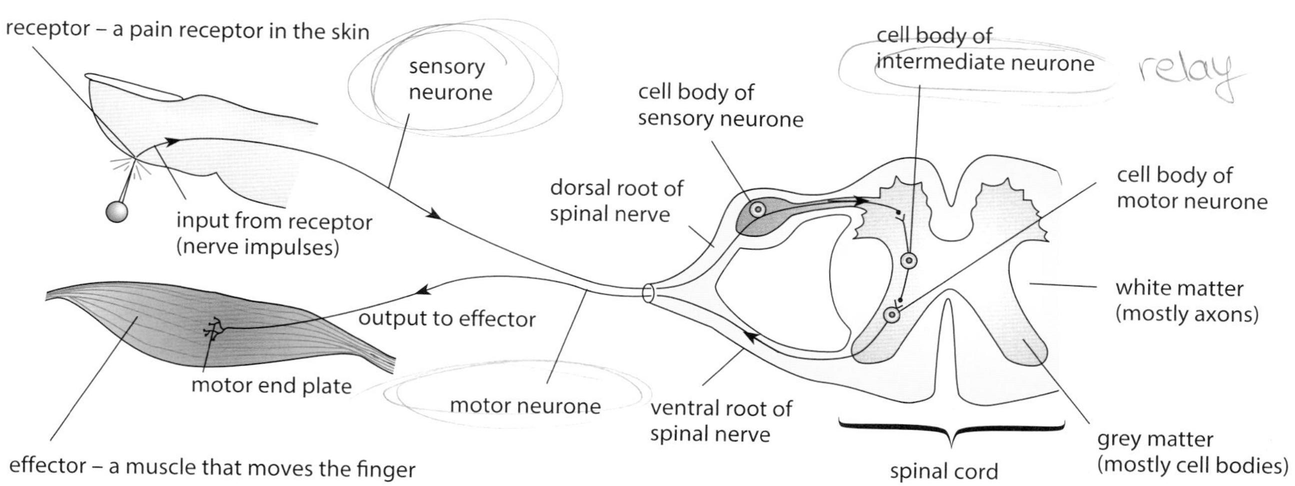

Figure 15.8 A reflex arc. The spinal cord is shown in transverse section.

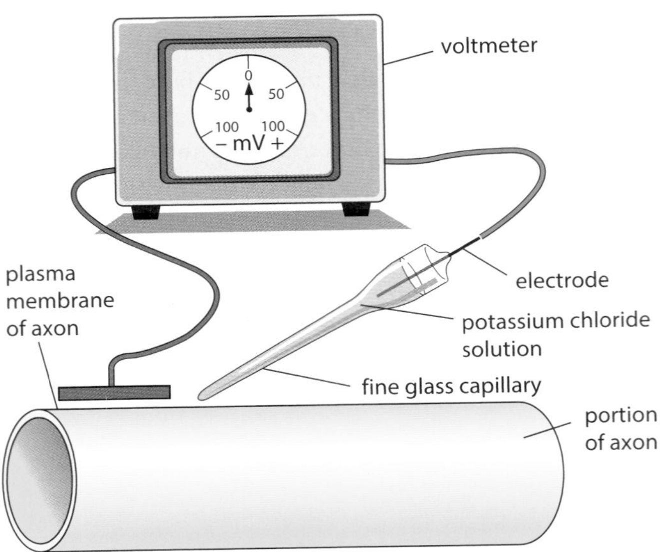

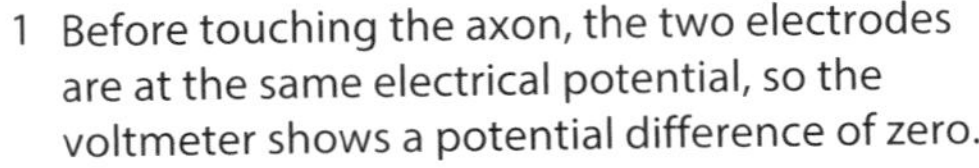

1 Before touching the axon, the two electrodes are at the same electrical potential, so the voltmeter shows a potential difference of zero.

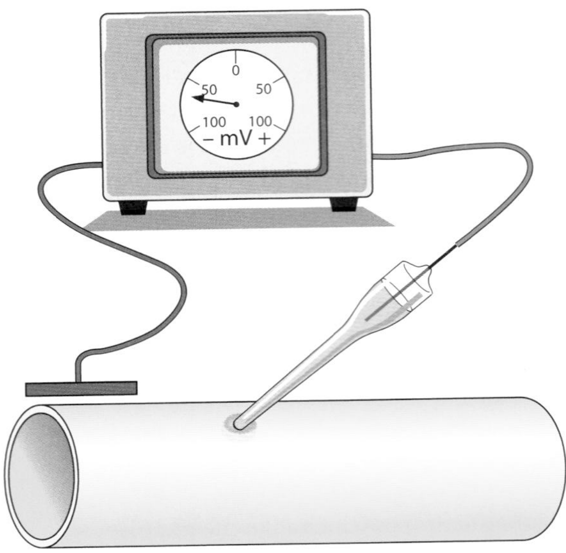

2 When one electrode is pushed inside the axon, the voltmeter shows that there is a potential difference between the inside and outside of about −70 mV inside with respect to outside.

Figure 15.9 Measuring the resting potential of an axon.

The membrane has protein channels for potassium and for sodium which are open all the time. There are far more of these for potassium than for sodium. Therefore, some potassium diffuses back out again much faster than sodium diffuses back in. In addition, there are many large, negatively charged molecules inside the cell that attract the potassium ions reducing the chance that they will diffuse out. The result of these effects is an overall excess of negative ions inside the membrane compared with outside. The membrane is relatively impermeable to sodium ions but there are two things that influence the inward movement of sodium ions during an action potential. There is a steep concentration gradient, and also the inside of the membrane is negatively charged, which attracts positively charged ions. A 'double' gradient like this is known as an electrochemical gradient.

a

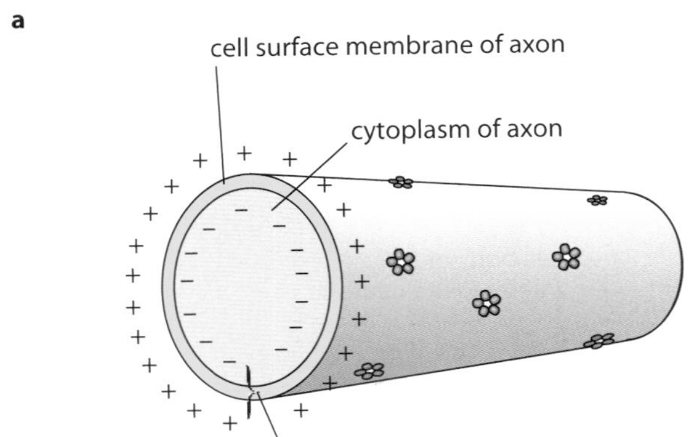

The inside of the axon is negatively charged in comparison with the outside. The difference is about −70 mV.

b

Sodium ions are constantly pumped out and potassium ions in.

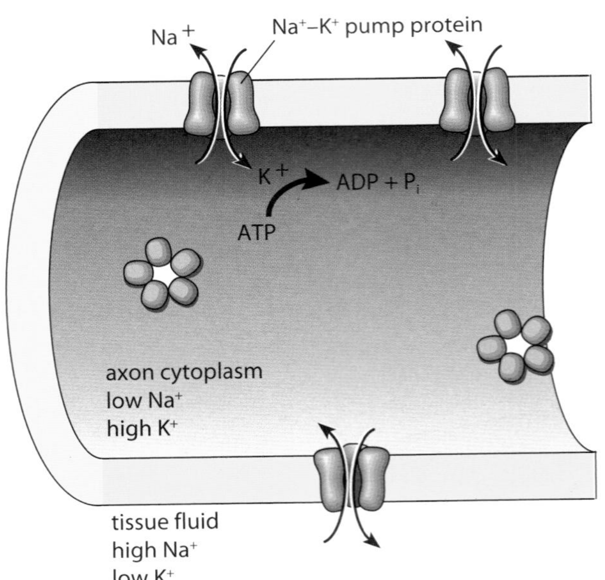

Figure 15.10 **a** At rest, an axon has negative electrical potential inside. **b** The sodium–potassium pump maintains the resting potential by keeping more sodium ions outside than there are potassium ions inside.

Action potentials

With a small addition to the apparatus shown in Figure 15.9, it is possible to stimulate the axon with a very brief, small electric current (Figure 15.11). If the axon is stimulated in this way, the steady trace on the computer screen suddenly changes. The potential difference across the cell surface membrane of the axon suddenly switches from −70 mV to +30 mV. It swiftly returns to normal after a brief 'overshoot' (Figure 15.12). The whole process takes about 3 milliseconds (ms).

This rapid, fleeting change in potential difference across the membrane is the action potential. It is caused by changes in the permeability of the cell surface membrane to sodium ions and potassium ions.

As well as the channels that are open all the time, there are other channels in the cell surface membrane that allow sodium ions or potassium ions to pass through. They open and close depending on the electrical potential (or voltage) across the membrane, and are therefore said to be **voltage-gated channels**. When the membrane is at its resting potential, these channels are closed.

First, the electric current used to stimulate the axon causes the opening of the voltage-gated channels in the cell surface membrane which allow sodium ions to pass through. As there is a much greater concentration of sodium ions outside the axon than inside, sodium ions enter through the open channels. To begin with only a few channels open. This changes the potential difference across the membrane, which becomes less negative on the inside. This **depolarisation** triggers some more channels to open so that more sodium ions enter. There is more depolarisation. If the potential difference reaches about −50 mV, then many more channels open and the inside reaches a potential of +30 mV compared with the outside. This is an example of positive feedback

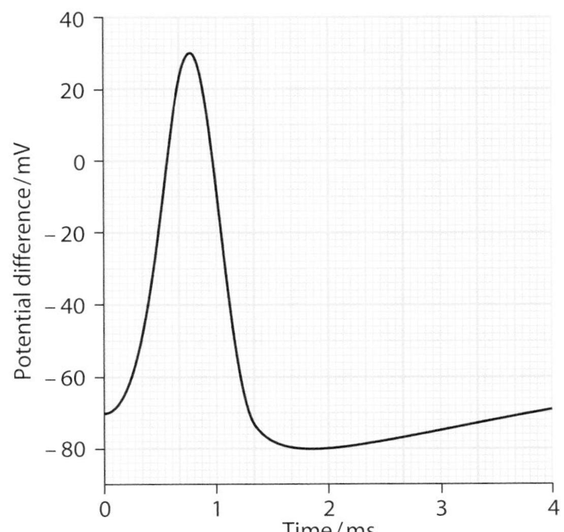

Figure 15.12 An action potential.

(page 303) because a small depolarisation leads to a greater and greater depolarisation. Action potentials are only generated if the potential difference reaches a value between −60 mV and −50 mV. This value is the **threshold potential**. If it is **less** than this, then an action potential does not occur.

After about 1 ms, all the sodium ion voltage-gated channels close, so sodium ions stop diffusing into the axon. At the same time, the potassium ion channels open. Potassium ions therefore diffuse **out** of the axon, down their concentration gradient. The outward movement of potassium ions removes positive charge from inside the axon to the outside, thus returning the potential difference to normal (−70 mV). This is called **repolarisation**. In fact, the potential difference across the membrane briefly

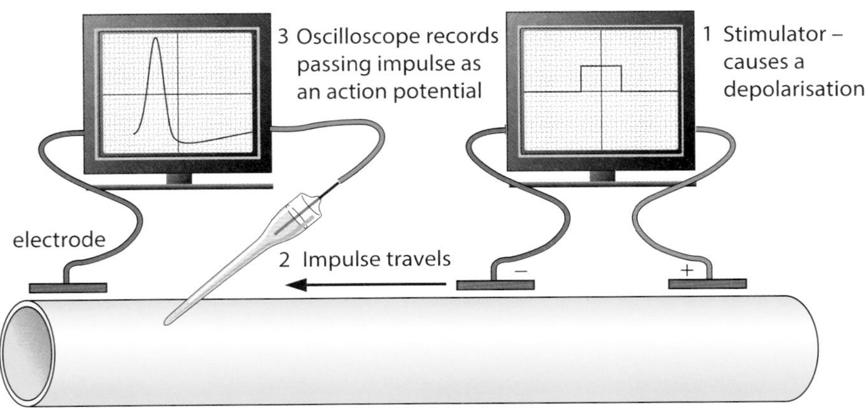

Figure 15.11 Recording of an action potential.

becomes even more negative than the normal resting potential. The potassium ion channels then close and the sodium ion channels become responsive to depolarisation again. The sodium–potassium pump continues to pump sodium ions out and potassium ions in, and this helps to maintain the distribution of sodium ions and potassium ions across the membrane so that many more action potentials can occur.

Transmission of action potentials

Figure 15.12 shows what happens at one particular point in an axon membrane. However, the function of a neurone is to transmit information **along** itself. How do action potentials help to transmit information along a neurone?

An action potential at any point in an axon's cell surface membrane triggers the production of an action potential in the membrane on either side of it. Figure 15.13 shows how it does this. The temporary depolarisation of the membrane at the site of the action potential causes a 'local circuit' to be set up between the depolarised region and the resting regions on either side of it. These local circuits depolarise the adjoining regions and so generate action potentials in them.

In practice, this only happens in an experimental situation when a stimulus is applied somewhere along an axon. In the body, action potentials begin at one end and 'new' action potentials are generated **ahead** and not behind. This is because the region behind it will still be recovering from the action potential it has just had, and the sodium ion voltage-gated channels are 'shut tight' and cannot be stimulated to open, however great the stimulus. This period of recovery is the **refractory period** when the axon is unresponsive. There are several consequences of there being refractory periods.

- Action potentials are discrete events; they do not merge into one another.
- There is a minimum time between action potentials occurring at any one place on a neurone.
- The length of the refractory period determines the maximum frequency at which impulses are transmitted along neurones.

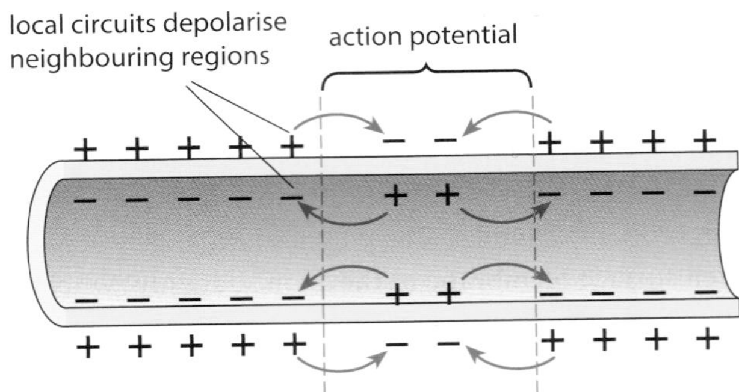

Figure 15.13 How local circuits cause an action potential to move along an axon.

QUESTION

15.3 Make a copy of Figure 15.12.

a On your graph, draw a horizontal line right across it to represent the resting potential.

b The resting potential is said to be –70 mV inside. What does this mean?

c Describe how a neurone maintains this resting potential.

d As an action potential begins, the potential difference changes from –70 mV to +30 mV inside.

  i Why is this called **depolarisation**?

  ii Annotate your graph to describe what is happening in the axon membrane to cause this depolarisation.

e Annotate your graph to describe what is happening between 1 ms and 2 ms.

f If the action potential starts at time 0, how long does it take for the resting potential to be restored?

g Indicate the refractory period on your graph.

How action potentials carry information

Action potentials do not change in size as they travel, nor do they change in size according to the intensity of the stimulus. However long an axon is, the action potential will continue to reach a peak value of +30 mV inside all the way along. A very strong light shining in your eyes will produce action potentials of precisely the same size as a dim light. The speed at which action potentials travel does not vary according to the size of the stimulus. In any one axon, the speed of axon potential transmission is always the same.

What **is** different about the action potentials resulting from a strong and a weak stimulus is their **frequency**. A strong stimulus produces a rapid succession of action potentials, each one following along the axon just behind its predecessor. A weak stimulus results in fewer action potentials per second (Figure 15.14).

Moreover, a strong stimulus is likely to stimulate more neurones than a weak stimulus. Whereas a weak stimulus might result in action potentials passing along just one or two neurones, a strong stimulus could produce action potentials in many more.

The brain can therefore interpret the **frequency** of action potentials arriving along the axon of a sensory neurone, and the **number** of neurones carrying action potentials, to get information about the **strength** of the stimulus being detected. The **nature** of the stimulus, whether it is light, heat, touch or so on, is deduced from the **position** of the sensory neurone bringing the information. If the neurone is from the retina of the eye, then the brain will interpret the information as meaning 'light'. If for some reason a different stimulus, such as pressure, stimulates a receptor cell in the retina, the brain will still interpret the action potentials from this receptor as meaning 'light'. This is why rubbing your eyes when they are shut can cause you to 'see' patterns of light.

Speed of conduction

In unmyelinated neurones, the speed of conduction is slow, being as low as $0.5\,\mathrm{m\,s^{-1}}$ in some cases. In a myelinated human neurone, action potentials travel at speeds of up to $100\,\mathrm{m\,s^{-1}}$. Myelin speeds up the rate at which action potentials travel, by insulating the axon membrane. Sodium and potassium ions cannot flow through the myelin sheath, so it is not possible for depolarisation or action potentials to occur in parts of the axon which are surrounded by the myelin sheath. Action potentials can only occur at the nodes of Ranvier, where all the channel proteins and pump proteins are concentrated.

Figure 15.15 shows how an action potential is transmitted along a myelinated axon. The local circuits exist from one node to the next. Thus action potentials 'jump' from one node to the next, a distance of 1–3 mm. This is called saltatory conduction. In a myelinated axon, saltatory conduction can increase the speed of transmission by up to 50 times that in an unmyelinated axon of the same diameter.

Diameter also affects the speed of transmission (Figure 15.16). Thick axons transmit impulses faster than thin ones, as their resistance is much less. Earthworms, which have no myelinated axons, have a small number of very thick unmyelinated ones that run all along their body.

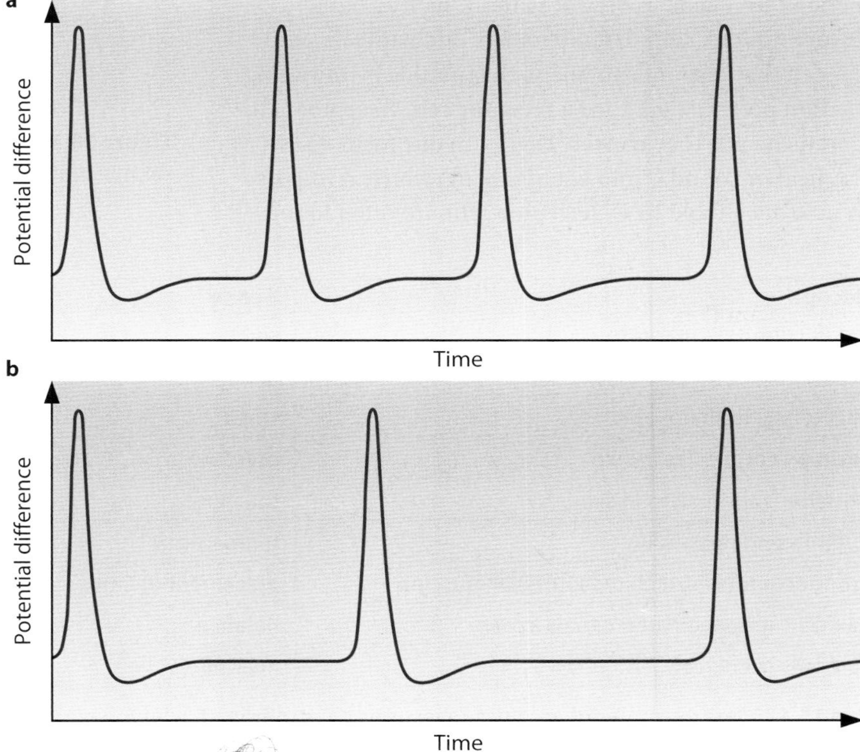

Figure 15.14 Action potentials resulting from **a** a strong stimulus and **b** a weak stimulus. Note that the size of each action potential remains the same, only its frequency changes.
a A high frequency of impulses is produced when a receptor is given a strong stimulus. This high frequency carries the message 'strong stimulus'.
b A lower frequency of impulses is produced when a receptor is given a weaker stimulus. This lower frequency carries the message 'weak stimulus'.

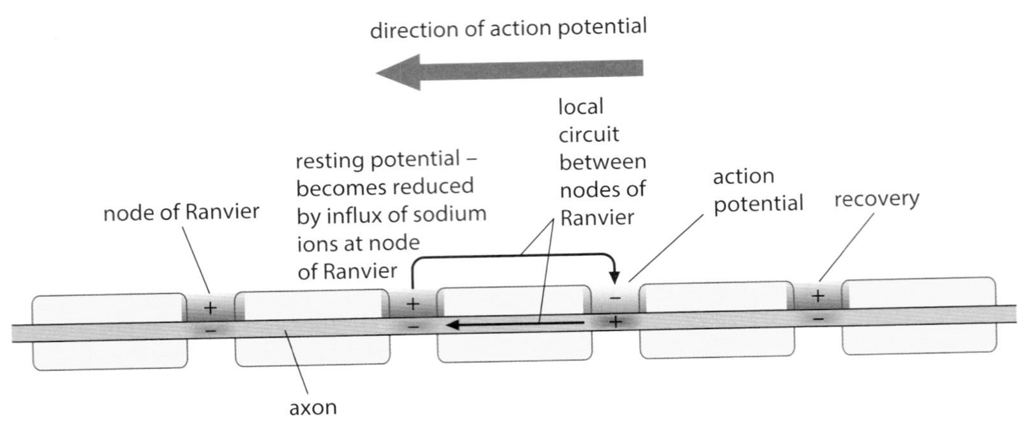

Figure 15.15 Transmission of an action potential in a myelinated axon. The myelin sheath acts as an insulator, preventing differences in potential across the parts of the axon membrane surrounded by the sheath. Potential differences can only occur at the nodes of Ranvier. The action potential therefore 'jumps' from one node to the next, travelling much more swiftly than in an unmyelinated axon.

A bird pecking at an earthworm that has just emerged from its burrow acts as a stimulus to set up impulses in these giant axons, which travel very quickly along the length of the body, stimulating muscles to contract. The rapid response may help the earthworm to escape down into its burrow.

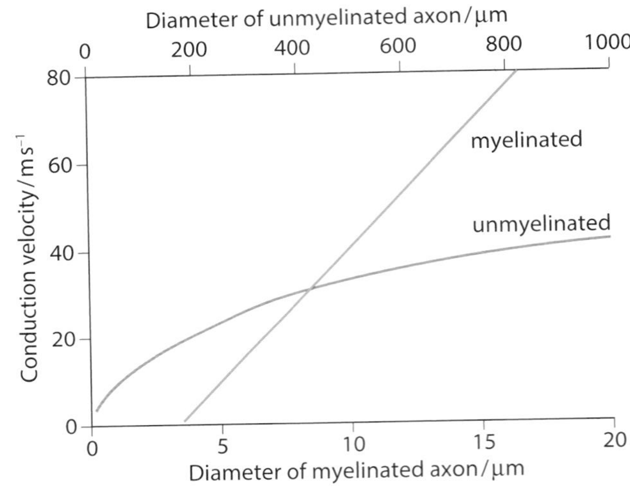

Figure 15.16 Speed of transmission in myelinated and unmyelinated axons of different diameters.

What starts off an action potential?

In the description of the generation of an action potential on page 335, the initial stimulus was a small electric current. In normal life, however, action potentials are generated by a wide variety of stimuli, such as light, pressure (touch), sound, temperature or chemicals.

A cell that responds to one such stimulus by initiating an action potential is called a receptor cell. Receptor cells are transducers: they convert energy in one form – such as light, heat or sound – into energy in an electrical impulse in a neurone (Table 15.1). Receptor cells are often found

Receptor	Sense	Form in which energy is received
rod or cone cells in retina	sight	light
taste buds on tongue	taste	chemical potential
olfactory cells in nose	smell	chemical potential
Pacinian corpuscles in skin	pressure	movement and pressure
Meissner's corpuscles in skin	touch	movement and pressure
Ruffini's endings in skin	temperature	heat
proprioceptors (stretch receptors) in muscles	placement of limbs	mechanical displacement – stretching
hair cells in semicircular canals in ear	balance	movement
hair cells in cochlea	hearing	sound

Table 15.1 Some examples of energy conversions by receptors. Each type of receptor converts a particular form of energy into electrical energy – that is, a nerve impulse. All of the receptors in the table except for stretch receptors respond to external stimuli. Stretch receptors respond to changes inside the muscles. There are many receptors that respond to other internal stimuli.

in sense organs; for example, light receptor cells are found in the eye, and sound receptor cells are found in the ear. Some receptors, such as light receptors in the eye and chemoreceptors in the taste buds, are specialised cells which detect a specific type of stimulus and influence the electrical activity of a sensory neurone. Other receptors, such as some kinds of touch receptors, are simply the ends of the sensory neurones themselves (Figure 15.17).

The tongue is covered in many small bumps or papillae. Each papilla has many taste buds over its surface (Figure 15.18). Within each taste bud are between 50 and 100 receptor cells that are sensitive to chemicals in the liquids that we drink or chemicals from our food that dissolve in saliva. Each chemoreceptor is covered with receptor proteins that detect these different chemicals. There are several types of receptor proteins, each detecting a different type of chemical and giving us a different sensation. There are five tastes: sweet, sour, salt, bitter and umami (savoury).

Chemoreceptors in the taste buds that detect salt are directly influenced by sodium ions. These ions diffuse through highly selective channel proteins in the cell surface membrane of the microvilli and this leads to depolarisation of the membrane. The increase in positive charge inside the cell is the receptor potential. If there is sufficient stimulation by sodium ions in the mouth then the receptor potential becomes large enough to stimulate the opening of voltage-gated calcium ion channels. Calcium ions enter the cytoplasm and lead to exocytosis of vesicles containing neurotransmitter from the basal membrane. The neurotransmitter stimulates an action potential in the sensory neurone (page 341) that transmits impulses to the taste centre in the cerebral cortex of the brain.

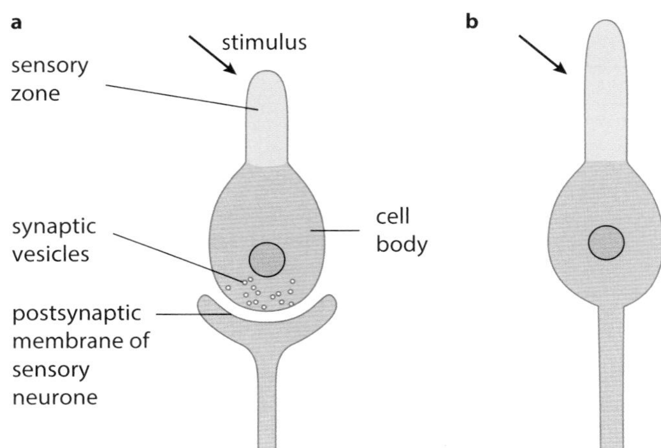

Figure 15.17 Receptors can be specialised cells **a** or simply the end of a sensory neurone **b**.

Other chemoreceptors in the taste buds use different methods. For example, the cells that are sensitive to sweet have protein receptors that stimulate a G protein, which activates an enzyme to produce many molecules of cyclic AMP. Cyclic AMP acts as a second messenger activating a cascade to amplify the signal leading to the closure of potassium ion channels (page 334). This also depolarises the membrane.

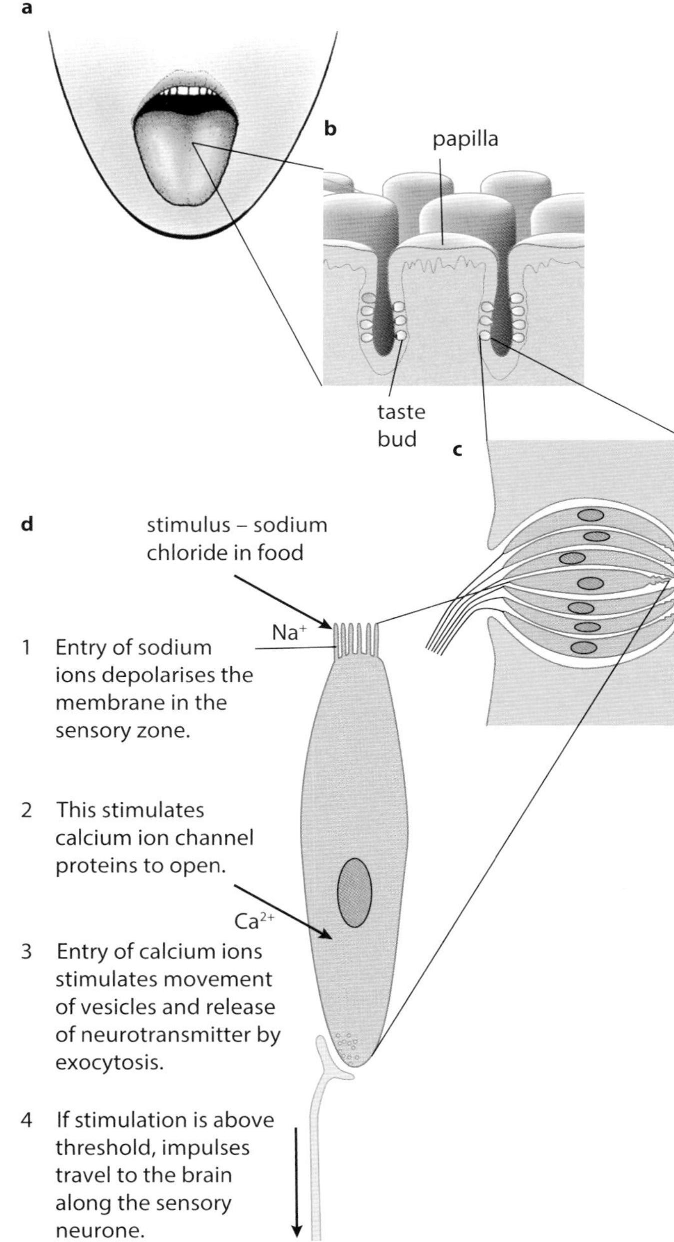

Figure 15.18 **a** Taste buds are in papillae that are distributed across the tongue. **b** A cross section through a papilla showing the distribution of taste buds. **c** A taste bud. **d** Details of one chemoreceptor cell.

When receptors are stimulated they are depolarised. If the stimulus is very weak, the cells are not depolarised very much and the sensory neurone is not activated to send impulses (Figure 15.19). If the stimulus is stronger, then the sensory neurone is activated and transmits impulses to the CNS. If the receptor potential is below a certain threshold, the stimulus only causes local depolarisation of the receptor cell. If the receptor potential is above the threshold, then the receptor cell stimulates the sensory neurone to send impulses. Above this threshold, action potentials are initiated in the sensory neurone. This is an example of the **all-or-nothing law**: neurones either transmit impulses from one end to the other or they do not. As we have already seen, the action potentials always have the same amplitude. As the stimuli increase in intensity, the action potentials are produced more frequently. The action potentials do not become bigger; they have the same amplitudes, not larger ones.

Threshold levels in receptors rarely stay constant all the time. With continued stimulation, they often increase so that it requires a greater stimulus before receptors send impulses along sensory neurones.

QUESTION

15.4 Use Figure 15.19 to answer these questions.

a Explain what is meant by the terms:
   i receptor potential
   ii threshold receptor potential
   iii the all-or-nothing law.

b Describe the relationship between the strength of the stimulus and the size of the receptor potential that is generated.

c Describe the relationship between the strength of the stimulus applied and the frequency of action potentials generated in the sensory neurone.

d What determines the maximum frequency of action potentials in a neurone?

e Threshold potentials in receptor cells can increase and decrease. Suggest the likely advantages of this.

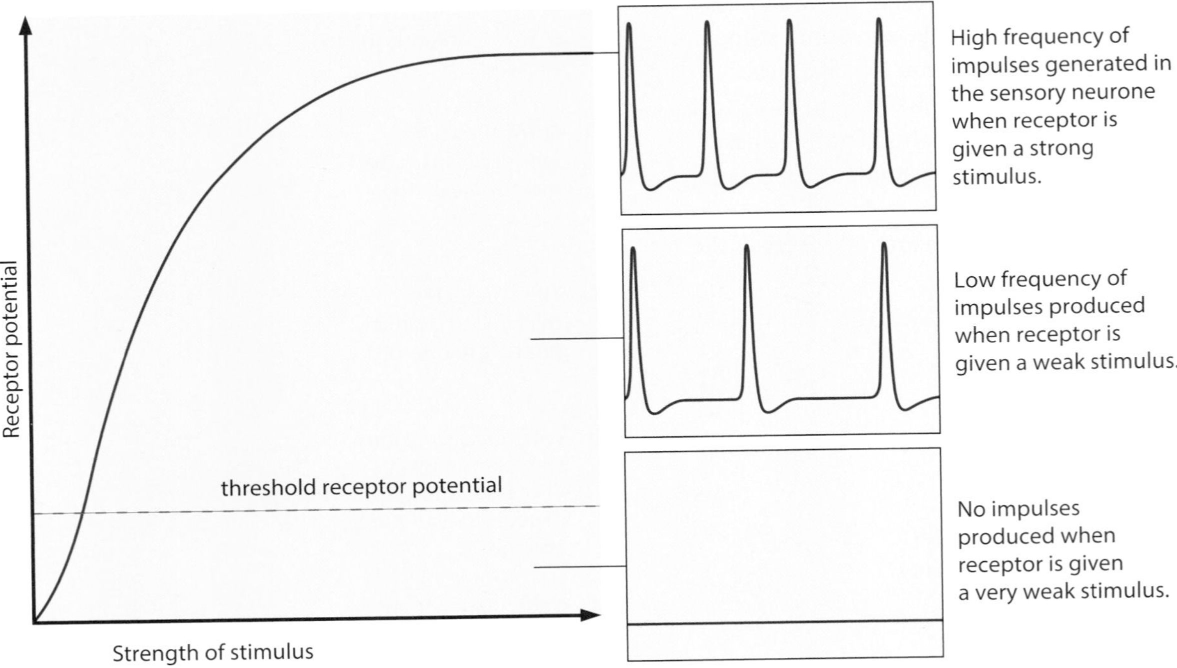

Figure 15.19 As the strength of a stimulus increases, the receptor potential also increases. If the receptor potential reaches the threshold, then impulses are sent along the sensory neurone at low frequency. Increasing the strength of the stimulus above the threshold increases the frequency of the impulses; it does not change their amplitude.

Synapses

Where two neurones meet, they do not quite touch. There is a very small gap, about 20 nm wide, between them. This gap is called the synaptic cleft. The parts of the two neurones near to the cleft, plus the cleft itself, make up a synapse (Figure 15.20).

The mechanism of synaptic transmission

Impulses cannot 'jump' across the type of synapse shown in Figure 15.20. Instead, molecules of a transmitter substance, or neurotransmitter, are released to stimulate the next neurone. This is an outline of the sequence of events that occurs.

- An action potential occurs at the cell surface membrane of the first neurone, or presynaptic neurone.
- The action potential causes the release of molecules of transmitter substance into the cleft.
- The molecules of transmitter substance diffuse across the cleft and bind temporarily to receptors on the postsynaptic neurone.
- The postsynaptic neurone responds to all the impulses arriving at any one time by depolarising; if the overall depolarisation is above its threshold, then it will send impulses.

Let us look at these processes in more detail. The cytoplasm of the presynaptic neurone contains vesicles of transmitter substance (Figure 15.21). More than 40 different transmitter substances are known; noradrenaline and acetylcholine (ACh) are found throughout the nervous system, whereas others such as **dopamine**, **glutamic acid** and gamma-aminobutyric acid (GABA) occur only in the brain. We will concentrate on the synapses that use acetylcholine as the transmitter substance. These are known as cholinergic synapses.

You will remember that, as an action potential occurs at one place on an axon, local circuits depolarise the next piece of membrane, stimulating the opening of sodium ion voltage-gated channels and so propagating the action potential. In the part of the membrane of the presynaptic neurone that is next to the synaptic cleft, the arrival of the action potential also causes **calcium ion voltage-gated channels** to open (Figure 15.22). Thus, the action potential causes not only sodium ions but also calcium ions to diffuse into the cytoplasm of the presynaptic neurone. There are virtually no calcium ions in the cytoplasm, but many in the tissue fluid surrounding the synapse. This means that there is a very steep electrochemical gradient for calcium ions.

The influx of calcium ions stimulates vesicles containing ACh to move to the presynaptic membrane and fuse with it, emptying their contents into the synaptic cleft (Figure 15.22). Each action potential causes just a few vesicles to do this, and each vesicle contains up to 10 000 molecules of ACh. The ACh diffuses across the synaptic cleft, usually in less than 0.5 ms.

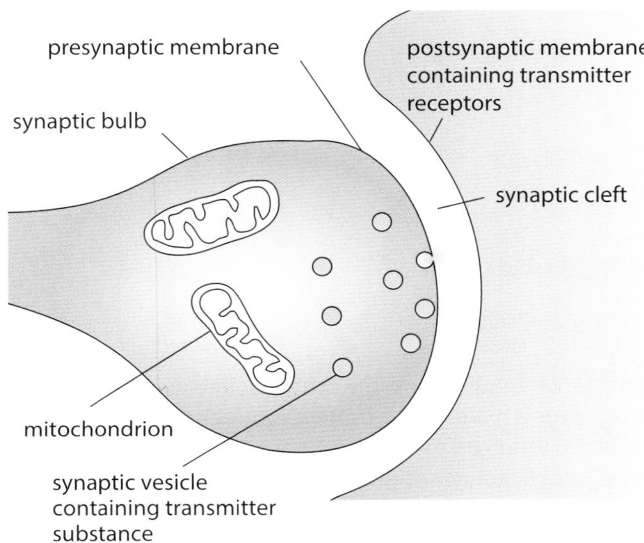

Figure 15.20 A synapse.

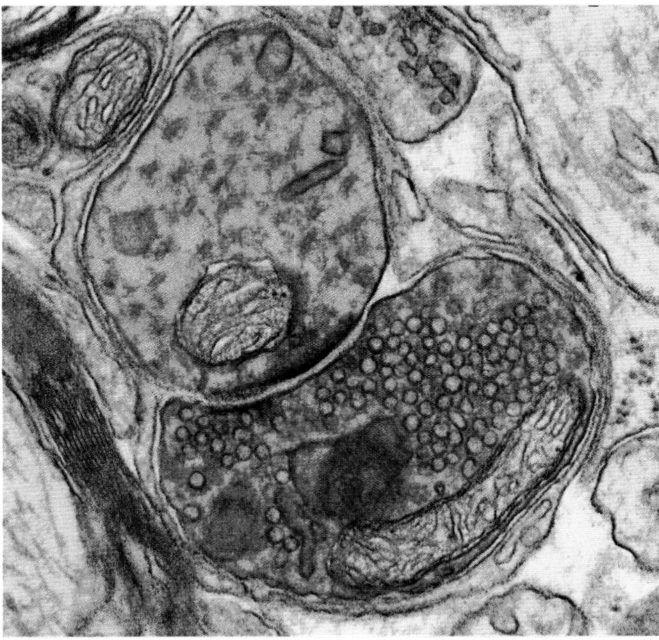

Figure 15.21 False-colour transmission electron micrograph of a synapse (×52 000). The presynaptic neurone (at the bottom) has mitochondria (shown in green) and numerous vesicles (blue), which contain the transmitter substance.

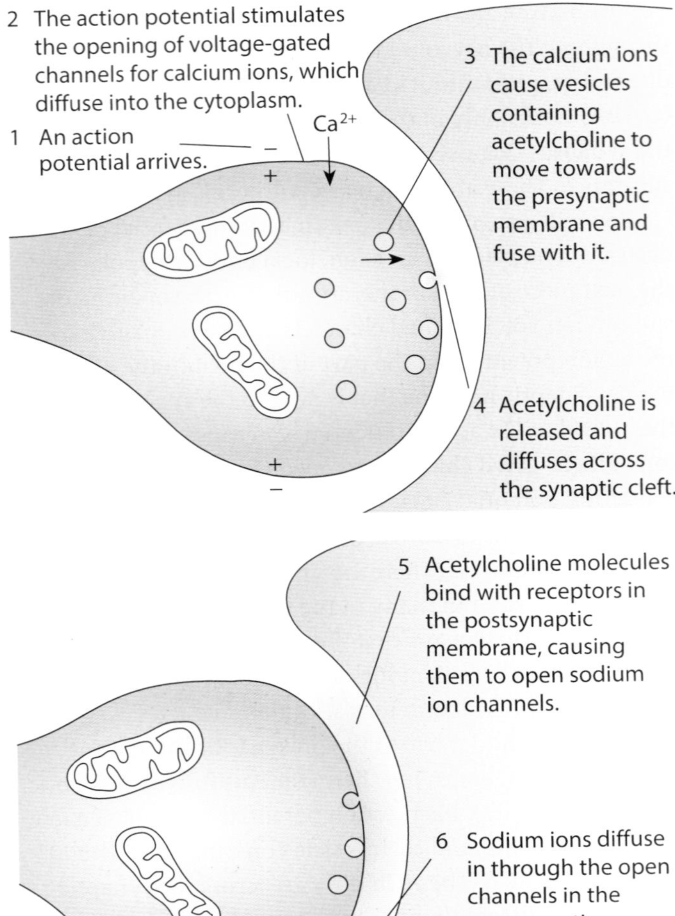

Figure 15.22 Synaptic transmission.

The cell surface membrane of the postsynaptic neurone contains **receptor proteins.** Part of the receptor protein molecule has a complementary shape to part of the ACh molecule, so that the ACh molecules can temporarily bind with the receptors. This changes the shape of the protein, opening channels through which sodium ions can pass (Figure 15.23). Sodium ions diffuse into the cytoplasm of the postsynaptic neurone and depolarise the membrane.

These receptor proteins with their channels are chemically gated ion channels as they are stimulated to open by chemicals (neurotransmitters) and not by a voltage change.

If the ACh remained bound to the postsynaptic receptors, the sodium channels would remain open, and the postsynaptic neurone would be permanently depolarised. The ACh is recycled to prevent this from happening and also to avoid wasting it. The synaptic cleft contains an enzyme, **acetylcholinesterase**, which catalyses the hydrolysis of each ACh molecule into acetate and choline.

The choline is taken back into the presynaptic neurone, where it is combined with acetyl coenzyme A to form ACh once more. The ACh is then transported into the presynaptic vesicles, ready for the next action potential. The entire sequence of events, from initial arrival of the action potential to the re-formation of ACh, takes about 5–10 ms.

QUESTION

15.5 a Name the process by which vesicles release their contents at the presynaptic membrane.

b Describe the role of acetylcholinesterase.

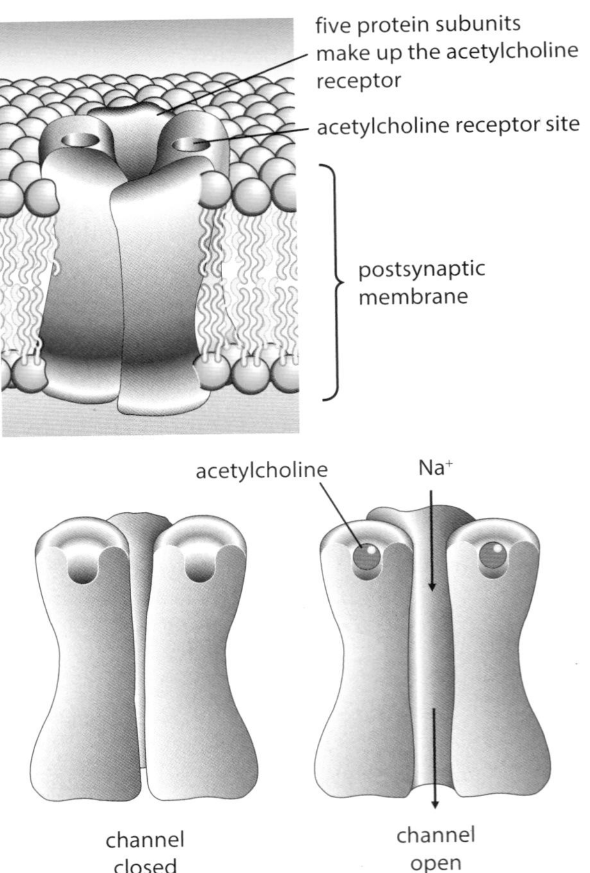

Figure 15.23 Detail of how the acetylcholine receptor works. The receptor is made of five protein subunits spanning the membrane arranged to form a cylinder. Two of these subunits contain acetylcholine receptor sites. When acetylcholine molecules bind with both of these receptor sites, the proteins change shape, opening the channel between the units. Parts of the protein molecules around this channel contain negatively charged amino acids, which attract positively charged sodium ions so they pass through the channel.

The depolarisation of the postsynaptic neurone only leads to the generation of an action potential if the potential difference is above the threshold for that neurone. If not, then there is no action potential. The chance that an action potential is generated and an impulse sent in the postsynaptic neurone is increased if more than one presynaptic neurone releases ACh at the same time or over a short period of time.

Much of the research on synapses has been done at synapses between a motor neurone and a muscle, not those between two neurones. The motor neurone forms a **motor end plate** with each muscle fibre and the synapse is called a **neuromuscular junction** (Figure 15.24). Such synapses function in the same way as described above. An action potential is produced in the muscle fibre, which may cause it to contract (pages 344–349).

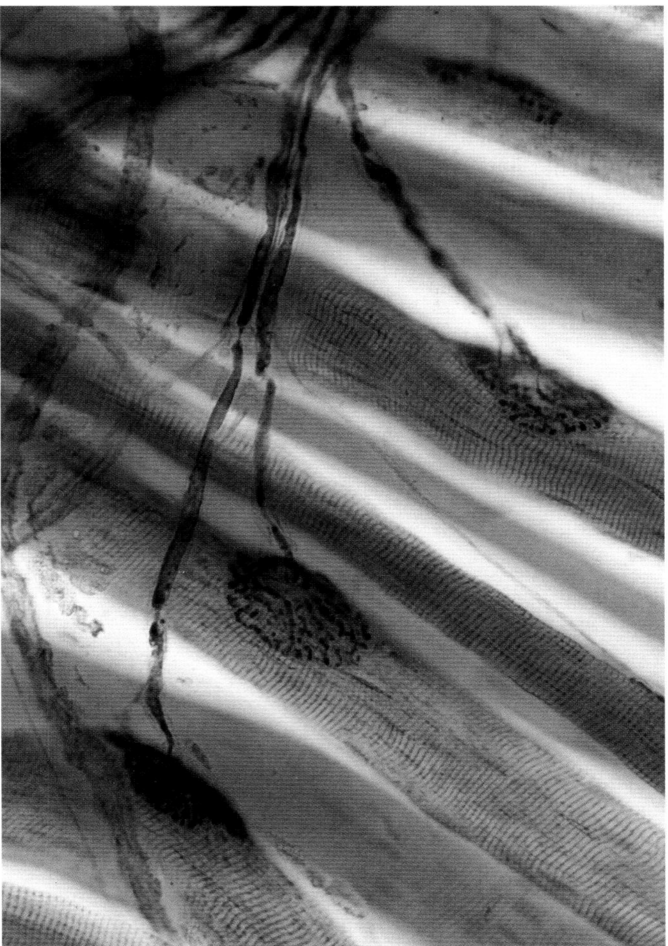

Figure 15.24 Photomicrograph of neuromuscular junctions (×200). The red tissue is muscle fibres, whereas the axons show as dark lines. The axons terminate in a number of branches on the surface of the muscle fibre, forming motor end plates. Action potentials are passed from the axon to the muscle, across a synaptic cleft, at these end plates.

The roles of synapses

Synapses slow down the rate of transmission of a nerve impulse that has to travel along two or more neurones. Responses to a stimulus would be much quicker if action potentials generated in a receptor travelled along an unbroken neuronal pathway from receptor to effector, rather than having to cross synapses on the way. So why have synapses?

- **Synapses ensure one-way transmission.** Impulses can only pass in one direction at synapses. This is because neurotransmitter is released on one side and its receptors are on the other. There is no way that chemical transmission can occur in the opposite direction.
- **Synapses allow integration of impulses.** Each sensory neurone has many branches at the end of its axon that form synapses with many relay (intermediate) neurones. The cell body of each motor neurone is covered with the terminations of many relay neurones. Motor neurones only transmit impulses if the net effect of the relay neurones is above the threshold at which it initiates action potentials. If the depolarisation of the postsynaptic membrane does not reach the threshold, no impulse is sent in that neurone. One advantage of this is that impulses with low frequencies do not travel from sensory neurones to reach the brain. This means that the brain is not overloaded with sensory information.
- **Synapses allow the interconnection of nerve pathways.** Synapses allow a wider range of behaviour than could be generated in a nervous system in which neurones were directly 'wired up' to each other. They do this by allowing the interconnection of many nerve pathways. This happens in two ways:
  - individual sensory and relay neurones have axons that branch to form synapses with many different neurones; this means that information from one neurone can spread out throughout the body to reach many relay neurones and many effectors as happens when we respond to dangerous situations
  - there are many neurones that terminate on each relay and motor neurone as they have many dendrites to give a large surface area for many synapses; this allows one neurone to integrate the information coming from many different parts of the body – something that is essential for decision-making in the brain.

- **Synapses are involved in memory and learning.** Despite much research, little is yet known about how memory operates. However, there is much evidence that it involves synapses. For example, if your brain frequently receives information about two things at the same time, say the sound of a particular voice and the sight of a particular face, then it is thought that **new** synapses form in your brain that link the neurones involved in the passing of information along the particular pathways from your ears and eyes. In future, when you hear the voice, information flowing from your ears along this pathway automatically flows into the other pathway too, so that your brain 'pictures' the face which goes with the voice. Compare how you respond when talking on the phone to someone you know well and someone you have never met.

QUESTION

15.6 Suggest why:

a impulses travel in only one direction at synapses

b if action potentials arrive repeatedly at a synapse, the synapse eventually becomes unable to transmit the impulse to the next neurone.

Muscle contraction

This section concerns the contraction of striated muscle. This type of muscle tissue makes up the many muscles in the body that are attached to the skeleton. Striated muscle only contracts when it is stimulated to do so by impulses that arrive via motor neurones. Muscle tissue like this is described as being neurogenic. You have already seen how the cardiac muscle in the heart is myogenic – it contracts and relaxes automatically, with no need for impulses arriving from neurones (Chapter 8 page 177). We have also mentioned a third type of muscle tissue, smooth muscle, which is found throughout the body in organs, such as in the gas exchange system (page 188), alimentary canal and in the walls of the arteries, arterioles and veins. Most smooth muscle only contracts when it receives impulses in motor neurones. However, smooth muscle in arteries also contracts when it is stretched by the pressure of blood surging through them. This happens without any input from the nervous system. This type of muscle is called smooth because, unlike the other two types of muscle tissue, it has no striations. Smooth muscle does **not** form smooth linings of tubular structures, such as the trachea and arteries; the lining of these structures is always formed by an epithelium. The structures and functions of the three types of muscle tissue are compared in Table 15.2.

	Type of muscle		
	striated	**cardiac**	**smooth**
Appearance in the light microscope	stripes (striations) at regular intervals	stripes (striations) at regular intervals	no striations
Cell structure	multinucleate (syncytium)	uninucleate cells joined by intercalated discs (Figure 8.23, page 173)	uninucleate cells
Shape of cells	long, unbranched cylinder	cells are shorter with branches that connect to adjacent cells	long, unbranched cells that taper at either end
Organisation of contractile proteins inside the cell	organised into parallel bundles of myofibrils	organised into parallel bundles of myofibrils	contractile proteins not organised into myofibrils
Distribution in the body	muscles attached to the skeleton	heart	tubular structures e.g. blood vessels (arteries, arterioles and veins), airways, gut, Fallopian tubes (oviducts), uterus
Control	neurogenic	myogenic	neurogenic

Table 15.2 Mammals have three types of muscle tissue: striated, cardiac and smooth.

It is obviously very important that the activities of the different muscles in our bodies are coordinated. When a muscle contracts, it exerts a force on a particular part of the body, such as a bone. This results in a particular response. The nervous system ensures that the behaviour of each muscle is coordinated with all the other muscles, so that together they can bring about the desired movement without causing damage to any parts of the skeletal or muscular system.

The structure of striated muscle

A muscle such as a biceps is made up of thousands of muscle fibres (Figure 15.25). Each muscle fibre is a very specialised 'cell' with a highly organised arrangement of contractile proteins in the cytoplasm, surrounded by a cell surface membrane. Some biologists prefer not to call it a cell, because it contains many nuclei. Instead they prefer the term **syncytium** to describe the multinucleate muscle fibre. The parts of the fibre are known by different terms. The cell surface membrane is the **sarcolemma**, the cytoplasm is **sarcoplasm** and the endoplasmic reticulum is **sarcoplasmic reticulum** (SR). The cell surface membrane has many deep infoldings into the interior of the muscle fibre, called **transverse system tubules** or **T-tubules** for short (Figure 15.26). These run close to the sarcoplasmic

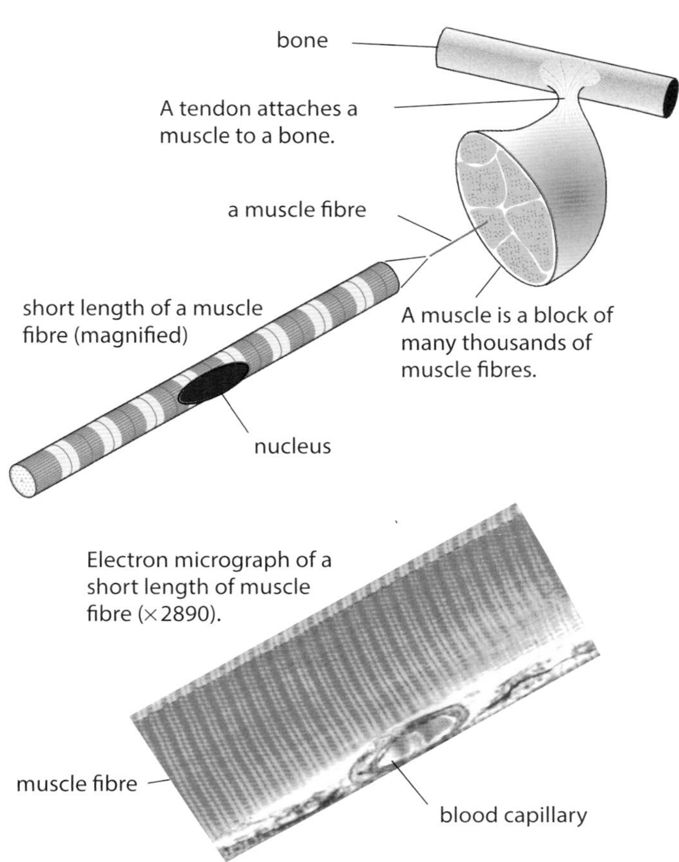

Figure 15.25 The structure of a muscle. As each muscle is composed of several tissues (striated muscle tissue, blood, nerves and connective tissue) it is an example of an organ.

Short length of muscle fibre

Highly magnified edge of muscle fibre

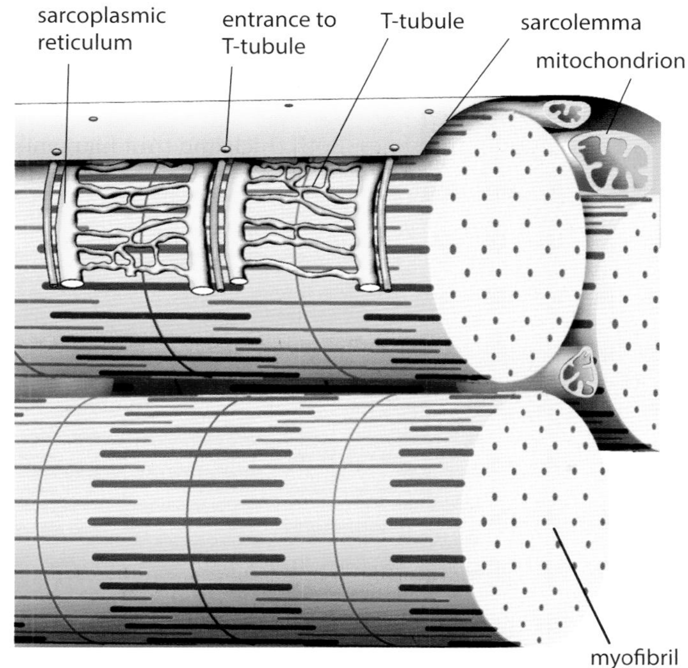

Electronmicrograph of muscle fibre (×27 000)

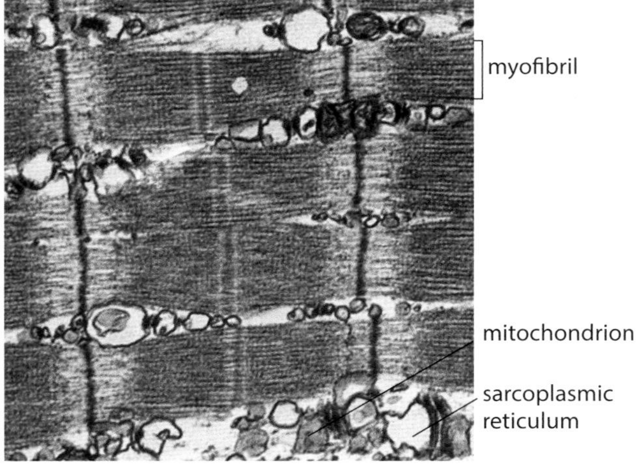

Figure 15.26 Ultrastructure of part of a muscle fibre.

reticulum. The membranes of the sarcoplasmic reticulum have huge numbers of protein pumps that transport calcium ions into the cisternae of the SR. The sarcoplasm contains a large number of mitochondria, often packed tightly between the **myofibrils**. These carry out aerobic respiration, generating the ATP that is required for muscle contraction.

The most striking thing about a muscle fibre is its stripes, or striations. These are produced by a very regular arrangement of many myofibrils in the sarcoplasm. Each myofibril is striped in exactly the same way, and is lined up precisely against the next one, so producing the pattern you can see in Figures 15.25 and 15.26.

This is as much as we can see using a light microscope, but with an electron microscope it is possible to see that each myofibril is itself made up of yet smaller components, called filaments. Parallel groups of thick filaments lie between groups of thin ones. Both thick and thin filaments are made up of protein. The thick filaments are made mostly of **myosin**, whilst the thin ones are made mostly of **actin**. Now we can understand what causes the stripes. The darker parts of the stripes, the A bands, correspond to the thick (myosin) filaments. The lighter parts, the I bands, are where there are no thick filaments, only thin (actin) filaments (Figure 15.27). The very darkest parts of the A band are produced by the overlap of thick and thin filaments, while the lighter area within the A band, known as the H band, represents the parts where only the thick filaments are present. A line known as the Z line provides an attachment for the actin filaments, while the M line does the same for the myosin filaments. The part of a myofibril between two Z lines is called a **sarcomere**. Myofibrils are cylindrical in shape, so the Z line is in fact a disc separating one sarcomere from another and is also called the Z disc.

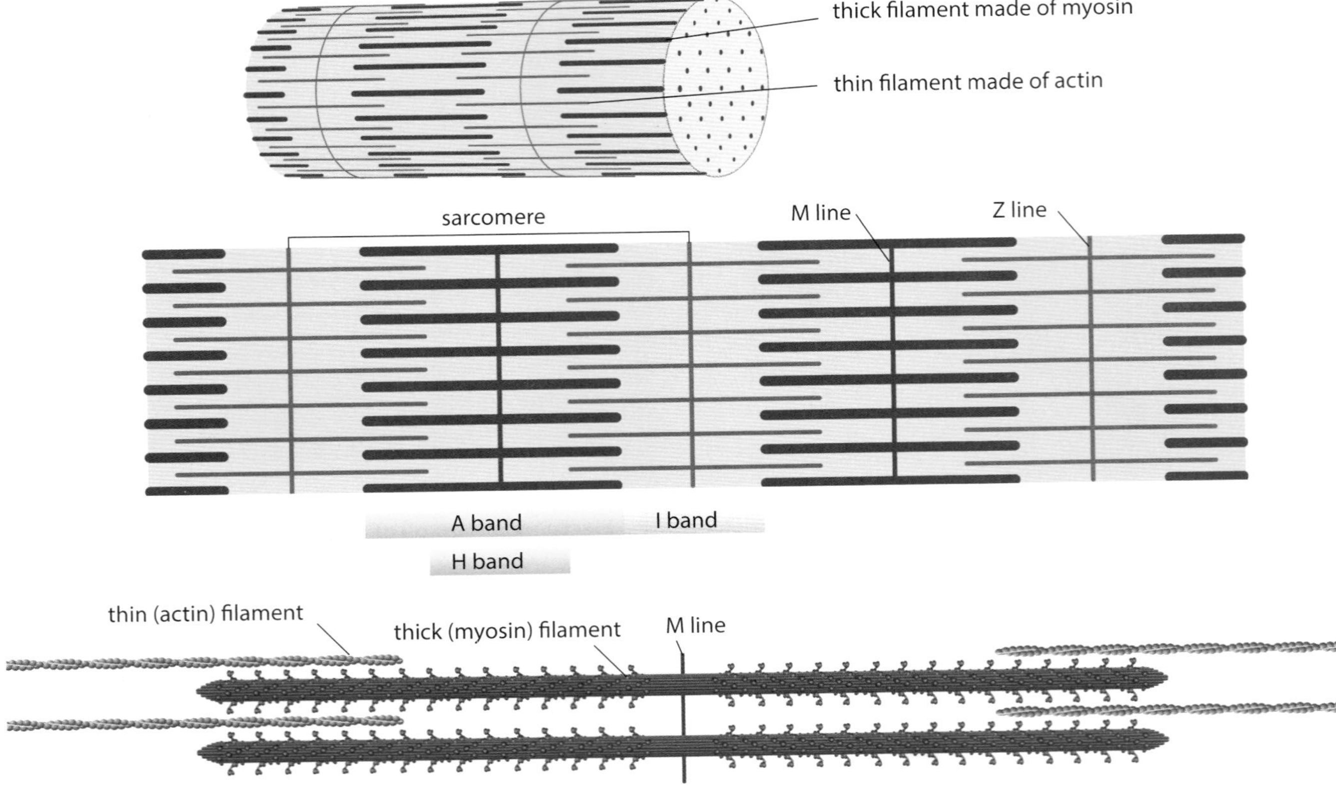

Figure 15.27 The structure of a myofibril.

QUESTION

15.7 a Use Figure 15.27 to make a simple diagram to show the arrangement of the thick and thin filaments in a sarcomere of a resting muscle. In your diagram, show and label the following: one thick filament, four thin filaments and two Z lines.

b Indicate and label the following on your diagram: A band, I band and H band.

c The average length of a sarcomere in a resting muscle is 2.25 µm. Use this figure to calculate the magnification of your diagram.

Structure of thick and thin filaments

Thick filaments are composed of many molecules of myosin, which is a fibrous protein with a globular head. The fibrous portion helps to anchor the molecule into the thick filament. Within the thick filament, many myosin molecules all lie together in a bundle with their globular heads all pointing away from the M line.

The main component of thin filaments, actin, is a globular protein. Many actin molecules are linked together to form a chain. Two of these chains are twisted together to form a thin filament. Also twisted around the actin chains is a fibrous protein called **tropomyosin**. Another protein, **troponin**, is attached to the actin chain at regular intervals (Figure 15.28).

How muscles contract

Muscles cause movement by contracting. The sarcomeres in each myofibril get shorter as the Z discs are pulled closer together. Figure 15.28 shows how this happens. It is known as the sliding filament model of muscle contraction.

The energy for the movement comes from ATP molecules that are attached to the myosin heads. Each myosin head is an ATPase.

When a muscle contracts, calcium ions are released from stores in the SR and bind to troponin. This stimulates troponin molecules to change shape (Figure 15.28). The troponin and tropomyosin proteins move to a different position on the thin filaments, so exposing parts of the actin molecules which act as binding sites for

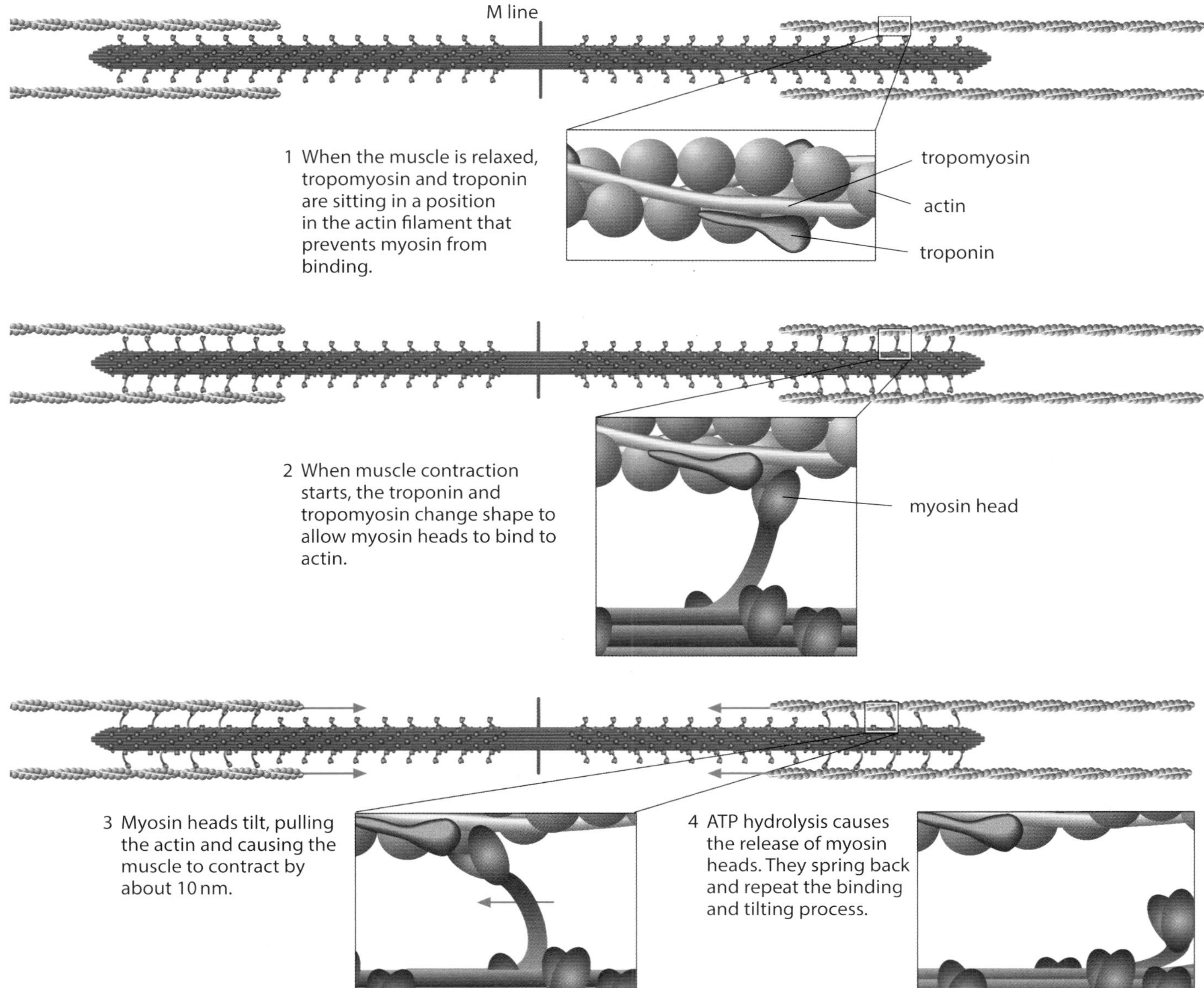

Figure 15.28 The sliding filament model of muscle contraction.

myosin. The myosin heads bind with these sites, forming cross-bridges between the two types of filament.

Next, the myosin heads tilt, pulling the actin filaments along towards the centre of the sarcomere. The heads then hydrolyse ATP molecules, which provide enough energy to force the heads to let go of the actin. The heads tip back to their previous positions and bind again to the exposed sites on the actin. The thin filaments have moved as a result of the previous power stroke, so myosin heads now bind to actin further along the thin filaments closer to the Z disc. They tilt again, pulling the actin filaments even further along, then hydrolyse more ATP molecules so that they can let go again. This goes on and on, so long as the troponin and tropomyosin molecules are not blocking the binding sites, and so long as the muscle has a supply of ATP.

Stimulating muscle to contract

Skeletal muscle contracts when it receives an impulse from a neurone. An impulse moves along the axon of a motor neurone and arrives at the presynaptic membrane (Figure 15.29). A neurotransmitter, generally acetylcholine, diffuses across the neuromuscular junction and binds to receptor proteins on the postsynaptic membrane – which is the sarcolemma (the cell surface membrane of the muscle fibre). The binding of acetylcholine stimulates the ion channels to open, so that sodium ions enter to depolarise the membrane and generate an action potential in the sarcolemma.

Impulses pass along the sarcolemma and along the T-tubules towards the centre of the muscle fibre. The membranes of the sarcoplasmic reticulum are very close to the T-tubules. The arrival of the impulses causes calcium ion channels in the membranes to open. Calcium ions diffuse out, down a very steep concentration gradient, into the sarcoplasm surrounding the myofibrils.

The calcium ions bind with the troponin molecules that are part of the thin filaments. This changes the shape of the troponin molecules, which causes the troponin and tropomyosin to move away and expose the binding sites for the myosin heads. The myosin heads attach to the binding sites on the thin filaments and form cross-bridges (Figures 15.28 and 15.29). When there is no longer any stimulation from the motor neurone, there are no impulses conducted along the T-tubules. Released from stimulation, the calcium ion channels in the SR close and the calcium pumps move calcium ions back into stores in the sarcoplasmic reticulum. As calcium ions leave their

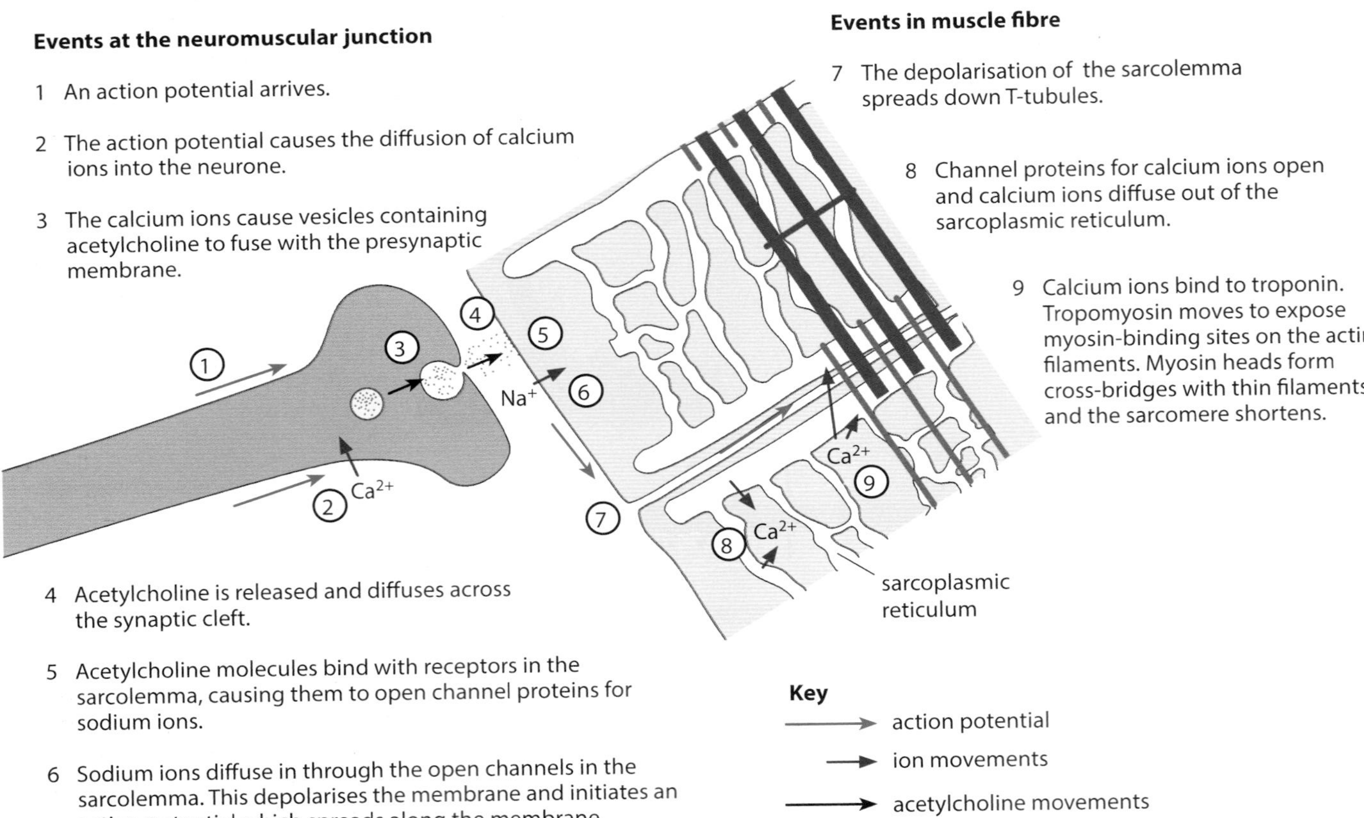

Figure 15.29 The sequence of events that follows the arrival of an impulse at a motor end plate.

binding sites on troponin, tropomyosin moves back to cover the myosin-binding sites on the thin filaments.

When there are no cross-bridges between thick and thin filaments, the muscle is in a relaxed state. There is nothing to hold the filaments together so any pulling force applied to the muscle will lengthen the sarcomeres so that they are ready to contract (and shorten) again. Each skeletal muscle in the body has an antagonist – a muscle that restores sarcomeres to their original lengths when it contracts. For example, the triceps is the antagonist of the biceps.

QUESTION

15.8 Interneuronal synapses are those found between neurones, such as those in a reflex arc (Figure 15.8). Describe the similarities and differences between the structure and function of interneuronal synapses and neuromuscular junctions.

Providing ATP for muscle contraction

A contracting muscle uses a lot of ATP. The very small quantity of ATP in the muscle fibres in a resting muscle is used up rapidly once the muscle starts to contract. More ATP is produced by respiration – both aerobic respiration inside the mitochondria and, when that cannot supply ATP fast enough, also by lactic fermentation in the sarcoplasm (Figure 15.30).

Muscles also have another source of ATP, produced from a substance called **creatine phosphate**. They keep stores of this substance in their sarcoplasm. It is their immediate source of energy once they have used the small quantity of ATP in the sarcoplasm. A phosphate group can quickly and easily be removed from each creatine phosphate molecule and combined with ADP to produce more ATP:

creatine phosphate + ADP → creatine + ATP

Later, when the demand for energy has slowed down or stopped, ATP molecules produced by respiration can be used to 'recharge' the creatine:

creatine + ATP → creatine phosphate + ADP

In the meantime, however, if energy is still being demanded by the muscles and there is no ATP spare to regenerate the creatine phosphate, the creatine is converted to creatinine and excreted in urine (page 304).

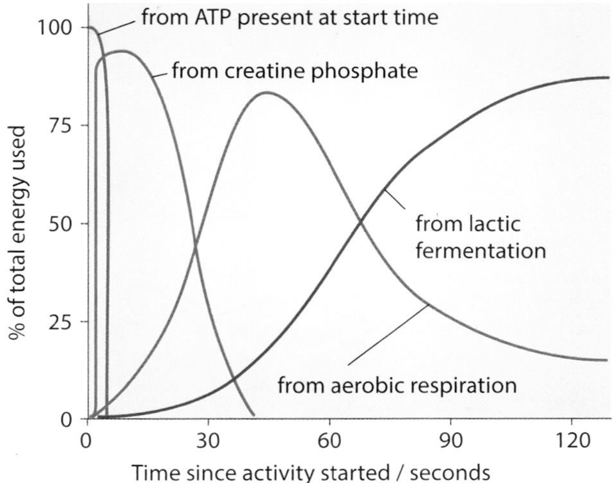

Figure 15.30 Energy sources used in muscle at high power output.

Hormonal communication

The control by the nervous system is very fast. However, it is also very expensive in terms of the energy needed for pumping sodium and potassium ions to maintain resting potentials and in protein synthesis to make all the channels and pump proteins. Energy is also needed to maintain all the neurones and other cells in the nervous system, such as Schwann cells. A much 'cheaper' alternative is to use hormones that are secreted in tiny quantities and dispersed around the body in the blood. Homeostatic functions, such as the control of blood glucose concentration and the water potential of the blood, need to be coordinated all the time, but they do not need to be coordinated in a hurry. Hormones are ideal for controlling these functions.

Hormones such as adrenaline, insulin, glucagon and ADH are made in **endocrine glands**. A **gland** is a group of cells that produces and releases one or more substances, a process known as **secretion**. Endocrine glands contain secretory cells that pass their products directly into the blood. As endocrine glands do not have ducts, they are often known as ductless glands. Hormones are cell signalling molecules.

The hormones we considered in Chapter 14 are peptides or small proteins. They are water soluble, so they cannot cross the phospholipid bilayer of cell surface membranes. These hormones bind to receptors on their target cells that in turn activate second messengers to transfer the signal throughout the cytoplasm.

The steroid hormones that we are about to consider are lipid soluble, so they can pass through the phospholipid bilayer. Once they have crossed the cell surface membrane, they bind to receptor molecules inside the cytoplasm or the nucleus and activate processes such as transcription (page 119).

Hormonal control of the human menstrual cycle

After puberty in women, the ovaries and the uterus go through a series of changes that recur approximately every 28 days – the **menstrual cycle**. Figure 15.31 shows the changes that occur during the cycle in the ovary. In the middle of this cycle the female gamete is released in the oviduct. If fertilisation occurs while the gamete is in the oviduct, the embryo that develops needs somewhere to embed itself to continue its development. There is a cycle of changes that occurs in the uterus so that the lining is ready to receive the embryo, allowing it to continue its development. The uterine cycle (Figure 15.32) is synchronised with the ovarian cycle so that the endometrium (lining of the uterus) is ready to receive the embryo at the right time.

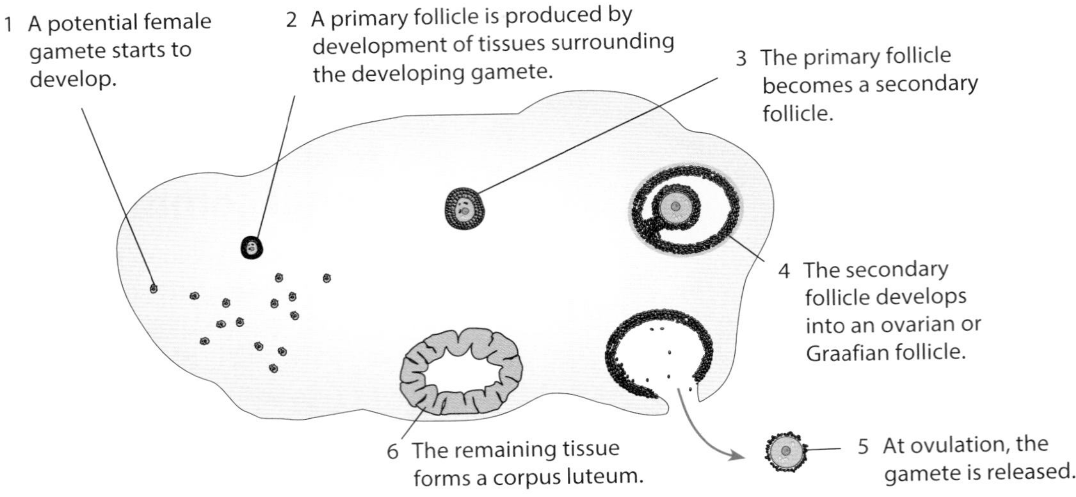

Figure 15.31 The ovary, showing the stages leading up to and following ovulation.

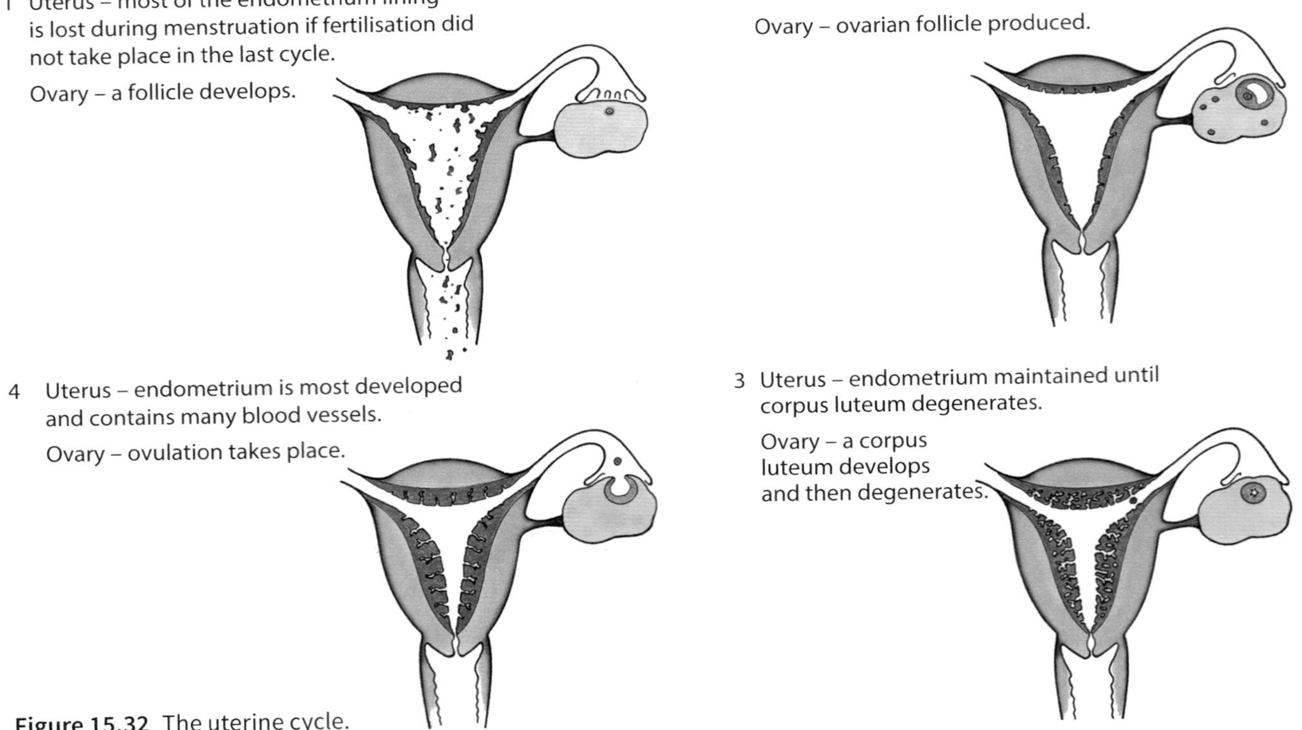

Figure 15.32 The uterine cycle.

The menstrual cycle is coordinated by glycoprotein hormones released by the anterior pituitary gland (Figure 14.16, page 312) and by the ovaries. The anterior pituitary gland secretes **follicle stimulating hormone** (FSH) and **luteinising hormone** (LH). These hormones control the activity of the ovaries. During the monthly ovarian cycle, follicles develop which secrete the steroid hormone **oestrogen**. After the female gamete is released from the ovary at ovulation, the remains of the follicle secretes **progesterone**, another steroid hormone. Figure 15.33 shows the changes in these hormones during the menstrual cycle.

The menstrual cycle is considered to begin with the onset of menstruation. Menstruation usually lasts for about four to eight days. During this time, the anterior pituitary gland secretes LH and FSH, and their concentrations increase over the next few days.

In the ovary, one follicle becomes the 'dominant' one. The presence of LH and FSH stimulates the secretion of oestrogen from the cells surrounding the follicle. The presence of oestrogen in the blood has a negative feedback effect (page 301) on the production of LH and FSH, so the concentrations of these two hormones decrease. The oestrogen stimulates the endometrium to grow, thicken and develop numerous blood capillaries.

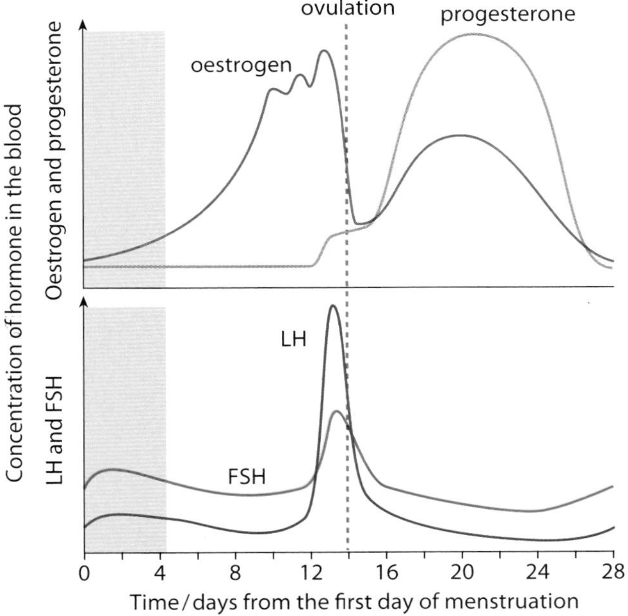

Figure 15.33 Changes in the concentrations of hormones in the blood during the menstrual cycle.

When the oestrogen concentration of the blood has reached a level of around two to four times its level at the beginning of the cycle, it stimulates a surge in the secretion of LH and, to a lesser extent, of FSH. The surge of LH causes the dominant follicle to burst and to shed its gamete into the oviduct. This usually happens about 14–36 hours after the LH surge. The follicle then collapses to form the **corpus luteum** (yellow body), which secretes progesterone and some oestrogen. Together these two hormones maintain the lining of the uterus, making it ready to receive the embryo if fertilisation occurs. Progesterone also inhibits the anterior pituitary from secreting FSH so no more follicles develop.

High levels of oestrogen and progesterone in the second half of the cycle inhibit the secretion of FSH and LH by the anterior pituitary. This means that there is less stimulation of the corpus luteum so that it begins to degenerate and secrete less oestrogen and progesterone. As the concentrations of these two hormones decrease, the endometrium is not maintained and menstruation begins. The decrease also releases the anterior pituitary from inhibition, so FSH is secreted to begin another cycle.

QUESTIONS

15.9 Explain why steroid hormones, such as progesterone and oestrogen, can pass easily through the cell surface membrane, whereas FSH and LH cannot.

15.10 Make a table to compare coordination in mammals by the nervous system and by the endocrine system.

Birth control

'Birth control' means taking control over if and when a couple have a child. It may involve contraception, which means preventing fertilisation when sexual intercourse takes place. There are also several methods of birth control that do not prevent conception, but rather prevent the tiny embryo from implanting into the lining of the uterus. These can be termed anti-implantation methods, and they include the use of intra-uterine devices (IUDs) and the 'morning after' pill. Here we look only at methods that use hormones to prevent pregnancy.

The birth control pill

At the moment, a birth control pill is available only for women, although considerable research is going into producing a pill that men could use as a contraceptive measure.

The 'pill' was developed in the 1960s, and its introduction had a huge impact on the freedom of women to have sexual intercourse without running the risk of becoming pregnant. While most would consider that this has been a great advance, it has also contributed to the rise in the incidence of sexually transmitted diseases, including HIV/AIDS, because more people have had unprotected sex with more than one partner.

The pill contains steroid hormones that suppress ovulation. Usually, synthetic hormones rather than natural ones are used, because they are not broken down so rapidly in the body and therefore act for longer. Some forms of the pill contain progesterone only, but most contain both progesterone and oestrogen, and are known as 'combined oral contraceptives'. There are many different types, with slightly different ratios of these hormones, because women are not all alike in the way their bodies respond to the pill.

With most types of oral contraceptive, the woman takes one pill daily for 21 days and then stops for seven days, during which time menstruation occurs. For some types, she continues to take a different coloured, inactive, pill for these seven days.

Both oestrogen and progesterone suppress the secretion of FSH and LH from the anterior pituitary gland. This is an example of negative feedback (page 301). Look again at Figure 15.32. You will see that, in the menstrual cycle, the highest concentrations of FSH and LH are produced when the concentration of oestrogen starts to fall and when progesterone concentration has only just started to rise. Taking the pill daily, starting at the end of menstruation, keeps oestrogen and progesterone concentrations high. This suppresses the secretion of FSH and LH, and prevents their concentrations from reaching the levels that would stimulate ovulation. This essentially mimics the natural situation during the second half of the menstrual cycle.

Stopping taking the pill after 21 days allows the concentrations of oestrogen and progesterone to fall to the point at which the uterine lining is no longer maintained. Menstruation occurs, and this reassures the woman taking the pill that she is **not** pregnant.

This combined oral contraceptive is very effective at preventing conception, but the women taking it have to be careful not to miss even a single day's pill as this might allow ovulation to take place, and lead to fertilisation if she has unprotected sex at that time.

The combination of oestrogen and progesterone can also be given by means of a skin patch from which the hormones are absorbed through the skin, by injection or by inserting an implant under the skin that is effective for several months. Obviously, with these methods, no menstruation takes place.

Pills containing only progesterone may allow ovulation to occur. They seem to work as contraceptives by reducing the ability of sperm to fertilise the egg and by making the mucus secreted by the cervix more viscous and so less easily penetrated by sperm.

QUESTION

15.11 The graphs in Figure 15.34 show part of a woman's 28-day oral contraceptive cycle. The top row shows the days on which she took a combined progesterone and oestrogen pill. The part of the graph below this illustrates the changes in concentrations of progesterone and oestrogen (steroids) in her blood. The bottom graph shows the activity of the follicles in her ovaries.

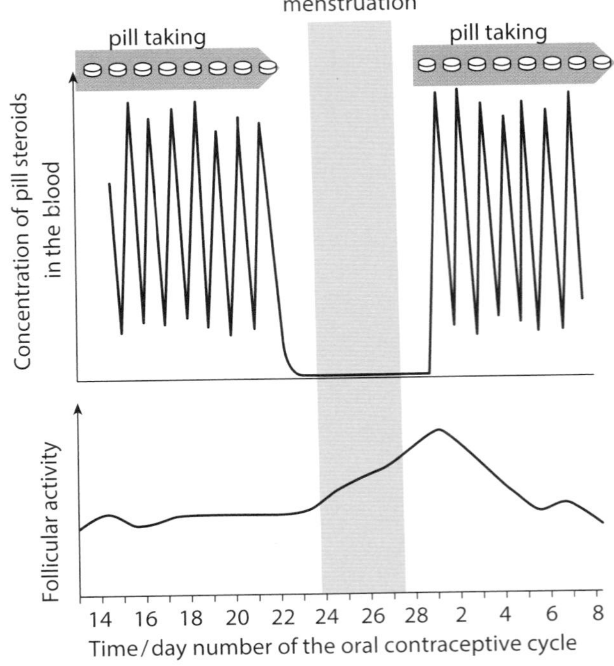

Figure 15.34

- **a** How many days of the cycle are shown in these graphs?
- **b** Describe the patterns shown by the concentration of steroids in the woman's blood, and relate these to her pill-taking schedule.
- **c** Describe the patterns shown by the level of follicular activity. Explain how the concentrations of steroids in the blood can cause the patterns you describe.

The morning-after pill

This form of birth control is intended to be taken **after** a woman has had unprotected sexual intercourse and thinks that she might be pregnant. It might be taken by a woman who forgot to take her oral contraceptive pill, or if a condom broke, or by someone who was raped, as well as by a woman who simply did not take any precautions to prevent pregnancy. It works for up to 72 hours after intercourse, not just the 'morning after'.

The pill contains a synthetic progesterone-like hormone. If taken early enough, it reduces the chances of a sperm reaching and fertilising an egg. However, in most cases, it probably prevents a pregnancy by stopping the embryo implanting into the uterus.

Control and coordination in plants

Plants, like animals, have communication systems that allow coordination between different parts of their bodies. They too must respond to changes in their external and internal environments, as we saw in Chapter 14. Most plant responses involve changing some aspect of their growth to respond to factors such as gravity, light and water availability. Plants can also respond fairly quickly to changes in carbon dioxide concentration, lack of water, grazing by animals and infection by fungi and bacteria. Some of these responses are brought about by quick changes in turgidity, as happens when stomata respond to changes in humidity, carbon dioxide concentration and water availability.

Electrical communication in plants

Plant cells have electrochemical gradients across their cell surface membranes in the same way as in animal cells. They also have resting potentials. As in animals, plant action potentials are triggered when the membrane is depolarised. In at least some species, some responses to stimuli are coordinated by action potentials. The 'sensitive plant', *Mimosa*, responds to touch by folding up its leaves.

Microelectrodes inserted into leaf cells detect changes in potential difference that are very similar to action potentials in animals. The depolarisation results not from the influx of positively charged sodium ions, but from the outflow of negatively charged chloride ions. Repolarisation is achieved in the same way by the outflow of potassium ions. Plants do not have specific nerve cells, but many of their cells transmit waves of electrical activity that are very similar to those transmitted along the neurones of animals. The action potentials travel along the cell membranes of plant cells and from cell to cell through plasmodesmata that are lined by cell membrane (Figure 1.28, page 20). The action potentials generally last much longer and travel more slowly than in animal neurones.

Many different stimuli trigger action potentials in plants. Chemicals coming into contact with a plant's surface trigger action potentials. For example, dripping a solution of acid of a similar pH to acid rain on soya bean leaves causes action potentials to sweep across them. In potato plants, Colorado beetle larvae feeding on leaves have been shown to induce action potentials. No-one knows what effect, if any, these action potentials have, but it is thought that they might bring about changes in the metabolic reactions taking place in some parts of the plant.

The Venus fly trap is a carnivorous plant that obtains a supply of nitrogen compounds by trapping and digesting small animals, mostly insects. Charles Darwin made the first scientific study of carnivorous plants describing Venus fly traps as 'one of the most wonderful plants in the world'. The specialised leaf is divided into two lobes either side of a midrib. The inside of each lobe is often red and has nectar-secreting glands around the edge to attract insects. Each lobe has three stiff sensory hairs that respond to being deflected. The outer edges of the lobes have stiff hairs that interlock to trap the insect inside. The surface of the lobes has many glands that secrete enzymes for the digestion of trapped insects. The touch of a fly or other insect on the sensory hairs on the inside of the folded leaves of the Venus fly trap stimulates action potentials that travel very fast across the leaf causing it to fold over and trap the insect (Figure 15.35).

The deflection of a sensory hair activates calcium ion channels in cells at the base of the hair. These channels open so that calcium ions flow in to generate a receptor potential. If two of these hairs are stimulated within a period of 20 to 35 seconds, or one hair is touched twice within the same time interval, action potentials travel across the trap. When the second trigger takes too long to occur after the first, the trap will not close, but a new time interval starts again. If a hair is deflected a third time then the trap will still close. The time between stimulus and response is about 0.5 s. It takes the trap less than 0.3 s to close and trap the insect.

The lobes of the leaf bulge upwards when the trap is open. They are convex in shape. No one is quite sure how the trap closes, but as Darwin noticed, the lobes rapidly change into a concave shape, bending downwards so the trap snaps shut. This happens too fast to be simply the result of water movement from the cells on the top of the

Figure 15.35 The leaves of the Venus fly trap, *Dionaea muscipula*, have a group of stiff, sensitive hairs in their centres. When these are touched, the leaves respond by closing, trapping whatever was crawling over them. Digestive juices are then secreted, and the soluble products absorbed into the leaf cells.

lobes to cells underneath. Instead, it is likely that the rapid change occurs as a result of a release of elastic tension in the cell walls.

However, the trap is not completely closed at this moment. To seal the trap, it requires ongoing activation of the trigger hairs by the trapped prey. Unless the prey is able to escape, it will further stimulate the inner surface of the lobes, thereby triggering further action potentials. This forces the edges of the lobes together, sealing the trap to form an external 'stomach' in which prey digestion occurs. Further deflections of the sensory hairs by the trapped insect stimulate the entry of calcium ions into gland cells. Here, calcium ions stimulate the exocytosis of vesicles containing digestive enzymes in a similar way to their role in synapses (page 341). The traps stay shut for up to a week for digestion to take place. Once the insect is digested, the cells on the upper surface of the midrib grow slowly so the leaf reopens and tension builds in the cell walls of the midrib so the trap is set again.

Venus fly traps have two adaptations to avoid closing unnecessarily and wasting energy. First, the stimulation of a single hair does not trigger closure. This prevents the traps closing when it rains or when a piece of debris falls into the trap. Second, the gaps between the stiff hairs that form the 'bars' of the trap allow very small insects to crawl out. The plant would waste energy digesting a very small 'meal'.

Chemical communication in plants

Chemicals known as **plant hormones** or plant growth regulators are responsible for most communication within plants. Unlike animal hormones, plant growth regulators are not produced in specialised cells within glands, but in a variety of tissues. They move in the plant either directly from cell to cell (by diffusion or active transport) or are carried in the phloem sap or xylem sap. Some may not move far from their site of synthesis and may have their effects on nearby cells.

Here we consider two types of plant growth regulator:

- **auxins**, which influence many aspects of growth including elongation growth which determines the overall length of roots and shoots
- **gibberellins**, which are involved in seed germination and controlling stem elongation.

Abscisic acid (ABA) is another plant hormone, which controls the response of plants to environmental stresses such as shortage of water (Chapter 14, page 323).

Plant hormones interact with receptors on the surface of cells or in the cytoplasm or nucleus. These receptors usually initiate a series of chemical or ionic signals that amplify and transmit the signal within the cell in much the same way that we saw in Chapters 4 and 14.

Auxins and elongation growth

Plants make several chemicals known as **auxins**, of which the principal one is **IAA** (indole 3-acetic acid, Figure 15.36). Here, we refer to this simply as 'auxin' in the singular. Auxin is synthesised in the growing tips (meristems) of shoots and roots, where the cells are dividing. It is transported back down the shoot, or up the root, by active transport from cell to cell, and also to a lesser extent in phloem sap.

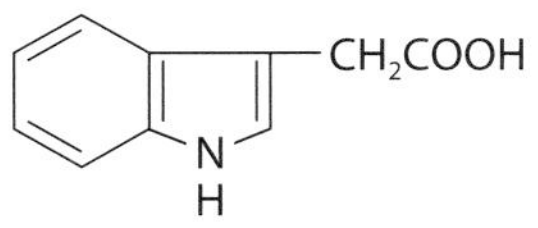

CH_2COOH (indole ring with N–H)

Figure 15.36 The molecular structure of indole 3-acetic acid, IAA.

Growth in plants occurs at meristems, such as those at shoot tips and root tips (Chapter 5). Growth occurs in three stages: cell division by mitosis, cell elongation by absorption of water, and cell differentiation. Auxin is involved in controlling growth by elongation (Figure 15.37).

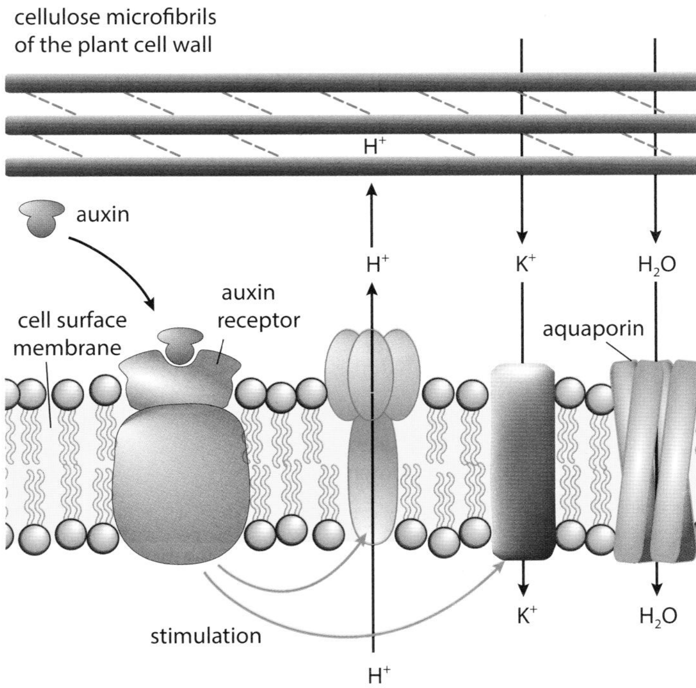

Figure 15.37 The binding of auxin to its receptor is thought to activate a membrane protein, which stimulates the pumping of protons out of the cell into the cell wall where they lower the pH and break bonds. Potassium ion channels are also stimulated to open leading to an increase in potassium ion concentration in the cytoplasm. This decreases the water potential so water enters through aquaporins.

Auxin stimulates cells to pump hydrogen ions (protons) into the cell wall. This acidifies the cell walls which leads to a loosening of the bonds between cellulose microfibrils and the matrix that surrounds them. The cells absorb water by osmosis and the pressure potential causes the wall to stretch so that these cells become longer, or elongate.

The details of this process are shown in Figure 15.37. Molecules of auxin bind to a receptor protein on the cell surface membrane. The binding of auxin stimulates ATPase proton pumps to move hydrogen ions across the cell surface membrane from the cytoplasm into the cell wall. In the cell walls are proteins known as expansins that are activated by the decrease in pH. The expansins loosen the linkages between cellulose microfibrils. It is not known exactly how they do this, but it is thought that expansins disrupt the non-covalent interactions between the cellulose microfibrils and surrounding substances, such as hemicelluloses, in the cell wall. This disruption occurs briefly so that microfibrils can move past each other allowing the cell to expand without losing much of the overall strength of the wall.

Gibberellins

Gibberellins are plant growth regulators that are synthesised in most parts of plants. They are present in especially high concentrations in young leaves and in seeds, and are also found in stems, where they have an important role in determining their growth.

Gibberellins and stem elongation

The height of some plants is partly controlled by their genes. For example, tallness in pea plants is affected by a gene with two alleles; if the dominant allele, **Le**, is present, the plants can grow tall, but plants homozygous for the recessive allele, **le**, always remain short. The dominant allele of this gene regulates the synthesis of the last enzyme in a pathway that produces an active form of gibberellin, GA_1. Active gibberellin stimulates cell division and cell elongation in the stem, so causing the plant to grow tall. A substitution mutation in this gene gives rise to a change from alanine to threonine in the primary structure of the enzyme near its active site, producing a non-functional enzyme. This mutation has given rise to the recessive allele, **le**. Homozygous plants, **lele**, are genetically dwarf as they do not have the active form of gibberellin. Applying active gibberellin to plants which would normally remain short, such as cabbages, can stimulate them to grow tall.

Gibberellins and seed germination

Gibberellins are involved in the control of germination of seeds, such as those of wheat and barley. Figure 15.38 shows the structure of a barley seed. When the seed is shed from the parent plant, it is in a state of **dormancy**; that is, it contains very little water and is metabolically inactive. This is useful because it allows the seed to survive in adverse conditions, such as through a cold winter, only germinating when the temperature rises in spring.

The seed contains an embryo, which will grow to form the new plant when the seed germinates. The embryo is surrounded by **endosperm**, which is an energy store containing the polysaccharide starch. On the outer edge of the endosperm is a protein-rich **aleurone layer**. The whole seed is covered by a tough, waterproof, protective layer (Figure 15.38).

The absorption of water at the beginning of germination stimulates the embryo to produce gibberellins. These gibberellins diffuse to the aleurone layer and stimulate the cells to synthesise amylase. The amylase mobilises energy reserves by hydrolysing starch molecules in the endosperm, converting them to soluble maltose molecules. These maltose molecules are converted to glucose and transported to the embryo, providing a source of carbohydrate that can be respired to provide energy as the embryo begins to grow.

Gibberellins cause these effects by regulating genes that are involved in the synthesis of amylase. In barley seeds, it has been shown that application of gibberellin causes an increase in the transcription of mRNA coding for amylase. It has this action by promoting the destruction of DELLA proteins that inhibit factors that promote transcription (Chapter 16 page 391). (DELLA stands for the first five amino acids in the primary sequence of these proteins.)

QUESTION

15.12 a i Explain the advantages to plants of having fast responses to stimuli.

ii The closure of a leaf of the Venus fly trap is an example of the all-or-nothing law. Explain why.

iii Suggest the advantage to Venus fly traps of digesting insects.

b Outline how auxin stimulates elongation growth.

c i What are the phenotypes of plants with the genotypes **LeLe**, **Lele** and **lele**?

ii When gibberellins are applied to dwarf pea plants, the plants grow in height. Explain what this tells us about dwarfness in pea plants.

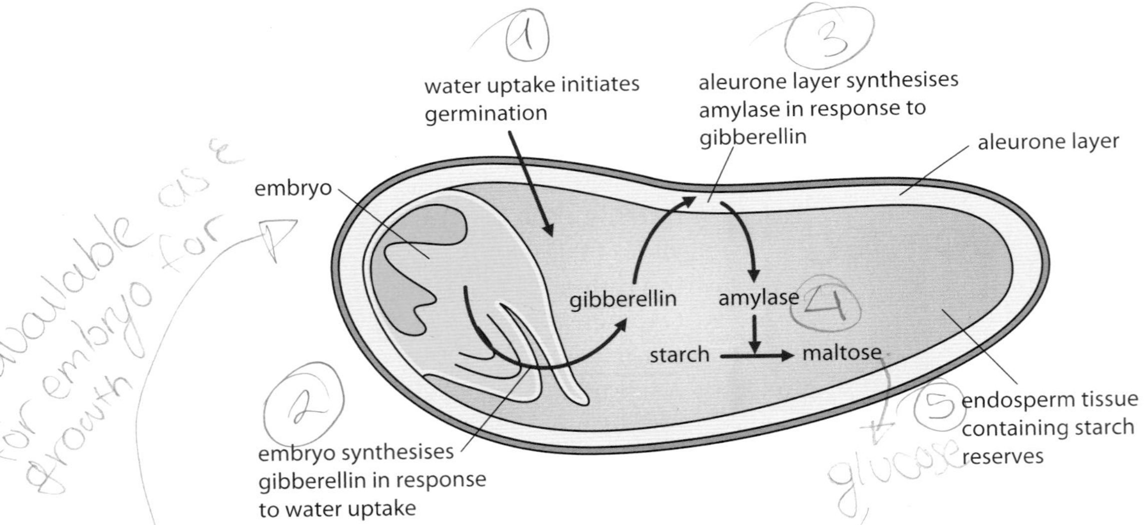

Figure 15.38 Longitudinal section through a barley seed, showing how secretion of gibberellins by the embryo results in the mobilisation of starch reserves during germination.

Summary

- Animals and plants have internal communication systems that allow information to pass between different parts of their bodies, and so help them to respond to changes in their external and internal environments.
- Neurones are cells adapted for the rapid transmission of electrical impulses; to do this, they have long thin processes called axons. Sensory neurones transmit impulses from receptors to the central nervous system (brain and spinal cord). Motor neurones transmit impulses from the central nervous system to effectors. Relay neurones transmit impulses within the central nervous system. Sensory, relay and motor neurones are found in series in reflex arcs that control fast, automatic responses to stimuli.
- Receptors are either specialised cells or the endings of sensory neurones; they act as transducers converting the energy of stimuli into electrical impulses. There are receptors that detect external stimuli, such as light and sound, and also receptors that detect internal changes, such as changes in blood pressure and carbon dioxide concentration of the blood.
- Neurones have a resting potential, which is a potential difference across their membranes, with the inside having a negative potential compared with the outside; this potential difference is about −70 mV. An action potential is a rapid reversal of this potential, caused by changes in permeability of the cell surface membrane to potassium and sodium ions. Action potentials are always the same size. Information about the strength of a stimulus is given by the frequency of action potentials produced. Action potentials are propagated along axons by local circuits that depolarise regions of membrane ahead of the action potential. This depolarisation stimulates sodium ion voltage-gated channels to open, so that the permeability to sodium increases and the action potential occurs further down the axon.
- Axons are repolarised by the opening of potassium ion voltage-gated channels that allow potassium ions to diffuse out of the axon. After a short refractory period when the sodium channels cannot open, the membrane is able to respond again. Refractory periods determine the maximum frequency of impulses. In vertebrates, the axons of many neurones are insulated by a myelin sheath, which speeds up the transmission of impulses.
- Action potentials may be initiated within the brain or at a receptor. Environmental changes result in permeability changes in the membranes of receptor cells, which in turn produce changes in potential difference across the membrane. If the potential difference is sufficiently great and above the threshold for the receptor cell, this will trigger an action potential in a sensory neurone.
- A synapse is a junction between two neurones or between a motor neurone and a muscle cell. At cholinergic synapses, a transmitter substance, acetylcholine, is released when action potentials arrive. Impulses pass in one direction only, because transmitter substances are released by exocytosis by the presynaptic neurone to bind to receptor proteins that are only found on the postsynaptic neurone.
- At least several hundred neurones are likely to have synapses on the dendrites and cell body of each neurone within the central nervous system. The presence of many synapses on the cell body of a neurone allows integration within the nervous system, resulting in complex and variable patterns of behaviour, and in learning and memory.
- Striated (skeletal) muscle is made of many multinucleate cells called muscle fibres, which contain many myofibrils. Myofibrils contain regularly arranged thick (myosin) and thin (actin) filaments which produce the striations seen in the muscle. Thick filaments are made of myosin molecules that have a globular head and fibrous tail; the head is an ATPase. Thin filaments are composed of actin, troponin and tropomyosin. Each myofibril is divided into sarcomeres by Z discs; the thin filaments are attached to the Z discs and the thick filaments can slide in between the thin filaments.
- The arrival of an action potential at a neuromuscular junction causes the release of acetylcholine which diffuses across the synaptic gap and binds to receptor proteins on the sarcolemma. The binding of acetylcholine causes the opening of sodium ion channels so the sarcolemma is depolarised by the entry of sodium ions. An action potential passes across the sarcolemma and down T-tubules where it leads to sarcoplasmic reticulum becoming permeable to calcium ions that are stored within it. Calcium ions diffuse out of the sarcoplasmic reticulum to bind to troponin causing tropomyosin to move so exposing binding sites on the actin in the thin filaments.

- Myosin heads bind to thin filaments to form cross-bridges; the myosin heads then tilt pulling the thin filaments together so that each sarcomere decreases in length as the filaments slide over each other. The myosin ATPase then hydrolyses ATP, providing the energy for myosin heads to detach from the thin filament and flip back ready to bind with actin again. This process is repeated many times during a contraction, but it can only be reversed by the relaxation of the muscle (with no cross-bridges) and the contraction of antagonist muscle that pulls the filaments further away lengthening each sarcomere.
- Muscles rely on energy from the small quantity of ATP in the sarcoplasm, small stores of creatine phosphate, aerobic respiration in mitochondria and, if oxygen is in short supply, lactate fermentation (anaerobic respiration) in the sarcoplasm.
- Hormones are chemicals that are made in endocrine glands and transported in blood plasma to their target cells, where they bind to specific receptors and so affect the behaviour of the cells.
- The menstrual cycle is controlled by secretions of FSH and LH from the anterior pituitary gland, and oestrogen and progesterone from the ovaries. In the first week of the cycle, while menstruation is taking place, LH and FSH are secreted, and these cause the secretion of oestrogen, which causes the endometrium to thicken. Negative feedback reduces the secretion of LH and FSH. Halfway through the cycle, a surge of LH causes ovulation to take place. After ovulation, the empty follicle becomes a corpus luteum and secretes progesterone, which maintains the endometrium. At the end of the cycle, the corpus luteum degenerates; the reduction in progesterone and oestrogen causes menstruation.
- Contraceptive pills contain oestrogen and/or progesterone; these hormones suppress the secretion of FSH and LH from the anterior pituitary gland to prevent ovulation.
- Plants make limited use of action potentials, but they are used to coordinate fast responses; for example, the closing of Venus fly traps.
- Plants produce several chemicals known as plant growth substances that are involved in the control of growth and responses to environmental changes. Auxin is synthesised mainly in growing tips of shoots and roots, and stimulates cells to pump protons into the cell wall to lower the pH. Proteins in the cell wall known as expansins respond to a low pH by loosening the links between cellulose microfibrils and the matrix of the cell wall so allowing microfibrils to slide apart. Plant cells absorb water by osmosis and the internal pressure potential causes the walls to stretch and the cells to elongate.
- Gibberellin is synthesised in young leaves and in seeds. It stimulates growth of stems and germination of seeds such as those of wheat and barley. In germinating seeds, gibberellins activate the synthesis of amylase enzymes for the breakdown and mobilisation of starch.

End-of-chapter questions

1 Which of the following is an incorrect statement about the endocrine system?

A All hormones bind to receptors on the cell surface of their target cells.

B Endocrine glands are ductless.

C Endocrine glands secrete hormones into the blood.

D Hormones are transported in the blood plasma. [1]

2 Which statement describes the biological action of the oestrogen/progesterone contraceptive pill?

A Increased concentrations of the hormones increase the thickness of the endometrium of the uterus.

B Increased concentrations of the hormones prevent the shedding of the endometrium of the uterus.

C Increased concentrations of the hormones stimulate a surge in the secretion of LH from the anterior pituitary gland.

D Increased concentrations of the hormones suppress the secretion of FSH and LH from the anterior pituitary gland. [1]

3 Which of the following is responsible for saltatory conduction in myelinated neurones?

A axon membranes
B nodes of Ranvier
C Schwann cells
D voltage-gated channel proteins [1]

4 Which of the following correctly identifies the effects of the three plant hormones, abscisic acid (ABA), auxin and gibberellin?

	Abscisic acid	Auxin	Gibberellin
A	elongation growth of roots and shoots	stomatal closure	stem elongation
B	stimulates synthesis of amylase in seed germination	stem elongation	acidification of cell walls
C	stomatal closure	acidification of cell walls	stimulates synthesis of amylase in seed germination
D	stem elongation	elongation growth of roots and shoots	stomatal closure

[1]

5 Which statements about the concentrations of hormones in the human menstrual cycle are correct?

1 Shortly before ovulation, the concentration of oestrogen is high and concentration of progesterone low.
2 During the last quarter of the cycle, the concentrations of oestrogen and progesterone fall.
3 At the end of menstruation, the concentration of oestrogen is low but rising, and the concentration of progesterone is low.
4 Just before ovulation, the concentrations of LH and FSH suddenly rise.

A 1, 2, 3 and 4 **B** 1, 2 and 4 only **C** 2 and 3 only **D** 3 and 4 only [1]

6 The figure shows the changes in potential difference across the membrane of a neurone over a period of time. The membrane was stimulated at time **A** and time **B** with stimuli of different intensities.

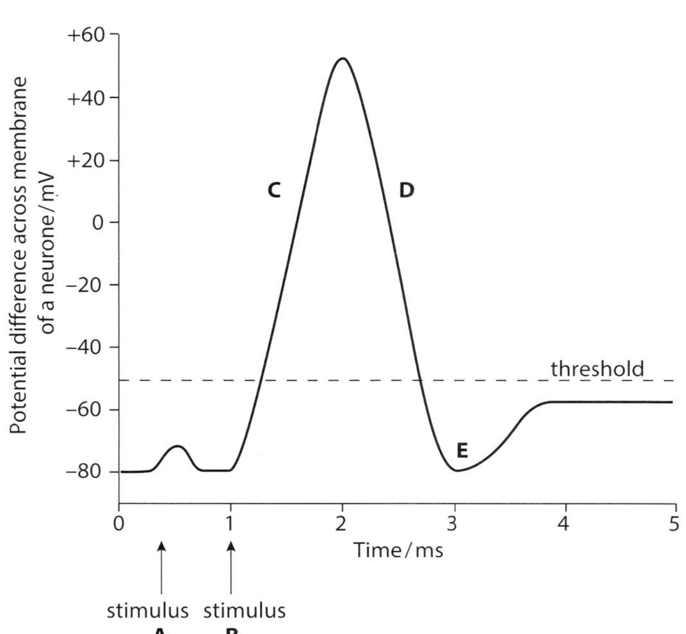

a Stimulus **B** resulted in an action potential. Describe what is occurring at **C**, **D** and **E**. [6]

b Suggest why stimulus **A** did not result in an action potential being produced whereas stimulus **B** did. [2]

[Total: 8]

Cambridge International AS and A Level Biology 9700/04, Question 8, October/November 2007

7 The electron micrograph shows parts of some myofibrils in a striated muscle that is in a relaxed state.

a i Name the parts labelled **K**, **L** and **M**. [3]

ii How many myofibrils are visible in the electron micrograph? Explain your answer. [2]

b i There are many glycogen granules and mitochondria visible in the electron micrograph. Explain why they are both there. [2]

ii Describe how you can tell that this electron micrograph is from relaxed muscle and not contracted muscle. [3]

c The electron micrograph is magnified 20 000 times. Calculate the actual length of the sarcomere which includes the region labelled **K**. Give your answer in micrometres (μm). [2]

[Total: 12]

8 The diagrams show a sarcomere in different states of contraction.

A

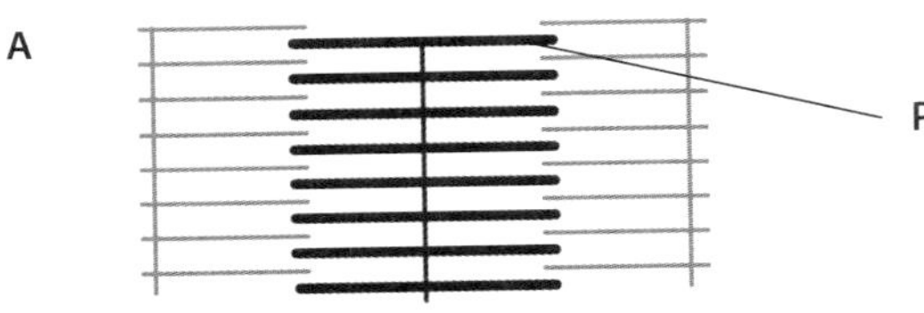

C

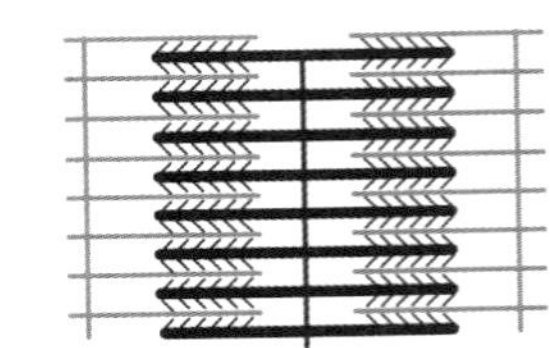

B

Q

R

D

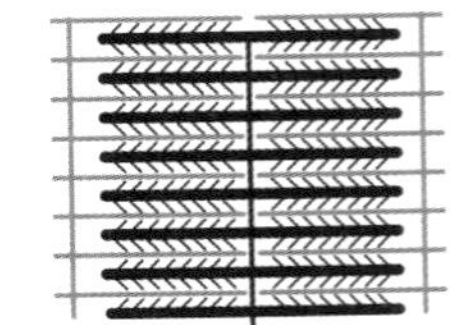

a Name the parts labelled **P**, **Q** and **R**. [3]

b Explain why there are no actin–myosin cross-bridges visible in diagram **A**. [2]

c Muscle fibres are able to contract with more force in some states of contraction than others. Suggest which of the diagrams shows the state that can develop the greatest force, and explain the reasons for your answer. [4]

d Explain why the muscle shown in diagram **D** would not be able to contract any further. [1]

e A muscle can contract with force, but it cannot pull itself back to its original relaxed length.

i With reference to the mechanism of muscle contraction, explain why this is so. [2]

ii Suggest how the muscle in diagram **D** could be returned to the state shown in diagram **A**. [2]

[Total: 14]

9 A biopsy was taken from a leg muscle of a healthy racehorse. The muscle fibres were teased apart and cross-sections were taken from one of the muscle fibres. These cross-sections were examined with a transmission electron microscope. The figure shows drawings made from three different cross-sections of a myofibril from the muscle fibre.

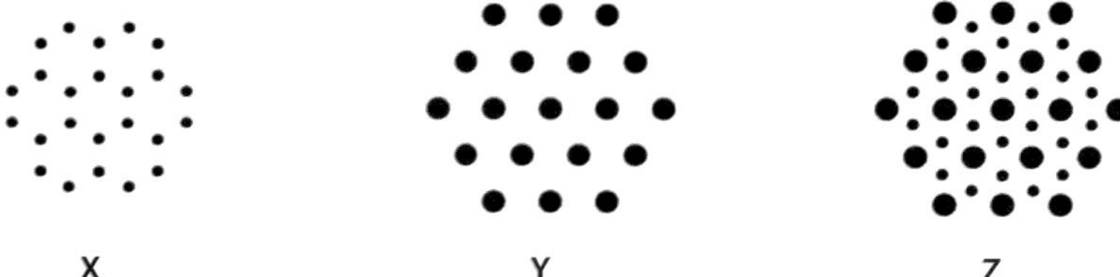

a Explain the differences between the sections **X**, **Y** and **Z**. You may draw a labelled diagram to illustrate your answer. [4]

b The sections were taken from a relaxed muscle fibre. Suggest how the sections would appear if taken from a fibre that had contracted to its maximum extent. Explain your answer. [3]

c Muscle weakness in racehorses may sometimes be related to a deficiency of calcium.
Outline the roles of calcium ions in the coordination of muscle contraction. [6]

[Total: 13]

10 Copy and complete the table to show, for each hormone, the precise site of its secretion, and its effects on the ovary or on the endometrium of the uterus.

Hormone	Site of secretion	Effect(s) of hormone	
		ovary	endometrium
FSH	A.P.G	follicle production	none
LH	A.P.G	ovulation	none
oestrogen		none	thickens
progesterone		none	keeps it thick

[8]

11 The Pacinian corpuscle is a type of receptor found in the dermis of the skin. Pacinian corpuscles contain an ending of a sensory neurone, surrounded by several layers of connective tissue called a capsule. The activity of a Pacinian corpuscle was investigated by inserting microelectrodes into the axon at the positions shown in the diagram below.

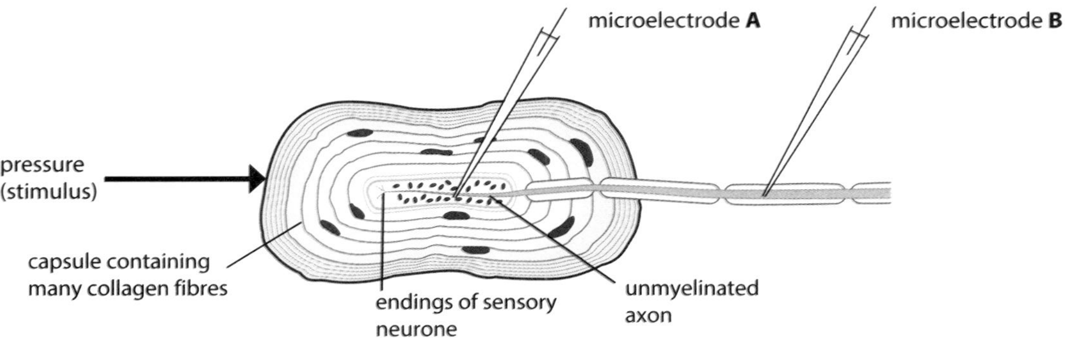

Pressure was applied to the Pacinian corpuscle and recordings made of the electrical activity in the axon at microelectrodes **A** and **B**. The results are shown in the diagram below.

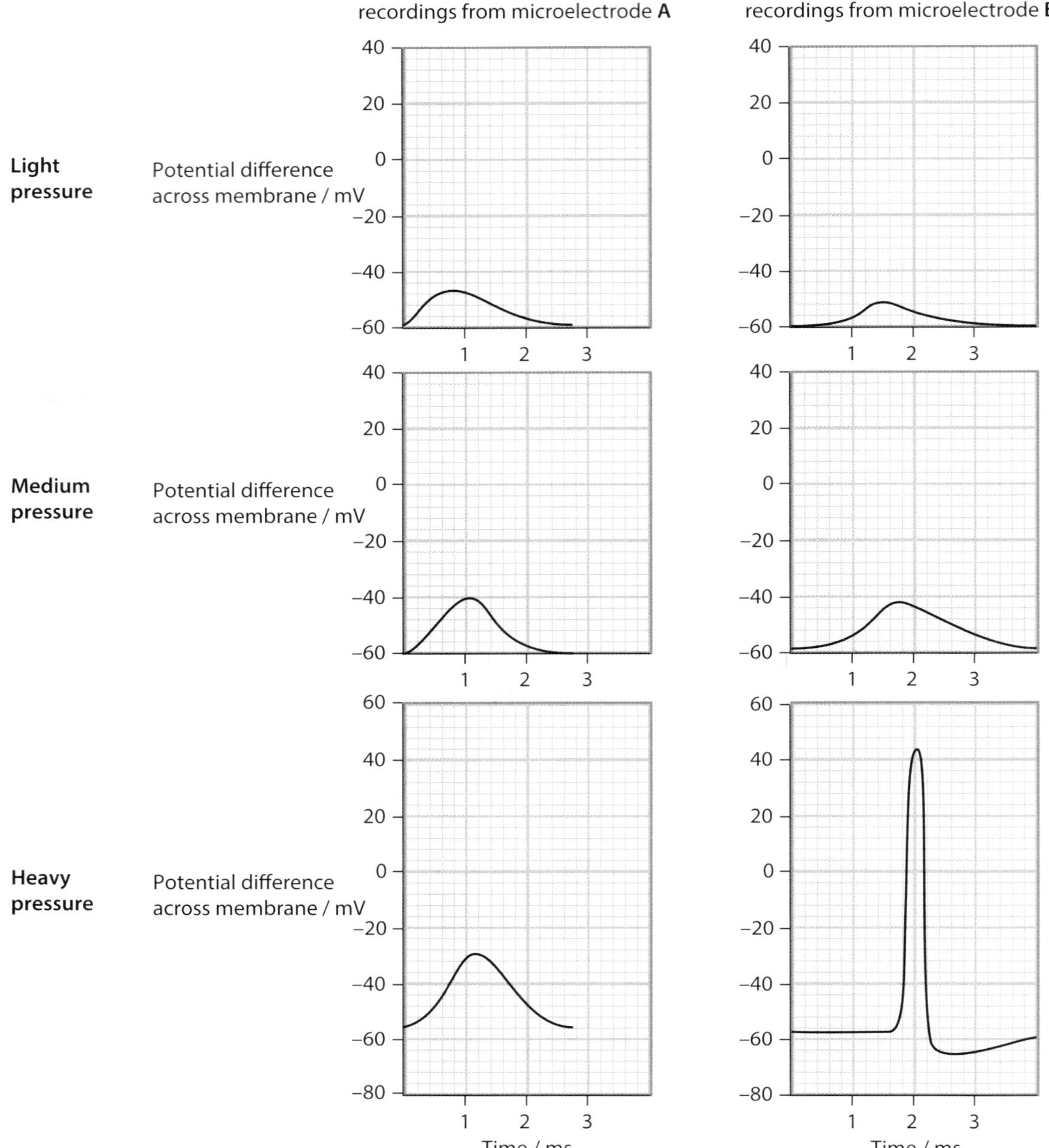

a Suggest what happened in the unmyelinated region of the axon as pressure was applied to the Pacinian corpuscle. [4]

b Explain the pattern of recordings from microelectrode **B** as the pressure applied to the corpuscle was increased. [4]

c Explain why sensory neurones from Pacinian corpuscles are myelinated and not unmyelinated. [3]

[Total: 11]

12 Gibberellin is a plant growth regulator.

a Outline the role of gibberellin in the germination of seeds such as those of wheat and barley. [5]

In an investigation of the effects of gibberellin, plants of short-stemmed and long-stemmed varieties of five cultivated species were grown from seed. The young plants of each species were divided into two groups. One group of plants was sprayed with a solution of gibberellin each day. A control group was sprayed with the same volume of water. After eight weeks, the stem length of each plant was measured and means calculated for each group of plants. A statistical test was carried out to determine whether the difference between the treatments for each species was significant.

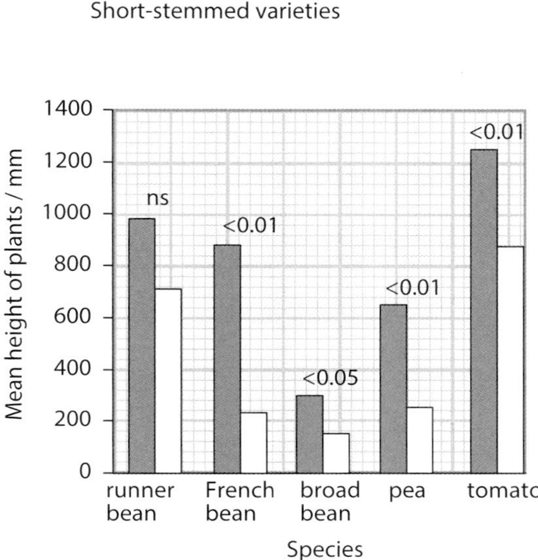

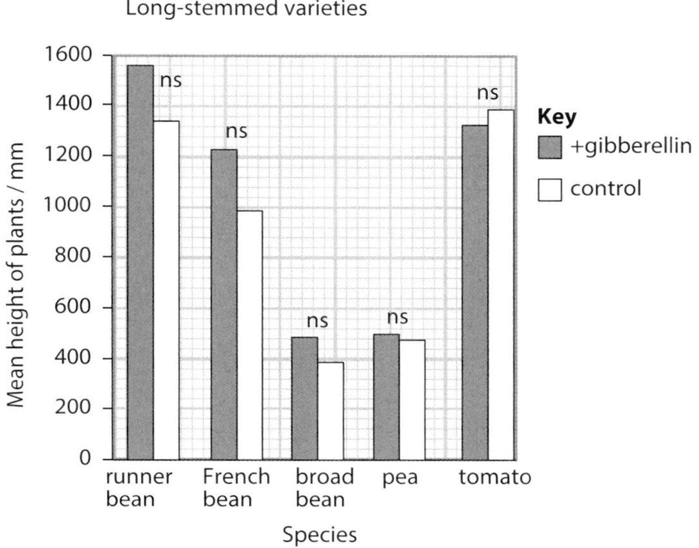

The results are shown in the figure. The *p* value for each species is given.

b Using the information in the figure, describe the effect of adding gibberellin solution to the two varieties of the five species. [5]

c Explain why the short-stemmed variety of pea showed a more significant growth in height when treated with gibberellin than the long-stemmed variety. [3]

d Suggest the advantages of cultivating crops of short-stemmed varieties of peas and beans rather than long-stemmed varieties. [3]

[Total: 16]

13 a Explain how a nerve impulse is transmitted along a motor neurone. [9]

b Describe how an impulse crosses a synapse. [6]

[Total: 15]

14 a Describe a reflex arc **and** explain why such reflex arcs are important. [7]

b Describe the structure of a myelin sheath **and** explain its role in the speed of transmission of a nerve impulse. [8]

[Total: 15]

Cambridge International AS and A Level Biology Paper 41, Question 10, October/November 2009

Chapter 16: Inherited change

Learning outcomes

You should be able to:

- explain what is meant by homologous pairs of chromosomes
- outline the role of meiosis in sexual reproduction
- describe the behaviour of homologous chromosomes during meiosis and explain how this leads to genetic variation
- use genetic diagrams, in both monohybrid and dihybrid crosses, to solve problems involving codominance, linkage, multiple alleles or gene interaction
- use the chi-squared (χ^2) test to assess the significance of differences between observed and expected results
- explain how gene mutation occurs, and describe the effect of mutant alleles in some human conditions
- describe the control of gene expression

Katydids

Oblong-winged katydids are normally green and well camouflaged. Occasionally, a bright pink katydid occurs (Figure 16.1).

The pink colour is caused by an allele (variety) of the gene that determines body colour. It has always been thought that the allele causing the pink colour was recessive, but breeding experiments in 2013 showed – surprisingly – that this is a dominant allele. A katydid that has one allele for pink colour and one for green colour is pink. As you read this chapter, think about how breeding experiments with katydids could have determined whether the pink colour is caused by a dominant allele or a recessive allele.

Figure 16.1 A rare pink katydid, *Amblycorypha oblongifolia*.

You have already learned that the nuclei of eukaryotic cells contain chromosomes and that the number of chromosomes is characteristic of the species. Look back to Chapter 5 to remind yourself about the structure of chromosomes.

Homologous chromosomes

Figure 16.2 shows the chromosomes from Figure 5.2 (page 95) rearranged into numbered pairs, and Figure 16.3 is a diagram of the chromosomes.

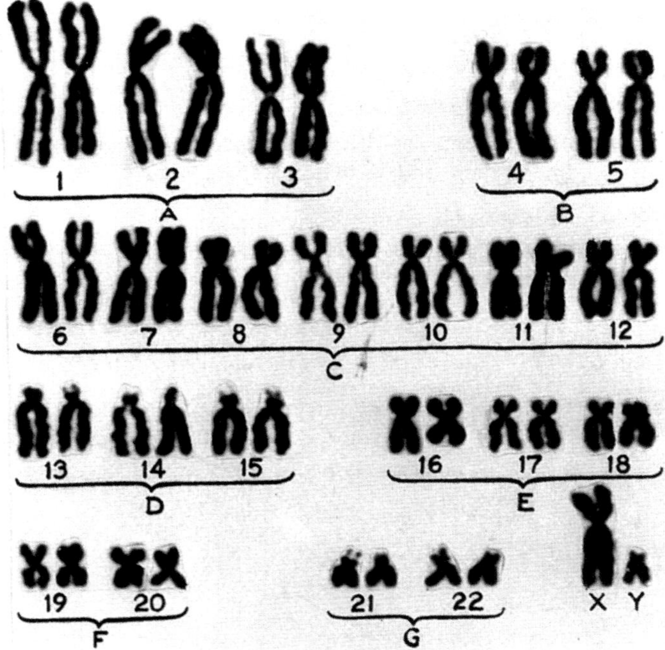

Figure 16.2 Karyogram of a human male. There are 22 homologous pairs of non-sex chromosomes (autosomes). The sex chromosomes (X, female; Y, male) are placed separately.

A photograph such as Figure 16.2 is called a karyogram. Karyograms are prepared by cutting out individual chromosomes from a picture like Figure 5.2 and rearranging them.

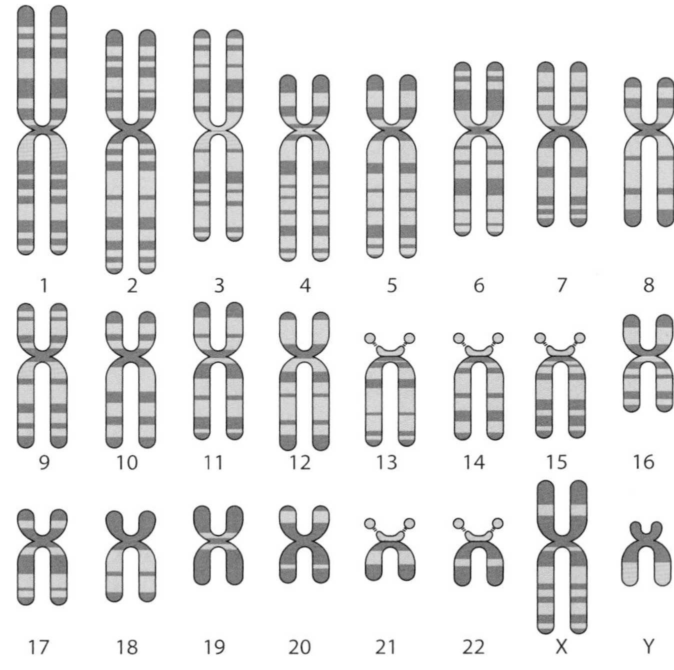

Figure 16.3 Diagram showing banding patterns of human chromosomes when stained. Green areas represent those regions that stain with ultraviolet fluorescence staining; orange areas are variable bands. Note that the number of genes is greater than the number of stained bands. Only one chromosome of each pair is shown, except for the sex chromosomes, which are both shown.

QUESTION

16.1 Look at Figures 16.2 and 16.3 and try to decide why the numbered chromosomes are arranged in the particular order shown.

Note the following in Figures 16.2 and 16.3.

- There are 22 matching pairs of chromosomes. These are called **homologous chromosomes**. (The word 'homologous' means similar in structure and composition.) Each pair is given a number. In the original zygote, one of each pair came from the mother, and one from the father. There is also a non-matching pair labelled X and Y. There are, therefore, two sets of 23 chromosomes – one set of 23 from the father and one set of 23 from the mother.
- The non-matching X and Y chromosomes are the **sex chromosomes**, which determine the sex of the individual. All the other chromosomes are called **autosomes**. It is conventional to position the two sex chromosomes to one side in a karyogram, so that the sex of the organism can be recognised quickly. In humans, females have two X chromosomes, and males have one X and one Y chromosome. The Y chromosome has portions missing and is therefore smaller than the X chromosome.
- The pairs of chromosomes can be distinguished not only by size and shape, but because each pair has a distinctive banding pattern when stained with certain stains, as shown in Figure 16.3.

Each chromosome has a characteristic set of **genes** which code for different features. Scientists are gradually identifying which genes are located on which chromosomes and what their precise functions are. For example, we now know that the gene for the genetic disorder cystic fibrosis is located on chromosome 7.

The gene for a particular characteristic is always found at the same position, or **locus** (plural: **loci**), on a chromosome. Figure 16.4 shows a map of some of the genes on the human female sex chromosome (X), which, if faulty, are involved in genetic diseases.

Each chromosome typically has several hundred to several thousand gene loci, many more than shown in Figure 16.4. The total number of different genes in humans is thought to be about 20 000–25 000.

Each member of a homologous pair possesses genes controlling the same characteristics (Figure 16.5). A gene for a given characteristic may exist in different forms (**alleles**) which are expressed differently. For example, the gene for eye colour has two forms, or alleles, one coding for blue eyes and one for brown eyes. An individual could possess both alleles, one on the maternal chromosome and the other on the paternal chromosome.

X chromosome

Normal expression	Disease or genetic disorder caused by faulty expression
enables kidneys to retain phosphate	a form of rickets known as hypophosphataemic rickets
controls production of a membrane protein found in muscle fibres	muscular dystrophy
controls production of cytochrome b in white blood cells	white blood cells unable to kill bacteria, leading to recurrent infections and death in childhood
controls production of testosterone receptor in fetus	interrupted development of testes, leading to partial physical feminisation of genetic males (testicular feminisation or androgen insensitivity syndrome)
centromere – no known genes	
controls production of factor IX protein, which is needed for blood clotting	haemophilia B
FMR1 (fragile-X mental retardation 1) gene makes a protein needed for normal brain development	a form of mental retardation known as fragile-X syndrome (FXS), the most common form of learning difficulties in boys
normal colour vision – produces pigment in retina	red–green colour-blindness
controls production of factor VIII protein, which is needed for blood clotting	haemophilia A

Figure 16.4 Locations of some of the genes on the human female sex chromosome (the X chromosome), showing the effects of normal and faulty expression. The chromosome has 153 million base pairs.

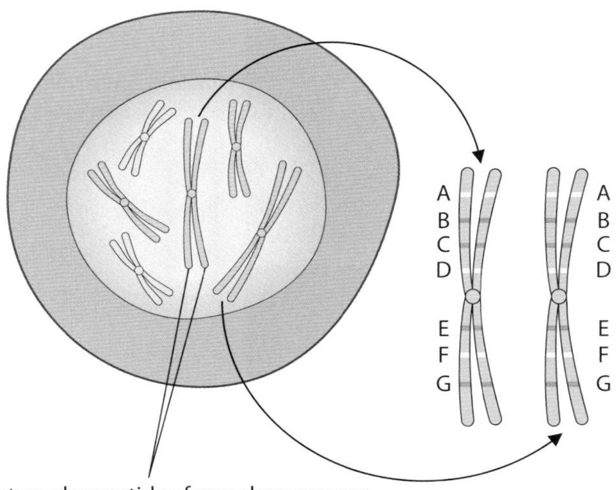

Figure 16.5 Homologous chromosomes carry the same genes at the same loci. Just seven genes, labelled A–G, are shown on this pair of chromosomes, but in reality there are often hundreds or thousands of genes on each chromosome. This is a diploid cell as there are two complete sets of chromosomes ($2n = 6$).

Homologous chromosomes are a pair of chromosomes in a diploid cell that have the same structure as each other, with the same genes (but not necessarily the same alleles of those genes) at the same loci, and that pair together to form a bivalent during the first division of meiosis.

A **gene** is a length of DNA that codes for a particular protein or polypeptide.

An **allele** is a particular variety of a gene.

A **locus** is the position at which a particular gene is found on a particular chromosome; the same gene is always found at the same locus.

Haploid and diploid cells

When animals other than humans are examined, we again find that cells usually contain two sets of chromosomes. Such cells are described as diploid. This is represented as $2n$, where n = the number of chromosomes in one set of chromosomes.

Not all cells are diploid. As we shall see, gametes have only one set of chromosomes. A cell which contains only one set of chromosomes is described as haploid. This is represented as n. In humans, therefore, normal body cells are diploid ($2n$), with 46 chromosomes, and gametes are haploid (n), with 23 chromosomes.

Two types of nuclear division

Figure 16.6 shows a brief summary of the life cycle of an animal such as a human. Two main stages are involved.

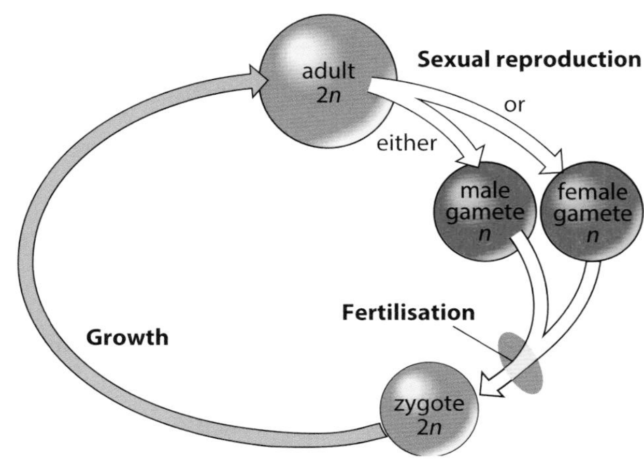

Figure 16.6 Outline of the life cycle of an animal.

- **Growth**. When a diploid zygote, which is one cell, grows into an adult with millions of cells, the new cells must be genetically identical, with the same number of chromosomes as the cells that divided to produce them. The type of nuclear division that achieves this is called mitosis.
- **Sexual reproduction**. For the life cycle to contain sexual reproduction, there must be a point before fertilisation takes place when the number of chromosomes is halved. This is explained in Figure 16.6. This results in the gametes containing only one set of chromosomes, rather than two sets. If there were no point in the life cycle when the number of chromosomes halved, then the number of chromosomes would double every generation, as shown in Figure 16.7a. The type of nuclear division that halves the chromosome number is called meiosis. Gametes are always haploid as a result of meiosis. Meiosis is sometimes described as a **reduction division**, because the number of chromosomes is reduced.

As you will see in this chapter, meiosis does more than halve the number of chromosomes in a cell. Meiosis also introduces genetic variation into the gametes and therefore the zygotes that are produced. Genetic variation may also arise as a result of **mutation**, which can occur at any stage in a life cycle. Such variation is the raw material on which natural selection has worked to produce the huge range of species that live on Earth (Chapter 17.)

a If chromosome number is not halved, the number of chromosomes doubles every generation:

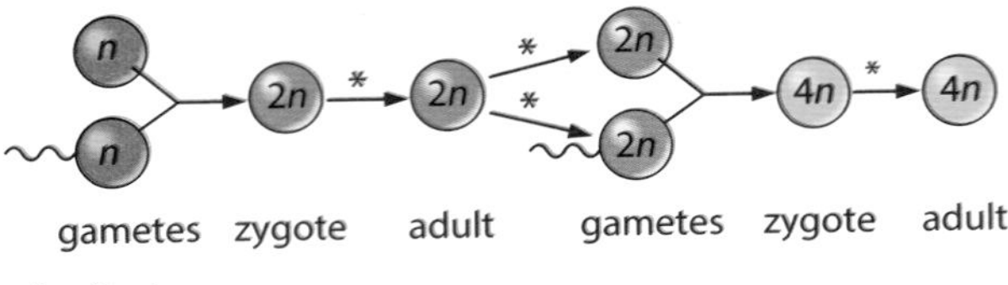

b If chromosome number is halved, the number of chromosomes stays the same every generation:

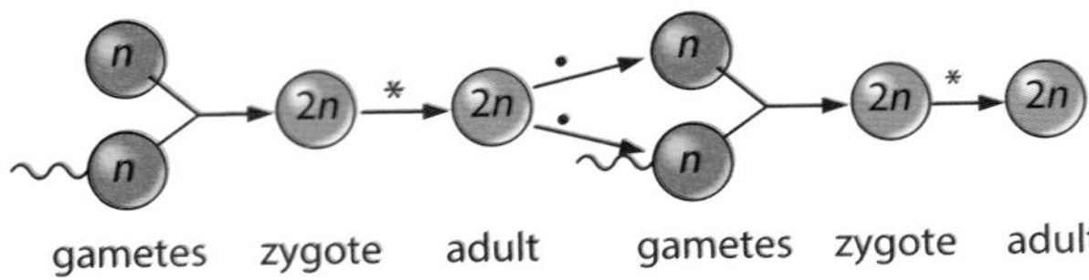

c

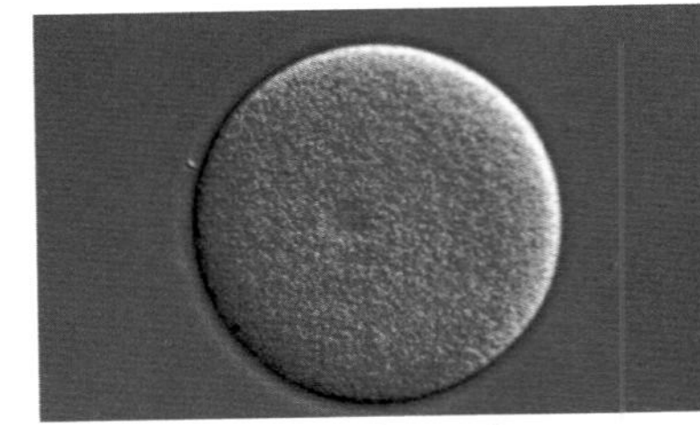

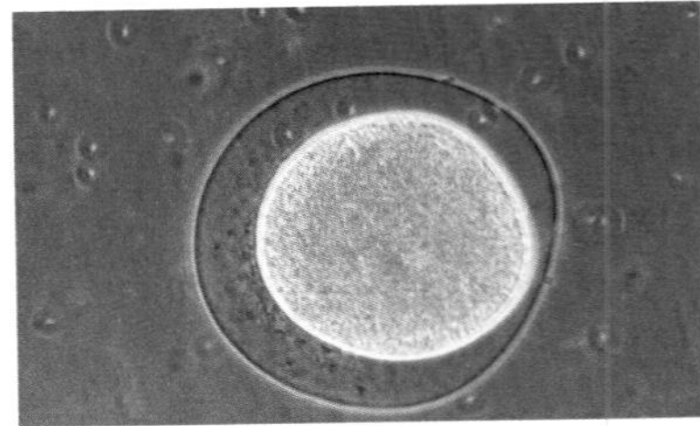

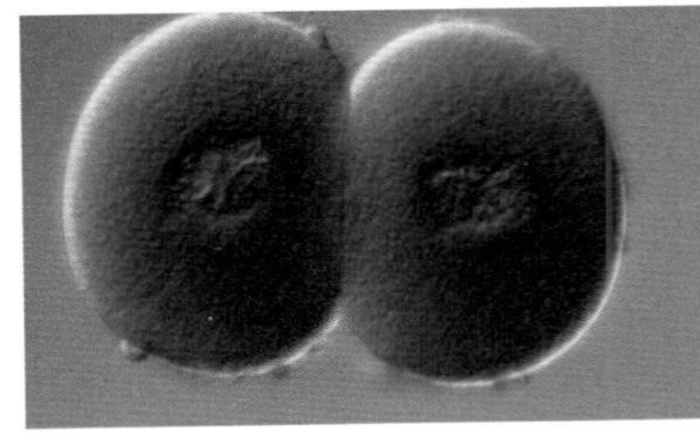

Figure 16.7 A life cycle in which the chromosome number is **a** not halved; **b** halved; **c** life cycle stages in a sea urchin .

A **diploid** cell is one that possesses two complete sets of chromosomes; the abbreviation for diploid is $2n$.

A **haploid** cell is one that possesses one complete set of chromosomes; the abbreviation for haploid is n.

Meiosis

The process of meiosis is best described by means of annotated diagrams (Figure 16.8). An animal cell is shown where $2n = 4$, and different colours represent maternal and paternal chromosomes. The associated behaviour of the nuclear envelope, cell surface membrane and centrosomes is also shown. Remember, each centrosome contains a pair of centrioles (Chapters 1 and 5).

Unlike mitosis (page 97), meiosis involves two divisions, called meiosis I and meiosis II. Meiosis I is a reduction division, resulting in two daughter nuclei with **half** the number of chromosomes of the parent nucleus. In meiosis II, the chromosomes behave as in mitosis, so that each of the two haploid daughter nuclei divides again. Meiosis therefore results in a total of four haploid nuclei. Note that it is the behaviour of the chromosomes in **meiosis I** that is particularly important and contrasts with mitosis.

Figure 16.8 summarises the process of meiosis diagrammatically. Figure 16.9 shows photographs of the process as seen with a light microscope.

Two of the events that take place during meiosis help to produce genetic variation between the daughter cells that are produced. These are independent assortment of the homologous chromosomes, and crossing over, which happens between the chromatids of homologous chromosomes. When these genetically different gametes fuse, randomly, at fertilisation, yet more variation is produced amongst the offspring. In order to understand how these events produce variation, we first need to consider the **genes** that are carried on the chromosomes, and the way in which these are passed on from parents to offspring. This branch of biology is known as genetics.

Meiosis I

1 Early prophase I
– as mitosis early prophase

2 Middle prophase I
Homologous chromosomes pair up. This process is called synapsis. Each pair is called a bivalent.

centrosomes moving to opposite ends of nucleus, as in mitosis

3 Late prophase I

nuclear envelope breaks up as in mitosis

crossing over of chromatids may occur

nucleolus 'disappears' as in mitosis

Bivalent showing crossing over:

chromatids may break and may reconnect to another chromatid

centromere

chiasma = point where crossing over occurs (plural; chiasmata)

one or more chiasmata may form, anywhere along length

At the end of prophase I a spindle is formed.

4 Metaphase I (showing crossing over of long chromatids)

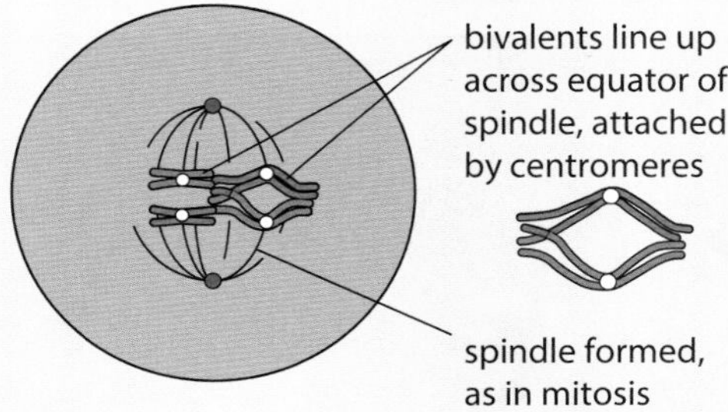

5 Anaphase I
Centromeres do not divide, unlike in mitosis.

Whole chromosomes move towards opposite ends of spindle, centromeres first, pulled by microtubules.

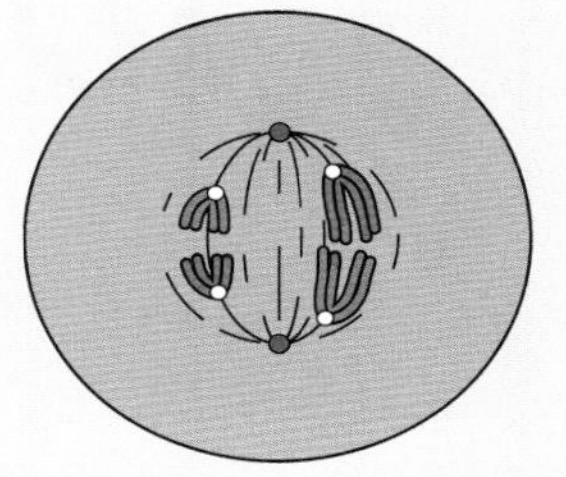

6 Telophase I

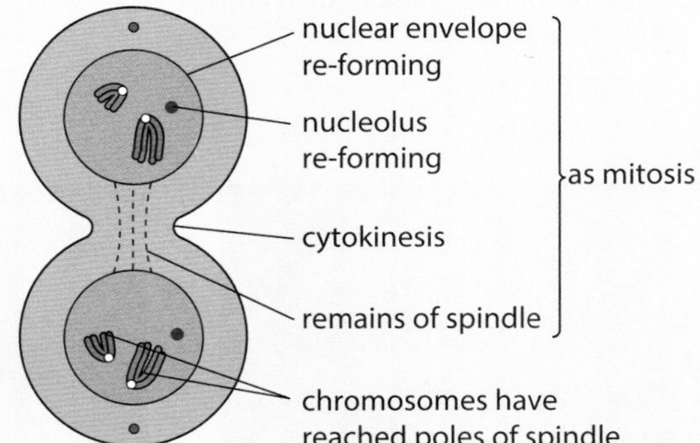

Animal cells usually divide before entering meiosis II. Many plant cells go straight into meiosis II with no reformation of nuclear envelopes or nucleoli. During meiosis II, chromatids separate as in mitosis.

Meiosis II

7 Prophase II

8 Metaphase II

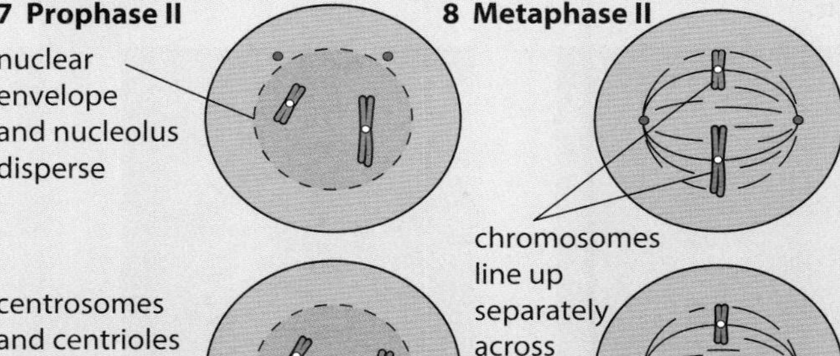

9 Anaphase II

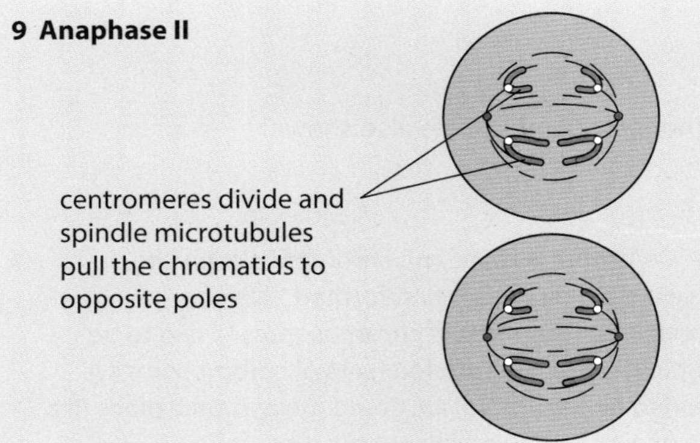

10 Telophase II

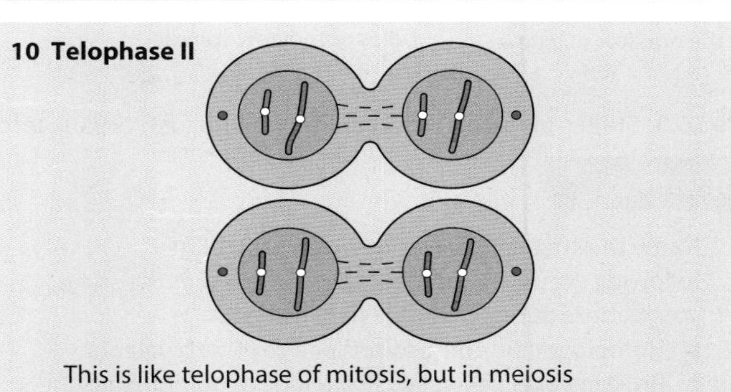

This is like telophase of mitosis, but in meiosis telophase II four haploid daughter cells are formed

Figure 16.8 Meiosis and cytokinesis in an animal cell. Compare this process with nuclear division by mitosis, shown in Figure 5.7 (page 98).

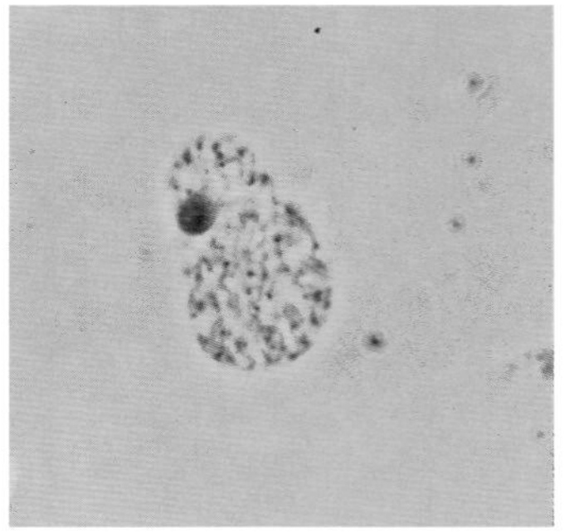

a interphase nucleus

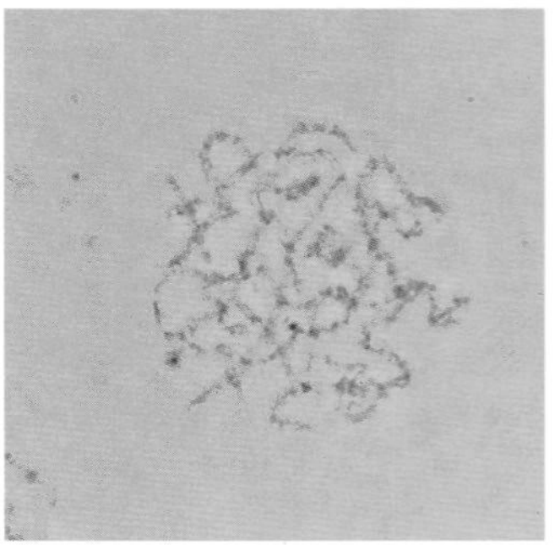

b meiosis I, early prophase I: chromosomes condensing and becoming visible

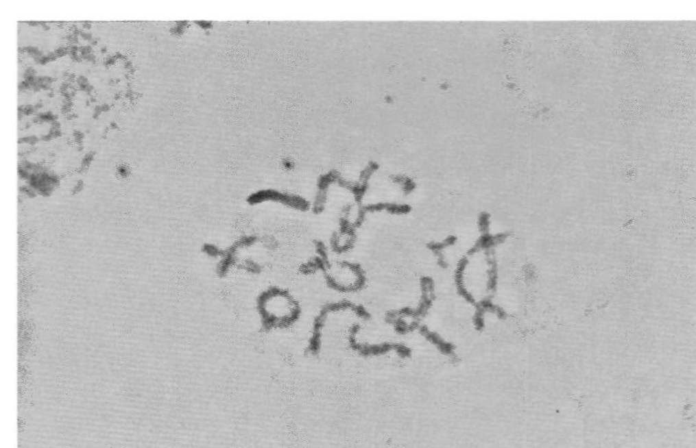

c prophase I: homologous chromosomes have paired up, forming bivalents, and crossing over of chromatids is occurring; members of each pair of chromosomes are repelling each other but are still held at the crossing-over points (chiasmata)

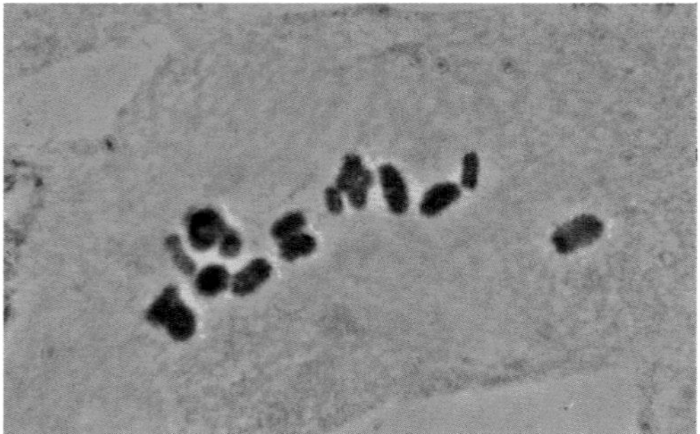

d metaphase I: bivalents line up across the equator of the spindle; the spindle is not visible in the photo; e anaphase I: homologous chromosomes move to opposite poles of the spindle

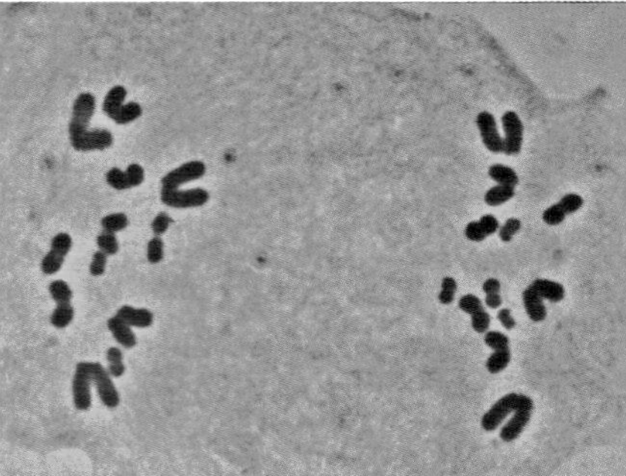

e anaphase I: homologous chromosomes move to opposite poles of the spindle

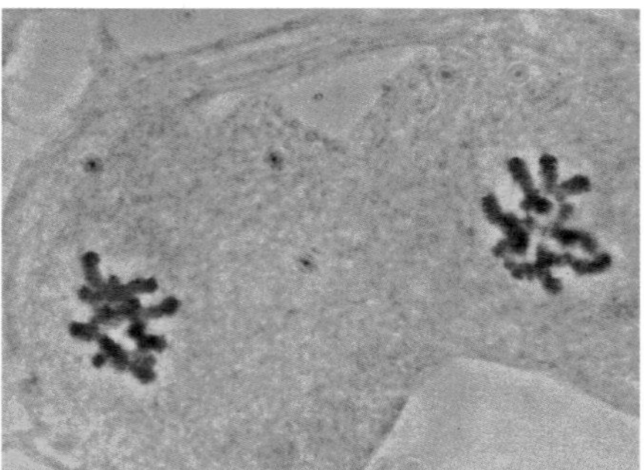

f telophase I and cytokinesis

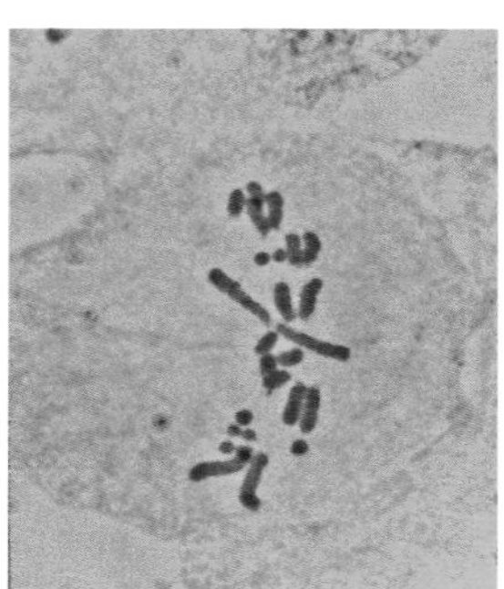

g meiosis II, metaphase II: single chromosomes line up across the equator of a new spindle

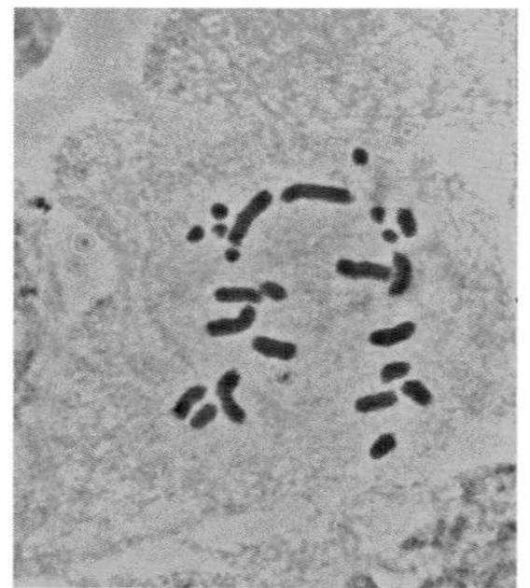

h anaphase II: chromatids separate and move to opposite poles of the new spindle

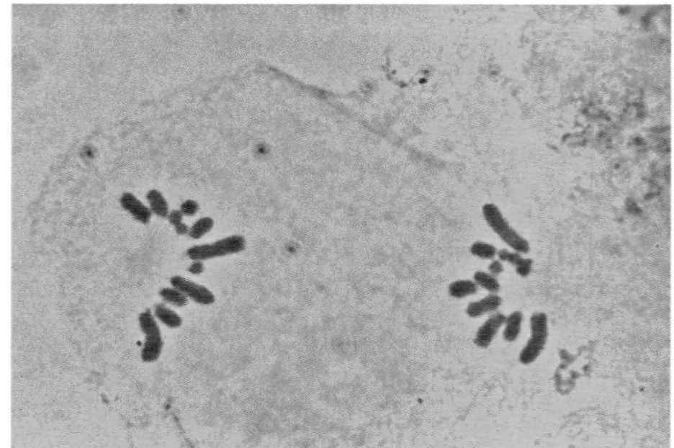

i late anaphase II

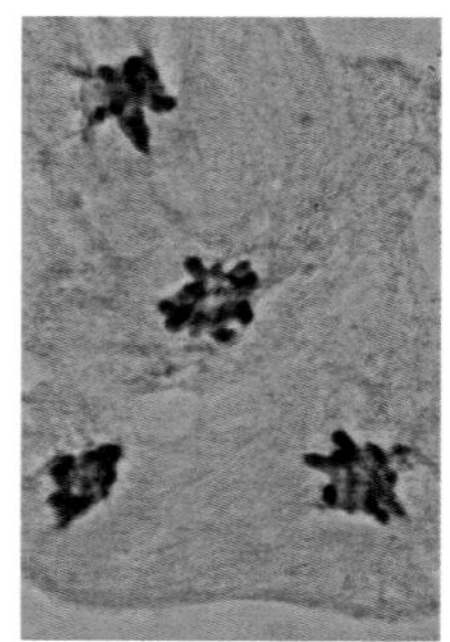

j telophase II

Figure 16.9 Stages of meiosis in an animal cell (locust) (×950). Interphase (not part of meiosis) is also shown.

QUESTIONS

16.2 Name the stage of meiosis at which each of the following occurs. Remember to state whether the stage you name is during division I or division II.

- a Homologous chromosomes pair to form bivalents.
- b Crossing over between chromatids of homologous chromosomes takes place.
- c Homologous chromosomes separate.
- d Centromeres split and chromatids separate.
- e Haploid nuclei are first formed.

16.3 A cell with three sets of chromosomes is said to be triploid, $3n$. A cell with four sets of chromosomes is said to be tetraploid, $4n$. Could meiosis take place in a $3n$ or a $4n$ cell? Explain your answer.

Gametogenesis in humans

In animals such as humans, meiosis occurs as gametes are formed inside the testes and ovaries. The formation of male gametes is known as spermatogenesis (Figure 16.10) and the formation of female gametes as oogenesis (Figure 16.11).

Sperm production takes place inside tubules in the testes. Here, diploid cells divide by mitosis to produce numerous diploid **spermatogonia**, which grow to form diploid **primary spermatocytes**. The first division of meiosis then takes place, forming two haploid **secondary spermatocytes**. The second division of meiosis then produces haploid **spermatids**, which mature into spermatozoa.

Oogenesis follows a similar pattern, but many fewer gametes are made than during spermatogenesis, and the process takes much longer, with long 'waiting stages'. It takes place inside the ovaries, where diploid cells divide by mitosis to produce many **oogonia**. These begin to divide by meiosis, but stop when they reach prophase I. At this stage, they are called **primary oocytes**, and they are, of course, still diploid. All of this happens before a baby girl is born, and at birth she has around 400 000 primary oocytes in her ovaries.

When she reaches puberty, some of the primary oocytes get a little further with their division by meiosis. They proceed from prophase I to the end of the first meiotic division, forming two haploid cells. However, the division is uneven; one cell gets most of the cytoplasm, and becomes a **secondary oocyte**, while the other is little more than a nucleus, and is called a **polar body**. The polar body can be thought of as simply a way of getting rid of half of the chromosomes, and has no further role to play in reproduction.

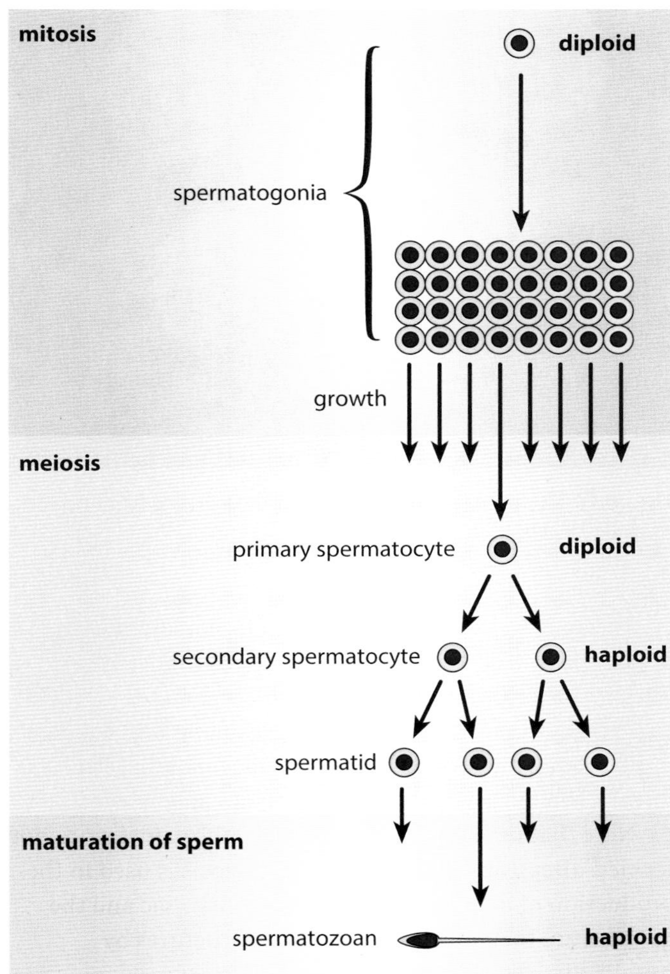

Figure 16.10 Spermatogenesis in humans.

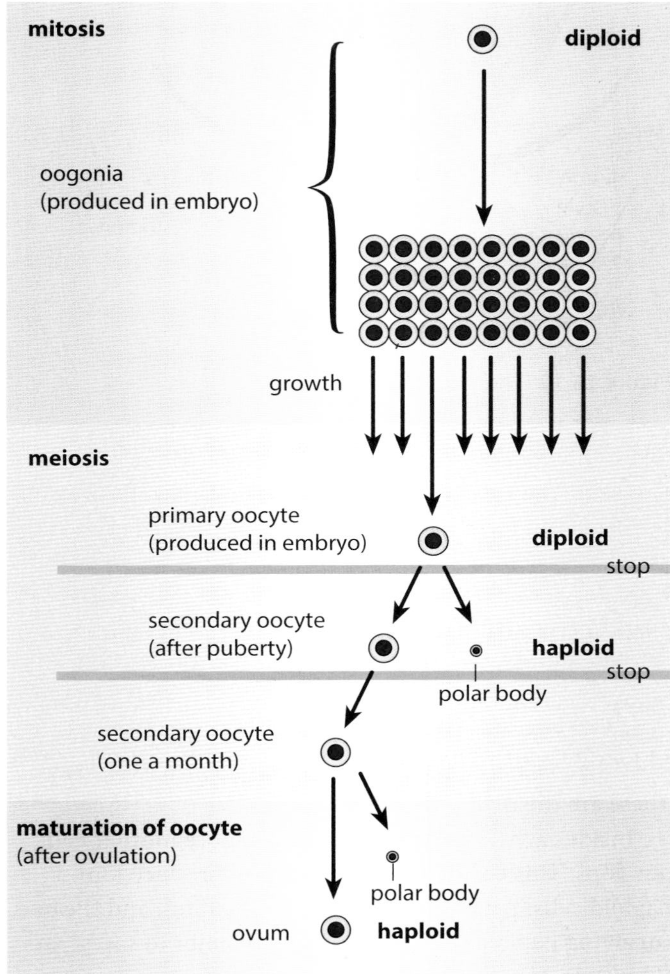

Figure 16.11 Oogenesis in humans.

Each month, one secondary oocyte is released into the oviduct from one of the ovaries. If it is fertilised, it continues its division by meiosis, and can now be called an **ovum**. The chromosomes of the spermatozoan and the ovum join together to form a single diploid nucleus, and the cell that is made by this process is called a **zygote**. The zygote can now divide repeatedly by mitosis to form first an embryo, and then a fetus.

Gametogenesis in flowering plants

Figure 16.12 shows the structure of a typical flower. Male gametes are produced in the anthers, and female gametes in the ovules.

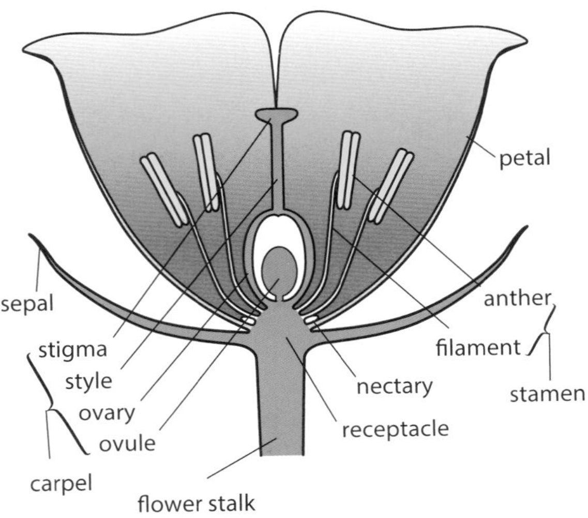

Figure 16.12 The structure of a flower.

Inside the anthers, diploid **pollen mother cells** divide by meiosis to form four haploid cells. The nuclei of each of these haploid cells then divide by mitosis, but the cell itself does not divide (cytokinesis does not take place), resulting in cells that each contain two haploid nuclei. These cells mature into **pollen grains**, each surrounded by a protective wall made up of a tough exine and thinner intine (Figure 16.13). One of the haploid nuclei is called the **tube nucleus**, and the other is the **generative nucleus**. These are the male gametes.

Inside each ovule, a large, diploid, **spore mother cell** develops. This cell divides by meiosis to produce four haploid cells. All but one of these degenerates, and the one surviving haploid cell develops into an **embryo sac** (Figure 16.14).

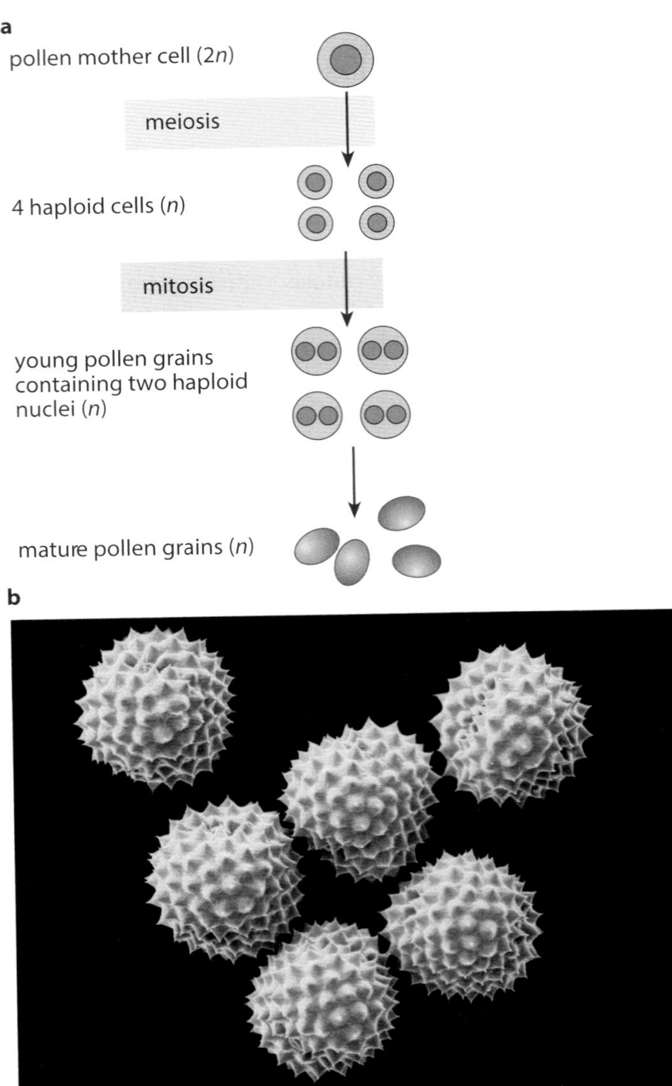

Figure 16.13 **a** The development of pollen grains from pollen mother cells. **b** Mature pollen grains (×1500).

The embryo sac grows larger, and its haploid nucleus divides by mitosis three times, forming eight haploid nuclei. One of these becomes the female gamete.

Fertilisation occurs when a male gamete from a pollen grain fuses with a female gamete inside an ovule. This forms a diploid zygote, which grows into an embryo plant.

Note that in plants, unlike animals, the gametes are not formed directly by meiosis. Instead, meiosis is used in the production of pollen grains and the embryo sac and the gametes are then formed inside these structures by mitotic divisions.

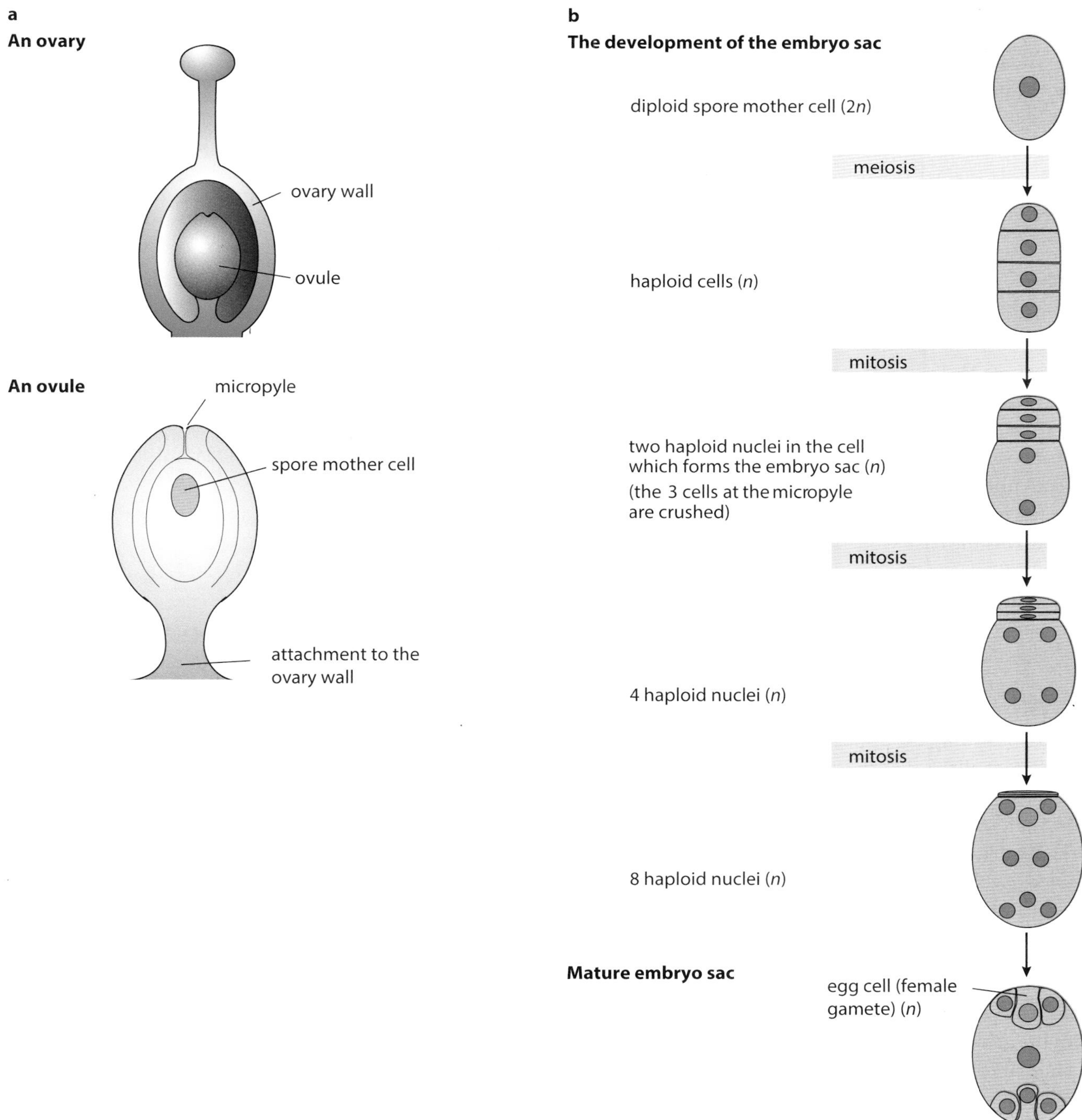

Figure 16.14 **a** Structure of an ovary and an ovule. **b** Development of the embryo sac.

Genetics

You will remember that a **gene** is a length of DNA that codes for the production of a polypeptide molecule. The code is held in the sequence of nucleotide bases in the DNA. A triplet of three bases codes for one amino acid in the polypeptide that will be constructed on the ribosomes in the cell (Chapter 6). One chromosome contains enough DNA to code for many polypeptides.

Alleles

The gene that codes for the production of the β-globin polypeptide of the haemoglobin molecule (pages 42–43) is on chromosome 11. Each cell contains two copies of this gene, one maternal in origin (from the mother) and one paternal (from the father).

There are several forms or varieties of this gene. One variety contains the base sequence GGACTTCTC and codes for the normal β-globin polypeptide. Another variety contains the base sequence GGACATCTC and codes for a different sequence of amino acids that forms a variant of the β-globin polypeptide known as the **sickle cell** β-globin polypeptide. Different varieties of the same gene are called **alleles**.

Genotype

Most genes, including the β-globin polypeptide gene, have several different alleles. For the moment, we will consider only the above two alleles of this gene.

For simplicity, the different alleles of a gene can be represented by symbols. In this case, they can be represented as follows:

$\mathbf{Hb^A}$ = the allele for the normal β-globin polypeptide

$\mathbf{Hb^S}$ = the allele for the sickle cell β-globin polypeptide

The letters **Hb** stand for the locus of the haemoglobin gene, whereas the superscripts ^{A} and ^{S} stand for particular alleles of the gene.

In a human cell, which is diploid, there are two copies of the β-globin polypeptide gene. The two copies might be:

$\mathbf{Hb^AHb^A}$ or $\mathbf{Hb^SHb^S}$ or $\mathbf{Hb^AHb^S}$.

The alleles that an organism has form its genotype In this case, where we are considering just two different alleles, there are three possible genotypes.

QUESTION

16.4 If there were three different alleles, how many possible genotypes would there be?

A genotype in which the two alleles of a gene are the same – for example, $\mathbf{Hb^AHb^A}$ – is said to be homozygous for that particular gene. A genotype in which the two alleles of a gene are different – for example, $\mathbf{Hb^AHb^S}$ – is said to be heterozygous for that gene. The organism can also be described as homozygous or heterozygous for that characteristic.

QUESTION

16.5 How many of the genotypes in your answer to Question 16.4 are homozygous, and how many are heterozygous?

A **genotype** is the alleles possessed by an organism.

Homozygous means having two identical alleles of a gene.

Heterozygous means having two different alleles of a gene.

Genotype affects phenotype

A person with the genotype $\mathbf{Hb^AHb^A}$ has two copies of the gene in each cell coding for the production of the normal β-globin polypeptide. All of the person's haemoglobin will be normal.

A person with the genotype $\mathbf{Hb^SHb^S}$ has two copies of the gene in each cell coding for the production of the sickle cell β-globin polypeptide. All of this person's haemoglobin will be sickle cell haemoglobin, which is inefficient at transporting oxygen. The person will have sickle cell anaemia. This is a very dangerous disease, in which great care has to be taken not to allow the blood to become short of oxygen, or death may occur (page 388). A person with the genotype $\mathbf{Hb^AHb^S}$ has one allele of the β-globin gene in each cell coding for the production of the normal β-globin, and one coding for the production of the sickle cell β-globin. Half of the person's haemoglobin will be normal, and half will be sickle cell haemoglobin. Such people have sickle cell trait, and are sometimes referred to as 'carriers'. They will probably be completely unaware that they have sickle cell trait, because they have enough normal haemoglobin to carry enough oxygen, and so will have no problems at all. They will appear to be perfectly healthy. Difficulties arise only very occasionally – for example, if a person with sickle cell trait does strenuous exercise at high altitudes, when oxygen concentrations in the blood might become very low (page 172).

The observable characteristics of an individual are called the person's phenotype. We will normally use the word 'phenotype' to describe just the one or two particular characteristics that we are interested in. In this case, we are considering the characteristic of having, or not having, sickle cell anaemia (Table 16.1).

Genotype	**Phenotype**
Hb^AHb^A	normal
Hb^AHb^S	normal, but with sickle cell trait
Hb^SHb^S	sickle cell anaemia

Table 16.1 Genotypes and phenotypes for sickle cell anaemia.

An organism's **phenotype** is its characteristics, often resulting from an interaction between its genotype and its environment.

Inheriting genes

In sexual reproduction, haploid gametes are made, following meiosis, from diploid body cells. Each gamete contains one of each pair of chromosomes. Therefore, each gamete contains only one copy of each gene.

Think about what happens when sperm are made in the testes of a man who has the genotype $\mathbf{Hb^AHb^S}$. Each time a cell divides during meiosis, four gametes are made, two of them with the $\mathbf{Hb^A}$ allele and two with the $\mathbf{Hb^S}$ allele. Of all the millions of sperm that are made in his lifetime, half will have the genotype $\mathbf{Hb^A}$ and half will have the genotype $\mathbf{Hb^S}$ (Figure 16.15).

Similarly, a heterozygous woman will produce eggs of which half have the genotype $\mathbf{Hb^A}$ and half have the genotype $\mathbf{Hb^S}$.

This information can be used to predict the possible genotypes of children born to a couple who are both heterozygous. Each time fertilisation occurs, either an $\mathbf{Hb^A}$ sperm or an $\mathbf{Hb^S}$ sperm may fertilise either an $\mathbf{Hb^A}$ egg or an $\mathbf{Hb^S}$ egg. The possible results can be shown like this:

		Genotypes of eggs	
		(Hb^A)	(Hb^S)
Genotypes of sperm	(Hb^A)	Hb^AHb^A normal	Hb^AHb^S sickle cell trait
	(Hb^S)	Hb^AHb^S sickle cell trait	Hb^SHb^S sickle cell anaemia

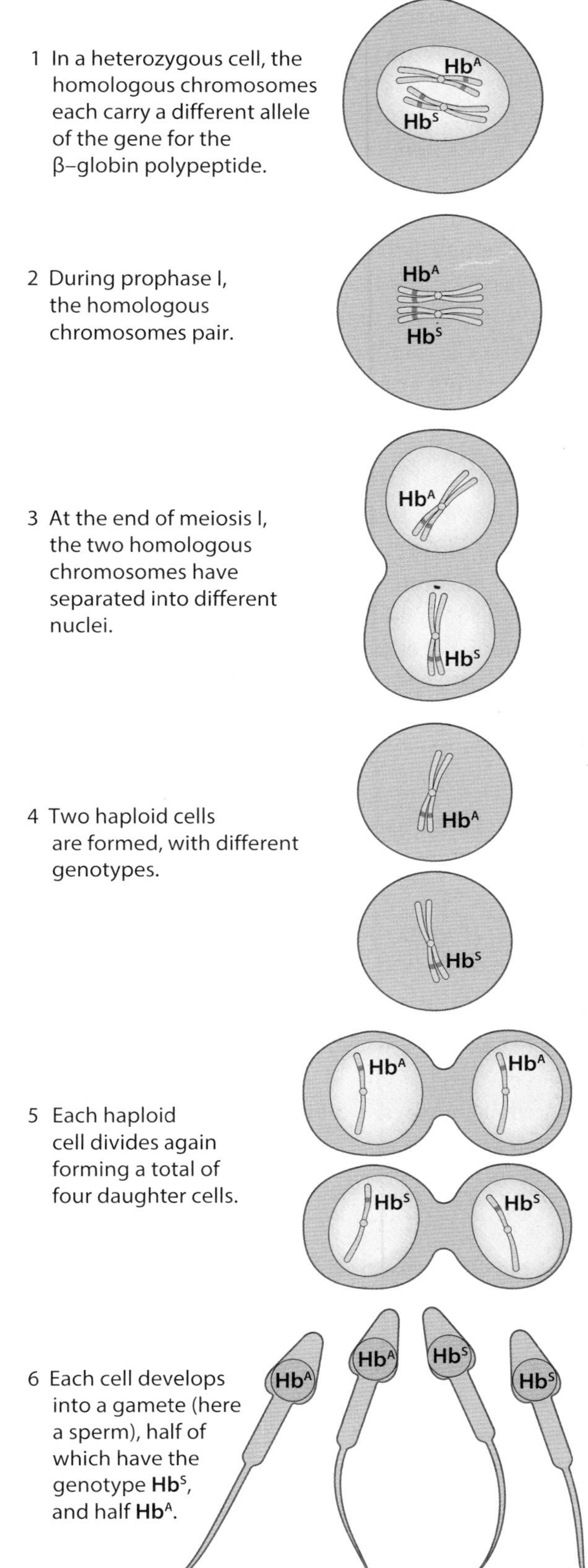

Figure 16.15 Meiosis of a heterozygous cell produces gametes of two different genotypes. Only one pair of homologous chromosomes is shown.

As there are equal numbers of each type of sperm and each type of egg, the chances of each of these four possibilities are also equal. Each time a child is conceived, there is a one in four chance that it will have the genotype $\mathbf{Hb^AHb^A}$, a one in four chance that it will be $\mathbf{Hb^SHb^S}$ and a two in four chance that it will be $\mathbf{Hb^AHb^S}$. Another way of describing these chances is to say that the probability of a child being $\mathbf{Hb^SHb^S}$ is 0.25, the probability of being $\mathbf{Hb^AHb^A}$ is 0.25, and the probability of being $\mathbf{Hb^AHb^S}$ is 0.5. It is important to realise that these are only **probabilities.** It would not be surprising if this couple had two children, both of whom had the genotype $\mathbf{Hb^SHb^S}$ and so suffered from sickle cell anaemia.

Genetic diagrams

A genetic diagram is the standard way of showing the genotypes of offspring that might be expected from two parents. To illustrate genetic diagrams, let us consider flower colour in snapdragons (*Antirrhinum*).

One of the genes for flower colour has two alleles, namely $\mathbf{C^R}$, which gives red flowers, and $\mathbf{C^W}$, which gives white flowers. The phenotypes produced by each genotype are:

Genotype	Phenotype
C^RC^R	red
C^RC^W	pink
C^WC^W	white

What colour flowers would be expected in the offspring from a red and a pink snapdragon?

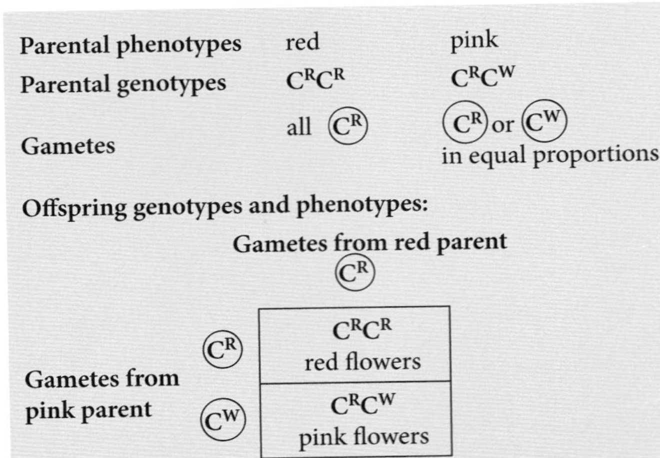

Thus, you would expect about half of the offspring to have red flowers and half to have pink flowers.

QUESTION

16.6 Red Poll cattle are homozygous for an allele that gives red coat colour. White Shorthorn are homozygous for an allele that gives white coat colour. When crossed, the offspring all have a mixture of red and white hairs in their coats, producing a colour called roan.

a Suggest suitable symbols for the two alleles of the coat colour gene.

b List the **three** possible genotypes for the coat colour gene and their phenotypes.

c Draw genetic diagrams to show the offspring expected from the following matings:

i a Red Poll with a roan **ii** two roans.

Dominance

In the examples used so far, both of the alleles in a heterozygous organism have an effect on the phenotype. A person with the genotype $\mathbf{Hb^AHb^S}$ has some normal haemoglobin and some sickle cell haemoglobin. A snapdragon with the genotype $\mathbf{C^RC^W}$ has some red colour and some white colour, so that the flowers appear pink. Alleles that behave like this are said to be codominant.

Frequently, however, only one allele has an effect in a heterozygous organism. This allele is said to be the dominant allele, whereas the one that has no effect is recessive. An example is stem colour in tomatoes. There are two alleles for stem colour, one of which produces green stems, and the other purple stems. In a tomato plant that has one allele for purple stems and one allele for green stems, the stems are exactly the same shade of purple as in a plant that has two alleles for purple stems. The allele for purple stems is dominant, and the allele for green stems is recessive.

When alleles of a gene behave like this, their symbols are written using a capital letter for the dominant allele and a small letter for the recessive allele. You are often free to choose the symbols you will use. In this case, the symbols could be **A** for the purple allele and **a** for the green allele. The possible genotypes and phenotypes for stem colour are:

Genotype	Phenotype
AA	purple stem
Aa	purple stem
aa	green stem

It is a good idea, when choosing symbols to use for alleles, to use letters where the capital looks very different from the small one. If you use symbols such as S and s or P and p, it can become difficult to tell them apart if they are written down quickly.

QUESTIONS

16.7 In mice, the gene for eye colour has two alleles. The allele for black eyes is dominant, whereas the allele for red eyes is recessive. Choose suitable symbols for these alleles, and then draw a genetic diagram to show the probable results of a cross between a heterozygous black-eyed mouse and a red-eyed mouse.

16.8 A species of poppy may have plain petals or petals with a large black spot near the base. If two plants with spotted petals are crossed, the offspring always have spotted petals. A cross between unspotted and spotted plants sometimes produces offspring that all have unspotted petals, and sometimes produces half spotted and half unspotted offspring. Explain these results.

A **dominant** allele is one whose effect on the phenotype of a heterozygote is identical to its effect in a homozygote.

A **recessive** allele is one that is only expressed when no dominant allele is present.

Codominant alleles both have an effect on the phenotype of a heterozygous organism.

F1 and F2 generations

The symbols F1 and F2 may be used in genetic diagrams. These symbols have specific meanings and should not be used in other circumstances.

The **F1 generation** is the offspring resulting from a cross between an organism with a homozygous dominant genotype, and one with a homozygous recessive genotype.

The **F2 generation** is the offspring resulting from a cross between two F1 (heterozygous) organisms

Test crosses

Where alleles show dominance, it is not possible to tell the genotype of an organism showing the dominant characteristic just by looking at it. A purple-stemmed tomato plant might have the genotype **AA**, or it might have the genotype **Aa**. To find out its genotype, it could be crossed with a green-stemmed tomato plant.

If the purple-stemmed tomato plant's genotype is **AA**:

Parental phenotypes	purple	green
Parental genotypes	AA	aa
Gametes	Ⓐ	ⓐ
Offspring	all Aa purple	

If its genotype is **Aa**:

Parental phenotypes	purple	green
Parental genotypes	Aa	aa
Gametes	Ⓐ or ⓐ	ⓐ
Offspring	Aa purple	aa green

So, from the colours of the offspring, you can tell the genotype of the purple parent. If any green offspring are produced, then the purple parent must have the genotype **Aa**.

This cross is called a **test cross**. A test cross always involves crossing an organism showing the dominant phenotype with one that is homozygous recessive. (You may come across the term 'backcross' in some books, but test cross is the better term to use.)

A **test cross** is a genetic cross in which an organism showing a characteristic caused by a dominant allele is crossed with an organism that is homozygous recessive; the phenotypes of the offspring can be a guide to whether the first organism is homozygous or heterozygous.

QUESTION

16.9 In Dalmatian dogs, the colour of the spots is determined by a gene that has two alleles. The allele for black spots is dominant, and the allele for brown spots is recessive.
A breeder wanted to know the genotype of a black-spotted female dog. She crossed her with a brown-spotted male dog, and a litter of three puppies was produced, all of which were black. The breeder concluded that the female was homozygous for the allele for black spots. Was she right?
Explain your answer.

Multiple alleles

So far, we have considered just two alleles, or varieties, of any one gene. Most genes, however, have more than two alleles. An example of this situation, known as multiple alleles, is the gene for human blood groups.

The four blood groups A, B, AB and O are all determined by a single gene. Three alleles of this gene exist, I^A, I^B, and I^o. Of these, I^A and I^B are codominant, whereas I^o is recessive to both I^A and I^B. As a diploid cell can carry only two alleles, the possible genotypes and phenotypes are as shown in Table 16.2.

Genotype	Blood group
I^AI^A	A
I^AI^B	AB
I^AI^o	A
I^BI^B	B
I^BI^o	B
I^oI^o	O

Table 16.2 Genotypes and phenotypes for blood groups.

QUESTIONS

16.10 A man of blood group B and a woman of blood group A have three children. One is group A, one group B and one group O. What are the genotypes of these five people?

16.11 Coat colour in rabbits is determined by a gene with four alleles. The allele for agouti (normal) coat is dominant to all of the other three alleles. The allele for albino coat is recessive to the other three alleles. The allele for chinchilla (grey) coat is dominant to the allele for Himalayan (white with black ears, nose, feet and tail) (Figure 16.16).

a Write down the **ten** possible genotypes for coat colour, and their phenotypes.

b Draw genetic diagrams to explain each of the following.

i An albino rabbit is crossed with a chinchilla rabbit, producing offspring that are all chinchilla. Two of these chinchilla offspring are then crossed, producing four chinchilla offspring and two albino.

ii An agouti rabbit is crossed with a Himalayan rabbit, producing three agouti offspring and three Himalayan.

iii Two agouti rabbits produce a litter of five young, three of whom are agouti and two chinchilla. The two chinchilla young are then crossed, producing four chinchilla offspring and one Himalayan.

Sex inheritance

In humans, sex is determined by one of the 23 pairs of chromosomes. These chromosomes are called the sex chromosomes. The other 22 pairs are called autosomes.

The sex chromosomes differ from the autosomes in that the two sex chromosomes in a cell are not always alike. They do not always have the same genes in the same position, and so they are not always homologous. This is because there are two types of sex chromosome, known as the X and Y chromosomes because of their shapes. The Y chromosome is much shorter than the X, and carries fewer genes. A person with two X chromosomes is female, whereas a person with one X and one Y chromosome is male.

QUESTION

16.12 Draw a genetic diagram to explain why there is always an equal chance that a child will be male or female. (You can do this in just the same way as the other genetic diagrams you have drawn, but using symbols to represent whole chromosomes, not genes.)

Figure 16.16 Colour variations in rabbits, caused by multiple alleles of a single gene: **a** agouti; **b** albino; **c** chinchilla; **d** Himalayan.

Sex linkage

The X chromosome contains many different genes. (You can see some of these in Figure 16.4, page 366.) One of them is a gene that codes for the production of a protein needed for blood clotting, called factor VIII. There are two alleles of this gene, the dominant one, **H**, producing normal factor VIII, and the recessive one, **h**, resulting in a lack of factor VIII. The recessive allele causes the disease haemophilia, in which the blood fails to clot properly.

The fact that the gene for haemophilia is on the X chromosome, and not on an autosome, affects the way that it is inherited. Females, who have two X chromosomes, have two copies of the gene. Males have only one X chromosome, and so have only one copy of the gene. Therefore, the possible genotypes for men and women are different. They are shown in Figure 16.17.

The factor VIII gene is said to be sex linked. A sex-linked gene is one that is found on a part of the X chromosome not matched by the Y, and therefore not found on the Y chromosome.

Genotypes including sex-linked genes are always represented by symbols that show that they are on an X chromosome. Thus the genotype of a woman who has the allele **H** on one of her X chromosomes, and the allele **h** on the other, is written as X^HX^h.

You can draw genetic diagrams to show how sex-linked genes are inherited, in exactly the same way as for other genes. For example, the following diagram shows the children that could be born to a couple where the man does not have haemophilia, while the woman is a carrier for the disease.

Parental phenotypes	normal man	carrier woman
Parental genotypes	X^HY	X^HX^h
Gametes	(X^H) or (Y)	(X^H) or (X^h)

Offspring genotypes and phenotypes:

Gametes from man	**Gametes from woman:** (X^H)	(X^h)
(X^H)	X^HX^H normal female	X^HX^h carrier female
(Y)	X^HY normal male	X^hY haemophiliac male

Each time this couple has a child, there is a 0.25 probability that it will be a normal girl, a 0.25 probability that it will be a normal boy, a 0.25 probability that it will be a carrier girl, and a 0.25 probability that it will be a boy with haemophilia.

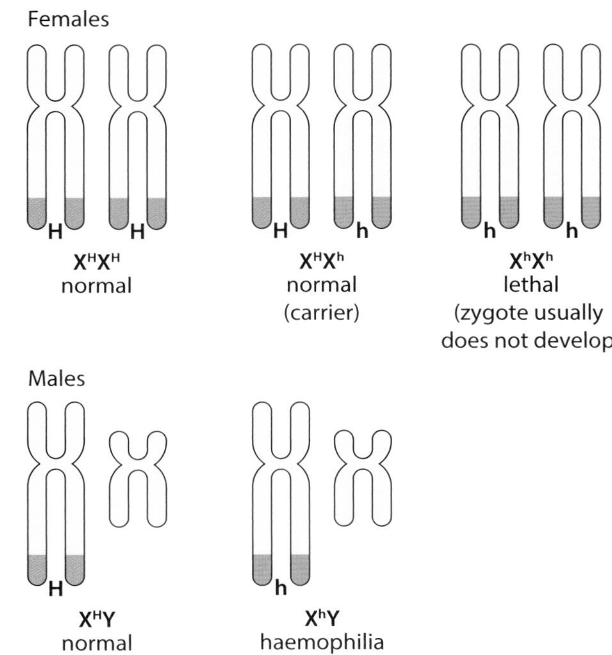

Figure 16.17 The possible genotypes and phenotypes for haemophilia.

QUESTIONS

16.13 Can a man with haemophilia pass on the disease to:

a his son?

b his grandson?

16.14 One of the genes for colour vision in humans is found on the X chromosome but not on the Y chromosome. The dominant allele of this gene gives normal colour vision, whereas a recessive allele produces red–green colour blindness.

a Choose suitable symbols for these alleles, and then write down all of the possible genotypes for a man and for a woman.

b A couple who both have normal colour vision have a child with colour blindness. Explain how this may happen, and state what the sex of the colour-blind child must be.

c Is it possible for a colour-blind girl to be born? Explain your answer.

QUESTION

16.15 One of the genes for coat colour in cats is sex linked. The allele $\mathbf{C^O}$ gives orange fur, whereas $\mathbf{C^B}$ gives black fur. The two alleles are codominant, and when both are present the cat has patches of orange and black, which is known as tortoiseshell.

a Explain why male cats cannot be tortoiseshell.

b Draw a genetic diagram to show the expected genotypes and phenotypes of the offspring from a cross between an orange male and a tortoiseshell female cat. (Remember to show the X and Y chromosomes, as well as the symbols for the alleles.)

Dihybrid crosses

So far, we have considered the inheritance of just one gene. Such examples are called **monohybrid crosses**. **Dihybrid crosses** look at the inheritance of two genes at once.

You have already seen that, in tomato plants, there is a gene that codes for stem colour. This gene has two alleles:

stem colour gene **A** = allele for purple stem
a = allele for green stem

where **A** is dominant and **a** is recessive.

A different gene, at a different locus on a different chromosome, codes for leaf shape. Again, there are two alleles:

leaf shape gene **D** = allele for cut leaves (jagged edges)
d = allele for potato leaves (smooth edges)

where **D** is dominant and **d** is recessive.

At metaphase of meiosis I, the pairs of homologous chromosomes line up on the equator independently of each other. For two pairs of chromosomes, there are two possible orientations (Figure 16.18).

At the end of meiosis II, each orientation gives two types of gamete. There are therefore four types of gamete altogether.

What will happen if a plant that is heterozygous for both of these genes (for stem colour and leaf shape) is crossed with a plant with a green stem and potato leaves?

Figure 16.18 shows the alleles in a cell of the plant that is heterozygous for both genes. When this cell undergoes meiosis to produce gametes, the pairs of homologous chromosomes line up independently of each other on the equator during metaphase I. There are two ways in which the two pairs of chromosomes can do this. If there are many such cells undergoing meiosis, then the chromosomes in roughly half of them will probably line up one way, and the other half will line up the other way. This is **independent assortment**. We can therefore predict that the gametes formed from these heterozygous cells will be of four types, **AD**, **Ad**, **aD** and **ad**, occurring in approximately equal numbers.

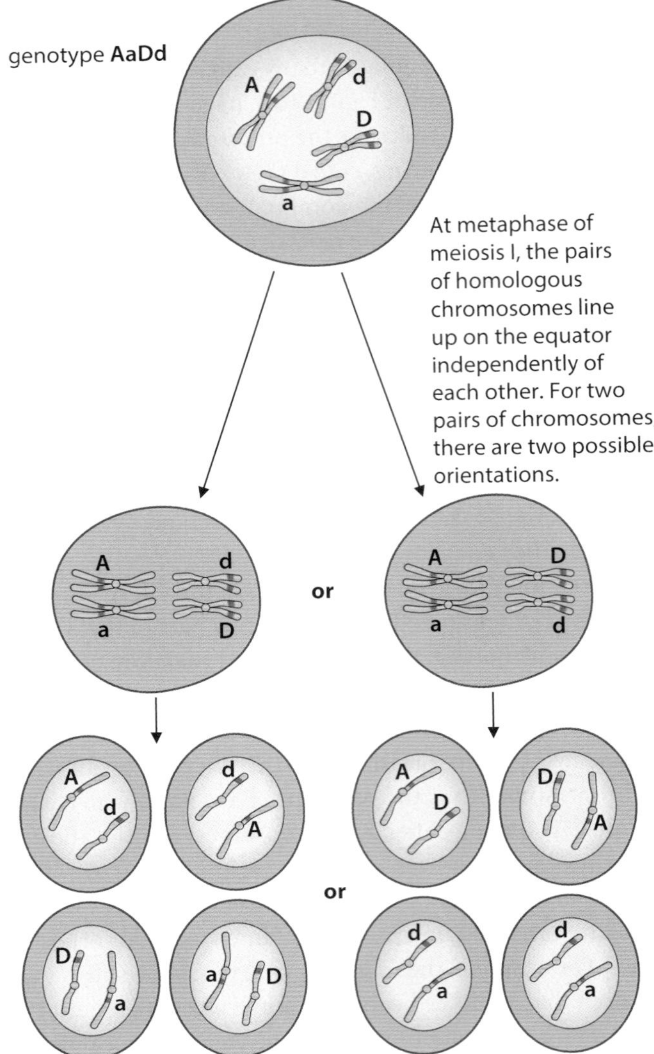

Figure 16.18 Independent assortment of homologous chromosomes during meiosis I results in a variety of genotypes in the gametes formed.

The plant with green stem and potato leaves must have the genotype **aadd**. Each of its gametes will contain one **a** allele and one **d** allele. All of the gametes will have the genotype **ad**.

Parental phenotypes	purple stem, cut leaves	green stem, potato leaves
Parental genotypes	**AaDd**	**aadd**
Gametes	(AD) or (Ad) or (aD) or (ad) in equal proportions	all (ad)

At fertilisation, any of the four types of gamete from the heterozygous parent may fuse with the gametes from the homozygous parent. The genotypes of the offspring will be:

		Gametes from green, potato plant (ad)
Gametes from purple, cut plant	(AD)	**AaDd** purple stem, cut leaves
	(Ad)	**Aadd** purple stem, potato leaves
	(aD)	**aaDd** green stem, cut leaves
	(ad)	**aadd** green stem, potato leaves

From this cross, therefore, we would expect approximately equal numbers of the four possible phenotypes. This 1 : 1 : 1 : 1 ratio is typical of a dihybrid cross between a heterozygous organism and a homozygous recessive organism where the alleles show complete dominance.

If both parents are heterozygous, then things become a little more complicated, because both of them will produce four kinds of gametes.

Parental phenotypes	purple stem, cut leaves	purple stem, cut leaves
Parental genotypes	**AaDd**	**AaDd**
Gametes	(AD) or (Ad) or (aD) or (ad) in equal proportions	(AD) or (Ad) or (aD) or (ad) in equal proportions

Offspring genotypes and phenotypes:

		Gametes from one parent (AD)	(Ad)	(aD)	(ad)
Gametes from other parent	(AD)	AADD purple, cut	AADd purple, cut	AaDD purple, cut	AaDd purple, cut
	(Ad)	AADd purple, cut	AAdd purple, potato	AaDd purple, cut	Aadd purple, potato
	(aD)	AaDD purple, cut	AaDd purple, cut	aaDD green, cut	aaDd green, cut
	(ad)	AaDd purple, cut	Aadd purple, potato	aaDd green, cut	aadd green, potato

If you sort out the numbers of each phenotype among these 16 possibilities, you will find that the offspring would be expected to occur in the following ratio:

9 purple, cut : 3 purple, potato : 3 green, cut : 1 green, potato

This 9 : 3 : 3 : 1 ratio is typical of a dihybrid cross between two heterozygous organisms where the two alleles show complete dominance and where the genes are on different chromosomes.

QUESTIONS

16.16 Explain the contribution made to the variation among the offspring of this cross by:

- **a** independent assortment
- **b** random fertilisation.

16.17 Draw genetic diagrams to show the genotypes of the offspring from each of the following crosses.

- **a** AABb × aabb
- **b** GgHh × gghh
- **c** TTyy × ttYY
- **d** eeFf × Eeff

QUESTIONS

16.18 The allele for grey fur in a species of animal is dominant to white, and the allele for long tail is dominant to short.

a Using the symbols **G** and **g** for coat colour, and **T** and **t** for tail length, draw a genetic diagram to show the genotypes and phenotypes of the offspring you would expect from a cross between a pure-breeding grey animal with a long tail and a pure-breeding white animal with a short tail.

b If this first generation of offspring were bred together, what would be the expected phenotypes in the second generation of offspring, and in what ratios would they occur?

16.19 In a species of plant, the allele for tall stem is dominant to short. The two alleles for leaf colour, giving green or white in the homozygous condition, are codominant, producing variegated leaves in the heterozygote.

A plant with tall stems and green leaves was crossed with a plant with short stems and variegated leaves. The offspring from this cross consisted of plants with tall stems and green leaves and plants with tall stems and variegated leaves in the ratio of 1:1. Construct a genetic diagram to explain this cross.

16.20 In a species of animal, it is known that the allele for black eyes is dominant to the allele for red eyes, and that the allele for long fur is dominant to the allele for short fur.

a What are the possible genotypes for an animal with black eyes and long fur?

b How could you find out which genotype this animal had?

Interactions between loci

You have already seen interactions between alleles at the *same* locus, namely:

- codominant alleles in flower colour in snapdragons
- dominant and recessive alleles in tomato plant stem colour
- multiple alleles in the inheritance of the ABO blood groups.

There are also cases where *different* loci interact to affect one phenotypic character.

In the inheritance of feather colour in chickens, there is an interaction between two gene loci, **I**/**i** and **C**/**c**. Individuals carrying the dominant allele, **I**, have white feathers even if they also carry the dominant allele, **C**, for coloured feathers. Birds that are homozygous recessive are also white.

QUESTION

16.21 List the genotypes that will result in coloured feathers in chickens.

White Leghorn chickens have the genotype **IICC**, while white Wyandotte chickens have the genotype **iicc**. A white Leghorn is crossed with a white Wyandotte.

Parental phenotypes	white	white
Parental genotypes	IICC	iicc
Gametes	(IC)	(ic)
Offspring (F1) genotypes	all IiCc	
Offspring phenotypes	all white	

These offspring are interbred to give another generation.

Parental phenotypes	white	white
Parental genotypes	IiCc	IiCc
Gametes	(IC) or (Ic) or (iC) or (ic)	(IC) or (Ic) or (iC) or (ic)
	in equal proportions	

Offspring (F2) genotypes and phenotypes:

Gametes from other parent \ **Gametes from one parent**	(IC)	(Ic)	(iC)	(ic)
(IC)	IICC white	IICc white	IiCC white	IiCc white
(Ic)	IICc white	IIcc white	IiCc white	Iicc white
(iC)	IiCC white	IiCc white	iiCC coloured	iiCc coloured
(ic)	IiCc white	Iicc white	iiCc coloured	iicc white

The usual 9 : 3 : 3 : 1 ratio expected in this generation has been modified to (9+3+1) : 3 giving 13 white : 3 coloured.

A different interaction is shown by the inheritance of flower colour in *Salvia*. A pure-breeding, pink flowered variety of *Salvia* was crossed with a pure-breeding, white-flowered variety. The offspring had purple flowers. Interbreeding these offspring to give another generation resulted in purple, pink and white-flowered plants in a ratio of 9 : 3 : 4. Two loci, **A**/**a** and **B**/**b**, on different chromosomes are involved:

Genotype	Phenotype
A–B–	purple
A–bb	pink
aaB–	white
aabb	white

(– indicates that either allele of the gene may be present)

The homozyote recessive **aa** affects the **B**/**b** locus. Neither the dominant allele, **B**, for purple flower colour, nor the recessive allele, **b**, for pink flower colour can be expressed in the absence of a dominant **A** allele.

QUESTION

16.22 Draw a genetic diagram of the *Salvia* cross described above to show the 9 : 3 : 4 ratio in the second generation.

Autosomal linkage

When two or more gene loci are on the same chromosome, they do not assort independently in meiosis as they would if they were on different chromosomes. The genes are said to be **linked**.

You have already learned about sex-linked genes (page 379). Genes on a chromosome other than the sex chromosomes are said to be autosomally linked.

Linkage is the presence of two genes on the same chromosome, so that they tend to be inherited together and do not assort independently.

The fruit fly, *Drosophila*, normally has a striped body and antennae with a feathery arista (Figure 16.19). The gene for body colour and the gene for antennal shape are close together on the same chromosome and so are linked.

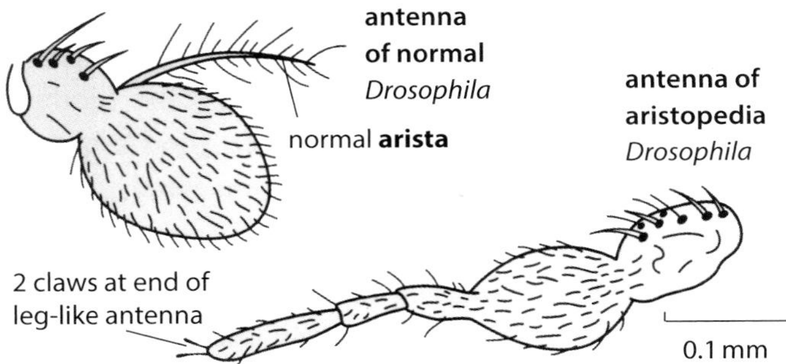

Figure 16.19 Normal and aristopedia *Drosophila* antennae.

A black body with no stripes results from a recessive allele called 'ebony'. A recessive allele for antennal shape, called 'aristopedia', gives an antenna looking rather like a *Drosophila* leg, with two claws on the end.

Body colour gene:
- **E** = allele for striped body
- **e** = allele for ebony body

Antennal shape gene:
- **A** = allele for normal antennae
- **a** = allele for aristopedia antennae

To help keep track of linked alleles in a genetic diagram, you can bracket each linkage group. In this case the genotype of the striped body fly with normal antennae is written (**EA**)(**EA**) and not **EEAA**, which would indicate that the genes were not on the same chromosomes.

A fly homozygous for striped body and normal antennae was crossed with a fly homozygous for ebony body and aristopedia antennae. All the offspring had striped bodies and normal antennae.

Parental phenotypes	striped body normal antennae	ebony body aristopedia antennae
Parental genotypes	(EA)(EA)	(ea)(ea)
Gametes	(EA)	(ea)
Offspring (F1) genotypes and phenotypes	all (EA)(ea) striped body normal antennae	

Male offspring were then test crossed with females homozygous for ebony body and aristopedia antennae, producing the two original parent types in equal numbers.

Parental phenotypes	male striped body normal antennae	female ebony body aristopedia antennae
Parental genotypes	(EA)(ea)	(ea)(ea)
Gametes	(EA) or (ea) in equal proportions	(ea)

Offspring genotypes and phenotypes:

		Gametes from female parent: (ea)
Gametes from male parent	(EA)	(EA)(ea) striped body, normal antennae
	(ea)	(ea)(ea) ebony body, aristopedia antennae

The test cross gives a 1 : 1 ratio of the two original parental types and not the 1 : 1 : 1 : 1 ratio expected from a dihybrid cross. (If you are uncertain about these ratios, repeat the cross above but assume that the genes are not linked. This should result in 1 **EeAa** : 1 **Eeaa** : 1 **eeAa** : 1 **eeaa**.) The dihybrid cross has behaved as a monohybrid cross. The alleles that went into the cross together remained together.

The genes of any organism fall into a number of linkage groups equal to the number of pairs of homologous chromosomes. Total linkage is very rare. Almost always, the linkage groups are broken by crossing over during meiosis. In the example discussed above, a male *Drosophila* was test crossed to show the effect of linkage because, unusually, there is no crossing over in male *Drosophila*.

Crossing over

During prophase I of meiosis, a pair of homologous chromosomes (a bivalent) can be seen to be joined by chiasmata (Figure 16.20). The chromatids of a bivalent may break and reconnect to another, non-sister chromatid. This results in an exchange of gene loci between a maternal and paternal chromatid (Figure 16.21).

Let us return to the *Drosophila* cross described above and test cross the female offspring. Figure 16.21 will help you follow what happens in the cross. Large numbers of the parental types of flies are produced. They are in a 1 : 1 ratio. Smaller numbers of **recombinant** flies are produced. These result from crossing over and 'recombine' the characteristics of the original parents into some flies that have a striped body and aristopedia antennae and others that have an ebony body with normal antennae. The two recombinant classes themselves are in a 1 : 1 ratio.

In this particular cross, we would typically find:

striped body, normal antennae	44%	parental classes
ebony body, aristopedia antennae	44%	
striped body, aristopedia antennae	6%	recombinant classes
ebony body, normal antennae	6%	

The **cross over value** is the percentage of offspring that belong to the recombinant classes. In this case it is 6% + 6% = 12%. This is a measure of the distance apart of the two gene loci on their chromosomes. The smaller the cross over value, the closer the loci are together. The chance of a cross over taking place between two loci is directly related to their distance apart.

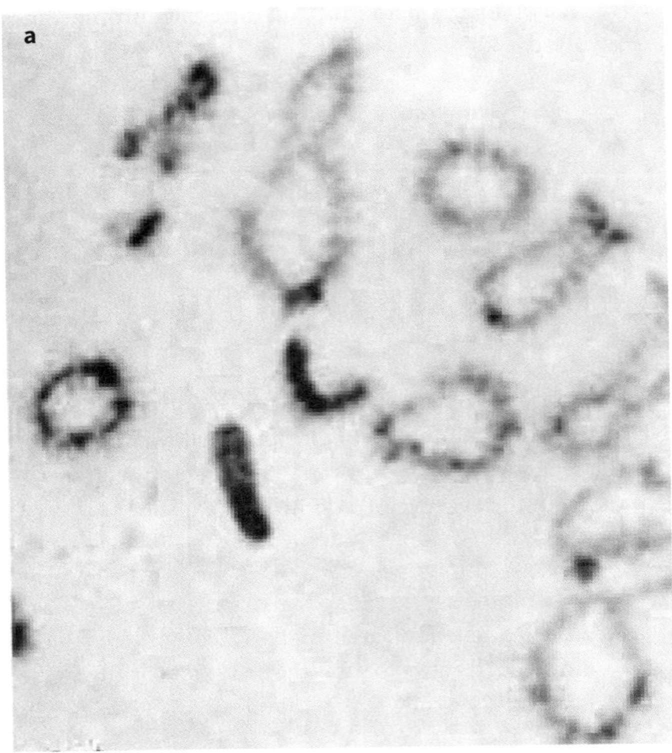

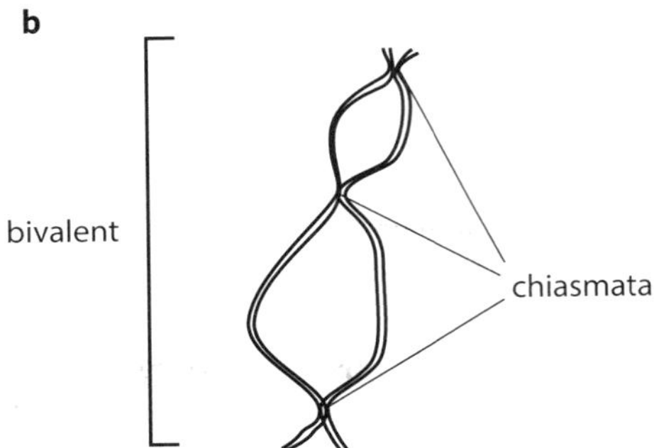

Figure 16.20 **a** Photomicrograph of bivalents in prophase I of meiosis, showing chiasmata. A chiasma shows that crossing over has occurred between two chromatids. **b** Interpretive drawing of one bivalent.

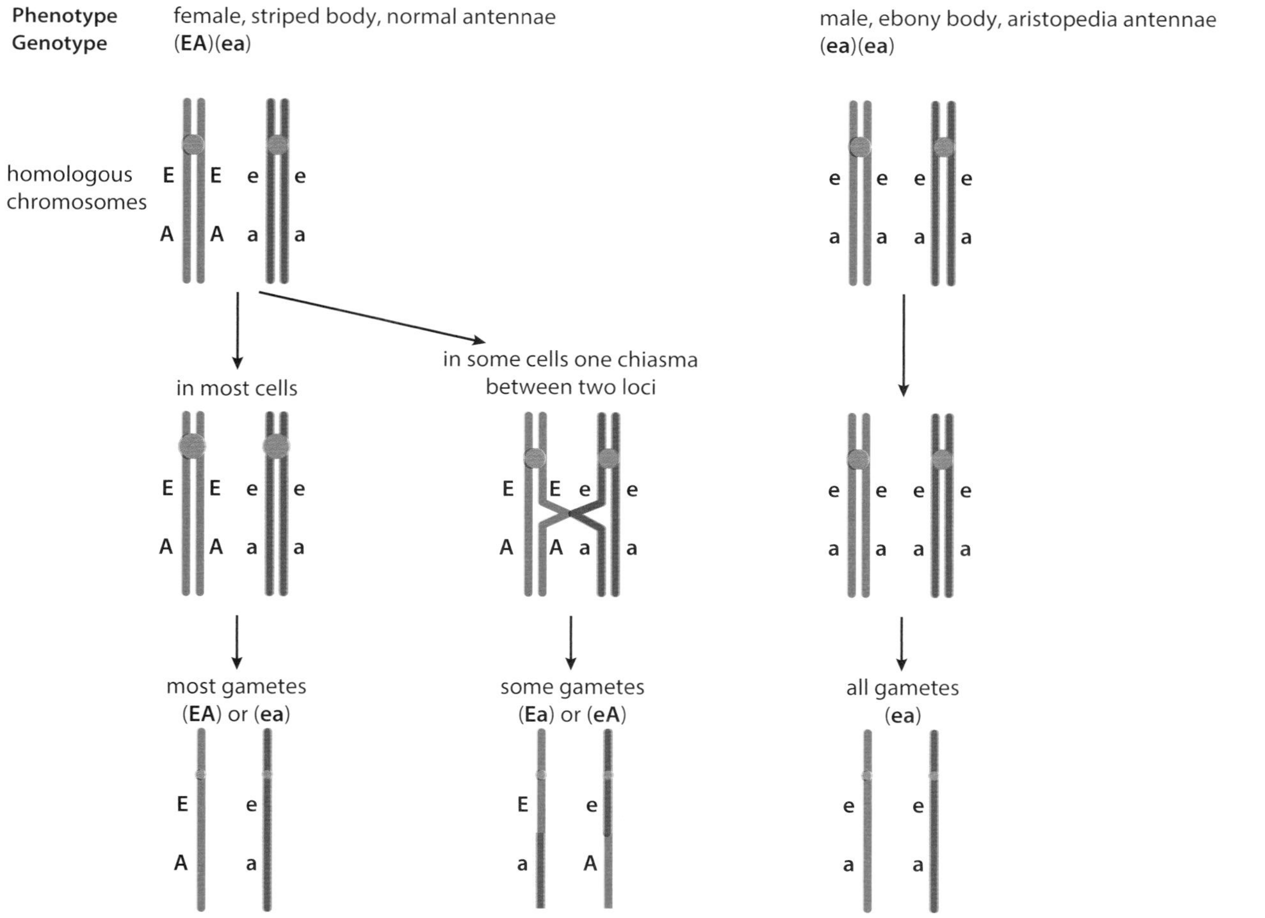

Phenotype female, striped body, normal antennae
Genotype (**EA**)(**ea**)

male, ebony body, aristopedia antennae
(**ea**)(**ea**)

homologous chromosomes

in most cells

in some cells one chiasma between two loci

most gametes (**EA**) or (**ea**)

some gametes (**Ea**) or (**eA**)

all gametes (**ea**)

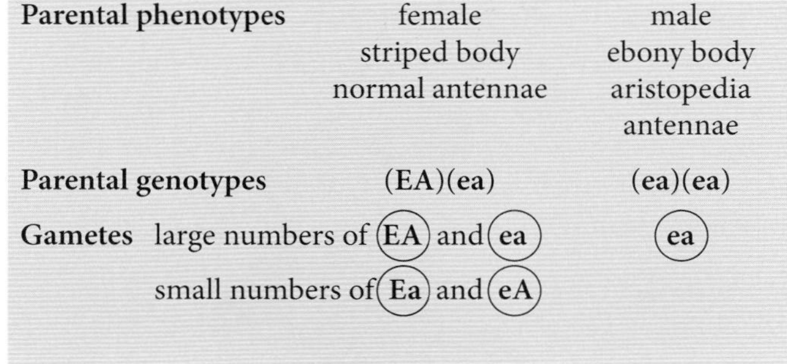

	female	male
Parental phenotypes	female striped body normal antennae	male ebony body aristopedia antennae
Parental genotypes	(EA)(ea)	(ea)(ea)
Gametes	large numbers of (EA) and (ea) small numbers of (Ea) and (eA)	(ea)

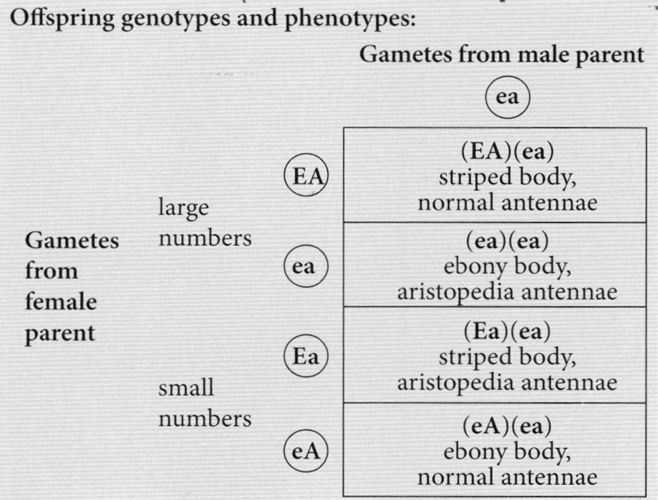

Offspring genotypes and phenotypes:

			Gametes from male parent (ea)
Gametes from female parent	large numbers	(EA)	(EA)(ea) striped body, normal antennae
		(ea)	(ea)(ea) ebony body, aristopedia antennae
	small numbers	(Ea)	(Ea)(ea) striped body, aristopedia antennae
		(eA)	(eA)(ea) ebony body, normal antennae

Figure 16.21 Crossing over in female *Drosophila*.

QUESTION

16.23 Pure-breeding *Drosophila* with straight wings and grey bodies were crossed with pure-breeding curled-wing, ebony-bodied flies. All the offspring were straight-winged and grey-bodied.
Female offspring were then test crossed with curled-wing, ebony-bodied males, giving the following results:

straight wing, grey body	113
straight wing, ebony body	30
curled wing, grey body	29
curled wing, ebony body	115

a State the ratio of phenotypes expected in a dihybrid test cross such as this.

b Explain the discrepancy between the expected result and the results given.

c Calculate the cross over value.

d Is the curled wing locus closer to the ebony locus than is the aristopedia locus?
Explain your answer.

The χ^2 (chi-squared) test

If you look back at the cross between the two heterozygous tomato plants on pages 380–381, you will see that we would expect to see a 9 : 3 : 3 : 1 ratio of phenotypes in the offspring. It is important to remember that this ratio represents the **probability** of getting these phenotypes, and we would probably be rather surprised if the numbers came out absolutely precisely to this ratio.

But just how much difference might we be happy with, before we began to worry that perhaps the situation was not quite what we had thought? For example, let us imagine that the two plants produced a total of 144 offspring. If the parents really were both heterozygous, and if the purple stem and cut leaf alleles really are dominant, and if the alleles really do assort independently, then we would expect the following numbers of each phenotype to be present in the offspring:

purple, cut $= \frac{9}{16} \times 144 = 81$

purple, potato $= \frac{3}{16} \times 144 = 27$

green, cut $= \frac{3}{16} \times 144 = 27$

green, potato $= \frac{1}{16} \times 144 = 9$

But imagine that, among these 144 offspring, the results we actually observed were as follows:

purple, cut	86	green, cut	24
purple, potato	26	green, potato	8

We might ask: are these results sufficiently close to the ones we expected that the differences between them have probably just arisen by chance, or are they so different that something unexpected must be going on?

To answer this question, we can use a statistical test called the chi-squared (χ^2) test. This test allows us to compare our observed results with the expected results, and decide whether or not there is a significant difference between them.

The first stage in carrying out this test is to work out the expected results, as we have already done. These, and the observed results, are then recorded in a table like the one below. We then calculate the difference between each set of results, and square each difference. (Squaring gets rid of any minus signs – it is irrelevant whether the differences are negative or positive.) Then we divide each squared difference by the expected value, and add up all of these answers:

$$\chi^2 = \sum \frac{(O - E)^2}{E}$$

where: Σ = sum of
O = observed value
E = expected value

Phenotypes of plants	purple stems, cut leaves	purple stems, potato leaves	green stems, cut leaves	green stems, potato leaves
Observed number (O)	86	26	24	8
Expected ratio	9 :	3 :	3 :	1
Expected number (E)	81	27	27	9
$O - E$	+5	−1	−3	−1
$(O - E)^2$	25	1	9	1
$(O - E)^2/E$	0.31	0.04	0.33	0.11
$\chi^2 = \sum \frac{(O - E)^2}{E} = 0.79$				

So now we have our value of χ^2. Next we have to work out what it means. To do this, we look in a table that relates χ^2 values to probabilities (Table 16.3). The probabilities given in the table are the probabilities that the differences between our expected and observed results are due to chance.

For example, a probability of 0.05 means that we would expect these differences to occur in five out of every 100 experiments, or one in 20, just by chance. A probability of 0.01 means that we would expect these differences to occur in one out of every 100 experiments, just by chance.

In biological experiments, we usually take a probability of 0.05 as being the critical one. If our χ^2 value represents a probability of 0.05 or larger, then we can be fairly certain that the differences between our observed and expected results are due to chance – the differences between them are **not significant**. However, if the probability is smaller than 0.05, then it is likely that the difference **is significant**, and we must reconsider our assumptions about what was going on in this cross.

There is one more aspect of our results to consider before we can look up our value of χ^2 in Table 16.3. This is the number of degrees of freedom in our results. The degrees of freedom take into account the number of comparisons made. (Remember that to get our value for χ^2, we added up all our calculated values, so obviously the larger the number of observed and expected values we have, the larger χ^2 is likely to be. We need to compensate for this.) To work out the number of degrees of freedom, simply calculate the number of classes of data minus 1. Here we have four classes of data (the four possible sets of phenotypes), so the degrees of freedom are: $4 - 1 = 3$.

Now, at last, we can look at Table 16.3 to determine whether our results show a significant deviation from what we expected. The numbers in the body of the table are χ^2 values. We look at the third row in the table (because that is the one relevant to 3 degrees of freedom), and find the χ^2 value that represents a probability of 0.05. You can see that this is 7.82. Our calculated value of χ^2 was 0.79. So our value is a much, much smaller value than the one we have read from the table. In fact, we cannot find anything like this number in the table – it would be way off the left-hand side, representing a probability of much more than 0.1 (1 in 10) that the difference in our results is just due to chance. So we can say that the difference between our observed and expected results is almost certainly due to chance, and there is **no significant difference** between what we expected and what we actually got.

Degrees of freedom	Probability greater than			
	0.1	0.05	0.01	0.001
1	2.71	3.84	6.64	10.83
2	4.60	5.99	9.21	13.82
3	6.25	7.82	11.34	16.27
4	7.78	9.49	13.28	18.46

Table 16.3 Table of χ^2 values.

QUESTION

16.24 Look back at your answer to Question 16.18b. In the actual crosses between the animals in this generation, the numbers of each phenotype obtained in the offspring were:

grey, long	54
grey, short	4
white, long	4
white, short	18

Use a χ^2 test to determine whether or not the difference between these observed results and the expected results is significant.

Mutations

You have seen that most genes have several different variants, called alleles. A gene is made up of a sequence of nucleotides, each with its own base. The different alleles of a gene contain slightly different sequences of bases.

These different alleles originally arose by a process called **mutation**. Mutation is an unpredictable change in the genetic material of an organism. A change in the structure of a DNA molecule, producing a different allele of a gene, is a **gene mutation**. Mutations may also cause changes in the structure or number of whole chromosomes in a cell, in which case they are known as **chromosome mutations** (or chromosome aberrations).

Mutations may occur completely randomly, with no obvious cause. However, there are several environmental factors that significantly increase the chances of a mutation occurring. All types of ionising radiation (alpha, beta and gamma radiation) can damage DNA molecules, altering the structure of the bases within them. Ultraviolet radiation has a similar effect, as do many chemicals – for example, mustard gas. A substance that increases the chances of mutation occurring is said to be a **mutagen**.

In gene mutations, there are three different ways in which the sequence of bases in a gene may be altered. These are:

- base substitution, where one base simply takes the place of another; for example, CCT GAG GAG may change to CCT GTG GAG
- base addition, where one or more extra bases are added to the sequence; for example, CCT GAG GAG may change to CCA TGA GGA G
- base deletion, where one or more bases are lost from the sequence; for example, CCT GAG GAG may change to CCG AGG AG.

Base additions or deletions usually have a very significant effect on the structure, and therefore the function, of the polypeptide that the allele codes for. If you look up in Appendix 2 the amino acids that are coded for by the 'normal' sequence shown above, you will see that it is Gly Leu Leu. But the new sequence resulting from the base addition codes for Gly Thr Pro, and that resulting from the base deletion is Gly Ser. Base additions or deletions always have large effects, because they alter every set of three bases that 'follows' them in the DNA molecule. Base additions or deletions are said to cause **frame shifts** in the code. Often, the effects are so large that the protein that is made is totally useless. Or the addition or deletion may introduce a 'stop' triplet part way through a gene, so that a complete protein is never made at all.

Base substitutions, on the other hand, often have no effect at all. A mutation that has no apparent effect on an organism is said to be a **silent mutation**. Base substitutions are often silent mutations because many amino acids have more than one triplet code (see Appendix 2 again), so even if one base is changed, the same amino acid is still coded for. You have seen above that a change from CCT to CCA or CCG makes no difference – the amino acid that will be slotted into the chain at that point will still be Gly.

However, base substitutions **can** have very large effects. Suppose, for example, the base sequence ATG (coding for Tyr) mutated to ATT. This is a 'stop' triplet, so the synthesis of the protein would stop at this point.

Sickle cell anaemia

One example of a base substitution that has a significant effect on the phenotype is the one involved in the inherited blood disorder, sickle cell anaemia. (We have already looked at the difference between the $\mathbf{Hb^A}$ and $\mathbf{Hb^S}$ alleles on page 374 and the inheritance of this disease on pages 374–376.)

You will remember that the gene that codes for the amino acid sequence in the β-globin polypeptide is not the same in everyone. In most people, the β-globin polypeptide begins with the amino acid sequence coded from the $\mathbf{Hb^A}$ allele:

Val-His-Leu-Thr-Pro-Glu-Glu-Lys-

But in people with the $\mathbf{Hb^S}$ allele, the base sequence CTT is replaced by CAT, and the amino acid sequence becomes:

Val-His-Leu-Thr-Pro-Val-Glu-Lys-

This small difference in the amino acid sequence makes little difference to the haemoglobin molecule when it is combined with oxygen. But when it is not combined with oxygen, the 'unusual' β-globin polypeptides make the haemoglobin molecule much less soluble. The molecules tend to stick to each other, forming long fibres inside the red blood cells. The red cells are pulled out of shape, into a half-moon or sickle shape. When this happens, the distorted cells become useless at transporting oxygen. They also get stuck in small capillaries, stopping any unaffected cells from getting through (Figure 2.25b, page 34).

A person with this unusual β-globin can suffer severe anaemia (lack of oxygen transported to the cells) and may die. Sickle cell anaemia is especially common in some parts of Africa and in India. You can read about the reasons for this distribution on pages 407–408.

Albinism

Albinism provides an example of the relationship between a gene, an enzyme and a human phenotype.

In albinism, the dark pigment melanin is totally or partially missing from the eyes, skin and hair. In humans this results in pale blue or pink irises in the eyes and very pale skin and hair (Figure 16.22). The pupils of the eyes appear red. The condition is often accompanied by poor

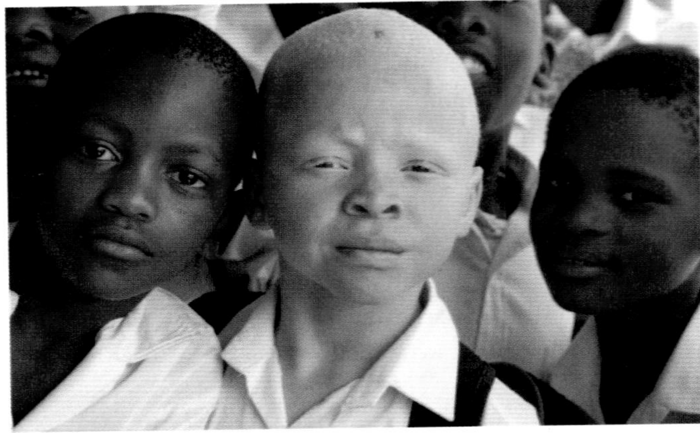

Figure 16.22 An albino boy with his classmates in South Africa.

vision, by rapid, jerky movements of the eyes and by a tendency to avoid bright light.

Mutations at several loci may be responsible for this condition, but in its classic form the mutation is an autosomal recessive and individuals that are homozygous for the recessive allele show albinism. Such individuals occur in about 1 in 17 000 births worldwide. However, the condition is relatively common in some populations such as the Hopi in Arizona and the Kuna San Blas Indians in Panama. A different form of albinism, which affects the eyes, but not the skin, is sex-linked.

A mutation in the gene for the enzyme tyrosinase results in either the absence of tyrosinase or the presence of inactive tyrosinase in the cells responsible for melanin production. In these melanocytes, the first two steps of the conversion of the amino acid, tyrosine into melanin cannot take place. Tyrosine cannot be converted into DOPA and dopaquinone.

$$\text{tyrosine} \xrightarrow{\text{tyrosinase}} \text{DOPA} \xrightarrow{\text{tyrosinase}} \text{dopaquinone} \rightarrow \text{melanin}$$

Tyrosinase is an oxidase and has two copper atoms in its active site which bind an oxygen molecule. It is a transmembrane protein and is found in the membrane of large organelles in the melanocytes called melanosomes. Most of the protein, including the active site, is inside the melanosome.

Tyrosinases occur in plant as well as in animal tissues. The action of the enzyme can be seen in the blackening of a slice of potato left exposed to the air.

Huntington's disease

So far, in the examples of the inheritance of human conditions, the mutations have been inherited as recessive alleles. Huntington's disease (HD) provides an example of a mutation that is inherited as a dominant allele. This means that most people with the condition are heterozygous and have a 1 in 2 chance of passing on the condition to a child.

HD is a neurological disorder resulting in involuntary movements (chorea) and progressive mental deterioration. Brain cells are lost and the ventricles of the brain become larger. The age of onset is variable, but occurs most commonly in middle age, so that individuals may have children before they know that they themselves have the condition.

The mutation is an unstable segment in a gene on chromosome 4 coding for a protein, huntingtin. In people who do not have HD, the segment is made up of a small number of repeats of the triplet of bases CAG. People with HD have a larger number of repeats of the CAG triplet. This is called a 'stutter'. There is a rough inverse correlation between the number of times the triplet of bases is repeated and the age of onset of the condition: the more stutters, the earlier the condition appears.

Gene control in prokaryotes

In both prokaryotes and eukaryotes, transcription of a gene is controlled by transcription factors. These are proteins that bind to a specific DNA sequence and control the flow of information from DNA to RNA by controlling the formation of mRNA.

To understand how gene expression in bacteria is controlled, you must distinguish between structural genes and regulatory genes.

- Genes that code for proteins required by a cell are called structural genes. Such proteins may literally form part of a cellular structure, but they may also have some other role, such as acting as an enzyme.
- Genes that code for proteins that regulate the expression of other genes are called regulatory genes.

You must also distinguish between repressible and inducible enzymes.

- The synthesis of a repressible enzyme can be prevented by binding a repressor protein to a specific site, called an operator, on a bacterium's DNA.
- The synthesis of an inducible enzyme occurs only when its substrate is present. Transcription of the gene occurs as a result of the inducer (the enzyme's substrate) interacting with the protein produced by the regulatory gene.

The different roles of structural and regulatory genes can be seen by looking at the control of gene expression in a prokaryote using the *lac* operon. An **operon** is a length of DNA making up a unit of gene expression in a bacterium. It consists of one or more structural genes and also control regions of DNA that are recognised by the products of regulatory genes.

The *lac* operon

The enzyme β-galactosidase hydrolyses the disaccharide lactose to the monosaccharides glucose and galactose.

In the bacterium, *Escherichia coli*, the number of molecules of this enzyme present in a bacterial cell varies according to the concentration of lactose in the medium in which the bacterium is growing. The bacterium has one copy of the gene coding for β-galactosidase and so, to alter the concentration of the enzyme in its cell, it must regulate the transcription of the gene.

The *lac* operon consists of a cluster of three structural genes and a length of DNA including operator and promoter regions. The three structural genes are:

- *lacZ*, coding for β-galactosidase
- *lacY*, coding for permease (which allows lactose to enter the cell)
- *lacA*, coding for transacetylase.

Close to the promoter, but not actually part of the operon, is its regulatory gene (Figure 16.23).

The sequence of events when there is no lactose in the medium in which the bacterium is growing is as follows:

- the regulatory gene codes for a protein called a repressor
- the repressor binds to the operator region, close to the gene for β-galactosidase
- in the presence of bound repressor at the operator, RNA polymerase cannot bind to DNA at the promoter region
- no transcription of the three structural genes can take place.

The repressor protein is allosteric. This means that it has two binding sites. When the protein binds to a molecule at one site, this affects its ability to bind to a different molecule at the other binding site. The site that binds to DNA is separate from the site that binds to lactose. When lactose binds to its site, the shape of the protein changes so that the DNA-binding site is closed. Compare this mechanism with enzyme inhibition in Chapter 3.

When lactose is present in the medium in which the bacterium is growing:

- lactose is taken up by the bacterium
- lactose binds to the repressor protein, distorting its shape and preventing it from binding to DNA at the operator site
- transcription is no longer inhibited and messenger RNA is produced from the three structural genes. The genes have been switched on and are transcribed together (Figure 16.23).

This mechanism allows the bacterium to produce β-galactosidase, permease and transacetylase only when lactose is available in the surrounding medium and to produce them in equal amounts. It avoids the waste of energy and materials in producing enzymes for taking up and hydrolysing a sugar that the bacterium may never meet. However, the sugar can be hydrolysed when it is available. The enzyme β-galactosidase is an inducible enzyme. Look back at the distinction between repressible and inducible enzymes and convince yourself that this is so.

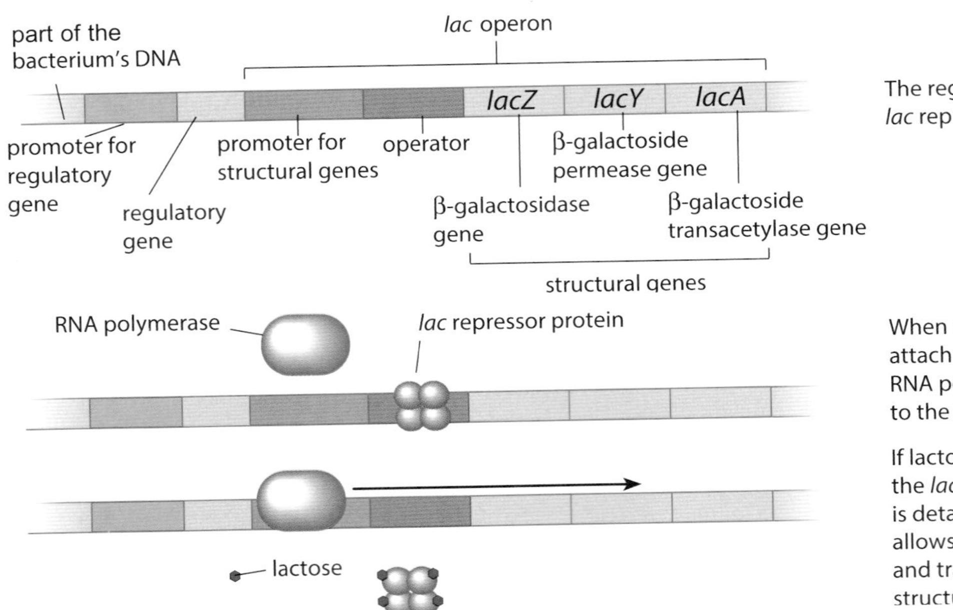

The regulatory gene codes for the *lac* repressor protein.

When the *lac* repressor protein is attached to the operator gene, RNA polymerase cannot attach to the DNA.

If lactose is present, it binds to the *lac* repressor protein, which is detached from the DNA. This allows RNA polymerase to bind and transcribe the operon's structural genes.

Figure 16.23 Regulation of gene expression by the *lac* operon.

The bacterium uses glucose in preference to other sugars. When a bacterium finds both glucose and lactose in the medium in which it is growing, it represses the use of lactose by suppressing the *lac* operon by means of a different transcription control factor.

Gene control in eukaryotes

In general, the number of different proteins that act as transcription factors increases with increasing size of the genome. This means that eukaryotes have many more ways of regulating gene expression than have prokaryotes. In humans, for example, about 10% of the genes code for proteins which act as transcription factors.

The factors may bind to the promoter region of a gene. They may increase or decrease the transcription of the gene. Whatever the mechanism, their role is to make sure that genes are expressed in the correct cell at the correct time and to the correct extent.

Effects of transcription factors include the following.

- General transcription factors are necessary for transcription to occur. They form part of the protein complex that binds to the promoter region of the gene concerned.
- Other factors activate appropriate genes in sequence, allowing the correct pattern of development of body regions.
- A transcription factor is responsible for the determination of sex in mammals.
- Transcription factors allow responses to environmental stimuli, such as switching on the correct genes to respond to high environmental temperatures.
- Some transcription factors, including the products of proto-oncogenes and tumour suppressor genes, regulate the cell cycle, growth and apoptosis (programmed cell death) (Chapter 5).
- Hormones have their effect through transcription factors.

The plant hormone, gibberellin, controls seed germination in plants such as wheat and barley by stimulating the synthesis of amylase (Chapter 15, page 356). It is a good example of how a hormone can influence transcription. It has been shown that, in barley seeds, application of gibberellin causes an increase in the transcription of mRNA coding for amylase.

Gibberellin has this effect by causing the breakdown of DELLA proteins (Figure 16.24). A DELLA protein inhibits the binding of a transcription factor, such as phytochrome-interacting protein (PIF), to a gene promoter. By causing the breakdown of the DELLA protein, gibberellin allows PIF to bind to its target promoter. Transcription of the gene can then take place, resulting in an increase in amylase production.

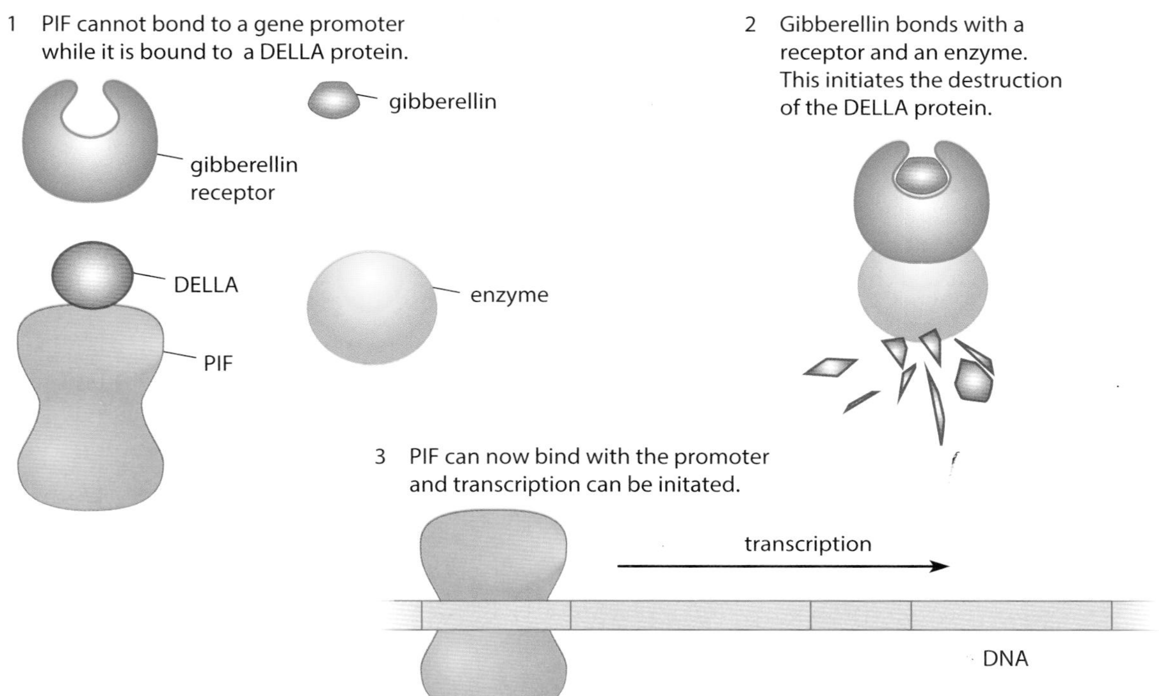

Figure 16.24 How gibberellin controls gene transcription.

Summary

- Homologous chromosomes are pairs of chromosomes in a diploid cell that have the same structure and the same genes, but not necessarily the same varieties of those genes.
- Meiosis consists of two divisions. The first division, meiosis I, separates the homologous chromosomes, so that each cell now has only one of each pair. The second division, meiosis II, separates the chromatids of each chromosome. Meiotic division therefore produces four cells, each with one complete set of chromosomes.
- Diploid organisms contain two copies of each gene in each of their cells. In sexual reproduction, gametes are formed containing one copy of each gene. Each offspring receives two copies of each gene, one from each of its parents.
- The cells produced by meiosis are genetically different from each other and from their parent cell. This results from independent assortment of the chromosomes as the bivalents line up on the equator during metaphase I, and also from crossing over between the chromatids of homologous chromosomes during prophase I.
- Genetic variation also results from random fertilisation, as gametes containing different varieties of genes fuse together to form a zygote.
- An organism's genetic constitution is its genotype. The observable expression of its genes is its phenotype. Different varieties of a gene are called alleles. Alleles may show dominance, codominance or recessiveness. An organism possessing two identical alleles of a gene is homozygous; an organism possessing two different alleles of a gene is heterozygous. If a gene has several different alleles, such as the gene for human blood groups, these are known as multiple alleles. The position of a gene on a particular chromosome is its locus.
- A gene found on the X chromosome but not on the Y chromosome is known as a sex-linked gene. Genes that are close together on a chromosome that is not a sex chromosome are said to be autosomally linked.
- The genotype of an organism showing dominant characteristics can be determined by looking at the offspring produced when it is crossed with an organism showing recessive characteristics. This is called a test cross. Monohybrid crosses consider the inheritance of one gene. Dihybrid crosses consider the inheritance of two different genes. Different genes may interact to affect the same phenotypic character. The chi-squared (χ^2) test can be used to find out whether any differences between expected results and observed results of a genetic cross are due to chance, or whether the difference is significant.
- Mutation can be defined as an unpredictable change in the base sequence in a DNA molecule (gene mutation) or in the structure or number of chromosomes (chromosome mutation). New alleles arise by gene mutation. Gene mutations include base substitutions, deletions or additions.
- The $\mathbf{Hb^S}$ (sickle cell) allele arose by base substitution. The allele responsible for Huntington's disease includes a repeated triplet of nucleotides called a 'stutter'. Albinism and haemophilia show the effect on the phenotype of missing or inactive polypeptides.
- 'Structural' genes code for the proteins required by a cell for its structure or metabolism, whereas 'regulatory' genes control the expression of other genes. A repressor protein can block the synthesis of a 'repressible' enzyme, by binding to the gene's operator site. An 'inducible' enzyme is synthesised only when its substrate is present.
- The *lac* operon provides an example of how a prokaryote can alter the transcription of a cluster of structural genes coding for enzymes concerned with lactose uptake and metabolism, depending on whether or not lactose is present.
- Transcription factors in eukaryotes make sure that genes are expressed in the correct cell, at the correct time and to the correct extent. In plants, gibberellins allow gene transcription by causing the breakdown of DELLA proteins which inhibit the binding of transcription factors.

End-of-chapter questions

1 A cell in the process of meiosis was seen to have a spindle with sister chromatids being drawn towards opposite poles of the cell. In what stage of meiosis was the cell?

A anaphase I

B anaphase II

C metaphase I

D metaphase II [1]

2 All the offspring of a cross between pure-bred red-flowered and pure-bred white-flowered snapdragons were pink. Two of these pink-flowered plants were interbred. What proportion of the offspring were pink?

A 25%

B 33%

C 50%

D 100% [1]

3 A man has haemophilia. Which statement correctly describes the inheritance of the gene causing his condition?

A He inherited the recessive allele from his mother.

B He inherited the dominant allele from his father.

C He can pass the recessive allele to a son.

D He can pass the dominant allele to a daughter. [1]

4 The diploid (2*n*) chromosome number of *Drosophila* is 8. Copy and complete the table to show the different outcomes of mitotic and meiotic division of a *Drosophila* cell.

	Mitosis	Meiosis
number of division cycles		
number of daughter cells		
number of chromosomes per nucleus in daughter cells		

[3]

5 Copy and complete the table to compare meiosis with mitosis.

Mitosis	Meiosis
maintains the chromosome number	
does not involve crossing over or independent assortment	
gives daughter nuclei that are genetically identical (apart from mutation) to one another and to the parent nucleus	

[3]

6 a Describe the essential difference between meiosis I and meiosis II. [2]

b State the similarity between meiosis II and mitosis. [1]

[Total: 3]

7 Diagram **1** shows chromosomes in the nucleus of a diploid cell.

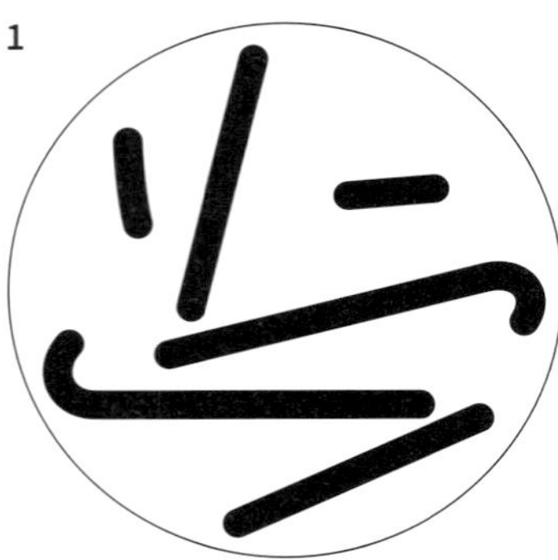

a Draw the nucleus of a gamete produced from this cell. [1]

b What type of nuclear division would be used in the production of the gamete? [1]

c Draw a diagram to show what the nucleus would look like in anaphase of **mitosis**. [3]

Diagrams **2** and **3** below show the same diploid nucleus as in diagram **1**. However, the chromosomes have been shaded.

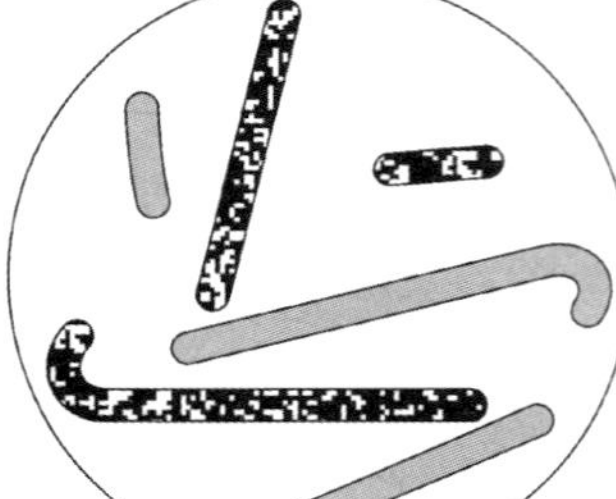

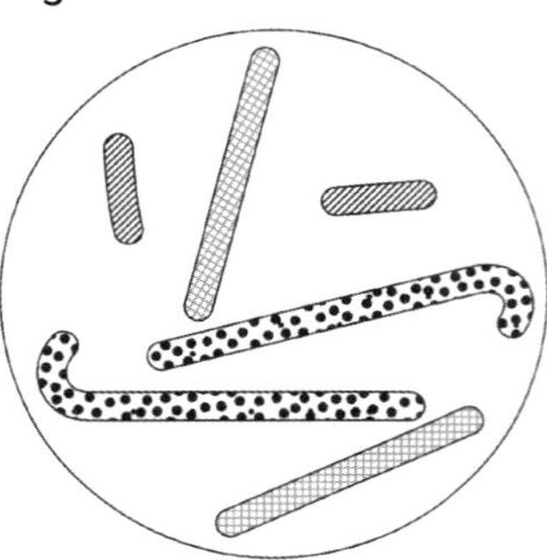

d State what the different types of shading represent in each nucleus. [2]

e Draw a karyogram based on the diploid nucleus shown in all three diagrams. [3]

[Total: 10]

8 There is no crossing over during meiosis in male *Drosophila.* Assuming that no mutation occurs, the only source of genetic variation is independent assortment. Given that the diploid (2*n*) chromosome number is 8, calculate the number of genetically different spermatozoa that can be produced. [1]

9 Distinguish between the following pairs of terms.

a **genotype** and **phenotype** [2]

b **homozygous** and **heterozygous** [2]

[Total: 4]

10 In sweet-pea plants, the gene **A**/**a** controls flower colour. The dominant allele gives purple flowers and the recessive allele red flowers.

A second gene, **B**/**b,** controls the shape of the pollen grains. The dominant allele gives elongated grains and the recessive allele spherical grains.

A plant with the genotype **AaBb** was test-crossed by interbreeding it with a plant with red flowers and spherical pollen grains.

Copy and complete the table to show the expected ratio of phenotypes of the offspring of this cross. The gametes from one parent are already in the table.

			Gametes of the other parent
Gametes of one parent	**AB**	genotype: phenotype:	
	Ab	genotype: phenotype:	
	aB	genotype: phenotype:	
	ab	genotype: phenotype:	

[5]

11 a The fruit fly, *Drosophila melanogaster,* feeds on sugars found in damaged fruits. A fly with normal features is called a wild type. It has a grey striped body and its wings are longer than its abdomen. There are mutant variations such as an ebony-coloured body or vestigial wings. These three types of fly are shown in the diagrams.

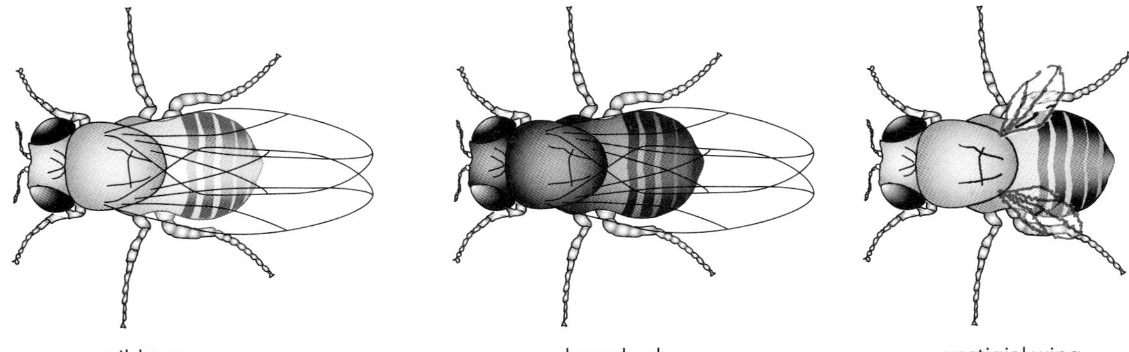

wild type ebony body vestigial wing

Wild-type features are coded for by dominant alleles: **A** for wild-type body and **B** for wild-type wings.

Explain what is meant by the terms *allele* and *dominant*. [2]

b Two wild-type fruit flies were crossed. Each had alleles **A** and **B** and carried alleles for ebony body and vestigial wings.

Draw a genetic diagram to show the possible offspring of this cross. [6]

c When the two heterozygous flies in **b** were crossed, 384 eggs hatched and developed into adult flies. A chi-squared (χ^2) test was carried out to test the significance of the differences between observed and expected results:

$$\chi^2 = \sum \frac{(O - E)^2}{E}$$

where:

$\sum$ = sum of

O = observed value

E = expected value

i Copy and complete the table.

	Phenotypes of *Drosophila melanogaster*			
	grey body long wing	grey body vestigial wing	ebony body long wing	ebony body vestigial wing
Observed number (O)	207	79	68	30
Expected ratio	9	3	3	1
Expected number (E)	216	72	72	24
$O - E$	−9		−4	6
$(O - E)^2$	81		16	36
$(O - E)^2/E$	0.38		0.22	1.50

[3]

ii Calculate the value for χ^2. [1]

The table below relates χ^2 values to probability values.

As four classes of data were counted, the number of degrees of freedom was 4 − 1 = 3. The table gives values of χ^2 where there are three degrees of freedom.

Probability greater than	0.50	0.20	0.10	0.05	0.01	0.001
Values for χ^2	2.37	4.64	6.25	7.82	11.34	16.27

iii Using your value for χ^2 and the table above, explain whether or not the observed results were significantly different from the expected results. [2]

[Total: 14]

Cambridge International AS and A Level Biology 9700/41, Question 7, October/November 2009

Chapter 17: Selection and evolution

Learning outcomes

You should be able to:

- describe the differences between continuous and discontinuous variation
- explain how the environment may affect phenotype
- explain the importance of variation as the basis for natural and artificial selection
- explain how natural selection occurs
- describe how humans use selective breeding to improve features of crop plants and domesticated animals
- explain how new species may arise, and how molecular evidence can provide information about relationships between species
- explain why organisms may become extinct

Deceptive flowers

Coevolution is the evolution of the adaptations of two species as a result of the selective pressures each exerts on the other. This has occurred in the adaptations of the flowers of various species of plants, such as orchids, and their specific pollinators.

The structure of an orchid flower consists of three outer sepals, two petals and a conspicuously shaped and coloured 'lip' (labellum), which is the third petal. In the centre of the flower is the 'column' which carries packages of pollen on its upper side and the surface of the stigma on its lower side. This arrangement encourages cross-pollination and avoids self-pollination.

In bee orchids and fly orchids, the lip both looks and smells like a female of the species of pollinating insect (Figure 17.1). Male insects, attempting to mate with the 'decoy female', collect pollen on their backs and deliver it to the stigma of another flower.

Over time, the lip has become more and more like a female insect. The orchid has a dedicated pollinator that is not attracted to the flowers of other species. The male insect receives his reward of nectar or wax so long as he is attracted by the decoy female.

Figure 17.1 A fly orchid.

Variation

In Chapter 16, you have seen how sexual reproduction produces **genetic variation** among the individuals in a population. Genetic variation is caused by:

- independent assortment of chromosomes, and therefore alleles, during meiosis
- crossing over between chromatids of homologous chromosomes during meiosis
- random mating between organisms within a species
- random fertilisation of gametes
- mutation.

The first four of these processes reshuffle existing alleles in the population. Offspring have combinations of alleles which differ from those of their parents and from each other. This genetic variation produces **phenotypic variation**.

Mutation, however, does not reshuffle alleles that are already present. Mutation can produce completely new alleles. This may happen, for example, if a mistake occurs in DNA replication, so that a new base sequence occurs in a gene. This is probably how the sickle cell allele of the gene for the production of the β-globin polypeptide first arose. Such a change in a gene, which is quite unpredictable, is called a **gene mutation**. The new allele is very often recessive, so it frequently does not show up in the population until some generations after the mutation actually occurred, when by chance two descendants of organisms in which the mutation happened mate and produce offspring.

Mutations that occur in body cells, or somatic cells, often have no effects at all on the organism. Somatic mutations cannot be passed on to offspring by sexual reproduction. However, mutations in cells in the ovaries or testes of an animal, or in the ovaries or anthers of a plant, may be inherited by offspring. If a cell containing a mutation divides to form gametes, then the gametes may also contain the mutated gene. If such a gamete is one of the two which fuse to form a zygote, then the mutated gene will also be in the zygote. This single cell then divides repeatedly to form a new organism, in which all the cells will contain the mutated gene.

Genetic variation, whether caused by the reshuffling of alleles during meiosis and sexual reproduction or by the introduction of new alleles by mutation, can be passed on by parents to their offspring, giving differences in phenotype. Genetic variation provides the raw material on which natural selection can act. Variation within a population means that some individuals have features that give them an advantage over other members of that population. Variation in phenotype is also caused by the **environment** in which organisms live. For example, some organisms might be larger than others because they had access to better quality food while they were growing. Variation caused by the environment is **not** passed on by parents to their offspring.

QUESTION

17.1 Explain why variation caused by the environment cannot be passed from an organism to its offspring.

Continuous and discontinuous variation

Phenotypic differences between you and your friends include qualitative differences such as blood groups and quantitative differences such as height and mass.

Qualitative differences fall into clearly distinguishable categories, with no intermediates – for example, you have one of four possible ABO blood groups: A, B, AB or O. This is **discontinuous variation.**

In contrast, the quantitative differences between your individual heights or masses may be small and difficult to distinguish. When the heights of a large number of people are measured, there are no distinguishable height classes. Instead there is a range of heights between two extremes (Figure 17.2). This is **continuous variation.**

The genetic basis of continuous and discontinuous variation

Both qualitative and quantitative differences in phenotype may be inherited. Both may involve several different genes. However, there are important differences between them.

In discontinuous (qualitative) variation:

- different alleles at a single gene locus have large effects on the phenotype
- different genes have quite different effects on the phenotype.

In continuous (quantitative) variation:

- different alleles at a single gene locus have small effects on the phenotype
- different genes have the same, often additive, effect on the phenotype
- a large number of genes may have a combined effect on a particular phenotypic trait; these genes are known as polygenes.

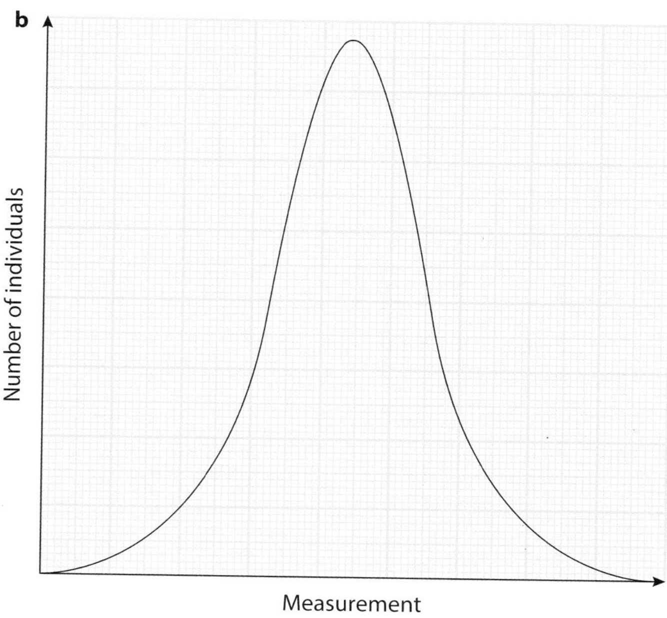

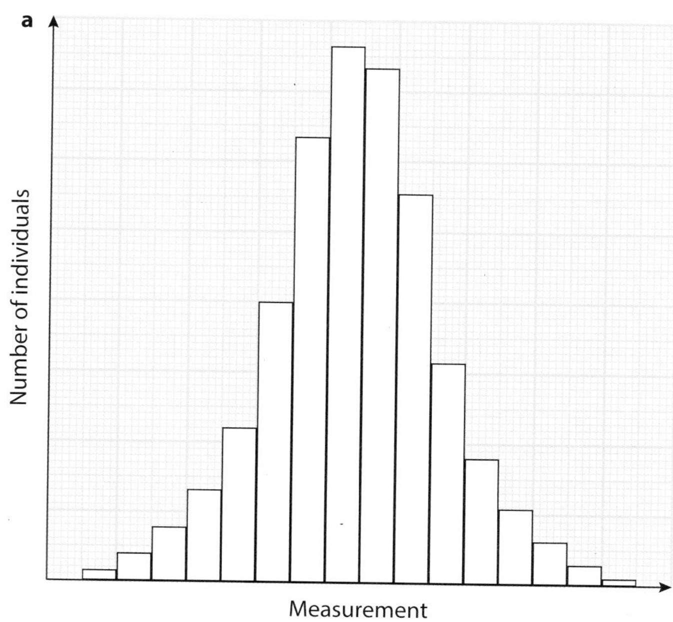

Figure 17.2 **a** Distribution curve and **b** a histogram, showing continuous variation.

In Chapter 16, you met a number of examples of discontinuous variation showing the large effect of the different alleles of a single gene. The inheritance of sickle cell anaemia and haemophilia are examples of discontinuous variation in humans. Flower colour in snapdragons, stem colour of tomato plants and feather colour of chickens are examples from other organisms. From these examples, you can see that dominance and gene interaction tend to reduce phenotypic variation.

Two of the typical effects of the inheritance of continuous variation, namely the small effects of the different alleles of one gene on the phenotype and the additive effect of different genes on the same phenotypic character, can be seen in a hypothetical example of the inheritance of an organism's height.

Suppose that the height of an organism is controlled by two unlinked (that is, on different chromosomes) genes: **A/a** and **B/b**. The recessive alleles of both genes (**a** and **b**) each contribute x cm to the height of the organism. The dominant alleles (**A** and **B**) each add $2x$ cm.

Since the effect of such genes is additive, the homozygote recessive (**aabb**) is therefore potentially $4x$ cm tall and the homozygote dominant (**AABB**) is potentially $8x$ cm tall. The other genotypes will fall between these extremes.

Parental phenotypes	$4x$ cm tall	$8x$ cm tall
Parental genotypes	**aabb**	**AABB**
Gametes	(**ab**)	(**AB**)
Offspring genotypes	all **AaBb**	
Offspring phenotypes	all $6x$ cm tall	

Interbreeding these potentially $6x$ cm tall offspring gives all possible genotypes and phenotypes among the 16 possibilities.

Parental phenotypes	$6x$ cm tall	$6x$ cm tall
Parental genotypes	**AaBb**	**AaBb**
Gametes	(**AB**) or (**Ab**) or (**aB**) or (**ab**)	(**AB**) or (**Ab**) or (**aB**) or (**ab**)
	in equal proportions	

continued ...

Offspring genotypes and phenotypes:

	Gametes from one parent: (**AB**)	(**Ab**)	(**aB**)	(**ab**)
Gametes from other parent (**AB**)	**AABB** $8x$ cm	**AABb** $7x$ cm	**AaBB** $7x$ cm	**AaBb** $6x$ cm
(**Ab**)	**AABb** $7x$ cm	**AAbb** $6x$ cm	**AaBb** $6x$ cm	**Aabb** $5x$ cm
(**aB**)	**AaBB** $7x$ cm	**AaBb** $6x$ cm	**aaBB** $6x$ cm	**aaBb** $5x$ cm
(**ab**)	**AaBb** $6x$ cm	**Aabb** $5x$ cm	**aaBb** $5x$ cm	**aabb** $4x$ cm

The number of offspring and their potential heights according to their genotypes are summarised in the histogram in Figure 17.3. These results fall approximately on a normal distribution curve.

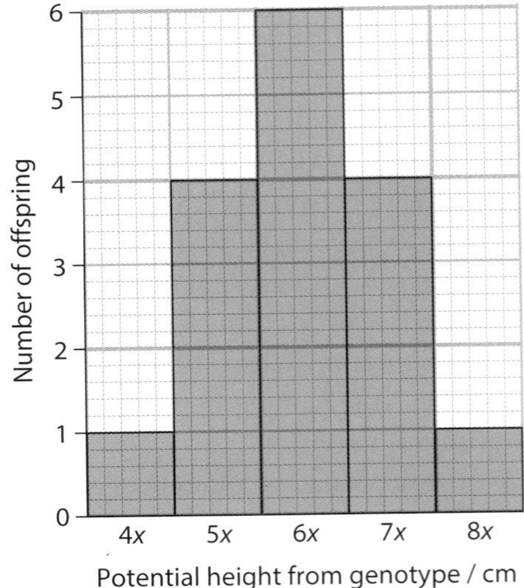

Figure 17.3 The additive effect of alleles.

These hypothetical results come from assuming that two unlinked genes, each with two alleles, contribute to the height of the organism. Think about what would happen to the quantitative character if more genes, each with an additive effect, were involved (polygenes). The genes may have more than two alleles. Suppose that all the genes affecting height are on different chromosomes: the number of discrete height classes increases as more genes are involved and the differences between these classes get less. Even if two or more of the genes are linked on the same chromosome (Chapter 16, page 383), potentially reducing the number of classes of offspring and increasing the difference between them, crossing over in meiosis (page 384) will restore the variation. The differences between different classes will be further smoothed out by environmental effects, as discussed in the next section.

QUESTION

17.2 **a** Distinguish between continuous and discontinuous variation.

b Explain the genetic basis of continuous variation.

Environmental effects on the phenotype

In our hypothetical example of continuous variation just given, the heights shown are those that would be expected from the genotype alone. If you were able to take a number of individuals, all with the same genotypic contribution to height, it would be most unlikely that their heights would be exactly the same when measured. Environmental effects may allow the full genetic potential height to be reached or may stunt it in some way.

One individual might have less food, or less nutritious food, than another with the same genetic contribution. A plant may be in a lower light intensity or in soil with fewer nutrients than another with the same genetic potential height. Other examples of the effect of the environment include the development of dark tips to ears, nose, paws and tail in the Himalayan colouring of rabbits (page 378) and of Siamese and Burmese cats (Figure 17.4). This colouring is caused by an allele which allows the formation of the dark pigment only at low temperature. The extremities are the coldest parts of the animal, so the colour is produced there. When an area somewhere else on its body is plucked of fur and kept cold, the new fur growing in this region will be dark.

In a classic experiment, the American geneticists Ralph Emerson and Edward East crossed two varieties of maize which differed markedly in cob length. Both of the parental varieties (Black Mexican and Tom Thumb) were pure-bred lines. The cob lengths of the plants used as parents and the first and second generations of offspring resulting from the cross were measured to the nearest centimetre. The number of cobs in each length category was counted. The results are shown in Table 17.1.

Both parental varieties were pure-bred and so were homozygous at a large number of loci. The first generation were genetically different from the parents but were genetically the same as one another. The phenotypic variation that you can see within the two parental varieties and also within the first generation of offspring shows the effect of the environment. The second generation of offspring shows a much wider variation in cob length. This is both genetic and environmental.

The variation of two populations, such as the pure-bred Black Mexican and Tom Thumb maize plants, can be compared using the *t*-test (page 500).

Just as genetic variation provides the raw material on which natural selection can act, so in selective breeding it is important to know how much of the phenotypic variation is genetic and how much is environmental in origin. There is no point in selecting parents for a breeding programme on the basis of environmental variation.

Figure 17.4 In Siamese cats, dark hair develops where the skin is at a relatively low temperature.

Cob length / cm	5	6	7	8	9	10	11	12	13	14	15	16	17	18	19	20	21
Number of Black Mexican parent cobs									3	11	12	14	26	15	10	7	2
Number of Tom Thumb parent cobs	4	21	24	8													
Number of F1 offspring cobs					1	12	12	14	17	9	4						
Number of F2 offspring cobs			1	10	19	26	47	73	68	68	39	25	15	9	1		

Table 17.1 Variation in cob length of two parental varieties of maize and of the first (F1) and second (F2) generations of a cross between them.

Natural selection

All organisms have the reproductive potential to increase their populations. Rabbits, for example, produce several young in a litter, and each female may produce several litters each year. If all the young rabbits survived to adulthood and reproduced, then the rabbit population would increase rapidly. Figure 17.5 shows what might happen.

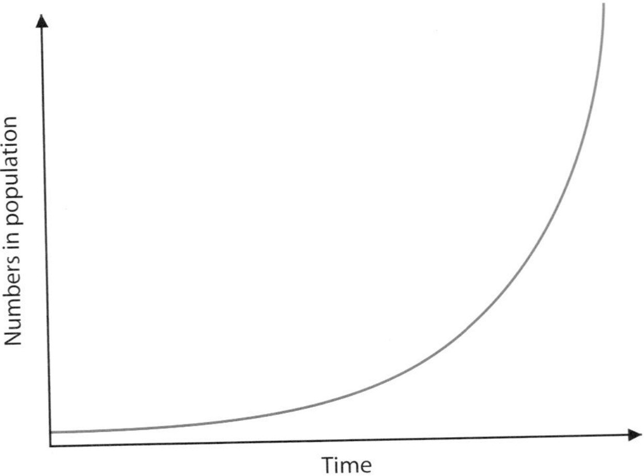

Figure 17.5 If left unchecked by environmental factors, numbers in a population may increase exponentially.

This sort of population growth actually did happen in Australia in the 19th century. In 1859, 12 pairs of rabbits from Britain were released on a ranch in Victoria, as a source of food. The rabbits found conditions to their liking. Rabbits feed on low-growing vegetation, especially grasses, of which there was an abundance. There were very few predators to feed on them, so the number of rabbits soared. Their numbers became so great that they seriously affected the availability of grazing for sheep (Figure 17.6).

Such population explosions are rare in normal circumstances. Although rabbit populations have the potential to increase at such a tremendous rate, they do not usually do so.

As a population of rabbits increases, various environmental factors come into play to keep down the rabbits' numbers. These factors may be **biotic** – caused by other living organisms such as through predation, competition for food, or infection by pathogens – or they may be **abiotic** – caused by non-living components of the environment such as water supply or nutrient levels in the soil. For example, the increasing number of rabbits eat an increasing amount of vegetation, until food is in short supply. The larger population of rabbits may allow the populations of predators such as foxes, stoats and weasels to increase. Overcrowding may occur, increasing the ease with which diseases such as myxomatosis may spread (Figure 17.7). This disease is caused by a virus that is transmitted by fleas. The closer together the rabbits live, the more easily fleas, and therefore viruses, will pass from one rabbit to another.

Figure 17.6 Attempts to control the rabbit population explosion in Australia in the mid to late 19th century included 'rabbit drives', in which huge numbers were rounded up and killed. Eventually, myxomatosis brought numbers down.

Figure 17.7 Rabbits living in dense populations are more likely to get myxomatosis than those in less crowded conditions.

These environmental factors act to reduce the rate of growth of the rabbit population. Of all the rabbits born, many will die from lack of food, or be killed by predators, or die from myxomatosis. Only a small proportion of young will grow to adulthood and reproduce, so population growth slows.

If the pressure of the environmental factors is sufficiently great, then the population size will decrease. Only when the numbers of rabbits have fallen considerably will the numbers be able to grow again. Over a period of time, the population will oscillate about a mean level. Figure 17.8 shows this kind of pattern in a lemming population over 11 years. The oscillations in lemming populations are particularly marked; in other species, they are usually less spectacular.

This type of pattern is shown by the populations of many organisms. The number of young produced is far greater than the number which will survive to adulthood. Many young die before reaching reproductive age.

What determines which will be the few rabbits to survive, and which will die? It may be just luck. However, some rabbits will be born with a better chance of survival than others. Variation within a population of rabbits means that some will have features which give them an advantage in the 'struggle for existence'.

One feature that may vary is coat colour. Most rabbits have alleles which give the normal agouti (brown) colour. A few, however, may be homozygous for the recessive allele which gives white coat. Such white rabbits will stand out distinctly from the others, and are more likely to be picked out by a predator such as a fox. They are less likely to survive than agouti rabbits. The chances of a white rabbit reproducing and passing on its alleles for white coat to its offspring are very small, so the allele for white coat will remain very rare in the population. The term 'fitness' is often used to refer to the extent to which organisms are adapted to their environment. **Fitness** is the capacity of an organism to survive and transmit its genotype to its offspring.

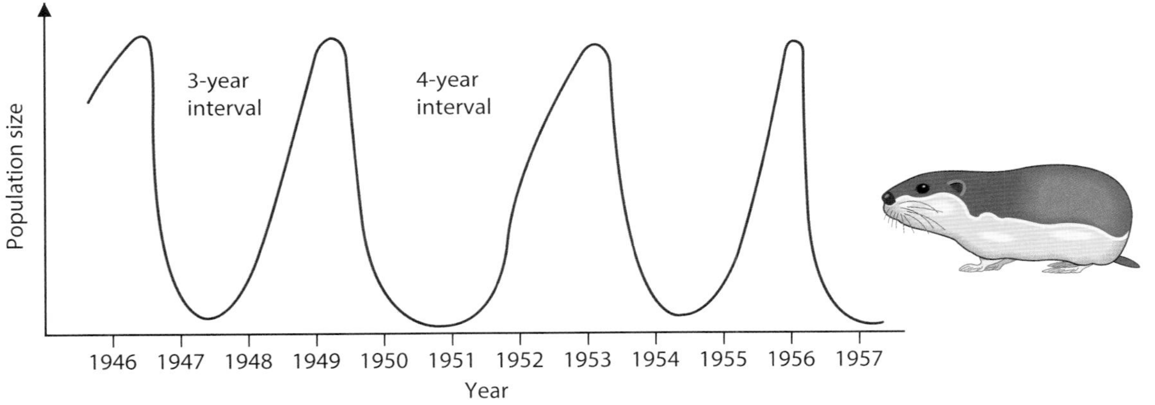

Figure 17.8 Lemming populations are famous for their large increases and decreases. In some years, populations become so large that lemmings may emigrate in one group from overcrowded areas. The reason for the oscillating population size is not known for certain, although it has been suggested that food supply or food quality may be the main cause. As the population size rises, food supplies run out, so the population size 'crashes'. Once the population size has decreased, food supplies begin to recover, and the population size rises again.

Predation by foxes is an example of a selection pressure. Selection pressures increase the chances of some alleles being passed on to the next generation, and decrease the chances of others. In this case, the alleles for agouti coat have a selective advantage over the alleles for white. The alleles for agouti will remain the commoner alleles in the population, while the alleles for white will remain very rare. The alleles for white coat may even disappear completely.

The effects of such selection pressures on the frequency of alleles in a population is called natural selection. Natural selection raises the frequency of alleles conferring an advantage, and reduces the frequency of alleles conferring a disadvantage.

QUESTION

17.3 Skomer is a small island off the coast of Wales. Rabbits have been living on the island for many years. There are no predators on the island.

- **a** Rabbits on Skomer are not all agouti. There are quite large numbers of rabbits of different colours, such as black and white. Suggest why this is so.
- **b** What do you think might be important selection pressures acting on rabbits on Skomer?

Evolution

The general theory of evolution is that organisms have changed over time. Usually, natural selection keeps things the way they are. This is stabilising selection (Figure 17.9a and b, and Figure 17.10). Agouti rabbits are the best-adapted rabbits to survive predation, so the agouti allele remains the most common coat colour allele in rabbit populations. Unless something changes, then natural selection will ensure that this continues to be the case.

However, if a **new environmental factor** or a **new allele** appears, then allele frequencies may also change. This is called directional selection (Figure 17.9c).

A third type of selection, called disruptive selection, can occur when conditions favour both extremes of a population. This type of selection maintains different phenotypes (polymorphism) in a population (Figure 17.9d).

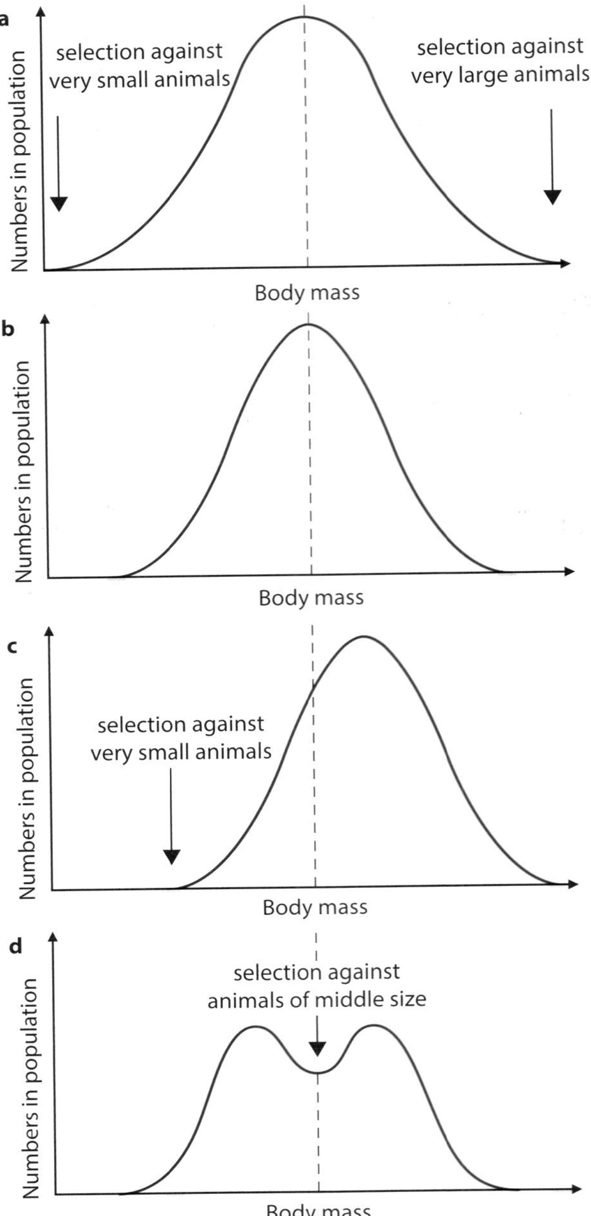

Figure 17.9 If a characteristic in a population, such as body mass, shows wide variation, selection pressures often act against the two extremes (graph **a**). Very small or very large individuals are less likely to survive and reproduce than those whose size lies nearer the centre of the range. This results in a population with a narrower range of body size (graph **b**). This type of selection, which tends to keep the variation in a characteristic centred around the same mean value, is called **stabilising selection.** Graph **c** shows what would happen if selection acted against smaller individuals but not larger ones. In this case, the range of variation shifts towards larger size. This type of selection, which results in a change in a characteristic in a particular direction, is called **directional selection**. Graph **d** shows the result of selection that favours both large and small individuals, but acts against those whose size is in the middle of the range. This is **disruptive selection.**

Figure 17.10 The tuatara, *Sphenodon punctatus*, is a lizard-like reptile that lives in New Zealand. Fossils of a virtually identical animal have been found in rocks 200 million years old. Natural selection has acted to keep the features of this organism the same over all this time.

Figure 17.11 The white winter coat of a mountain hare provides excellent camouflage from predators when viewed against snow.

A new environmental factor

Imagine that we are plunged into a new Ice Age. The climate becomes much colder, so that snow covers the ground for almost all of the year. Assuming that rabbits can cope with these conditions, white rabbits now have a selective advantage during seasons when snow lies on the ground, as they are better camouflaged (like the hare in Figure 17.11). Rabbits with white fur are more likely to survive and reproduce, passing on their alleles for white fur to their offspring. The frequency of the allele for white coat increases at the expense of the allele for agouti. Over many generations, almost all rabbits will come to have white coats rather than agouti.

A new allele

Because they are random events, most mutations that occur produce features that are harmful. That is, they produce organisms that are less well adapted to their environment than 'normal' organisms. Other mutations may be neutral, conferring neither an advantage nor a disadvantage on the organisms within which they occur. Occasionally, mutations may produce useful features.

Imagine that a mutation occurs in the coat colour gene of a rabbit, producing a new allele which gives a better-camouflaged coat colour than agouti. Rabbits possessing this new allele will have a selective advantage. They will be more likely to survive and reproduce than agouti rabbits, so the new allele will become more common in the population. Over many generations, almost all rabbits will come to have the new allele.

Such changes in allele frequency in a population are the basis of **evolution**. Evolution occurs because natural selection gives some alleles a better chance of survival than others. Over many generations, populations may gradually change, becoming better adapted to their environments. Examples of such change are the development of antibiotic resistance in bacteria (described in the next section) and industrial melanism in the peppered moth, *Biston betularia* (page 406).

In contrast, the role of malaria in the global distribution of sickle cell anaemia is an example of how the interaction of two strong selection pressures can maintain two alleles within certain populations (page 407).

Antibiotic resistance

Antibiotics are chemicals produced by living organisms, which inhibit or kill bacteria but do not normally harm human tissue. Most antibiotics are produced by fungi. The first antibiotic to be discovered was penicillin, which was first used during the Second World War to treat a wide range of diseases caused by bacteria. Penicillin stops cell wall formation in bacteria, so preventing cell reproduction.

When someone takes penicillin to treat a bacterial infection, bacteria that are sensitive to penicillin die. In most cases, this is the entire population of the disease-causing bacteria. However, by chance, there may be among them one or more individual bacteria with an allele giving resistance to penicillin. One example of such an allele occurs in some populations of the bacterium *Staphylococcus*, where some individual bacteria produce an enzyme, penicillinase, which inactivates penicillin.

As bacteria have only a single loop of DNA, they have only one copy of each gene, so the mutant allele will have an immediate effect on the phenotype of any bacterium possessing it. These individuals have a tremendous selective advantage. The bacteria without this allele will be killed, while those bacteria with resistance to penicillin can survive and reproduce. Bacteria reproduce very rapidly in ideal conditions, and even if there was initially only one resistant bacterium, it might produce ten thousand million descendants within 24 hours. A large population of a penicillin-resistant strain of *Staphylococcus* would result.

Such antibiotic-resistant strains of bacteria are continually appearing (Figure 17.12). By using antibiotics, we change the environmental factors which exert selection pressures on bacteria. A constant race is on to find new antibiotics against new resistant strains of bacteria.

Alleles for antibiotic resistance often occur on plasmids (Figure 1.30, page 21). Plasmids are quite frequently transferred from one bacterium to another, even between different species. Thus it is even possible for resistance to a particular antibiotic to arise in one species of bacterium, and be passed on to another. The more we use antibiotics, the greater the selection pressure we exert on bacteria to evolve resistance to them.

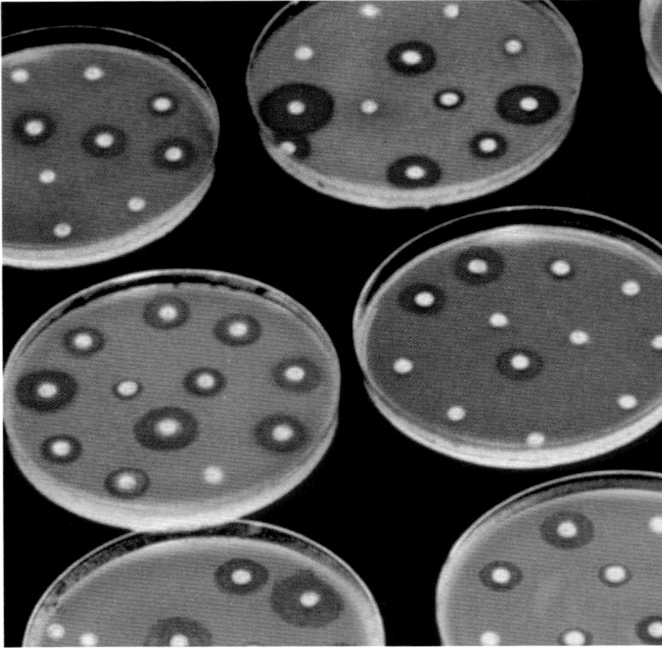

Figure 17.12 The red gel in each of these Petri dishes has been inoculated with bacteria. The small light blue circles are discs impregnated with antibiotics. Bacteria that are resistant to an antibiotic are able to grow right up to the disc containing it. See also Figure 10.16, page 215.

QUESTION

17.4 Suggest how each of the following might decrease the chances of an antibiotic-resistant strain of bacteria developing:

a limiting the use of antibiotics to cases where there is a real need

b regularly changing the type of antibiotic that is prescribed for a particular disease

c using two or more antibiotics together to treat a bacterial infection.

Industrial melanism

One well-documented case of the way in which changing environmental factors may produce changes in allele frequencies is that of the peppered moth, *Biston betularia* (Figure 17.13), in the UK and Ireland. This is a night-flying moth which spends the day resting underneath the branches of trees. It relies on camouflage to protect it from insect-eating birds that hunt by sight. Until 1849, all specimens of this moth in collections had pale wings with dark markings, giving a speckled appearance. In 1849, however, a black (melanic) individual was caught near Manchester (Figure 17.14). During the rest of the 19th century, the numbers of black *Biston betularia* increased dramatically in some areas, whereas in other parts of the country the speckled form remained the more common.

Figure 17.13 Dark form of peppered moth on dark and pale tree bark.

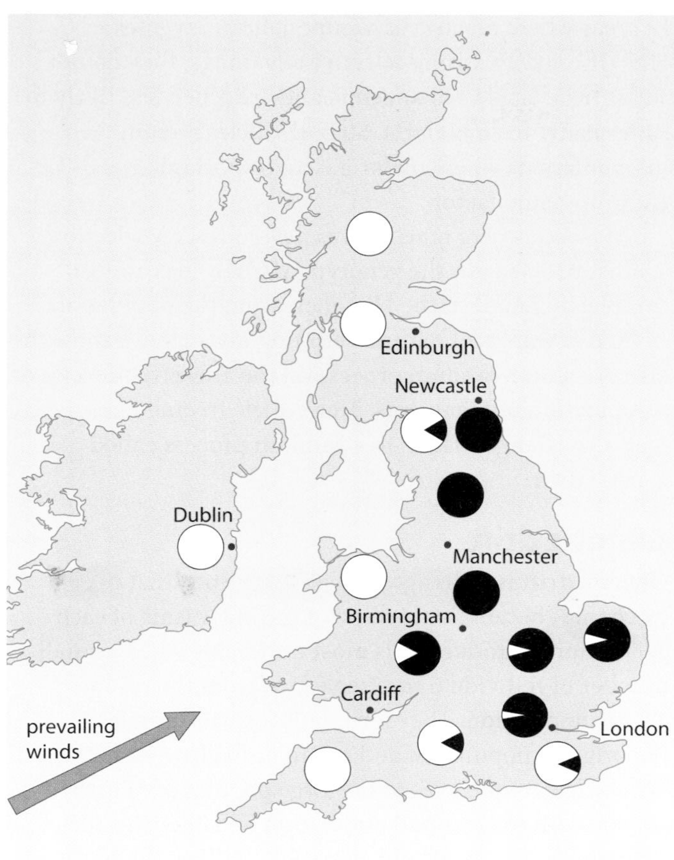

Figure 17.14 The distribution of the pale and dark forms of the peppered moth, *Biston betularia*, in the UK and Ireland during the early 1960s. The ratio of dark to pale areas in each circle shows the ratio of dark to pale moths in that part of the country.

The difference in the black and speckled forms of the moth is caused by a single gene. The normal speckled colouring is produced by a recessive allele of this gene, **c**, while the black colour is produced by a dominant allele, **C**. Up until the late 1960s, the frequency of the allele **C** increased in areas near to industrial cities. In non-industrial areas, the allele **c** remained the more common allele.

The selection pressure causing the change of allele frequency in industrial areas was predation by birds. In areas with unpolluted air, tree branches are often covered with grey, brown and green lichen. On such tree branches, speckled moths are superbly camouflaged.

However, lichens are very sensitive to pollutants such as sulfur dioxide, and do not grow on trees near to or downwind of industries releasing pollutants into the air. Trees in these areas therefore have much darker bark, against which the dark moths are better camouflaged. Experiments have shown that pale moths have a much higher chance of survival in unpolluted areas than dark moths, while in polluted areas the dark moths have the selective advantage. As air pollution from industry is reduced, the selective advantage swings back in favour of the speckled variety. So we would expect the proportion of speckled moths to increase if we succeeded in reducing the output of certain pollutants. This is, in fact, what has happened since the 1970s.

It is important to realise that mutations to the **C** allele have probably always been happening in *B. betularia* populations. The mutation was not caused by pollution. Until the 19th century there was such a strong selection pressure against the **C** allele that it remained exceedingly rare. Mutations of the **c** allele to the **C** allele may have occurred quite frequently, but moths with this allele would almost certainly have been eaten by birds before they could reproduce. Changes in environmental factors only affect the likelihood of an allele surviving in a population; they do **not** affect the likelihood of such an allele arising by mutation.

Sickle cell anaemia

In Chapter 16, we saw how an allele, $\mathbf{Hb^S}$, of the gene that codes for the production of the β-globin polypeptide can produce sickling of red blood cells. People who are homozygous for this allele have sickle cell anaemia. This is a severe form of anaemia that is often lethal.

The possession of two copies of this allele obviously puts a person at a great selective disadvantage. People who are homozygous for the sickle cell allele are less likely to survive and reproduce. Until recently, almost everyone with sickle cell anaemia died before reaching reproductive age. Yet the frequency of the sickle cell allele is very high in some parts of the world. In some parts of East Africa, almost 50% of babies born are carriers for this allele, and 14% are homozygous, suffering from sickle cell anaemia. How can this be explained?

The parts of the world where the sickle cell allele is most common are also the parts of the world where malaria is found (Figure 17.15). Malaria is caused by a protoctist parasite, *Plasmodium*, which can be introduced into a person's blood when an infected mosquito bites (Figure 17.16 and pages 202–205). The parasites enter the red blood cells and multiply inside them. Malaria is the major source of illness and death in many parts of the world.

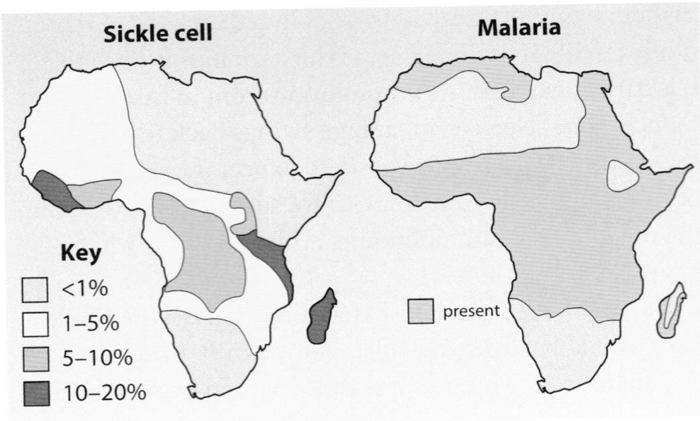

Figure 17.15 The distribution of people with at least one copy of the sickle cell allele, and the distribution of malaria, in Africa.

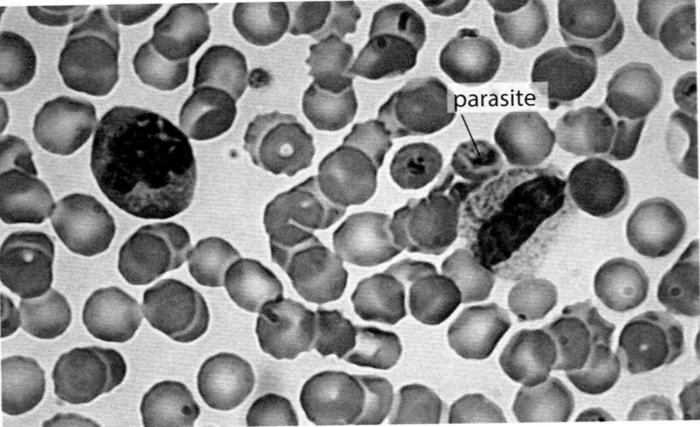

Figure 17.16 Red blood cells infected with malarial parasite. Some cells have multiple parasites.

In studies carried out in some African states, it has been found that people who are heterozygous for the sickle cell allele (**Hb**^{S}) are much less likely to suffer from a serious attack of malaria than people who are homozygous for the **Hb**^{A} allele. Heterozygous people with malaria only have about one-third the number of *Plasmodium* in their blood as do **Hb**^{A}**Hb**^{A} homozygotes. In one study, of a sample of 100 children who died from malaria, all except one were **Hb**^{A}**Hb**^{A} homozygotes, although within the population as a whole, 20% of people were heterozygotes.

There are, therefore, two strong selection pressures acting on these two alleles.

- Selection against people who are homozygous for the sickle cell allele, **Hb**^{S}**Hb**^{S}, is very strong, because they become seriously anaemic.
- Selection against people who are homozygous **Hb**^{A}**Hb**^{A}, is also very strong, because they are more likely to die from malaria.

In areas where malaria is common, heterozygotes, **Hb**^{A}**Hb**^{S}, have a strong selective advantage; they do not suffer from sickle cell anaemia and are much less likely to suffer badly from malaria. So both alleles remain in populations where malaria is an important environmental factor.

In places where malaria was never present, selection against people with the genotype **Hb**^{S}**Hb**^{S} has almost completely removed the **Hb**^{S} allele from the population.

The examples of natural selection given above show the effect of a **non-random** process on the allele frequencies of a population of organisms. These allele frequencies may also change thanks to a **random** process called genetic drift.

Genetic drift

Genetic drift is a change in allele frequency that occurs by chance, because only some of the organisms of each generation reproduce. It is most noticeable when a small number of individuals are separated from the rest of a large population. They form only a small sample of the original population and so are unlikely to have the same allele frequencies as the large population. Further genetic drift in the small population will alter the allele frequencies still more and evolution of this population may take a different direction from that of the larger parent population. This process, occurring in a recently isolated small population, is called the **founder effect**.

The Hardy–Weinberg principle

When a particular phenotypic trait is controlled by two alleles of a single gene, **A**/**a**, the population will be made up of three genotypes: **AA**, **Aa** and **aa**. Calculations based on the **Hardy–Weinberg principle** allow the proportions of each of these genotypes in a large, randomly mating population to be calculated.

The frequency of a genotype is its proportion of the total population. The total is the whole population (that is 1) and the frequencies are given as decimals (e.g. 0.25) of the total.

We use the letter p to represent the frequency of the dominant allele, **A**, in the population and the letter q to represent the frequency of the recessive allele, **a**. Then, since there are only two alleles of this gene:

$$p + q = 1 \qquad \textbf{(Equation 1)}$$

- the chance of an offspring inheriting a dominant allele from both parents $= p \times p = p^2$
- the chance of an offspring inheriting a recessive allele from both parents $= q \times q = q^2$
- the chance of an offspring inheriting a dominant allele from the father and a recessive allele from the mother $= p \times q = pq$
- the chance of an offspring inheriting a dominant allele from the mother and a recessive allele from the father $= p \times q = pq$

So, $p^2 + 2pq + q^2 = 1$ **(Equation 2)**

WORKED EXAMPLE 1

Calculating genotype frequency

The homozygous recessives in a population can be recognised and counted. Suppose that the incidence of the **aa** genotype is 1 in 100 individuals (1%).

Then, $q^2 = 0.01$ and $q = \sqrt{0.01} = 0.1$

So, using **Equation 1**:

$p = 1 - 0.1 = 0.9$ and $p^2 = (0.9)^2 = 0.81$

That is 81% of the population are homozygous **AA**.

And, $2pq = 2 \times 0.9 \times 0.1 = 0.18$

Or, using **Equation 2**:

$2pq = 1 - (0.01 + 0.81) = 0.18$

That is, 18% of the population are heterozygous **Aa**.

These Hardy–Weinberg calculations do not apply when the population is small or when there is:

- significant selective pressure against one of the genotypes
- migration of individuals carrying one of the two alleles into, or out of, the population
- non-random mating.

What is the use of these calculations? When the ratios of the different genotypes in a population have been determined, their predicted ratios in the next generation can be compared with the observed values. Any differences can be tested for significance using the χ^2 test (page 386). If the differences are significant and migration and non-random mating can be discounted, then there is evidence that directional selection is occurring in the population.

QUESTION

17.5 a A phenotypic trait is controlled by two alleles of a single gene **A/a**. Explain why only the homozygous recessives, **aa**, can be recognised.

b Calculate the proportions of homozygous dominant and of heterozygous individuals in a population in which the proportion of homozygous recessives is 20%.

Artificial selection

Selective breeding of dairy cattle

Sometimes, the most important selection pressures on organisms are those applied by humans. When humans purposefully apply selection pressures to populations, the process is known as **artificial selection**.

Consider, for example, the development of modern breeds of cattle. Cattle have been domesticated for a very long time (Figure 17.17).

Figure 17.17 The original wild cattle from which individuals were first domesticated are thought to have looked very much like **a**, the modern Chillingham White breed. Selective breeding over many centuries has produced many different breeds, such as **b**, the Guernsey. Guernseys have been bred for the production of large quantities of fat-rich milk. Notice the large udder compared with the Chillingham.

For thousands of years, people have tried to 'improve' their cattle. Desired features include docility (making the animal easier to control), fast growth rates and high milk yields. Increases in these characteristics have been achieved by **selective breeding**. Individuals showing one or more of these desired features to a larger degree than other individuals are chosen for breeding. Some of the alleles conferring these features are passed on to the individuals' offspring. Again, the 'best' animals from this generation are chosen for breeding. Over many generations, alleles conferring the desired characteristics increase in frequency, while those conferring characteristics not desired by the breeder decrease in frequency. In many cases, such 'disadvantageous' alleles are lost entirely.

Such selective breeding of dairy cattle presents the breeder with problems. The animals are large and take time to reach maturity. The gestation period is long and the number of offspring produced is small. A bull cannot be assessed for milk production since this a sex-limited trait (note that this is not the same as sex-linked). Instead, the performance of the bull's female offspring is looked at to see whether or not to use the bull in further crosses. This is called progeny testing and is a measure of the bull's value to the breeder.

It is important to realise that selective breeders have to consider the whole genotype of an organism, not just the genes affecting the desired trait, such as increased milk yield. Within each organism's genotype are all the alleles of genes that adapt it to its particular environment. These genes are called background genes.

Suppose that the chosen parents come from the same environment and are from varieties that have already undergone some artificial selection. It is likely that such parents share a large number of alleles of background genes, so the offspring will be adapted for the same environment.

But suppose instead that one of the chosen parents comes from a different part of the world. The offspring will inherit appropriate alleles from only one parent. It may show the trait being selected for, but it may not be well-adapted to its environment.

Crop improvement

The same problem is seen when a cross is made between a cultivated plant and a related wild species. Although most species will not breed with a different species, some can be interbred to give fertile offspring. Such species are often those that do not normally come into contact with one another, because they live in different habitats or areas. The wild parent will have alleles that are not wanted and which have probably been selected out of the cultivated parent.

Farmers have been growing cereal crops for thousands of years. There is evidence that wheat was being grown in the so-called 'fertile crescent' – land that was watered by the rivers Nile, Tigris and Euphrates – at least 10 000 years ago (Figure 17.18). In South and Central America, maize was being farmed at least 7000 years ago.

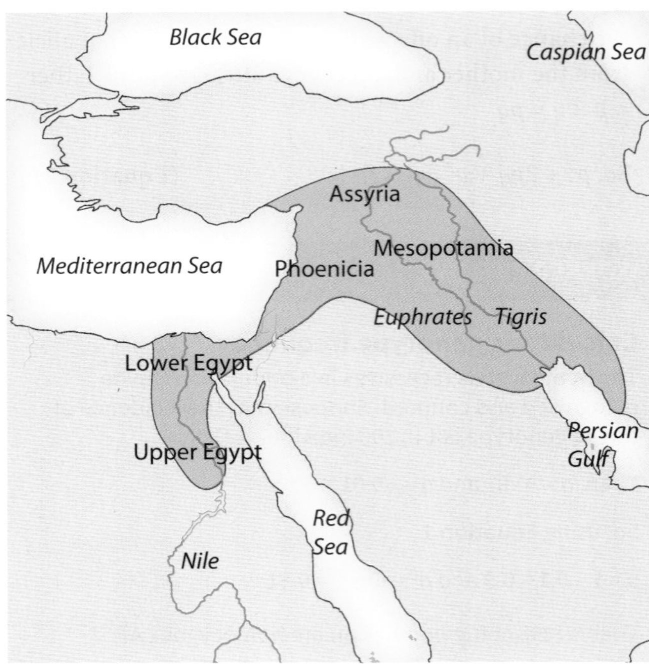

Figure 17.18 Wheat was first farmed in the 'fertile crescent' (shown in green) around 10 000 years ago.

It was not until the 20th century that we really understood how we can affect the characteristics of crop plants by artificial selection and selective breeding. But, although these early farmers knew nothing of genes and inheritance, they did realise that characteristics were passed on from parents to offspring. The farmers picked out the best plants that grew in one year, allowing them to breed and produce the grain for the next year. Over thousands of years, this has brought about great changes in the cultivated varieties of crop plants, compared with their wild ancestors.

Today, selective breeding continues to be the main method by which new varieties of crop plants are produced. In some cases, however, gene technology is being used to alter or add genes into a species in order to change its characteristics.

Most modern varieties of wheat belong to the species *Triticum aestivum*. Selective breeding has produced many different varieties of wheat. Much of it is grown to produce grains rich in gluten, which makes them good for making bread flour. For making other food products such as pastry, varieties that contain less gluten are best.

Breeding for resistance to various fungal diseases, such as head blight, caused by *Fusarium*, is important, because of the loss of yield resulting from such infections. Successful introduction of an allele giving resistance takes many generations, especially when it comes from a wheat grown in a different part of the world. To help with such selective breeding, the Wheat Genetic Improvement Network was set up in the UK in 2003 to bring together research workers and commercial plant breeders. Its aim is to support the development of new varieties by screening seed collections for plants with traits such as disease resistance, or climate resilience (Figure 17.19), or efficient use of nitrogen fertilisers. Any plant with a suitable trait is grown in large numbers and passed to the commercial breeders.

Wheat plants now have much shorter stems than they did only 50 years ago. This makes them easier to harvest and means they have higher yields (because they put more energy into making seeds rather than growing tall) (Figure 17.20). The shorter stems also make the plants less susceptible to being knocked flat by heavy rains, and means they produce less straw, which has little value and costs money to dispose of.

Figure 17.19 Wheat breeders are attempting to produce new varieties of wheat that will be able to grow in the higher temperatures that global warming is expected to bring.

Figure 17.20 Harvesting wheat.

Most of the dwarf varieties of wheat carry mutant alleles of two reduced height (Rht) genes. These genes code for DELLA proteins which reduce the effect of gibberellins on growth (page 355). The mutant alleles cause dwarfism by producing more of, or more active forms of, these transcription inhibitors. A mutant allele of a different gene, called 'Tom Thumb', has its dwarfing effect because the plant cells do not have receptors for gibberellins and so cannot respond to the hormone.

Rice, *Oryza sativa*, is also the subject of much selective breeding. The International Rice Research Institute, based in the Philippines, holds the rice gene bank and together with the Global Rice Science Partnership coordinates research aimed at improving the ability of rice farmers to feed growing populations.

The yield of rice can be reduced by bacterial diseases such as bacterial blight, and by a range of fungal diseases including various 'spots' and 'smuts'. The most significant fungal disease is rice blast, caused by the fungus *Magnaporthe*. Researchers are hoping to use selective breeding to produce varieties of rice that show some resistance to all these diseases.

Inbreeding and hybridisation in maize

Maize, *Zea mays*, is also known as corn in some parts of the world. It is a sturdy, tall grass with broad, strap-shaped leaves (Figure 17.21). Maize grows best in climates with long, hot summers, which provide plenty of time for its cobs (seed heads) to ripen. It was originally grown in Central and South America, but now it forms the staple crop in some regions of Africa, and is grown as food for people or animals in Europe, America, Australia, New Zealand, China and Indonesia. If maize plants are inbred (crossed with other plants with genotypes like their own), the plants in each generation become progressively smaller and weaker (Figure 17.22). This **inbreeding depression** occurs because, in maize, homozygous plants are less vigorous than heterozygous ones. Outbreeding – crossing with other, less closely related plants – produces heterozygous plants that are healthier, grow taller and produce higher yields.

Figure 17.21 Maize plants in flower.

Figure 17.22 The effects of inbreeding depression in maize over eight generations.

However, if outbreeding is done at random, the farmer would end up with a field full of maize in which there was a lot of variation between the individual plants. This would make things very difficult. To be able to harvest and sell the crop easily, a farmer needs the plants to be uniform. They should all be about the same height and all ripen at about the same time.

So the challenge when growing maize is to achieve both heterozygosity and uniformity. Farmers buy maize seed from companies that specialise in using inbreeding to produce homozygous maize plants, and then crossing them. This produces F1 plants that all have the same genotype. There are many different homozygous maize varieties, and different crosses between them can produce a large number of different hybrids, suited for different purposes. Every year, thousands of new maize hybrids are trialled, searching for varieties with characteristics such as high yields, resistance to more pests and diseases, and good growth in nutrient-poor soils or where water is in short supply.

QUESTIONS

17.6 Use a genetic diagram to show why all the plants produced by crossing two different homozygous strains of maize will have the same genotype.

17.7 Explain why farmers need to buy maize seed from commercial seed producers each year, rather than saving their own seed to plant.

The Darwin–Wallace theory of evolution by natural selection

The original theory that natural selection might be a mechanism by which evolution could occur was put forward independently by both Charles Darwin and Alfred Russel Wallace in 1856. They knew nothing of genes or mutations, so did not understand how natural variation could arise or be inherited. Nevertheless, they realised the significance of variation. Their observations and deductions can be summarised briefly as follows.

Observation 1 Organisms produce more offspring than are needed to replace the parents.

Observation 2 Natural populations tend to remain stable in size over long periods.

Deduction 1 There is competition for survival (a 'struggle for existence').

Observation 3 There is variation among the individuals of a given species.

Deduction 2 The best adapted variants will be selected for by the natural conditions operating at the time. In other words, natural selection will occur. The 'best' variants have a selective advantage; 'survival of the fittest' occurs.

As you can see, this theory, put forward well over a century ago, hardly differs from what we now know about natural selection and evolution. The major difference is that we can now think of natural selection as selecting particular **alleles** or groups of alleles.

The title of Darwin's most famous and important book contained the words *On the Origin of Species*. Yet, despite his thorough consideration of how natural selection could cause evolution, he did not attempt to explain how **new species** could be produced. This process is called speciation.

Species and speciation

In this chapter, you have seen how natural selection can act on variation within a population to bring about changes in allele frequencies. There is now a wealth of evidence to support the idea that natural selection is the force that has produced all of the different species of organisms on Earth. Yet in the examples of directional selection described on pages 404–407 (the evolution of antibiotic resistance in bacteria, and changes in the frequency of wing colour in peppered moths) no **new** species have been produced. How can natural selection produce new species? Before we can begin to answer this question, we must answer another: exactly what is a species? This proves to be an extremely difficult question, with no neat answer.

One definition of a species that is quite widely accepted by biologists is: a group of organisms, with similar morphological, physiological, biochemical and behavioural features, which can interbreed to produce fertile offspring, and are reproductively isolated from other species.

'Morphological' features are structural features, while 'physiological' features are the way that the body works. 'Biochemical' features include the sequence of bases in DNA molecules and the sequence of amino acids in proteins.

Thus all donkeys look and work like donkeys, and can breed with other donkeys to produce more donkeys, which themselves can interbreed. All donkeys belong to the same species. Donkeys can interbreed with organisms of another similar species, horses, to produce offspring called mules. However, mules are infertile; they cannot breed and are effectively a 'dead-end'. Thus, donkeys and horses belong to different species.

When a decision needs to be made as to whether two organisms belong to the same species or to two different species, the organisms should ideally be tested to find out if they can interbreed successfully, producing fertile offspring. However, as you can imagine, this is not always possible. Perhaps the organisms are dead; they may even be museum specimens or fossils. Perhaps they are both of the same sex. Perhaps the biologist making the decision does not have the time or the facilities to attempt to interbreed them. Perhaps the organisms will not breed in captivity. Perhaps they are not organisms which reproduce sexually, but only asexually. Perhaps they are immature, and not yet able to breed.

As a result of all of these problems, it is quite rare to test the ability of two organisms to interbreed. Biologists frequently rely only on morphological, biochemical, physiological and behavioural differences to decide whether they are looking at specimens from one or two species. In practice, it may only be morphological features which are considered, because physiological and biochemical ones, and to some extent behavioural ones, are more time-consuming to investigate. Sometimes, however, detailed studies of DNA sequences may be used to assess how similar two organisms are to each other.

It can be extremely difficult to decide when these features are sufficiently similar or different to define two organisms as belonging to the same or different species.

This leads to great uncertainties and disagreements about whether to lump many slightly different variations of organisms together into one species, or whether to split them up into many different species.

Despite the problems described above, many biologists would agree that the feature which really decides whether or not two organisms belong to different species is their inability to interbreed successfully. In explaining how natural selection can produce new species, therefore, we must consider how a group of interbreeding organisms (that is, all of the same species) can produce another group of organisms which cannot interbreed successfully with the first group. The two groups must undergo reproductive isolation.

Reproductive isolation can take very different forms. **Prezygotic** (before a zygote is formed) isolating mechanisms include:

- individuals not recognising one another as potential mates or not responding to mating behaviour
- animals being physically unable to mate
- incompatibility of pollen and stigma in plants
- inability of a male gamete to fuse with a female gamete.

Postzygotic isolating mechanisms include:

- failure of cell division in the zygote
- non-viable offspring (offspring that soon die)
- viable, but sterile offspring.

Obviously, the last is the most wasteful of energy and resources.

Investigating how reproductive isolation can arise is not easy. The main difficulty is that this process **takes time**. A speciation experiment in a laboratory would have to run for many years. The evidence that we have for the ways in which speciation can occur is almost all circumstantial evidence. We can look at populations of organisms at one moment in time, that is now, and use the patterns we can see to suggest what might have happened, and might still be happening, over long periods of time.

Figure 17.23 *Hibiscus clayi* is found only on Hawaii, where it is in danger of extinction.

Allopatric speciation

One picture that emerges from this kind of observation is that geographical isolation has played a major role in the evolution of many species. This is suggested by the fact that many islands have their own unique groups of species. The Hawaiian and Galapagos islands, for example, are famous for their spectacular arrays of species of all kinds of animals and plants found nowhere else in the world (Figure 17.23).

Geographical isolation requires a barrier of some kind to arise between two populations of the same species, preventing them from mixing. This barrier might be a stretch of water. We can imagine that a group of organisms, perhaps a population of a species of bird, somehow arrived on one of the Hawaiian islands from mainland America; the birds might have been blown off course by a storm. Here, separated by hundreds of miles of ocean from the rest of their species on mainland America, the group interbred. The selection pressures on the island were very different from those on the mainland, resulting in different alleles being selected for. Over time, the morphological, physiological and behavioural features of the island population became so different from the mainland population that the two populations could no longer interbreed. A new species had evolved.

You can probably think of many other ways in which two populations of a species could be physically separated. A species living in dense forest, for example, could become split if large areas of forest are cut down, leaving 'islands' of forest in a 'sea' of agricultural land. Very small or immobile organisms can be isolated by smaller-scale barriers.

Speciation which happens like this, when two populations are separated from each other geographically, is called allopatric speciation. 'Allopatric' means 'in different places'.

However, it is also possible for new species to arise without the original populations being separated by a geographical barrier. This is known as sympatric speciation.

Sympatric speciation

Perhaps the commonest way in which sympatric speciation can occur is through polyploidy.

A polyploid organism is one with more than two complete sets of chromosomes in its cells. This can happen if, for example, meiosis goes wrong when gametes are being formed, so that a gamete ends up with two sets of chromosomes instead of one set. If two such gametes fuse, then the zygote gets four complete sets of chromosomes. It is said to be **tetraploid**.

Tetraploids formed in this way are often sterile. As there are four of each kind of chromosome, all four try to 'pair' up during meiosis I, and get in a terrible muddle. It is very difficult for the cell to divide by meiosis and produce new cells each with complete sets of chromosomes. However, the cell may well be able to grow perfectly well, and to reproduce asexually. There is nothing to stop mitosis happening absolutely normally. (Remember that chromosomes do not need to pair up in mitosis – they each behave quite independently.) This does quite often happen in plants but only rarely in animals, largely because most animals do not reproduce asexually.

Just occasionally, this tetraploid plant may manage to produce gametes. They will be diploid gametes. If one of these should fuse with a gamete from the normal, diploid, plant, then the resulting zygote will be **triploid**. Once again, it may be able to grow normally, but it will certainly be sterile. There is no way in which it can produce gametes, because it cannot share the three sets of chromosomes out evenly between the daughter cells.

So, the original diploid plant and the tetraploid that was produced from it cannot interbreed successfully. They can be considered to be different species. A new species has arisen in just one generation.

The kind of polyploid just described contained four sets of chromosomes all from the same species. It is said to be an **autopolyploid**. ('Auto' means 'self'.) Polyploids can also be formed that contain, say, two sets of chromosomes from one species and two sets from another closely related species. They are called **allopolyploids**. ('Allo' means 'other' or 'different'.) Meiosis actually happens more easily in an allotetraploid than in an autotetraploid, because the chromosomes from each species are not quite identical. So the two chromosomes from one species pair up with each other, while the two chromosomes from the other species pair up. This produces a much less muddled situation than in an autopolyploid, where the chromosomes try to get together in fours, so it is much more likely that meiosis can come to a successful conclusion. The allopolyploid may well be able to produce plenty of gametes. It is fertile.

Once again, however, the allopolyploid cannot interbreed with individuals from its parent species, for the same reasons as the autopolyploid. It is a new species.

One well-documented instance of speciation through allopolyploidy is the cord grass *Spartina anglica*. This is a vigorous grass that grows in salt marshes.

Before 1830, the species of *Spartina* that grew in these places in England was *S. maritima*. Then, in 1829, a different species called *S. alterniflora* was imported from America (Figure 17.24). *S. maritima* and *S. alterniflora* hybridised (interbred), producing a new species called *S. townsendii* (Figure 17.25). This is a diploid plant, with one set of chromosomes from *S. maritima* and one set from *S. alterniflora*. It is sterile, because the two sets of chromosomes from its parents cannot pair up, so it cannot undergo meiosis successfully. Nor can *S. townsendii* interbreed with either of its two parents, which is what makes it a different species. Although it is sterile, it has been able to spread rapidly, reproducing asexually by producing long underground stems called rhizomes, from which new plants can grow.

Figure 17.24 *Spartina alterniflora.*

Figure 17.25 *Spartina townsendii.* 

At some later time, probably around 1892, faulty cell division in *S. townsendii* somehow produced cells with double the number of chromosomes. A tetraploid plant was produced, probably from the fusion of two abnormal diploid gametes from *S. townsendii*. So this tetraploid has two sets of chromosomes that originally came from *S. maritima*, and two sets from *S. alterniflora*. It is an allotetraploid. These chromosomes can pair up with each other, two and two, during meiosis, so this tetraploid plant is fertile. It has been named *S. anglica*. It is more vigorous than any of the other three species, and has spread so widely and so successfully that it has practically replaced them in England.

Molecular comparisons between species

Molecular evidence from comparisons of the amino acid sequences of proteins and of the nucleotide sequences of mitochondrial DNA can be used to reveal similarities between related species.

Comparing amino acid sequences of proteins

As you saw in Chapter 16, changing a single amino acid in the primary structure of a protein may cause a dramatic change in its structure and function. However, for many proteins, small changes in the amino acid sequence leave the overall structure and the function of the protein unaltered. Typically, the part of the molecule essential for its function (such as the active site of an enzyme) remains the same, but other parts of the molecule may show changes. When the amino acid sequence of a particular protein is compared in different species, the number of differences gives a measure of how closely related the species are.

Let us take cytochrome c as an example. Cytochrome c is a component of the electron transfer chain in oxidative phosphorylation in mitochondria (pages 273–275). A protein with such an important function is expected to have a similar sequence of amino acids in different species since a poorly adapted cytochrome c molecule would result in the death of the organism.

When the sequences of cytochrome c from humans, mice and rats were compared, it was found that:

- all three molecules consist of 104 amino acids
- the sequences of mouse and rat cytochromes c are identical
- nine amino acids in human cytochrome c are different from the mouse or rat sequence
- most of these substitutions in human cytochrome c are of amino acids with the same type of R group (Appendix 1).

This comparison suggests that mice and rats are closely related species, sharing a recent common ancestor, and that humans are more distantly related, sharing a common ancestor with mice and rats less recently.

When the sequences of cytochrome c from other species, such as a fruit fly or a nematode worm, are also examined, the number of differences from the human sequence increases. These organisms are less closely related to humans.

Comparing nucleotide sequences of mitochondrial DNA

Differences in the nucleotide sequences of mitochondrial DNA (mtDNA) can be used to study the origin and spread of our own species, *Homo sapiens*. Human mitochondrial DNA is inherited through the female line. A zygote contains the mitochondria of the ovum, but not of the sperm. Since the mitochondrial DNA is circular (page 276) and cannot undergo any form of crossing over, changes in the nucleotide sequence can only arise by mutation.

Mitochondrial DNA mutates faster than nuclear DNA, acquiring one mutation every 25 000 years. Unlike nuclear DNA, mitochondrial DNA is not protected by histone proteins (page 95) and oxidative phosphorylation in the mitochondria (page 273) can produce forms of oxygen that act as mutagens.

Different human populations show differences in mitochondrial DNA sequences. These provide evidence for the origin of *H. sapiens* in Africa and for the subsequent migrations of the species around the world. These studies have led to the suggestion that all modern humans, of whatever race, are descendants from one woman, called Mitochondrial Eve, who lived in Africa between 150 000 and 200 000 years ago. This date is derived from the 'molecular clock' hypothesis, which assumes a constant rate of mutation over time and that the greater the number of differences in the sequence of nucleotides, the longer ago those individuals shared a common ancestor. The 'clock' can be calibrated by comparing nucleotide sequences of species whose date of speciation can be estimated from fossil evidence.

Analysis of mitochondrial DNA of the different species of anole lizards that are found throughout the Caribbean and the adjacent mainland provides evidence of their relationships (Figure 17.26). Each island species of lizard is found only on one island or a small group of islands. Table 17.2 shows the results of comparing part of the mitochondrial DNA of four of the species. These results show that the three species *Anolis brunneus*, *A. smaragdinus* and *A. carolinensis* are more closely related to *A. porcatus* than they are to each other. This suggests that these species have each originated from separate events in which a few individuals of *A. porcatus* spread from Cuba to three different places. The mitochondrial DNA analysis shows that allopatric speciation has occurred.

Extinctions

Species may become extinct, perhaps as a result of a change in climate or increased competition from a better-adapted species.

The International Union for Conservation of Nature (IUCN) annually publishes a Red List of threatened species. The 2013 list contained 21 286 species. (Visit the IUCN website at www.iucn.org and search for Red List.)

The species in the Red List are all under threat of extinction – of disappearing forever from the Earth. Of course, millions of species have become extinct in the past, sometimes huge numbers at one time in so-called mass extinctions. However, these events were all natural, and at least some are thought to have been caused by sudden changes in the environment, such as a large asteroid colliding with the Earth.

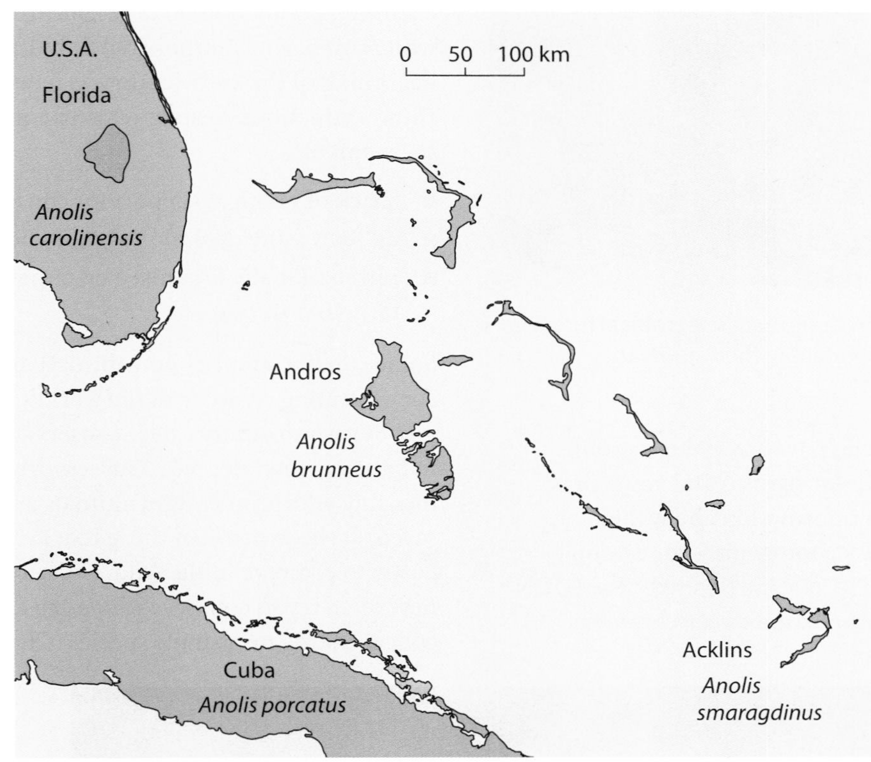

Figure 17.26 Anole lizard species in the Caribbean.

	A. brunneus	*A. smaragdinus*	*A. carolinensis*	*A. porcatus*
A. brunneus				
A. smaragdinus	12.1			
A. carolinensis	16.7	15.0		
A. porcatus	11.3	8.9	13.2	

Table 17.2 The results of comparing part of the mitochondrial DNA of four of the species of anole lizards. The smaller the number, the smaller the differences between the base sequences of the two species.

We are currently facing the likelihood of another mass extinction, this time caused by us. The main reason for this is the loss of habitats. Many species are adapted for survival in a particular habitat with a particular range of environmental conditions. Humans are destroying habitats by draining wetlands, cutting down rainforests (Figure 17.27), and polluting the air, water and soil. Another reason for a species to become extinct is if we kill too many of them, perhaps for sport or for food.

Figure 17.27 Orang-utans live only in dense tropical forest in Borneo. Deforestation is threatening their survival.

Some endangered species have a very high profile – for example, pandas, rhinos or tigers – and you will be able to find a great deal of information about them on the internet. Others are less photogenic – for example, the Kerry slug (Figure 17.28). Not surprisingly, the IUCN Red List has a very high proportion of vertebrates as opposed to invertebrates, and green plants as opposed to protoctists. There are no prokaryotes on the list. We have absolutely no idea how many of these are threatened.

Despite the high profile of some mammalian species, extinctions continue. Estimates in 2011 suggest that the global population of all tiger species is about 5000 individuals. In India, which is expected soon to overtake China as the nation with the largest human population, the pressure on the remaining tiger populations is intense. In China, where tiger products are thought to cure a variety of ills, poaching is still common and is highly organised.

In 2011, the western black rhino of Africa was declared extinct by the IUCN, which also described the northern white rhino of central Africa as 'probably extinct in the wild'. It is thought that the last Javan rhino in Vietnam was killed by poachers in 2010, leaving only the small and declining population on Java. Only the African southern white rhino, which numbered only 100 individuals at the beginning of the 20th century, is now flourishing. These rhino extinctions, despite years of conservation efforts, are the result of:

- a lack of political support for conservation
- an increasing demand for rhino horn
- internationally organised criminal groups targeting rhinos.

Some conservationists now think that it is time to stop concentrating on some of the world's high-profile species and to turn to others where conservation efforts are likely to have a greater degree of success. This would involve focusing efforts on certain animals and plants that **can** be saved, at the expense of those that are too difficult or too costly to preserve in the wild. Conservation programmes now often try to conserve whole ecosystems, rather than concentrating on a single species (Chapter 18).

Figure 17.28 Kerry slugs are found only in the south west of Ireland.

Summary

- Phenotypic variation may be continuous (as in the height or mass of an organism) or discontinuous (as in the human ABO blood groups). The genotype of an organism gives it the potential to show a particular characteristic. In many cases, the degree to which this characteristic is shown is also influenced by the organism's environment. Genetic variation within a population is the raw material on which natural selection can act.
- Meiosis, random mating and the random fusion of gametes produce genetic variation within populations of sexually reproducing organisms. Variation is also caused by the interaction of the environment with genetic factors, but such environmentally induced variation is not passed on to an organism's offspring. The only source of new alleles is mutation.
- All species of organisms have the reproductive potential to increase the sizes of their populations but, in the long term, this rarely happens. This is because environmental factors come into play to limit population growth. Such factors decrease the rate of reproduction or increase the rate of mortality so that many individuals die before reaching reproductive age.
- Within a population, certain alleles may increase the chance that an individual will survive long enough to be able to reproduce successfully. These alleles are, therefore, more likely to be passed on to the next generation than others. This is known as natural selection. Normally, natural selection keeps allele frequencies as they are; this is stabilising selection. However, if environmental factors that exert selection pressures change, or if new alleles appear in a population, then natural selection may cause a change in the frequencies of alleles; this is directional selection. Over many generations, directional selection may produce large changes in allele frequencies. This is how evolution occurs.
- The evolution of antibiotic resistance in bacteria and the spread of industrial melanism in moths are examples of changes in allele frequencies. The role of malaria in the global distribution of sickle cell anaemia is an example of how two strong opposing selection pressures can counterbalance each other in maintaining two alleles within certain populations.
- Allele frequencies in a small population may change thanks to a random process called genetic drift. The allele frequencies and the proportions of genotypes of a particular gene in a population can be calculated using the Hardy–Weinberg principle.
- A species can be defined as a group of organisms with similar morphology, behaviour, physiology and biochemistry that are capable of interbreeding to produce fertile offspring. However, in practice, it is not always possible to determine whether or not organisms can interbreed.
- Artificial selection involves the choice by humans of which organisms to allow to breed together, in order to bring about a desirable change in characteristics. Thus, artificial selection, like natural selection, can affect allele frequencies in a population.
- New species arise by a process called speciation. In allopatric speciation, two populations become isolated from one another, perhaps by some geographical feature, and then evolve along different lines until they become so different that they can no longer interbreed. In sympatric speciation, new species may arise through polyploidy.
- Species can become extinct (no longer existing on Earth) through a variety of mechanisms. Most are the result of human activity and include climate change, competition, habitat loss and killing.

End-of-chapter questions

1 Which of the following gives rise to genetic variation in a population?

1 crossing over and independent assortment in meiosis
2 different environmental conditions
3 random mating and fertilisation
4 mutation

A 1, 2, 3 and 4
B 1, 2 and 3 only
C 1, 3 and 4 only
D 2, 3 and 4 only [1]

2 A species of finch living on an isolated island shows variation in beak size. Birds with larger beaks can eat larger seeds.

After a period of drought on the island, large seeds were more plentiful than small seeds and the average size of the finches' beaks increased.

What explains this increase in size of beak?

A artificial selection acting against finches with small beaks
B directional selection acting against finches with small beaks
C increased rate of mutation resulting in finches with larger beaks
D stabilising selection acting against finches with the smallest and largest beaks [1]

3 Which effect of natural selection is likely to lead to speciation?

A Differences between populations are increased.
B The range of genetic variation is reduced.
C The range of phenotypic variation is reduced.
D Favourable alleles are maintained in the population. [1]

4 There are three genotypes of the gene for the β-globin polypeptide: **Hb^AHb^A**, **Hb^AHb^S** and **Hb^SHb^S**.

Copy and complete the table to show which genotypes have a selective advantage or disadvantage in different regions of the world.

	Region with no malaria	Region with malaria
Genotype(s) with selective advantage		
Genotype(s) with selective disadvantage		

[4]

5 The wings of butterflies are covered with microscopic scales that give them their colour and also provide waterproofing. The wings of some species have large transparent areas through which the colour of the vegetation on which the butterfly has settled can be seen. Because they lack scales, these areas have poor waterproofing. The butterflies are eaten by birds.

a Describe two selection pressures that are likely to control the size of the transparent areas of the wings of these butterflies. [2]

b In what circumstances might there be selection for larger transparent areas in the wings? [1]

[Total: 3]

6 Rearrange the order of the following statements to give a flow diagram showing the evolution of resistance to the antibiotic streptomycin by the bacterium *Escherichia coli*.

1 Most of the population of *E. coli* is resistant to streptomycin.
2 A mutation in a DNA triplet of a plasmid, changing TTT to TTG, gives an *E. coli* bacterium resistance to streptomycin.
3 The resistant bacterium divides and passes copies of the R plasmid (plasmid with gene for resistance to antibiotic) to its offspring.
4 Sensitive bacteria die in the presence of streptomycin as a selective agent.
5 The frequency of the mutated gene in the population increases.
6 The resistant bacterium has a selective advantage and survives. [3]

7 Hybrids produced by crossing two different inbred (homozygous) varieties are often more vigorous in their growth than either of their parents. Copy and complete the flow diagram to show how breeders of maize produce plants that show hybrid vigour.

Inbred line 1

Genotype: homozygous

Phenotype: uniform

Yield: low

Inbred line 2

Genotype: ....

Phenotype: ....

Yield: ....

Hybrid

Genotype: ....

Phenotype: ....

Yield: .... [6]

8 Copy and complete the table to compare artificial selection with natural selection.

Natural selection	Artificial selection
the selective agent is the total environment of the organism	
adaptations to the prevailing conditions are selected	
many different traits contributing to fitness are selected	

[3]

9 Pale and dark peppered moths were collected and placed on pale and dark areas of bark on trees in a park in Liverpool, England (Figure 17.13, page 406). Some of the moths were predated by birds. The results of the investigation are shown in the table.

Colour of moth	Percentage of moths taken by birds	
	from pale bark	from dark bark
pale	20	44
dark	40	15

a 40 dark moths were placed on dark bark. Calculate the number of moths taken by birds. Show your working. [2]

b Suggest an explanation for the differences in the numbers of moths taken by birds. [4]

[Total: 6]

10 The snail *Cepaea nemoralis* may have a yellow, pink or brown shell. Each colour shell may have up to five dark bands, or have no bands. Both shell colour and number of bands are genetically controlled. The snails are eaten by birds such as thrushes, which hunt by sight.

The following observations were made:

- Most snails living on a uniform background, such as short grass, have no bands.
- Most snails living on a green background, such as grass, are yellow.
- Most snails living on a non-uniform background, such as rough vegetation, have bands.

a Suggest an explanation for these observations. [4]

b Predict the phenotype of snails living on a dark background of dead leaves. [2]

c Suggest what will happen, during the course of a year, to the frequencies of the different alleles controlling shell colour and banding in a snail population living in deciduous woodland. (Deciduous trees shed their leaves in autumn. The background for the snails will be made up of dead leaves in the autumn and winter, and green vegetation in the spring and summer.) [4]

[Total: 10]

11 The heliconid butterflies of South America have brightly coloured patterns on their wings. A hybrid between two species, *Heliconius cydno* and *H. melpomene*, has wing patterns that are different from both parental species.

An investigation was carried out to see whether the hybrid was a new species.

Separate groups of four butterflies, each consisting of a male and female of one of the parental species and a male and female of the hybrid, were placed together and their choices of mates recorded. The results are shown in the table.

	Number of matings	
	H. melpomene **male**	**hybrid male**
H. melpomene **female**	15	0
hybrid female	0	15
	H. cydno **male**	**hybrid male**
H. cydno **female**	5	3
hybrid female	0	5

a With reference to the information in the table, explain whether or not the results of the investigation suggest that the hybrid butterfly is a separate species. [4]

b Suggest how the hybrid could be reproductively isolated from the two parent species of butterfly. [2]

c Briefly describe how allopatric speciation can occur. [4]

[Total: 10]

Chapter 18: Biodiversity, classification and conservation

Learning outcomes

You should be able to:

- explain the importance of biodiversity in terms of ecosystems, habitats and species
- investigate ecosystems using techniques for assessing the occurrence, abundance and distribution of species
- use statistical methods to analyse the relationships between the distribution and abundance of species and abiotic or biotic factors
- calculate an index of diversity and use it to compare the biodiversity of different areas
- describe and explain how species are classified
- outline the characteristic features of the three domains and of the four kingdoms of eukaryotic organisms
- discuss the threats to the biodiversity of aquatic and terrestrial ecosystems and the reasons for the need to maintain biodiversity
- discuss some of the methods used to protect endangered species including the control of invasive or alien species
- discuss the roles of non-governmental organisations in local and global conservation

Mockingbirds

Mockingbirds are found throughout the Americas. There are four species of mockingbird in the Galapagos Islands in the Pacific of which the Floreana mockingbird is the rarest (Figure 18.1). It became extinct on the island of Floreana in the second half of the 19th century, mainly because people colonised the island and introduced species, such as rats, that destroyed its nests. People also cut down the prickly pear cactus that is the bird's source of food and a favourite nesting site. Two small populations remain on nearby tiny islands. Conservation programmes have had some success at increasing the numbers of this species and there are plans to reintroduce the species to Floreana. One of the rarest birds in the world, it is worth saving from extinction not least because when Darwin visited Floreana in 1835, he noticed that the mockingbirds on different islands within the Galapagos were not all alike. The rest, as they say, is history!

Figure 18.1 Floreana mockingbird, *Nesomimus trifasciatus*.

So far in this book, we have presented evidence for the unity of life. In this chapter we look at the diversity of life on Earth. At its simplest, diversity can be thought of as a catalogue of all the different species.

You will have noticed that some species are given both their common and scientific names. When a new species is discovered, it is given a scientific name using the **binomial system** that was developed by the Swedish scientist Linnaeus in the 18th century. Each species is given two names: the first is its genus and the two names together indicate its species.

For animals, most plants and fungi, the term species is often used to mean a group of organisms that appear similar to each other, and that can breed together successfully.

> A **species** is a group of organisms with similar morphology and physiology, which can breed together to produce fertile offspring and are reproductively isolated from other species.

New species are discovered and named all the time. No-one knows for certain how many different species there are on Earth. More than 1.5 million species of animals have been described and named to date. Around a quarter of a million flowering plant species and thousands of other plants, such as ferns and mosses, are known. Add to this the fungi, bacteria and other single-celled organisms, and we are looking at a very large number of species. Even though amateur and professional biologists have been cataloguing species for hundreds of years, there are still places on Earth that have not been explored fully, such as the oceans and tropical forests. One tree in a tropical rainforest can contain as many as 1200 species of beetles; almost every time biologists do a thorough count of beetles in a small area of rainforest, they find many previously unknown species. Scientists also do not agree on the species that have been described and often subdivide them or group them together. Estimates of the total number of species range up to 100 million species.

Species do not live in isolation; they share their living space with others to form communities like that in a forest (Figure 18.2). These communities of organisms interact with each other and with their environment.

Figure 18.2 Trees in tropical forests, as here in Costa Rica, are often covered with epiphytes – plants that grow on other plants. These provide far more habitats for small animals, such as beetles, than does the bare bark of trees in temperate forests.

Ecosystems

An ecosystem is a relatively self-contained, interacting community of organisms, and the environment in which they live and with which they interact. A forest ecosystem includes not only the community of organisms, but also the soil, the dead leaves that form leaf litter, water in the rain and in streams, the air, the rocks and all the physical and chemical factors which influence organisms. No ecosystem is entirely self-contained; organisms in one ecosystem interact with those from others. Many birds, for example, migrate from one ecosystem to another often over very great distances to find food and to breed.

You can think of ecosystems on different scales: a small pond is an ecosystem as the water provides a very different environment from the soil and air around it. The organisms in the pond community are therefore very different from those of the surrounding ecosystem. At the other end of the scale is the open ocean which is a huge ecosystem. Some ecosystems are very complex, such as tropical rainforests and coral reefs, and others very simple, such as a sandy desert.

An **ecosystem** is a relatively self-contained, interacting community of organisms, and the environment in which they live and with which they interact.

A habitat describes the place where a species lives within an ecosystem. The niche occupied by a species is more than a physical description of a place. A niche is the role of an organism in an ecosystem; it is how the organism 'fits into' the ecosystem (Figure 18.3). An organism's niche describes where it is, how it obtains energy, how it interacts with both its physical environment and with other species. Organisms require resources from their surroundings and they have special adaptations for obtaining them. In many ecosystems, there are similar niches that may be occupied by the same species, or, more likely, by different ones. In Australia, the niche for large herbivorous mammals is filled by kangaroos. On the savannah of Africa, there are zebra, antelope, wildebeest, elephants and many more species filling 'large herbivore' niches. They coexist because there are differences in such things as the type of vegetation that they eat. It is almost impossible to provide a complete description of the niche of any organism, because there are so many ways in which one organism interacts with other components of the ecosystem of which it is a part.

A **niche** is the role of an organism in an ecosystem.

Figure 18.3 The niche of this great egret includes the freshwater ecosystem where it spends much of its time feeding. It also includes the nearby trees where it roosts and nests.

QUESTION

18.1 Distinguish between the terms **ecosystem** and **niche**.

Biodiversity

We live in a time of great biodiversity, perhaps the greatest there has ever been in the history of the Earth. Tropical forests and coral reefs (Figure 18.24, page 440) are two of the most species-rich areas on Earth. We also live in a time when our activities are causing severe problems for many ecosystems and driving species to extinction (page 441)

Biodiversity can be defined as the degree of variation of life forms in an ecosystem. This is usually taken to include diversity at three levels:

- the variation in ecosystems or habitats
- the number of different species in the ecosystem and their relative abundance
- the genetic variation within each species.

Some areas of the world have very high biodiversity. Examples of these include the Congo basin in Africa, South-East Asia, the Caribbean and Central America, Amazonia and south-west Australia. These areas have many endemic species – that is, species that are only found in these areas and nowhere else (Figure 18.4).

Figure 18.4 New Zealand was isolated for many millions of years so that it has many endemic species and also some, like this tuatara, *Sphenodon* sp., that have become extinct elsewhere.

Species diversity

The number of species in a community is known as **species richness**. Species diversity takes species richness into account, but also includes a measure of the evenness of the abundance of the different species. The more species there are, and the more evenly the number of organisms are distributed among the different species, the greater the species diversity. Coral reefs have a very high biodiversity; such an ecosystem offers many different ecological niches, which are exploited by different species. Species diversity is considered important because ecosystems with high species diversity tend to be more stable than ones with limited diversity; they are more able to resist changes.

Some ecosystems are dominated by one or two species and other species may be rare. This is the case in the natural pine forests in Florida and temperate forests in Canada which are dominated by a few tree species.

The tropics are important centres for biodiversity possibly because living conditions are not too extreme (no frost, snow or ice), there is light of high intensity all year round and birds and mammals do not need to expend energy keeping warm. For example, there are about 1500 species of bird in Central America, but only 300 in the Northwest Territories of Canada.

Genetic diversity

Genetic diversity is the diversity of alleles within the genes in the genome of a single species. All the individuals of a species have the same genes, but they do not all have the same alleles of those genes. Genetic diversity within a species can be assessed by finding out what proportion of genes have different alleles and how many alleles there are per gene.

The genetic diversity that exists between varieties of cultivated plants and domesticated animals is obvious because we can see the differences (consider types of garden rose or breeds of dog). Similar genetic diversity, although not always so obvious, exists in natural populations (Figure 18.5). The genetic differences between populations of the same species exist because populations may be adapted slightly differently in different parts of their range. There is also genetic diversity within each population. This diversity is important in providing populations with the ability to adapt to changes in biotic and abiotic factors, such as competition with other species, evading new predators, resisting new strains of disease and changes in temperature, salinity, humidity and rainfall.

Figure 18.5 These snails all belong to the same species, *Cepaea nemoralis*. The differences between them are the result of different alleles for shell colour and banding.

Assessing species diversity

Collecting organisms and making species lists

Imagine you are in an ecosystem like that in Figure 18.6. The most obvious species are the large plants and maybe some of the larger animals, particularly bird species. The first task when assessing species diversity is to identify and catalogue the types of organism and build a species list.

Figure 18.6 How would you start to investigate and catalogue the biodiversity of an area like this?

QUESTION

18.2 a The snails in Figure 18.5 look very different from one another. Explain why they are all members of the same species.

b Explain the term **genetic diversity**.

c Suggest and explain the effect of the following on genetic diversity: artificial selection (selective breeding), habitat destruction and the release of farmed fish into the wild.

Biologists use identification keys to name the organisms that they find. There are different forms of key – some have drawings or photographs with identifications; others ask a series of questions. The most common of these is a dichotomous key; in your own fieldwork you may use one of these to identify some plant species. Identification requires good skills of observation.

At first, it is a good idea to do a timed search throughout the area you are studying to see how many species you can collect and then identify. If you cannot identify particular species, take photographs of them and name them as species **A**, species **B**, and so on. Some animals will be hard to find and collect, especially small ones such as tiny beetles. A pooter is a simple piece of apparatus that is used to collect these animals (Figure 18.7). Breathing air into the mouth sucks up small animals into a plastic container. They can then be removed and studied and identified using a hand lens and then returned to their habitat.

Figure 18.7 This ecologist is using a pooter to collect small animals from high in the tree canopy in a Yungas forest, along the eastern slopes of the Andes in Argentina.

It is rare to try to catalogue everything. In an area of grassland or woodland, you might choose to concentrate on just one or two groups, such as flowering plants and insects. On a rocky shore, a study may involve the most obvious organisms such as seaweeds and molluscs (Worked example 3, page 432).

There are now two questions to ask: how are the different species spread throughout the ecosystem and how many individuals of each species are there? The answers to these two questions describe what we call **distribution** and **abundance**.

QUESTION

18.3 Some students surveyed the species diversity of an area of woodland and some grassland nearby. They used a 20-minute timed search.

a Explain why they used the same technique for the two areas.

b They found 56 species in the woodland and 12 in the grassland. What other data could they have collected to compare the biodiversity of the two ecosystems?

Sampling

To find out which species are present in an ecosystem, and the size of the population of each of them, the ideal method would be to find, identify and count every single organism that lives there. We can sometimes do this if the area is very small or the species are very large. But it is only rarely possible. Instead, we take **samples** from the area we are interested in, and use these to make an estimate of the total numbers in the area.

Sampling can be **random** or **systematic**. If an area looks reasonably uniform, or if there is no clear pattern to the way species are distributed, then it is best to use random sampling.

Random sampling using quadrats

A quadrat is a square frame that marks off an area of ground, or water, where you can identify the different species present and/or take a measurement of their abundance. You need to decide on a suitable size for the quadrat and how many samples you will take.

Samples must be taken randomly to avoid any bias. For example, you might choose to take all of your samples from the place with fewest species simply because it is the easiest to do. This would not be representative of the whole area you are surveying. The usual way to ensure that a sample is random is to mark out an area with measuring tapes and use a random number generator, such as an app on a mobile phone. The random numbers give you the coordinates of the sampling points in relation to the two tapes you have used to mark out the area (Figure 18.8).

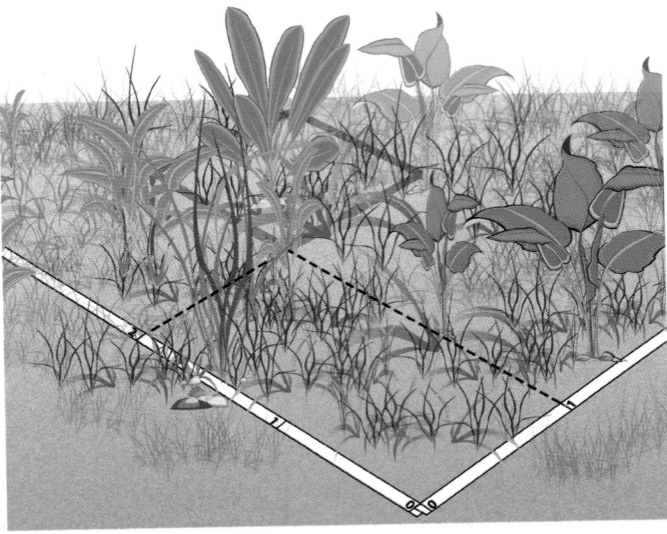

Figure 18.8 In random sampling, quadrats are positioned randomly in an area marked off by measuring tapes. This reduces the chances of bias in sampling the ecosystem.

You can use your results in two different ways: to calculate **species frequency** and **species density**. Species frequency is a measure of the chance of a particular species being found within any one quadrat. You simply record whether the species was present in each quadrat that you analyse. For example, if you placed your quadrat 50 times, and found daisy plants in 22 of your samples, then the species frequency for daisies is:

$$\frac{22}{50} \times 100 = 44\%$$

Species density is a measure of how many individuals there are per unit area – for example, per square metre. The number of individuals that you have counted is divided by the total area of all your quadrats.

It is not always possible to count individual plants and animals because of the way that they grow. For example, many animals and plants grow over surfaces forming a covering and it is almost impossible to count individuals. How do you decide how many grass plants there are in a quadrat that you have placed on a lawn? In this case, you can estimate the **percentage cover** of the species within your quadrat (Figure 18.9). To help with this, you can use a 100 cm × 100 cm quadrat with wires running across it at 10 cm intervals in each direction, dividing the quadrat into

100 smaller squares. You then decide approximately what percentage of the area inside the quadrat is occupied by each species. These percentages may not add up to 100%. For example, there might be bare ground in the quadrat, so the numbers will come to less than 100%. Or there may be plants overlying one another, in which case the numbers may add up to more than 100%. An alternative to estimating percentage cover of each species is to use an abundance scale, such as the Braun–Blanquet scale for number and plant cover (Table 18.1).

Figure 18.9 Estimating percentage cover. This 1 m² quadrat is divided into 100 small squares to make it easier to make the estimation for each species.

Braun–Blanquet Cover Scale	
Description	**Value**
Very few plants, cover is less than 1%	+
Many plants, but cover is 1–5%	1
Very many plants or cover is 6–25%	2
Any number of plants; cover is 26–50%	3
Any number of plants; cover is 51–75%	4
Cover is greater than 75%	5

Table 18.1 The Braun–Blanquet scale for recording vegetation within quadrats.

QUESTIONS

18.4 A survey gave the following results for a species of the red sea anemone, *Isactinia tenebrosa* (Figure 18.10), on a rocky shore in New Zealand, using a quadrat with an area of 0.25 m².

Quadrat	1	2	3	4	5	6	7	8	9	10
Number of red sea anemones	0	3	0	1	0	0	5	2	0	1

a Calculate **i** the species frequency, and **ii** the species density of *I. tenebrosa* from the results of this survey.

b Suggest when it might be more appropriate to use species frequency rather than species density to record the abundance of a species.

18.5 A survey was made of Benghal dayflower, *Commelina benghalensis*, growing on a lawn and in a field of young soybean plants. Ten 1.0 m² quadrats were placed randomly in each area, and the number of dayflower plants in each quadrat was counted. The results are shown in the table.

Quadrat	1	2	3	4	5	6	7	8	9	10
Number of dayflowers on lawn	0	0	4	3	0	1	2	4	0	3
Number of dayflowers in field	0	0	0	2	5	0	0	1	0	0

a Calculate **i** the species frequency, and **ii** the species density of dayflower plants in each of the two areas.

b Explain why it is important to use randomly placed quadrats.

c Suggest two disadvantages with calculating percentage cover or using an abundance scale, such as the Braun–Blanquet scale.

Figure 18.10 A red sea anemone that lives between the tides on rocky shores.

Estimating numbers of mobile animals

Quadrats are obviously no use for finding or counting mobile animals, so different methods have to be used for these.

Small mammals, such as mice and voles, can be caught in traps that are filled with hay for bedding and suitable food as bait. Insects and other invertebrates, such as spiders, can be captured by sweep netting. Pond nets are used for sampling aquatic organisms. The techniques for this vary according to the size of the body of water, and whether it is still or moving. Single birds can be counted quite easily, although this does become more difficult in dense forest. Flocks of birds are much more difficult to count, although it can be done by counting a group of ten birds and estimating how many such groups there are.

A good method of estimating the population size of mobile organisms, if used with care, is the mark–release–recapture technique. First, as many individuals as possible are caught. Each individual is marked, in a way that will not affect its future chance of survival. The marked individuals are counted, returned to their habitat and left to mix randomly with the rest of the population.

When enough time has elapsed for the mixing to take place, another large sample is captured. The number of marked and unmarked individuals is counted. The proportion of marked to unmarked individuals is then used to calculate an estimate of the total number in the population (Worked example 1). For example, if you find that one-tenth of the second sample was marked, then you presume that you originally caught one-tenth of the population in your first sample. Your best estimate is therefore that the number in the population is ten times the number you caught and marked in your first sample.

WORKED EXAMPLE 1

The mark–release–recapture technique

Brown planthoppers are a serious insect pest of rice. Some students used sweep nets to catch a large sample of planthoppers in a field of rice. Each animal was marked with a very small spot of non-toxic waterproof paint and then they were released across the field. The next day, a second large sample was caught.

Number caught and marked in first sample = 247
Number caught in second sample = 259
Number in the second sample that had been marked = 16

So the estimated number in the population $= \frac{247 \times 259}{16}$

$= 3998$

Simpson's Index of Diversity

When you have collected information about the abundance of the species in the area you are studying, you can use your results to calculate a value for the species diversity in that area. We can do this using Simpson's Index of Diversity, *D*. One formula for this is:

$$D = 1 - \left(\sum \left(\frac{n}{N}\right)^2\right)$$

where n is the total number of organisms of a particular species, and *N* is the total number of organisms of all species (Worked example 2).

Values of *D* range from 0 to 1. A value near 0 represents

WORKED EXAMPLE 2

Simpson's Index of Diversity

A sample was made of the animals living on two rocky shores. 10 quadrats were placed on each shore, and the number of animals of each species in each quadrat was counted. The results are shown in the table.

Species	Number of individuals, *n*	
	Shore A	Shore B
painted topshells	24	51
limpets	367	125
dogwhelks	192	63
snakelocks anemones	14	0
beadlet anemones	83	22
barnacles	112	391
mussels	207	116
periwinkles	108	93
total number of individuals, *N*	1107	861

To determine Simpson's Index for shore **A**, calculate $\frac{n}{N}$ for each species, square each value, add them up and subtract from 1. Repeat the procedure for shore **B**.

Species	Shore A		
	n	$\frac{n}{N}$	$\left(\frac{n}{N}\right)^2$
painted topshells	24	0.022	0.000
limpets	367	0.332	0.110
dogwhelks	192	0.173	0.030
snakelocks anemones	14	0.013	0.000
beadlet anemones	83	0.075	0.006
barnacles	112	0.101	0.010
mussels	207	0.187	0.035
periwinkles	108	0.098	0.010
total number of individuals, *N*	1107	$\sum \left(\frac{n}{N}\right)^2 = 0.201$	

For shore **A**, Simpson's Index of Diversity (*D*)
$= 1 - 0.201 = 0.799$

QUESTION

18.6 **a** Calculate *D* for shore **B** in Worked example 18.2. Show all your working as in the table. The easiest way to do this is to use a spreadsheet. Once you have set up a spreadsheet you can use it for calculating this index of diversity for other data.

b Compare the diversity of the two shores.

a very low species diversity. A value near 1 represents a very high species diversity.

One advantage of this method is that you do not need to identify all, or even any, of the organisms present to the level of species. You can, for example, just decide to call the species of anemone that has short tentacles 'anemone A', and the species that has few long tentacles 'anemone B'. So long as you can recognise that they are different species, you do not need to find their scientific names. But beware: some species have many phenotypic forms (Figure 18.5).

The higher the number we get for *D*, the greater the diversity. You can probably see that the diversity depends on the number of different species there are, and also the abundance of each of those species. A community with 10 species, but where only one species is present in large numbers and the other 9 are very rare, is less diverse than one with the same number of species but where several different species have a similar abundance.

Comparisons using this diversity index should be on a 'like for like' basis, so the communities should be similar and the organisms chosen should also be similar. For example, it should not be used to compare the diversity of fish in a lake with the diversity of moths in a forest.

Systematic sampling

Random sampling is not suitable for every place that you may wish to survey. You might want to investigate how species are distributed in an area where the physical conditions, such as altitude, soil moisture content, soil type, soil pH, exposure or light intensity change. For example, suppose you want to investigate the change at the edge of a field where it becomes very marshy. In this case, you should randomly select a starting point in the field and lay out a measuring tape in a straight line to the marshy area. You then sample the organisms that are present along the line, which is called a transect. The simplest way to do this is to record the identity of the organisms that touch the line at set set distances – for example, every two metres. This **line transect** will give you qualitative data that can be presented as in Figure 18.11. You can also use the **belt transect** technique by placing a quadrat at regular intervals along the line and recording the abundance of each species within the quadrat. Data from a line transect can be shown as a drawing. Data from a belt transect can be plotted as a set of bar charts or as a kite diagram (Worked example 3).

Line transect – a line across one or more habitats
The organisms found at regular points along a line are noted. Transects are used to detect changes in community composition along a line across one or more habitats.

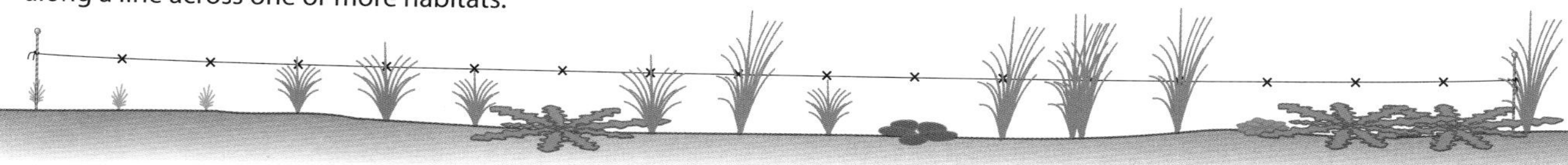

Interrupted belt transect
The abundance of organisms within quadrats placed at regular points along a line is noted.

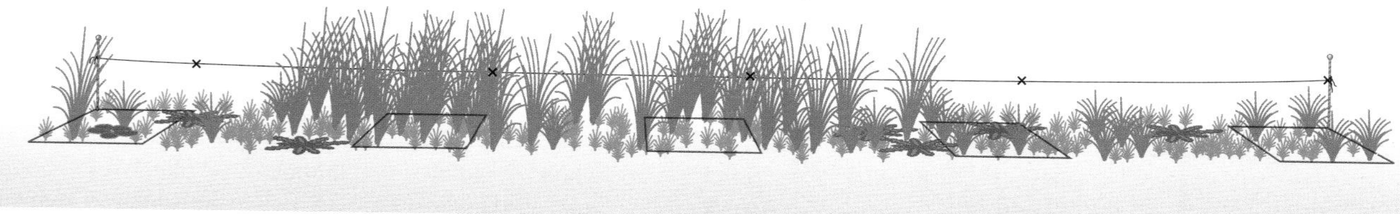

Figure 18.11 Systematic sampling using transects: **a** a line transect, and **b** an interrupted belt transect.

WORKED EXAMPLE 3

Drawing kite diagrams to display data on species abundance

Some students used an interrupted belt transect to investigate the distribution and abundance of seaweeds and molluscs on a rocky shore placing a quadrat every 20 m. The students began sampling at 20 m below mean sea level, when the tide was out, and worked their way up the shore. They used two different scales for assessing the abundance of seaweeds and molluscs. On these scales, 5 indicates the highest level of abundance and 1 the lowest. If a species is not present in a quadrat it is obviously recorded as 0. The table shows the data collected from an interrupted belt transect on a rocky shore.

Distance on the shore / m	Seaweeds (algae)				Molluscs				
	Kelp	Serrated wrack	Bladder wrack	Spiral wrack	Dog whelk	Edible periwinkle	Limpet	Flat periwinkle	Rough periwinkle
200	0	0	0	0	0	0	0	0	0
180	0	0	0	0	0	0	0	0	5
160	0	0	0	0	0	0	0	0	5
140	0	0	5	5	0	0	0	0	0
120	0	0	1	4	0	0	0	0	0
100	0	5	5	0	0	0	0	0	0
80	0	5	5	0	0	0	5	4	5
60	0	5	5	0	0	1	4	0	4
40	0	5	5	0	0	4	4	0	0
20	0	4	5	0	1	4	5	0	0
0	3	5	0	0	0	0	0	0	0

The data collected by the students is shown in the kite diagram in Figure 18.12.

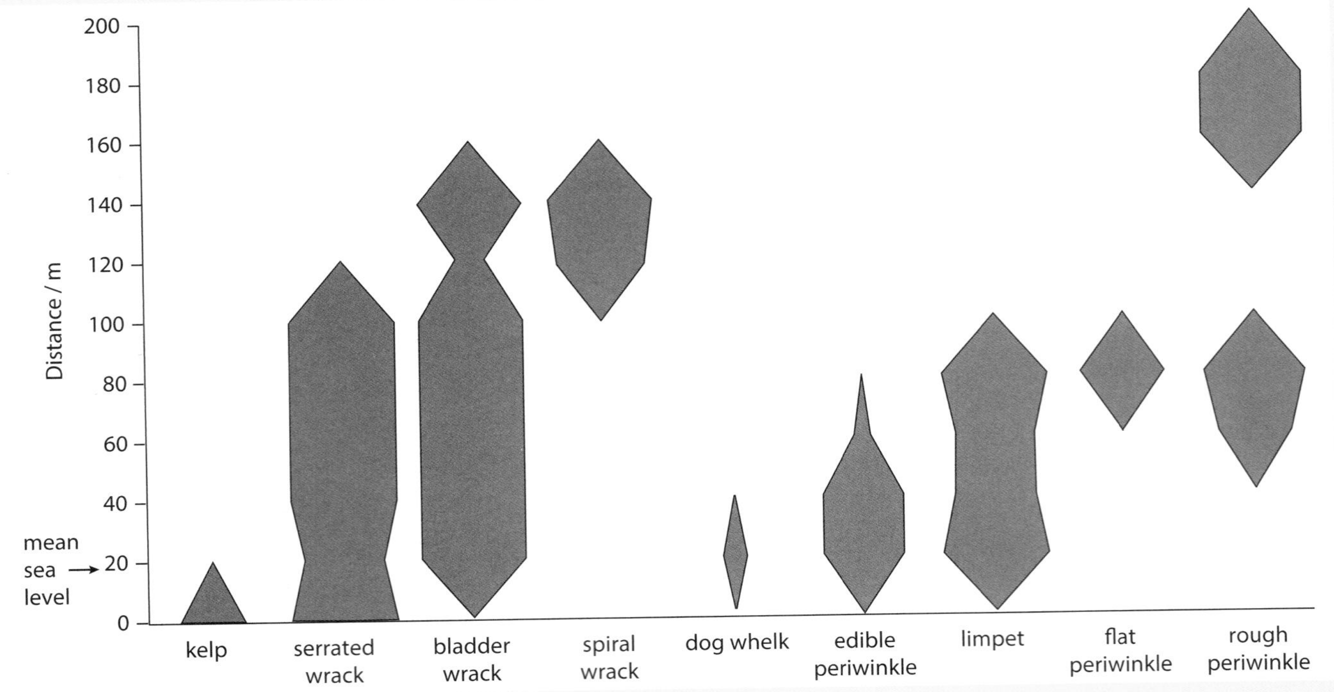

Figure 18.12 A kite diagram shows the data from a belt transect. The distribution of seaweeds and molluscs from low tide to the top of the rocky shore is shown by the lengths of the 'kites' and the abundance by their widths. This gives a representation of the ecosystem in a way that is easy to see.

QUESTION

18.7 In a survey of trees in a dry tropical forest, some students identified five tree species (**A** to **E**). They counted the numbers of trees in an area 100 m × 100 m with these results:

Tree species	Number
A	56
B	48
C	12
D	6
E	3

a Calculate the Simpson's Index of Diversity for the trees within the area sampled.

b Explain the advantage of using data on species diversity and abundance when calculating an index of diversity.

c The Simpson's Index of Diversity for the vegetation in an area of open grassland was 0.8; for a similar sized area of vegetation beneath some conifer trees it was 0.2. What do you conclude from these results?

Correlation

While doing random sampling or carrying out a belt transect you may observe that two plant species always seem to occur together. Is there in fact an association between them? Or you might want to know if there is any relationship between the distribution and abundance of a species and an abiotic factor, such as exposure to light, temperature, soil water content or salinity (saltiness). To decide if there is an association, you can plot scatter graphs and make a judgment by eye. Alternatively, you can calculate a correlation coefficient (r) to assess the strength of any correlation that you suspect to exist. Figure 18.13 shows the sorts of relationships that you may find.

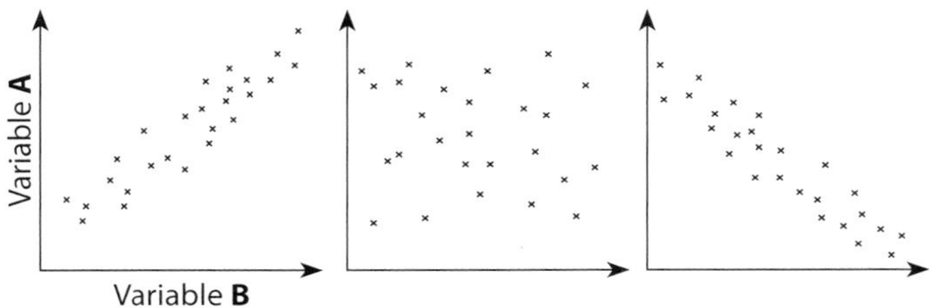

Figure 18.13 Three types of association: **a** a positive linear correlation, **b** no correlation, and **c** a negative linear correlation.

The strongest correlation you can have is when all the points lie on a straight line: there is a **linear correlation**. This is a correlation coefficient of 1. If as variable **A** increases so does variable **B**, the relationship is a positive correlation. If as variable A increases, variable **B** decreases then the relationship is a negative correlation. A correlation coefficient of 0 means there is no correlation at all (Figure 18.13).

You can calculate a correlation coefficient to determine whether there is indeed a linear relationship and also to find out the strength of that relationship. The strength means how close the points are to the straight line.

Pearson's correlation coefficient can only be used where you can see that there might be a linear correlation (**a** and **c** in Figure 18.13) and when you have collected quantitative data as measurements (for example, length, height, depth, light intensity, mass) or counts (for example, number of plant species in quadrats). The data must be distributed normally, or you must be fairly sure that this is the case.

Sometimes you may not have collected quantitative data, but used an abundance scale (Table 18.1) or you may not be sure if your quantitative data is normally distributed. It might also be possible that a graph of your results shows that the data is correlated, but not in a linear fashion. If so, then you can calculate **Spearman's rank correlation coefficient**, which involves ranking the data recorded for each variable and assessing the difference between the ranks.

You should always remember that correlation does not mean that changes in one variable cause changes in the other variable. These correlation coefficients are ways for you to test a relationship that you have observed and recorded to see if the variables are correlated and, if so, to find the strength of that correlation.

Before going any further, you should read pages 501–504 in Chapter P2 which show you how to calculate these correlation coefficients.

Spearman's rank correlation

An ecologist was studying the composition of vegetation on moorland following a reclamation scheme. Two species – common heather, *Calluna vulgaris*, and bilberry, *Vaccinium myrtillus* appeared to be growing together. He assessed the abundance of these two species by recording the percentage cover in 11 quadrats as shown in Table 18.3.

To find out if there is a relationship between the percentage cover of these two species, the first task is

Quadrat	Percentage cover	
	C. vulgaris	*V. myrtillus*
1	30	15
2	37	23
3	15	6
4	15	10
5	20	11
6	9	10
7	3	3
8	5	1
9	10	5
10	25	17
11	35	30

Table 18.2 Results from study of composition of vegetation on moorland.

to make a **null hypothesis** that there is **no correlation** between the percentage cover of the two species.

The equation to calculate the Spearman rank correlation, r_s, is:

$$r_s = 1 - \left(\frac{6 \times \Sigma D^2}{n^3 - n}\right)$$

where n is the number of pairs of items in the sample and *D* is the difference between each pair of ranked measurements and Σ is the 'sum of'.

The next step is to draw a scatter graph to see if it looks as if there is a correlation between the abundance of the two species. This can be done very quickly using the graphing facility of a spreadsheet program. You can now follow the steps shown on page 503 to calculate the value for r_s. Again, the quickest way to do this is to set up a spreadsheet that will do the calculations.

The ecologist calculated the value of r_s to be 0.930. A correlation coefficient of +0.930 is very close to +1, so we can conclude that there is a positive correlation between the two species and that the strength of the association is very high. The ecologist was also able to reject the null hypothesis and accept the alternative hypothesis that there is a correlation between the abundance of *C. vulgaris* and *V. myrtillus* on this reclaimed moorland.

QUESTION

18.8 a Draw a scatter graph to show the data in Table 18.2.

b Follow the worked example on page 503 to calculate the Spearman rank correlation coefficient (r_s) for the data in Table 18.2. Show all the steps in your calculation.

c Explain why the ecologist was able to reject the null hypothesis.

d State the conclusion that the ecologist could make from this investigation.

e The ecologist thinks that the relationship is the result of habitat preference, that both species prefer drier soil. Describe an investigation that he could carry out to test this hypothesis.

Pearson's linear correlation

In many investigations, the data collected are for two continuous variables and the data within each variable show a normal distribution. When this is the case, Pearson's correlation coefficient can be used. This also calculates numbers between +1 and –1 and the result is interpreted in the same way. The method of calculation looks more complex, but the test can be done easily with a spreadsheet.

The first step is to check whether the relationship between the two continuous variables appears to be linear by drawing a scatter graph. The correlation coefficient should not be calculated if the relationship is not linear. For correlation only purposes, it does not really matter on which axis the variables are plotted. The nearer the scatter of points is to a straight line, the higher the strength of association between the variables. Before using this test you must also be satisfied that the data for the variables you are investigating show a normal distribution.

As trees grow older, they tend to get cracks in their bark. A student measured the width of cracks on many pine trees in a plantation and found that they varied considerably. The data she collected showed a normal distribution. She noticed that the larger, and presumably older, trees tended to have wider cracks in their bark than the smaller trees. She wanted to see if there was a correlation between the size of the trees and the size of these cracks. She chose to measure the circumference of each tree as a measure of their overall size. She measured the width of the cracks in the bark. This means that she collected continuous data for each of the two variables – tree circumference and crack width. She investigated this by selecting twelve trees at random and measuring the circumference of each trunk and the widths of three cracks on the bark at head height. Her results are in Table 18.3.

Number of tree	Circumference of tree / metres	Mean width of crack / mm
1	1.77	50
2	1.65	28
3	1.81	60
4	0.89	24
5	1.97	95
6	2.15	51
7	0.18	2
8	0.46	15
9	2.11	69
10	2.00	64
11	2.42	74
12	1.89	69

Table 18.3 Widths of cracks on pine trees in a plantation.

The student plotted these results on a scatter graph and found that they look as if there might be a linear correlation between them. She then used the following formula to calculate Pearson's correlation coefficient, *r*.

$$r = \frac{\Sigma xy - n\bar{x}\bar{y}}{ns_x s_y}$$

The student calculated the correlation coefficient as $r = 0.79$.

QUESTION

18.9 a Draw a scatter graph of the data in Table 18.4.

b Use the worked example on page 501 to help you calculate the Pearson's correlation coefficient (*r*) for the data.

c Explain why the student was able to reject the null hypothesis.

d State the conclusion that the student could make from this investigation.

Classification

You will have noticed in the table in Worked example 3 that the students working on the rocky shore studied two groups of organisms: algae and molluscs. With such a huge number of different kinds of organisms living on Earth, biologists have always wanted to arrange them into groups, a process called **classification**. We find it difficult to memorise or absorb information about thousands of different unrelated objects. By grouping them into different categories, it is much easier to understand them and to remember their key features.

Taxonomy is the study and practice of classification, which involves placing organisms in a series of taxonomic units, or taxa (singular: taxon). In biological classification, these taxa form a **hierarchy**. Each kind of organism is assigned to its own **species**, and similar species are grouped into a **genus** (plural: genera). Similar genera are grouped into a **family**, families into an **order**, orders into a **class**, classes into a **phylum** (plural: phyla) and phyla into a **kingdom**. The domain is at the top of this hierarchical system. Table 18.4 shows how African bush elephants and hibiscus plants (Figure 18.14) are classified.

Taxon	African bush elephant	Hibiscus
domain	Eukarya	Eukarya
kingdom	Animalia	Plantae
phylum	Chordata	Angiosperms
class	Mammalia	Dicotyledonae
order	Proboscidea	Malvales
family	Elephantidae	Malvaceae
genus	*Loxodonta*	*Hibiscus*
species	*Loxodonta africana*	*Hibiscus rosa-sinensis*

Table 18.4 The classification of African bush elephants and hibiscus plants.

Figure 18.14 *Hibiscus rosa-sinensis* is a plant that has spread from Asia to much of the tropics and sub-tropics. Flower colour is a good example of the genetic diversity in this species.

QUESTION

18.10 a The giraffe, *Giraffa camelopardalis*, is a mammal that belongs to the order Artiodactyla and the family Giraffidae. Make a table to show how the giraffe is classified.

b Use examples from Table 18.4 to explain

i the term **taxon**

ii why the classification system is hierarchical.

Three domains

Biologists used to divide organisms into two large groupings based on their cell structure. In Chapter 1, you saw that prokaryotes and eukaryotes have significantly different cellular structures. In the 1970s, prokaryotes were discovered living in extreme environments, such as hot springs where temperatures often exceed 100 °C. These extremophiles, as they are known, were not like typical bacteria. Studies revealed that the genes coding for the RNA that makes up their ribosomes were more like those of eukaryotes. They were found to share features with both typical bacteria and eukaryotes. At this time, studies of molecular biology assumed a much greater significance in taxonomy. This meant that a new taxon, the **domain**, had to be introduced to reflect the differences between these extremophiles and typical bacteria. The domain is the taxon at the top of the hierarchy (Table 18.4). The prokaryotes are divided between the domains Bacteria and Archaea and all the eukaryotes are placed into the domain Eukarya.

Many Archaea live in extreme environments, such as hot springs, around deep volcanic vents (black smokers) in the oceans (Figure 18.15) and in lakes where there is a very high concentration of salt. Some of them produce methane, cannot survive where there is oxygen and have many unusual enzymes. Since they were discovered in extreme environments they have been found in many less extreme environments; for example, they form an important part of the plankton in the oceans.

In several ways, the Archaea appear to have more in common with the Eukarya than with Bacteria. It is thought that Bacteria and Archaea separated from each other very early in the evolution of life. The Archaea and Eukarya probably diverged later.

Figure 18.15 A deep sea hydrothermal vent (black smoker) surrounded by giant tubeworms. The community relies on energy made available by bacteria and archaeans.

Domain Bacteria

Bacteria are prokaryotic as their cells have no nucleus. They are all small organisms that vary in size between that of the largest virus and the smallest single-celled eukaryote. The characteristic features of bacteria are:

- cells with no nucleus
- DNA exists as a circular 'chromosome' and does not have histone proteins associated with it
- smaller circular molecules of DNA called plasmids are often present
- no membrane-bound organelles (such as mitochondria, endoplasmic reticulum, Golgi body, chloroplasts) are present
- ribosomes (70 S) are smaller than in eukaryotic cells
- cell wall is always present and contains peptidoglycans (not cellulose)
- cells divide by binary fission, not by mitosis
- usually exist as single cells or small groups of cells.

Look at Figure 1.30 on page 21 to remind yourself of these features and to see how prokaryotic cell structure differs from that of eukaryotes. There are electron micrographs of two pathogenic species of bacteria in Chapter 10. Figure 18.16 shows the cyanobacterium *Nostoc*.

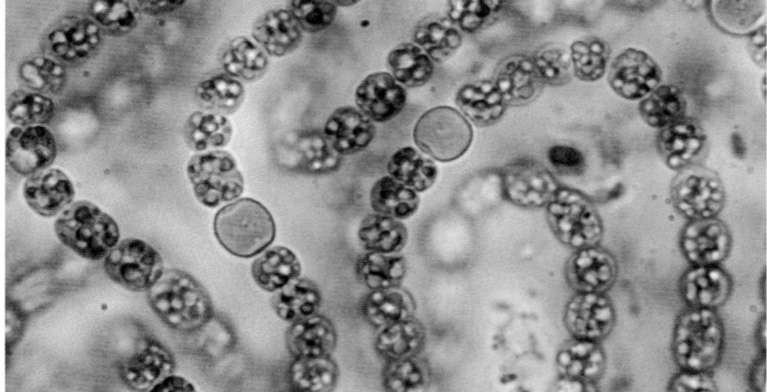

Figure 18.16 The filamentous cyanobacterium, *Nostoc*. This species fixes carbon dioxide in photosynthesis; it also fixes nitrogen by converting N_2 into organic forms of nitrogen in the wider, lighter green cells in its filaments. (× 600)

Domain Archaea

Archaeans are also prokaryotic as their cells have no nucleus. Their range of size is similar to that of bacteria. Many inhabit extreme environments (Figure 18.17). The characteristic features of archaeans are:

- cells with no membrane-bound organelles
- DNA exists as a circular 'chromosome' and does have histone proteins associated with it
- smaller circular molecules of DNA called plasmids are often present
- ribosomes (70 S) are smaller than in eukaryotic cells, but they have features that are similar to those in eukaryotic ribosomes, not to bacterial ribosomes
- cell wall always present, but does not contain peptidoglycans
- cells divide by binary fission, not by mitosis
- usually exist as single cells or small groups of cells.

The metabolism of archaeans is similar to that of bacteria, but the way in which transcription occurs has much in common with eukaryotes.

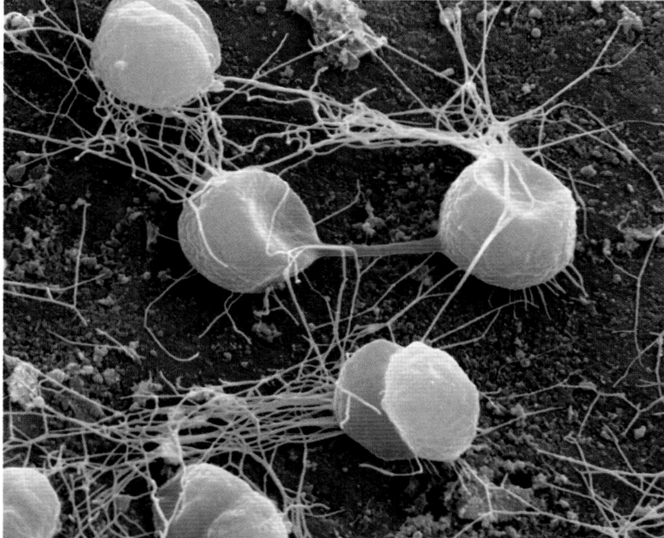

Figure 18.17 A scanning electron micrograph of the archaean *Pyrococcus furiosus* (× 12 500). Although they look like bacteria, archaeans have differences in metabolism and genetics. The flagella seen here are also different. *P. furiosus* is only found in near-boiling water; if the temperature falls below 70 °C it freezes and dies. It respires anaerobically using sulfur instead of oxygen as the final electron acceptor.

Domain Eukarya

All the organisms classified into this domain have cells with nuclei and membrane-bound organelles. Their characteristic features are:

- cells with a nucleus and membrane-bound organelles
- DNA in the nucleus arranged as linear chromosomes with histone proteins
- ribosomes (80 S) in the cytosol are larger than in prokaryotes; chloroplasts and mitochondria have 70S ribosomes, like those in prokaryotes.
- chloroplast and mitochondrial DNA is circular as in prokaryotes
- a great diversity of forms: there are unicellular (Figure 18.18), colonial (Figure 18.19) and multicellular organisms
- cell division is by mitosis
- many different ways of reproducing – asexually and sexually.

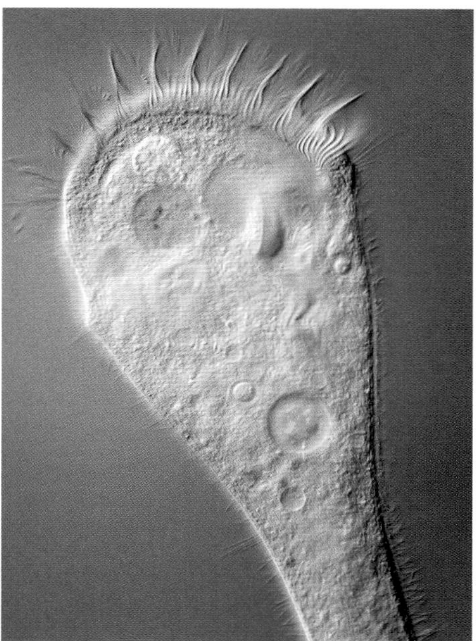

Figure 18.18 *Stentor roseli*, a protoctistan covered in many cilia which it uses for movement and for feeding. Although unicellular, it has considerable specialisation of regions within its body (× 240).

QUESTION

18.11 Make a table to compare the features of the three domains: Bacteria, Archaea and Eukarya.

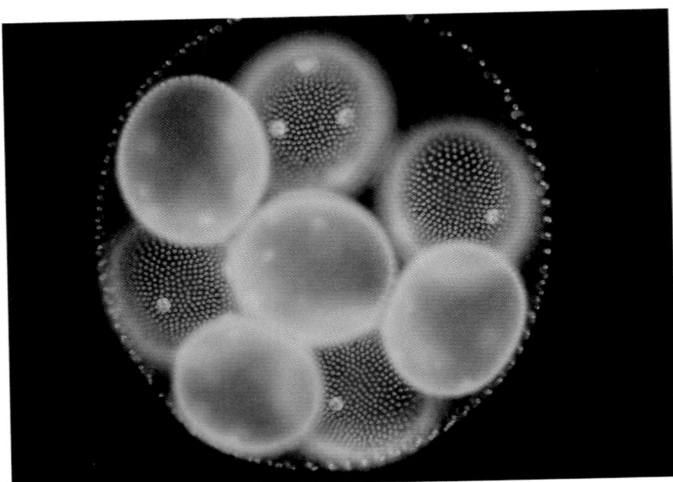

Figure 18.19 *Volvox globator* is a spherical colonial green protoctist. The body is composed of thousands of cells with flagella. These cells work together in a coordinated way but there is little specialisation of cells. Cells at one pole detect light so the colony swims towards the light. Cells at the other pole are specialised for reproduction. Inside there are new colonies that are just about to be released (× 60).

The three domains are each divided into kingdoms. The following section describes the features of the four kingdoms of the Eukarya.

Kingdom Protoctista

The Protoctista is made up of a very diverse range of eukaryotic organisms, which includes those that are often called protozoans ('simple animals') and algae, such as seaweeds. Any eukaryote that is not a fungus, plant or animal is classified as a protoctist (Figures 18.18 and 18.19). The characteristic features of protoctists are:

- eukaryotic
- mostly single-celled, or exist as groups of similar cells
- some have animal-like cells (no cell wall) and are sometimes known as protozoa
- others have plant-like cells (with cellulose cell walls and chloroplasts) and are sometimes known as algae.

Many organisms in this kingdom may actually be more closely related to organisms in other kingdoms than they are to each other. For example, there are strong arguments for classifying algae as plants.

Kingdom Fungi

Fungi have some similarities with plants, but none of them is able to photosynthesise. They are all heterotrophic, obtaining energy and carbon from dead and decaying matter or by feeding as parasites on living organisms. There is a vast range in size from the microscopic yeasts to what may be the world's largest organism. A specimen of the honey fungus, *Armillaria bulbosa*, grows in a forest in Wisconsin, USA and spreads over 160 000 m^2. Not only is it possibly the largest organism in the world, but it may also be the oldest at 1500 to 10 000 years old; its estimated mass is 100 tonnes.

Characteristic features of fungi are:

- eukaryotic
- do not have chlorophyll and do not photosynthesise
- heterotrophic nutrition – they use organic compounds made by other organisms as their source of energy and source of molecules for metabolism
- reproduce by means of spores (Figure 18.20)
- simple body form, which may be unicellular or made up of long threads called hyphae (with or without cross walls) (Figure 18.21); large fungi such as mushrooms produce large compacted masses of hyphae known as 'fruiting bodies' to release spores
- cells have cell walls made of chitin or other substances, not cellulose
- never have cilia or flagella.

Figure 18.20 A puffball fungus, *Lycoperdon* sp., releasing millions of microscopic spores. Their method of feeding on dead and decaying matter means that eventually the food is all used up. A few of these spores may land on a suitable food source and be able to grow.

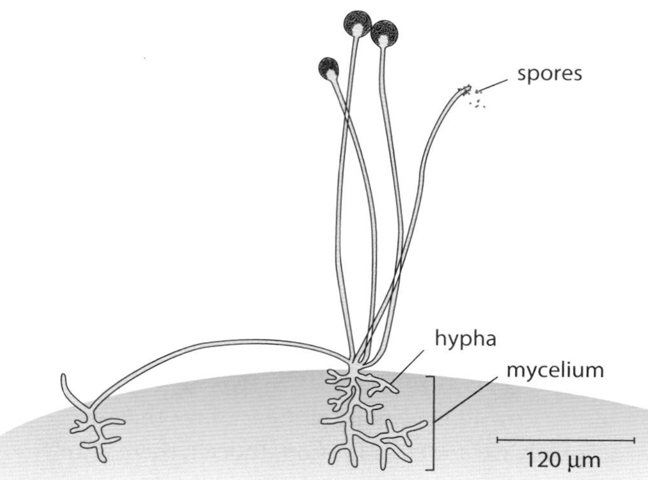

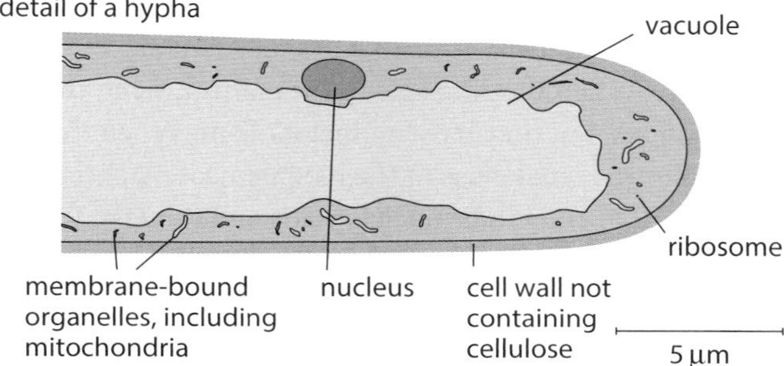

Figure 18.21 The bread mould fungus, *Rhizopus nigricans,* and a detail of the end of one hypha.

Figure 18.22 Tree ferns, *Cyathea* sp., growing in Whirinaki Conservation Park, New Zealand.

Kingdom Plantae

Plantae (plants) are all multicellular photosynthetic organisms (Figures 18.22 and 18.23). They have complex bodies that are often highly branched both above and below ground.

Characteristic features of plants are:

- multicellular eukaryotes with cells that are differentiated to form tissues and organs
- few types of specialised cells
- some cells have chloroplasts and photosynthesise (Figures 1.28 and 1.29, pages 20 and 21)
- cells have large, often permanent vacuoles for support
- autotrophic nutrition
- cell walls are always present and are made of cellulose
- cells may occasionally have flagella – for example, male gametes in ferns.

Figure 18.23 Bristlecone pines are some of the oldest trees on Earth, many estimated to be 2000 to 3000 years old. These grow at over 2700 metres in the Great Basin National Park in Nevada.

Kingdom Animalia

The Animalia are multicellular organisms that are all heterotrophic with many ways of obtaining their food. There is a great diversity of forms within this kingdom (Figure 18.24). The nervous system is unique to the animal kingdom.

Characteristic features of animals are:

- multicellular eukaryotes with many different types of specialised cells
- cells that are differentiated to form tissues and organs
- cells do not have chloroplasts and cannot photosynthesise (although some, such as coral polyps have photosynthetic protoctists living within their tissues)
- cell vacuoles are small and temporary (for example, lysosomes and food vacuoles)
- heterotrophic nutrition
- cells do not have cell walls
- communication is by the nervous system (Chapter 15)
- cells sometimes have cilia or flagella.

Figure 18.24 The crown-of-thorns starfish, *Acanthaster planci*, feeds on coral in the Great Barrier Reef. It has been through several population explosions over recent years causing destruction of much of the coral on parts of the reef (page 443).

QUESTION

18.12 a Which eukaryotic kingdoms contain:
- i autotrophic organisms
- ii heterotrophic organisms?

b Make a table to compare the features of the four kingdoms of eukaryotes.

Viruses

Viruses are microorganisms whose structure is only visible with electron microscopes (Figure 18.25). Viruses are acellular – they do not have a cellular structure like bacteria and fungi. In Chapter 1 we looked, briefly, at their structure (Figure 1.31, page 22); in Chapters 10 and 11 we discussed some human viral pathogens, and in Chapter 19 you will see that they play important roles in gene technology.

You will have noticed that viruses are not in the classification system we have discussed so far. This is because viruses have none of the features that we traditionally use for classification. Indeed, there is an argument that they should not be considered to be living organisms at all. They do, however, have particles made of proteins and nucleic acids that are found in all cellular organisms. When they are free in the environment, they are infectious, but they have no metabolism. When they infect cells, they make use of the biochemical machinery of the host cell to copy their nucleic acids and to make their proteins, often leading to destruction of the host cells. The energy for these processes is provided by respiration in the host cell.

The taxonomic system for classifying viruses is based on the diseases which they cause, the type of nucleic acid they contain (DNA or RNA) and whether the nucleic acid is single-stranded or double-stranded. In cellular organisms, DNA is double-stranded and RNA is single-stranded but in viruses both can be either single-stranded or double-stranded. Some examples of these four groups of viruses are shown in Table 18.5.

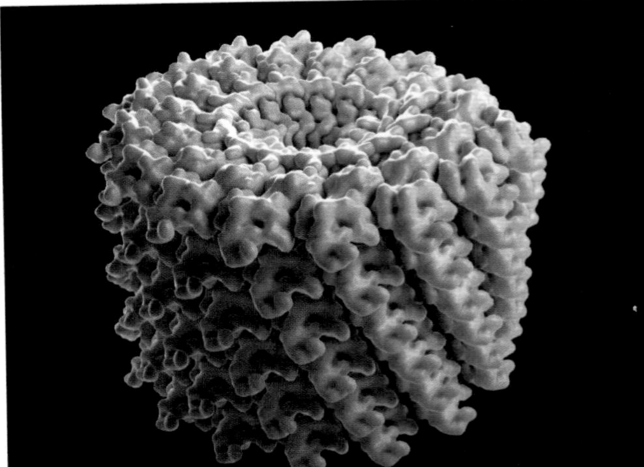

Figure 18.25 A model showing the structure of part of the tobacco mosaic virus, TMV. The blue strand is the single strand of RNA and the rest is composed of proteins. In a survey of plant pathologists in 2011, TMV was voted the most important plant virus.

Nucleic acid	Number of strands	Example	Host organism	Disease
DNA	1	canine parvovirus type 2	dogs	canine parvovirus
		African cassava mosaic virus	cassava plants	mosaic disease
	2	varicella zoster virus (VZV)	humans	chickenpox
RNA	1	rotavirus	humans	gastroenteritis
		morbillivirus	humans	measles
		tobacco mosaic virus (TMV)	tobacco, tomato, pepper	mosaic disease
	2	human immunodeficiency virus	humans	HIV/AIDS

Table 18.5 Viruses are classified into four groups using the type and structure of their nucleic acid as the criteria for grouping them.

Figure 18.26 Viruses are the ultimate in parasitism. This tobacco plant is showing signs of infection by the tobacco mosaic virus. TMV spreads rapidly through crops of plants, such as tobacco and tomato, severely reducing their growth.

QUESTION

18.13 a Explain why viruses are not included in the three domain classification system described on pages 436–437.

b State the features that are used in the classification of viruses.

Threats to biodiversity

Biodiversity is under threat in many aquatic and terrestrial ecosystems as the human population continues to increase and we take more resources from the environment and produce increasing quantities of waste. Ecosystems and species are being lost at an alarming rate, not just by the direct action of humans, but also indirectly as a result of climate change.

There are five major threats to biodiversity:

- habitat loss and the degradation of the environment
- climate change
- excessive use of fertilisers and industrial and domestic forms of pollution
- the overexploitation and unsustainable use of resources
- the effects of invasive alien species on native species, especially endemics.

The destruction of the natural environment leads to **habitat loss**. The clearing of land for agriculture, housing, transport, leisure facilities and industry removes vegetation. Consequently, many species of plant and animal either lose their habitats completely or their habitats become divided into small areas; this is known as **habitat fragmentation**. Most at risk of extinction are endemic species on small islands (Figure 18.1).

Deforestation has had a devastating effect on the biodiversity of some countries. Madagascar, famed for its unique plant and animal life, has lost almost all of its natural forest. Much of the forest in the northern hemisphere has been cleared by humans to grow food – a process that started about 10 000 years ago. Vast tracts of forests remained in the southern hemisphere until well into the 20th century, but much of those in South-East Asia, Africa, Amazonia and Central America have now been cut down and have often been replaced with cattle ranches and plantations of oil palm, which have much lower biodiversity. Deforestation can lead to severe land degradation as a result of soil erosion once the vegetation is removed.

Although agriculture provides most of our food, we still rely on taking wild fish from the environment. It is very difficult to know whether fish stocks are sustainable, but the history of the fishing industry suggests that many species have been driven to near extinction by overfishing. Many fisheries, such as those for cod on the Grand Banks in the North Atlantic, herring in the North Sea and a variety of species in the East China Sea have declined or collapsed. The response to the steep decrease in large, predatory species is to fish further down the food chain taking smaller fish that other animals, such as marine mammals and sea birds, depend upon. Fishing is just one example of the **overexploitation** of resources. Another example is the removal by logging companies of valuable trees, such as teak and mahogany, at a rate faster than they can regenerate.

The loss of a single species can have devastating effects on the rest of its community. The Pacific sea otter, *Enhydra lutris*, is a predator of sea urchins in kelp forests (Figures 18.27 and 18.28). In the 19th century sea otters were hunted for their fur and there was a striking change to the whole of the food web as urchins exploded in numbers and ate their way through the stipes ('stems') of the kelp forests (Figure 18.28). The loss of one species, the sea otter, led to catastrophic loss of many others. Organisms like the sea otter that play a central role in an ecosystem are known as keystone species. Sea otters are now protected and their numbers increased in the latter half of the 20th century but now they are being predated by killer whales that may have less prey to hunt because of overfishing.

Figure 18.27 A sea otter, *Enhydra lutris*.

Figure 18.28 A kelp forest in the Pacific Ocean off the coast of California. The giant seaweeds provide habitats for many species including sea urchins, fish and sea otters.

The African bush elephant, *Loxodonta africana*, is a keystone species of the savannah grasslands of East and Southern Africa. Bush elephants are very destructive of vegetation as they push over and eat many tree species. This extreme form of grazing helps to maintain this ecosystem, which is renowned for its diversity of large mammals as well as many other species. Elephant dung provides a very rich habitat – in fact almost an ecosystem in itself – for many organisms including fungi and dung beetles. Elephants were once hunted widely for their ivory and their populations decreased considerably. Now protected by international agreements, they are still at risk of poaching to supply the illegal trade in ivory (Figure 18.29).

Pollution is a major threat to many ecosystems. In many countries, industrial and domestic waste is processed to reduce its impact on the environment. For example, sewage is treated before it reaches aquatic ecosystems, such as rivers and the sea. Much toxic industrial waste is collected and disposed of so that it cannot leak into the environment. However, this does not happen everywhere with the result that many ecosystems are polluted, often with substances that animals' bodies are unable to metabolise or excrete. Polychlorinated biphenyls (PCBs) were used in various manufacturing processes. Waste from factories used to flow directly into rivers without any form of treatment. Even though PCBs are no longer used, the substance persists in the environment and has entered food chains. Among its effects are weakening of immune systems and reduction in fertility in birds and mammals. PCBs were one of the factors contributing to the deaths of seals in the North Sea from a viral infection. Non-biodegradable plastic is a major marine pollutant. Animals, such as dolphins and turtles, get caught in discarded fishing nets and die. Turtles eat plastic bags mistaking them for jelly fish.

Much of the world's farmland is in low-lying land near coasts. Fertilisers that have not been absorbed by crop plants drain into rivers and then into the sea. The extra nutrients that become available to river and marine ecosystems cause growth of producers, such as algae. This often occurs faster than herbivorous organisms, such as fish, can feed on them to keep their growth under control. Many of these algae produce toxic substances and their growth often unbalances food webs. Excess growth of algae has catastrophic effects on coral reefs and hugely reduces biodiversity.

Pollution of the air leads to problems for aquatic and terrestrial ecosystems. The combustion of fuel with a high sulfur content, such as coal, leads to high concentrations of sulfur dioxide in the atmosphere. This reacts with water vapour to fall as acid rain. Acid rain has destroyed vegetation and led to the acidification of aquatic ecosystems in parts of the world downwind of highly industrialised areas. Few animals can survive and/or breed in waters of low pH, so the biodiversity has decreased markedly. Many ecosystems are still at risk from acid rain.

Industrialisation and the extraction and combustion of fossil fuels have also led to an increase in the concentrations of carbon dioxide and methane in the atmosphere. These are both **greenhouse gases**. High emissions of methane are associated with cattle and rice farming and the breakdown, under anaerobic conditions, of organic waste in landfill sites. The build-up of greenhouse gases is leading to **climate change**. Global warming is likely to produce changes in the distribution of terrestrial ecosystems. Organisms are expected to migrate north or south to cooler latitudes and also to higher altitudes. There will be competition between migrating organisms and species in existing communities. The acidification of the oceans may spell catastrophe for coral reefs and those species, such as many molluscs, that make their skeletons and shells from calcium carbonate. Additionally, corals are very sensitive to temperature increases. The algae that live inside the polyps tend to leave the animals if the temperature remains high for a period of time. This leads to coral turning white. Coral bleaching, as it is called, can lead to the death of the coral. Coral reefs are one of the most diverse ecosystems on Earth. We rely on them to provide protection for many coastlines. A large proportion of coral reefs have already been destroyed or degraded by overfishing, mining and fertiliser run-off.

The rise in sea levels associated with global warming will bring many problems for coastal ecosystems which are some of the most productive on Earth. Some ecosystems will become even more restricted in their range than is currently the case and some will become even more fragmented. For example, ecosystems associated with high altitude will retreat higher up mountains. As the world warms, where will they go then?

The frequency of natural catastrophes, such as hurricanes, typhoons, severe storms and flooding is thought to be on the increase. Following typhoons in the Pacific, flooding increases the concentration of nutrients in coastal waters. This encourages growth of phytoplankton which provides food for the larvae of the crown-of-thorns starfish, *Acanthaster planci*. Huge numbers of adults then eat the coral (Figure 18.24). Eventually the numbers of starfish decrease and the coral regrows. If these population explosions happen every ten years or so, the coral has time to recover; if they are more frequent than this, then it may not.

Why does biodiversity matter?

Moral and ethical reasons

For many people, the loss of biodiversity is a simple moral or ethical issue: we share our planet with a huge range of other organisms and we have no right to drive them to extinction. Some people believe that humans have custody of the Earth and should therefore value and protect the organisms that share the planet with us.

Ecological reasons

There are ecological reasons why biodiversity matters. In general, the higher the diversity of an ecosystem, the less likely it is to be unbalanced by changes in conditions or threats such as pollution. All the organisms in an ecosystem interact in many different ways, and, as we have seen, if one key species disappears, this can affect the whole community. We are part of many ecosystems and rely on them in many ways.

Ecosystems are of direct value to humans. Many of the drugs that we use originate from living organisms. Antibiotics are isolated from fungi and bacteria; anti-cancer drugs have been isolated from plants such as the Madagascan periwinkle, *Catharanthus roseus*, and the Pacific yew tree, *Taxus brevifolia*, which is the source of the drug paclitaxel (better known as Taxol®). The natural habitat of the Madagascan periwinkle has been almost lost, largely because of slash-and-burn agriculture; fortunately, the plant is able to survive in artificial habitats, such as along roadsides, and is now cultivated in many countries. The Himalayan yew, which is also a source of paclitaxel, is threatened by over-harvesting for medicinal use and collection for fuel.

There is currently much interest in cataloguing plants used in traditional Chinese and Indian medicines to see if they can provide drugs that can be mass-produced. There are doubtless many more that we do not know about. If we allow tropical forests with their great biodiversity to disappear, then we are undoubtedly losing species that could be beneficial to us.

Aesthetic reasons

There is an aesthetic argument for maintaining biodiversity. Many people gain pleasure from studying or just appreciating the natural world, which continues to provide much inspiration for artists, photographers, poets, writers and other creative people.

Wildlife is a source of income for many countries as ecotourism has increased in popularity. Countries such as Belize, Malaysia, the Maldives and Costa Rica encourage tourists to visit their National Parks. This form of tourism provides employment and contributes to the economies of these nations.

Social and commercial reasons

Our crop plants do not have as much genetic diversity as their wild relatives, because it has been lost by selective breeding for uniform, high-yielding crops. The wild relatives of maize grow in the states of Oaxaca and Puebla in Mexico; they can provide the genetic resources we might need to widen the genetic diversity of cultivated maize if it is affected by disease or by other catastrophes. Many of these wild relatives are threatened by climate change, habitat destruction and perhaps the spread of genetically modified crops. A species of rice, *Oryza longistaminata*, which grows wild in Mali in North Africa, is not suitable for cultivation as a crop plant because of its low yield and poor taste. However, it is resistant to a large number of different strains of the disease of rice known as bacterial blight. It has been successfully interbred with cultivated rice, *O. sativa* (Chapter 17), to give varieties of rice with resistance to the disease. Another plant example is the potato. There are about 150 species of potato growing in the Andes, but outside that region the world's crop comes from a single species, *Solanum tuberosum*. This means that the crop is vulnerable to diseases, such as potato blight. The International Potato Center (CIP) in Peru has used the Andean species as a source of alleles for resistance. These alleles have been introduced into the crop species both by interbreeding and by gene technology (Chapter 19). Obviously it is important to conserve all the Andean potato species.

Few give any thought to the contribution of microorganisms, which are the source of many useful products, not least antibiotics. The heat-stable enzyme *Taq* polymerase was discovered in a thermophilic bacterium, *Thermus aquaticus*, from a hot spring in Yellowstone National Park in the USA. This enzyme is mass produced by genetically modified bacteria for use in the polymerase chain reaction (page 471) which is used by forensic and other scientists to increase quantities of DNA for analysis. There are likely to be many other such compounds, especially enzymes, in archaeans that live in extreme conditions not unlike those in some industrial processes.

Other services

Ecosystems provide services for us. Forests and peat bogs absorb carbon dioxide and may help to reduce the effect of increases in carbon dioxide in the atmosphere.

Organic waste material added to waters is broken down by microorganisms. The transpiration of plants contributes to the water cycle providing us with drinking and irrigation water. Termites and ants along with many species of fungi and bacteria recycle elements, such as carbon, nitrogen, sulfur and phosphorus. Without this recycling, the supply of nitrates, sulfates and phosphates for plants would become limiting. Plant growth would slow and there would be less food available for organisms in other trophic levels.

QUESTION

18.14 a List some of the threats to aquatic and terrestrial ecosystems.

b State **five** reasons why it is important to maintain biodiversity.

Protecting endangered species

An endangered species is one that is threatened with extinction (Chapter 17, page 417). There are a variety of ways to protect endangered species.

The best way to conserve any species is to keep it in its natural habitat. Maintaining the natural habitat means that all the 'life support systems' are provided. In the public mind, conservation tends to concentrate on individual species or groups of species. High-profile programmes have centred on mammals, such as giant pandas and whales. Equally important is the protection of whole ecosystems threatened by development; the most popular of these is the tropical rainforest although there are many other, less well-known, rare ecosystems that should be conserved, such as karst limestone, because they are very vulnerable to pollution.

National parks

Most countries now set aside areas where wildlife and the environment have some form of protection, and where the activities of humans are limited. For example, conservation areas may be set up where there are strict limits on building, grazing farm animals, hunting or other activities that might adversely affect animals and plants that live there.

National parks are areas of land that are controlled by the government of a country and protected by legislation. Agriculture, building, mining and other industrial activities are strictly controlled. In some countries, such as Kenya, national parks act as conservation areas where populations of wild animals are maintained (Figure 18.29).

Figure 18.29 Elephants in the Amboseli National Park, Kenya. Elephants throughout Africa are exposed to numerous threats, not least poaching. Biodiversity suffers if their numbers increase or decrease. There is a delicate balance to achieve and this needs careful management by park authorities.

The world's first national park, the Yellowstone National Park in the USA, was set up in 1872. It is the last remaining nearly intact ecosystem of the northern temperate climatic zone and serves as both a recreational and a conservation area. Three of the animal species that can be found there are listed in the USA as threatened (bald eagle, grizzly bear and lynx) and two as endangered (whooping crane and grey wolf).

Much of the Galapagos Islands, which belong to Ecuador, is a national park. Since its establishment over 50 years ago, the park authorities have done much active conservation. They have restricted access to the uninhabited islands and limited access to other areas which are sensitive to human interference. A marine reserve of 133 000 km^2 has been set up to protect the environment from the destructive activities of fishing. The reserve is cared for by local people as well as by conservation organisations. Alien animal species, such as rats and goats, are being removed and invasive plants, such as elephant grass, dug up and destroyed. There are captive breeding and reintroduction programmes, notably for giant tortoises, coordinated by the Charles Darwin Research Station.

There are restrictions on human activities in national parks. Tourism brings in money to pay for the maintenance of the parks, and also helps to inform people about how conservation takes place. This raises awareness of important issues and can elicit support from the public. This works best if local people are involved in some way, so they feel that the park is 'theirs' and can obtain benefits from it. This may involve allowing them to use some areas of the park for herding their animals or growing crops, employing them as wardens or rangers, or using some of the money raised from tourism to improve local health or education facilities.

Marine parks, like that in the Galapagos Islands, have been set up in many places to conserve fragile ecosystems and areas at risk of overfishing, dredging and pollution. The marine reserve off the coast of Little Cayman in the Caribbean is a 'no-take' reserve that protects one of the last spawning grounds of an endangered fish, the Nassau grouper, *Epinephelus striatus*. The establishment of marine parks and reserves around the coast of New Zealand has increased biodiversity and also led to an increase in fish catches.

Some conservation areas are designated by international bodies. Wetland habitats, such as estuaries, salt marshes, blanket bogs, ponds and mangrove forests, are ecosystems with high biodiversity. Ramsar sites are wetlands considered to be important for the conservation of wildlife. They are designated under an international treaty signed at Ramsar in Iran in 1971. Under the terms of the convention, a designated site must be 'used wisely'. This gives protection against such threats as building development and extraction of minerals. The ecologically important Okavango Delta in Botswana is a Ramsar site.

The standard of management of parks and reserves varies throughout the world. Some countries have the resources and the national will to provide excellent protection and careful management. Others do not.

Even though species and habitats are protected, the threats remain so great that some species have to be removed from their natural environment and placed somewhere safer. The mountain forests of Panama in Central America were the habitat of the golden frog, *Atelopus zeteki*. It faced threats from habitat loss, over-collection for the pet trade and the disease chytridiomycosis, which has caused the collapse of many amphibian populations across the world. In 2005, the Houston Zoo in Texas established a conservation centre in Panama so that the golden frog could be protected. The Society Islands of French Polynesia in the Pacific, including Tahiti and Moorea, have a rich biodiversity including many endemic species. Studies of the different species of the land snail genus *Partula* revealed how much diversity can evolve on islands. However, by the end of the 20th century, the number of species of *Partula* on the Society Islands had fallen from 61 to 5. Individuals of the remaining species were transferred to zoos around the world for their protection and conservation.

QUESTION

18.15 Explain the importance of management of national parks and other, similar protected areas.

Zoos

Zoos have a variety of functions in addition to providing enjoyment and interest for visitors who can see and study animals that they would not otherwise be able to see.

Zoos provide protection for endangered and vulnerable species, and have had success with captive breeding programmes, often with the long-term aim of reintroducing the animals to their natural habitat. For example, The Durrell Wildlife Conservation Trust at the Jersey Zoo has been involved with the captive breeding of various species of tamarins from Brazil (Figure 18.30).

A problem with breeding animals from small populations is inbreeding. The cheetah, *Acinonyx jubatus*, is a species classified by the IUCN as vulnerable (page 417). Genetic diversity among cheetahs is very low because they nearly became extinct 10 000 years ago and only a few survived. Maintaining the genetic diversity is an aim in the conservation of many species, including the cheetah. In the wild, female cheetahs tend to mate with many different males, which helps to increase genetic diversity in the population.

Figure 18.30 Golden lion tamarin, *Leontopithecus rosalia*, from the coastal forests of Brazil. As their habitat has been destroyed they have been rescued, bred in captivity and reintroduced to protected reserves.

Figure 18.31 A conservation success story: the scimitar-horned oryx, *Oryx dammah*, saved from extinction and bred in captivity, is now protected in reserves in Senegal, Tunisia and Morocco.

Zoos also have an important role in research, especially in trying to gain a better understanding of breeding habits, habitat requirements and ways to increase genetic diversity. The Zoological Society of London (ZSL), like many large zoos, has an important programme of research.

The major goal of captive breeding is to reintroduce animals to their natural habitat. This can prove extremely difficult as there are many factors that affect the success of these schemes. The Emperor Valley Zoo in Trinidad, in collaboration with Cincinnati Zoo in the USA, has successfully reintroduced captive-bred blue-and-gold macaws, *Ara ararauna*, to the nearby Nariva Swamp. The scimitar-horned oryx, *Oryx dammah*, was driven almost to extinction in its habitat, the semi-deserts in northern Africa, by hunting for its meat and skin (Figure 18.31). During the 1960s and 1970s, it was recognised that if nothing was done, the oryx would become extinct. A few oryx were caught and transported to zoos in several places around the world. A captive breeding programme was successful and breeding herds of these animals have been established in reserves in North Africa.

However, not every conservation attempt has been a success story. Some animals simply refuse to breed in captivity. Often, it is not possible to create suitable habitats for them, so they cannot be returned to the wild. Sometimes, even if a habitat exists, it is very difficult for the animals to adapt to living in it after being cared for in a zoo. The giant panda is a good example. In 2011, there were over 300 pandas in zoos and research centres in their native China, and others on loan to zoos around the world. A captive breeding programme was started in 1963, and since then about 300 pandas have been born in captivity, but so far no panda has successfully been returned to the wild (Figure 18.32). The first captive-bred panda to be released was killed, aged five years, probably by other pandas. Female golden lion tamarins released into forest reserves in Brazil often die before breeding because they do not have the climbing and foraging skills that they need to survive. Some captive-bred animals do not know how to avoid predators, find food or rear their own young.

Figure 18.32 The giant panda cub, Yuan Zai, born in the Taipei Zoo in Taiwan on 6 July 2013. Is she destined to remain in captivity her whole life or will she be released into the wild?

Assisted reproduction

Assisted reproduction is a solution to the problem of inbreeding. Zoos used to transport large mammals between them as part of their captive breeding programmes. Movement of large mammals is difficult and expensive and breeding did not always happen. A much cheaper option is to collect semen and keep it frozen in a **sperm bank**. Samples are collected from males, checked for sperm activity and then diluted with a medium containing a buffer solution and albumen. Small volumes of semen are put into thin tubes known as straws, which are stored in liquid nitrogen at −196 °C.

Different methods of assisted reproduction solve the problem of males and females who do not show any courtship behaviour and will not mate. In **artificial insemination** (AI), a straw is placed into warm water so that sperm become active and then put into a catheter, which is inserted into the vagina, through the cervix and into the uterus. This may happen when the female is naturally 'on heat', but also may follow hormone treatment so she ovulates at the time of AI. The hormone treatment can stimulate the female to 'superovulate' to produce a large number of follicles (page 350). Following AI, the resulting embryos may be 'flushed out' of the uterus and transferred to other females that have had hormonal treatment to prepare them for pregnancy. These females need not be of the same species – they could be a related, but not endangered species. This process of **embryo transfer** protects the endangered animal from the risks of pregnancy and means that she can be a source of many offspring. Females that receive embryos like this are **surrogate** mothers. This technique of embryo transfer has been used for many species including wild ox and different species of African antelope.

In order to carry out **in vitro fertilisation** (IVF), oocytes are collected by inserting a needle into the ovaries and withdrawing some mature follicles. The oocytes are kept in a culture medium for a short time and then mixed with semen. The resulting zygotes divide to form embryos, which are cultured for several days and then placed into the mother or into several females of the same or different species.

Eggs (oocytes) and embryos can also be stored in much the same way as sperm. Eggs are more difficult to freeze as they are more likely to be damaged by the freezing or thawing processes. Eggs are large cells with lots of water which tends to form ice crystals that damage internal membranes. Eggs are fertilised in vitro and then frozen until such time as a surrogate mother becomes available. A '**frozen zoo**', such as the one at the San Diego Zoo, holds genetic resources in the form of sperm, eggs and embryos from many endangered and vulnerable species until they might be needed. Frozen zoos can hold much more genetic diversity than a normal zoo and the material can be kept for very long periods of time.

Problems of successful conservation

Sometimes, conservation practices have been too successful and the organism saved from extinction has increased in numbers beyond the capacity of the ecosystem to sustain such numbers. **Culling** is often used to reduce numbers although it is a practice that arouses much emotion, especially when it is used to control the numbers of elephants. Between 1966 and 1994 more than 16 000 elephants were culled in the Kruger National Park, South Africa, to limit the growth of the population. Transferring animals to places where there are small populations is one option, but that is not easy over large distances and is expensive. An alternative is to use methods of birth control. San Diego Zoo has experimented with sedating male wild mammals and cutting their sperm ducts (vasectomy). Chemical contraceptives are available, but – unlike the contraceptives that women may use – these are not steroid hormones (pages 352–353). Instead, a vaccine is used which targets the region surrounding the layer of glycoproteins around the egg – the zona pellucida. When the vaccine is injected into a female animal, it stimulates an immune response that produces antibodies against these glycoproteins (Chapter 11). These antibodies attach to the glycoproteins around the female's own eggs, so blocking sperm from fertilising the egg. This method has a 90% success rate in mammals.

QUESTIONS

18.16 a Suggest why some animals do not breed in captivity.

b Explain the following terms: artificial insemination, in vitro fertilisation, sperm bank, embryo transfer, surrogacy and 'frozen zoo'.

c Explain why it is important that zoos do not keep small, reproductively isolated populations of animals for breeding purposes.

18.17 Many people think that keeping animals in zoos should be banned. Outline the arguments for zoos.

Botanic gardens

Botanic gardens play similar roles to zoos for endangered plants. Seeds or cuttings are collected from species in the wild and then used to build up a population of plants from which, one day, some plants may be reintroduced to their natural habitats. It is also possible to take small samples of cells and grow them on agar in sterile conditions. The cells divide by mitosis to give a mass of cells that can be cloned by subdividing them. When the cells are transferred to a medium containing an appropriate mixture of plant hormones, they grow stems and roots and can then be transferred to grow in soil. These techniques of tissue culture and cloning are used to produce large numbers of plants from a few original specimens.

The roles of botanic gardens are to:

- protect endangered plant species; the world's botanic gardens already cultivate around one-third of the world's known plant species, many of which are increasingly threatened in the wild by environmental degradation and climate change
- research methods of reproduction and growth so that species cultivated in botanic gardens can be grown in appropriate conditions and be propagated
- research conservation methods so plants can be introduced to new habitats if their original habitat has been destroyed
- reintroduce species to habitats where they have become very rare or extinct
- educate the public in the many roles of plants in ecosystems and their economic value.

It often takes a long time to reintroduce a plant species and ensure its survival. This is especially true with slow-growing plants, such as Sargent's cherry palm, *Pseudophoenix sargentii*. Specimens of this species were grown at the Fairchild Tropical Botanical Gardens in Miami before being reintroduced to its natural habitat in the Florida Keys.

As well as cultivating plants, botanic gardens may store seeds in a seed bank. The Royal Botanic Gardens at Kew, UK, runs a hugely ambitious project called the Millennium Seed Bank, which began in 2000 (Figures 18.33 and 18.34). The bank's ambition is to collect and store seeds from at least 25% of the world's plants by 2025, so that even if the plants become extinct in the wild there will still be seeds from which they can be grown. If possible, seeds of the same species are collected from different sites, so that the stored samples contain a good proportion of the total gene pool for that species.

Figure 18.33 The Millennium Seed Bank, Wakehurst Place, UK. Seeds arriving at the seed bank are checked for pests and diseases, assessed for viability, dried, and then stored in airtight jars (Figure 18.34) and kept in the seed-storage vault at −20 °C.

Figure 18.34 A botanist with one of the seed collections in a cold vault at the Millennium Seed Bank.

The Svalbard Global Seed Vault is a seed bank run by the government of Norway. The vault is at the end of a 120 metre tunnel cut into the rock of a mountain on the island of Spitsbergen, within the Arctic Circle. The vault is thought to have ideal storage conditions, and it is only opened in winter when the environmental temperature falls close to the operating temperature of −18 °C. Its first seed samples went into storage in January 2008, and by 2013 the vault held over 770 000 different seed samples of crop varieties from all over the world.

National and international organisations collect and store seeds so that the genetic diversity in our crop plants is not lost. Seeds from each sample are stored in the vault at Svalbard. If seeds stored elsewhere are lost for any reason, such as an environmental disaster that destroys a seed bank, then there will always be duplicate samples available from Svalbard. Such problems could range from mismanagement or loss of funding to accidents or failure of equipment. The depositing seed bank owns its seeds and alone has access to them. Any research organisation wishing to use any of the seeds has to apply to the original seed bank for such seeds. The storage of the seeds is free of charge, as the costs of the upkeep of the vault are covered by the Norwegian government and by the Global Crop Diversity Trust. This trust also helps developing countries to select and package seeds that can then be sent to Svalbard to be stored.

Another important seed bank is that of the International Rice Institute in the Philippines, which holds all rice varieties.

Many plants produce seeds, known as orthodox seeds, that remain viable for at least 15 years if they are carefully dehydrated until they contain only about 5% water, and then stored at around −15 to −20 °C. With this small water content, there is little danger that cells in the seed will be damaged by ice crystals during freezing and thawing.

The only way to find out whether or not stored seeds are still viable is to try to germinate them. Seed banks carry out germination tests at five-year intervals. When fewer than 85% of the seeds germinate successfully, then plants are grown from these seeds so that fresh seed can be collected and stored.

When such plants are grown from samples of stored seed, there is the possibility of altering the genetic diversity that was originally stored. Small samples of seeds from rare plants present a particular problem, as even smaller samples of the original are taken to test for viability or to grow into plants to increase the number of seeds in store. Such samples are unlikely to contain all the genetic diversity of the original sample. The only answer to this problem is to put as large and diverse a sample as possible into store in the first place.

Most seeds are easy to store, but some plants have seeds that cannot be dried and frozen. These 'recalcitrant seeds', as they are called, include seeds of economically important tropical species, such as rubber, coconut palm, coffee and cocoa (Figure 18.35). The only ways to keep the genetic diversity of these species are to collect seeds and grow successive generations of plants or to keep them as tissue culture. Cocoa is banked as trees. The International Cocoa Genebank in Trinidad has about 12 000 trees – examples of all the cocoa varieties found in Latin America and the Caribbean. Selected material from the Trinidad collection is distributed to other cocoa-producing countries after it has been through quarantine at the University of Reading in the UK.

Coconut palms are particularly difficult to bank (Figure 18.36). The seed (the coconut) is very large, and the embryo is too large to freeze successfully. Collectors remove the embryos from the seeds, culture them in sterile tubes and eventually plant them.

Figure 18.35 The future of cocoa, *Theobroma cacao*, is threatened by diseases, climate change, natural disasters, limited genetic diversity and the failure to manage plantations by replacing old trees. 30–40% of the world's production is lost to pests and disease.

Figure 18.36 Germinating coconut seeds ready to be planted in nursery plots. There are gene banks for coconut in Karnataka, India and in Papua New Guinea.

QUESTIONS

18.18 **a** Why is it important to keep seeds in seed banks?
b Explain why it is not possible to keep the seeds of all species in seed banks, such as the Millennium Seed Bank and the Svalbard Seed Bank.

18.19 It has been suggested that seed banks put selection pressures on the seeds that are different from those that the plants would experience in the wild.
a How might these selection pressures differ?
b How might this affect the chances of success in returning the plants to the wild?
c Suggest some particular problems that face gene banks for plants with recalcitrant seeds, such as cocoa and coconut.

Controlling alien species

Alien or invasive species are those that have moved from one ecosystem to another where they were previously unknown. People have been responsible for the movement of species about the globe by trading animals and plants or unwittingly carrying them on ships. Some species have been introduced as biological control agents to control pests. The small Indian mongoose, *Herpestes auropunctatus*, was introduced to Jamaica in 1872 and proved so successful at controlling rats in the cane fields that it was introduced elsewhere. Unfortunately, it then became a predator of other animals. Perhaps the most notorious introduction is the cane toad, introduced to Queensland, Australia from Hawaii in 1935 to control an insect pest of sugar cane. In Australia, the cane toad has become a pest as it breeds rapidly and has spread across the eastern, northern and western parts of the country. The cane toad has few predators in Australia, mainly because it produces a powerful toxin that kills most animals that eat it. The species known to be most at risk from invasion by cane toads is the northern quoll, *Dasyurus hallucatus*, which tries to eat the toads. Numbers of this endangered marsupial carnivore decrease steeply after cane toads invade its habitat. Cane toads probably compete with some other amphibian species for food and are known to eat the chicks of a ground-nesting bird, the rainbow bee-eater, *Merops ornatus*.

Other alien species are escapees or animals introduced for sport. It is thought that the rabbit, introduced to Australia in the 19th century, has been responsible for more loss of biodiversity there than any other factor. Burmese pythons have invaded the Everglades National Park in Florida, probably because pet owners found that they could not look after them anymore and just let them go into the wild. The pythons feed on a wide variety of mammals and birds so competing with native predators. This makes it difficult to conserve endangered species in the national park. Humans are the python's only predator, although efforts to remove pythons completely from Florida have failed and it looks as if trying to stop the population increasing too much is the only option available.

The red lionfish, *Pterois volitans*, is native to the seas of South-East Asia. No-one knows how it came to invade the waters of the Caribbean, but it seems likely to have escaped from aquaria in the United States. It has spread throughout the Caribbean, eating its way through many local species on coral reefs. Again, there is no natural predator of the animal in its new environment. In Belize, divers are encouraged to spear them to reduce their populations. The Jamaican government believes that if it encourages people to develop a taste for them, fishermen will catch more of them and their population will soon decrease (Figure 18.37).

Invasive species have a variety of effects on their new environments. As well as being successful predators with few controls, they may compete effectively with native organisms that occupy the same niche, pushing them to extinction. They may also introduce diseases that spread to similar organisms that have never been exposed to the pathogens. Some invasive plants grow so successfully that they cover huge areas of land or water. The water hyacinth, *Eichhornia crassipes*, is a floating aquatic plant that spreads rapidly when introduced to new habitats. It blocks sunlight from reaching native aquatic plants and reduces the oxygen concentration of the water, so killing fish. Water hyacinth also provides a habitat for mosquito larvae so its control is important for the sake of human health too. Japanese knotweed, *Fallopia japonica*, has a very vigorous root system and its growth is so strong that it can force its way through concrete and damage buildings, roads and walls. It also outcompetes native species simply by reducing the space where they can grow.

Figure 18.37 Red lionfish, *Pterois volitans*, an alien species that escaped into the Caribbean and is causing havoc on coral reefs as it eats many reef animals and has no natural predator.

QUESTION

18.20 a Explain the damage that alien species may have on an ecosystem.

b Suggest how you might investigate the effect of an alien plant species on the biodiversity of an ecosystem.

International conservation organisations

CITES

In 1973, 145 countries signed an agreement to control the trade in endangered species and any products from them, such as furs, skins and ivory. More countries have joined since. This agreement is called the Convention on International Trade in Endangered Species of Wild Flora and Fauna, CITES for short.

CITES considers the evidence presented to it about endangered and vulnerable species and assigns them to one of three Appendices as shown in Table 18.6.

The species listed in the CITES appendices are reviewed by expert committees and the list is growing. However, there is concern that a CITES listing does not always benefit a species (Figure 18.38). If trade in a species or its products becomes illegal, then the price that can be obtained for those products rises, and this is likely to make it worthwhile for people to break the law. Particular problems arise when it is announced in advance that a species will go on the list; in the months between the announcement and the introduction of the new law, trade in that species tends to increase.

You can learn a lot more about CITES and see photographs and information about the species that are listed on their website.

Figure 18.38 Policemen in Malaysia examine dead sea turtles from a boat stopped on its way to the souvenir market. All sea turtles are listed in CITES Appendix 1, and poaching remains a serious threat to sea turtle populations, along with fishing nets, pollution, and habitat destruction.

CITES Appendix	Criteria	Trading regulations	Animal examples	Plant examples
I	Species that are the most endangered and threatened with extinction	All trade in species or their products is banned	Orang-utans, *Pongo abelii* and *P. pygmaeus* (Borneo, Indonesia)	Kinabulu pitcher plant, *Nepenthes kinabaluensis* (Malaysia)
II	Species that are not threatened with extinction, but will be unless trade is closely controlled	Trade is only allowed if an export permit is granted by the countries concerned	Sir David's long-beaked echidna, *Zaglossus attenboroughi* (Papua, Indonesia)	All species in the genus *Nepenthes*; Venus fly trap, *Dionaea muscipula*
III	Species included at the request of a country that regulates trade in the species and needs the cooperation of other countries to prevent unsustainable or illegal exploitation	Trade in these species is regulated; permits are required, but they are easier to obtain than for species in Appendix II	Mauritian pink pigeon, *Columba mayeri*	Spur tree from Nepal, *Tetracentron sinense*

Table 18.6 CITES Appendices I, II and III.

World Wide Fund for Nature

The World Wide Fund for Nature (WWF) is one of the best known campaigning groups for wildlife. Established in 1961, WWF is the largest international non-governmental organisation (NGO) specialising in conservation. Its mission statement is 'to stop the degradation of the planet's natural environment and to build a future in which humans live in harmony with nature'. To this end, it funds conservation projects, publicises environmental issues and campaigns to save ecosystems from degradation and species from extinction. One of its recent campaigns is to stop prospecting for oil in Africa's oldest national park. The Virunga National Park in the Democratic Republic of the Congo (DRC) is one of the last refuges of the mountain gorilla, *Gorilla beringei beringei*, in the wild.

Restoring degraded habitats

Conservation involves restoring areas that have been degraded by human activity or by natural catastrophes, such as flood, fire, hurricane, typhoon and earthquake. This can be done on a small scale when a farmer decides to plant trees on land that is no longer needed for food production or has become degraded by overuse.

Mangrove forests are found throughout the tropics. This is an extremely rich ecosystem that provides valuable protection to coastlines from storms (Figure 18.39). Many mangrove forests have been cut down to make way for coastal developments. One place where replanting mangroves is important is in the Sunderbans delta region which covers some 26 000 square kilometres in India and Bangladesh and now has the largest mangrove forest in the world. This area with high diversity is at risk of rising sea levels and planting mangrove is thought likely to offer some protection against the effects of this. Mangrove forests provide important 'ecosystem services'. They reduce coastal erosion by reducing the effects of strong waves during storms and they act as a barrier to rising sea levels by trapping sediment. They are also important nurseries for young fish.

Figure 18.39 Young people planting mangrove seedlings on the island of Bali in Indonesia. In the Philippines, a group of students led a community effort to replant 100 000 mangroves in seven months after learning how mangroves protect coastal areas from storms yet were being cut down for charcoal.

After centuries of deforestation, soil erosion and severe land degradation in Haiti, efforts are now being made to restore some of the forest that used to cover the country. About 70% of the country's land is not suitable for agriculture any more and there is a severe shortage of firewood. Numerous NGOs are working with community groups in tree planting projects in what is in effect a rescue mission. Haiti is a lesson to the world as to the effects of neglecting the welfare of trees.

A special reclamation project was undertaken in a disused clay mine in Cornwall in the UK. The area became the site of the Eden Project, which is dedicated to educating people in plant biodiversity and the need for conservation (Figure 18.40). Beneath the domes are communities of plants from different regions of the world. Visitors learn about the major roles that plants play in our economy and the extent to which we depend on them. We ignore their biology and their welfare at our peril.

Figure 18.40 The Eden Project, near St Austell in Cornwall, UK. This is an example of reclamation, and also of education in the importance of sustainable development and the conservation of the world's plant life.

Summary

- A species is a group of organisms that have many features in common and are able to interbreed and produce fertile offspring. An ecosystem is a relatively self-contained, interacting community of organisms, and the environment in which they live and with which they interact. A niche is the role of an organism in an ecosystem; it is how it 'fits into' the ecosystem.
- Biodiversity is considered at three different levels: variation in ecosystems or habitats, the number of species and their relative abundance, and genetic variation within each species.
- Random sampling in fieldwork is important to collect data that is representative of the area being studied and to avoid bias in choice of samples. Random sampling is appropriate for habitats where species are distributed uniformly across the whole area and all are exposed to the same environmental conditions.
- Frame quadrats are used to assess the abundance of organisms. Abundance can be recorded as species frequency, species density and percentage cover. Line and belt transects are used to investigate the distribution of organisms in a habitat where conditions are not uniform (e.g. from low to high altitude). Line transects show qualitative changes in species distribution; belt transects show changes in abundance as well as distribution. Mark–release–recapture is a method used to estimate the numbers of mobile animals in a particular place.
- Simpson's Index of Diversity (D) is used to calculate the biodiversity of a habitat. The range of values is 0 (low biodiversity) to 1 (high biodiversity). Spearman's rank correlation and Pearson's linear correlation are used to see whether there is a relationship between two features.
- Species are classified into a taxonomic hierarchy comprising: domain, kingdom, phylum, class, order, family, genus and species. There are three domains: Bacteria, Archaea and Eukarya. Viruses are not included in this classification as they are acellular; they have no cells and are composed of only nucleic acid (DNA or RNA) surrounded by a protein coat. Viruses are classified according to the type of nucleic acid that they have and whether it is single-stranded or double-stranded.
- There are many threats to the biodiversity of aquatic and terrestrial ecosystems; these include overpopulation (by humans), deforestation, overexploitation, pollution and climate change. There are moral and ethical reasons for maintaining biodiversity, and also more practical ones; for example, plants may be sources of future medicines and animals may provide alleles to use in animal breeding.

- Conservation of an endangered animal species may involve captive breeding programmes, in which viable populations are built up in zoos and wildlife parks. These programmes try to ensure that the gene pool is maintained and inbreeding is avoided. At the same time, attempts are made to provide a suitable habitat in the wild, so that captive-bred animals can eventually be reintroduced to the wild; it is important to involve local people in such projects to increase the level of acceptance and the chances of success.
- Methods of assisted reproduction, such as IVF, embryo transfer and surrogacy are used in the conservation of endangered mammals. Some populations of protected animals increase beyond the resources available in parks so that culling and/or contraceptive methods are used to prevent their overpopulation. One successful method of contraception is using a vaccine against the zona pellucida that surrounds the egg; this prevents fertilisation occurring.
- Botanic gardens and seed banks help to conserve threatened plant species by breeding them for reintroduction into an appropriate habitat. Seed banks provide suitable conditions to keep different types of seeds alive for as long as possible. Samples of the seeds are grown into adult plants every now and then, so that fresh seed can be collected.
- Many countries have protected areas called national parks, which often cover large areas. National parks are set up to conserve rare and endangered species and to maintain their habitats; often legislation is passed to ensure their protection. Within national parks, activities such as agriculture, building, mining and other industries are forbidden or strictly regulated; access may be limited but not forbidden, as one aim of most such parks is to educate people about the importance of conservation.
- Marine parks have been created in many parts of the world to protect the biodiversity of endangered ecosystems such as coral reefs; these parks usually prohibit fishing. One benefit of marine parks is that populations of fish in the area have increased. Other, smaller, conservation areas may be created to protect particular species and habitats.
- Alien species are those that have been introduced or have invaded ecosystems where they did not exist before; alien or invasive species need to be controlled because they pose threats to native species. Alien species may be effective predators and/or competitors with native species.
- Non-governmental organisations, such as the World Wide Fund for Nature (WWF) and the Convention on International Trade in Endangered Species of Wild Fauna and Flora (CITES), play important roles in local and global conservation.
- An important part of conservation is restoring degraded habitats, so that they may support a flourishing community with high biodiversity. Mangrove forest is being replanted in many parts of the world to provide protection against storm damage, flooding and rising sea levels; mangrove forests are also important nursery grounds for young fish.

End-of-chapter questions

1 Biodiversity includes:

A ecosystem and habitat diversity
B species diversity and ecosystem diversity
C species diversity, genetic diversity and ecosystem diversity
D species diversity only [1]

2 Which of the following is the definition of the term **niche**?

A all the environmental factors that determine where an organism lives
B all the food webs in an ecosystem
C the place where an organism lives
D the role that a species fulfills in a community [1]

3 Which of the statements about Protoctista are correct?

1 A eukaryote that is not a fungus, plant or animal is a protoctist.
2 An organism with cellulose cell walls and chloroplasts may be a protoctist.
3 An organism existing as a group of similar cells may be a protoctist.
4 A single-celled heterotrophic eukaryote is a protoctist.

A 1, 2, 3 and 4
B 1, 2 and 4
C 2 and 3
D 3 and 4 only [1]

4 Some students investigated what size of quadrat they should use to assess the abundance of plant species in an old field ecosystem. They used quadrats of side 10, 25, 50, 75 and 100 cm and recorded how many plant species were present. They repeated their investigation five times and calculated mean numbers of species per quadrat. Their results are in the graph below.

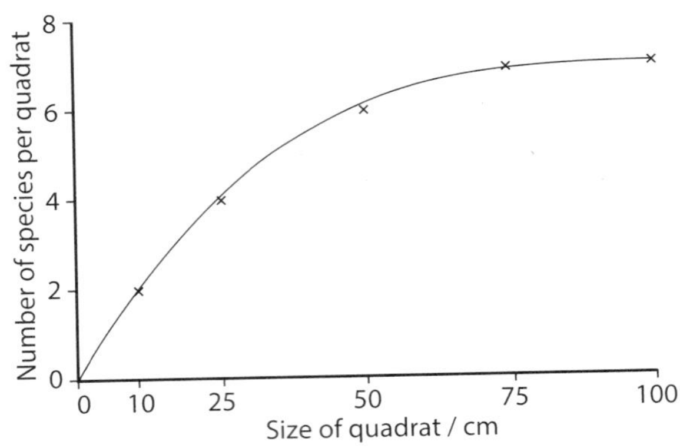

a Calculate the area of each quadrat that they used. [2]

b Explain why the students took five results for each quadrat. [2]

c Based on their results, the students decided to use the 50 cm quadrat to study the old field. Why did they choose the 50 cm quadrat? [2]

d Explain how they would use the 50 cm quadrat to estimate the abundance of the different plant species in the field. [3]

[Total: 9]

5 Five 0.25 m^2 quadrats were placed randomly in an area of grassland in the UK. The percentage of each quadrat occupied by each species of plant was estimated to the nearest 5% and recorded in the following table.

Species	Percentage cover of each plant species in each quadrat to the nearest 5%					Mean percentage cover
	1	2	3	4	5	
Timothy grass	60	30	35	70	25	
Yorkshire fog grass	25	70	30	15	40	
plantain	0	5	0	5	0	2
meadow buttercup	0	0	15	0	10	5
dock	5	10	5	0	5	5
cowslip	0	0	0	5	0	1
white clover	15	0	25	25	10	15
bare ground	0	15	15	5	20	11

a Calculate the mean percentage cover for the first two species in the table. [1]

b Explain why the percentage cover for all the species in each quadrat adds up to more than 100%. [1]

c Suggest why the percentage cover was recorded to the nearest 5%. [1]

d Could these results be used to obtain a valid estimate of the species density for each species? Explain your answer. [4]

e State **four** factors that influence plant biodiversity in a field ecosystem. [4]

[Total: 11]

6 A sample of 39 ground beetles was captured from an area of waste ground measuring 100 × 25 metres. Each animal was marked and then released. A second sample of 35 was caught on the following day. Of these, 20 had been marked.

a Use these results to estimate the number of ground beetles in the population. Show your working. [2]

b State **three** assumptions that must be made in order to make this estimate. [3]

c Describe a method that you could use to check that the mark–release–recapture method gives a valid estimate of the ground beetle population in the area of waste ground. [4]

[Total: 9]

7 Light is an important abiotic factor that determines the distribution and abundance of plants. Some plants are adapted to grow in areas of low light intensity. They are known as shade plants. Some students investigated the abundance of dog's mercury, *Mercurialis perennis*, under forest trees in eight different locations. At each location, they used ten randomly positioned quadrats to measure the percentage cover of this shade plant.

Location	Light intensity at ground level / percentage of full sun	Mean percentage cover of *M. perennis*
A	90	10
B	50	65
C	30	90
D	20	80
E	80	34
F	88	20
G	93	10
H	100	0

a Draw a scatter graph to see if there is a correlation between light intensity beneath the canopy and abundance of *M. perennis*. [2]

b State a null hypothesis for this investigation. [1]

c Use the data in the table to calculate the Spearman rank correlation, r_s. Show all your working. [4]

d What can you conclude from your calculation? You will need to consult Table P2.8 on page 504 to help you answer this question. [3]

[Total: 10]

8 Stoneflies, *Plecoptera* spp., lay eggs in freshwater streams and rivers. The eggs hatch into nymphs which live in the water for several years before changing to adults. Stonefly nymphs are known as good indicators of pollution since they are very sensitive to a decrease in the oxygen concentration of the water.

A biologist wanted to find out whether stonefly nymphs would be suitable as an indicator of water hardness. She collected samples from 12 streams and obtained values of calcium carbonate concentration from the local water authority. The number of stonefly nymphs and the concentration of calcium carbonate for each of the 12 streams are in the table below.

Stream	Number of stonefly nymphs	Concentration of calcium carbonate / arbitrary units
1	42	17
2	40	20
3	30	22
4	7	28
5	12	42
6	10	55
7	8	55
8	7	75
9	3	80
10	7	90
11	5	145
12	2	145

a Draw a scatter graph to see if there is a correlation between the number of stonefly nymphs and the hardness of the water. [2]

b State a null hypothesis for this investigation. [1]

c Use the data in the table to calculate the Spearman's rank correlation, r_s. Show your working. [4]

d What can you conclude from your calculation? You will need to consult Table P2.8 on page 504 to help you answer this question. [3]

[Total: 10]

9 Woodlice are small terrestrial crustaceans that feed on organic matter in leaf litter and in the soil. While carrying out an ecological survey in woodland, a student noticed that there seemed to be more woodlice in areas where there was plenty of leaf litter and other organic matter. She also noticed that there appeared to be more leaf litter towards the middle of the wood than at the edge. To find out if there is a relationship between the number of woodlice and the organic content of the soil, she used a line transect. At each point along the transect, she:

- took a sample of soil.
- counted the number of woodlice in three separate samples of leaf litter from the surface of the soil.

She took the samples of soil to a laboratory to find out their percentage organic content. Her results are in the table below.

Distance along transect / m	Percentage organic matter in the soil	Number of woodlice in sample			
		1	2	3	mean
1	5.42	2	0	0	0.6
4	10.02	3	4	0	2.3
8	15.56	3	2	2	2.3
16	8.25	0	1	3	1.3
20	9.62	5	7	6	6.0
24	11.73	9	9	9	9.0
28	10.67	8	14	15	12.3
32	9.36	14	16	3	11.6
36	11.35	12	17	6	11.6
40	15.11	17	9	7	11.0
44	20.87	20	2	9	10.3
48	20.30	20	1	12	11.0

a i Explain why the student has taken three samples of woodlice at each point along the transect. [2]

ii Suggest **three** further pieces of information you would need if you were to repeat the student's investigation. [3]

b Draw a scatter graph to see if there is any relationship between the number of woodlice and the mass of organic matter in the soil. [2]

c State the null hypothesis for this investigation. [1]

d Carry out the Spearman's rank test on the data that you have used to draw the scatter graph. [4]

e What conclusions can you make? You will need to consult Table P2.8 on page 504 to help you answer this question. [4]

[Total: 16]

10 Researchers investigated aspects of the anatomy of the Melina tree, *Gmelina arborea*, which has been introduced into wet and dry forests in Costa Rica. The researchers wanted to know if the diameter of the trees at head height was correlated with the overall height of the trees. The table shows the data for seven trees from each type of forest.

Tree number	Wet forest		Dry forest	
	diameter / cm	height / m	diameter / cm	height / m
1	38.0	25.5	31.9	20.8
2	31.5	22.0	29.8	19.5
3	33.2	18.5	32.5	24.7
4	32.0	23.5	30.5	24.0
5	30.7	18.0	33.2	27.1
6	25.0	19.0	19.7	19.5
7	28.5	22.5	22.5	23.3

The researchers calculated the Pearson's correlation coefficient for the two areas as:

Type of forest	Pearson's correlation coefficient
wet forest	0.47
dry forest	0.42

The critical value at $p = 0.05$ for samples of this size is 0.75.

a Explain what conclusions the researchers should make from these results. [4]

b Explain why the researchers chose to calculate the Pearson's correlation coefficient and not the Spearman's rank correlation coefficient. [2]

c The researchers next measured the diameters of trees of the same heights in the two forests to see if their growth in width was influenced by the rainfall in the two forests.

i State the independent and dependent variables in the investigation. [2]

ii Describe the method that the researchers would follow to gain results. [5]

iii State how the researchers would find out if the difference between the two populations of trees is significant. [1]

[Total: 14]

11 Sharks are an important part of the biodiversity of marine ecosystems. Many species are endangered as a result of overfishing. In 2004, the great white shark, *Carcharodon carcharias*, was added to Appendix II of CITES. Two other sharks, the porbeagle, *Lamna nasus*, and the scalloped hammerhead shark, *Sphyrna lewini,* were added to Appendix III in 2013.

a Explain what is meant by the term **biodiversity**. [3]

b i The great white shark is an endangered species. Explain what is meant by the term **endangered**. [1]

ii Explain how the addition of species to the CITES lists provides protection for endangered species, such as sharks. [3]

c List **four** practical reasons why humans should maintain biodiversity. [4]

[Total: 11]

12 The numbers of elephants in Hwange National Park in Zimbabwe were counted each year between 1980 and 2001. In the 1980s, some of the elephants were culled. This practice stopped in 1986. The graph shows the changes in the numbers of elephants ($\pm S_M$, see page 499), the bars indicate the numbers of elephants that were culled.

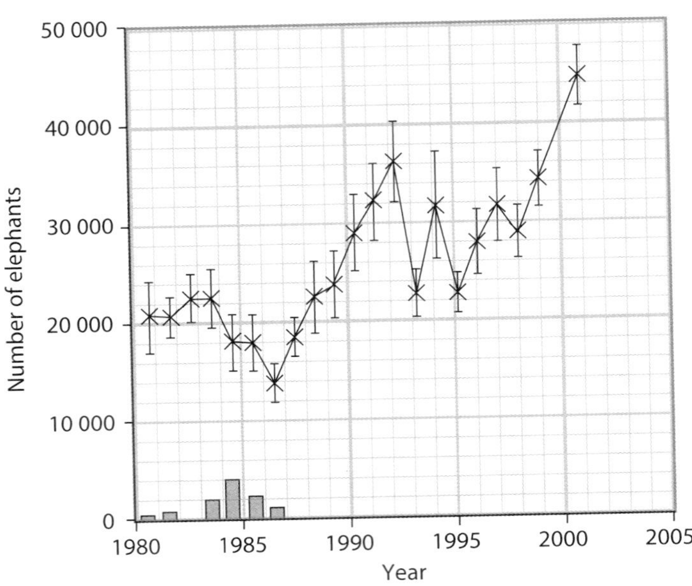

a Explain the importance of showing the standard error (S_M) on the graph. [2]

b Describe the changes in the numbers of elephants between 1980 and 2001. [4]

c Elephants feed by browsing on woody vegetation and other plants. They strip off bark and push over some trees, keeping patches of grassland open, but leaving much woodland.

Explain how both high and low densities of elephants could result in reduced species diversity in national parks in Africa. [4]

d Explain the reasons for culling animals, such as elephants, in national parks, other than to prevent loss of biodiversity. [4]

e Populations of animals can also be controlled by the use of contraceptives.

Explain how this method is used. [4]

[Total: 18]

13 a There are many seed banks in different parts of the world. Explain the reasons for depositing samples of seeds in cold storage. [6]

b Explain what is meant by the term **frozen zoo**. [3]

c Przewalski's horse, *Equus ferus przewalskii*, is a species of horse that became extinct in the wild in the 1980s. Zoos throughout the world already held large numbers of this species. A captive breeding programme was so successful that these horses have been returned to the Dzungarian Gobi Strictly Protected Area in Mongolia.

i Explain how captive breeding of animals, such as Przewalski's horse, can help to conserve an endangered species. [3]

ii Explain how genetic diversity is maintained during a captive breeding programme. [3]

iii Suggest how the success of a reintroduction scheme may be assessed. [3]

[Total: 18]

Chapter 19:
Genetic technology

Learning outcomes

You should be able to:

- explain the principles of genetic technology
- describe the tools and techniques available to the genetic engineer
- describe the uses of the polymerase chain reaction (PCR) and gel electrophoresis
- explain the use of microarrays
- outline the use of bioinformatics in sequencing genomes
- describe examples of the uses of genetic technology in medicine
- describe some uses of genetically modified organisms in agriculture

Crispr

Until recently, genetic technology has had to rely on inaccurate methods of editing DNA. For example, a modified virus can be used to insert DNA into the human genome, but we have had no control about just where the DNA is inserted. It might be inserted in the middle of another gene, with unpredictable consequences.

A technique called Crispr is changing this. Crispr has been developed from a mechanism used by some bacteria to defend themselves against viruses. It was first discovered in 1987, but its details were worked out only in 2012. Since then, it has been used successfully to edit DNA in a range of organisms, including human cells. A Crispr-associated (CAS) enzyme cuts DNA at a point that is determined by a short strand of RNA. Since the RNA can be made 'to order' to match any unique sequence of DNA, the genetic engineer can dictate precisely where a cut is made in human DNA to remove a 'faulty' allele or to insert a therapeutic allele as in gene therapy (Figure 19.1). The genetic engineer now has a tool kit that can 'cut and paste' genes with much greater precision.

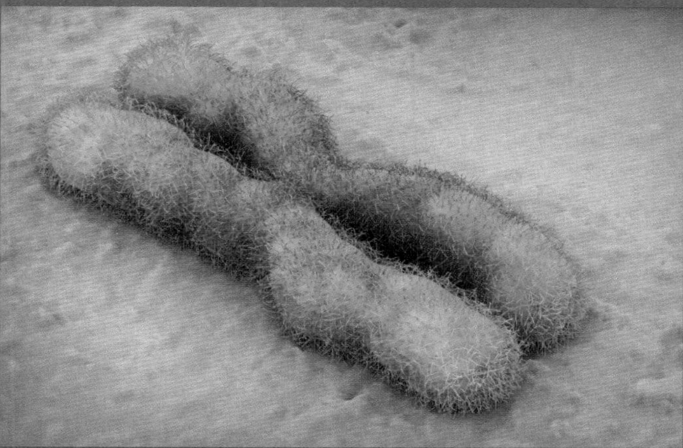

Figure 19.1 Crispr makes it possible to insert DNA at a precisely determined point in a chromosome.

The structure of DNA, and the way in which it codes for protein synthesis, was worked out during the 1950s and 1960s. Since then, this knowledge has developed so much that we can change the DNA in a cell, and thereby change the proteins which that cell synthesises. Not only that, but we can sequence the nucleotides in DNA and compare nucleotide sequences in different organisms. It is also possible to carry out genetic tests to see if people are carriers of genetic diseases and, in a few cases, use gene therapy to treat those who have these diseases. Gene technology has brought huge benefits to many people, but may have consequences that we cannot foresee. These technologies raise social, economic and ethical issues, that we must confront.

Genetic engineering

The aim of genetic engineering is to remove a gene (or genes) from one organism and transfer it into another so that the gene is expressed in its new host. The DNA that has been altered by this process and which now contains lengths of nucleotides from two different organisms is called recombinant DNA (rDNA). The organism which now expresses the new gene or genes is known as a transgenic organism or a genetically modified organism (GMO).

Recombinant DNA is DNA made by joining pieces from two or more different sources.

Genetic engineering provides a way of overcoming barriers to gene transfer between species. Indeed the genes are often taken from an organism in a different kingdom, such as a bacterial gene inserted into a plant or a human gene inserted into a bacterium. Unlike selective breeding, where whole sets of genes are involved, genetic engineering often results in the transfer of a single gene.

We will look first at the general principles involved in genetic engineering and then at some of the techniques in more detail.

An overview of gene transfer

There are many different ways in which a GMO may be produced, but these steps are essential.

1 The gene that is required is identified. It may be cut from a chromosome, made from mRNA by reverse transcription or synthesised from nucleotides.

2 Multiple copies of the gene are made using the technique known as the polymerase chain reaction (PCR).

3 The gene is inserted into a vector which delivers the gene to the cells of the organism. Examples of vectors are plasmids, viruses and liposomes.

4 The vector takes the gene into the cells.

5 The cells that have the new gene are identified and cloned.

To perform these steps, the genetic engineer needs a 'toolkit' consisting of:

- enzymes, such as restriction endonucleases, ligase and reverse transcriptase
- vectors, including plasmids and viruses
- genes coding for easily identifiable substances that can be used as markers.

Tools for the gene technologist

Restriction enzymes

Restriction endonucleases are a class of enzymes from bacteria which recognise and break down the DNA of invading viruses known as bacteriophages (phages for short). Bacteria make enzymes that cut phage DNA into smaller pieces. These enzymes cut the sugar–phosphate backbone of DNA at specific places within the molecule. This is why they are known as endonucleases ('endo' means within). Their role in bacteria is to restrict a viral infection, hence the name restriction endonuclease or restriction enzyme.

Each restriction enzyme binds to a specific target site on DNA and cuts at that site. Bacterial DNA is protected from such an attack either by chemical markers or by not having the target sites. These target sites, or restriction sites, are specific sequences of bases. For example, the restriction enzyme called *Bam*HI always cuts DNA where there is a GGATCC sequence on one strand and its complementary sequence, CCTAGG, on the other. You will notice that this sequence reads the same in both directions: it is a palindrome. Many, but not all, restriction sites are palindromic. Restriction enzymes either cut straight across the sugar-phosphate backbone to give blunt ends or they cut in a staggered fashion to give sticky ends (Figure 19.2).

Sticky ends are short lengths of unpaired bases. They are known as sticky ends because they can easily form hydrogen bonds with complementary sequences

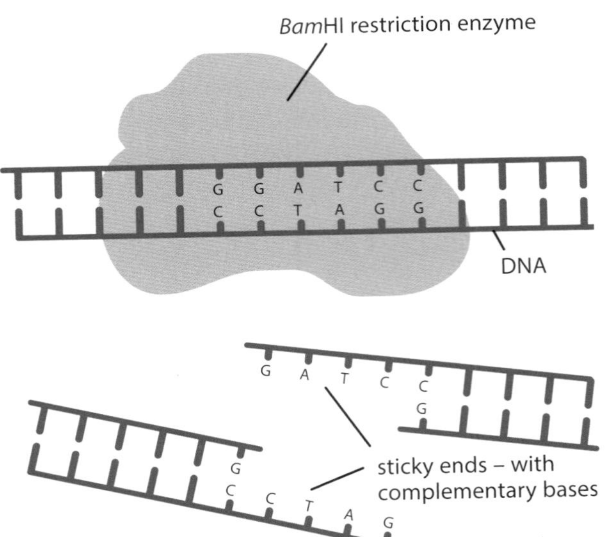

Figure 19.2 The restriction enzyme, *Bam*HI, makes staggered cuts in DNA to give sticky ends.

of bases on other pieces of DNA cut with the same restriction enzyme. When long pieces of DNA are cut with a restriction enzyme, there will be a mixture of different lengths. To find the specific piece of DNA required involves separating the lengths of DNA using gel electrophoresis and using gene probes. These techniques are described on pages 470–471. Multiple copies of the required piece of DNA can be made using the polymerase chain reaction (PCR) which is described on page 471.

QUESTION

19.1 a Explain the term **sticky end**.

b Four different lengths of DNA, A to D, were cut with the restriction enzyme *Eco*RI. These fragments had the following numbers of restriction sites for this enzyme:
A – 5, **B** – 7, **C** – 0 and **D** – 3.
State the number of fragments that will be formed from each length of DNA after incubation with *Eco*RI.

Restriction enzyme	Restriction site	Site of cut across DNA	Source of enzyme
*Eco*RI	5′–GAATTC–3′ 3′–CTTAAG–5′	–G ⌊ AATTC– –CTTAA ⌉ G–	*Escherichia coli*
*Bam*HI	5′–GGATCC–3′ 3′–CCTAGG–5′	–G ⌊ GATCC– –CCTAG ⌉ G–	*Bacillus amyloliquefaciens*
*Hind*III	5′–AAGCTT–3′ 3′–TTCGAA–5′	–A ⌊ AGCTT– –TTCGA ⌉ A–	*Haemophilus influenzae*
*Hae*III	5′–GGCC–3′ 3′–CCGG–5′	–GG | CC– –CC | GG–	*Haemophilus aegyptius*

Table 19.1 Four restriction enzymes and their target sites.

Restriction enzymes are named by an abbreviation which indicates their origin (Table 19.1). Roman numbers are added to distinguish different enzymes from the same source. For example, *Eco*RI comes from *Escherichia coli* (strain RY13), and was the first to be identified from this source.

Now that many proteins have been sequenced, it is possible to use the genetic code (page 513) to synthesise DNA artificially from nucleotides rather than cutting it out of chromosomal DNA or making it by reverse transcription. Genes, and even complete genomes, can be made directly from DNA nucleotides without the need for template DNA. Scientists can do this by choosing codons for the amino acid sequence that they need. The sequence of nucleotides is held in a computer that directs the synthesis of short fragments of DNA. These fragments are then joined together to make a longer sequence of nucleotides that can be inserted into plasmids for use in genetic engineering. This method is used to generate novel genes that are used, for example, in the synthesis of vaccines (page 233) and they have even been used to produce the genomes of bacteria consisting of a million base pairs.

Vectors

Inserting a gene into a plasmid vector

In order to get a new gene into a recipient cell, a go-between called a vector often has to be used. One type of vector is a plasmid (Figure 19.3). These are small, circular pieces of double-stranded DNA (page 114). Plasmids occur naturally in bacteria and often contain genes for antibiotic resistance. They can be exchanged between bacteria – even between different species of bacteria. If a genetic engineer inserts a piece of DNA into a plasmid, then the plasmid can be used to take the DNA into a bacterial cell.

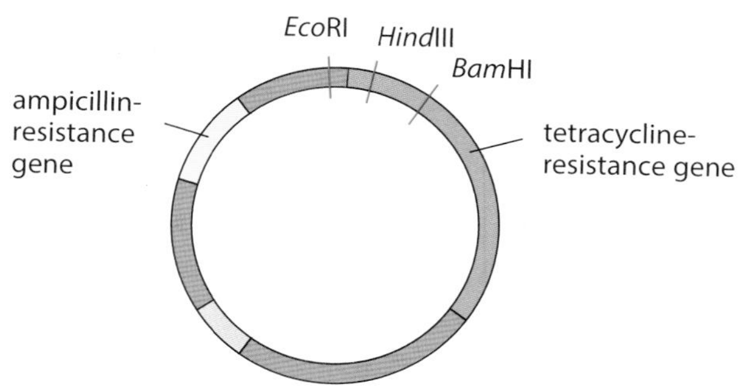

Figure 19.3 Plasmid pBR322 was used in the production of human insulin.

To get the plasmids, the bacteria containing them are treated with enzymes to break down their cell walls. The 'naked' bacteria are then spun at high speed in a centrifuge, so that the relatively large bacterial chromosomes are separated from the much smaller plasmids.

The circular DNA of the plasmid is cut open using a restriction enzyme (Figure 19.4). The same enzyme as the one used to cut out the gene should be used, so that the sticky ends are complementary. If a restriction enzyme

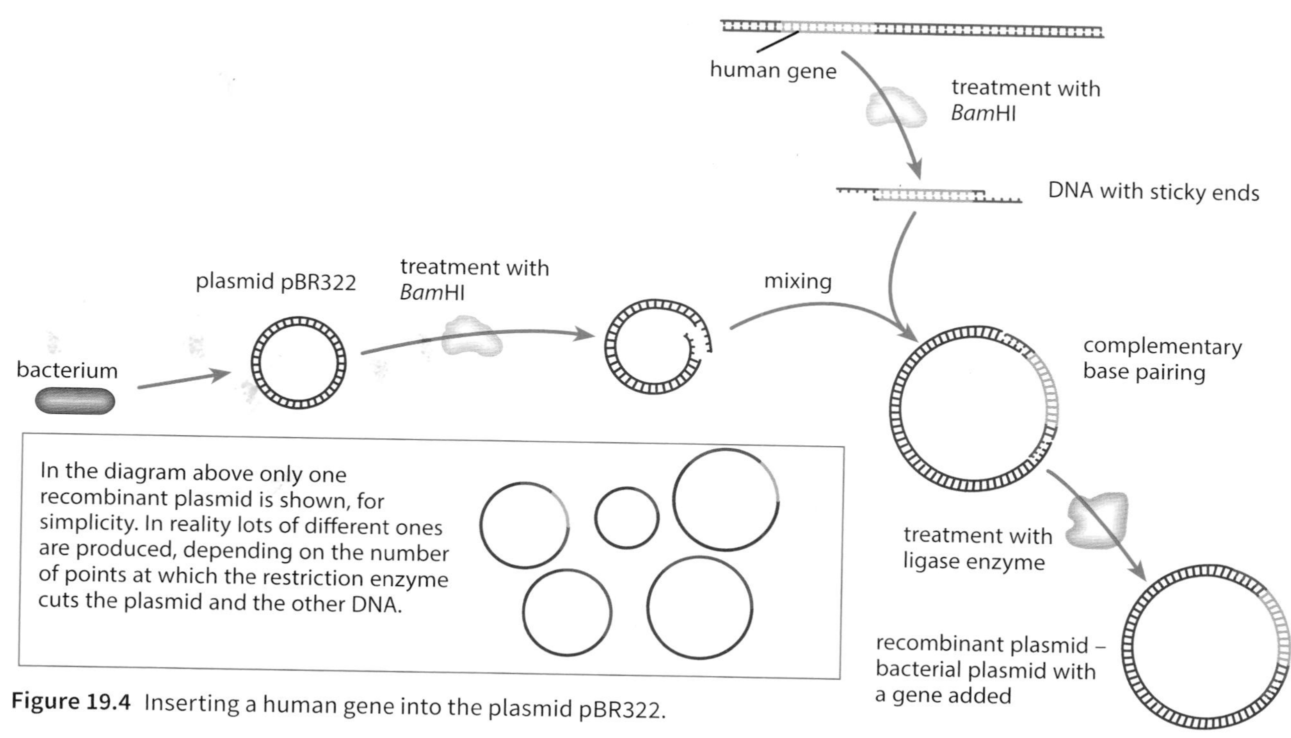

Figure 19.4 Inserting a human gene into the plasmid pBR322.

is used that gives blunt ends, then sticky ends need to be attached to both the gene and the plasmid DNA.

The opened plasmids and the lengths of DNA are mixed together. Some of the plasmid sticky ends pair up with the sticky ends on the new gene. The enzyme DNA ligase is used to link together the sugar–phosphate backbones of the DNA molecule and the plasmid, producing a closed circle of double-stranded DNA containing the new gene. This is now recombinant DNA.

Bacterial plasmids can be modified to produce good vectors. Plasmids can also be made artificially. For example, the pUC group of plasmids have:

- a low molecular mass, so they are readily taken up by bacteria
- an origin of replication so they can be copied
- several single target sites for different restriction enzymes in a short length of DNA called a polylinker
- one or more marker genes, allowing identification of cells that have taken up the plasmid.

Plasmids are not the only type of vector that can be used. Viruses can also be used as vectors. A third group of vectors are liposomes, which are tiny spheres of lipid containing the DNA. There is more about these other vectors later in this chapter.

Getting the plasmids into bacteria

The next step in the process is to get bacteria to take up the plasmids. The bacteria are treated by putting them into a solution with a high concentration of calcium ions, then cooled and given a heat shock to increase the chances of plasmids passing through the cell surface membrane. A small proportion of the bacteria, perhaps 1%, take up plasmids with the gene, and are said to be transformed. The rest either take up plasmids that have closed without incorporating a gene or do not take up any plasmids at all.

Identifying bacteria with recombinant DNA

It is important to identify which bacteria have been successfully transformed so that they can be used to make the gene product. This used to be done by spreading the bacteria on agar plates each containing an antibiotic. So if, for example, the insulin gene had been inserted into the plasmid at a point in the gene for tetracycline resistance in pBR322, then any bacteria which had taken up plasmids with the recombinant DNA would not be able to grow on agar containing tetracycline. However, this technique has fallen out of favour, and has largely been replaced by simpler methods of identifying transformed bacteria (page 468).

DNA polymerase in bacteria copies the plasmids; the bacteria then divide by binary fission so that each daughter cell has several copies of the plasmid. The bacteria transcribe the new gene and may translate it to give the required gene product, such as insulin.

QUESTION

19.2 Summarise the advantages of using plasmids as vectors in genetic engineering.

Insulin production

One form of diabetes mellitus is caused by the inability of the pancreas to produce insulin (Chapter 14). Before insulin from GM bacteria became available, people with this form of diabetes were treated with insulin extracted from the pancreases of pigs or cattle. In the 1970s, biotechnology companies began to work on the idea of inserting the gene for human insulin into a bacterium and then using this bacterium to make insulin. They tried several different approaches, finally succeeding in the early 1980s. This form of human insulin became available in 1983. The procedure involved in the production of insulin is shown in Figure 19.5.

There were problems in locating and isolating the gene coding for human insulin from all of the rest of the DNA in a human cell. Instead of cutting out the gene from the DNA in the relevant chromosome, researchers extracted mRNA for insulin from pancreatic β cells, which are the only cells to express the insulin gene. These cells contain large quantities of mRNA for insulin as they are its only source in the body. The mRNA was then incubated with the enzyme reverse transcriptase which comes from the group of viruses called retroviruses (Chapter 10, page 206). As its name suggests, this enzyme reverses transcription, using mRNA as a template to make single-stranded DNA. These single-stranded DNA molecules were then converted to double-stranded DNA molecules using DNA polymerase to assemble nucleotides to make the complementary strand. The genetic engineers now had insulin genes that they could insert into plasmids to transform the bacterium *Escherichia coli*.

The main advantage of this form of insulin is that there is now a reliable supply available to meet the increasing demand. Supplies are not dependent on factors such as availability through the meat trade.

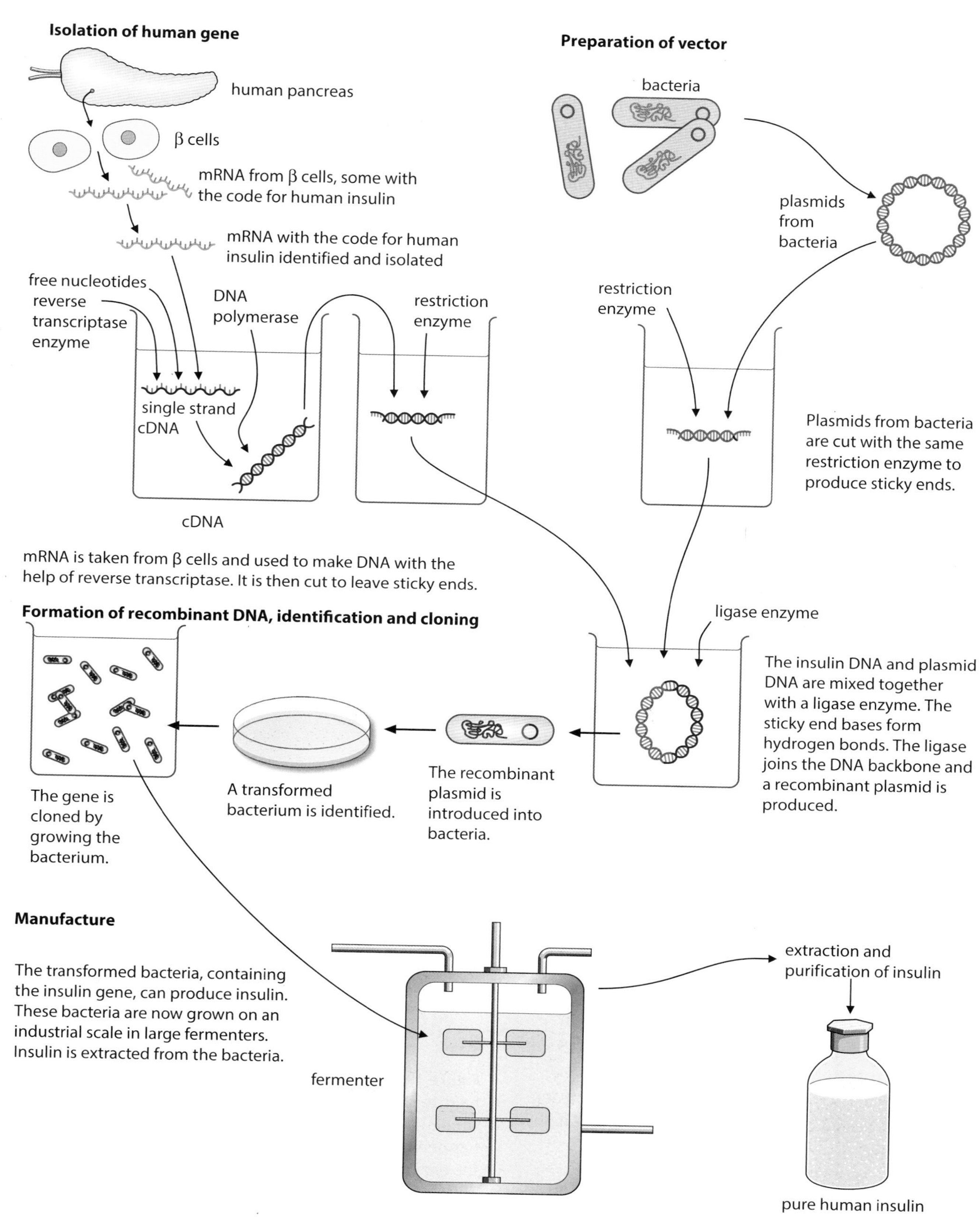

Figure 19.5 Producing insulin from genetically modified bacteria.

Genetic engineers have changed the nucleotide sequence of the insulin gene to give molecules with different amino acid sequences. These insulin analogues have different properties; for example, they can either act faster than animal insulin (useful for taking immediately after a meal) or more slowly over a period of between 8 and 24 hours to give a background blood concentration of insulin. Many diabetics take both these forms of recombinant insulin at the same time.

Other genetic markers

There is some concern about using antibiotic resistance genes as markers. Could the antibiotic resistance genes spread to other bacteria, producing strains of pathogenic (disease-causing) bacteria that we could not kill with antibiotics? In insulin production, the risk is probably very small, because the genetically modified bacteria are only grown in fermenters and not released into the wild. But now there are many different kinds of genetically modified bacteria around, some of which are used in situations in which their genes might be passed on to other bacteria. If these bacteria were pathogens, then we might end up with diseases that are untreatable.

Because of the risk of creating pathogenic antibiotic-resistant bacteria, there is now much less use of antibiotic resistance genes in this way, and other ways have been developed in which the successfully transformed bacteria can be identified. One method uses enzymes that produce fluorescent substances. For example, enzymes obtained from jellyfish make a protein called GFP (green fluorescent protein) that fluoresces bright green in ultraviolet light. The gene for the enzyme is inserted into the plasmids. So all that needs to be done to identify the bacteria that have taken up the plasmid is to shine ultraviolet light onto them. The ones that glow green are the genetically modified ones. The same marker gene can be used in a range of organisms (Figure 19.6).

Another marker is the enzyme β-glucuronidase (known as GUS for short), which originates from *E. coli*. Any transformed cell that contains this enzyme, when incubated with some specific colourless or non-fluorescent substrates, can transform them into coloured or fluorescent products. This is especially useful in detecting the activity of inserted genes in plants, such as the sundew in Figure 19.7.

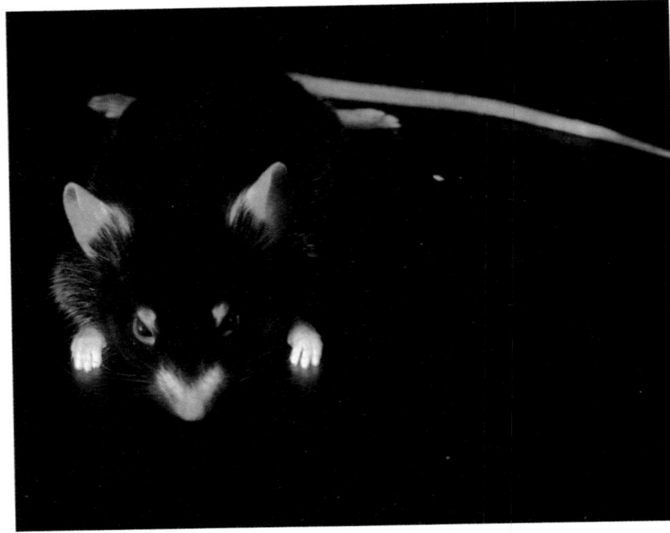

Figure 19.6 A transgenic mouse expressing a gene for a fluorescent protein.

Figure 19.7 Sundews are carnivorous plants that use sticky hairs to catch insects. On the left is a leaf of a transgenic sundew plant which is expressing the gene for GUS. The leaf has placed in a solution of a colourless substance and the enzyme GUS has converted it into this dark blue colour. This indicates that the plant has been genetically modified successfully. On the right is a normal sundew leaf.

QUESTION

19.3 Would all the bacteria that fluoresce definitely have taken up the gene that it is hoped was inserted into them? Explain your answer.

Promoters

Bacteria contain many different genes, which make many different proteins. But not all these genes are switched on at once. The bacteria make only the proteins that are required in the conditions in which they are growing. For example, as we saw in Chapter 16, *E. coli* bacteria make the enzyme β-galactosidase only when they are growing in a medium containing lactose and there is no glucose available.

The expression of genes, such as those in the *lac* operon, is controlled by a promoter – the region of DNA to which RNA polymerase binds as it starts transcription. If we want the gene that we are going to insert into a bacterium to be expressed, then we also have to insert an appropriate promoter. When bacteria were first transformed to produce insulin, the insulin gene was inserted next to the β-galactosidase gene so they shared a promoter. The promoter switched on the gene when the bacterium needed to metabolise lactose. So, if the bacteria were grown in a medium containing lactose but no glucose, they synthesised both β-galactosidase and human insulin.

The promoter not only allows RNA polymerase to bind to DNA but also ensures that it recognises which of the two DNA strands is the template strand. Within the sequence of nucleotides in the promoter region is the transcription start point – the first nucleotide of the gene to be transcribed. In this way, the promoter can be said to control the expression of a gene and can ensure a high level of gene expression. In eukaryotes, various proteins known as transcription factors are also required to bind to the promoter region or to RNA polymerase before transcription can begin (Chapter 16, page 391).

Gel electrophoresis

Gel electrophoresis is a technique that is used to separate different molecules. It is used extensively in the analysis of proteins and DNA. This technique involves placing a mixture of molecules into wells cut into agarose gel and applying an electric field. The movement of charged molecules within the gel in response to the electric field depends on a number of factors. The most important are:

- net (overall) charge – negatively charged molecules move towards the anode (+) and positively charged molecules move towards the cathode (–); highly charged molecules move faster than those with less overall charge
- size – smaller molecules move through the gel faster than larger molecules
- composition of the gel – common gels are polyacrylamide for proteins and agarose for DNA; the size of the 'pores' within the gel determines the speed with which proteins and fragments of DNA move.

Electrophoresis of proteins

The charge on proteins is dependent on the ionisation of the R groups on the amino acid residues. You will remember from Chapter 2 that some amino acids have R groups that can be positively charged ($-NH^{3+}$) and some have R groups that can be negatively charged ($-COO^{-}$). Whether these R groups are charged or not depends on the pH. When proteins are separated by electrophoresis, the procedure is carried out at a constant pH using a buffer solution. Usually proteins have a net negative charge (Figure 19.8).

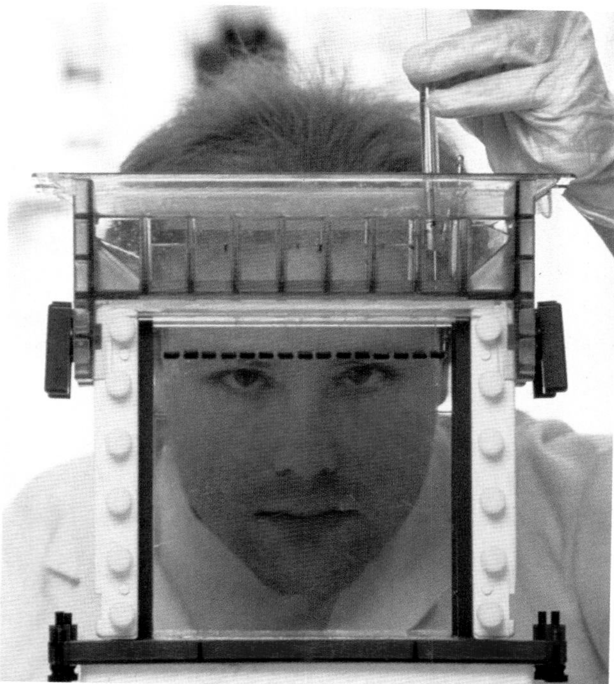

Figure 19.8 Gel electrophoresis of proteins. The gel was placed in the tank containing a suitable buffer solution. Protein samples stained red have been added to wells along the top of the gel. They are migrating downwards towards the anode.

Gel electrophoresis has been used to separate the polypeptides produced by different alleles of many genes. For example, allozymes are variant forms of enzymes produced by different alleles of the same gene.

There are also many variants of haemoglobin. Adult haemoglobin is composed of four polypeptides: 2 α-globins and 2 β-globins (Figure 2.23, page 42, and Chapter 16, page 374). In sickle cell anaemia, a variant of β-globin has an amino acid with a non-polar R group instead of one with an R group that is charged. These two variants of the β-globin can be separated by electrophoresis because they have different net charges. This means that haemoglobin molecules in people who have sickle cell anaemia have a slightly lower negative charge than normal haemoglobin and so the molecules do not move as far through the gel as molecules of normal haemoglobin. The test to find out whether someone carries the sickle cell allele makes use of this difference.

Figure 19.9 shows the separation of different forms of haemoglobin in samples of blood taken from the members of a family which has a child with sickle cell anaemia.

QUESTION

19.4 Identify the genotypes of the family whose haemoglobin was analysed in Figure 19.9.

Electrophoresis of DNA

DNA fragments carry a small charge thanks to the negatively charged phosphate groups. In DNA electrophoresis, these fragments move through the gel towards the anode. The smaller the fragments, the faster they move. We will look at an example: the use of genetic profiling (fingerprinting) in forensic science. Figure 19.10 shows how genetic profiling is carried out.

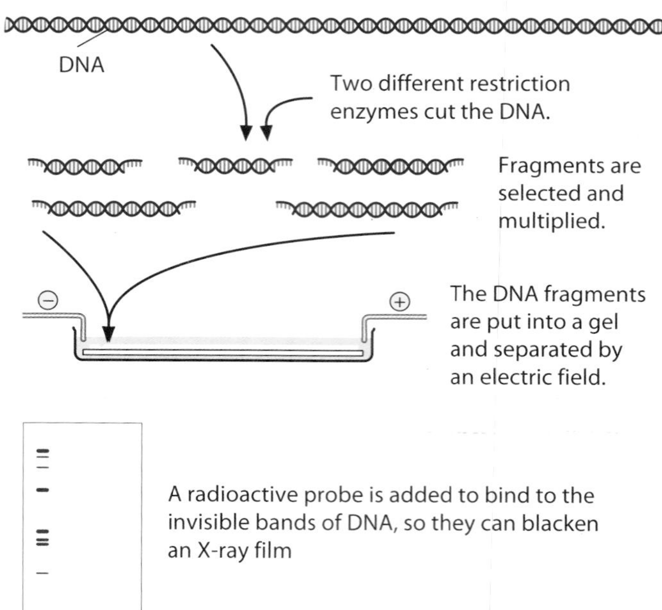

Figure 19.10 Genetic profiling.

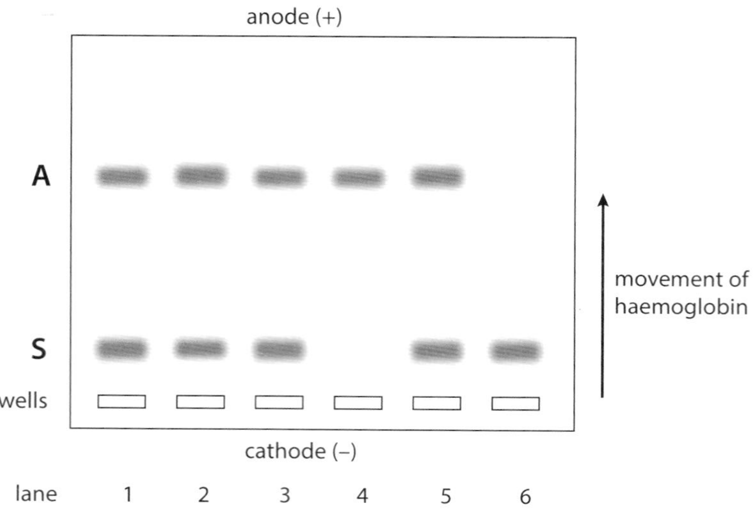

Figure 19.9 Separation of haemoglobin by gel electrophoresis. This analysis was carried out on a family in which one child has sickle cell anaemia. Lane 1 contains haemoglobin standards, A = normal haemoglobin, S = sickle cell haemoglobin; lanes 2 and 3 are the haemoglobin samples from parents; lanes 4, 5 and 6 are haemoglobin samples from their children.

A region of DNA that is known to vary between different people is chosen. These regions often contain variable numbers of repeated DNA sequences and are known as **variable number tandem repeats (VNTRs)**. Only identical twins share all their VNTR sequences.

DNA can be extracted from almost anything that has come from a person's body – the root of a hair, a tiny spot of blood or semen at a crime scene, or saliva where someone has drunk from a cup. Usually the quantity of DNA is increased by using the polymerase chain reaction (PCR), which makes many copies of the DNA that has been found.

The DNA is then chopped into pieces using restriction enzymes known to cleave it close to the VNTR regions. Now the DNA is ready for electrophoresis. When the current is turned off, the gel contains DNA fragments that have ended up in different places. These fragments are not visible straight away.

To make the fragments visible, they are carefully transferred onto absorbent paper, which is placed on top of the gel. The paper is then heated just enough to make the two strands in each DNA molecule separate from one another. Short sequences of single-stranded DNA called **probes** are added; they have base sequences complementary to the VNTR regions. The probes also contain a radioactive phosphorus isotope so when the paper is placed on an X-ray film, the radiation emitted by the probes (which are stuck to the DNA fragments) make the film go dark. So, we end up with a pattern of dark stripes on the film matching the positions that the DNA fragments reached on the agarose gel (Figures 19.11 and 19.12). Alternatively, the probes may be labelled with a fluorescent stain that shows up when ultraviolet light is shone onto them.

Polymerase chain reaction

The polymerase chain reaction, generally known as PCR, is used in almost every application of gene technology. It is a method for rapid production of a very large number of copies of a particular fragment of DNA. Virtually unlimited quantities of a length of DNA can be produced from the smallest quantity of DNA (even one molecule). Figure 19.13 shows the steps involved in PCR.

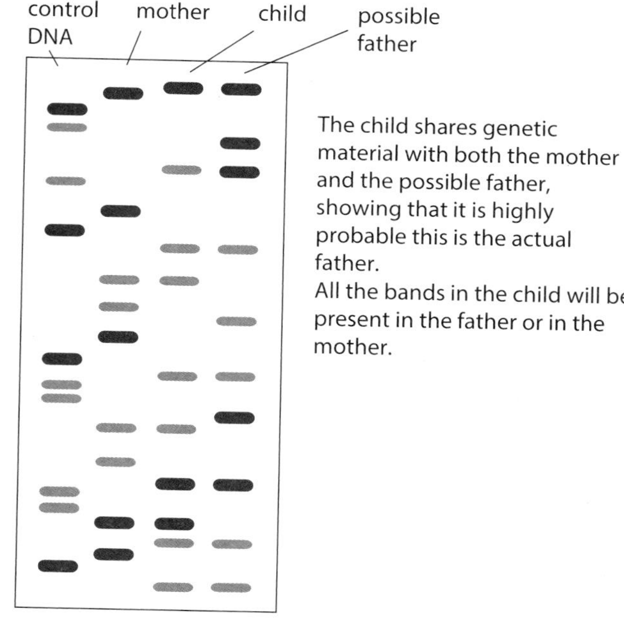

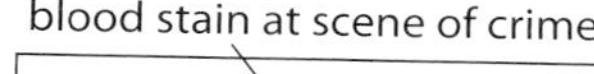

Figure 19.11 Using DNA profiling in paternity testing.

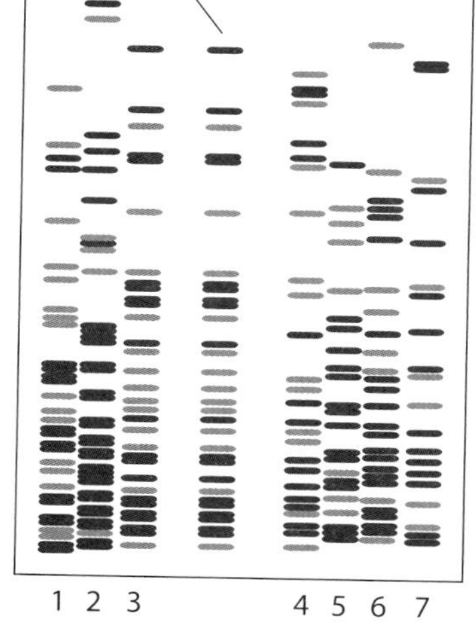

Figure 19.12 Using DNA profiling in crime scene analysis.

QUESTION

19.5 Look at Figure 19.12. Which suspect was at the scene of the crime? What is the evidence supporting this?

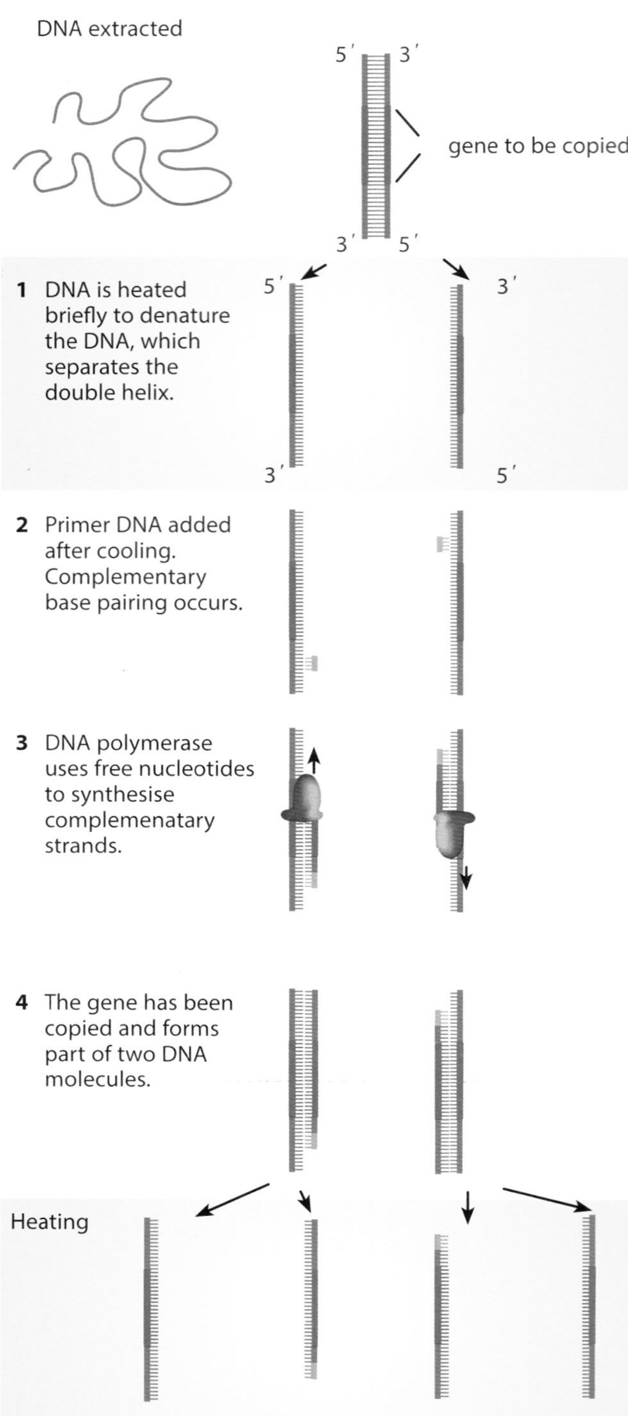

Figure 19.13 The polymerase chain reaction.

First, the DNA is denatured, usually by heating it. This separates the DNA molecule into its two strands, leaving bases exposed.

The enzyme DNA polymerase is then used to build new strands of DNA against the exposed ones. However, DNA polymerase cannot just begin doing this with no 'guidance'. A primer is used to begin the process. This is a short length of DNA, often about 20 base pairs long, that has a base sequence complementary to the start of the part of the DNA strand that is to be copied. The primer attaches to the start of the DNA strand, and then the DNA polymerase continues to add nucleotides all along the rest of the DNA strand.

Once the DNA has been copied, the mixture is heated again, which once more separates the two strands in each DNA molecule, leaving them available for copying again. Once more, the primers fix themselves to the start of each strand of unpaired nucleotides, and DNA polymerase makes complementary copies of them.

The three stages in each round of copying need different temperatures.

- Denaturing the double-stranded DNA molecules to make single-stranded ones requires a high temperature, around 95 °C.
- Attaching the primers to the ends of the single-stranded DNA molecules (known as annealing) requires a temperature of about 65 °C.
- Building up complete new DNA strands using DNA polymerase (known as elongation) requires a temperature of around 72 °C. The DNA polymerases used for this process come from microorganisms that have evolved to live in hot environments.

Most laboratories that work with DNA have a machine that automatically changes the temperature of the mixture. The DNA sample is placed into a tube together with the primers, free nucleotides, a buffer solution and the DNA polymerase. The PCR machine is switched on and left to work. The tubes are very small (they hold about 0.05 cm^3) and have very thin walls, so when the temperature in the machine changes, the temperature inside the tubes changes very quickly.

QUESTION

19.6 Explain the difference between a primer and a probe.

You can see that theoretically this could go on forever, making more and more copies of what might originally have been just a tiny number of DNA molecules. A single DNA molecule can be used to produce literally billions of copies of itself in just a few hours. PCR has made it possible to get enough DNA from a tiny sample – for example, a microscopic portion of a drop of blood left at a crime scene.

Taq polymerase was the first heat-stable DNA polymerase to be used in PCR. It was isolated from the thermophilic bacterium, *Thermus aquaticus*, which is found in hot springs in Yellowstone Park in the USA (Figure 3.10, page 60). It is valuable for PCR for two reasons. First, is not destroyed by the denaturation step, so it does not have to be replaced during each cycle. Second, its high optimum temperature means that the temperature for the elongation step does not have to be dropped below that of the annealing process, so efficiency is maximised.

PCR is now routinely used in forensic science to amplify DNA from the smallest tissue samples left at the scene of a crime. Many crimes have been solved with the help of PCR together with analysis of DNA using gel electrophoresis.

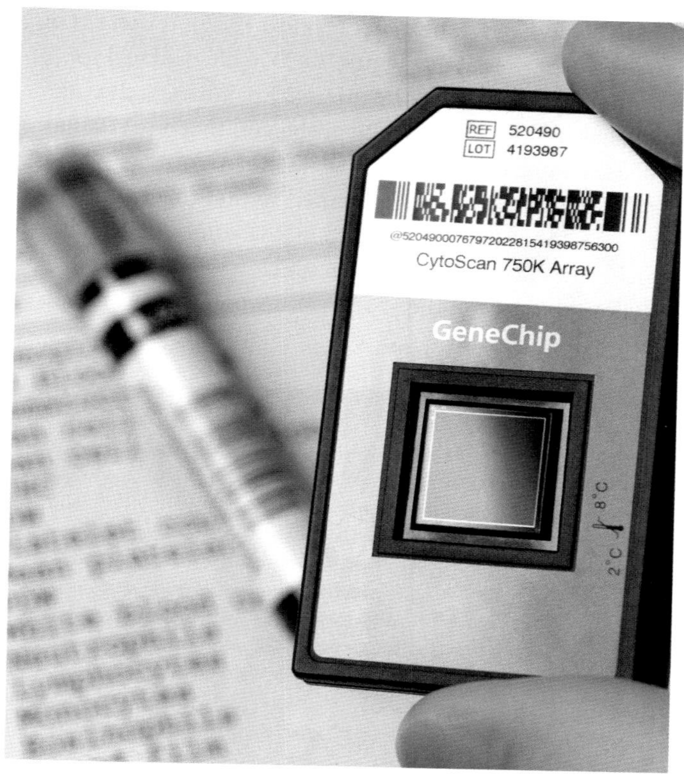

Figure 19.14 A microarray, also known as a DNA chip.

QUESTION

19.7 a How many molecules of DNA are produced from one double-stranded starting molecule, after eight cycles of PCR?

b Explain why it is not possible to use PCR to increase the number of RNA molecules in the same way as it is used to increase the number of DNA molecules.

Microarrays

Microarrays have proved a valuable tool to identify the genes present in an organism's genome and to find out which genes are expressed within cells. They have allowed researchers to study very large numbers of genes in a short period of time, increasing the information available.

A microarray is based on a small piece of glass or plastic usually 2 cm^2 (Figure 19.14). Short lengths of single-stranded DNA are attached to this support in a regular two-dimensional pattern, with 10 000 or more different positions per cm^2. Each individual position has multiple copies of the same DNA probe. It is possible to search databases to find DNA probes for a huge range of genes. Having selected the gene probes required, an automated process applies those probes to the positions on the microarray.

When microarrays are used to analyse genomic DNA, the probes are from known locations across the chromosomes of the organism involved and are 500 or more base pairs in length. A single microarray can even hold probes from the entire human genome.

Microarrays can be used to compare the genes present in two different species. DNA is collected from each species and cut up into fragments and denatured to give lengths of single-stranded DNA. The DNA is labelled with fluorescent tags so that – for example – DNA from one species may be labelled with green tags and DNA from the other species labelled with red tags. The labelled DNA samples are mixed together and allowed to hybridise with the probes on the microarray. Any DNA that does not bind to probes on the microarray is washed off. The microarray is then inspected using ultraviolet light, which causes the tags to fluoresce. Where this happens, we know that hybridisation has taken place because the DNA fragments are complementary to the probes. Green and red fluorescent spots indicate where DNA from one species only has hybridised with the probes. Where DNA from both species hybridise with a probe, a yellow colour is seen. Yellow spots indicate that the two species have DNA with exactly the same base sequence. This suggests that they have the same genes (Figure 19.15). If there is no colour (or, in the case of Figure 19.15, a blue colour) for a particular position on the microarray it means that no DNA has hybridised with the probe and that a particular gene is not present in either species.

The microarray is then scanned so that the data can be read by a computer. Data stored by the computer indicate which genes are present in both species, which genes are only found in one of the species and which genes are not present in either species.

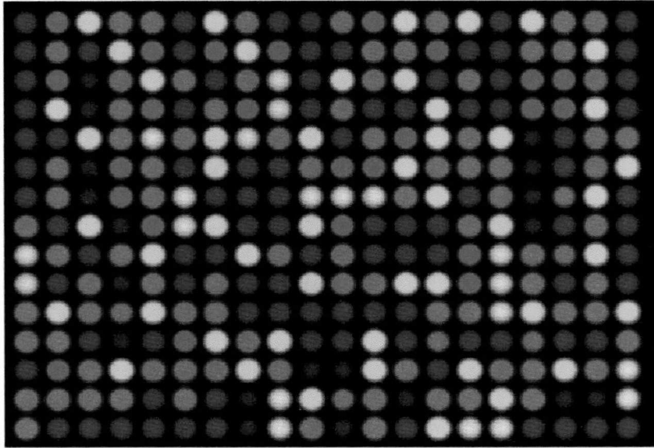

Figure 19.15 A DNA microarray as viewed with a laser scanner. The colours are analysed to show which genes or alleles are present.

Microarrays also make it possible to detect which genes are being expressed at any specific time in each cell in the body. For example, the genes that are expressed in a cancer cell are different from those active in non-cancerous cells. Microarrays are used to compare which genes are active by identifying the genes that are being transcribed into mRNA (Figure 19.16). The mRNA from the two types of cell is collected and reverse transcriptase is used to convert mRNA to cDNA (page 467). As the quantity of mRNA in a cell at one time is quite small, the quantity of cDNA may need to be increased by PCR. The cDNA is labelled with fluorescent tags, denatured to give single-stranded DNA and allowed to hybridise with probes on the microarray. Spots on the microarray that fluoresce indicate the genes that were being transcribed in the cell. The intensity of light emitted by each spot indicates the level of activity of each gene. A high intensity indicates that many mRNA molecules were present in the sample, while a low intensity indicates that there were very few. The results therefore not only show which genes are acive, but also their level of activity. This information is changing the way in which cancers are treated.

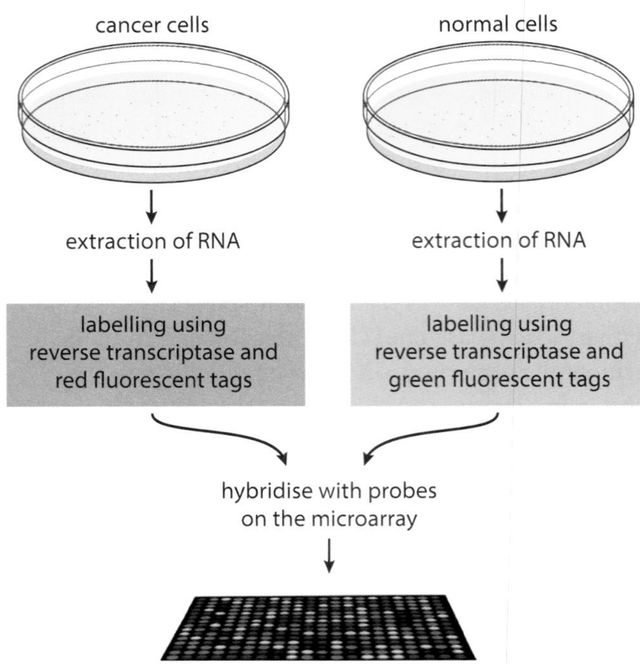

Figure 19.16 How to use a microarray to compare the mRNA molecules present in cancerous and non-cancerous cells. The results identify which genes in the cancerous cells that are not normally expressed are being transcribed.

QUESTION

19.8 The latest estimate of the number of genes in the human genome is 21 000. Before the invention of microarrays, it was very time consuming to find out which genes were expressed in any particular cell.

- **a** Explain how it is possible to find out which genes are active in a cell at a particular time in its development.
- **b** Why is it not possible to use the same technique to find out which genes are active in red blood cells?

Bioinformatics

Research into the genes that are present in different organisms and the genes that are expressed at any one time in an organism's life generates huge quantities of data. As we have seen, one DNA chip alone may give 10 000 pieces of information about the presence and absence of genes in genomes or the activity of genes within cells.

Gene sequencing has also generated huge quantities of data. This technique establishes the sequence of base pairs in sections of DNA. Sequencing DNA is now a fully automated process and the genomes of many species have been published. There is also a vast quantity of data about the primary structures, shapes and functions of proteins.

Bioinformatics combines biological data with computer technology and statistics. It builds up databases and allows links to be made between them. The databases hold gene sequences, sequences of complete genomes, amino acid sequences of proteins and protein structures. Computer technology facilitates the collection and analysis of this mass of information and allows access to it via the internet.

There are databases that specialise in holding different types of information; for example, on DNA sequences and the primary structures of proteins. The information needs to be in a form that can be searched, so software developers play an important role in developing systems that allow this. In 2014, these databases held over 6×10^{11} base pairs or 600 Gbp (Gigabasepairs), equivalent to 200 human genome equivalents or huges. A huge is 3×10^{9} base pairs. Databases that hold the coordinates required to show 3D models hold details of over 100 000 different proteins and nucleic acids. The quantity of data is vast and growing at an exponential rate.

The database Ensembl holds data on the genomes of eukaryotic organisms. Among others, it holds the human genome and the genomes of zebra fish and mice that are used a great deal in research. UniProt (universal protein resource) holds information on the primary sequences of proteins and the functions of many proteins, such as enzymes. The search tool BLAST (basic local alignment search tool) is an algorithm for comparing primary biological sequence information, such as the primary sequences of different proteins or the nucleotide sequences of genes. Researchers use BLAST to find similarities between sequences that they are studying and those already saved in databases.

When a genome has been sequenced, comparisons can be made with other known genomes. For example, the human genome can be compared to the genomes of the fruit fly, *Drosophila*, the nematode worm, *Caenorhabditis*, or the malarial parasite, *Plasmodium*. Sequences can be matched and degrees of similarity calculated. Similarly, comparisons can be made between amino acid sequences of proteins or structures of proteins. Close similarities indicate recent common ancestry.

Human genes, such as those that are concerned with development, may be found in other organisms such as *Drosophila*. This makes *Drosophila* a useful model for investigating the way in which such genes have their effect. Microarrays can be used to find out when and where genes are expressed during the development of a fruit fly. Researchers can then access information about these genes and the proteins that they code for. For example, they can search databases for identical or similar base sequences in other organisms, compare primary structures of proteins and visualise the 3D structure of the proteins.

Caenorhabditis elegans was the first multicellular organism to have its genome fully sequenced. It has fewer than 1000 cells in its body, of which about 300 are nerve cells. It is conveniently transparent, allowing the developmental fate of each of its cells to be mapped. Because of its simplicity, it is used as a model organism for studying the genetics of organ development, the development of neurones into a nervous system and many other areas of biology such as cell death, ageing and behaviour.

All the information we have about the genome of *Plasmodium* is now available in databases. This information is being used to find new methods to control the parasite. For example, being able to read gene sequences is providing valuable information in the development of vaccines for malaria.

Bioinformatics is the collection, processing and analysis of biological information and data using computer software.

Genetic technology and medicine

Genetic technology allows products specific to humans to be made. We have already looked at the advantages of producing human insulin by recombinant DNA techniques (page 466).

Other human proteins are produced by similar techniques – for example:

- human growth hormone
- thyroid stimulating hormone
- factor VIII – a blood clotting protein.

There are advantages in using bacteria, yeasts and cultures of mammalian cells to produce these proteins. These cells have simple nutritional requirements; large volumes of product are produced; the production facilities do not require much space and the processes can be carried out almost anywhere in the world. There are few practical and ethical problems, because proteins do not have to be extracted from animal sources or by collecting blood from many donors. The disadvantage of using bacteria to produce human proteins is that bacteria do not modify their proteins in the same way that eukaryotes do. It is much better, therefore, to use eukaryotic cells to produce human proteins.

Genetically modified hamster cells are used by several companies to produce factor VIII. This protein is essential for blood clotting, and people who cannot make it are said to have haemophilia. The human gene for making factor VIII has been inserted into hamster kidney and ovary cells which are then cultured in fermenters. The cells constantly produce factor VIII which is extracted and purified before being used to treat people with haemophilia. These people need regular injections of factor VIII which, before the availability of recombinant factor VIII, came from donated blood. Using donated blood carried risks of infection – for example, from HIV (Chapter 10). Recombinant factor VIII avoids such problems.

High yields of the enzyme adenosine deaminase (ADA), which is used to treat severe combined immunodeficiency disease (SCID), are made by a genetically modified insect larva, the cabbage looper moth caterpillar. This enzyme is administered to patients while they are waiting for gene therapy or when gene therapy is not possible (page 477).

Some proteins are even produced by transgenic animals. Sheep and goats have been genetically modified to produce human proteins in their milk:

- human antithrombin is produced by goats – this protein is used to stop blood clotting
- human alpha-antitrypsin is produced by sheep – this is used to treat people with emphysema.

Genetic screening

Genetic screening is the analysis of a person's DNA to check for the presence of a particular allele. This can be done in adults, in a fetus or embryo in the uterus, or in a newly formed embryo produced by in vitro fertilisation.

An adult woman with a family history of breast cancer may choose to be screened for the faulty alleles of the genes *Brca-1* and *Brca-2*, which considerably increase an individual's chance of developing breast cancer. Should the results be positive, the woman may elect to have her breasts removed (elective mastectomy) before such cancer appears.

In 1989, the first 'designer baby' was created. Officially known as pre-implantation genetic diagnosis (PGD), the technique involved mixing the father's sperm with the mother's eggs (oocytes) in a dish – that is, a 'normal' IVF procedure. It was the next step that was new. At the eight-cell stage, one of the cells from the tiny embryo was removed. The DNA in the cell was analysed and used to predict whether or not the embryo would have a genetic disease for which both parents were carriers. An embryo that was not carrying the allele that would cause the disease was chosen for implantation, and embryos that did have this allele were discarded.

Since then, many babies have been born using this technique. It has been used to avoid pregnancies in which the baby would have had Duchenne muscular dystrophy, thalassaemia, haemophilia, Huntington's disease and others. In 2004, it was first used in the UK to produce a baby that was a tissue match with an elder sibling, with a view to using cells from the umbilical cord as a transplant into the sick child.

For some time, genetic testing of embryos has been leaving prospective parents with very difficult choices to make if the embryo is found to have a genetic condition such as Down's syndrome or cystic fibrosis. The decision about whether or not to have a termination is very difficult to make. Now, though, advances in medical technology have provided us with even more ethical issues to consider.

The ethics of genetic screening

In 2004, UK law allowed an embryo to be chosen that did not have an allele for a genetic disease, and also one that **did** have a tissue type that would allow a successful transplant into a sick elder brother or sister. But it did not allow the addition of an allele to an egg, sperm or zygote.

A line has to be drawn somewhere, but feelings can run high. Many people believe that the law is allowing too much, while others think that it should allow more. Other countries have different attitudes and different regulations.

There is still controversy over other, long-established outcomes of genetic screening. For example, a fetus can now be screened for a genetic disease while in the uterus, using amniocentesis or chorionic villus sampling. The parents may then decide to have the pregnancy terminated if the embryo is found to have a genetic disease. However, there have been cases where this decision has been made even though the 'defect' has been a relatively minor one with which the child could be expected to lead a fairly normal life. Some parents have decided to terminate pregnancies simply because the child is not the sex that they want. They have also used PGD to select the sex of the embryo that they choose to implant. Many think that this sex preselection, as it is called, is totally unethical.

Amniocentesis is used to obtain a sample of amniotic fluid at 15 to 16 weeks of pregnancy. Various tests can be carried out on this sample to check the health of the fetus. Most amniocentesis samples, however, are to look for chromosomal mutations (page 387).

Ultrasound scanning is used to visualise the fetus and to locate the position of the placenta, fetus and umbilical cord. A suitable point for the insertion of the hypodermic syringe needle is chosen and this is marked on the abdominal skin surface. Generally, this position is away from the fetus, umbilical cord and placenta.

Chorionic villus sampling can be carried out between 10 and 13 weeks of pregnancy, so it allows parents to get an earlier warning of any genetic abnormalities in the fetus than is possible with amniocentesis.

A small sample of part of the placenta called the chorion is removed by a needle. The needle is narrow (less than 0.8 mm in diameter). The procedure is monitored by ultrasound scanning (Figure 19.17).

Like amniocentesis, chorionic villus sampling has a small increased risk of miscarriage. It has been estimated that miscarriage rate is increased by about 1–2%. (The typical miscarriage rate for all women is about 2–3% at 10 to 12 weeks of pregnancy.) This is a slightly greater risk than for amniocentesis, but, before 15 weeks, chorionic villus sampling is probably less risky than amniocentesis.

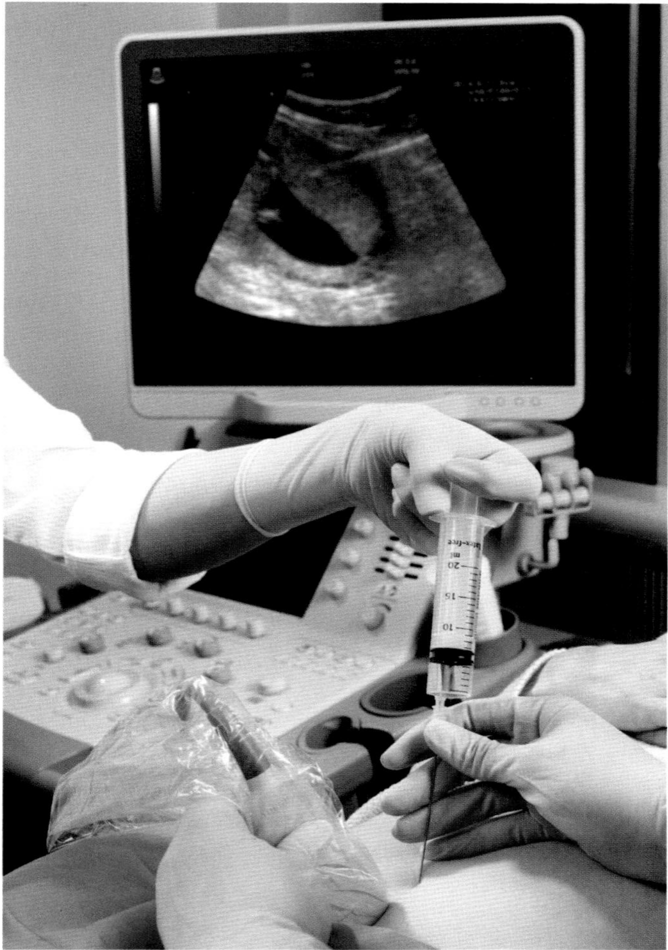

Figure 19.17 In chorionic villus sampling, an ultrasound scanner is used to guide the needle to the placenta to remove a small sample of the fetal chorionic villi which are embedded in the placenta. A small sample of the fetal blood is removed for analysis.

Thalassaemia is a blood disease similar to sickle cell anaemia. It used to be a common genetic disease in countries around the Mediterranean: Cyprus, Greece and southern Italy. The incidence of the disease has decreased significantly over recent years as a result of genetic screening and giving advice to couples who are identified as carriers of the mutant allele. Testing during pregnancy was also carried out. When a fetus was identified as having inherited the disorder, couples received advice about the possibility of terminating the pregnancy. Terminating pregnancies for a medical reason, rather than for any other, is known as therapeutic abortion.

Huntington's disease is a late-onset disease – symptoms do not usually appear until middle age, by which time people have usually already had children. There is no cure for this disease and the treatments available can only alleviate the symptoms. People in families with Huntington's face a dilemma: should they have the genetic test to find out whether or not they have the dominant allele for the disease? This also poses ethical dilemmas: would you rather be told that you are at high risk of developing this disease, even though nothing can be done about it, or live with the uncertainty of not knowing? Is it a good idea to have this information before you start a family? Decisions are made even more difficult by the possibility that a person with the dominant allele for Huntington's may live their whole life completely free of the disorder as it sometimes does not develop.

QUESTION

19.9 Explain the advantages of genetic screening.

Gene therapy

Gene technology and our rapidly increasing knowledge of the positions of particular genes on our chromosomes have given us the opportunity to identify many genes that are responsible for genetic disorders such as **sickle cell anaemia** (page 388) and **cystic fibrosis**. When genetic engineering really began to get going in the 1990s, it was envisaged that it would not be long before gene technology could cure these disorders by inserting 'normal' alleles of these genes into the cells. This process is called gene therapy. But gene therapy has proved to be far more difficult than was originally thought. The problems lie in getting normal alleles of the genes into a person's cells and then making them work properly when they get there.

The most common vectors that are used to carry the normal alleles into host cells are viruses (often retroviruses or lentiviruses) or small spheres of phospholipid called liposomes. Occasionally 'naked' DNA is used.

The first successful gene therapy was performed in 1990 on a four-year-old girl from Cleveland, Ohio. She suffered from the rare genetic disorder known as severe combined immunodeficiency (SCID). In this disorder, the immune system is crippled and sufferers die in infancy from common infections. Children showing the condition are often isolated inside plastic 'bubbles' to protect them from infections.

The defect in SCID involves the inability to make an enzyme, adenosine deaminase (ADA) which is vital for the functioning of the immune system. Some of the child's T-lymphocytes (page 230) were removed and normal alleles of the ADA gene were introduced into them, using a virus as a vector. The cells were then replaced. This was not a permanent cure. Regular transfusions (every three to five months) were necessary to keep the immune system functioning.

Two years later, gene therapy using stem cells harvested from bone marrow was successful, but in France, in 2000, four children who had received gene therapy for X-linked SCID developed leukaemia as a result of using a retrovirus as vector. Retroviruses insert their genes into the host's genome, but they do so randomly. This means that they may insert their genes within another gene or, more dangerously, into the regulatory sequence of a gene, which may then activate a nearby gene causing cancer.

Since then, researchers have used lentiviruses as vectors. These also insert their genes randomly into the host genome, but they can be modified to inactivate replication. HIV has been disabled in this way to act as a vector. The adeno-associated virus (AAV) is also now used as a vector. This virus does not insert its genes into the host genome and so they are not passed on to daughter cells when a cell divides. This is a problem when the host cells are short-lived (such as lymphocytes), but the virus has been used successfully with long-lived cells such as liver cells and neurones.

This work on vectors has led to increasingly successful gene therapies in the last few years, including the following.

- The eyesight of young men with a form of hereditary blindness, Leber congenital amaurosis, in which retinal cells die off gradually from an early age, has been improved.
- The normal allele of the β-globin gene has been successfully inserted into blood stem cells to correct the disorder, β-thalassaemia.
- Six people with haemophilia B (in which factor IX is missing) have at least seen their symptoms reduced.
- Five children were successfully treated for SCID in 2013.

We will look in more detail at the genetic disorder, cystic fibrosis, to illustrate some of the problems facing gene therapy.

Cystic fibrosis

Cystic fibrosis is a genetic disorder in which abnormally thick mucus is produced in the lungs and other parts of the body. A person with cystic fibrosis is very prone to bacterial infections in the lungs because it is difficult for the mucus to be removed, allowing bacteria to breed in it. People with cystic fibrosis need daily therapy to help them to cough up this mucus (Figure 19.18). The thick mucus adversely affects many other parts of the body. The pancreatic duct may become blocked, and people with cystic fibrosis often take pancreatic enzymes by mouth to help with digestion. Around 90% of men with cystic fibrosis are sterile, because thick secretions block ducts in the reproductive system.

Cystic fibrosis is caused by a recessive allele of the gene that codes for a transporter protein called CFTR. This protein sits in the cell surface membranes of cells in the alveoli (and also elsewhere in the body) and allows

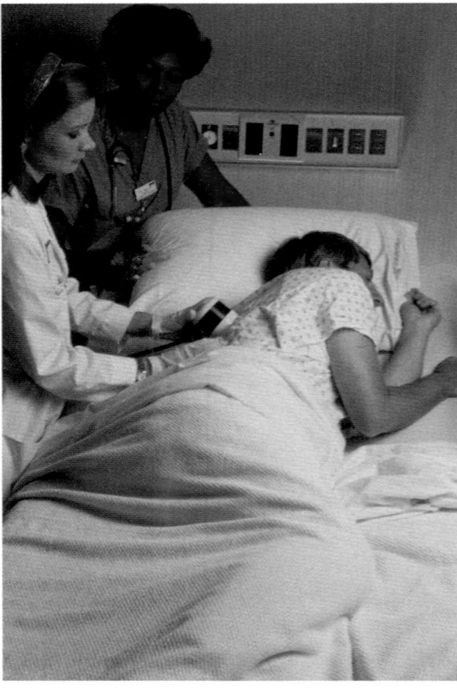

Figure 19.18 A person with cystic fibrosis is often treated with 'percussion therapy' – pummelling against the back to loosen the thick mucus so that it can be coughed up.

chloride ions to pass out of the cells. The recessive allele codes for a faulty version of this protein that does not act properly as a chloride ion transporter.

Normally, the cells lining the airways and in the lungs pump out chloride ions (Cl^-) through the channel in the cell surface membrane formed by CFTR. This results in a relatively high concentration of chloride ions outside the cells. This reduces the water potential below that of the cytoplasm of the cells. So water moves out of the cells by osmosis, down the water potential gradient. It mixes with the mucus there, making it thin enough for easy removal by the sweeping movements of cilia (Figure 19.19).

However, in someone with cystic fibrosis, much less water moves out of the cells, so the mucus on their surfaces stays thick and sticky. The cilia, or even coughing, can't remove it all.

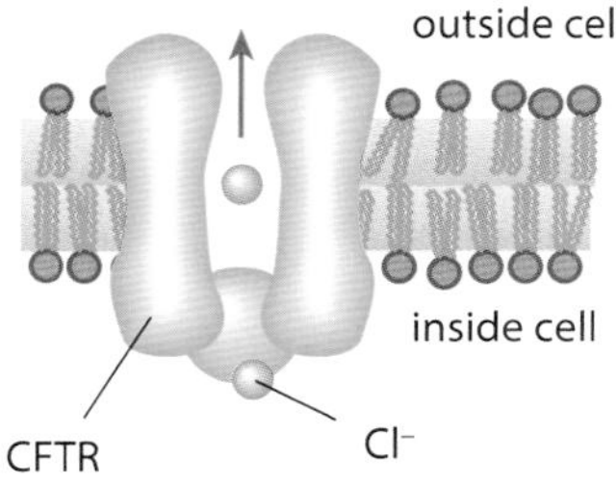

Figure 19.19 The CFTR protein forms channels for chloride ions in the cell surface membrane.

The *CFTR* gene

The *CFTR* gene is found on chromosome 7 and consists of about 250 000 bases. Mutations in this gene have produced several different defective alleles. The commonest of these is the result of a deletion of three bases. The CFTR protein made using the code on this allele is therefore missing one amino acid. The machinery in the cell recognises that this is not the right protein and does not place it in the cell surface membrane.

Because the faulty *CFTR* alleles are recessive, someone with one faulty allele and one normal allele is able to make enough of the CFTR protein to remain healthy. The person is a symptom-free carrier of the disease. Each time two heterozygous people have a child, there is a one in four chance that their child will have the disease.

Because it is caused by a single gene, cystic fibrosis could be a good candidate for gene therapy.

If the normal dominant allele could be inserted into cells in the lungs, the correct CFTR should be made. In theory, there is no reason why this should not happen. In practice, there have been major problems in getting the allele into the cells.

In the UK, trials began in 1993. The normal allele was inserted into liposomes (tiny balls of lipid molecules), which were then sprayed as an aerosol into the noses of nine volunteers. This succeeded in introducing the allele into a few cells lining the nose, but the effect only lasted for a week, because these cells have a very short natural lifespan.

Researchers in the USA tried a different vector. In a trial involving several people with cystic fibrosis, they introduced the normal allele into normally harmless viruses and then used these to carry the allele into the passages of the gas exchange system. The allele did indeed enter some cells there, but some of the volunteers experienced unpleasant side-effects as a result of infection by the virus. As a result, the trials were stopped.

To be used as a treatment, the allele really needs to get into many cells throughout the respiratory system, including the ones that divide to form new surface cells. This has so far not been achieved.

More recently, a different approach has been taken. In some people with cystic fibrosis, the mutation in the gene has simply replaced one base with another. This has created a 'stop' codon in the middle of the gene. The gene is transcribed (that is, mRNA is made from it) in the normal way, but translation on the ribosomes stops when this codon is reached. This means that only a short length of the CFTR protein is made. A drug called PTC124 has been found to allow translation to just keep going across this stop codon, so the entire protein is made, albeit with a wrong or missing amino acid in the middle of it. Clinical trials have shown hopeful signs that this may allow enough CFTR to be made to significantly relieve the symptoms of some people with cystic fibrosis. It is much easier to do than 'classic' gene therapy, because it only involves the patient taking a pill each day.

Occasionally, DNA has been inserted directly into tissues without the use of any vector. This so-called naked DNA has been used in trials of gene therapy for skin, muscular and heart disorders. The advantage of using this method is that it removes the problems associated with using vectors.

QUESTION

19.10 Gene therapy for cystic fibrosis would be successful if only **one** copy of the normal allele of the gene was successfully inserted into the cells. Explain why this is so.

Somatic and germ cell gene therapy

Gene therapy involves introducing a 'correct' allele into a person's cells as a treatment for a genetic disease. So far, all attempts to do this in humans have involved placing the allele in body cells, otherwise known as **somatic cells**. However, another possibility would be to insert the allele into germ cells – that is, cells that are involved in sexual reproduction, such as gametes or an early embryo. For example, in theory, a woman with cystic fibrosis could opt to try to conceive a baby using IVF. Eggs (oocytes) would be harvested from her in the normal way. Then the 'correct' allele of the *CFTR* gene could be injected into an egg and this egg fertilised by a sperm to produce a zygote.

At present, this is illegal in humans. However, it has been successfully done in other animals, although it looks as though there is a high chance of offspring produced in this way having other, unpredicted, diseases.

The problem that many people see with germ cell therapy is that **all** the cells of the child are produced from this genetically engineered zygote, and therefore will all carry the gene that has been inserted. When the child grows up and produces eggs or sperm, these gametes will also contain the allele and therefore it will be passed on to their children. We say that the allele is in the 'germ line', being passed on from generation to generation.

Genetic technology and agriculture

Genetically modified plants

Proteins for use in medicine can be produced from genetically modified plants, so avoiding any problem of contamination by animal proteins. Examples include vaccines, albumin and the proteins found in breast milk that are used to treat diarrhoea in infants. However, the vast bulk of genetically modified plants grown around the world are crop plants modified to be resistant to herbicides, such as glufosinate and glyphosate, or crops that are resistant to insect pests. These modifications increase crop yield. A few crops, such as vitamin A-enhanced rice, provide improved nutrition.

Herbicide-resistant crops

Oil seed rape, *Brassica napus*, is grown in many parts of the world as a source of vegetable oil which is used as biodiesel fuel, as a lubricant and in human and animal foods (Figure 19.20). Natural rape seed oil contains substances (erucic acid and glucosinolates) that are undesirable in oil that is to be used in human or animal food. A hybrid, bred in Canada to produce low concentrations of these undesirable substances, was called canola (Canadian oilseed low acid), and this name is now often used to mean any variety of oil seed rape.

Gene technology has been used to produce herbicide-resistant strains. Growing a herbicide-resistant crop allows fields to be sprayed with herbicide after the crop has germinated, killing any weeds that would otherwise compete with the crop for space, light, water or ions. This increases the yield of the crop. Oil seed rape that is resistant to the herbicide glyphosate, or to the related glufosinate, is grown in a number of countries.

Glyphosate inhibits an enzyme involved in the synthesis of three amino acids: phenylalanine, tyrosine and tryptophan. Glyphosate is absorbed by a plant's leaves and is transported to the growing tips. The amino acids are needed for producing essential proteins, so the plant dies. Various microorganisms have versions of the enzyme involved in the synthesis of phenylalanine, tyrosine and tryptophan that are not affected by glyphosate. The gene that was transferred into crop plants came from a strain of the bacterium *Agrobacterium*.

Tobacco has been made resistant to two different herbicides: sulfonylurea and dinitroaniline. In both cases the genes were taken from other species of plant.

Figure 19.20 Oil seed rape **a** in flower and **b** in seed.

The most likely detrimental effects on the environment of growing a herbicide-resistant crop are that:

- the genetically modified plant will become an agricultural weed
- pollen will transfer the gene to wild relatives, producing hybrid offspring that are invasive weeds
- herbicide-resistant weeds will evolve because so much of the same herbicide is used.

In 1993, an investigation to compare invasiveness of normal and genetically modified oil seed rape plants was carried out. Three genetic lines were compared: non-engineered oilseed rape and two different genetically engineered versions of the same cultivar. The rates of population increase were compared in plants grown in a total of 12 different environments. The environments differed in, for example, the presence and absence of cultivated and uncultivated background vegetation, and presence and absence of various herbivores and pathogens. There was no evidence that genetic engineering increased the invasiveness of oil seed rape plants. Where differences between normal and genetically modified plants existed, the genetically engineered plants were slightly less invasive than the unmodified plants.

The risk of pollen transfer, by wind or by insects, is real. Oil seed rape interbreeds easily with two related species: wild radish and wild turnip. Its flowers are adapted for insect pollination, but are also pollinated by wind. Although 'safe' planting distances are specified for trials of genetically modified plants (for example, 200 m for oil seed rape), pollen from various plants has been found between 1000 and 1500 m away from those plants. Bees visiting some flowers have been found to forage at distances of more than 4000 m. Safe planting distances should be increased to allow the organic farming industry to maintain its 'GM-free' certification.

Experimental crosses between glufosinate-resistant oil seed rape and both wild radish and wild turnip have shown that resistance can be passed to the hybrid offspring and that it persists through several further generations of their offspring. However, there is as yet little evidence of this occurring outside the laboratory.

Herbicide-resistant mutant plants of various species have been found growing near fields where glyphosate has been much used. However, the herbicide is not only used on resistant crop species. Gene technology is not directly responsible for this evolution of resistance, which may arise in the absence of any genetically modified crop.

Insect-resistant crops

Another important agricultural development is that of genetically modified plants protected against attack by insect pests. Maize is protected against the corn borer, which eats the leaves of the plants and then burrows into the stalk, eating its way upwards until the plant cannot support the ear. Cotton is protected against pests such as the boll weevil (Figure 19.21). In both plants, yield is improved.

Insect-resistant tobacco also exists, and is protected against the tobacco bud worm, but as yet it has not been grown commercially.

The most likely detrimental effects on the environment of growing an insect-resistant crop are:

- the evolution of resistance by the insect pests
- a damaging effect on other species of insects
- the transfer of the added gene to other species of plant.

However, less pesticide is used, reducing the risk of spray carrying to and affecting non-target species of insects in other areas. Remember also that only insects that actually eat the crop are affected.

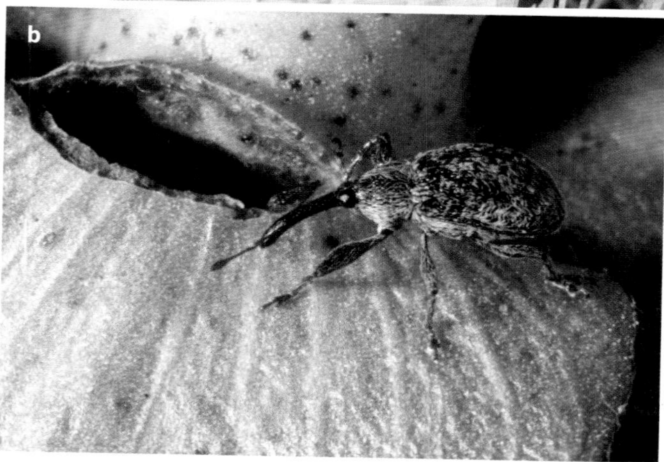

Figure 19.21 **a** Corn borer, **b** boll weevil.

A gene for a toxin, **Bt toxin**, which is lethal to insects that eat it but harmless to other animals, has been taken from a bacterium, *Bacillus thuringiensis*. Different strains of *B. thuringiensis* produce different toxins that can be used against different insect species. Crop plants that contain the Bt toxin gene from *B. thuringiensis* produce their own insecticides. However, insect populations can evolve resistance to toxins. Large numbers of crop plants containing the genes for Bt toxin may accelerate the evolution of resistance to it.

Many populations of corn borers in the USA are now resistant to Bt toxin. From the outset, growers have been encouraged to plant up to 50% of their maize as non-genetically modified maize in so called 'refuges'. Bt resistance in corn borers happens to be a recessive allele. Adult corn borers in the refuges are mostly homozygous dominant or heterozygous. These insects supply the dominant alleles to counteract resistance when adult corn borers from fields and refuges mate.

The pollen of Bt maize (corn) expresses the gene and has been found to disperse at least 60 m by wind. In the USA, milkweed frequently grows around the edge of maize fields and is a food source for the caterpillars of the monarch butterfly. Half of the summer population of monarch butterflies is found in the maize-growing areas of the USA. An experiment was set up in which caterpillars were fed milkweed leaves dusted with pollen from Bt maize, pollen from unmodified maize or no pollen at all. Caterpillar survival after four days of feeding on leaves dusted with pollen from Bt maize was 56%, whereas no caterpillars died after eating leaves dusted with pollen from unmodified maize or leaves with no pollen. However, further studies have shown that this laboratory-based experiment does not reflect the situation in the field, where the butterflies and caterpillars are not normally present at the time when pollen is shed.

Various aquatic insect larvae live in the streams in the maize-growing areas of the USA. Leaves from genetically modified Bt plants end up in the streams and may be eaten by, for example, caddis larvae. Experiments showed a small reduction in growth of larvae fed on Bt leaves. Another experiment, in which caddis larvae were fed on material containing different concentrations of Bt toxin, found a significant effect on larval growth at a concentration twice that actually found in the streams. This is, as yet, a potential rather than an actual problem, but one that needs careful monitoring.

There is thought to be a danger, if Bt maize is grown in Mexico, of it pollinating its wild 'parent' species, teosinte, and transferring genes to it. However, it has been found that maize pollen, once released from the anthers and exposed to the air, is not viable after two hours. This requires a two-hour wind-drift distance between a genetically modified crop and any teosinte habitats.

There is some evidence of reduced populations of microorganisms in soil in which Bt maize has been growing, but no practical detriment has been seen.

It must not be forgotten that genetically modified crop seed is expensive and that its cost may remove any advantage of growing resistant crops. Growers need to buy seed each season, which again keeps costs high when compared with those of traditional varieties. In parts of the world where a great deal of a genetically modified crop is grown, there is the danger of losing biodiversity (Chapter 18).

Golden Rice

Rice is a staple food in many parts of the world. Where people are poor and rice forms the major part of their diet, deficiency of vitamin A is a common and serious problem. Vitamin A deficiency can cause blindness. The World Health Organization estimates that as many as 500 000 children go blind each year as a result of vitamin A deficiency. Even more importantly, lack of vitamin A can cause an immune deficiency syndrome, and this is a significant cause of mortality in some parts of the world, particularly in children. It is estimated that, in 2010, more than two and half million children died of vitamin A deficiency.

Vitamin A is a fat-soluble vitamin found in oily fish and animal products such as eggs, milk, cheese and liver. It is also made in our bodies from carotene, the orange carotenoid pigment found in carrots. Pro-vitamin A carotenoids are also present in the aleurone layer of rice grains, but not in the endosperm, the energy storage tissue in the seed that humans eat. The aleurone layer is removed from rice when it is polished to produce white rice. Brown rice still contains the aleurone layer. The aleurone layer goes rancid if the rice is stored for any length of time, which is why white rice is produced and usually eaten instead. Children of families living in poverty often lack animal products in their diets as they are too expensive. Even if such children have a diet containing a wide range of vegetables rich in carotenoids, it is still difficult for them to avoid vitamin A deficiency.

In the 1990s, a project was undertaken to produce a variety of rice that contained carotene in its endosperm. Genes for carotene production were taken from daffodils and a common soil bacterium, now named *Pantoea ananatis*, and inserted into rice. Further research showed that substituting the gene from daffodil with one from maize gave even higher quantities of carotene, and the single transformation with these genes is the basis of all current Golden Rice. Figure 19.22 shows how the

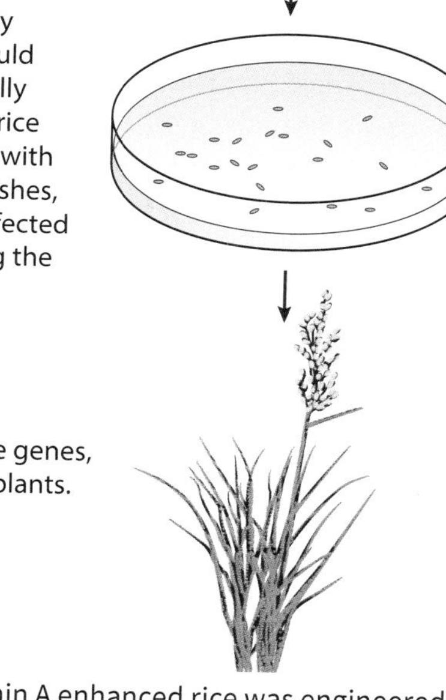

Figure 19.22 Pro-vitamin A enhanced rice was engineered using genes from the maize and a bacterium.

Figure 19.23 Normal rice on the left; Golden Rice on the right.

research was carried out. The genetically modified rice is called Golden Rice, because it contains a lot of the orange pigment carotene (Figure 19.23).

The genetically modified rice is being bred into other varieties of rice to produce varieties that grow well in the conditions in different parts of the world, with the same yield, pest resistance and eating qualities as the original varieties. For example, the International Rice Research Institute (IRRI) has worked with researchers in Bangladesh to produce a pro-vitamin A enhanced ('Golden') variety of Bangladesh's most popular rice variety. Research with children in China has shown that Golden Rice may be as useful as a source of vitamin A from vitamin A capsules, eggs, or milk to overcome vitamin A deficiency in rice-consuming populations.

There has been quite a lot of controversy over Golden Rice. Several non-governmental organisations, opposed to the use of genetic engineering in any crops, have condemned Golden Rice as being the wrong way to solve the problem of people eating diets that are short of vitamin A. One of their arguments is that the main reason that people eat diets that are short of vitamin A is poverty, and that the way to solve the problem is to help them out of poverty so that they have access to a more varied diet. Others say that, although it would be better if we could somehow lift these people out of poverty, this cannot be quickly achieved.

Despite the research, development and evaluation of Golden Rice that has taken place over the last ten years, it is not yet available to farmers and consumers because it has to be approved by national authorities in each country first.

With the help of the scientists who initially donated their technology invention, an international network of public sector rice research institutes and funding from bodies such as the Bill and Melinda Gates Foundation, Golden Rice seed will made available in developing countries at no greater cost than white rice seed. Everyone agrees that we need to solve the root causes of poor diets – which include numerous political, cultural and economic issues – but, meanwhile, pro-vitamin A enhanced rice could help millions of people to avoid blindness or death.

QUESTION

19.11 a Explain why promoters were inserted into the plasmids (Figure 19.22).

b In the production of bacteria that synthesise human insulin (page 467), plasmids acted as vectors to introduce the gene into the bacterial cells. What were the vectors used in the production of vitamin A enhanced rice? Explain your answer.

Genetically modified animals

Genetically modified animals for food production are much rarer than crop plants. An example is the GM Atlantic salmon, developed in the USA and Canada (Figure 19.24). A growth-hormone regulating gene from a Pacific Chinook salmon and a promoter from another species of fish, an ocean pout, were injected into a fertilised egg of an Atlantic salmon. By producing growth hormone throughout the year, the salmon are able to grow all year, instead of just in spring and summer. As a result, fish reach market size in about eighteen months, compared with the three years needed by an unmodified fish. It is proposed to rear only sterile females and to farm them in land-based tanks. The characteristics of the GM salmon reduce their ability to compete with wild salmon in a natural environment. This has led the US Food and Drug Administration (FDA) to declare that they are 'highly unlikely to have any significant effects on the environment' and 'as safe as food as conventional Atlantic salmon'.

In 2013, Canada approved the production of GM salmon eggs on a commercial scale, but neither Canada nor the USA FDA had yet given permission for GM salmon to enter the human food chain.

Social implications of using genetically modified organisms in food production

Some genetically modified plants are grown in strict containment in glasshouses, but a totally different set of problems emerges when genetically engineered organisms such as crop plants and organisms for the biological control of pests are intended for use in the general environment. Can such organisms be used safely?

It might seem likely that few countries would object to the growth of genetically modified crops that produce vaccines for human or animal use, yet there are people who object to the growth of pro-vitamin A enhanced rice (page 482). However, most objections are raised against the growth of herbicide-resistant or insect-resistant crops. The concerns about these genetically modified crops are as follows.

- The modified crop plants may become agricultural weeds or invade natural habitats.
- The introduced gene(s) may be transferred by pollen to wild relatives whose hybrid offspring may become more invasive.
- The introduced gene(s) may be transferred by pollen to unmodified plants growing on a farm with organic certification.
- The modified plants may be a direct hazard to humans, domestic animals or other beneficial animals, by being toxic or producing allergies.
- The herbicide that can now be used on the crop will leave toxic residues in the crop.
- Genetically modified seeds are expensive, as is herbicide, and their cost may remove any advantage of growing a resistant crop.
- Growers mostly need to buy seed each season, keeping costs high, unlike for traditional varieties, where the grower kept seed from one crop to sow for the next.

Figure 19.24 A GM salmon and non-GM salmon of the same age. Note that the GM fish do not grow larger than non-GM salmon, but attain their maximum size more quickly.

- In parts of the world where a lot of genetically modified crops are grown, there is a danger of losing traditional varieties with their desirable background genes for particular localities and their possibly unknown traits that might be useful in a world where the climate is changing. This requires a programme of growing and harvesting traditional varieties and setting up a seed bank to preserve them.

Despite these concerns, there are now millions of hectares of genetically modified crops and trees growing across the world. In the USA in 2011, half the cotton crop and more than half the maize and soya crops were genetically modified. Significant areas of China, Brazil and India are used for these crops, and farmers in developing countries are adopting the products of gene technology with enthusiasm. The exception is Europe, with its careful, but strict, controls. But Europe also has well-organised groups of protesters. Almost all of the field trials of genetically modified crops that have taken place in the UK during the last ten years have been vandalised (Figure 19.25).

But are there any damaging effects on human societies of genetic technology? Have any of the theoretical hazards had an actual effect on human societies?

There is little evidence of genes 'escaping' into the wild. No 'superweed' has appeared to reduce crop growth. There are no examples of foods produced from genetically modified organisms unexpectedly turning out to be toxic or allergenic. Unless the known effects of genetically modified crops become much greater than have so far been measured, the effect on human societies may be said to be small, but positive. There are, though, possible effects that cannot yet be measured, such as the future consequences of any loss of biodiversity from growing genetically modified crops.

Figure 19.25 This genetically modified maize, growing in Shropshire in the UK, is protected by an electric fence.

Summary

- Genetic technology involves using a variety of techniques to investigate the sequence of nucleotides in DNA and alter an organism's DNA. Genetic engineering involves the extracting of genes from one organism and placing them into the DNA of another to form recombinant DNA (rDNA). The gene(s) need to be inserted in such a way that they will be expressed in the genetically modified organism (GMO).
- Restriction enzymes cut across DNA at specific sites, known as restriction sites: these can be staggered cuts that give rise to short lengths of unpaired bases known as sticky ends or straight cuts to give blunt ends. Pieces of DNA with sticky ends that are complementary to each other are able to join together by forming hydrogen bonds. The enzyme ligase joins the sugar–phosphate backbones of pieces of DNA.
- In genetic engineering, vectors are used to carry pieces of DNA into cells: typical examples are plasmids, viruses and liposomes. Plasmids are small circles of double-stranded DNA; they are useful for genetic engineering because they can be cut with restriction enzymes and have promoters and gene markers (e.g. genes for antibiotics, GFP or GUS) inserted into them alongside the gene(s) to transform the host cell. A promoter must be inserted alongside the gene because organisms will not transcribe and express a gene unless there is a binding site for RNA polymerase.
- Cells that have taken up plasmids with the desired gene can be identified by detecting fluorescence (GFP) or appropriate staining (GUS).
- Lengths of DNA for genetic modification can be synthesised directly from mRNA by using the enzyme reverse transcriptase. Specific lengths of DNA can also be synthesised from nucleotides using knowledge of the genetic code.

- Electrophoresis is used to separate proteins and fragments of DNA of different lengths; the material to be tested is placed in wells cut in a gel and a voltage applied across the gel.
- The polymerase chain reaction (PCR) is a method of making very large numbers of copies of DNA from very small quantities (even one molecule). In PCR, DNA is denatured by heat to separate the strands; a short length of DNA known as a primer attaches to one end of each strand so that DNA polymerase can start synthesising a complementary strand using free nucleotides (as in replication). The double-stranded copies of DNA are separated again and the process repeated many times to 'bulk up' the DNA. Heat-stable DNA polymerases are used in PCR; the first was *Taq* polymerase that is found in a thermophilic bacterium, *Thermus aquaticus*.
- A gene probe is a length of single-stranded DNA, which has a known base sequence and is used to hybridise with lengths of DNA which have the complementary sequence; probes are labelled in some way to make them 'visible' (e.g. with radioactive phosphorus). PCR and gene probes are used in forensic investigations to look for matches between the DNA left at crime scenes and the DNA of suspects.
- Microarrays contain many thousands of gene probes and are used in two ways: to analyse the presence or absence of genes in different genomes and detect the presence of mRNA from cells to detect the genes that are being expressed at any one time.
- Bioinformatics deals with the storage and analysis of biological data; more specifically, nucleotide sequences and amino acid sequences of proteins.
- Many human proteins are now produced by GMOs, such as bacteria and yeasts and even animals, such as goats and sheep. This makes available drugs to treat diseases (e.g. human growth hormone) and secures the supply of others that were available from other sources (e.g. insulin).
- Genetic screening involves testing people to find out if they carry any faulty alleles for genes that can cause disease; there are genetic tests for many genetic diseases including breast cancer associated with *Brca*-1 and *Brca*-2, haemophilia, sickle cell anaemia, Huntington's disease and cystic fibrosis. Genetic counsellors may help people, who find that they or their unborn child have a disease-causing allele, to make a decision about how to act on this information.
- Gene therapy involves the addition of genes to human or animal cells that can cure, or reduce the symptoms of genetic diseases, such as SCID and some inherited eye diseases. Successful gene therapy involves selecting a suitable vector, such as viruses or liposomes, or inserting DNA directly into cells (naked DNA). Several attempts have been made to insert normal alleles of the *CFTR* gene into people with cystic fibrosis; so far there has been only limited success, because it is difficult to get the alleles into the cells that express the gene. Even when successful, it will have to be repeated at frequent intervals as the cells have a short lifespan.
- Genetic technology can provide benefits in, for example, agriculture and medicine, but has the associated risk of the escape of the gene concerned into organisms other than the intended host. The risk is seen to be particularly high for genetically modified crops that are released into the environment to grow.
- Genetic engineering is used to improve the quality and yield of crop plants and livestock in ways designed to solve the demand for food across the world; examples are Bt maize (corn) and pro-vitamin A enhanced rice (Golden Rice). Crops, such as maize, cotton, tobacco and oil seed rape, have been genetically modified for herbicide resistance and insect resistance to decrease losses and increase production.
- The social implications of genetic technology are the beneficial or otherwise effects of the technology on human societies. Ethics are sets of standards by which a particular group of people agree to regulate their behaviour, distinguishing an acceptable from an unacceptable activity. Each group must decide, first, whether research into gene technology is acceptable, and then whether or not it is acceptable to adopt the successful technologies.

End-of-chapter questions

1 Different enzymes are used in the various steps involved in the production of bacteria capable of synthesising a human protein. Which step is catalysed by a restriction enzyme?

A cloning DNA
B cutting open a plasmid vector
C producing cDNA from mRNA
D reforming the DNA double helix [1]

2 What describes a promoter?

A a length of DNA that controls the expression of a gene
B a piece of RNA that binds to DNA to switch off a gene
C a polypeptide that binds to DNA to switch on a gene
D a triplet code of three DNA nucleotides that codes for 'stop' [1]

3 Which statement correctly describes the electrophoresis of DNA fragments?

A Larger fragments of DNA move more rapidly to the anode than smaller fragments.
B Positively charged fragments of DNA move to the anode.
C Small negatively charged fragments of DNA move rapidly to the cathode.
D Smaller fragments of DNA move more rapidly than larger fragments. [1]

4 The table shows enzymes that are used in gene technology. Copy and complete the table to show the role of each enzyme.

Enzyme	Role
DNA ligase	
DNA polymerase	
restriction enzymes	
reverse transcriptase	

[4]

5 Rearrange the statements below to produce a flow diagram showing the steps involved in producing bacteria capable of synthesising a human protein such as human growth hormone (hGH).

1 Insert the plasmid into a host bacterium.
2 Isolate mRNA for hGH.
3 Insert the DNA into a plasmid and use ligase to seal the 'nicks' in the sugar–phosphate chains.
4 Use DNA polymerase to clone the DNA.
5 Clone the modified bacteria and harvest hGH.
6 Use reverse transcriptase to produce cDNA.
7 Use a restriction enzyme to cut a plasmid vector. [4]

6 a Genetic fingerprinting reveals the differences in variable number tandem repeats (VNTRs) in the DNA of different individuals. Explain what is meant by a VNTR. [3]

b Examine the figure, which shows diagrammatic DNA profiles of a mother, her child and a possible father of the child. Decide, giving your reasons, whether the possible father is the biological father of the child. [3]

[Total: 6]

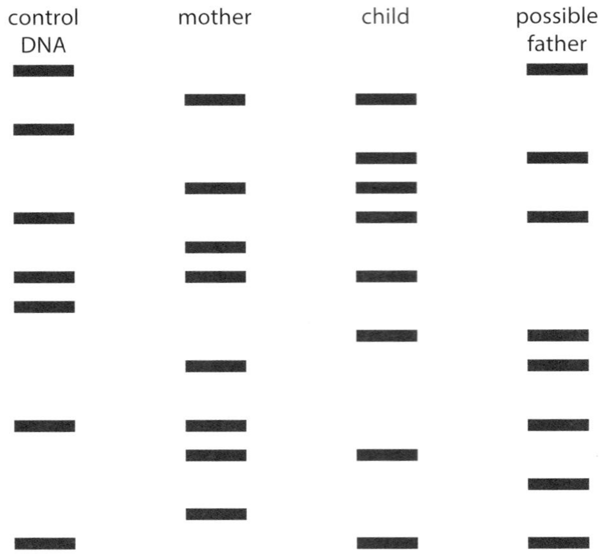

7 a Explain what is meant by:

i gene therapy [1]

ii genetic screening. [1]

b Explain why it is easier to devise a gene therapy for a condition caused by a recessive allele than for one caused by a dominant allele. [5]

[Total: 7]

8 a Draw a genetic diagram to show how two heterozygous parents may produce a child with cystic fibrosis. Use the symbols **A**/**a** in your diagram. [3]

b State the probability of one of the children of these parents suffering from cystic fibrosis. [1]

[Total: 4]

9 The figure shows the CFTR (cystic fibrosis transmembrane conductance regulator) protein in a cell surface membrane.

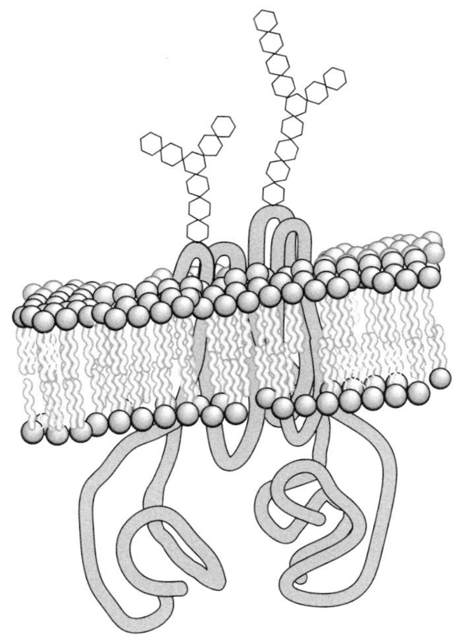

a i Describe the normal function of the CFTR protein. [2]

ii Use the letter **E** to indicate the external face of the membrane. State how you identified this face. [1]

b Cystic fibrosis is caused by a recessive allele of the *CFTR* gene.

i Explain the meaning of the term recessive allele. [2]

ii Explain how cystic fibrosis affects the function of the lungs. [3]

c As cystic fibrosis is caused by a recessive allele of a single gene, it is a good candidate for gene therapy. Trials were undertaken in the 1990s, attempting to deliver the normal allele of the *CFTR* gene into cells of the respiratory tract, using viruses or liposomes as vectors. Explain how viruses deliver the allele into cells. [2]

d In some people with cystic fibrosis, the allele has a single-base mutation which produces a 'nonsense' (stop) codon within the gene.

i Explain how this mutation would prevent normal CFTR protein being produced. [2]

ii A new type of drug, PTC124, enables translation to continue through the nonsense codon. Trials in mice homozygous for a *CFTR* allele containing the nonsense codon have found that animals treated with PTC124 produce normal CFTR protein in their cells. The drug is taken orally and is readily taken up into cells all over the body.

Using your knowledge of the progress towards successful gene therapy for cystic fibrosis, suggest why PTC124 could be a simpler and more reliable treatment for the disease. [3]

[Total: 15]

Cambridge International AS and A Level Biology 9700/04, Question 2, October/November 2008

Chapter P2:
Planning, analysis and evaluation

Learning outcomes

You should be able to:

- construct a testable, falsifiable hypothesis
- plan an investigation to test a hypothesis or to investigate a problem, identifying the independent variable and the dependent variable, and listing key variables to control
- describe how the independent variable would be altered, how the dependent variable would be measured, and how the key variables would be controlled
- describe how any control experiments would be used
- describe a logical sequence of steps for the investigation, including identification of any risks and how they can be minimised
- describe how to make up solutions in per cent (mass per volume) and in mol dm^{-3}
- find the mean, median and mode of a set of data
- find the interquartile range of a set of data
- calculate and use the standard deviation of a set of data
- calculate standard error, and use it to draw error bars on graphs and to interpret apparent differences between two sets of data
- use a t-test to find the probability that differences between two sets of data are due to chance
- use Pearson's linear correlation to find out if two sets of data have a linear relationship
- use Spearman's rank correlation to find out if two sets of data are correlated
- evaluate data, including identifying and dealing with anomalous results, assessing the adequacy of the range of data and the control of key variables, and making informed judgements about the degree of confidence that can be put in conclusions

The practical skills that you will develop during your A level course are tested by a written paper. In this examination, you will write answers on a question paper, but the questions actually test skills that you will have developed as you carried out practical work during your AS and A level course. The skills that are tested build on those that are tested at AS level. They fall into two main groups:

- planning
- analysis, conclusions and evaluation.

Each time you carry out an experiment or analyse data collected through experiment, you will use a variety of these skills. In this chapter, we will look at the different components of the skills in detail, and consider what you must be able to do in order to demonstrate these skills most effectively.

Some of the skills are the same as those that were tested at AS level. However, the questions are usually a little more demanding at A level.

Planning an investigation

As you progress through your A level biology course, you should develop the ability to plan your own experiments, rather than simply following instructions on a worksheet provided by someone else. In a practical examination, you are likely to be asked to plan an experiment that investigates the effect of one factor (the independent variable) on another (the dependent variable).

The question will usually give you a scenario to start you off. Sometimes, this will be familiar – you will be able to remember a similar experiment that you have done yourself. Sometimes it will be completely new, and you will have to use your experience in other experiments to think about the best way to design the particular one you have been asked to plan.

It is very important to read the instructions in the question extremely carefully. For example, the question might tell you what apparatus you should use in your experimental design. If you do not use that apparatus in your plan – even if you do not think it the best apparatus for that particular investigation – then it is unlikely that you are answering the question fully.

Constructing a hypothesis

Once you have been given a scenario, the question may ask you to construct a **hypothesis**. A hypothesis is a prediction of how you think the two variables are related to one another. It must be stated in a way that you can test by experiment, through the collection of quantitative results. (Quantitative results are ones that involve numerical data rather than descriptions.)

For example, you might be asked to plan an experiment to investigate the relationship between temperature and the rate of respiration of yeast. So what will your hypothesis be? Here are two possibilities.

- The rate of respiration of yeast increases with temperature.
- As temperature increases, the rate of respiration of yeast will increase up to a maximum temperature, above which it will fall.

Either of those hypotheses is fine. Both of them can be tested by changing the temperature and measuring the rate of respiration.

Note, however, that you cannot 'prove' that a hypothesis is correct just by doing one experiment. Your results may **support** your hypothesis, but they cannot **prove** it. You would need to do many more experiments before you can be sure that your hypothesis really is correct in all situations.

But you can **disprove** a hypothesis more easily. If you found that the rate of respiration did not increase as temperature increases, then this suggests that the hypothesis is incorrect. Nevertheless, it would be a good idea to do the experiment two or three times more, to make sure that the results can be repeated.

Sometimes, you could be asked to sketch a graph of your predicted results, if your hypothesis is supported. A sketch graph relating to the first hypothesis is shown in Figure P2.1a, and one relating to the second hypothesis is shown in Figure P2.1b.

Using the right apparatus

Imagine the question has asked you to plan your experiment using the apparatus shown in Figure P2.2.

As the yeast respires, it produces carbon dioxide gas. This collects above the liquid in the syringe, increasing the pressure and causing the meniscus to move down the capillary tubing.

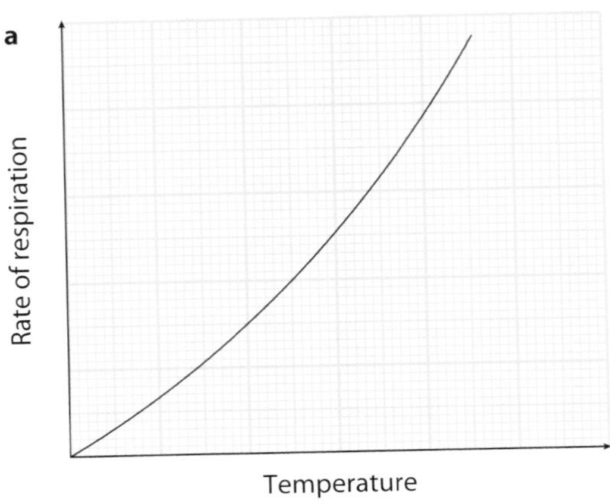

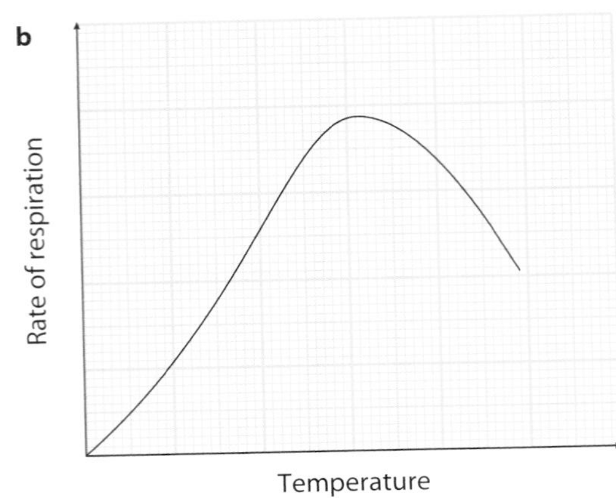

Figure P2.1 Using sketch graphs to show predicted results: **a** rate of respiration of yeast will increase with temperature; **b** as temperature increases, the rate of respiration will increase up to a maximum temperature, above which it will fall.

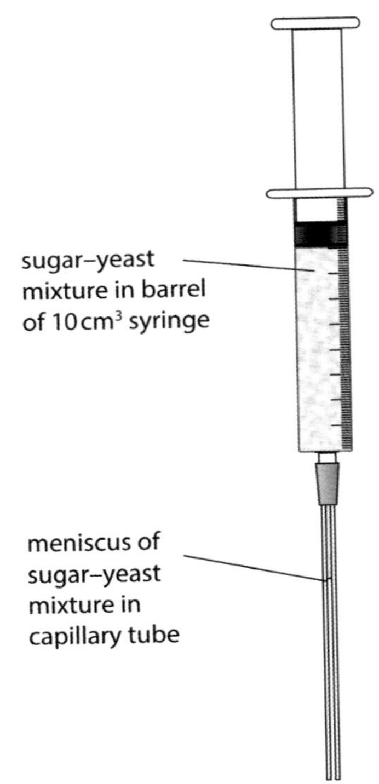

Figure P2.2 Apparatus for measuring the rate of respiration of yeast.

You may have done an experiment to investigate the relationship between temperature and rate of respiration of yeast, using a respirometer, like the experiment on page 280, and you may think that this would be a much better way of doing this investigation. You would be right – but if the question asks you to use the syringe apparatus, then that is what you must do.

Identifying variables

Both the hypotheses on the previous page include reference to the independent and dependent variables. Make sure that you know which is which. It is usually a good idea to make a clear statement about these as part of your plan.

You also need to think about the controlled variables – the ones that you will keep constant as you change the independent variable. It is very important to include only variables that really might have an effect on your results – the key variables – and not just give a long list including unimportant ones.

Changing the independent variable

You should be prepared to describe how you will change the independent variable. In this experiment, this means changing the temperature, and you should already know how to do this using a water bath. You should describe how you would do this, and how you would measure the different values of temperature.

There are many other possible independent variables that a question could involve. For example, the independent variable could be pH, in which case you would use buffer solutions. It could be concentration. By now, you should be familiar with using a stock solution to make up solutions with lower concentrations, using the serial dilution technique – this is described on page 248 in Chapter P1. At A level, however, you could also be asked

QUESTION

P2.1 Describe how you could use the apparatus in Figure P2.2 to measure the rate of respiration of yeast.

how to make up the initial stock solution with a particular concentration.

For example, you may need to describe how to make up a 1% sugar solution. This means a solution containing 1 g of sugar for every 100 g of water. Remember that 1 cm^3 of water has a mass of 1 g. So, to make up a 1% sugar solution you would:

- use a top pan balance to measure out 1 g of sugar
- place it into a 100 cm^3 volumetric flask (a kind of flask that enables the volume to be measured very accurately)
- add a small amount of distilled water and dissolve the sugar thoroughly
- add distilled water to make up to exactly 100 cm^3.

You could also be asked to make up a 1 mol dm^{-3} solution. This means a solution containing 1 mole of the solute in 1 dm^3 of solution. A **mole** is the relative molecular mass of a substance in grams. To work this out, you need to know the molecular formula of the substance, and the relative atomic masses of each atom in the formula. You would be given these in the question. For example, for sucrose:

- molecular formula of sucrose is $C_{12}H_{22}O_{11}$
- relative atomic masses are: carbon 12, hydrogen 1, oxygen 16.

So the relative molecular mass of sucrose is:

$$(12\times12)+(22\times1)+(11\times16) = 144+22+176 = 342$$

QUESTION

P2.2 Which of these variables should be controlled (kept constant) when investigating the effect of temperature on the rate of respiration of yeast?

a the concentration of sugar solution
b the initial volume of the sugar–yeast mixture inside the syringe
c the length of the capillary tubing
d the diameter of the capillary tubing
e the type of sugar (e.g. glucose, sucrose)
f the light intensity
g the pH of the sugar–yeast mixture

To make a 1 mol dm^{-3} solution:

- use a top pan balance to measure out 342 g of sugar (sucrose)
- put the sugar into a 1 dm^3 volumetric flask
- add a small amount of distilled water and shake until the sugar has completely dissolved
- add more distilled water until the meniscus of the liquid is exactly on the 1 dm^3 mark (Figure P2.3).

You also need to think about a suitable **range** and **interval** of the independent variable. This is described in Chapter P1 on page 248. Read the question carefully, as it may give you some clues about this. If not, use your biological knowledge to help you make your decision.

1 Add 342 g of sucrose to a beaker. Take the mass of the beaker into account or tare the beaker to zero.

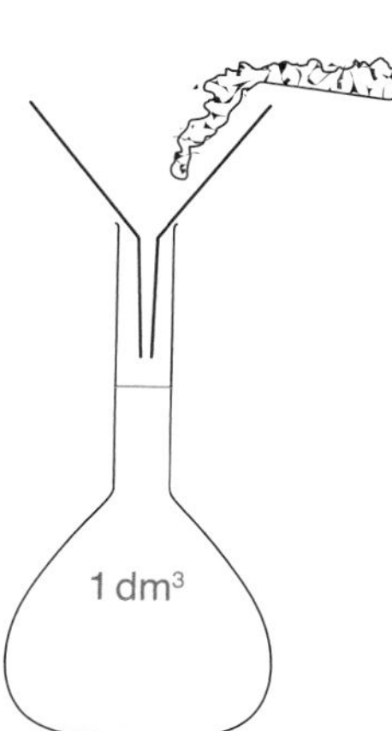

2 Pour the sucrose into a volumetric flask. Wash any sugar crystals left behind into the flask.

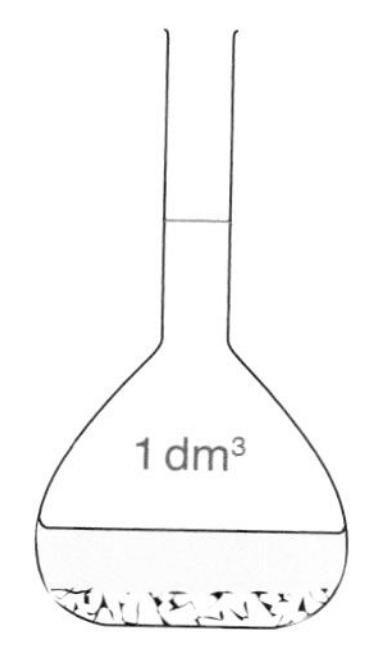

3 Add water and shake the flask until all the sucrose is dissolved.

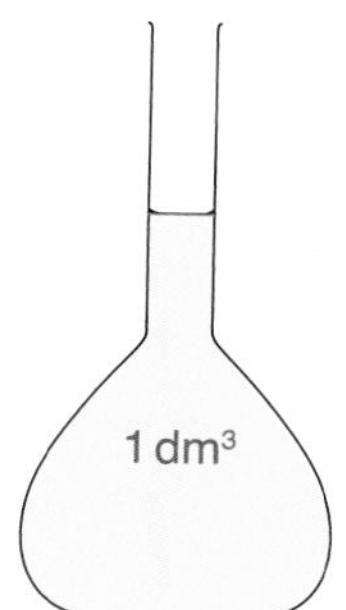

4 Top up the solution with water until the meniscus reaches the line.

Figure P2.3 Making up a 1 mol dm^{-3} solution of sucrose.

QUESTION

P2.3 The molecular formula of glucose is $C_6H_{12}O_6$.

- **a** Describe how you would make up 100 cm^3 of a 1% solution of glucose.
- **b** Describe how you would make up 250 cm^3 of a 1% solution of glucose.
- **c** Describe how you would make up a 1 mol dm^{-3} solution of glucose.
- **d** Describe how you would use the 1 mol dm^{-3} solution to make up a 0.5 mol dm^{-3} solution.

Measuring the dependent variable

Now let us go back to the yeast respiration rate experiment, and the apparatus shown in Figure P2.2.

The dependent variable is the rate of respiration of the yeast. This can be measured by recording the rate of movement of the meniscus. Remember that, if you are investigating **rate**, then **time** must come into your measurements. You would need to record the position of the meniscus at time 0, and then perhaps continue to do this at regular time intervals for, say, ten minutes. Alternatively, you could just record the position at time 0 and again at ten minutes. You would need to do this at each temperature that you have decided to test.

Whatever the experiment that you are describing, take care to describe exactly how you would measure the dependent variable. Say what measuring instruments you would use, and what you would do to make sure that your measurements are made accurately. Say exactly what you would measure and when you would measure it.

Identifying different types of variable

Often, the data about the dependent variable that you collect in your experiment (your results) are numerical. These are called **quantitative** data.

These quantitative data may be **continuous** or **discrete**. If the variable is continuous, then each measurement, count or reading can be any value between two extremes. Your results will not necessarily be whole numbers. The results for the yeast respiration rate experiment will be a quantitative and continuous variable.

If the variable is discrete, then each measurement, count or reading can only be one of a set number of discrete values. For example, you might be asked to count the number of prickles on each leaf in a sample of holly leaves. The number of prickles will always be a whole number – you cannot have half a prickle on a leaf.

Sometimes, the data you collect about the dependent variable are not numerical. These are called **qualitative** data.

Qualitative data can be **ordered** or **categoric**. Ordered (ordinal) variables are those that – although you do not have actual numerical values for them – can be organised into an order or sequence. For example, you might do a series of Benedict's tests on a set of glucose solutions of unknown concentration, and decide on the relative depth of colour of each one. You can sort these into an order – from the one that is least brick-red to the one that is darkest brick-red – but you can't assign an actual numeric value for the colour of any of them.

Categoric variables are completely discrete, and you can't put them into order. Each observation fits into a particular, clearly defined category. For example, in a sample of dead leaves taken from a forest floor, you might record the species of tree from which each leaf comes. Each leaf comes from one species – there are no 'overlaps' between categories, and there is no way of ordering or ranking them.

Controlling the controlled variables

You will be expected to describe how to control each of the key variables in your plan. This is described on pages 253–254 in Chapter P1.

You might also need to think about doing a control experiment as part of your investigation. The purpose of a control is to check that it is the factor you are investigating that is affecting the dependent variable, and not some other factor. For example, in the yeast respiration experiment, it would be a good idea to set up at least one syringe with sugar solution but no yeast. This would be a check that it is something that the yeast is doing that is causing the change in position of the meniscus.

Describing the sequence of steps

When you are describing your planned experiment, make sure that you describe a logical sequence of steps that you would follow. It is always worth jotting these down roughly first (perhaps on a spare page on the examination paper), to make sure that the steps follow one another in a sensible order, and that everything is fully explained. You might like to draw labelled or annotated diagrams to explain some of the steps – it is sometimes quicker and easier to do it that way. Diagrams are very often the best way to describe how you would assemble the apparatus.

Risk assessment

An extremely important part of planning any experiment is to think about the potential hazards involved. In biology experiments, there often are not any significant risks, and if that is the case then you should say so. Do not invent risks when there are none. Do, though, always mention risk, even if it is just to say that you do not think there is any.

If you do identify any significant risks, then you should explain how you would minimise them. For example, in the yeast experiment, you might decide to use temperatures up to 80 °C. This is hot enough to burn the skin, so you would need to take precautions when handling the apparatus in a water bath at that temperature. You should lift the apparatus in and out using tongs, or use heatproof and waterproof gloves.

Recording and displaying results

You may be asked to construct a results table that you could use to fill in your results. This is described on pages 255–256 in Chapter P1.

You may also want to describe how you would use the data to plot a graph, explaining what you would put on each axis and the type of graph you would draw.

Using the data to reach a conclusion

You may be asked how you would use your collected results to reach a conclusion. Usually, the conclusion will be whether or not the results support your hypothesis. (Remember that they will never prove it.) This is described on page 259 in Chapter P1.

Analysis, conclusions and evaluation

In a written practical examination, generally at least one question will provide you with sets of data. You could be asked how to process the data (for example, by doing calculations) in order to be able to use them to make a conclusion. You could be asked to assess the reliability of the investigation. This should be familiar from your AS work, but there is one big step up at this stage – you need to be able to use **statistics** to assess the variability of the data, or the significance of your results.

Note that you do **not** have to learn any of the formulae for the statistical tests. They will always be provided for you in the question.

Mean, median and mode

A large lemon tree has several hundred fruits on it. Imagine that you have measured the masses of 40 lemons from this tree. These are the results, recorded to the nearest gram and arranged in order of increasing mass.

57, 60, 67, 72, 72, 76, 78, 79, 81, 83, 84, 86, 87,
88, 88, 90, 92, 92, 93, 94, 95, 97, 98, 99, 100,
101, 101, 103, 105, 106, 107, 109, 111, 113,
119, 120, 125, 128, 132, 135

To calculate the **mean**, add up all the readings and divide by the total number of readings.

$$\text{mean} = \frac{\text{sum of individual values}}{\text{total number of values}} = \frac{3823}{40} = 96$$

When you do this calculation, your calculator will read 95.575. However, you should not record this value. You need to consider the appropriate number of decimal places to use. Here, the masses of the lemons were recorded to the nearest whole number, so we can record the mean either to the nearest whole number, which is 96, or to one more decimal place, which is 95.6.

To find the median and mode, you need to plot your results as a frequency histogram (Figure P2.4).

The mode (modal class) is the most common class in the set of results. For these results, the modal class is 90–94 g.

The median is the middle value of all the values in the data set. In this case, the median class is 95–99 g.

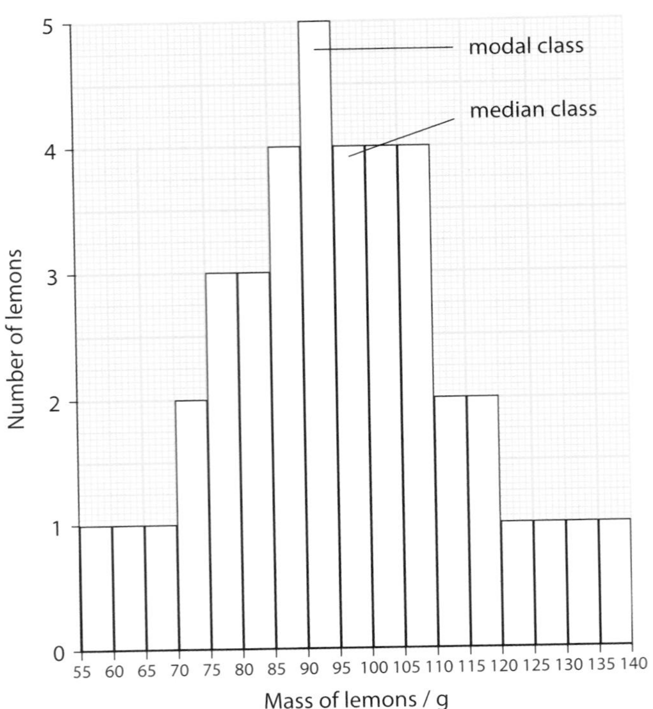

Figure P2.4 Lemon mass data plotted as a histogram.

Range and interquartile range

We have already met the term **range**, in the context of the range of the independent variable. It means exactly the same thing when applied to the results – it is the spread between the smallest number and the largest. For the fruit masses, the range is from 57 g to 135 g.

The interquartile range is the range into which the middle 50% of your data fall. For the fruits, we had 40 fruits in total. One-quarter of 40 is ten. The ten smallest fruits had masses between 57 g and 83 g. The ten largest fruits had masses between 106 g and 135 g. The interquartile range is therefore the range of mass shown by the remaining 20 fruits, which is 84 g to 105 g. This can also be expressed as 21 g, the difference between 105 g and 84 g.

Why would we want to know the interquartile range? It is sometimes useful if we want to compare two sets of data. Imagine that you have collected another set of fruits from a different tree, and want to compare the two sets. By concentrating on the middle 50% of the range, you eliminate the fruits with extreme masses at either end of the range. Comparing the interquartile ranges of both sets rather than comparing the complete ranges may give you a better idea of how similar or different the sets are.

Normal distribution

Many sets of data produce a symmetrical pattern when they are plotted as a frequency diagram. This is called a normal distribution (Figure P2.5a).

The data in a frequency diagram can also be plotted as a line graph (Figure P2.5b). If the data show a normal distribution, then this curve is completely symmetrical.

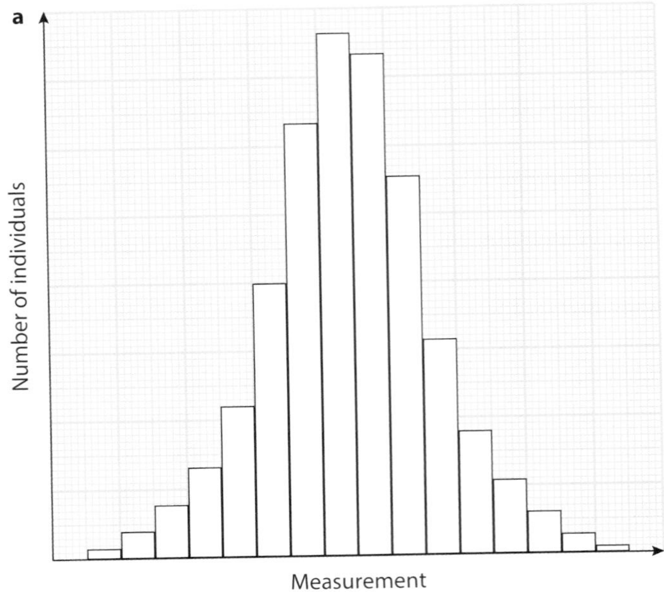

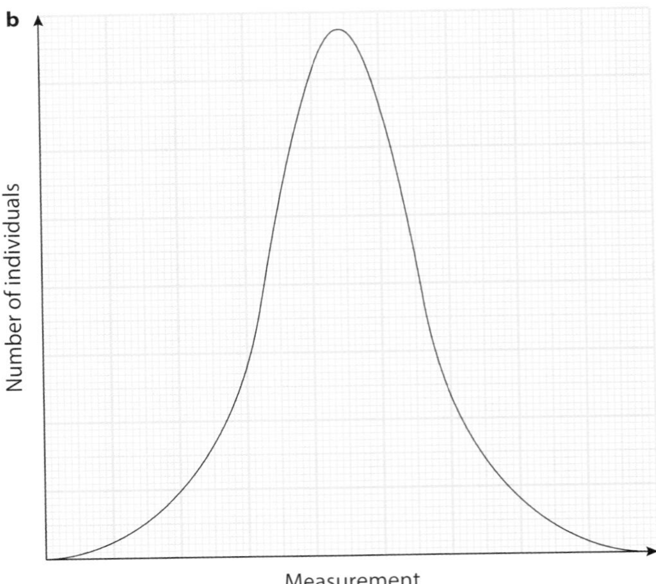

Figure P2.5 Normal distribution curves shown as **a** frequency diagram and **b** a line graph.

QUESTION

P2.4 In a perfect normal distribution curve, what will be the relationship between the mean, median and mode?

Standard deviation

A useful statistic to know about data that have an approximately normal distribution is how far they spread out on either side of the mean value. This is called the **standard deviation**. The larger the standard deviation, the wider the variation from the mean (Figure P2.6).

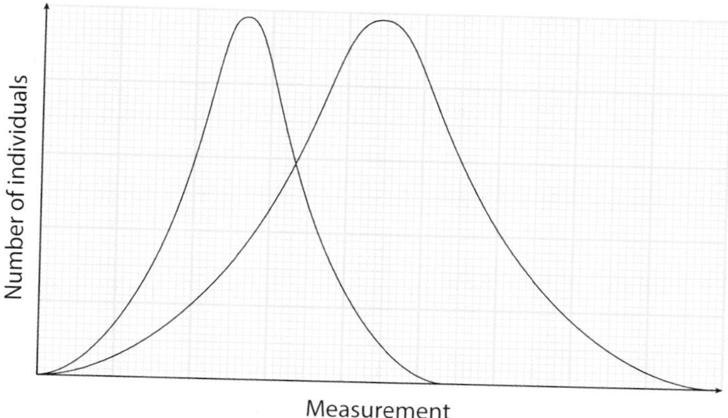

Figure P2.6 Normal distribution curves with small and large standard deviations.

A student measured the length of 21 petals from flowers of a population of a species of plant growing in woodland. These were the results:

Petal lengths in woodland population / mm						
3.1	3.2	2.7	3.1	3.0	3.2	3.3
3.1	3.1	3.3	3.3	3.2	3.2	3.3
3.2	2.9	3.4	2.9	3.0	2.9	3.2

The formula for calculating standard deviation is:

$$s = \sqrt{\frac{\Sigma(x - \bar{x})^2}{n - 1}}$$

where:

$\bar{x}$ is the mean
Σ stands for 'sum of'
x refers to the individual values in a set of data
n is the total number of observations (individual values, readings or measurements) in one set of data
s is standard deviation
$\sqrt{}$ is the symbol for square root

You may have a calculator that can do all the hard work for you – you just key in the individual values and it will calculate the standard deviation. However, you do need to know how to do the calculation yourself. The best way is to set your data out in a table, and work through it step by step.

1 List the measurement for each petal in the first column of a table like Table P2.1.
2 Calculate the mean for the petal length by adding all the measurements and dividing this total by the number of measurements.
3 Calculate the difference from the mean for each observation. This is $(x - \bar{x})$.
4 Calculate the squares of each of these differences from the mean. This is $(x - \bar{x})^2$.
5 Calculate the sum of the squares. This is $\Sigma(x - \bar{x})^2$.
6 Divide the sum of the squares by $n - 1$.
7 Find the square root of this. The result is the standard deviation, s, for that data set.

Table P2.1 shows the calculation of the standard deviation for petals from plants in the woodland.

QUESTION

P2.5 The student measured the petal length from a second population of the same species of plant, this time growing in a garden. These are the results:

Petal lengths in garden population / mm						
2.8	3.1	2.9	3.2	2.9	2.7	3.0
2.8	2.9	3.0	3.2	3.1	3.0	3.2
3.0	3.1	3.3	3.2	2.9		

Show that the standard deviation for this set of data is 0.16.

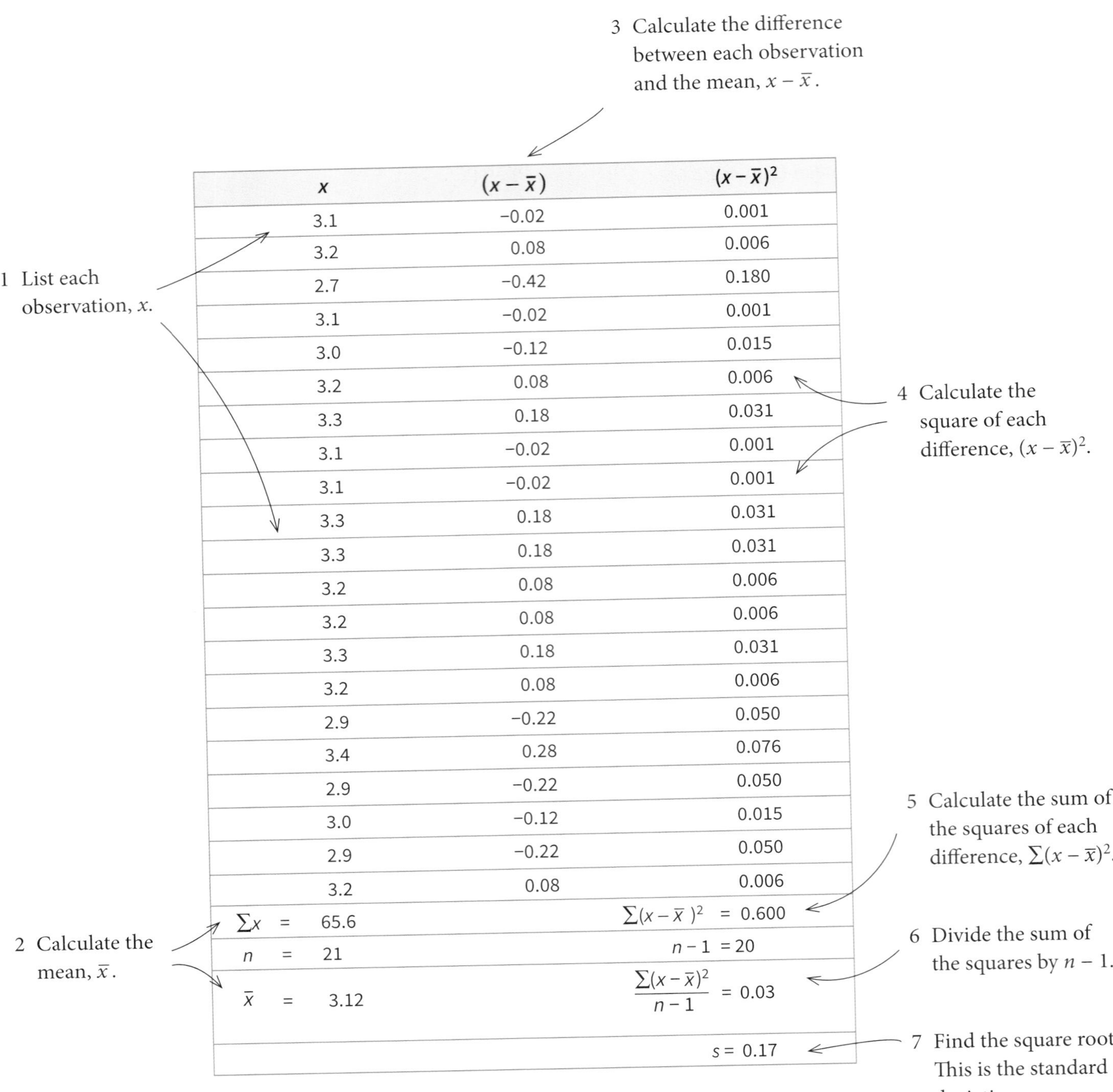

x	$(x - \bar{x})$	$(x - \bar{x})^2$
3.1	−0.02	0.001
3.2	0.08	0.006
2.7	−0.42	0.180
3.1	−0.02	0.001
3.0	−0.12	0.015
3.2	0.08	0.006
3.3	0.18	0.031
3.1	−0.02	0.001
3.1	−0.02	0.001
3.3	0.18	0.031
3.3	0.18	0.031
3.2	0.08	0.006
3.2	0.08	0.006
3.3	0.18	0.031
3.2	0.08	0.006
2.9	−0.22	0.050
3.4	0.28	0.076
2.9	−0.22	0.050
3.0	−0.12	0.015
2.9	−0.22	0.050
3.2	0.08	0.006
Σx = 65.6		$\Sigma(x - \bar{x})^2$ = 0.600
n = 21		$n - 1$ = 20
$\bar{x}$ = 3.12		$\frac{\Sigma(x - \bar{x})^2}{n - 1}$ = 0.03
		s = 0.17

Table P2.1 Calculation of standard deviation for petal length in a sample of plants from woodland. All lengths are in mm.

Standard error

The 21 petals measured were just a sample of all the thousands of petals on the plants in the wood and in the garden. If we took another sample, would we get the same value for the mean petal length? We cannot be certain without actually doing this, but there is a calculation that we can do to give us a good idea of how close our mean value is to the true mean value for all of the petals in the wood. The calculation works out the **standard error** (S_M) for our data.

Once you have worked out the standard deviation, s, then the standard error is very easy to calculate. The formula is:

$$S_M = \frac{s}{\sqrt{n}}$$

where S_M = standard error
s = standard deviation
n = the sample size (in this case, the number of petals in the sample)

So, for the petals in woodland:

$$S_M = \frac{0.17}{\sqrt{21}} = \frac{0.17}{4.58} = 0.04$$

What does this value tell us?

The standard error tells us how certain we can be that our mean value is the true mean for the population that we have sampled.

We can be 95% certain that – if we took a second sample from the same population – the mean for that second sample would lie within $2 \times$ our value of S_M from the mean for our first sample.

So here, we can be 95% certain that the mean petal length of a second sample would lie within 2×0.04 mm of our mean value for the first sample.

QUESTION

P2.6 Show that the standard error for the lengths of the sample of petals taken from the garden (Question P2.5) is also 0.04. Show each step in your working.

Error bars

The standard error can be used to draw error bars on a graph. Figure P2.7 shows the means for the two groups of petals, plotted on a bar chart.

The bars drawn through the tops of the plotted bars are called error bars.

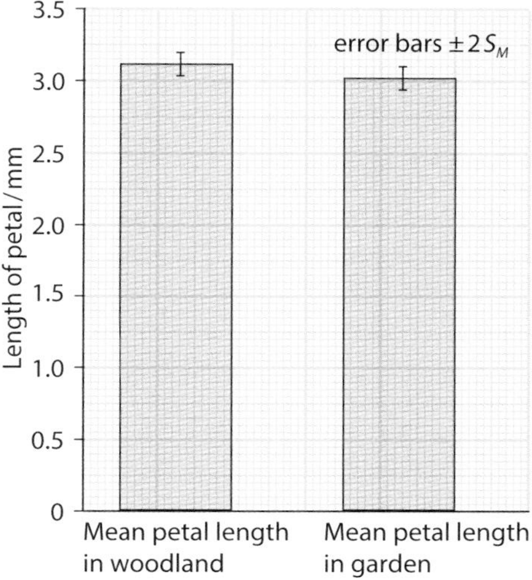

Figure P2.7 Mean petal length of plants in woodland and garden.

QUESTION

P2.7 From the data in Figure P2.7, is there strong evidence that the lengths of the petals in the woodland are significantly different from the lengths of the petals in the garden? Explain your answer.

If we draw an error bar that extends two standard errors above the mean and two standard errors below it, then we can be 95% certain that the true value of the mean lies within this range.

We can use these error bars to help us to decide whether or not there is a significant difference between the petal length in the woodland and the garden. If the error bars overlap, then the difference between the two groups is definitely not significant. If the error bars do not overlap, we still cannot be sure that the difference is significant – but at least we know it is possible that it is. You can also add error bars to line graphs, where your individual points represent mean values.

To find out whether the difference is significant, we can do a further statistical calculation, called a t-test.

The *t*-test

The t-test is used to assess whether or not the means of two sets of data with roughly normal distributions, are significantly different from one another.

For this example, we will use data from another investigation.

The corolla (petal) length of two populations of gentian were measured in mm.

Corolla lengths of population A:

13, 16, 15, 12, 18, 13, 13, 16, 19, 15, 18, 15, 15, 17, 15,

Corolla lengths of population B:

16, 14, 16, 18, 13, 17, 19, 20, 17, 15, 16, 16, 19, 21, 18,

The formula for the *t*-test is:

$$t = \frac{|\bar{x}_1 - \bar{x}_2|}{\sqrt{\left(\frac{s_1^2}{n_1} + \frac{s_2^2}{n_2}\right)}}$$

$\bar{x}_1$ is the mean of sample 1

$\bar{x}_2$ is the mean of sample 2

s_1 is the standard deviation of sample 1

s_2 is the standard deviation of sample 2

n_1 is the number of individual measurements in sample 1

n_2 is the number of individual measurements in sample 2

1 For each set of data, calculate the mean.
2 Calculate the differences from the mean of all observations in each data set. This is $x - \bar{x}$.
3 Calculate the squares of these. This is $(x - \bar{x})^2$.
4 Calculate the sum of the squares. This is $\sum(x - \bar{x})^2$.
5 Divide this by $n_1 - 1$ for the first set and $n_2 - 1$ for the second set.
6 Take the square root of this. The result is the standard deviation for each set of data.
For population A, $s_1 = 4.24$.
For population B, $s_2 = 4.86$.
7 Square the standard deviation and divide by the number of observations in that sample, for both samples.
8 Add these values together for the two samples and take the square root of this.
9 Divide the difference in the two sample means with the value from step 8. This is *t* and, in this case, is 1.93.
10 Calculate the total degrees of freedom for all the data (v).
$v = (n_1 - 1) + (n_2 - 1) = 28$
11 Refer to the table of *t* values for 28 degrees of freedom and a value of $t = 1.93$ (Table P2.2).

Degrees of freedom	**Value of *t***			
1	6.31	12.7	63.7	63.6
2	2.92	4.30	9.93	31.6
3	2.35	3.18	5.84	12.9
4	2.13	2.78	4.60	8.61
5	2.02	2.57	4.03	6.87
6	1.94	2.45	3.71	5.96
7	1.90	2.37	3.50	5.41
8	1.86	2.31	3.36	5.04
9	1.83	2.26	3.25	4.78
10	1.81	2.23	3.17	4.59
11	1.80	2.20	3.11	4.44
12	1.78	2.18	3.06	4.32
13	1.77	2.16	3.01	4.22
14	1.76	2.15	2.98	4.14
15	1.75	2.13	2.95	4.07
16	1.75	2.12	2.92	4.02
17	1.74	2.11	2.90	3.97
18	1.73	2.10	2.88	3.92
19	1.73	2.09	2.86	3.88
20	1.73	2.09	2.85	3.85
22	1.72	2.07	2.82	3.79
24	1.71	2.06	2.80	3.75
26	1.71	2.06	2.78	3.71
28	1.70	2.05	2.76	3.67
30	1.70	2.04	2.75	3.65
>30	1.64	1.96	2.58	3.29
Probability that chance could have produced this value of *t*	0.10	0.05	0.01	0.001
Confidence level	10%	5%	1%	0.1%

Table P2.2 Values of *t*.

Using the table of probabilities in the *t*-test

In statistical tests that compare samples, it is the convention to start off by making the assumption that there is no significant difference between the samples. You assume that they are just two samples from an identical population. This is called the null hypothesis. In this case, the null hypothesis would be:

There is no difference between the corolla length in population A and population B.

The probabilities that we look up in the t-test table are **probabilities that the null hypothesis is correct** and there is no significant difference in the samples. The probability you find is the probability that any difference between the samples is just due to chance.

You may remember that this idea is also used in the χ^2 test (pages 386–387). In both the χ^2 test and the t-test, we take a probability of 0.05 as being the critical one. This is sometimes called the **5% confidence level**.

If our t-test value represents a probability of 0.05 or more, then we assume that the differences between the two sets of data are due only to chance. The differences between them are not significant.

If the probability is less than 0.05, then this is strong evidence that something more than chance is causing the difference between the two sets of data. We can say that the difference is significant.

If the total number of observations (both samples added together) is below 30, error due to chance is significant and the table of t makes an adjustment to critical values to take this into account, which is why you need to calculate the value of degrees of freedom. However, above 30 observations, the number of observations makes little or no difference to critical values of t.

For our data sets of corolla length, the value of t is 1.93. This is below the critical value of t at the 5% confidence level for 28 degrees of freedom, which is 2.05. The chance of getting this difference in the means of the two sets of data through random error is greater than 5%. So, we can say that the two means are not significantly different.

The χ^2 test

You have already met the χ^2 test, on pages 386–387. This statistical test is used to determine whether any differences between a set of observed values are significantly different from the expected values. It is most likely to be used in genetics or ecology.

Pearson's linear correlation

Sometimes, you want to see if there is a relationship between two variables – are they correlated? For example, perhaps we have measured the numbers of species **P** and species **Q** in ten different 1 m^2 quadrats. Table P2.3 shows what we found. Are the numbers of these two species related to each other?

Quadrat	Number of individuals of species P	Number of individuals of species Q
1	10	21
2	9	20
3	11	22
4	7	17
5	8	16
6	14	23
7	10	20
8	12	24
9	12	22
10	9	19

Table P2.3 Numbers of species **P** and species **Q** in ten quadrats.

The first thing to do is to plot a scatter graph of your data. This is shown in Figure P2.8. Note that it does not matter which set of values goes on the x-axis or the y-axis.

Looking at this graph, we can see that the data look as though they might lie approximately on a straight line. It looks as though there might be a **linear correlation** between them.

To find out whether this is so, we can carry out Pearson's linear correlation test. This test can only be used if the data are interval data and are normally distributed, which is the case for these data. We can see that because they do not appear to be skewed in any one direction, and there are no obvious outliers.

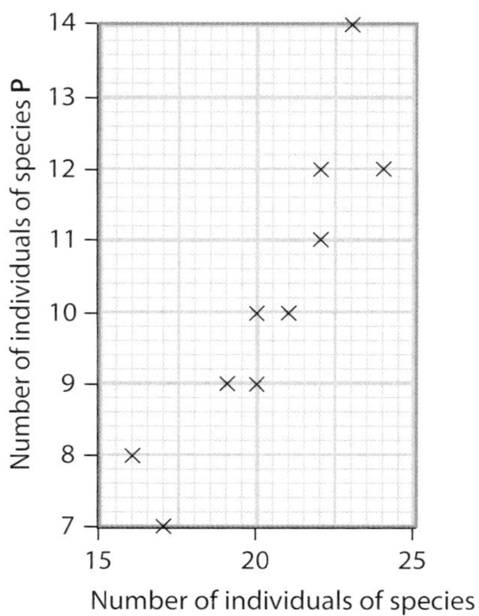

Figure P2.8 Scatter graph of the data in Table P2.3.

The formula for this test is:

$$r = \frac{\sum xy - n\bar{x}\bar{y}}{ns_x s_y}$$

where:

r is the correlation coefficient

x is the number of species **P** in a quadrat

y is the number of species **Q** in the same quadrat

n is the number of readings (in this case, 10)

$\bar{x}$ is the mean number of species **P**

$\bar{y}$ is the mean number of species **Q**

s_x is the standard deviation for the numbers of **P**

s_y is the standard deviation for the numbers of **Q**

Table P2.4 shows you the steps to follow in order to calculate the value of r, using the data in Table P2.3.

The value of r, 0.81, is the correlation coefficient. This value should always work out somewhere between –1 and +1. (If it doesn't check your calculation!)

A value of +1 means total positive correlation between your two sets of figures.

A value of –1 means total negative correlation between your two sets of figures.

A value of 0 means there is no correlation.

Here, we have a value of r that lies quite close to 1. We can say that there is a positive, linear correlation between the numbers of species **P** and the numbers of species **Q**.

1 Calculate $x \times y$ for each set of values.

2 Calculate the means for each set of figures, $\bar{x}$ and $\bar{y}$.

3 Calculate $n\bar{x}\bar{y}$. Here, $n = 10$, $\bar{x} = 10.2$ and $\bar{y} = 20.4$, so $n\bar{x}\bar{y} = 10 \times 10.2 \times 20.4$

4 Add up all the values of xy, to find $\sum xy$.

5 Now calculate the standard deviation, s, for each set of figures. The method for doing this is shown in Table P2.1 on page 498.

6 Now substitute your numbers into the formula and calculate r.

Quadrat	**Number of species P, x**	**Number of species Q, y**	***xy***
1	10	21	210
2	9	20	180
3	11	22	242
4	7	17	119
5	8	16	128
6	14	23	322
7	10	20	200
8	12	24	288
9	12	22	264
10	9	19	171
mean	$\bar{x} = 10.2$	$\bar{y} = 20.4$	$\sum xy = 2124$
$n\bar{x}\bar{y}$	$10 \times 10.2 \times 20.4 = 2080.8$		
standard deviation	$s_x = 2.10$	$s_y = 2.55$	

$$r = \frac{\sum xy - n\bar{x}\bar{y}}{ns_x s_y}$$

$$= \frac{2124 - (10 \times 10.2 \times 20.4)}{10 \times 2.10 \times 2.55}$$

$$= \frac{2124 - 2080.8}{53.55}$$

$$= \frac{43.2}{53.55}$$

$$= 0.81$$

Table P2.4 Calculating Pearson's linear correlation for the data in Table P2.3.

Spearman's rank correlation

Spearman's rank correleation is used to find out if there is a correlation between two sets of variables, when they are **not** normally distributed.

As with Pearson's linear correlation test, the first thing to do is to plot your data as a scatter graph, and see if they look as though there may be a correlation. Note that, for this test, the correlation need not be a straight line – the correlation need not be linear.

Let's say that you have counted the numbers of species **R** and species **S** in 10 quadrats. Table P2.5 shows your results, and Figure P2.9 shows these data plotted as a scatter graph.

Quadrat	Number of species R	Number of species S
1	38	24
2	2	5
3	22	8
4	50	31
5	28	27
6	8	4
7	42	36
8	13	6
9	20	11
10	43	30

Table P2.5 Numbers of species **R** and species **S** found in 10 quadrats.

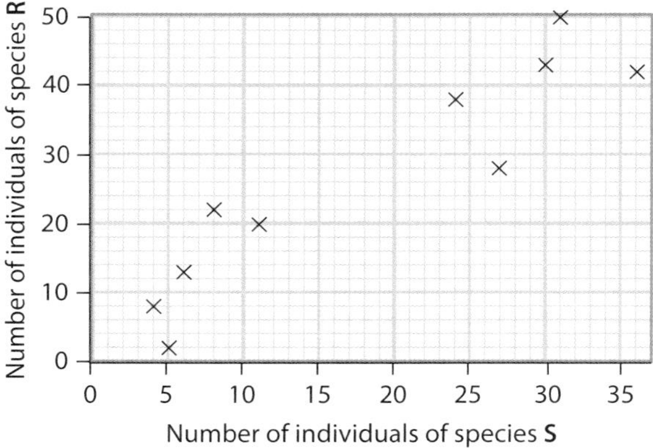

Figure P2.9 Scatter graph of the data in Table P2.5.

Now rank each set of data. For example, for the number of species **R**, Quadrat **4** has the largest number, so that is ranked as number 1. This is shown in Table P2.6.

Quadrat	Number of species R	Rank for species R	Number of species S	Rank for species S
1	38	7	24	6
2	2	1	5	2
3	22	5	8	4
4	50	10	31	9
5	28	6	27	7
6	8	2	4	1
7	42	8	36	10
8	13	3	6	3
9	20	4	11	5
10	43	9	30	8

Table P2.6 Ranked data from Table 2.5.

Once you have ranked both sets of results, you need to calculate the differences in rank, *D*, by subtracting the rank of species **S** from the rank of species **R**. Then square each of these values. Add them together to find $\sum D^2$. This is shown in Table P2.7.

Quadrat	Rank for species R	Rank for species S	Difference in rank, *D*	D^2
1	7	6	1	1
2	1	2	–1	1
3	5	4	1	1
4	10	9	1	1
5	6	7	–1	1
6	2	1	1	1
7	8	10	–2	4
8	3	3	0	0
9	4	5	–1	1
10	9	8	1	1
				$\sum D^2 = 12$

Table P2.7 Calculating $\sum D^2$ for the data in Table P2.5.

The formula for calculating Spearman's rank correlation coefficient is:

$$r_s = 1 - \left(\frac{6 \times \sum D^2}{n^3 - n}\right)$$

where:

r_s is Spearman's rank coefficient

$\sum D^2$ is the sum of the differences between the ranks of the two samples

n is the number of samples

Substituting into this formula, we can calculate Spearman's rank correlation coefficient for the distribution of species **R** and species **S**.

$$r_s = 1 - \left(\frac{(6 \times 12)}{(10^3 - 10)}\right)$$
$$= \frac{1 - 72}{1000 - 10}$$
$$= 1 - \frac{72}{990}$$
$$= 1 - 0.072$$
$$= 0.928$$
$$= 0.93 \text{ (to 2 decimal places)}$$

This number is the correlation coefficient. The closer the value is to 1, the more likely it is that there is a genuine correlation between the two sets of data.

Our value is very close to 1, so it certainly looks as though there is strong correlation. However, as with the t-test and the χ^2 test, we need to look up this value in a table, and compare it against a critical value. As in most statistical tests used in biology, we use a probability of 0.05 as our baseline; if our value indicates a probability of 0.05 or less, then we can say that there is a significant correlation between our two samples. Another way of saying this is that the null hypothesis, that there is no correlation between the two samples, is not supported.

Table P2.8 shows these critical values for samples with different numbers of readings.

Notice that, the smaller the number of the readings we have taken, the larger our value of r_s needs to be in order to say that there is a significant correlation. This makes logical sense – if we have taken only 5 readings, then we would really need them all to be ranked identically to be able to say they are correlated. If we have taken 16, then we can accept a smaller value of r_s.

Remember that showing there is a correlation between two variables does not indicate a causal relationship – in other words, we can't say that the numbers of species **R** have an effect on the numbers of species **S**, or vice versa. There could well be other variables that are causing both of their numbers to vary.

For our data, we have 10 quadrats, so $n = 10$. The critical value is therefore 0.65. Our value is much greater than this, so we can accept that there is a significant correlation between the numbers of species **R** and the numbers of species **S**.

Table P2.9 summarises the circumstances in which you would use each of the statistical tests that help you to decide whether or not there is a relationship between two sets of data.

Evaluating evidence

It is important to be able to assess how much trust you can have in the data that you have collected in an experiment, and therefore how much confidence you can have in any conclusions you have drawn. Statistical tests are a big help in deciding this. But you also need to think about the experiment itself and any sources of error that might have affected your results.

Sources of error were discussed in Chapter P1, on page 263. You will remember that sources of error stem from two main sources – limitations in the apparatus and measuring instruments, and difficulty in controlling key variables.

In a written practical examination at A level, you will often be analysing data that you have not collected yourself, so you will have only the information provided in the question and your own experience of carrying out experiments to help you to decide how much confidence you can have in the reliability of the data. Important things to think about include the following questions.

- How well were the key variables controlled?
- Was the range and interval of the independent variable adequate?
- Do any of the results appear to be anomalous? If so, what could have caused these anomalous readings?
- Have the provided readings been replicated sufficiently?

Another aspect of the experiment that needs to be considered is its validity. A valid experiment really does test the hypothesis or question that is being investigated.

n	**5**	**6**	**7**	**8**	**9**	**10**	**11**	**12**	**14**	**16**
Critical value of r_s	1.00	0.89	0.79	0.76	0.68	0.65	0.60	0.54	0.51	0.51

Table P2.8 Critical values of r_s at the 0.05 probability level.

A key feature to consider is whether you really were measuring the dependent variable that you intended to measure. For example, if you were doing a transpiration experiment using a potometer, were you really measuring the rate of loss of water vapour from the shoot's leaves? The measurement you were making was actually the rate of uptake of water by the shoot from the potometer. You should be able to explain why you think (or do not think!) that this measurement can be relied upon to give you valid data about the rate of transpiration.

It is important to be able to bring all of this information together, and to be able to make an informed judgement about the overall validity of the investigation and how much it can be trusted for testing the hypothesis. As at AS level, you should be able to suggest improvements to the experiment that would increase its reliability.

Statistical test	When to use it	Criteria for using the test	Examples of use	How to interpret the value you calculate
t-test	You want to know if two sets of continuous data are significantly different from one another.	• You have two sets of continuous, quantitative data (page 494). • You have more than 10 but less than 30 readings for each set of data. • Both sets of data come from populations that have normal distributions. • The standard deviations for the two sets of data are very similar.	Are the surface areas of the leaves on the north-facing side of a tree significantly different from the surface areas on the south-facing side? Are the reaction times of students who have drunk a caffeine-containing drink significantly different from students who have drunk water?	Use a *t*-test table to look up your value of *t*. If this value is greater than the *t* value for a probability of 0.05 (the critical value), then you can say that your two populations are significantly different.
χ^2 test	You want to know if your observed results differ significantly from your expected results.	• You have two or more sets of quantitative data, which belong to two or more discontinuous categories (i.e. they are nominal data – page 494)	Are the numbers of offspring of different phenotypes obtained in a genetic cross significantly different from the expected numbers?	Use a χ^2 table to look up your value of χ^2. If this value is greater than the χ^2 value for a probability of 0.05, then you can say that your observed results differ significantly from your expected results.
Pearson's linear correlation	You want to know if there is a linear correlation between two paired sets of data.	• You have two sets of interval data. • You have at least 5 pairs of data, but preferably 10 or more. • A scatter graph suggests there might be a linear relationship between them. • Both sets of data have an approximately normal distribution.	Is there a linear correlation between the rate of an enzymic reaction and the concentration of an inhibitor? Is there a linear correlation between the numbers of limpets and the numbers of dog whelks on a sea shore?	A value close to +1 indicates a positive linear correlation. A value close to –1 indicates a negative linear correlation. A value close to 0 indicates no correlation.
Spearman's rank correlation	You want to know if there is a correlation (not necessarily linear) between two paired sets of data.	• You have quantitative data that can be ranked. • The samples for each set of data have been made randomly. • You have at least 5 pairs of data, but preferably between 10 and 30. • A scatter graph suggests there might be a relationship between the two sets of data (not necessarily linear). Note: you cannot use this test if the scatter graph is U-shaped, i.e. the correlation is positive for some values and negative for others.	Is there correlation between the surface area of a fruit and the time it takes to fall to the ground? Is there a correlation between the numbers of limpets and the numbers of dog whelks on a sea shore?	Use a correlation coefficient table to look up your value of r_s. If your value of r_s is greater than the r_s value for a probability of 0.05, you can say there is a significant correlation between your two values.

Table P2.9 Summary of four statistical tests.

Conclusions and discussion

The construction of a simple conclusion was described in Chapter P1 on page 259. In a written practical examination at A level, you will often be making conclusions from data that you have not collected yourself, which is a bit more difficult. The more experience you have of doing real practical work, the better equipped you will be to understand how to make conclusions from data provided to you.

Your conclusion should begin with a simple statement about whether or not the hypothesis that was being tested is supported. This would also be the point at which you could mention the results of any statistical tests, and how they have helped you to make your conclusion.

You may also be asked to discuss the data and your conclusion in more depth. You should be prepared to give a description of the data, pointing out key features. This might involve looking for trends or patterns in the data, and identifying points on a graph where there is a marked change in gradient (Chapter P1, page 260).

You could be asked to use the data to make further predictions, perhaps suggesting another hypothesis that could be tested. For example, for the petal length investigation (page 497), we could start to think about why the petal length in the woodland is greater than in the garden. A new hypothesis could be: petals grow longer in lower light intensity.

As well as describing the data, you could be asked to use your scientific knowledge to attempt to explain them. It is important to remember that the data in an A level question could relate to anything from either the first or second year of your course, so you need to revise all of your work from both years in preparation for these examination papers.

Summary

- See also the summary for Chapter P1.
- A hypothesis about the relationship between two variables predicts how one variable affects the other. It should be testable and falsifiable by experiment.
- To make up a 1% (mass/volume) solution, dissolve 1 g of the solute in a small amount of water, then make up to a total volume of 1 dm^3. To make up a 1 mol dm^{-3} solution, dissolve 1 mole of the solute in a small amount of water, then make up to a total volume of 1 dm^3.
- The mean of a set of data is calculated by adding up all the individual values and dividing by the total number of readings. The median is the middle value in the set of results. The mode is the most common value in the set of results. The interquartile range is the range into which the middle 50% of the data fall. Standard deviation is a measure of how much the data are spread on either side of the mean.
- Standard error is a measure of the likelihood of the mean of your sample being the true mean of the whole population. There is a 95% probability that the true mean lies within ±2 standard errors of the mean you have calculated. This can be shown by drawing error bars on a bar chart, where the error bar extends 2 standard errors above and below the plotted value.
- If the error bars for two sets of data overlap, then there is no significant difference between the two sets of data. If the error bars do not overlap, it is possible that there is a significant difference between them, but this is not necessarily so.
- The *t*-test is used to determine whether or not two sets of quantitative data, each with an approximately normal distribution, are significantly different from one another.
- The χ^2 test is used to determine whether or not observed results differ significantly from expected results.
- The Pearson linear correlation test is used to determine whether or nor there is a linear correlation between two sets of quantitative data.
- Spearman's rank correlation test is used to determine whether or nor there is a correlation between two paired sets of data that can be ranked.
- When discussing an experiment you need also to consider possible sources of error and validity of the experiment, and be able to suggest ways in which the experiment's reliability could be improved. You should be able to reach a conclusion about the data and use of statistical tests, and perhaps make suggestions for further experimental work.

End-of-chapter questions

1 a Calculate the standard deviation, s, for the fruit mass data on page 495. Use the formula:

$$s = \sqrt{\frac{\Sigma(x - \bar{x})^2}{n - 1}}$$

Show each step in your working. [4]

b Calculate the standard error, S_M, for this set of data. Use the formula:

$$S_M = \frac{s}{\sqrt{n}}$$

Show each step in your working. [2]

c A sample of lemon fruits was taken from a different population of lemon trees. The mean mass of this sample of fruits was 84 g. The standard error was 0.52.

Draw a bar chart to show the means of the fruit masses for these two populations.

Draw error bars on your bar chart, and add a key to explain what the error bars represent. [5]

d Are the differences between the masses of the fruits in the two different populations significant? Explain your answer. [3]

[Total: 14]

2 Using the data on page 497, carry out a t-test to determine whether the difference between petal lengths in woodland and in a garden is significant. (The standard errors for these two data sets have already been calculated, and are given on page 499.)

Show all of your working, and explain your conclusions fully. [10]

3 The standard deviations for the data in Table P2.4 have been calculated for you. Do these calculations for yourself, showing your working fully. [5]

4 The scatter graphs show values for x plotted against values for y.

A

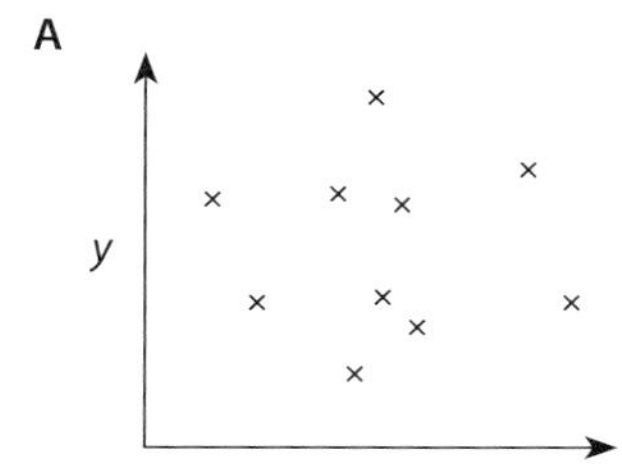

B

y

x

C

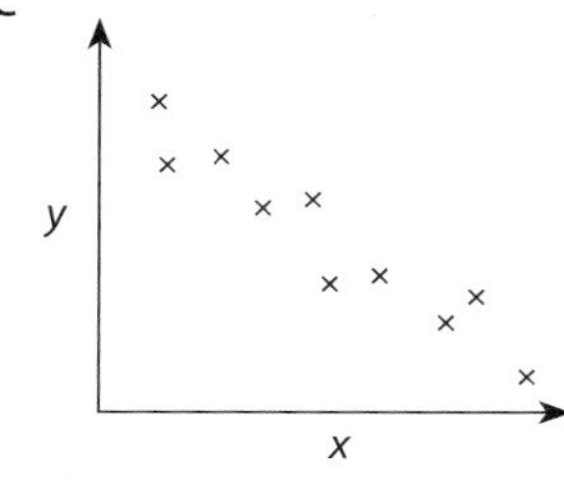

D

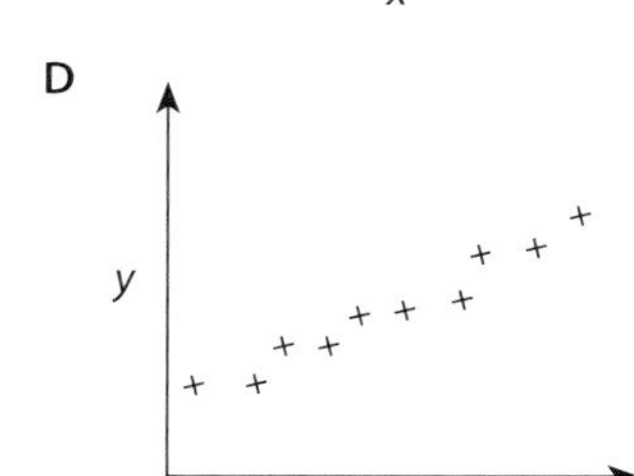

E

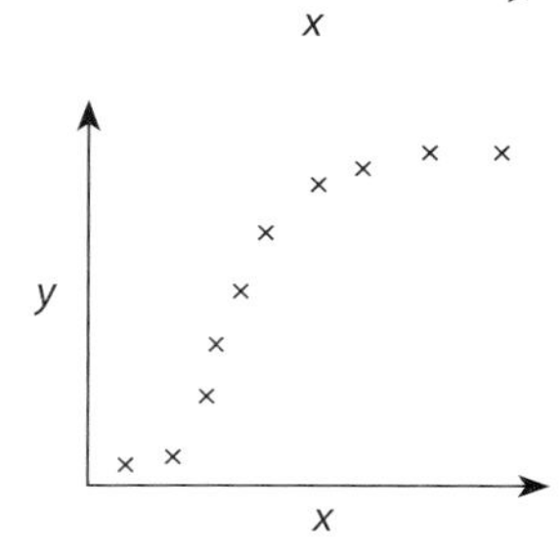

State the letter of any graph or graphs that:

a indicate there could be a positive correlation [1]

b indicate that there is probably no correlation [1]

c indicate that there is a linear correlation [1]

d indicate that Spearman's rank correlation could be used [1]

e indicate that Pearson's linear correlation test could be used [1]

[Total: 5]

5 A student measured the length of 8 randomly selected bean pods and the total mass of the seeds inside them. The table shows her results.

Pod	Length / mm	Mass of seeds / g
1	134	35
2	71	18
3	121	30
4	83	21
5	99	23
6	107	29
7	82	17
8	119	34

a Draw a scatter graph of these results. [3]

b Describe the relationship that is suggested by your scatter graph. [1]

c Use Pearson's linear correlation test to analyse these results. Show all of your working, and explain what your calculated value of *r* suggests about the relationship between the length of a bean pod and the mass of the seeds. [10]

[Total: 14]

6 The table shows the numbers of two species, **F** and **G**, found in in eight randomly placed 10 × 10 m quadrats.

Quadrat	Number of species F	Number of species G
1	34	8
2	8	22
3	47	8
4	19	21
5	6	38
6	41	3
7	22	15
8	38	10

a Draw a scatter graph of these results. [3]

b Describe the relationship that is suggested by your scatter graph. [1]

c Explain why it is better to use Spearman's rank correlation, rather than Pearson's linear correlation test, to find out if there is a correlation between the numbers of species **F** and species **G**. [1]

d Use Spearman's rank correlation to analyse these results. (Note: if two values are equal, then you must give them equal rank.) Show all of your working, and explain what your calculated value indicates. [8]

[Total: 13]

7 A student noticed that the leaves on a plant growing close to a wall had two sorts of leaves. The leaves next to the wall were in the shade and looked different from the leaves on the side away from the wall that were exposed to the sun. The length of the internodes on the stem also looked different.

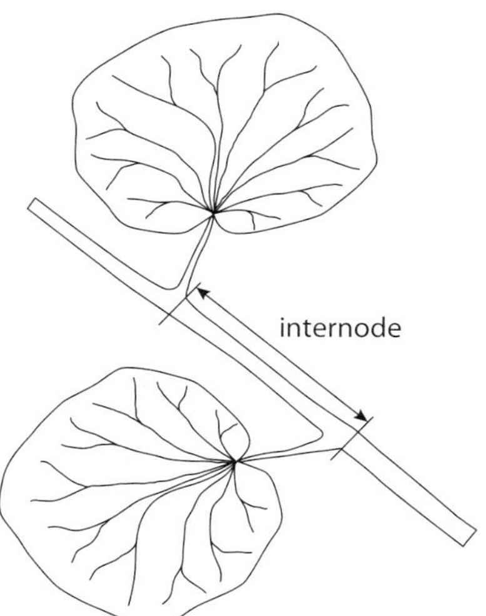

The student decided to investigate the differences by measuring some features of 30 leaves and internodes from each side of the plant.

The figure on the left shows the leaf shape. The figure on the right shows an internode.

	Shaded leaves	Exposed leaves
mean internode length / mm	23±4	15±3
mean surface area of leaves / mm²	2750±12	1800±15
mean mass of leaves / mg	50±8	60±10
mean leaf surface area : leaf mass ratio	55±9	30±6
rate of water loss / mg mm⁻² h⁻¹	50±11	65±12

a i State the independent variable being investigated. [1]

ii Outline the procedures the student could use to obtain these results. [8]

The student carried out *t*-tests for the leaf surface area : leaf mass ratio and for internode length.

The leaf surface area : leaf mass ratio gave the value ***t* = 12.6**.

The formula for the *t*-test is:

$$t = \frac{|\bar{x}_1 - \bar{x}_2|}{\sqrt{\left(\frac{s_1^2}{n_1} + \frac{s_2^2}{n_2}\right)}}$$

b i Copy and complete the calculation to find the value of *t* for the internode length. [3]

$$t = \frac{\quad - \quad}{\sqrt{\frac{4^2}{30} + \frac{\quad}{\quad}}}$$

$$t = \frac{\quad}{0.9}$$

$$t =$$

The table shows the critical values at $p = <0.05$ for the *t*-test.

Degrees of freedom	18	20	21	22	23	24	25	26	27	28	29	30	40	60	∞
Critical value	2.10	2.09	2.08	2.07	2.06	2.06	2.06	2.06	2.05	2.05	2.04	2.04	2.02	2.00	1.96

The number of degrees of freedom is 58.

ii State how the number of degrees of freedom was calculated. [1]

iii State and explain the meaning of these results. [2]

[Total: 15]

Cambridge International AS and A Level Biology 9700/51, Question 1, November 2010

8 The figure shows one type of potometer used by a student to investigate transpiration.

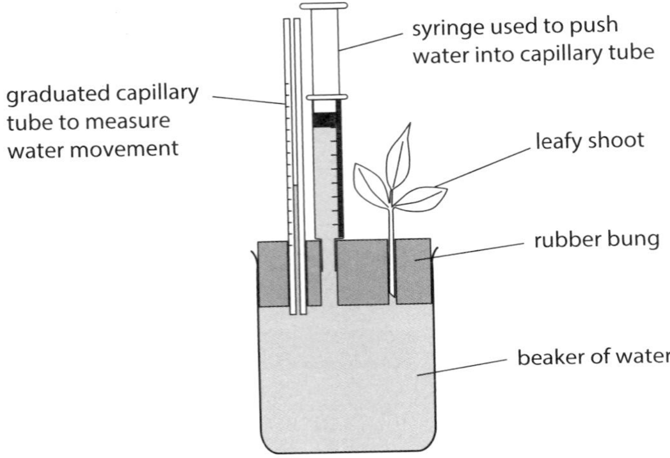

a i Suggest a hypothesis the student could test about the transpiration of a mesophyte (a plant adapted to a moist environment) and a xerophyte (a plant adapted to a dry environment). [1]

ii Outline a procedure involving the potometer that the student could use to test this hypothesis. [8]

iii The capillary tube measures the distance moved by the water. Explain how the actual volume of water lost can be calculated. [2]

b Sketch a graph to predict the expected results of the investigation. [2]

c i The student then measured the surface area of the leaves by tracing the outline on a grid and counting the number of squares covered by the leaves. This area was doubled.

Mesophyte:

surface area of leaves $= 36\,cm^2$

water loss in 30 minutes $= 0.018\,cm^3$

Calculate the rate of water loss in $cm^3\,m^{-2}\,min^{-1}$. Show all the steps in your calculation. [3]

ii State a statistical test that the student could use to find out if the difference in water loss between the two types of leaf is significant. State a reason for your choice. [2]

d In a further investigation, the student measured the loss in mass of each type of leaf.
The figure shows the experimental set-up.
The table shows the results of this investigation.

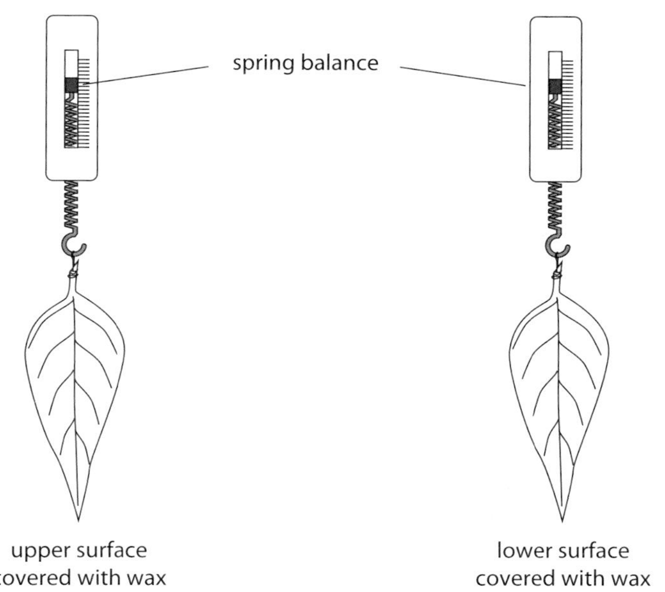

Day	Loss in mass / g per day			
	Upper side covered		Lower side covered	
	Mesophyte	Xerophyte	Mesophyte	Xerophyte
1	4.25	0.55	1.15	0.05
2	3.20	0.35	1.00	0.05
3	1.55	0.20	0.75	0.00
4	0.50	0.10	0.95	0.05
5	0.05	0.04	1.00	0.00
Total loss in mass / g	9.55	1.24	4.85	0.15

State **three** conclusions that can be drawn from these results. [3]

[Total: 21]

Cambridge International AS and A Level Biology 9700/52, Question 1, June 2010

Appendix 1: Amino acid R groups

The general formula for an amino acid is shown in Figure 2.16a (page 40). In the list below, only the R groups are shown; the rest of the amino acid molecule is represented by a block.

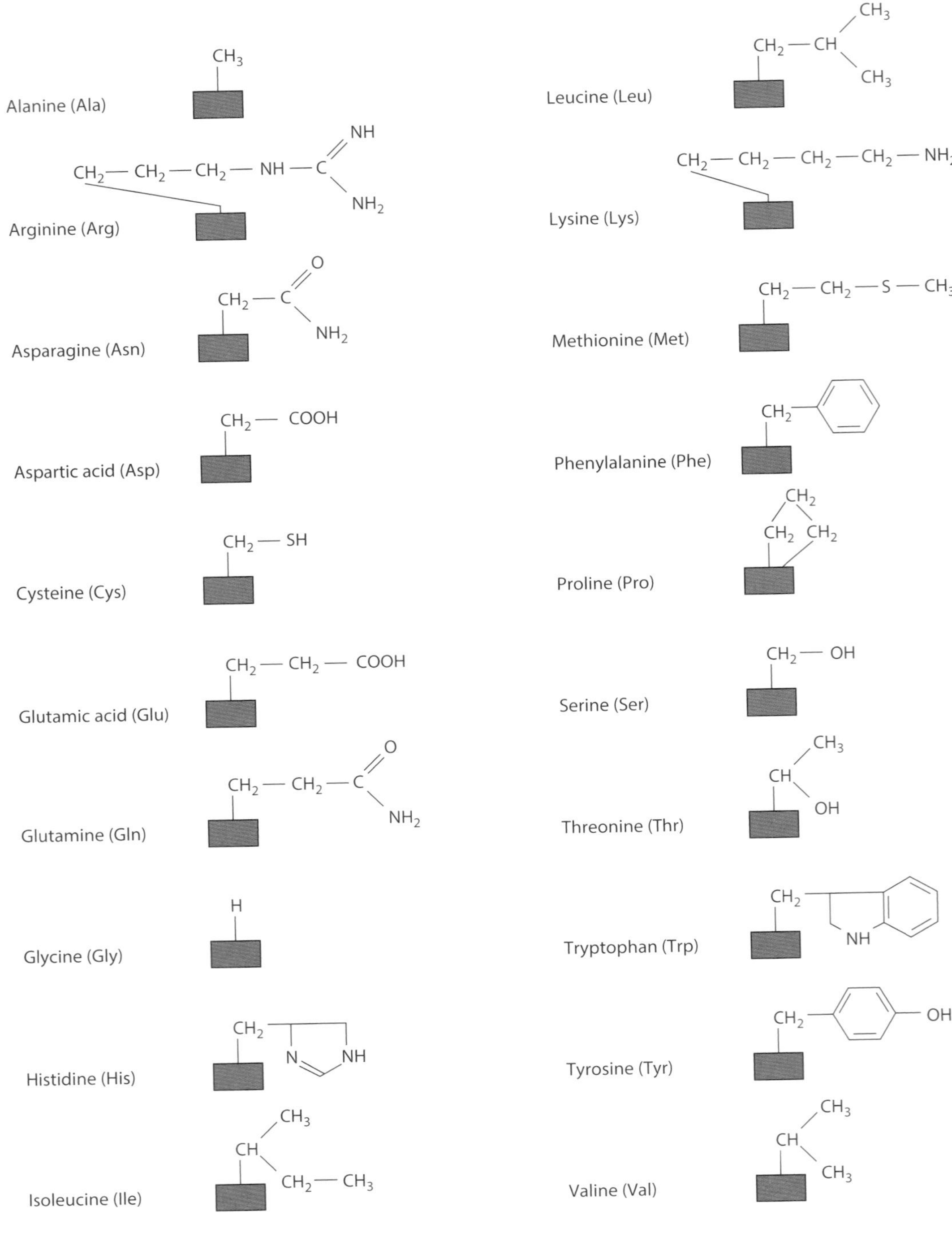

Appendix 2: DNA and RNA triplet codes

The table below shows all the possible triplets of bases in a DNA molecule and what each codes for. The three-letter abbreviation for each amino acid is, in most cases, the first three letters of its full name (Appendix 1).

First position	Second position				Third position
	A	**G**	**T**	**C**	
A	Phe	Ser	Tyr	Cys	A
	Phe	Ser	Tyr	Cys	G
	Leu	Ser	STOP	STOP	T
	Leu	Ser	STOP	Trp	C
G	Leu	Pro	His	Arg	A
	Leu	Pro	His	Arg	G
	Leu	Pro	Gln	Arg	T
	Leu	Pro	Gln	Arg	C
T	Ile	Thr	Asn	Ser	A
	Ile	Thr	Asn	Ser	G
	Ile	Thr	Lys	Arg	T
	Met	Thr	Lys	Arg	C
C	Val	Ala	Asp	Gly	A
	Val	Ala	Asp	Gly	G
	Val	Ala	Glu	Gly	T
	Val	Ala	Glu	Gly	C

The table below shows all the possible triplets of bases in a mRNA molecule and what each codes for.

First position	Second position				Third position
	U	**C**	**A**	**G**	
U	Phe	Ser	Tyr	Cys	U
	Phe	Ser	Tyr	Cys	C
	Leu	Ser	STOP	STOP	A
	Leu	Ser	STOP	Trp	G
C	Leu	Pro	His	Arg	U
	Leu	Pro	His	Arg	C
	Leu	Pro	Gln	Arg	A
	Leu	Pro	Gln	Arg	G
A	Ile	Thr	Asn	Ser	U
	Ile	Thr	Asn	Ser	C
	Ile	Thr	Lys	Arg	A
	Met	Thr	Lys	Arg	G
G	Val	Ala	Asp	Gly	U
	Val	Ala	Asp	Gly	C
	Val	Ala	Glu	Gly	A
	Val	Ala	Glu	Gly	G

Glossary

α cell a cell in the islets of Langerhans in the pancreas that senses when blood glucose levels are low and secretes glucagon in response

α-helix a helical structure formed by a polypeptide chain, held in place by hydrogen bonds; an α-helix is an example of secondary structure in a protein

abiotic factor a physical characteristic of a habitat, such as temperature, light intensity and soil pH

abscisic acid (ABA) an inhibitory plant growth regulator that causes closure of stomata in dry conditions

absorption spectrum a graph of the absorbance of different wavelengths of light by a compound such as a photosynthetic pigment

accessory pigment a pigment that is not essential to photosynthesis but which absorbs light of different wavelengths and passes the energy to chlorophyll *a*

acetylcholine (ACh) a transmitter substance found, for example, in the presynaptic neurone at neuromuscular junctions

acetylcholinesterase an enzyme that rapidly breaks down acetylcholine at synapses

actin the protein that makes up the thin filaments in striated muscle

action potential a brief change in the potential difference across cell surface membranes of neurones and muscle cells caused by the inward movement of sodium ions followed by the outward movement of potassium ions; it rapidly travels along the length of the neurone

action spectrum a graph showing the effect of different wavelengths of light on a process, e.g. on the rate of photosynthesis

activation energy the energy that must be provided to make a reaction take place; enzymes reduce the activation energy required for a substrate to change into a product

active immunity immunity gained when an antigen enters the body, an immune response occurs and antibodies are produced by plasma cells

active site an area on an enzyme molecule where the substrate can bind

active transport the movement of molecules or ions through transport proteins across a cell membrane, against their concentration gradient, using energy from ATP

adenine (A) nitrogen-containing purine base found in DNA and RNA

ADP adenosine diphosphate

adrenaline a hormone secreted by the adrenal glands in times of stress or excitement

aerenchyma plant tissue containing air spaces

afferent leading towards; e.g. the afferent blood vessel leads towards a glomerulus

agarose gel a jelly made from agar, used in electrophoresis

alcoholic fermentation anaerobic respiration in which glucose is converted to ethanol

aleurone layer a layer of tissue around the endosperm in a cereal seed that synthesises amylase during germination

alien species a species that has been moved into a new ecosystem where it was previously unknown; also known as invasive species

allele a particular variety of a gene

allopatric speciation speciation that takes place as a result of two populations living in different places and having no contact with each other

allopolyploid possessing more than two sets of chromosomes, where the chromosomes come from two different species

all-or-nothing law a law that states neurones and muscle cells only transmit impulses if the initial stimulus is sufficient to increase the membrane potential above a threshold potential

amniocentesis taking a small sample of amniotic fluid during the 15th or 16th week of pregnancy, so that the genotype of the fetus can be determined

amylopectin a polymer of α-glucose monomers linked by both 1, 4 and 1, 6 linkages, forming a branched chain; amylopectin is a constituent of starch

amylose a polymer of α-glucose monomers linked by 1, 4 linkages, forming a curving chain; amylose is a constituent of starch

Animalia one of the four kingdoms of the domain Eukarya; eukaryotic organisms which are multicellular and heterotrophic and have a nervous system

antibiotic a substance produced by a living organism that is capable of killing or inhibiting the growth of a microorganism

antibody a glycoprotein (immunoglobulin) made by plasma cells derived from B-lymphocytes, secreted in response to an antigen; the variable region of the antibody molecule is complementary in shape to its specific antigen

anticodon sequence of three unpaired bases on a tRNA molecule that binds with a codon on mRNA

antidiuretic hormone (ADH) a hormone secreted from the pituitary gland that increases water reabsorption in the kidneys and therefore reduces water loss in urine

antigen a substance that is foreign to the body and stimulates an immune response

apoplastic pathway the non-living system of interconnected cell walls extending throughout a plant, used as a transport pathway for the movement of water and mineral ions

Archaea a domain of prokaryotic organisms that resemble bacteria but share some features with eukaryotes

arteriole small blood vessel that carries blood from arteries to capillaries; arterioles determine how much blood flows through capillaries

artery a blood vessel that carries blood away from the heart; it has a relatively thick wall and contains large amounts of elastic fibres

artificial immunity immunity gained either by vaccination (active) or by injecting antibodies (passive)

artificial insemination (AI) injection of semen collected from a male into the uterus

artificial selection the selection by humans of organisms with desired traits

ATP adenosine triphosphate – the universal energy currency of cells

ATP synthase the enzyme catalysing the phosphorylation of ADP to ATP

atrioventricular node a patch of tissue in the septum of the heart, through which the wave of electrical excitation is passed from the atria to the Purkyne tissue

autoimmune disease a type of disease in which there is a mistaken immune response to a self antigen leading to the production of antibodies and the destruction of body tissues; myasthenia gravis is an example

autopolyploid possessing more than two sets of chromosomes, where all the sets are from the same species

autosomes all the chromosomes except the X and Y (sex) chromosomes

autotroph an organism that can trap an inorganic carbon source (carbon dioxide) using energy from light or from chemicals

auxin a plant growth regulator (plant hormone) that stimulates cell elongation

axon a long cytoplasmic process of a neurone

β cell a cell in the islets of Langerhans in the pancreas that senses when blood glucose levels are high and secretes insulin in response

β-galactosidase an enzyme which catalyses the hydrolysis of lactose to glucose and galactose

β-lactamase an enzyme made by antibiotic-resistant bacteria that breaks down penicillin molecules

β-pleated sheet a loose, sheet-like structure formed by hydrogen bonding between parallel polypeptide chains; a β-pleated sheet is an example of secondary structure in a protein

B-lymphocyte a type of lymphocyte that gives rise to plasma cells and secretes antibodies

Bacteria the domain that contains all prokaryotic organisms except those classified as Archaea

Benedict's test a test for the presence of reducing sugars; the unknown substance is heated with Benedict's reagent, and a change from a clear blue solution to the production of a yellow, red or brown precipitate indicates the presence of reducing sugars such as glucose

biodiversity the variety of ecosystems and species in an area and the genetic diversity within species

bioinformatics the collection, processing and analysis of biological information and data using computer software

biosensor a device that uses a biological material such as an enzyme to measure the concentration of a chemical compound

biotic factor a factor which affects a population or an environment, that is caused by living organisms; examples include competition, predation and parasitism

biuret test a test for the presence of amine groups and thus for the presence of proteins; biuret reagent is added to the unknown substance, and a change from pale blue to purple indicates the presence of proteins

birth control pill a pill that contains oestrogen and/or progesterone to prevent ovulation

Bohr effect the decrease in affinity of haemoglobin for oxygen that occurs when carbon dioxide is present

Bowman's capsule the cup-shaped part of a nephron that surrounds a glomerulus and collects filtrate from the blood

bronchitis a disease in which the airways in the lungs become inflamed and congested with mucus; chronic bronchitis is often associated with smoking

Bt toxin insecticidal toxin produced by the bacterium *Bacillus thuringiensis*

bundle sheath cells a ring of cells in the leaves of C4 plants in which the Calvin cycle takes place out of contact with the air spaces in the leaf

C3 plant a plant in which the first product of photosynthesis is a three-carbon compound

C4 plant a plant in which the first product of photosynthesis is a four-carbon compound; most tropical grasses (but not rice) are C4 plants

calorimeter the apparatus in which the energy value of a compound can be measured by burning it in oxygen

Calvin cycle a cycle of reactions in photosynthesis in which carbon dioxide is fixed into carbohydrate

cancer a disease, often but not always treatable, that results from a breakdown in the usual control mechanisms that regulate cell division; certain cells divide uncontrollably and form tumours, from which cells may break away and form secondary tumours in other areas of the body (metastasis)

capillary the smallest type of blood vessel, whose function is to facilitate exchange of substances between the blood and the tissues; capillary walls are made up of a single layer of squamous epithelium, and their internal diameter is only a little larger than that of a red blood cell

carcinogen a substance that can cause cancer

cardiac cycle the sequence of events taking place during one heart beat

cardiac muscle the muscle tissue that makes up most of the heart; cardiac muscle cells are striated

cardiovascular diseases degenerative diseases of the heart and circulatory system, e.g. coronary heart disease or stroke

carotenoid a yellow, orange or red plant pigment used as an accessory pigment in photosynthesis

carrier someone who is infected with a pathogen but does not show any symptoms of the disease; people like this can be sources of infection that are hard to trace

carrier protein a type of membrane protein which changes shape to allow the passage into or out of the cell of specific ions or molecules by facilitated diffusion or active transport

cell the basic unit of all living organisms; it is surrounded by a cell surface membrane (also known as the plasma membrane) and contains cytoplasm and organelles

cell cycle the sequence of events that takes place from one cell division until the next; it is made up of interphase, mitosis and cytokinesis

cell signalling the molecular mechanisms by which cells detect and respond to external stimuli, including communication between cells

cell surface membrane a very thin membrane (about 7 nm diameter) surrounding all cells; it is partially permeable and controls the exchange of materials between the cell and its environment

cell wall a wall surrounding prokaryote, plant and fungal cells; the wall contains a strengthening material which protects the cell from mechanical damage, supports it and prevents it from bursting by osmosis if the cell is surrounded by a solution with a higher water potential

centriole one of two small, cylindrical structures, made from microtubules, found just outside the nucleus in animal cells, in a region known as the centrosome; they are also found at the bases of cilia and flagella

CFTR a transmembrane protein that controls the transport of chloride ions out of cells; CFTR written in italics refers to the gene for the CFTR protein

channel protein a membrane protein of fixed shape which has a water-filled pore through which selected hydrophilic ions or molecules can pass see facilitated diffusion

chemiosmosis the synthesis of ATP using energy stored as a difference in hydrogen ion concentration across a membrane in a chloroplast or mitochondrion

chemoautotroph autotroph such as a nitrifying bacterium that uses a chemical energy source

chemoreceptor a receptor cell that responds to chemical stimuli; chemoreceptors are found in taste buds on the tongue, in the nose and in blood vessels where they detect changes in oxygen and carbon dioxide concentrations

chiasma (plural: chiasmata) the visible effect(s) of crossing over

chi-squared (χ^2) test a statistical test that can be used to determine whether any difference between observed results and expected results is significant or is due to chance

chlorophyll a green pigment responsible for light capture in photosynthesis in algae and plants

chloroplast the photosynthetic organelle in eukaryotes

cholesterol a small, lipid-related molecule with a hydrophilic head and a hydrophobic tail which is an essential constituent of membranes, particularly in animal cells, conferring fluidity, flexibility and stability to the membrane

cholinergic synapse a synapse at which the transmitter substance is acetylcholine

chorionic villus sampling taking a small sample from one of the villi in the placenta, between the tenth and 13th week of pregnancy, so that the genotype of the fetus can be determined

chromatid one of two identical parts of a chromosome, held together by a centromere, formed during interphase by the replication of the DNA strand

chromatin the loosely coiled form of chromosomes during interphase of the cell cycle; chromatin is made of DNA and proteins and is visible as loosely distributed patches or fibres within the nucleus when stained

chromosome a structure made of DNA and histones, found in the nucleus of a eukaryotic cell; the term bacterial chromosome is now commonly used for the circular strand of DNA present in a prokaryotic cell

chromosome mutation a random and unpredictable change in the structure or number of chromosomes in a cell

chronic obstructive pulmonary disease (COPD) a disease of the lungs characterised by bronchitis and emphysema

codominant two alleles are said to be codominant when both alleles have an effect on the phenotype of a heterozygous organism

codon sequence of three bases on an mRNA molecule that codes for a specific amino acid or for a stop signal

collecting duct the last section of a nephron, from which water can be absorbed back into the bloodstream before the urine flows into the ureter

collenchyma a modified form of parenchyma in which the corners of the cells have extra cellulose thickening, providing extra support, as in the midrib of leaves and at the corners of square stems; in three dimensions the tissue occurs in strands (as in celery petioles)

community all of the living organisms, of all species, that are found in a particular ecosystem at a particular time

companion cell a cell with an unthickened cellulose wall and dense cytoplasm that is found in close association with a phloem sieve element to which it is directly linked via many plasmodesmata; the companion cell and the sieve element form a functional unit

competitive inhibition when a substance reduces the rate of activity of an enzyme by competing with the substrate molecules for the enzyme's active site; increasing the concentration of the substrate reduces the degree of inhibition

complementary base pairing the hydrogen bonding of A with T or U and of C with G in nucleic acids

condensation reaction a chemical reaction involving the joining together of two molecules by removal of a water molecule

control experiment an experiment in which the factor whose effect is being investigated (the independent variable) is absent; it is used as a standard of comparison

controlled variables see standardised variables

coronary heart disease a disease of the heart caused by damage to the coronary arteries

corpus luteum the 'yellow body' that develops from an ovarian follicle after ovulation

counter-current multiplier an arrangement in which fluid in adjacent tubes flows in opposite directions, allowing relatively large differences in concentration to be built up

creatinine a nitrogenous excretory substance produced from the breakdown of creatine
crista (plural: cristae) a fold of the inner membrane of the mitochondrial envelope on which are found stalked particles of ATP synthase and electron transport chains associated with aerobic respiration
crossing over an event that occurs during meiosis I, when chromatids of two homologous chromosomes break and rejoin so that a part of one chromatid swaps places with the same part of the other
cystic fibrosis a genetic disease caused by a recessive allele of the *CFTR* (cystic fibrosis transmembrane regulator) gene
cytokinins plant growth regulators that stimulate cell division
cytoplasm the contents of a cell, excluding the nucleus
cytosine (C) nitrogen-containing pyrimidine base found in DNA and RNA
deamination the breakdown of excess amino acids in the liver, by the removal of the amine group; ammonia and eventually urea are formed from the amine group
dendrite a short cytoplasmic process of a neurone that receives nerve impulses from other neurones
deoxyribose a pentose (5C) sugar found in DNA
dependent variable in an experiment, the variable that changes as a result of changing the independent variable
depolarisation the reversal of the resting potential across the cell surface membrane of a neurone or muscle cell, so that the inside becomes positively charged compared with the outside
diabetes an illness in which the pancreas does not make sufficient insulin, or where cells do not respond appropriately to insulin
diffusion the net movement of molecules or ions from a region of higher concentration to a region of lower concentration down a gradient, as a result of the random movements of particles
dihybrid cross a genetic cross in which two different genes are considered
diploid cell one that possesses two complete sets of chromosomes; the abbreviation for diploid is $2n$
directional selection a type of selection in which the most common varieties of an organism are selected against, resulting in a change in the features of the population
disaccharide a sugar molecule consisting of two monosaccharides joined together by a glycosidic bond
disruptive selection natural selection that favours the survival of individuals at two different points within the range of variation, resulting in two different phenotypes
DNA ligase an enzyme that catalyses the linkage of nucleotides during the formation of a DNA molecule
domain one of the three major groups into which all organisms are classified
dominant an allele is said to be dominant when its effect on the phenotype of a heterozygote is identical to its effect in a homozygote
dormancy a state of 'suspended animation', in which metabolism is slowed right down, enabling survival in adverse conditions
double circulation a circulatory system in which the blood travels twice through the heart on one complete circuit of the body; the pathway from heart to lungs and back to the heart is known as the pulmonary circulation, and that from heart to the rest of the body and back to the heart as the systemic circulation
ecosystem all the living organisms of all species (the biotic component) and all the non-living components (the abiotic component) that are found together in a defined area and that interact with one another
effector an organ or tissue that carries out an action in response to a stimulus; muscles and glands are effectors
efferent leading away from
electrochemical gradient a gradient across a cell surface membrane that involves both a difference in concentrations of ions and a potential difference e.g. the entry of sodium ions into neurones
electron transport chain chain of adjacently arranged carrier molecules in the inner mitochondrial membrane, along which electrons pass by redox reactions
embryo transfer embryos are removed from a female mammal and transferred to one or more surrogate mothers to complete development until birth
emphysema a disease in which alveoli are destroyed, giving large air spaces and decreased surface area for gaseous exchange; it is often associated with chronic bronchitis
endemic of disease, a disease that is always in a population; of species, a species that is only found in a certain area and nowhere else
endocrine gland a gland that secretes its products, which are always hormones, directly into the blood
endocytosis the bulk movement of liquids (pinocytosis) or solids (phagocytosis) into a cell, by the infolding of the cell surface membrane to form vesicles containing the substance; endocytosis is an active process requiring ATP
endoplasmic reticulum (ER) a network of flattened sacs running through the cytoplasm of eukaryotic cells; molecules, particularly proteins, can be transported through the cell inside the sacs separate from the rest of the cytoplasm; ER is continuous with the outer membrane of the nuclear envelope
endosperm a tissue that stores food in some seeds, such as cereal grains
endothelium a tissue that lines the inside of a structure, such as the inner surface of a blood vessel
enzyme a protein produced by a living organism that acts as a catalyst in a specific reaction by reducing activation energy
epithelium a tissue that covers the outside of a structure
Eukarya domain that contains all eukaryotic organisms: protoctists, fungi, plants and animals
eukaryotic cell a cell containing a nucleus and other membrane-bound organelles
evolution changes in a population that may lead to speciation or extinction
excretion the removal of toxic or waste products of metabolism from the body

exocytosis the bulk movement of liquids or solids out of a cell, by the fusion of vesicles containing the substance with the cell surface membrane; exocytosis is an active process requiring ATP

expansins proteins in the cell walls of plants that loosen the attachment of microfibrils of cellulose during elongation growth

F1 the offspring resulting from a cross between an organism with a homozygous dominant genotype, and one with a homozyogus recessive genotype

F2 the offspring resulting from a cross between two F1 (heterozygous) organisms

facilitated diffusion the diffusion of a substance through transport proteins in a cell membrane; the proteins provide hydrophilic areas that allow the molecules or ions to pass through the membrane which would otherwise be less permeable to them

factor VIII one of several substances that must be present in blood in order for clotting to occur (see haemophilia)

fibrous protein a protein whose molecules have a relatively long, thin structure that is generally insoluble and metabolically inactive, and whose function is usually structural, e.g. keratin and collagen

fitness the probability of an organism surviving and reproducing in the environment in which it is found

fluid mosaic model the currently accepted basic model of membrane structure, proposed by Singer and Nicolson in 1972, in which protein molecules are free to move about in a fluid bilayer of phospholipid molecules

follicle stimulating hormone (FSH) glycoprotein hormone secreted by anterior pituitary gland to stimulate the development of follicles in the ovary

Fungi one of the four kingdoms of the domain Eukarya; eukaryotic organisms which do not photosynthesise and have cell walls but without cellulose

gamete a haploid cell specialised for fertilisation

gase exchange the movement of gases between an organism and its environment, e.g. the intake of oxygen and the loss of carbon dioxide; gas exchange often takes place across a specialised surface such as the alveoli of the lungs

gene a length of DNA that codes for a particular protein or polypeptide

gene mutation a change in the base sequence in part of a DNA molecule

gene technology a procedure by which one or more selected genes are removed from one organism and inserted into another, which is then said to be transgenic

gene therapy treatment of a genetic disorder by altering a person's genotype

genetic diversity all the alleles of all the genes in the genome of a species

genetic profiling (fingerprinting) sequencing a length of DNA from an organism, to compare with the sequence of the same DNA from other organisms

genetic screening testing an embryo, fetus or adult to find out whether a particular allele is present

genetic variation the occurrence of genetic differences between individuals

genetics the study of heritability and variation

genotype the alleles possessed by an organism

geographical isolation the separation of two populations of the same species by a geographic barrier that prevents them from meeting

germ cells cells that can divide to form gametes

gibberellin a plant growth regulator (plant hormone) that regulates plant height (stem growth) and stimulates seed germination; a lack of gibberellin causes dwarfness

globin a protein which forms the basic unit of a group of proteins that bind oxygen, namely myoglobin and different forms of haemoglobin; globin has a characteristic tertiary structure; it exists in several closely similar forms, such as α-globin, β-globin and γ-globin

globular protein a protein whose molecules are folded into a relatively spherical shape, and which is often water-soluble and metabolically active, e.g. insulin and haemoglobin

glomerular filtration rate the rate at which fluid passes from the glomerular capillaries into the Bowman's capsules in the kidneys

glomerulus a group of capillaries within the 'cup' of a Bowman's capsule in the cortex of the kidney

glucagon a small peptide hormone secreted by the α cells in the islets of Langerhans in the pancreas that brings about an increase in the blood glucose level

gluconeogenesis formation of glucose in the liver from non-carbohydrate sources such as amino acids, pyruvate, lactate and glycerol

glucose oxidase enzyme that converts glucose and oxygen to gluconolactone and hydrogen peroxide

glycogen a polysaccharide made of many glucose molecules linked together, that acts as a glucose store in liver and muscle cells

glycolysis the splitting (lysis) of glucose

glycosidic bond a C–O–C link between two monosaccharide molecules, formed by a condensation reaction

Golgi body (Golgi apparatus, Golgi complex) an organelle found in eukaryotic cells; the Golgi apparatus consists of a stack of flattened sacs, constantly forming at one end and breaking up into Golgi vesicles at the other end; Golgi vesicles carry their contents to other parts of the cell, often to the cell surface membrane for secretion; the Golgi apparatus chemically modifies the molecules it transports; for example, sugars may be added to proteins to make glycoproteins

granum (plural: grana) a stack of circular thylakoids in a chloroplast

guanine (G) nitrogen-containing purine base found in DNA and RNA

guard cell a sausage-shaped epidermal cell found with another, in a pair bounding a stoma and controlling its opening or closure

habitat the place where an organism, a population or a community lives

haemoglobin the red pigment found in red blood cells, whose molecules contain four iron atoms within a globular protein made up of four polypeptides, and that combines reversibly with oxygen

haemophilia a genetic disease in which there is an insufficient amount of a clotting factor such as factor VIII in the blood

haploid cell one that possesses one complete set of chromosomes; the abbreviation for haploid is n

heterotroph an organism needing a supply of organic molecules as its carbon source

heterozygous having two different alleles of a gene

homeostasis maintaining a relatively constant environment for the cells within the body

homologous chromosomes a pair of chromosomes in a diploid cell that have the same structure as each other, with the same genes (but not necessarily the same alleles of those genes) at the same loci, and that pair together to form a bivalent during the first division of meiosis

homozygous having two identical alleles of a gene

hormone a substance secreted by an endocrine gland, that is carried in blood plasma to another part of the body where it has an effect (plant hormone – see plant growth regulator)

hybridoma a cell formed by the fusion of a plasma cell and a cancer cell; it can both secrete antibodies and divide to form other cells like itself

hydrogen bond a relatively weak bond formed by the attraction between a group with a small positive charge on a hydrogen atom and another group carrying a small negative charge, e.g. between two $-O^{\delta-}H^{\delta+}$ groups

hydrolysis a reaction in which a complex molecule is broken down to simpler ones, involving the addition of water

hypha (plural: hyphae) a long threadlike structure surrounded by a cell surface membrane and cell wall that makes up part of the body of a fungus; hyphae may or may not have cross-walls

IAA a type of auxin

immobilised enzymes enzymes that have been fixed to a surface or within a bead of agar gel

immune response the complex series of responses of the body to the entry of a foreign antigen; it involves the activity of lymphocytes and phagocytes

immune system the body's defence system

immunity protection against infectious diseases, gained either actively or passively

immunoglobulin an antibody

immunological memory the ability of the immune system to respond quickly to antigens that it recognises as having entered the body before

in vitro fertilisation (IVF) the addition of sperm to eggs in a Petri dish or other container, in which fertilisation takes place

inbreeding depression increase in the proportion of debilitated offspring as a result of breeding between closely related organisms of the same species

independent assortment the way in which different alleles of genes on different chromosomes may end up in any combination in gametes, resulting from the random alignment of bivalents on the equator during meiosis I

independent variable the variable (factor) that is deliberately changed in an experiment

induced fit hypothesis a model for enzyme action; the substrate is a complementary shape to the active site of the enzyme, but not an exact fit – the enzyme or sometimes substrate can change shape slightly to ensure a perfect fit

infectious disease a disease caused by an organism such as a bacterium or virus

insulin a small peptide hormone secreted by the β cells in the islets of Langerhans in the pancreas that reduces blood glucose levels

interquartile range the range of values into which the middle 50% of a set of data falls

islet of Langerhans a group of cells in the pancreas which secrete insulin and glucagon

isotopes atoms of the same element with different atomic mass numbers

K_m see Michaelis–Menten constant

karyogram a photograph or diagram of a set of chromosomes from an individual; the chromosomes are normally arranged in their homologous pairs in order of size; sex chromosomes may be shown separately

kinetochore a protein structure found at the centromere of a chromatid to which microtubules attach during cell division

Krebs cycle a cycle of reactions in aerobic respiration in the matrix of a mitochondrion in which hydrogens pass to hydrogen carriers for subsequent ATP synthesis and some ATP is synthesised directly

lactate (or lactic acid) the end-product of lactic fermentation, often produced by muscles during exercise

lactic fermentation respiration in which, in the absence of oxygen, glucose is converted to lactic acid (lactate)

leukaemia cancer of the stem cells that give rise to white blood cells

light dependent reactions reactions in photosynthesis for which light energy is needed

light independent reactions reactions in photosynthesis for which light energy is not needed

limiting factor the one factor, of many affecting a process, that is nearest its lowest value and hence is rate-limiting

linkage the presence of two genes on the same chromosome, so that they tend to be inherited together and do not assort independently

link reaction decarboxylation and dehydrogenation of pyruvate, resulting in the formation of acetyl coenzyme A, linking glycolysis with the Krebs cycle

lock and key hypothesis a model for enzyme action; the substrate is a complementary shape to the active site of the enzyme, and fits exactly into the site

locus the position at which a particular gene is found on a particular chromosome; the same gene is always found at the same locus

loop of Henle the part of the nephron between the proximal and distal convoluted tubules; in humans, some of the loops of Henle are long and reach down into the medulla of the kidney
luteinising hormone (LH) a glycoprotein hormone that is secreted by the anterior pituitary gland to stimulate ovulation and the development of the corpus luteum
lymph an almost colourless fluid, very similar in composition to blood plasma but with fewer plasma proteins, that is present in lymph vessels
lymphocyte a type of white blood cell that is involved in the immune response; unlike phagocytes they become active only in the presence of a particular antigen that 'matches' their specific receptors or antibodies
lysosome a spherical organelle found in eukaryotic cells; a lysosome contains digestive (hydrolytic) enzymes and has a variety of destructive functions, such as removal of old cell organelles
macromolecule (see also polymer) a large biological molecule such as a protein, polysaccharide or nucleic acid
macrophage phagocytic cell found in tissues throughout the body; they act as antigen-presenting cells (APCs)
magnification the number of times greater that an image is than the actual object; magnification = image size ÷ actual (real) size of the object
mark–release–recapture a method of estimating the numbers of individuals in a population of mobile animals

median the middle value of all the values in a set of data showing a normal distribution
meiosis the type of nuclear division that results in a halving of chromosome number and a reshuffling of alleles; in humans, it occurs in the formation of gametes
memory cells lymphocytes which develop during an immune response and retain the ability to respond quickly when an antigen enters the body on a second or any subsequent occasion
menstrual cycle the changes that occur in the ovary and the uterus approximately every 28 days involving ovulation and the breakdown and loss of the lining of the uterus (menstruation)
mesophyll the internal tissue of a leaf blade with chloroplasts for photosynthesis and consisting of an upper layer of palisade mesophyll (the main photosynthetic tissue) and a lower layer of spongy mesophyll with large air spaces for gas exchange
messenger RNA (mRNA) a single-stranded RNA molecule that carries the genetic code from DNA to a ribosome
Michaelis–Menten constant (K_m) the substrate concentration at which an enzyme works at half its maximum rate ($\frac{1}{2}V_{max}$), used as a measure of the efficiency of an enzyme; the lower the value of K_m, the more efficient the enzyme
microtubules tiny tubes made of a protein called tubulin and found in most eukaryotic cells; microtubules have a large variety of functions, including cell support and determining cell shape; the 'spindle' on which chromosomes separate during nuclear division is made of microtubules
mitochondrion (plural: mitochondria) the organelle in eukaryotes in which aerobic respiration takes place
mitosis the division of a nucleus into two so that the two daughter cells have exactly the same number and type of chromosomes as the parent cell
mode (modal class) the value that occurs most frequently in a set of data showing a normal distribution
monoclonal antibodies many identical antibodies, made by hybridoma cells formed by the fusion of a plasma cell and a cancer cell
monohybrid cross a cross in which the inheritance of one gene is considered
monomer a relatively simple molecule which is used as a basic building block for the synthesis of a polymer; many monomers are joined together to make the polymer, usually by condensation reactions; common examples of molecules used as monomers are monosaccharides, amino acids and nucleotides
monosaccharide a molecule consisting of a single sugar unit with the general formula $(CH_2O)_n$
motor end plate the ending of an axon of a motor neurone, where it forms a synapse with a muscle
motor neurone a neurone whose cell body is in the brain or spinal cord, and that transmits action potentials to an effector such as a muscle or gland
multiple alleles the existence of three or more alleles of a gene, as, for example, in the determination of A,B,O blood groups
multiple resistance possession by a bacterium of a plasmid carrying resistance genes for several antibiotics or other drugs
murein see peptidoglycan
mutagen a substance that can cause mutation
mutation an unpredictable change in the structure of DNA, or in the structure and number of chromosomes
myelin a substance that surrounds many axons, made up of many layers of the cell surface membranes of Schwann cells
myofibril one of many cylindrical bundles of thick and thin filaments inside a muscle fibre
myosin the contractile protein that makes up the thick filaments in striated muscle
natural immunity immunity gained by being infected (active) or by receiving antibodies from the mother across the placenta or in breast milk (passive)
natural selection the way in which individuals with particular characteristics have a greater chance of survival than individuals without those characteristics, and are therefore more likely to breed and pass on the genes for these characteristics to their offspring

negative feedback a process in which a change in some parameter, such as blood glucose level, brings about processes which move its level back towards normal again

nephron a kidney tubule

nerve a bundle of numerous axons of many different neurones, surrounded by a sheath called the perineurium

neuromuscular junction a synapse between the axon of a motor neurone and a muscle

neurone a nerve cell; a cell which is specialised for the conduction of action potentials

neutrophil a phagocytic white blood cell

niche the role of an organism in an ecosystem

nicotine a chemical found in tobacco smoke that can bind with acetylcholine receptors on the postsynaptic membrane of cholinergic synapses

nitrogenous excretory product an unwanted product of metabolism that contains nitrogen, e.g. ammonia, urea or uric acid

node of Ranvier a short gap in the myelin sheath surrounding an axon

non-competitive inhibition when a substance reduces the rate of activity of an enzyme, but increasing the concentration of the substrate does not reduce the degree of inhibition; many non-competitive inhibitors bind to areas of the enzyme molecule other than the active site itself

non-infectious disease any disease not caused by a pathogen

non-self any substance or cell that is recognised by the immune system as being foreign and will stimulate an immune response

noradrenaline a neurotransmitter substance

normal distribution a pattern shown by a set of data that are distributed symmetrically on both sides of the mean value

nuclear envelope the two membranes, situated closely together, that surround the nucleus; the envelope is perforated with nuclear pores

nuclear pores pores found in the nuclear envelope which control the exchange of materials, e.g. mRNA, between the nucleus and the cytoplasm

nucleolus a small structure, one or more of which is found inside the nucleus; the nucleolus is usually visible as a densely stained body; its function is to manufacture ribosomes using the information in its own DNA

nucleosome a bead-like structure made of eight histone molecules, around which DNA is wrapped; nucleosomes are the fundamental subunits of chromatin

nucleotide a molecule consisting of a nitrogen-containing base, a pentose sugar and a phosphate group

nucleus a relatively large organelle found in eukaryotic cells, but absent from prokaryotic cells; the nucleus contains the cell's DNA and therefore controls the activities of the cell

null hypothesis a hypothesis that assumes there is no relationship between two variables, or that there is no significant difference between two samples

oestrogen steroid hormone secreted by follicles in the ovary; used in some contraceptive pills

oogenesis the production of egg cells

operon a group of structural genes headed by a non-coding sequence of DNA called the operator

organelle a functionally and structurally distinct part of a cell, e.g. a ribosome or mitochondrion

organic molecule a compound containing carbon and hydrogen

osmoreceptor a receptor cell that is sensitive to the water potential of the blood

osmoregulation the control of the water content of the fluids in the body

osmosis the net movement of water molecules from a region of higher water potential to a region of lower water potential, through a partially permeable membrane, as a result of their random motion (diffusion)

oxidative phosphorylation the synthesis of ATP from ADP and Pi using energy from oxidation reactions in aerobic respiration (compare photophosphorylation)

oxygen debt the volume of oxygen that is required at the end of exercise to metabolise the lactate that accumulates as a result of anaerobic respiration in muscles

palisade mesophyll see mesophyll

pancreas an organ lying close to the stomach that functions both as an exocrine gland (secreting pancreatic juice) and an endocrine gland (secreting insulin and glucagon)

parenchyma a basic plant tissue typically used as packing tissue between more specialised structures; it is metabolically active and may have a variety of functions such as food storage, support and transport via symplast and apoplast pathways

passive immunity immunity gained without an immune response; antibodies are injected (artificial) or pass from mother to child across the placenta or in breast milk (natural)

pathogen an organism that causes infectious disease

Pearson's linear correlation a statistical test used to determine whether two variables show a linear correlation

PEP carboxylase an enzyme found in C4 plants that catalyses the combination of carbon dioxide with PEP

PEP phosphoenolpyruvate, the initial carbon dioxide acceptor in a C4 plant

peptide bond a C–N link between two amino acid molecules, formed by a condensation reaction

peptidoglycan a substance, also known as murein, whose molecules are made of amino acid chains to which sugars are attached; bacterial cell walls contain peptidoglycans; they make the wall more rigid, preventing the cell from bursting by osmosis when it is surrounded by a solution with a higher water potential

phagocyte a type of cell that ingests and destroys pathogens or damaged body cells by the process of phagocytosis; some phagocytes are white blood cells

phagocytosis see endocytosis
phenotype the characteristics of an organism, often resulting from an interaction between its genotype and its environment
phloem tissue containing sieve tubes and other types of cell, responsible for the translocation of assimilates such as sucrose through a plant
phospholipid a substance whose molecules are made up of a glycerol molecule, two fatty acids and a phosphate group; a bilayer of phospholipids forms the basic structure of all cell membranes
phosphorylation the transfer of a phosphate group to an organic compound
photoautotroph autotroph that uses light energy in photosynthesis
photolysis the splitting of water using light energy: $H_2O \rightarrow 2H^+ + 2e^- + O_2$
photophosphorylation the synthesis of ATP from ADP and P_i using light energy in photosynthesis (compare oxidative phosphorylation)
photorespiration a wasteful reaction in which RuBP combines with oxygen rather than carbon dioxide; it is favoured by high temperatures and high light intensities
photosystem a cluster of light-harvesting accessory pigments surrounding a primary pigment or reaction centre
pinocytosis see endocytosis
plant growth regulator (plant hormone) any chemical produced in plants that influences their growth and development, e.g. auxins, gibberellins, cytokinins and abscisic acid
Plantae one of the four kingdoms of the domain Eukarya; eukaryotic organisms which are multicellular, have cell walls that contain cellulose and can photosynthesise
plasmid a small, circular piece of DNA in a bacterium (not its main 'chromosome'); plasmids often contain genes that confer resistance to antibiotics
plasmodesma (plural: plasmodesmata) a pore-like structure found in plant cell walls; plasmodesmata of neighbouring plant cells line up to form tube-like pores through the cell walls, allowing the controlled passage of materials from one cell to the other; the pores contain ER and are lined with the cell surface membrane
podocyte one of the cells that makes up the lining of Bowman's capsule surrounding the glomerular capillaries
polymer a giant molecule made from many similar repeating subunits joined together in a chain; the subunits are much smaller and simpler molecules known as monomers; polymers may also be referred to as macromolecules; examples of biological polymers are polysaccharides, proteins and nucleic acids; see monomer
polynucleotide a chain of nucleotides
polypeptide a long chain of amino acids formed by condensation reactions between the individual amino acids; proteins are made of one or more polypeptide chains; see peptide bond
polyploidy possessing more than two complete sets of chromosomes
polyribosome the group of ribosomes working on the same mRNA molecule
polysaccharide a polymer whose subunits are monosaccharides joined together by glycosidic bonds
population all of the organisms of the same species present in the same place and at the same time that can interbreed with one another
positive feedback a process in which a change in some parameter brings about processes that move its level even further in the direction of the initial change
postsynaptic neurone the neurone on the opposite side of a synapse to the neurone in which the action potential arrives
potometer a piece of apparatus used to measure the rate of uptake of water by a plant
presynaptic neurone a neurone ending at a synapse from which neurotransmitter is secreted when an action potential arrives
primary pigment see reaction centre
primary structure the sequence of amino acids in a polypeptide or protein
probe a length of DNA that has a complementary base sequence to another piece of DNA that you are trying to detect
progesterone a steroid hormone secreted by the corpus luteum in the ovary after ovulation; used in contraceptive pills
prokaryote see prokaryotic cell
prokaryotic cell a cell that does not contain a nucleus or any other membrane-bound organelles; bacteria and archeans are prokaryotes
promoter a length of DNA that controls the expression of a gene
prosthetic group a non-protein component that is attached to a protein to form a complex molecule; it aids the protein in its function; an example is the iron-containing haem group, which is attached to α-globin and β-globin to form haemoglobin
Protoctista one of the five kingdoms; eukaryotic organisms which are single-celled, or made up of groups of similar cells
protoctist a member of the Protoctista kingdom
protoplast the living contents of a plant cell, including the cell surface membrane, but excluding the cell wall
Purkyne tissue an area of tissue in the septum of the heart that conducts the wave of excitation from the atria to the base of the ventricles
quadrat a square frame which is used to mark out an area for sampling of organisms
quaternary structure the three-dimensional arrangement of two or more polypeptides, or of a polypeptide and a non-protein component such as haem, in a protein molecule

reaction centre a molecule of chlorophyll *a* that receives energy from the light absorbed by surrounding accessory pigments in a photosystem

receptor cell a cell which is sensitive to a change in the environment and that may generate an action potential as a result of a stimulus

receptor potential a change in the normal resting potential across the membrane of a receptor cell, caused by a stimulus

recessive an allele that is only expressed when no dominant allele is present

recombinant DNA DNA made by joining pieces from two or more different sources

redox reaction an oxidation–reduction reaction involving the transfer of electrons from a donor to an acceptor

reflex action a fast, automatic response to a stimulus; reflex actions may be innate (inborn) or learned (conditioned)

reflex arc the pathway taken by an action potential, leading to a reflex action; the action potential is generated in a receptor, passes along a sensory neurone into the brain or spinal cord, and then along a motor neurone to an effector

refractory period a period of time during which a neurone is recovering from an action potential, and during which another action potential cannot be generated

relay neurone a neurone whose cell body and many dendrites are all within the brain or spinal cord; it receives action potentials from a sensory neurone and transmits action potentials to a motor neurone

repolarisation returning the potential difference across the cell surface membrane of a neurone or muscle cell to normal following the depolarisation of an action potential

reproductive isolation the inability of two groups of organisms of the same species to breed with one another, e.g. because of geographical separation or because of behavioural differences

resolution the ability to distinguish between two objects very close together; the higher the resolution of an image, the greater the detail that can be seen

respiration enzymatic release of energy from organic compounds in living cells

respiratory quotient (RQ) the ratio of the volume of carbon dioxide given out in respiration to that of oxygen used

respirometer the apparatus for measuring the rate of oxygen consumption in respiration or for finding the respiratory quotient

resting potential the difference in electrical potential that is maintained across a neurone when it is not transmitting an action potential; it is normally about −70 mV inside and is maintained by the sodium–potassium pump

restriction enzyme an enzyme, originally derived from bacteria, that cuts DNA molecules; each type of restriction enzyme cuts only at a particular sequence of bases

ribose a pentose (5C) sugar found in RNA

ribosome a tiny organelle found in large numbers in all cells; prokaryotic ribosomes are smaller (20 nm diameter) than eukaryotic ribosomes (25 nm diameter); ribosomes are made of protein and ribosomal RNA and consist of two subunits; they are the sites of protein synthesis in cells

saltatory conduction conduction of an action potential along a myelinated axon, in which the action potential jumps from one node of Ranvier to the next

sarcolemma the cell surface membrane of a muscle fibre

sarcomere the part of a myofibril between two Z discs

sarcoplasm the cytoplasm of a muscle fibre

sarcoplasmic reticulum the endoplasmic reticulum of muscle fibre

Schwann cell a cell which is in close association with a neurone, whose cell surface membrane wraps around and around the axon of the neurone to form a myelin sheath

secondary structure the structure of a protein molecule resulting from the regular coiling or folding of the chain of amino acids, e.g. an α-helix or β-pleated sheet

secretion the release of a useful substance from a cell or gland

seed bank facility where seeds are dried and kept in cold storage to conserve plant biodiversity

selection pressure an environmental factor that confers greater chances of survival and reproduction on some individuals than on others in a population

selective breeding choosing only organisms with desirable features, from which to breed

selective reabsorption movement of certain substances from the filtrate back into the blood in the kidney nephron

self refers to substances produced by the body that the immune system does not recognise as foreign, so they do not stimulate an immune response

semi-conservative replication the method by which a DNA molecule is copied to form two identical molecules, each containing one strand from the original molecule and one newly synthesised strand

sensory neurone a neurone that transmits action potentials from a receptor to the central nervous system

set point the ideal value of a physiological factor that the body controls in homeostasis

sex chromosomes the pair of chromosomes that determine the sex of an individual; in humans, they are the X and Y chromosomes

sex-linked gene a gene that is carried on an X chromosome but not on a Y chromosome

sexual reproduction reproduction involving the fusion of gametes (fertilisation) to produce a zygote

sickle cell anaemia a genetic disease caused by a faulty gene coding for haemoglobin, in which haemoglobin tends to precipitate when oxygen concentrations are low

sickle cell trait a person who is heterozygous for the sickle cell allele is said to have sickle cell trait; there are normally no symptoms, except occasionally in very severe conditions of oxygen shortage

sieve element or sieve tube element a cell found in phloem tissue, with non-thickened cellulose walls, very little cytoplasm, no nucleus and end walls perforated to form sieve plates, through which sap containing sucrose is transported

sieve tube tubes formed by sieve elements lined up end to end

silent mutation a mutation in which the change in the DNA has no discernible effect on an organism

Simpson's index of diversity a method to assess the biodiversity of an ecosystem

sinoatrial node a patch of muscle in the wall of the right atrium of the heart, whose intrinsic rate of rhythmic contraction is faster than that of the rest of the cardiac muscle, and from which waves of excitation spread to the rest of the heart to initiate its contraction during the cardiac cycle

smooth muscle type of muscle tissue found in walls of blood vessels (except capillaries), trachea, bronchi and bronchioles, alimentary canal and ureter; the muscle cells are not striated

sodium–potassium pump a membrane protein (or proteins) that moves sodium ions out of a cell and potassium ions into it, using ATP

somatic cells cells that are not involved in gamete formation

Spearman's rank correlation a statistical test to determine whether two variables are correlated

speciation the production of new species

species a group of organisms with similar morphology and physiology, which can breed together to produce fertile offspring and which is reproductively isolated from other species

species diversity all the species in an ecosystem

spongy mesophyll see mesophyll

stabilising selection a type of natural selection in which the status quo is maintained because the organisms are already well adapted to their environment

standard deviation a measure of how widely a set of data is spread out on either side of the mean

standard error a measure of how likely it is that a mean calculated from a sample represents the true mean for the whole population

standardised variables variables (factors) that are kept constant in an experiment; only the independent variable should be changed

stem cell a relatively unspecialised cell that retains the ability to divide an unlimited number of times

stimulus a change in the environment that is detected by a receptor, and which may cause a response

stoma (plural: stomata) a pore in the epidermis of a leaf, bounded by two guard cells and needed for efficient gas exchange

striated muscle type of muscle tissue in skeletal muscles; the muscle fibres have regular striations that can be seen under the light microscope

stroke damage to the brain caused by bursting or blockage of an artery

stroma the matrix of a chloroplast in which the light independent reactions of photosynthesis occur

surrogacy see surrogate

surrogate female mammal which receives an embryo from another female; the embryo develops inside this surrogate mother which is not its biological parent

sympatric speciation the emergence of a new species from another species where the two are living in the same place; it can happen, for example, as a result of polyploidy

symplast the living system of interconnected protoplasts extending throughout a plant, used as a transport pathway for the movement of water and solutes; individual protoplasts are connected via plasmodesmata

synapse a point at which two neurones meet but do not touch; the synapse is made up of the end of the presynaptic neurone, the synaptic cleft and the end of the postsynaptic neurone

synaptic cleft a very small gap between two neurones at a synapse

T-lymphocyte a lymphocyte that does not secrete antibodies; T helper lymphocytes stimulate the immune system to respond during an infection, and killer T lymphocytes destroy human cells that are infected with pathogens such as bacteria and viruses

taxon (plural taxa) one of the groups used in the hierarchical classification system for organisms, e.g. species, genus, family, order, class, phylum, kingdom and domain

telomeres repetitive sequences of DNA at the end of chromosomes that protect genes from the chromosome shortening that happens at each cell division

tertiary structure the compact structure of a protein molecule resulting from the three-dimensional coiling of the already-folded chain of amino acids

test cross a genetic cross in which an organism showing a characteristic caused by a dominant allele is crossed with an organism that is homozygous recessive; the phenotypes of the offspring can be a guide to whether the first organism is homozygous or heterozygous

tetraploid possessing four complete sets of chromosomes

thermoregulation the control of core body temperature

thylakoid a flattened, membrane-bound, fluid-filled sac, which is the site of the light-dependent reactions of photosynthesis in a chloroplast

thymine (T) nitrogen-containing pyrimidine base found in DNA

tissue fluid the almost colourless fluid that fills the spaces between body cells; tissue fluid forms from the fluid that leaks from blood capillaries, and most of it eventually collects into lymph vessels where it forms lymph

tonoplast the partially permeable membrane that surrounds plant vacuoles

transcription production of an mRNA molecule on DNA

transect a line along which samples are taken, either by noting the species at equal distances (line transect) or placing quadrats at regular intervals (belt transect)

transfer RNA (tRNA) a folded, single-stranded RNA molecule that carries an amino acid to a ribosome for protein synthesis

transformed describes an organism that has taken up DNA introduced from a different species of organism

translation production of a polypeptide at a ribosome from the code on mRNA

translocation the transport of assimilates such as sucrose through a plant, in phloem tissue; translocation requires the input of metabolic energy; the term is sometimes used more generally to include transport in the xylem

transmission the transfer of a pathogen from one person to another

transmitter substance a chemical that is released from a presynaptic neurone when an action potential arrives, and that then diffuses across the synaptic cleft and may initiate an action potential in the postsynaptic neurone

transpiration the loss of water vapour from a plant to its environment, by diffusion down a water potential gradient; most transpiration takes place through the stomata in the leaves

transverse system tubule (also known as T-system tubule or T-tubule) infolding of the sarcolemma that goes deep into a muscle fibre

triglyceride a lipid whose molecules are made up of a glycerol molecule and three fatty acids

triploid possessing three complete sets of chromosomes

tropomyosin a fibrous protein that is part of the thin filaments in myofibrils in striated muscle

troponin a calcium-binding protein that is part of the thin filaments in myofibrils in striated muscle

***t*-test** a statistical procedure used to determine whether the means of two samples differ significantly

ultrafiltration filtration on a molecular scale e.g. the filtration that occurs as blood flows through capillaries, especially those in glomeruli in the kidney

uracil (U) nitrogen-containing pyrimidine base found in RNA

urea a nitrogenous excretory product produced in the liver from the deamination of amino acids

ureter a tube that carries urine from a kidney to the bladder

urethra a tube that carries urine from the bladder to the outside

uric acid a nitrogenous excretory product, made by the breakdown of purines

V_{max} the theoretical maximum rate of an enzyme-controlled reaction, obtained when all the active sites are occupied

vaccination giving a vaccine containing antigens for a disease, either by injection or by mouth; vaccination confers artificial active immunity

vacuole an organelle found in eukaryotic cells; a large, permanent central vacuole is a typical feature of plant cells, where it has a variety of functions, including storage of biochemicals such as salts, sugars and waste products; temporary vacuoles, such as phagocytic vacuoles (also known as phagocytic vesicles), may form in animal cells (see endocytosis)

variable number tandem repeat (VNTR) a length of DNA that contains different numbers of repeated base sequences in different individuals in a species

vector a means of delivering genes into a cell, used in gene technology; e.g. plasmids and viruses; see also disease vector

vein a blood vessel that carries blood back towards the heart; it has relatively thin walls and contains valves

vessel element see xylem vessel element

virus very small (20–300 nm) infectious particle which can replicate only inside living cells and consists essentially of a simple basic structure of a genetic code of DNA or RNA surrounded by a protein coat

VNTR see variable number tandem repeat

voltage-gated channel a channel protein through a cell membrane that opens or closes in response to changes in electrical potential across the membrane

water potential a measure of the tendency of water to move from one place to another; water moves from a solution with higher water potential to one with lower water potential; water potential is decreased by the addition of solute, and increased by the application of pressure; symbol is ψ or ψ_w

xerophyte a plant adapted to survive in conditions where water is in short supply

xylem tissue containing xylem vessel elements and other types of cells, responsible for support and the transport of water through a plant

xylem vessel a dead, empty tube with lignified walls, through which water is transported in plants; it is formed by xylem vessel elements lined up end to end

xylem vessel element a dead, lignified cell found in xylem specialised for transporting water and support; the ends of the cells break down and join with neighbouring elements to form long tubes called xylem vessels

zygote a cell formed by the fusion of two gametes; normally the gametes are haploid and the zygote is diploid

Index

Acknowledgements

The authors and publishers acknowledge the following sources of copyright material and are grateful for the permissions granted. While every effort has been made, it has not always been possible to identify the sources of all the material used, or to trace all copyright holders. If any omissions are brought to our notice, we will be happy to include the appropriate acknowledgements on reprinting.

Cover Colin Varndell/SPL; Chapter 1 Steve Gschmeissner/SPL; 1.1 Getty Images; 1.2 Dr Jeremy Burgess/SPL; 1.5, 1.18, 1.21 Dr Gopal Murti/SPL; 1.7 ISM/Phototake Inc – all rights reserved; 1.9a,b, 1.27 Biomedical Imaging Unit, University of Southampton; 1.13 Eye of Science/SPL; 1.15 Cultura Science/Matt Lincoln/SPL; 1.16Medimage/SPL; 1.19 NIBSC/SPL; 1.20 Dennis Kenkel Microscopy Inc/Visuals Unlimited/Corbis; 1.22 Bill Langcore/SPL; 1.23, 1.26 Biophoto Associates; 1.24b Dr Torsten Wittman/SPL; 1.29 Dr Kari Lounatmaa/SPL; EoC1Q9c Steve Gschmeissner/SPL; Chapter 2 lumpynoodles/iStock; 2.1 SSPL via Getty Images; 2.8 Dr Jeremy Burgess/SPL; 2.10 Bipohoto Associates; 2.11a, 2.11b Laguna Design/SPL; 2.14 Tom McHugh/SPL; 2.21 Dr Arthur Lesk/SPL; 2.25a,b Omikron/SPL; 2.26d J. Gross Biozentrum/SPL; 2.26e Steve Gschmeissner/SPL; 2.28 Claude Nuridsany & Marie Perennou/SPL; Chapter 3 Wolfgang Baumeiser/SPL; 3.1a National Geographic Images/Getty Images; 3.4b Div of Computer Research and Technology, National Institute of Health/SPL; 3.10 Simon Fraser/SPL; Chapter 4 Science Photo Library/Getty Images; 4.1 David McCarthy/SPL; 4.3 Visuals Unlimited/Corbis; 4.15 J.C. Revy, ISM/SPL; 4.21 Don W. Fawcett/SPL; Chapter 5 Science Photo Library/Getty Images; 5.1 Will Brown/Chemical Heritage Foundation/ SPL; 5.2 Biophoto Associates; 5.5 Arturo Londono, ISM/SPL; 5.8 *all,* 5.11a Eric Grave/SPL; 5.9 Manfred Kage/SPL; 5.12, 5.13a, 5.14b SPL; 5.13b James Stevenson/SPL; 5.14a Eye of Science/SPL; Chapter 6 Stocktrek Images/Superstock; 6.1 Power & Syred/SPL; 6.6 A. Barrington Brown/SPL; 6.7 SPL; 6.8 Photo Researchers/Alamy; 6.15 Professor Oscar Miller/SPL; Chapter 7 Moment/Getty Images; 7.1 © Copyright CSIRO Australia, from article 'Gilding the gumtree - scientists strike gold in leaves', October 2013; 7.3 Visuals Unlimited/SPL; 7.5, 7.13a,c, 7.22 *detail*, 7.23b, 7.31b Dr Keith Wheeler/SPL; 7.7, 7.11, 7.23e 7.31d Biophoto Associates/SPL; 7.17, 7.21e, 7.22 Geoff Jones; 7.18 Martyn F. Chillmaid/SPL; 7.21a,c,d 7.23c Power & Syred/SPL; 7.21b Sinclair Stammers/SPL; 7.29a Dr Jeremy Burgess/SPL; 7.33 Bill Brooks/Alamy; Chapter 8 E+/Getty Images; 8.1 dpa picture alliance/SPL; 8.6, 8.24 CNRI/SPL; 8.7 Ed Reschke/Peter Arnold/Getty Images; 8.9 Image Source Plus/Alamy; 8.13a Steve Allen/SPL; 8.14 Phototake Inc/Alamy; 8.16 Steve Gscmeissner/SPL; 8.20 Harry Kikstra/Getty Images; 8.21 SPL; EoC8Q2 Phototake Inc/Alamy; Chapter 9 LightRocket via Getty Images; 9.1a RGB Ventures LLC dba SuperStock/Alamy; 9.1b BSIP SA/Alamy; 9.3a,b John Adds; 9.3c, 9.7 Biophoto Associates/SPL; 9.4 Dr Yorgos Nikas/SPL; 9.5 CNRI/SPL; 9.8 GCA/SPL; 9.9 Dr Tony Brain/SPL; EoC9Q11a,b Ed Reschke/Peter Arnold/Getty Images; Chapter 10 David Scharf/SPL; 10.1 Karen Kasmauski/Science Faction/Corbis; 10.2 Eye of Science/SPL; 10.5 Cecil H Fox/SPL; 10.6 Omikron/SPL; 10.8 NIBSC/SPL; 10.10 Kwangshin Kim/SPL; 10.11epa european pressphoto agency b.v./Alamy; 10.16 John Durham/SPL; Chapter 11 Eye of Science/SPL; 11.1 My Planet/Alamy; 11.2 Biophoto Associate/SPL; 11.3 Biophoto Associates; 11.7 Steve Gschmeissner/SPL, 11.9 J.C. Revy, ISM/SPL; 11.16 Jesco Von Puttkamer/Eye Ubiquitous; 11.17 Karen Kasmauski/Science Faction/Corbis; 11.20 Dr M. A. Ansary/SPL; 11.22 Hank Mergan/SPL; Practice Test 1 Photothek via Getty Images; PT1.16 Geoff Jones; EoCPT1Q4*l* Dr Keith Wheeler; EoCPT1Q4*r* Visuals Unlimited/SPL; Chapter 12 Steve Gschmeissner/SPL; 12.1 Jochan Tack/Alamy; 12.13 Dr Keith Porter/SPL; 12.14 from *Cell Membranes* ed. G. Weissmann and R. Claiborne (1975); 12.20 Geoff Jones; 12.21 UCLES; Chapter 13 Frank Krahmer/Getty Images; 13.1 Ashley Cooper/Alamy; 13.6 Biophoto Associates; 13.10 Martyn F. Chillmaid/SPL; 13.11 Scenics & Science/Alamy; 13.12 Garry Delong/SPL; Chapter 14 Steve Gschmeissner/Science Photo Library/Getty Images; 14.1 Michio Hoshino/Minden Pictures/FLPA; 14.3b Scipro/SPL; 14.07a,b, EoC14Q6 Manfred Kage/SPL; 14.11 Phototake Inc/Alamy; 14.23 Biomedical Imaging Unit, University of Southampton; 14.26 Martin Riedl/SPL; 14.28 Jim Varney/SPL; 14.29 Geoff Jones; 14.31 Power & Syred/SPL; Chapter 15 Alfred Pasieka/SPL; 15.1 William Taufic/Corbis; 15.5 , 15.25 Manfred Kage/SPL; 15.6 Biophoto Associates; 15.21Tomas Deerinick, NCMIR/SPL; 15.24 Ed Reschke/Photolibrary/Getty Images; 15.26 Biology Medic/SPL; 15.34 Nigel Cattlin/Holt Studios; EoC15Q7 Dr Rosalind King/SPL; Chapter 16 Equinox Graphics/SPL; 16.1Kim Taylor/naturepl.com; 16.2, 16.9 Biophoto Associates; 16.7*ct* Gerrard Peaucellier, ISM/SPL; 16.7*cc* Biology Media/SPL; 16.7*cb* Manfred Kage/SPL; 16.13 Medical-on-line/Alamy; 16.16a Wayne Hutchison/Holt Studios; 16.16b Ammit/Alamy; 16.16c John Daniels/Alamy; 16.16d Arco ImagesGmbH/Alamy; 16.20a from "The Use of Grasshopper Chromosomes to Demonstrate

Meiosis" in *Tuatara: Journal of the Biological Society of Victoria University of Wellington,* 18 (1), 1970, by J. M. Martin http://nzetc.victoria.ac.nz/tm/scholarly/Bio18Tuat01-fig-Bio18Tuat01_007a.html; 16.22 Freidrich Stark/Alamy; Chapter 17 Don Johnston, All Canada Images/Getty Images; 17.1 Richard Becker/Alamy; 17.4 Juniors Bildarchiv GmbH/Alamy; 17.6 Popperfoto/Getty Images; 17.7 Mark Boulton/Alamy; 17.10 Dick Roberts/Holt Studios; 17.10 Dick Roberts/Holt Studios; 17.11 Eric Dragesco/ardea; 17.12 Stan Levy/SPL; 17.13a,b John Mason/ardea; 17.16 Cecil H Fox/SPL; 17.17a Pat Morris/ardea; 17.17b J B Bottomley/ardea; 17.19 Nigel Cattlin/Alamy; 17.20, 17.22 The Connecticut Agricultural Experiment Station; 17.21 David Ionut/Shutterstock; 17.23 florapix/Alamy; 17.24 John W. Bova/SPL; 17.25 age fotostock/Alamy; 17.27 Martin Harvey/NHPA; 17.28 Robert Thompson/NHPA; Chapter 18 Dennis Frates/Alamy; 18.1, 18.6, 18.40 Richard Fosbery; 18.2 LOOK Die Bildagentur de Fotografen GmbH/Alamy; 18.3 Sarah8000/iStock; 18.4 David Wall Photo/Lonely Planet Images/Getty Images; 18.5 GlobalIP/iStock; 18.7 Philippe Psaila/SPL; 18.9 Martyn F Chillmaid/SPL; 18.10 goldenangel/iStock; 18.14 irabel8/Shutterstock; 18.15 NOAA PMEL Vents Program/SPL; 18.16 Visuals Unlimited/Getty Images; 18.17 Eye of Science/SPL; 18.18, 18.19 Wim Van Egmond/Visuals Unlimited Inc/SPL; 18.21 Chris Howe/Wild Places Photography/Alamy; 18.22 Frans Landting Studio/Alamy; 18.23 Fuse/Thinkstock; 18.24 Rainer von Brandis/iStock; 18.25 Science Picture Co/Getty Images; 18.26 Nigel Cattlin/Visuals Unlimited/Corbis; 18.27 jgareri/iStock; 18.28 Mark Conlin/Alamy; 18.29 oversnap/iStock; 18.30 Jaroslav Frank/iStock; 18.31 Ariadne Van Zandbergen/Alamy; 18.32 Imaginechina/Corbis; 18.33 James King-Holmes/SPL; 18.34 Frans Lanting, Mint Images/SPL; 18.35 YinYang/iStock; 18.36 commablack/iStock; 18.37 Atthapol Saita/Thinkstock; 18.38 scubazoo/SuperStock/Corbis; 18.39 AFP/Getty Images; 18.42 Biophoto Assocatiates, Chapter 19 Peter Menzel/SPL; 19.1 Science Picture Co/SPL; 19.6 Dr Charles Mazel/Visuals Unlimited, Inc/SPL; 19.7 Philippe Psaila/SPL; 19.8 Hank Morgan/SPL; 19.14 TEK Images/SPL; 19.15 Alfred Pasieka/SPL; 19.17 Saturn Stills/SPL; 19.18 Will & Denise McIntyre/SPL; 19.20a optimarc/Shutterstock; 19.20b Chris Leachman/Shutterstock; 19.21a Dr William Weber/SPL; 19.21b Nigel Cattlin/Alamy; 19.22 Geoff Jones; 19.23 Courtesy Golden Rice Humanitarian Board www.goldenrice.org; 19.24 AquaBounty Technologies, used with permission; 19.25 Andrew Fox/Alamy; Practice Test 2 Blend Image/Alamy

SPL = Science Photo Library

EoCQ = End of chapter question

t = top, *c* = centre, *b* = bottom

Page layout and artwork by Greenhill Wood Studios

CD-ROM: Terms and Conditions of use

This is a legal agreement between 'You' (which for individual purchasers means the individual customer and, for network purchasers, means the Educational Institution and its authorised users) and Cambridge University Press ('the Licensor') for Cambridge International AS and A Level Biology Fourth edition Course book with CD-ROM. By placing this CD in the CD-ROM drive of your computer You agree to the terms of this licence.

1. Limited licence

(a) You are purchasing only the right to use the CD-ROM and are acquiring no rights, express or implied to it or to the software ('Software' being the CD-ROM software, as installed on your computer terminals or server), other than those rights granted in this limited licence for not-for-profit educational use only.

(b) Cambridge University Press grants the customer the licence to use one copy of this CD-ROM either (i) on a single computer for use by one or more people at different times, or (ii) by a single person on one or more computers (provided the CD-ROM is only used on one computer at any one time and is only used by the customer), but not both.

(c) You shall not: (i) copy or authorise copying of the CD-ROM, (ii) translate the CD-ROM, (iii) reverse-engineer, alter, adapt, disassemble or decompile the CD-ROM, (iv) transfer, sell, lease, lend, profit from, assign or otherwise convey all or any portion of the CD-ROM or (v) operate the CD-ROM from a mainframe system, except as provided in these terms and conditions.

2. Copyright

(a) All original content is provided as part of the CD-ROM (including text, images and ancillary material) and is the copyright of, or licensed by a third party to, the Licensor, protected by copyright and all other applicable intellectual property laws and international treaties.

(b) You may not copy the CD-ROM except for making one copy of the CD-ROM solely for backup or archival purposes. You may not alter, remove or destroy any copyright notice or other material placed on or with this CD-ROM.

(c) The CD-ROM contains Adobe® Flash® Player.

Adobe® Flash® Player Copyright © 1996–2010 Adobe Systems Incorporated. All Rights Reserved.

Protected by U.S. Patent 6,879,327; Patents Pending in the United States and other countries. Adobe and Flash are either trademarks or registered trademarks in the United States and/or other countries.

3. Liability and Indemnification

(a) The CD-ROM is supplied 'as-is' with no express guarantee as to its suitability. To the extent permitted by applicable law, the Licensor is not liable for costs of procurement of substitute products, damages or losses of any kind whatsoever resulting from the use of this product, or errors or faults in the CD-ROM, and in every case the Licensor's liability shall be limited to the suggested list price or the amount actually paid by You for the product, whichever is lower.

(b) You accept that the Licensor is not responsible for the persistency, accuracy or availability of any urls of external or third party internet websites referred to on the CD-ROM and does not guarantee that any content on such websites is, or will remain, accurate, appropriate or available. The Licensor shall not be liable for any content made available from any websites and urls outside the Software.

(c) Where, through use of the CD-ROM and content you infringe the copyright of the Licensor you undertake to indemnify and keep indemnified the Licensor from and against any loss, cost, damage or expense (including without limitation damages paid to a third party and any reasonable legal costs) incurred by the Licensor as a result of such infringement.

4. Termination

Without prejudice to any other rights, the Licensor may terminate this licence if You fail to comply with the terms and conditions of the licence. In such event, You must destroy all copies of the CD-ROM.

5. Governing law

This agreement is governed by the laws of England, without regard to its conflict of laws provision, and each party irrevocably submits to the exclusive jurisdiction of the English courts.